Tomatoes and Tomato Products

Nutritional, Medicinal and Therapeutic Properties

Tomatoes and Tomato Products

Nutritional, Medicinal and Therapeutic Properties

Editors

Victor R. Preedy
Department of Nutrition and Dietetics
King's College London
London
UK

Ronald R. Watson
Division of Health Promotion Sciences
School of Medicine
University of Arizona
Tuscon, Arizona
USA

CRC Press
Taylor & Francis Group
Boca Raton London New York

CRC Press is an imprint of the
Taylor & Francis Group, an **informa** business
A SCIENCE PUBLISHERS BOOK

First published 2008 by Science Publishers Inc.

Published 2019 by CRC Press
Taylor & Francis Group
6000 Broken Sound Parkway NW, Suite 300
Boca Raton, FL 33487-2742

First issued in paperback 2019

ISBN 13: 978-0-367-45275-9 (pbk)
ISBN 13: 978-1-57808-534-7 (hbk)

Visit the Taylor & Francis Web site at
http://www.taylorandfrancis.com

and the CRC Press Web site at
http://www.crcpress.com

Library of Congress Cataloging-in-Publication Data

Tomatoes and tomato products: nutritional, medicinal and therapeutic properties/editors, Victor R. Preedy, Ronald R. Watson.

p. cm.

Includes bibliographical references and index.

ISBN 978-1-57808-534-7 (hardcover)

1. Tomatoes. 2. Tomatoes—Health aspects. 3. Tomato products—Health aspects. I. Preedy, Victor R. II. Watson, Ronald R. (Ronald Ross)

SB349.T679 2008

613.2′6--dc22

2008001809

Foreword

The tomato (*Lycopersicon esculentum*) is a vegetable crop of the Solanaceae family that originated in the Andean region of South America. It increased in popularity and was rapidly expanded into large-scale cultivation during the last half-century. During this period, considerable progress was made in tomato breeding and use, such as advances in increased yields, improved quality, better handling and storage durability, improved pest resistance, expanded processing techniques, and the development of a plethora of new tomato-based products. Tomato production has been transformed into a large agricultural industry, and tomato is one of the most important commercially produced vegetables in the world. Technology has revolutionized tomato production through cultural practices and new cultivars and new hybrids that can be grown virtually throughout the world.

Tomato quality is a function of several factors including the choice of cultivar, cultural practices, harvest time and method, storage, and handling procedures. The tomato processing industry has made tremendous advances, developing many forms of tomato-based foods, such as sauces, catsup (ketchup), puree, pastes, soups, juices and juice blends, and canned tomatoes either whole or in diced, sliced, quartered or stewed form. The tomato's attractive color and flavor have made it a dietary staple in many parts of the world. Nutritional considerations also bring the tomato to the forefront. Tomatoes and tomato products are rich sources of vitamin C and A, lycopene, β-carotene, lutein, lectin, and a variety of phenolic compounds such as flavonoids and phenolic acids. They are rich in folates, potassium, fiber, and protein, but low in fat and calories, as well as being cholesterol-free. With many consumers becoming increasingly aware of its nutritional and health benefits, the future of tomato production looks exceedingly bright.

Prof. Victor Preedy and Prof. Ronald Watson, both internationally known scholars, have edited this present volume, *Tomatoes and Tomato Products: Nutritional, Medicinal and Therapeutic Properties*. The book consists of three sections: characterization and composition of the tomato plant and fruit, cellular and metabolic effects of tomato and related products or components, analysis and methods. The book offers readers a broad and comprehensive coverage through its 29 chapters.

In the 12 chapters covering characterization and composition of the tomato plant and fruit, the authors discuss the origin of tomato, cultivation techniques and germplasm resources, cultivar and agricultural management on lycopene and vitamin C contents in tomato fruits, field trials of genetically modified tomato, postharvest ripening of tomato, tomato flavor, antioxidant activity in carotenoids in tomato plants, tomato lectin, zingiberene and curcumene in wild tomato leaves, the role of intracellular and secreted purple acid phosphatases in tomato phosphate nutrition, nutritional characterization of tomato juices, and acrylamide in tomato products.

In the 12 chapters covering cellular and metabolic effects of tomato and related products or components, the authors discuss tomato leaf crude extracts for insects and spider mite control, bioactive polysaccharides from tomato, ingestion of tomato products and lycopene isomers in plasma, comparative studies of antioxidant compounds in different tomato cultivars, antioxidant activity of fresh-cut tomatoes, effects of tomato on anticancer properties of saliva, DNA strand breaks and tomatoes, tomato carotenoids and the IGF system in cancer, consumption of tomato paste and benign prostate hyperplasia, experimental prostate cancer and administration of tomato and its constituents, and effect of tomato juice on the prevention and management of prostate cancer, adenosine deaminase and lung diseases.

In the 5 chapters covering analysis and methods, the authors discuss proteomics of tomato seed and pollen, gene transfer in tomato and detection of transgenic tomato products, assaying vitamins and micronutrients in tomato, methods for PCR and gene expression studies in tomato plants, and DNA analysis of tomato seeds in forensic evidence.

The contributing authors are among the world's experts in the plant science of tomatoes, the chemistry and biochemistry of tomato and products, and the area of human nutrition. Drawing on the wide range of expertise from scientists in various interdisciplinary research areas, including chemists, biochemists, molecular biologists, agronomists, food technologists, medicinal and health professionals, the multidisciplinary approach offers a comprehensive review of new research findings and provides timely information focused on tomato products and their potential benefits for human health. This monograph will serve as a critical reference book by providing a deeper understanding of the role of agronomy practice for tomato production, chemical and biochemical properties of tomato, and product development in promoting human health.

John Shi, Ph.D.
Senior Research Scientist
Guelph Food Research Center
Agriculture and Agri-Food Canada
Canada

Preface

The tomato is a humble fruit and has been part of the heritage of mankind for hundreds of years. In various cultures, tomatoes form an integral part of the diet and "cuisine". Moreover, as well as the attributes of a pleasant taste, epidemiological studies have shown that the tomato confers beneficial health effects. This may be due to the fact that the tomato contains an abundance of antioxidants and protective compounds, such as carotenoids, bioactive exopolysaccharides, flavonoids, phenolic acids, ascorbic acid and vitamin E.

Gaining an insight into the tomato plant, from its genetics to its role in the diet is one of the preludes to understanding public health nutrition in its broadest sense. But finding this knowledge in a single coherent volume is presently problematical. However, Tomatoes and tomato products: Nutritional, medicinal and therapeutic properties addresses this.

The contributors to Tomatoes and tomato products: Nutritional, medicinal and therapeutic properties are authors of international and national standing, leaders in the field and trend-setters. Emerging fields of science and important discoveries relating to tomatoes and related products will also be incorporated in these books. This represents one stop shopping of material related to tomatoes. This book will be essential reading for plant scientists, nutritionists, dieticians, pharmacologists, health care professionals and research scientists.

Professor Victor R Preedy, Professor Ronald Ross Watson

Contents

Foreword *v*
Preface *ix*
List of Contributors *xv*

PART 1
CHARACTERIZATION AND COMPOSITION OF TOMATO PLANT AND FRUIT

1. **Tomatoes: Origin, Cultivation Techniques and Germplasm Resources** **3**
   Derly José Henriques da Silva, Flávia Barbosa Abreu, Fabiano Ricardo Brunele Caliman, Adilson Castro Antonio and Vinood B. Patel
2. **Cultivar and Agricultural Management on Lycopene and Vitamin C Contents in Tomato Fruits** **27**
   Tiequan Zhang, John Shi, YutaoWang and Sophia Jun Xue
3. **Field Trials of Genetically Modified Tomato: Fruit Quality and Productivity** **47**
   G.L. Rotino, T. Pandolfini, R. Lo Scalzo, E. Sabatini, M. Fibiani and A. Spena
4. **Post-harvest Ripening of Tomato** **67**
   María Serrano, Pedro Javier Zapata, Fabián Guillén, Domingo Martínez-Romero, Salvador Castillo and Daniel Valero
5. **Tomato and Flavour** **85**
   Christian Salles

6. **Antioxidant Activity in Tomato: A Function of Genotype** 111
Charanjit Kaur and H.C. Kapoor

7. **Carotenoids in Tomato Plants** 133
B. Stephen Inbaraj and B.H. Chen

8. **Tomato Lectin** 165
Willy J. Peumans and Els J.M. Van Damme

9. **Presence of Zingiberene and Curcumene in Wild Tomato Leaves** 193
George F. Antonious

10. **The Role of Intracellular and Secreted Purple Acid Phosphatases in Tomato Phosphate Nutrition** 215
Gale G. Bozzo and William C. Plaxton

11. **Nutritional Characterization of Tomato Juices** 235
Concepción Sánchez-Moreno, Begoña de Ancos, Lucía Plaza, Pedro Elez-Martínez and M. Pilar Cano

12. **Acrylamide in Tomato Products** 259
Fernando Tateo and Monica Bononi

PART-2
CELLULAR AND METABOLIC EFFECTS OF TOMATO AND RELATED PRODUCTS OR COMPONENTS

13. **Tomato Leaf Crude Extracts for Insects and Spider Mite Control** 269
George F. Antonious and John C. Snyder

14. **Bioactive Polysaccharides from Tomato** 299
Giuseppina Tommonaro, Alfonso De Giulio, Giuseppe Strazzullo, Salvatore De Rosa, Barbara Nicolaus and Annarita Poli

15. **Ingestion of Tomato Products and Lycopene Isomers in Plasma** 317
Volker Böhm and Kati Fröhlich

16. **Antioxidant Compound Studies in Different Tomato Cultivars** 333
Giuseppina Tommonaro, Barbara Nicolaus, Rocco De Prisco, Alfonso De Giulio, Giuseppe Strazzullo and Annarita Poli

17. Antioxidant Activity of Fresh-cut Tomatoes: Effects of Minimal Processing and Maturity Stage at Harvest 345
Milza Moreira Lana

18. Tomato and Anticancer Properties of Saliva 377
Masahiro Toda and Kanehisa Morimoto

19. DNA Strand Breaks and Tomatoes 385
Karlis Briviba

20. Tomato Carotenoids and the IGF System in Cancer 395
Joseph Levy, Shlomo Walfisch, Yossi Walfisch, Amit Nahum, Keren Hirsch, Michael Danilenko and Yoav Sharoni

21. Tomato Paste and Benign Prostate Hyperplasia 411
Magda Edinger de Souza, Walter José Koff and Tania Weber Furlanetto

22. Tomatoes and Components as Modulators of Experimental Prostate Carcinogenesis 429
Elizabeth Miller Grainger, Kirstie Canene-Adams, John W. Erdman, Jr. and Steven K. Clinton

23. Tomato Juice, Prostate Cancer and Adenosine Deaminase Enzyme 457
Aslihan Avci and Ilker Durak

24. Effect of Tomato Juice on Prevention and Management of Lung Diseases: Cigarette Smoke-induced Emphysema in the Senescence-accelerated Mouse and Bronchial Asthma in Human 475
Kuniaki Seyama, Naoaki Tamura, Takahiro Inakuma and Koichi Aizawa

PART-3
ANALYSIS AND METHODS

25. Proteomics of Tomato Seed and Pollen 501
Vipen K. Sawhney and Inder S. Sheoran

26. Gene Transfer in Tomato and Detection of Transgenic Tomato Products 515
Theodoros H. Varzakas, Dimitris Argyropoulos and Ioannis S. Arvanitoyannis

27. Assaying Vitamins and Micronutrients in Tomato **537**
A.I. Olives, M.A. Martin, B. del Castillo and M.E. Torija

28. Methods for PCR and Gene Expression Studies in Tomato Plants **585**
Dimitris Argyropoulos, Theodoros H. Varzakas, Charoula Psallida and Ioannis S. Arvanitoyannis

29. DNA Analysis of Tomato Seeds in Forensic Evidence **617**
Henry C. Lee and Cheng-Lung Lee

Index *637*

List of Contributors

Abreu Flávia Barbosa

Universidade Federal de Mato Grosso do Sul, Campus de Chapadão do Sul, Caixa Postal 112, Cep: 79560-000, Chapadão do Sul, MS, Brazil.

Aizawa Koichi

Biogenics Research Department, KAGOME Co. Ltd., Research Institute, 17 Nishitomiyama, Nasusiobara-Shi Tochigi, 329-2762, Japan.

Ancos Begoña de

Department of Plant Foods Science and Technology, Instituto del Frío, Consejo Superior de Investigaciones Científicas (CSIC), José Antonio Novais 10, Ciudad Universitaria, E-28040 Madrid, Spain.

Antonio Adilson Castro

Universidade Federal de Viçosa, Departamento de Fitotecnia, Avenue P.H. Rolfs s/n, Viçosa, MG 36571-000, Brazil.

Antonious George F.

Department of Plant and Soil Science, Water Quality/Environmental Toxicology, Land Grant Program, 218 Atwood Research Center, Kentucky State University, Frankfort, Kentucky 40601, USA.

Argyropoulos Dimitris

Institute of Biotechnology, National Agricultural Research Foundation, Sofokli Venizelou 1 Lykovrisi 14123, Attiki Greece.

Arvanitoyannis Ioannis S.

Department of Agriculture, Ichthyology and Aquatic Environment, Agricultural Sciences, University of Thessaly, Fytokou Street, Nea Ionia Magnesias, 38446 Volos, Hellas, Greece.

Avci Aslihan

Ankara University, Faculty of Medicine, Department of Biochemistry, 06100 Sihhiye, Ankara, Turkey.

Böhm Volker

Institute of Nutrition, Friedrich Schiller University Jena, Dornburger Str. 25-29, 07743 Jena, Germany.

Bononi Monica

Analytical Research Laboratories - Food and Env - Di.Pro.Ve., Faculty of Agriculture, University of Milan, 2, Via Celoria, 20133 Milan, Italy.

Bozzo Gale G.

Horticultural Sciences Dept., Univ. of Florida, P.O. Box 110690, Gainesville, FL 32611-690, USA.

Briviba Karlis

Institute of Nutritional Physiology, Federal Research Centre for Nutrition and Food, Haid-und-Neu-Str. 9, 76131 Karlsruhe, Germany.

Caliman Fabiano Ricardo Brunele

Universidade Federal de Viçosa, Departamento de Fitotecnia, Avenue P.H. Rolfs s/n, Viçosa, MG 36571-000, Brazil.

Cano M. Pilar

Department of Plant Foods Science and Technology, Instituto del Frío, Consejo Superior de Investigaciones Científicas (CSIC), José Antonio Novais 10, Ciudad Universitaria, E-28040 Madrid, Spain.

Canene-Adams Kirstie

905 S. Goodwin Ave, 448 Bevier Hall University of Illinois Urbana, Illinois, 61801, USA.

Castillo Benito Del

S.D. Quimica Analitica, Facultad de Farmacia, Universidad Complutense de Madrid, Pza. Ramon y Cajal s/n, 28040 - Madrid, Spain.

Castillo Salvador

Dept. Food Technology, University Miguel Hernández, Ctra. Beniel km. 3.2, 03312, Orihuela Alicante, Spain.

Chen B.H.

Department of Food Science, Fu Jen University, Taipei 242, Taiwan.

Clinton Steven K.

A 434 Starling Loving Hall 320 West 10th Ave, The Ohio State University Columbus, Ohio 43210, USA.

da Silva Derly José Henriques

Universidade Federal de Viçosa, Departamento de Fitotecnia, Avenue P.H. Rolfs s/n, Viçosa, MG 36571-000, Brazil.

Danilenko Michael

Department of Clinical Biochemistry, Faculty of Health Sciences, Ben-Gurion University of the Negev, P.O. Box 653, Beer-Sheva, Israel.

de Souza Magda Edinger

Rua Demétrio Ribeiro, 244/603, Centro Porto Alegre, RS, Brazil, 90010-312.

Durak Ilker

Ankara University, Faculty of Medicine, Department of Biochemistry, 06100 Sihhiye, Ankara, Turkey.

Elez-Martínez Pedro

Department of Plant Foods Science and Technology, Instituto del Frío, Consejo Superior de Investigaciones Científicas (CSIC), José Antonio Novais 10, Ciudad Universitaria, E-28040 Madrid, Spain.

Erdman John W., Jr.

905 S. Goodwin Ave. 455 Bevier Hall University of Illinois Urbana, Illinois, 61801, USA.

Fibiani Marta

CRA-IAA, Agricultural Research Council, Research Unit for Agro Food Industry, Via Venezian, 26 20133, Milan, Italy.

Fröhlich Kati

Institute of Nutrition, Friedrich Schiller University Jena, Dornburger Str. 25-29, 07743 Jena, Germany.

Furlanetto Tania Weber

Hospital de Clínicas de Porto Alegre Rua Ramiro Barcelos, 2350/700 Porto Alegre, RS, Brazil 90035-903.

Giulio Alfonso De

Istituto di Chimica Biomolecolare, Consiglio Nazionale delle Ricerche (C.N.R.), Via Campi Flegrei, 34, 80078 Pozzuoli, Napoli, Italy.

Grainger Elizabeth Miller

A 434 Starling Loving Hall 320 West 10th Ave, The Ohio State University Columbus, Ohio 43210, USA.

Guillén Fabián

Dept. Food Technology, University Miguel Hernández, Ctra. Beniel km. 3.2, 03312, Orihuela Alicante, Spain.

Hirsch Keren

Department of Clinical Biochemistry, Faculty of Health Sciences, Ben-Gurion University of the Negev, P.O. Box 653, Beer-Sheva, Israel.

Inakuma Takahiro

Biogenics Research Department, KAGOME Co. Ltd., Research Institute, 17 Nishitomiyama, Nasusiobara-Shi Tochigi, 329-2762, Japan.

Inbaraj B. Stephen

Department of Food Science, Fu Jen University, Taipei 242, Taiwan.

Kapoor H.C.

Division of Biochemistry, Indian Agricultural Research Institute, New Delhi 110 012, India.

Kaur Charanjit

Division of Post Harvest Technology, Indian Agricultural Research Institute, New Delhi 110 012, India.

Koff Walter José

Hospital de Clínicas de Porto Alegre Rua Ramiro Barcelos, 2350/835 Porto Alegre, RS, Brazil, 90035-903.

Lana Milza Moreira

Embrapa Hortaliças (Embrapa Vegetables), Caixa Postal 218, 70359970, Brasília-DF, Brazil.

Lee Cheng-Lung

Head of Forensic Science Section, Hsin-Chu City Police Bureau, No. 1 Chung-Shan Rd., Hsin-chu, Taiwan 300.

Lee Henry C.

Division of Scientific Services, Department of Public Safety, 278 Colony Street, Meriden, CT 06451 USA.

Levy Joseph

Department of Clinical Biochemistry, Faculty of Health Sciences, Ben-Gurion University of the Negev, P.O. Box 653, Beer-Sheva, Israel.

Martin M. Antonia

S.D. Quimica Analitica, Facultad de Farmacia, Universidad Complutense de Madrid, Pza. Ramon y Cajal s/n, 28040 - Madrid, Spain.

Martínez-Romero Domingo

Dept. Food Technology, University Miguel Hernández, Ctra. Beniel km. 3.2, 03312, Orihuela Alicante, Spain.

Morimoto Kanehisa

Department of Social and Environmental Medicine, Osaka University Graduate School of Medicine, 2-2 Yamada-oka, Suita, Osaka 565-0871, Japan.

Nahum Amit

Department of Clinical Biochemistry, Faculty of Health Sciences, Ben-Gurion University of the Negev, P.O. Box 653, Beer-Sheva, Israel.

Nicolaus Barbara

Istituto di Chimica Biomolecolare, Consiglio Nazionale delle Ricerche (C.N.R.), Via Campi Flegrei, 34, 80078 Pozzuoli, Napoli, Italy.

Olives Ana I.

S.D. Quimica Analitica, Facultad de Farmacia, Universidad Complutense de Madrid, Pza. Ramon y Cajal s/n, 28040 - Madrid, Spain.

Pandolfini Tiziana

Faculty of Science, University of Verona, Strada Le Grazie 15, 37134 Verona, Italy.

Patel Vinood B.

University of Westminster, School of Biosciences, 115 New Cavendish Street, London, UK.

Peumans Willy J.

Nievelveldweg 9, 9310 Aalst, Belgium.

Plaxton William C.

Dept. of Biology, Queen's Univ., Kingston, Ontario, Canada K7L 3N6.

Plaza Lucía

Department of Plant Foods Science and Technology, Instituto del Frío, Consejo Superior de Investigaciones Científicas (CSIC), José Antonio Novais 10, Ciudad Universitaria, E-28040 Madrid, Spain.

Poli Annarita

Istituto di Chimica Biomolecolare, Consiglio Nazionale delle Ricerche (C.N.R.), Via Campi Flegrei, 34, 80078 Pozzuoli, Napoli, Italy.

Prisco Rocco De

Istituto di Chimica Biomolecolare, Consiglio Nazionale delle Ricerche (C.N.R.), Via Campi Flegrei, 34, 80078 Pozzuoli, Napoli, Italy.

Psallida Charoula

Institute of Biotechnology, National Agricultural Research Foundation, Sofokli Venizelou 1 Lykovrisi 14123, Attiki, Greece.

Rosa Salvatore De

Istituto di Chimica Biomolecolare, Consiglio Nazionale delle Ricerche (C.N.R.), Via Campi Flegrei, 34, 80078 Pozzuoli, Napoli, Italy.

Rotino Giuseppe Leonardo

CRA-ORL,Agricultural Research Council, Research Unit for Vegetable Crops, Via Paullese 28, 26836 Montanaso Lombardo (LO), Italy.

Sabatini Emidio

CRA-ORL, Agricultural Research Council, Research Unit for Vegetable Crops, Via Paullese 28, 26836 Montanaso Lombardo (LO), Italy.

Salles Christian

UMR 1129 FLAVIC, ENESAD, INRA, University of Burgundy, 17 rue Sully, BP86510, F21065 Dijon, France.

Sánchez-Moreno Concepción

Department of Plant Foods Science and Technology, Instituto del Frío, Consejo Superior de Investigaciones Científicas (CSIC), José Antonio Novais 10, Ciudad Universitaria, E-28040 Madrid, Spain.

Sawhney Vipen K.

Department of Biology, 112 Science Place, University of Saskatchewan, Saskatoon, Saskatchewan S7N 5E2, Canada.

Scalzo Roberto Lo

CRA-IAA, Agricultural Research Council, Research Unit for Agro Food Industry, Via Venezian, 26 20133, Milan, Italy.

Serrano María

Dept. Applied Biology, University Miguel Hernández, Ctra. Beniel km. 3.2, 03312, Orihuela Alicante, Spain.

Seyama Kuniaki

Department of Respiratory Medicine, Juntendo University School of Medicine, 2-1-1 Hongo, Bunkyo-Ku, Tokyo 113-8421, Japan.

Sharoni Yoav

Department of Clinical Biochemistry, Faculty of Health Sciences, Ben-Gurion University of the Negev, P.O. Box 653, Beer-Sheva, Israel.

Sheoran Inder S.

Department of Biology, 112 Science Place, University of Saskatchewan, Saskatoon, Saskatchewan S7N 5E2, Canada.

Shi John

Guelph Food Research Center, Agriculture and Agri-Food Canada, Guelph, Ontario N1G 5C9, Canada.

Snyder John C.

Department of Horticulture, University of Kentucky, Lexington, Kentucky 40546, USA.

Spena Angelo

Faculty of Science, University of Verona, Strada Le Grazie 15, 37134 Verona, Italy.

Strazzullo Giuseppe

Istituto di Chimica Biomolecolare, Consiglio Nazionale delle Ricerche (C.N.R.), Via Campi Flegrei, 34, 80078 Pozzuoli, Napoli, Italy.

Tamura Naoaki

Department of Respiratory Medicine, Koto Hospital, 6-8-5 Ojima, Koto-ku, 136-0072, Tokyo, Japan.

Tateo Fernando

Analytical Research Laboratories - Food and Env - Di.Pro.Ve., Faculty of Agriculture, University of Milan, 2, Via Celoria, 20133 Milan, Italy.

Tiziana Pandolfini

Faculty of Science, University of Verona, Strada Le Grazie 15, 37134 Verona, Italy.

Toda Masahiro

Department of Social and Environmental Medicine, Osaka University Graduate School of Medicine, 2-2 Yamada-oka, Suita, Osaka 565-0871, Japan.

Tommonaro Giuseppina

Istituto di Chimica Biomolecolare, Consiglio Nazionale delle Ricerche (C.N.R.), Via Campi Flegrei, 34, 80078 Pozzuoli, Napoli, Italy.

Torija M. Esperanza

Dept. Nutricion y Bromatologia II: Bromatologia, Facultad de Farmacia, Universidad Complutense de Madrid, Pza. Ramon y Cajal s/n, 28040 – Madrid, Spain.

Valero Daniel

Dept. Food Technology, University Miguel Hernández, Ctra. Beniel km. 3.2, 03312, Orihuela Alicante, Spain.

Van Damme Els J.M.

Department of Molecular Biotechnology, Lab. Biochemistry and Glycobiology, Ghent University, Coupure Links 653, B-9000 Gent, Belgium.

Varzakas Theodoros H.

Department of Technology of Agricultural Products, School of Agricultural Technology, Technological Educational Institute of Kalamata, Antikalamos 24100, Kalamata, Hellas, Greece.

Walfisch Shlomo

Colorectal Unit, Faculty of Health Sciences, Ben-Gurion University of the Negev, P.O. Box 653, Beer-Sheva, Israel.

Walfisch Yossi

Department of Clinical Biochemistry, Faculty of Health Sciences, Ben-Gurion University of the Negev, P.O. Box 653, Beer-Sheva, Israel.

Wang Yutao

Greenhouse and Processing Crops Research Center, Agriculture and Agri-Food Canada, Harrow, Ontario N0R 1G0, Canada.

Xue Sophia Jun

Guelph Food Research Center, Agriculture and Agri-Food Canada, Guelph, Ontario N1G 5C9, Canada.

Zapata Pedro Javier

Dept. Food Technology, University Miguel Hernández, Ctra. Beniel km. 3.2, 03312, Orihuela Alicante, Spain.

Zhang Tiequan

Greenhouse and Processing Crops Research Center, Agriculture and Agri-Food Canada, Harrow, Ontario N0R 1G0, Canada.

PART 1

Characterization and Composition of Tomato Plant and Fruit

1

Tomatoes: Origin, Cultivation Techniques and Germplasm Resources

Derly José Henriques da Silva[1*], Flávia Barbosa Abreu[2], Fabiano Ricardo Brunele Caliman[1], Adilson Castro Antonio[1] and Vinood B. Patel[3]

[1]Universidade Federal de Viçosa, Departamento de Fitotecnia, Avenue P.H. Rolfs s/n, Viçosa, MG 36571-000, Brazil

[2]Universidade Federal de Mato Grosso do Sul, Campus de Chapadão do Sul, Caixa Postal 112, Cep: 79560-000, Chapadão do Sul, MS, Brazil

[3]University of Westminster, School of Biosciences, 115 New Cavendish Street, London, UK

ABSTRACT

The tomato originated from South America and its most widely known scientific name is *Lycopersicon esculentum,* but it can also be identified as *Solanum lycopersicon*, as originally classified by Linnaeus in 1753, because of the similarity between tomatoes and potatoes. This fruit contains a variety of micro-components, such as lycopene (an antioxidant), potassium, vitamins (A, C, E and K), sucrose and folic acid.

Among several climatic factors affecting tomato cultivation, the temperature should be emphasized. This is because the rate of liquid assimilation, i.e., efficiency of growth, is observed when the temperature is 18-28°C. In this temperature range optimum tomato production is carried out using appropriate techniques, such as the production of seedlings, soil preparation, application of organic matter, transplanting, planting and coverage, chemical nutrition, staking, pruning and pest control.

*Corresponding author

Factors such as temperature, relative air humidity, luminosity and genetics can affect the tomato fruit content of organic acids (citric and malic acids), sugars (glucose, fructose and sucrose), solids insoluble in alcohol (proteins, celluloses, pectin and polysaccharides), carotenes, and lipids, among others. Finally, many physiological, phytopathological and entomological disturbances can be resolved by genes that are in germplasm banks.

INTRODUCTION

In legends and stories from pre-Columbian America the tomato was known as a fruit eaten by wolves and werewolves. It was later domesticated in Mexico and introduced in Europe, where initially it was considered a poisonous fruit. The tomato is now acquiring more and more consumers because of its versatility of use, in domestic cooking and in commercial production of processed and packaged soups, sauces and condiments.

The use of tomato in the modern diet is so extensive that it is almost impossible to dissociate it from the menus of fast foods and pizza parlours. Its per capita consumption in fresh and processed form surpasses 20 kg/yr (Jones et al. 1991). The fruit is a source of potassium, vitamin C, folic acid and carotenoids, with lycopene (antioxidant) being predominant. It also contains vitamin E, vitamin K and flavonoids. It has a low calorie content of around 20 kcal/100 g of fruit. Its flavour comes mainly from its sugars (fructose, glucose, sucrose) and organic acids (malic and citric). As the fruit ripens the content of fructose and glucose increases and the content of acids decreases (Jones et al. 1991).

BOTANICS AND ORIGIN

Officially the cultivated tomato belongs to the order Scrophulariales, suborder Solanineae, family Solanaceae, tribe Solaneae, genus Lycopersicon, subgenus Eulycopersicon, species *Lycopersicon esculentum* (lycopersicon = wolf peach, esculentum = edible).

The following subspecies are recognized:

Lycopersicon esculentum var. *esculentum* is the main plant of commercial interest, with great variability in form, colour and size of fruits. *Lycopersicon esculentum* var. *cerasiforme* is typically known as cherry tomato; the fruits are small, normally with diameters from 2 to 5 cm. *Lycopersicon esculentum* var. *pyriforme* has a pear-shaped fruit, with an average length of 4 cm. *Lycopersicon esculentum* var. *grandifolium* is known as "potato leaves"; leaves are large with few follicles. *Lycopersicon esculentum* var. *validium* plants are erect and compact with very short internodes. At present the scientific name is undergoing evaluation because Linnaeus (1753) was the first author of the scientific name that related the tomatoes to the genus *Solanum*, which includes the potato (*S. tuberosum* L.). One year later, Miller (1754) classified the tomato in the genus *Lycopersicon*. Today, Miller's classification is still used by the majority of botanists and plant breeders.

However, according to Peralta and Spooner (2000), a minority of authors (Spooner et al. 1993, Bohs and Olmstead 1999, Knapp and Spooner 1999, Olmstead at al. 1999) still name tomatoes *Solanum*. Supporting this molecular data is DNA from chloroplasts (Spooner et al. 1993, Bohs and Olmstead 1997, 1999, Olmstead and Palmer 1997, Olmstead et al. 1999), which reinforces the close genetic relationships of tomatoes and potatoes, justifying Linnaeus's original denomination of tomato as *Solanum* (Peralta and Spooner 2000). The continuation of the name *Lycopersicon* has been justified by the convenience and stability of the nomenclature. For this reason, Spooner et al. (1993) transferred *Lycopersicon* to *Solanum* sect. *Lycopersicon* and proposed new combinations of names for many of the species of *Lycopersicon* (Table 1).

As to the origin, the tomato plant has a possible centre of variability along the Pacific coast between the Equator and the north of Chile, including the Galapagos Islands, to an elevation of approximately 2,000 m in the Andes mountain range. However, the domestication and cultivation

Table 1 List of species of the genus *Solanum*, with the equivalents of species previously designated *Lycopersicon*.

Genus Solanum	*Equivalent Lycopersicon*
S. arcanum	*L. peruvianum*
S. cheesmaniae	*L. cheesmaniae*
S. chilense	*L. chilense*
S. chmielewskii	*L. chmeilewskii*
S. corneliomuelleri	*L. peruvianum*
S. galapagense	*L. cheesmaniae*
S. habrochaites	*L. hirsutum*
S. huaylasense	*L. peruvianum*
S. lycopersicum	*L. esculentum*
S. neorickii	*L. parviflorum*
S. pennellii	*L. pennellii*
S. peruvianum	*L. peruvianum*
S. pimpinellifolium	*L. pimpinellifolium*

were carried out by Indian tribes that inhabited Mexico (Giordano and Silva 2000), where the name tomato originated.

CLIMATIC DEMANDS AND PLANTING SEASONS

The crop can be grown under different conditions, but the most suitable are high altitudes, with low humidity and high luminosity. In regions with altitudes of 500-900 m, the tomato plant may be cultivated for the whole year. At altitude less than 300 m it is preferably cultivated in the winter and at altitude above 1200 m it is best cultivated in the summer (Fontes and Silva 2002). Among the diverse climatic factors, temperature merits emphasis, because higher net assimilation rate, that is, greater growth efficiency, is observed when the temperature is between 18 and 28°C (Jones et al. 1991).

The temperature requirement of the tomato plant varies with the development of the plant. In the germination phase the ideal temperature is 16-29°C. Close to 5°C or 40°C the germination is inhibited. For plant development the ideal average temperature is 21 to 24°C and for fruit set the optimum temperature is 24°C during the day and 14-17°C at night. With temperatures above 35°C at day and 30°C at night, fruit abortion occurs, resulting in the production of small fruits with few seeds, and low liberation and germination of the pollen grain. The negative effect of this temperature is more intense from 8 to 13 d before anthesis and is enhanced with increasing temperatures (Jones et al. 1991).

The range of temperatures ideal for the formation of lycopene, the main carotenoid responsible for the intense red colour of the fruit, is 20-24°C during the day and around 18°C at night. Temperatures above 30°C inhibit the formation of lycopene and favour the formation of other carotenoids, which gives a yellow-orange colour to the fruit.

SEEDLING PRODUCTION

Commercial production of tomato for the fresh market is highly regulated, beginning with the seedlings, which should originate from high quality seeds. The seedlings are produced in trays and in a substrate free of pests and diseases. However, for domestic production less rigorous formulae may be used for the production of seedlings.

Containers

The containers may be perforated polyethylene bags or made from newspapers that will be filled with substrate. To manufacture a bag out of newspaper, a strip of newspaper is cut to a width of 12 cm and a length of 60 cm and rolled around a PVC tube of 6 cm diameter. The lower part of the strip is bent into the interior of the PVC tube, making up the bottom. In this way a bag is produced to hold 170 cm^3 of substrate. The bags are filled with the appropriate substrate. A sample formula for substrate contains 60 L subsoil, 10 L sand, 20 L cured farmyard manure, 700 g dolomite and 300 g 2-20-2 fertilizer. Alternatively, the subsoil and sand mixture can be 2 parts soil and 1 part cured manure (Fontes and Silva 2002).

Trays

Normally trays made of polystyrene are used for the production of seedlings with cells shaped like a cone or inverted pyramid and the bottom open so as to allow for trimming of the roots by air. It is common to use trays with 128 cells, with cell volumes of 35 cm^3. Alternatively, trays with 72 cells might be used with volumes of 113 cm^3. After receiving the substrate and the seeds, the trays should not be in contact with the ground, but be held up by supports. The cells are planted and irrigated manually. Before trays are reused, they should be washed, immersed in 1% sodium hypochlorite solution for 30 min, washed again in clean water and allowed to air dry.

The seedlings should be produced in sites with abundant illumination in a protected environment covered with new plastic free of dust, fenced by anti-aphid lateral screens to reduce insect transmitters of viruses, and irrigated with pathogen-free water (Fontes and Silva 2002). The seedlings must be irrigated daily and are sprayed weekly with a protective insecticide. Currently, seedlings are vaccinated against virus-transmitting insects by means of systemic insecticides. However, such a practice is not officially recognized.

The nutrients in the substrate may not be sufficient for full development of the seedling. In this case the plant may be sprayed weekly with a nutrient solution. For example, the following nutrients may be mixed in 1 L water: 100 mg super simple fertilizer, 100 mg potassium chloride, 100 mg magnesium sulphate, 5 mg iron sulphate, 4 mg borax, 2 mg manganese sulphate, 2 mg zinc sulphate, 0.4 mg copper sulphate, 0.2 mg sodium molybdate and eventually 150 mg nitrocalcium.

PLANTING METHODS

Site

The ideal site will be easily accessible for labourers and equipment such as tractors or ploughs. The other requirements are ample sunshine, a good quality water supply, and a slightly inclined topography to facilitate drainage.

Soil Preparation

The tomato plant needs a structured soil, medium texture and good drainage so as to allow for good development of roots, thus avoiding phytosanitary problems. The most widespread procedure of soil preparation is the conventional one, which consists of ploughing twice and one passage of roto tiller. Direct planting of the tomato plant has been investigated: however, the results are still being assessed.

Liming

The tomato plant adapts well to soil with a pH of 5.5 to 6.0. This is because in these conditions it is possible to reduce the prejudicial effects of high concentrations of aluminium and manganese ions, as well as increase the availability of phosphorus and molybdenum. On the other hand, the tomato plant is demanding of calcium and magnesium, such that the application of calcium to the soil is indispensable.

Application of Organic Matter

In soils that are poor in organic matter (< 0.4 g/kg soil), it is difficult to obtain productivity around 140 t/ha without the addition of organic matter (manure). Although it is difficult to recommend a fixed quantity of organic matter, about 30 t/ha of well-matured farmyard manure is a good rule of thumb.

With respect to nitrogen and potassium, it is suggested that only 5% of the total recommended dose be used during the planting. The rest should be divided into weekly doses applied on the soil.

Transplanting

The seedlings grown in the newspaper cups or trays are transplanted in furrows. Normally the furrows are 1 to 1.20 m apart, with a width of approximately 15 cm and depth made by a tractor tractioned furrower. The furrows should be made in a 0.5% slope. Planting fertilizer is applied in the furrows, containing organic matter and macro- and micronutrients, and, if necessary, insecticide is applied in the soil. After these applications the planting furrows should be closed.

Seedlings are transplated when they have three to four definitive leaves. If the seedlings are in newspaper cups, it is not necessary to remove the cups at the time of transplanting. If they are in Styrofoam trays the trays should be removed, maintaining the substrate adhering to the roots. The plants should be placed in the furrows and a slight pressure exerted on the soil in the direction of the seedlings, to assure adhesion of the soil to the roots.

As to spacing, the most usual in Brazil is 1 to 1.2 m between furrows, and 0.30 to 0.60 m between plants in rows. Such spacing allows for 13,900 to 33,300 plants per hectare. In other countries a density of only 8,000 plants per hectare is recommended. The Mexican spacing is commonly used, in which plants are settled with a supporting stake at a spacing of 1 m between furrows and 0.4 m between plants in rows. In this fashion it is possible to grow 25,000 plants per hectare (Fontes and Silva 2002).

Staking and Pruning

Staking

An assembly constructed so as to leave the plant shoot free from contact with the soil should be set up 10 to 15 d after the seedlings are transplanted, before they start to fall. Pruning is also initiated by that time. The plants may be tied with ribbon, plastic tape or other materials.

Crossed Fence

The most common staking method for the tomato plant is the crossed fence or inverted V, although it is quite expensive. Bamboo stakes are used, supported by a set of posts and wire. The bamboo that is used to make the stakes should be fresh, 8-10 cm in diameter and longitudinally split. Bamboo posts of 15 cm diameter should be placed about every 8 m. These should be 2.30 m in length and inserted 40 cm into the soil. The upper part of each post, at the height of ± 1.80 m, is united with #14 wire, stretched parallel to the soil. At the end of each planting row a thicker post should be sunk as reinforcement. Then, 10 cm from the base of each plant, a bamboo stake 2.30 cm long is inserted in the soil to a depth of 5-10 cm. This stake should be bent so as to reach the wire above between the planting rows. This is the stake that will support the plant. After the staking structure is installed, the plant stem is pulled in the direction of the stake and tied to it.

Individual Vertical Staking, with Ribbon

As in a greenhouse, the tomato plant may be staked with plastic ribbon.

Posts of 15-20 cm diameter and 2.30 m length are affixed in the soil to a depth of 40 cm, every 4-5 m along the planting row. On these post two #12 wires are stretched parallel to the soil, one at the top and another at the base, close to the ground along the planting row. One strengthened wedge-shaped stake at a 45° angle should be stuck in the soil at the end of each planting row to hold up the post.

Almost always, depending on the weight of the clusters, it is necessary to affix thinner stakes (whole bamboo) between the posts, after every two or three plants, to sustain the weight of the fruits and to help keep the upper wire tight, with slight arching. In this method of staking the plants are tied before they fall over. One end of the ribbon is tied to the lower wire. Then the ribbon is looped once or twice around the plant and tied to the upper wire (or to the bamboo). The plants are trained as they grow, that is, at the appearance of each cluster they should be twisted around the ribbon (Fontes and Silva 2002).

Trellises or Horizontal Fence

The tomato plant may be staked in trellises made of 5 to 6 rows of #14 wire or bamboo, arranged horizontally with respect to the soil, at every 25-30 cm and tied to reinforcement stakes or posts affixed in the soil at every 5 m along the planting row. The tomato plant is tied to the wire or to the bamboo. If wire is used, it is necessary to put in a vertical bamboo stake after every three plants to support the weight of the plant. This staking can be used single or double.

Pruning

To prune a plant means to manage the development of the shoots. Tomato plants can be grown with one or two stems. The maintenance of two stems allows for the development of the first sprouting below the first cluster. However, studies at the Federal University of Viçosa (UFV) have concluded that growing tomato plants with one stem increases the production of large fruits without reducing the yield (Marim et al. 2005). The plants should be pruned weekly, eliminating the original shoot from the axilla of each leaf and allowing only the gemma of the apex to grow. This is when the plant is grown with one stem. Some producers grow the tomato plant with two stems. In this case, in the shoot, one axial gemma is left besides the apex gemma. The chosen shoot almost always is the one that sprouts in the axilla of the leaf situated immediately below the first inflorescence (Fontes and Silva 2002).

The grower should decide how many clusters should be left on the plant. The smaller the number, the larger will be the fruits. However, this results in lower production, although the contrary has been observed, that is, plants with many clusters tend to produce more small and medium-size fruits than larger fruits. At UFV, several investigations concluded that between six and seven clusters is optimal to obtain high productivity of large fruits.

Irrigation

For tomato plants, irrigation methods that wet the leaves should not be used because the crop is highly sensitive to leaf diseases. The furrow, dripping and micro-aspersion methods should be preferred. In evaluating the different methods it should be kept in mind that drip irrigation reduces labour requirement over the crop cycle.

Integrated Pest Management of Tomato

Key pest fruit borers are tomato pinworm (*Tuta absoluta*), small tomato borer (*Neoleucinodes elegantalis*) and tomato fruitworm (*Helicoverpa zea*). Secondary pests are the vectors of viruses, including whitefly (*Bemisia tabaci*), the green peach aphid (*Myzus persicae*), potato aphid (*Macrosiphum euphorbia*) and the thrips (*Frankliniella schulzei*). Other pests are the leaf miners (*Liriomyza* spp.), tomato russet mite (*Aculops lycopersici*), two-spotted spider mite (*Tetranychus urticae*) and red-legged spider mite (*Tetranychus ludeni*).

Control Decision Making

Sampling: The plantation should be divided into uniform plots according to the planted cultivar, age of the transplant, spacing, staking and slope of the terrain. Forty samples per plot should be collected. This is determined by taking 10 points of sampling, so as to cover the whole plot, and then evaluating four sequential plants in the planting row per sampling point, thus giving 40 samples per plot. In addition, weekly samples should be taken.

In the sampling for a virus vector, a leaf is shaken from the median third of the median canopy into a white plastic tray and the number of insects are counted (i.e., thrips, green peach aphid and whitefly). In the sampling of miners the presence of active miners in the two leaves between the third apical and medium of the canopy should be evaluated. As for the pests, the fruits should be evaluated from the second and third clusters from the apex for the presence of eggs of small tomato borers, tomato pinworm and tomato fruitworm.

Levels of Control

Artificial measures of control (i.e., chemical, applied or behavioural) should be employed only when the intensity of attack of the pests is equal to or larger than the levels of control, as stated in Table 2.

Table 2 Levels of control for pests of the tomato plant.

Pest group	*Levels of control*
Fruit borers	4% of attack
Virus vectors	1 virus vector per plant
Leaf miners	10% of attack

Control Techniques

Crop Control

Various measures of crop control can be used. These include netting in the nurseries and the seedbed to avoid transmission of viruses, planting of barriers (sorghum or corn) around the plantation area, physical barriers against pests, destruction of crop residues, use of dead covering (rice rind or straw), prevention of colonization by solar light reflection, apical clipping, which facilitates the penetration of the insecticide into the foliage, reduction of the crop period so as to break the pest cycle, and management of invading plants, which serve as shelter to natural enemies.

Biological Control

Bacillus thuringiensis with vegetable oil as an adhesive is used to control caterpillars and eggs of microhymenopter parasitoid of *Trichogramma pretiosum* to control the small tomato borers, tomato pinworm and tomato fruitworm.

Behavioural Control

The synthetic sexual pheromone of the tomato moth is already available on the market and tests have been carried out for the pheromone of the small tomato borer (Heuvelink Dorais 2005).

Chemical Control

A protective suit should be worn at all times and only insecticides registered with state and federal agencies should be employed on the crop in accordance with the agronomical guidelines. Indiscriminate use of fungicides should be avoided as it has a deleterious effect on entomopathogenic fungi. Chemical control should be employed only when the intensity of attack is equal to or surpasses the level of control.

Around 0.25% of mineral oil should be added to the insecticide broth for the control of leaf miners and fruit borers. The active constituents should be rotated for the management of resistance to insecticides in accordance with the recommendations of the Resistance toward Insecticides Committee. Selective insecticides should be employed in favour of the natural enemies and should be sprayed preferentially at the end of the afternoon when the activity of the natural enemies in the crop is lower (Tomato News 2000).

TOMATO PLANT DISEASES

Root, Pith and Vascular System

Pith and root rotting may be caused by fungus and bacteria capable of infecting a large number of plants. These diseases are difficult to control and normally cause total loss of the plant. The fungi *Sclerotium* rolfsi Sacc, *Rhizoctonia solani* and *Sclerotinia sclerotiorum* are inhabitants of the soil and form sclerotia (resting bodies that stay in the soil for long periods of time). The signs of infection are normally rotting, with necrosis of the tissues at the base of the stem. Depending on the age of the lesion and the environmental conditions, it is possible to see the sclerotia (dense, dark-coloured bodies) next to the mycelium (white and cotton-like) associated with the affected part.

Species of bacteria *Erwinia,* mainly *E. carotovora* ssp. *carotovora,* cause rotting of the soft tissues in the basal portion of the stem and normally destroy the plant cortex. This feature is popularly known as "hollow stem". The bacteria penetrates the tissues of the plant through injuries. Therefore, care is recommended during weeding, pruning and other maintenance activities.

Root galls are common in tomato crop. The nematode *Meloidogyne incognita* is abundant and the most common feature is alteration in the growth of root points where the galls form. Many species of plants are hosts to *M. incognita;* thus, before planting, it is necessary to investigate the occurrence of nematodes in the area.

Vascular wilt is caused by the fungi *Fusarium oxysporum* f.sp. *lycopersici* and *Verticillium dahliae* and *V. albo-atrum* and by the bacterium *Ralstonia solanacearum.* The fungus *F. oxysporum* f.sp. *lycopersici* affects only the tomato plant, whereas *V. dahliae, V. albo-atrum* and *R. solanacearum* may affect other hosts. The main features are vascular darkening and wilting of the leaves. Wilt caused by fungi is managed by the planting of resistant species. However, commercial species resistant to bacterial wilt are still not available.

Shoot Diseases: Powdery Mildew, Late Blight, Spots and Mosaic

Powdery mildew is the most serious disease in protected plantations but can also occur in the field, especially in the dry season. The signs are discoloured areas in the leaf that may progress to chlorosis. However, the structures of the pathogen, the fungus *Oidiopsis sicula,* grow externally to the host and are easily identified. Mycelium, conidiophores and conidium form a cottony whitish growth. The colonies are most frequently observed on the adaxial face of the leaf. Sulphur-based fungicides are in many situations efficient in controlling the spread of this fungus (Fontes and Silva 2000).

Late blight is one of the most destructive diseases of the tomato plant. The causal agent, *Phytophthora infestans,* may complete its life cycle in less than 5 d. Thus, under favourable environmental conditions (mild temperatures and high humidity), the intensity of the disease increases rapidly. The disease may manifest in leaves, stems and fruits and varies according to the affected organ. In leaves, the lesions initially appear waterlogged and later become necrosed. In stems and fruits dark brown spots are observed. Late blight is controlled almost exclusively by application of fungicide.

Leaf spot may be caused by fungi or bacteria. The affected areas are necrosed and in general are well delimited. The most frequently occurring fungus-originated spots are caused by *Alternaria solani, Septoria lycopersici* and *Stemphylium solani,* that is black spot, septoriosis and stenophylla spot, respectively. Hot and rainy seasons are most favourable to the development of spots. Several fungicides are used for the control of these diseases. Some plant varieties show resistance to *S. solani.*

The most common bacterial leaf spots are bacterial spot (*Xanthomonas campestris* pv. *vesicatoria)* and bacterial speck on tomato (*Pseudomonas syringae* pv. *tomato)* caused, respectively, by *Xanthomonas campestris* pv. *vesicatoria* and *Pseudomonas syringae* pv. *tomato.* Bacterial spot is favoured by high temperatures (> 24°C), whereas bacterial speck is favoured by mild temperatures (< 24°C), both needing high humidity to develop. The pathogens may be transmitted by seeds. Thus, the control should start with the acquisition of healthy seeds. The application of cupric fungicides helps in the control of spots of bacterial origin (Fontes and Silva, 2000).

The tomato plant is affected by many viruses. The most common diseases are caused by geminivirus, tobamovirus and tospovirus. The affected plants commonly exhibit subnormal growth, leaf deformation (e.g., curling, thinning, wrinkling) and variation in the green colour of leaves (mosaic appearance). Geminivirus and tospovirus are transmitted by vectors such as the whitefly (*Bemisia tabaci*) and thrips (*Frankliniella* spp.). Tobacco mosaic virus, a tobamovirus, is transmitted mechanically from a sick plant to a healthy one. Use of resistant varieties is the main method of control for these diseases (Jones et al. 1991).

These are only some of the several important diseases that affect the crop. It is important to emphasize that the adoption of a varied set of control measures is the most efficient way to manage diseases and reduce plant loss. The following practices are recommended:

- Choose a planting area free of pathogens and rotate crops.
- Use resistant varieties.
- Use healthy and treated seeds or seedlings.
- Prepare the soil well.
- Supply nutrients in a form adequate to satisfy the plant's demands.
- Execute crop treatments carefully so as to reduce plant injury.
- Irrigate adequately and effectively (e.g., ensure adequate quality and quantity of water and frequency of irrigation).
- Use products listed for the crop, conforming to the recommendation of the manufacturer.
- Integrate a larger number of practices to reduce the intensity of diseases.

HARVESTING, CLASSIFICATION AND PACKING

Harvesting starts 85 to 125 d after seeding. It may take 30 to 70 d depending on the number of clusters per plant and the variety. At this stage, the fruits may be cut without affecting the seeds, and this stage can be recognized by the colour of the fruit, which changes from intense green to green-yellow. If the consumer marker is nearby the fruit can be harvested ripe, that is totally red, the tomato will have more taste and flavour. The harvested tomato should be kept in the shade, cleaned of pesticide residues and checked for defects. Tomatoes should be classified by diameter and each class should be packed separately.

NUTRITIONAL COMPOSITION, COLOUR, FLAVOUR AND TASTE OF TOMATO AND RELATIONSHIP TO CLIMATE

Tomatoes are composed of 93-95% water. The remaining constituents include 5-7% inorganic compounds, organic acids (citric and malic), sugars (glucose, fructose and sucrose), solids insoluble in alcohol (proteins, cellulose, pectin, polysaccharides), carotenoids and lipids (Giordano and Ribeiro 2000). Although not rich in essential nutrients for human beings, the tomato possesses many valuable properties including abundance of potassium (Grierson and Kader 1986). Potassium is important in the control of the osmotic pressure of the blood, kidney function and control of heart muscle contractions (Anderson et al. 1998).

Carotenoids are also important to humans because of their nutraceutic property. The carotenoid lycopene is responsible for the red colour of the fruit and constitutes 75-83% of the total carotenoids. The β-carotene pigment is responsible for the yellowish colour and represents 3-7% (Dorais et al. 2001). Lycopene is found in great concentrations, but only in a restricted number of vegetables. The tomato is the main source of these, containing high amounts, which, however, vary as a function of time of harvest, geographic location and plant genotype. The highest concentrations are found in wild cultivars, up to double the concentrations found in commercial cultivars (Dorais et al. 2001). In the human organism the lycopene is present in high concentrations in the blood plasma, seemingly an essential fraction, acting as a natural defence pathway and acting as an antioxidant and antimutagenic agent.

The content of β-carotene determines the activity of vitamin A, which has been cited as important in the prevention of coronary diseases and cancer (Abdulnabi et al. 1996). The concentration of β-carotene varies considerably among species, cultivars or lineages. According to Stevens and Rick (1986), the concentrations are up to 100 times as high in progeny obtained from the crossing of the cultivated species of the tomato plant *L. esculentum* and the wild species *L. hirsutum*. Besides vitamin A, the most important vitamins in the constitution of the fruit are vitamin B_1 (thiamine), B_2 (riboflavin), B_3 (pantothenic acid), B_6, niacin, folic acid, biotin, C (ascorbic acid) and E (α-tocopherol). Among these, several studies have reported the importance of vitamin C in the human diet, cited mainly for its antioxidant activity (Abdulnabi et al. 1996).

Vitamin E is present exclusively in the seeds (Tomato News 2000) and as these are eliminated in most of the processes to which the fruits are subjected its contribution to the human diet becomes irrelevant. Processing also reduces levels of the remaining vitamins.

In addition to the substances mentioned, other characteristics should be observed aiming to obtain cultivars with high acceptance in the market, for consumption in fresh as well as processed form, with emphasis on the contents of soluble solids, pH, total acidity and colour. The content of soluble solids is one of the main characteristics of the fruits. It is in this fraction that sugars and acids, the main compounds related to flavour, are found. They are also indicators of the quality of the fruits for processing. The larger the content of soluble solids, the greater will be the industrial yield. In practical terms, for each increased amount of °Brix (percentage of soluble solids as flavour indicator) in fruit, there is an increase of 20% in the industrial yield. This content, besides being a genetic characteristic of the cultivar, is influenced by the edaphoclimatic conditions of the cultivation region (Giordano and Ribeiro 2000).

Reducing sugars such as glucose and sucrose represent 50% of the dry matter content of the fruit, the remaining being proteins, pectin, cellulose, hemicelluloses, organic acids, minerals, pigments, vitamins and lipids. The concentration of sugars varies greatly as a function of the cultivar and cultivation conditions. According to Dorais et al. (2001), the concentration of sugars may vary from 1.66 to 3.99% and 3.05 to 4.65% of the fresh matter, as a function of the cultivar and cultivation conditions, respectively.

As with the sugars, the organic acids are crucial to the flavour of the fruits and characteristics related to processing. There is a continuous variation in the acidity of the fruit during its development and maturation, increasing with the growth of the fruit until reaching its maximum with the development of colouration and diminishing with the advance of maturation (Stevens 1972).

Among the acids present in the tomato fruit, the main acids are citric and malic, with citric acid being predominant over malic acid (Bertin et al. 2000). Any change in the contents of citric and/or malic acid will alter the content of titratable acid and will change the degree of acidity of the fruit, altering its flavour. It is important to emphasize that the potassium present in the fruits is positively related to the reduction of maturing disorders and to the increase in acid concentration in the fruits (Ho 1996). In general, a tomato pulp with a pH lower than 4.5 is desirable. Under these conditions the development of microorganisms harmful to the conservation of the processed products is inhibited (Ho 1986). The pH also has an important effect on the time of heating required for the sterilization of tomato byproducts. The more alkaline pH implies a greater heating time is required in processing (Tigchelaar 1986).

The colour of the fruits is an essential parameter for consumption of fresh fruit as well as for the industry. Consumers associate the colour characteristics of foods with other quality attributes such as flavour and nutritional value. The colouring of the tomato is due to chlorophylls (green pigments), carotenoids (mainly lycopene and β-carotene) and xanthophylls (yellow pigments). The green colour of the immature fruits is due to a mixture of chlorophylls, levels of which decrease with the maturing of the fruit. When the fruit begins to mature yellow (β-carotene) is produced, which becomes more apparent with the decrease in the content of chlorophyll. Afterwards there is a rapid increase in the concentration of lycopene, which gives the fruit a red colour (Ho 1986).

The influence of environmental factors such as temperature, relative air moisture and luminosity on the attributes of the composition of the fruits has been extensively investigated. Temperature may influence the distribution of photo assimilates between the fruits and the vegetative part of the plant (Heuvelink 2005). High temperatures favour the distribution of photo assimilates to the fruits at the expense of vegetative growth (De Koning 1989). Besides that, there is a positive correlation between the temperature of cultivation of the fruit and import of photo assimilates due to the increase of the drainage force of the fruit (Ho and Hewitt 1986). Similarly, the water movement to the fruit increases with temperature, as long as the availability of water in the plant is not a limiting factor. Temperatures equal to or lower than 15°C considerably reduce the absorption of water by the fruit (Dorais et al. 2001).

The influence of temperature on the metabolism is direct, affecting cellular structures and other components that determine the quality of the fruit such as colour, size, and organoleptic properties. High temperatures accelerate the development of the fruit, reduce the time necessary for its maturing and its final size (Dorais et al. 2001). The colour develops best in environments with temperatures between 12 and 21°C. Temperatures below 10°C or above 30°C inhibit the normal development of the fruit and the synthesis of lycopene. The biosynthesis of β-carotene apparently is less sensitive to temperature. As a consequence of maturing under suboptimal conditions of temperature fruits acquire an orange colour (Stevens and Rick 1986). Studies have evidenced that temperatures near 23°C improve the quality of the fruit and also increase the content of dry matter (Dorais et al. 2001).

Relative humidity of the air also affects the composition of the fruits. Under conditions of very low relative humidity (15-22%), the photosynthetic ratio is significantly reduced because of the closing of the stomata. There is also a reduction in plant growth, fruit size and total production. Under conditions of low vapour pressure deficit (high relative humidity), there is a decrease in plant transpiration and a decrease in the absorption of nutrients (Dorais et al. 2001).

Light intensity also affects the composition of tomato fruits. Of the total luminous radiation that reaches the plant in a greenhouse, about 10% is reflected, 10% transmitted and approximately 80% absorbed (Dorais et al. 2001). Of the absorbed radiation, a small proportion (5%) is used in biological reactions such as photosynthesis and the largest proportion is dissipated by transpiration or convection.

High light intensity on the plant canopy may affect the contents of soluble sugars (Davies and Hobson 1981), ascorbic acid (Giovanelli et al. 1998), pigments (lycopene) and the quantity of photo assimilates available to the fruit. High light intensity is associated with the content of vitamin C in the fruits (Venter 1977). Thus, cultivations in greenhouse exhibit differentiated vitamin C content when compared to field plantations. Low light intensity reduces the synthesis of pigments, resulting in plants with uneven colouration. Although the formation of carotenoids in mature fruits does not require induction by light, shaded fruits have lower content of carotenoids (Dorais et al. 2001).

The interception of light by the plant canopy may be altered by the density of the plant. Increase in density is positively correlated with fruit production (kg/m^2) but negatively correlated with fruit size because of insufficient supply of assimilates to the fruits (Dorais et al. 2001). The choice of density should be based on the vigour of the plant, the cultivar and the levels of light intensity.

All of these factors are influenced by genetics. That is, new genes can be used as needed to mitigate the impact of diseases, pests and environment on the composition and flavour of the tomato. Several of these genes may be found in germplasm banks.

USEFUL GENES IN GERMPLASM BANKS

A survey of the Tomato Crop Germplasm Committee (Tomato CGC) identified a series of tomato problems that could be resolved (or minimized) by genetic methods using genes from germplasm banks (Tomato CGC 2003). Fontes and Silva (2002), Jones (1999) and Jones et al. (1991) analysed problems in tomato culture. These issues can be elucidated from genetic features conserved in germplasm banks. Some of these problems are diseases (Table 3), pests (Table 4) and abiotic stresses (Table 5). The solutions to some problems lie in wild species of *Lycopersicon* (Table 6).

Several genes of interest in the improvement of the tomato plant have been identified from accessions conserved in germplasm banks ranging from those important to nutrition or human health to genes resistant to biotic and abiotic factors that limit production in the field. The

Table 3 Diseases of tomato plants.

Common name	Scientific name
Verticillium wilt race 2	*Veticillium albo atrum*
Bacterial canker	*Corynbacterium michiganense*
Late blight	*Phytophthora infestans*
Geminiviruses	Begomovirus
Powdery mildew	*Leveillula taurica*
Fruit rots	*Erwinia carotovora*
Bacterial spot	*Xanthomonas vesicatoria*
CMV	Cucumber mosaic virus
Fusarium wilt race 3	*Fusarium oxysporium* f. *lycopersici*
PVY	Potato virus Y

Table 4 Resistant pests of tomato plants.

Common name	Scientific name
Silverleaf, whitefly	*Bemisia* spp.
Potato aphid	*Macrosiphum euphorbiae*
Spider mites	*Tetranichus* spp.
Pinworm	*Tuta absoluta*
Small tomato borer	*Neoleucinodes elegantalis*
Tomato fruitworm	*Helicoverpa zea*
Green peach aphid	*Myzus persicae*
Thrips	*Frankliniella schulze*
Tomato russet mite	*Aculops lycopersici*

Table 5 Abiotic stress resistances of interest in tomato breeding from tomato germplasm banks.

Characteristics	Description
Aluminium sensitivity	Sensitivity to content of aluminium in the soil solution, determined by the degree of root suppression
High temperature fruit set	Subjective evaluation based on fruit set at high temperatures that normally inhibit set in an area of evaluation
Salt sensitivity	Sensitivity to salt concentration determined by the degree of top growth suppression

characterization of wild parents of the tomato conserved in several germplasm banks has revealed potential with respect to characteristics of interest (Table 6).

Table 6 Characteristic features of conserved wild type tomato species found in germplasm banks around the world.

Source	*Character of interest*
L. peruvianum	Resistance to several pests; rich source of vitamin C
L. pennellii	Resistance to drought; increased contents of vitamins A and C and sugars
L. pimpinellifolium	Resistance to diseases, lower acidity, intense colour, greater content of vitamins and soluble solids
L. esculentum var. *cerasiforme*	Resistance to high temperatures and humidity and to fungi that attack leaves and roots
L. chmielewskii	Intensity of fruit colour and sugar content
L. chilense	Resistance to drought
L. chesmanii	Tolerance to sea water and pedicles without articulations
L. hirsutum	Resistance to insects, spider mites, viruses and other diseases; tolerance to cold
L. parviflorum	Intensity of fruit colour and high contents of soluble solids

In the United States the accessions to the germplasm are denominated by PI or LA. Several studies have demonstrated the usefulness of some accessions in breeding programmes. Accession PI 134417 of *hirsutum* f. *glabratum* is resistant to pests such as *Tuta absoluta, Scrobipalpuloides absoluta* and *Tetranychus* sp. (Leite et al. 2001, Gonçalves et al. 1998, Giustolin and Vendramim, 1994).

Accession PI 126445 of *L. hirsutum* f. *typicum* is another important source of resistance to *T. absoluta* and *Spodoptera exigua* (Eigenbrode et al. 1994). Accessions PI 127826 of *L. hirsutum* f. *hirsutum* (Freitas et al. 1998, Campos et al. 1999) and LA 1777 of *L. hirsutum* f. *typicum* (Ecole et al. 1999) also exhibit compounds that confer resistance to the tomato plant moth. As for disease resistance, PI 126410 of *L. pimpinellifolium* presents a recessive gene that controls resistance to PVY (Nagai 1993) and PI 126445 of *L. hirsutum* presents resistance to *Septoria lycopersici*.

Studies undertaken in the germplasm bank of vegetables at UFV (BGH-UFV) revealed that accession BGH 6902 of *L. hirsutum* possesses elevated levels of resistance to *Phytophthora infestans* (Abreu 2005) and *Pepper yellow mosaic virus* (Juhász et al. 2006).

Even though germplasm of wild species have several interesting characteristics for plant improvement, many times the transference of these genes to the cultivated species *L. esculentum* presents some limitations. Thus, the search for useful genes in germplasm of cultivated species is a more strategic action, since it easier to carry out the intended transfer. As an example, LA 1421 of *L. esculentum* presents resistance to *Ralstonia solanacearum* bacteria (Mohamed et al. 1997), which although it is a quantitative characteristic, becomes easier to incorporate into commercial cultivars.

Accessions of *L. esculentum* from BGH-UFV have been characterized and some interesting characteristics are available for tomato plant breeders. Accessions BGH 4619 and BGH 4686 have elevated contents of total sugars; BGH 4350 and BGH 4474 exhibit high fruit firmness; BGH 1498 and BGH 3472 presented tolerance to *Phytophthora infestans*; BGH 4206 and BGH 4054 presented resistance to *Alternaria solani*; and BGH 1538 and BGH 225 possess resistance to *Ralstonia solanacearum* (Silva et al. 2006).

CONCLUSION

Originally used as an ornamental plant, the tomato has many applications in the food industry. Besides being a nutritious food with low calorific content the tomato is an excellent source of antioxidants and vitamins. The tomato plant originated in South America and is now cultivated in many parts of the world, under a range of temperature conditions. However, the main challenges it faces are diseases, pests and antibiotic resistance, though these may be overcome by modifying genes.

REFERENCES

Abdulnabi, A.A. and A.H. Emhemed, G.D. Hussein, and A.B. Peter. 1996. Determination of antioxidant vitamins in tomatoes. Food Chem. 60: 207-212.

Abreu, F.B. 2005. Herança da resistência a *Phytophthora infestans*, de características de frutos e seleção de genótipos resistentes na geração F5 de cruzamento interespecífico de tomateiro. PhD Thesis, Universidade Federal de Viçosa, Viçosa, Brazil.

Adegoroye, A.S. and P.A. Jolliffe. 1987. Some inhibitory effects of radiation stress on tomato fruit ripening. J. Sci. Food Agric. 39: 297-302.

Anderson, L. and M.V. Dible, and P.R. Tukki. 1998. Nutrição. Guanabara, Rio de Janeiro, Brazil.

Bertin, N. and N. Ghichard, C. Leonardi, J.J. Longuesse, D. Langlois, and B. Naves. 2000. Seasonal evolution quality of fresh glasshouse tomato under Mediterranean conditions, as affected by vapour pressure deficit and plant fruit load. Ann. Bot. 85: 741-750.

Bohs, L. and R.G. Olmstead. 1997. Phylogenetic relationships in S. (Solanaceae) base on sequences. Syst. Bot. 22: 5-17.

Bohs, L. and R.G. Olmstead. *S.* phylogeny inferred from chloroplast DNA sequence data. *In:* M. Nee, D.E. Symon, R.N. Lester and J.P. Jessop. [eds.] 1999. Solanaceae IV, Advances in Biology and Utilization. Royal Botanic Gardens, Kew, UK.

Campos, G.A. and W.R. Maluf, M.G. Cardoso, L.R. Braga. A.V. Teodoro, E.R. Guimarães, F.R.G. Benites, S.M. Azevedo, and J.T.V. Resende. 1999. Resistência de tomateiros com altos teores de zingibereno, ou 2-tridecanona oriundos de cruzamentos interespecíficos de *Lycopersicon* a ácaros do gênero *Tetranychus*. Horticultura Brasileira 17: 22.

Davies, J.N. and G.E. Robson. 1981. The constituents of tomato fruit - the influence of environment, nutrition and genotype. CRC Critical Reviews in Food Science and Nutrition. 15: 205-280.

De Koning, A.N.M. 1992. The effect of temperature on development rate and length increase of tomato, cucumber and sweet pepper. Acta Hortic. 305: 51-55.

Dorais, M. and A. Gosselin, and A.P. Papadopoulos. 2001. Greenhouse tomato fruit quality. Hortic. Rev. 26: 239-306.

Ecole, C.C. and M.C. Picanço, G.N. Jham, and R.N.C. Guedes. 1999. Variability of *Lycopersicon hirsutum* f. *typicum* and possible compounds involved in its resistance to *Tuta absoluta*. Agric. For. Entomol. 1: 249-254.

Eingenbrode, S.D. and J.T. Trumble, J.G. Milliar, and K.K. White. 1994. Topical toxicity of tomato sesquiterpenes to the beet armyworm and the role of these compounds in resistance derived from an accession of *Lycopersicon hirsutum* f. *typicum*. J. Agric. Food Chem. 42: 807-810.

Fontes, P.C.R. and D.J.H. Silva. 2002. Produção de Tomate de Mesa. Aprenda Fácil. Viçosa, Brazil.

Freitas, J.A. and M.G. Cardoso, E.R. Maluf, C.D. Santos, D.L. Nelson, J.T. Costa, E.C. Souza, and L. Spada. 1998. Identificação do sesquiterpeno zingibereno, aleloquímico responsável pela resistência à *Tuta absoluta* (Meyrick, 1917) na cultura do tomateiro. Ciência Agrotecnologia 22: 483-489.

Giordano, L.B. and J.B.C. Silva. 2000. Tomate para processamento industrial. Embrapa comunicação para transferência de tecnologia, Brasília, Distrito Federal, Brazil.

Giovanelli, G. and V. Lavelli, C. Peri, and L. Guidi. 1998. Variation of antioxidante content in tomato during ripening. Proc. Tomato and Health Seminar. Pamplona, Spain, May 25-28, 122-130.

Giustolin, T.A. and J.D. Vendramin. 1994. Efeito dos aleloquímicos 2-tridecanona e 2-undecanona na biologia de *Tuta absoluta* (Meyrick). Anais da Sociedade Entomológica do Brasil 25: 417-422.

Gonçalves, M.I.F. and W.R. Maluf, L.A.A. Gomes, and L.V. Barbosa. 1998. Variation of 2-tridecanona level in tomato plant leaflets and resistance to two mite species (*Tetranychus* sp.). Euphytica 104: 33-38.

Grierson, D. and A.A. Kader. Fruit ripening and quality. pp. 241-280. *In:* J.G. Atherton and J. Rudich. [eds.] 1986. The Tomato Crop: A Scientific Basis for Improvement. Chapman and Hall, London, UK.

Heuvelink, E. and M. Dorais. Crop growth and yield. pp. 85-145. *In:* E. Heuvelink. [ed.] 2005. Tomatoes. Cabi Publishing, Cambridge, USA.

Ho, L.C. 1996. The mechanism of assimilate partitioning and carbohydrate compartmentation in fruit in relation to the quality and yield of tomato. J. Exp. Bot. 47: 1239-1243.

Ho, L.C. and D.J. Hewitt. Fruit development. pp. 201-240. *In:* J.G. Atherton and J. Rudich. [eds.] 1986. The Tomato Crop: A Scientific Basis for Improvement. Chapman and Hall, London, UK.

Jones, J.B. 1999. Tomato Plant Culture: In the Field, Greenhouse and Home Garden. CRC Press, Boca Raton, Florida, USA.

Jones, J.B. and J.P. Jones, R.E. Stall, and T.A. Zitter. [eds.] 1991. Compendium of Tomato Diseases. APS Press, Saint Paul, Minnesota, USA.

Jones, P. and L.H. Allen, J.W. Jones, and R. Valle. 1985. Photosynthesis and transpiration responses of soybean canopies to short- and long-term CO_2 treatments. Agron. J. 77: 119-126.

Juhász, A.C.P. and D.J.H. Silva, F.M. Zerbini, B.O. Soares, and G.A.H. Aguilera. 2006. Screening of *Lycopersicon* sp. accessions for resistance to *Pepper yellow mosaic virus*. Scientia Agricola 63: 510-512.

Knapp, S. and D.M. Spooner. 1999. A new name for a common Ecuatorian and Peruvian wild tomato species. Novon, 9: 375-376.

Leite, G.L.D. and M.C. Picanço, R.N.C. Guedes, and J.C. Zanuncio. 2001. Role of plant age in the resistance of *Lycopersicon hirsutum* f. *glabratum* to the tomato leafminer *Tuta absoluta* (Lepidoptera: Gelechiidae). Scientia Horticulturae 89: 103-113.

Marim, B.G. and D.J.H. Silva, M.A. Guimaraes, and G. Belfort. 2005. Sistemas de Tutoramento e condução do tomateiro visando produção de frutos para consumo *in natura*. Horticultura Brasileira 23: 951-955.

Mohamed, M.E.S. and P. Umaharan, and R.H. Phelps. 1997. Genetic nature of bacterial wilt resistance in tomato (*Lycopersicon esculentum* Mill.) accession LA 1421. Euphytica 96: 323-326.

Nagai, H. Tomate. pp. 301-313. *In:* A.M.C. Furlani and G.P. Viégas. [eds.] 1993. Melhoramento de plantas no Instituto Agronômico. Campinas: IAC editora, São Paulo, Brazil.

Olmstead, R.G. and J.D. Palmer. 1997. Implications for phylogeny, classification, and biogeography of *S.* from cpDNA restriction site variation. Syst. Bot. 22: 19-29.

Olmstead, R.G. and J.A. Sweere, R.E. Spangler, L. Bohs, and J.D. Palmer. Phylogeny and provisional classification of the Solanaceae based on chloroplast DNA. pp. 111-137. *In:* M. Nee, D.E. Symon, R.N. Lester and J.P. Jessop. [eds.] 1999. Solanaceae IV, Advances in Biology and Utilization. Royal Botanic Gardens, Kew, UK.

Peralta, I.E.D. and M. Spooner. 2000. Classification of wild tomatoes: a review. Kurtziana 28: 45-54.

Silva, D.J.H. and M.C.C.L. Moura, B.G. Marim, F.B. Abreu, G.R. Moreira, A.C.P Juhász, A.P. Mattedi, N.B. Ribeiro, J.G. Aguilera, and M.P. Flores. 2006. Banco de Germoplasma de hortaliças - BGH - UFV caracterização, avaliação e pré-melhoramento. Magistra 18: 30-33.

Spooner, D.M. and G.J. Anderson, and R.K. Jansen. 1993. Chloroplast DNA evidence for the interrelationships of tomatoes, potatoes, and cucumber (Solanaceae). Am. J. Bot. 80: 676-688.

Spooner, D.M. and I. Peralta, and S. Knapp. 2005. Comparison of AFLPs with other markers for phylogenetic inference in wild tomatoes Solanum L. section Lycopersicon (Mill.) Westtst. Taxon 54: 43-61.

Stevens, A.M. and C.M. Rick. Genetics and breeding. pp. 35-110. *In:* J.G. Atherton and J. Rudich. [eds.] 1986. The Tomato Crop: A Scientific Basis for Improvement. Chapman and Hall, London, UK.

Stevens, M.A. 1972. Citrate malate concentrations in tomato fruits: genetic control and maturational effects. J. Am. Soc. Hortic. Sci. 97: 655-658.

Tigchelaar, E.C. Tomato breeding. pp 381-422. *In:* M.J. Basset. [ed.] 1986. Breeding Vegetable Crops. AVI Publishing Comp/Westport, Connecticut, USA.

Tomato-News. A European Commission Concerted Action. Web site: http://www.tomato-news.com, October, 2000.

Tomato Crop Germplasm Committee Report (Tomato CGC). 2003. USDA National Plant Germplasm System, Washington, DC, USA.

Venter, F. 1977. Solar radiation and vitamin C content of tomato fruits. Acta Hortic. 58: 121-127.

2

Cultivar and Agricultural Management on Lycopene and Vitamin C Contents in Tomato Fruits

Tiequan Zhang[1], John Shi[2], YutaoWang[1] and Sophia Jun Xue[2]

[1]Greenhouse and Processing Crops Research Center, Agriculture and Agri-Food Canada, Harrow, Ontario N0R 1G0, Canada

[2]Guelph Food Research Center, Agriculture and Agri-Food Canada, Guelph, Ontario N1G 5C9, Canada

ABSTRACT

Lycopene, constituting 80-90% of the total carotenoid content present in tomatoes and tomato products, has been believed to contribute to the reduced risks of some types of cancers. Vitamin C of tomato fruits accounts for up to 40% of the recommended dietary allowance for human beings. As a result, enhancing the levels of these two health chemicals in tomato fruits may form an efficient way to improve human health conditions. In response to this opportunity, numerous investigations have been conducted to identify the factors influencing the contents of lycopene and vitamin C in tomatoes. The results demonstrate consistent differences in lycopene and vitamin C content between tomato cultivars, which can be magnified by agricultural management. A relationship has been established associating electrical conductivity (EC) and light intensity with lycopene and vitamin C content in tomato fruits. Generally, moderate EC growing conditions enhance tomato health quality; solar radiation is favourable to lycopene and vitamin C accumulation, whereas strongly intense light exposure inhibits lycopene synthesis. Temperatures beyond the optimum temperature range may inhibit lycopene biosynthesis. However, the effects of temperature on vitamin C content are not always conclusive. The effects of nutrients (N, P, K, and Ca) and water availability have also been reviewed, but results are sometimes contradictory. Up-to-date studies dealing with cultivar and agricultural management on lycopene and vitamin C contents in tomato fruits are reviewed in this chapter.

INTRODUCTION

Tomato, *Lycopersicon esculentum,* is among the ten most important fruits and vegetables in terms of consumption, with an estimated 124.4 million tons of tomato fruits produced every year all over the world (Maul 1999, FAO 2004). Tomatoes are important not only because of the large amount consumed, but also because of their high health and nutritional contributions to humans. A survey conducted by the University of California at Davis ranked tomatoes as the single most important fruit or vegetable of western diets in terms of overall source of vitamins and minerals (Petro-Turza 1986). Most important, tomato consumption has been shown to reduce the risks of cardiovascular disease and certain types of cancer, such as cancers of prostate, lung, and stomach (Giovannucci 1999, Canene-Adams et al. 2005). The benefits of tomatoes and tomato products have been attributed mostly to the significant amount of lycopene contained, which constitutes 80-90% of the total carotenoid content present in tomatoes and tomato products (Table 1) and appears to be the most efficient quencher *in vitro* of singlet oxygen and free radicals among the common carotenoids, in addition to their role in fruit colour (Dorais et al. 2001). Offord (1998) reported that fresh tomatoes are an important source of vitamin C (220 kg^{-1}), providing up to 40% of the recommended dietary allowance. Increasing the levels of dietary lycopene and vitamin C through the consumption of fresh tomatoes and tomato products has been recommended by many health experts. In particular, trend surveys in North America have suggested over the past decade that many people have strong interests in diet and nutrition to improve their

Table 1 The contribution of carotenoid species in tomato fruits.

Carotenoid species	*Composition (%)*
Lycopene	80-90
α-carotene	0.03
β-carotene	3-5
γ-carotene	1-1.3
ξ-carotene	1-2
Phytoene	5.6-10
Phytofluene	2.5-3.0
Neurosporene	7-9
Lutein	0.011-1.1

Source: Shi and Le Maguer 2000.

health and are actively making food choices (American Diabetes Association 2002). As a result, enhancing the health value of tomatoes and tomato products may form an efficient way to improve health conditions and to reduce the risk of cancer, which is becoming an increasing burden upon society.

To satisfy this demand, people should be aware of the possibility and potential of improving the health value of tomato fruits. That is, basically, we must understand the factors and their effects on the contents of the health-promoting chemicals present in tomatoes and tomato products. For the present, in order to maximize the content of lycopene and vitamin C in tomato fruits, numerous investigations have been conducted to evaluate the influences of genotype, agricultural practices, and the environment. Here we review the effects of cultivars and some major pre-harvest factors on contents of lycopene and vitamin C, two important healthful and nutritional compounds in tomato fruits.

CULTIVARS

Tomatoes include several hundred cultivars and hybrids, developed in response to consumer demands, but most cultivars have been bred for characteristics such as consistency in colour, shape and size, long shelf-life, high yield, uniform ripening, and disease resistance, while their nutritional and flavour values have been neglected. As interest in the high health values of tomatoes and tomato products rises among consumers, insight into the effects of cultivars on the health and nutritional quality of tomatoes would not only aid in the selection of high added-health-value cultivars for growers, but also provide breeders with important information to develop new tomato cultivars for superior health qualities.

Lycopene

Existing evidence shows that lycopene content varies significantly among tomato cultivars. Kuti and Konuru (2005) investigated 40 tomato varieties, including cluster F1 hybrid tomatoes, round breeding line tomatoes, and cherry tomato types, and found that lycopene content ranged from 4.3 to 116.7 mg kg^{-1} on a fresh weight basis, with cherry tomato types having the highest lycopene content (Table 2). Martínez-Valverde et al. (2002) observed that lycopene content ranged from 18.60 to 64.98 mg kg^{-1} fresh weight among nine commercial varieties of tomatoes produced in Spain. George et al. (2004) found that the extent of variation arising from different cultivars in lycopene content was 48 to 141 mg kg^{-1} in peels and 20 to

Table 2 Lycopene contents in raw red-ripe fruits of 40 tomato varieties grown under greenhouse and field environments.

Varieties		*Lycopene content (mg kg^{-1} fresh weight)*	
		Greenhouse	*Field*
Round tomato type	Golden Jubilee	30.4	23.3
	Early Cascade	29.6	20.2
	First Lady	37.2	26.5
	Better Boy	19.9	14
	Early Girl	23.9	20.2
	Super Steak	8.4	6.2
	Early Pick	18	12.6
	Fantastic	35.1	16.2
	Monte Carlo	8.3	6.1
	Dona	5.7	4.3
	Italian Beefsteak	30.6	29.6
	Terrific	23.4	11.7
	Mar Globe Select	26.1	21.3
	Stupice	45.9	27.5
	Druzba	29.7	23.6
	Red Brandywine	37.3	17.6
	Miracle Sweet	30.6	28.8
	Sweet Cluster	34.2	16.2
	Big Beef	27.4	13.5
	Saint Pierre	39.6	27.2
	Boxcar Willie	31.5	19.8
	Goliath Bush	23.8	20.5
	Red Plum	47.8	31.5
	Husky Red	33.6	10.8
	Bonny Best	26.4	16.2
	Polish Giant	9.7	5.3
	Keepsake	27.2	18.3
	Kada	35.1	19.9
	Sun Master	27.8	20.1
Cluster tomato type	ALMA-01	34.4	27.9
	GS-111	21.9	15.3
	GS-114	33.3	12.6
	TMT-510	30.6	23.4
	TMT-521	35.4	21.6
	TMT-555	29.6	14.3
Cherry tomato type	Juliet Hybrid	54.7	94.5
	Gardener's Delight	48.9	73.8
	Sugar Lump	63.6	116.7
	Sun Cherry	57.3	82.5

Source: Kuti and Konuru 2005.

69 mg kg^{-1} in pulp, on a fresh weight base, and that cherry tomatoes usually contain higher amounts of lycopene. Similar variations ranging from 50 to 110 mg kg^{-1} of lycopene content in tomatoes have also been reported by Abushita et al. (2000) in Hungarian varieties. Additionally, a variation of 26 to 63 mg kg^{-1} has been reported by Rao and Yadav (1988) and Thakur and Lal Kaushal (1995) for Indian varieties. On average, the deep-red varieties contain more than 50 mg lycopene kg^{-1} fresh weight, whereas the yellow varieties contain about 5 mg kg^{-1} (Hart and Scott, 1995). Tomato cultivars that contain the Crimson gene are usually found to have a higher lycopene content (50.86 to 57.86 mg kg^{-1} fresh weight) than those lacking the gene (26.22 to 43.18 mg kg^{-1} fresh weight) (Thompson et al. 2000). The Crimson gene may open up the lycopene formation pathway, thereby producing tomatoes with higher lycopene content (Thompson et al. 1964). However, the expression may be tempered by climate, as lycopene synthesis is favoured at temperatures between 16 and 21°C and inhibited at temperatures above 30°C (Leoni, 1999). On the contrary, when examining tomatoes of several varieties grown for processing in the European Union, no such variations in the carotene concentration were observed (Dumas et al. 2003).

The potential of lycopene to control cancers has already driven tomato breeders at the University of Florida to produce high lycopene cultivars. *Lycopersicon esculentum*'s wild relative, *L. pimpinellifolium,* also known as the currant tomato, produces tiny fruits that contain over 40 times more lycopene than domesticated tomatoes. Since hybrids between the two are relatively simple to achieve, this source of genetic diversity is open for exploitation and will most likely become a sought-after hybrid trait (Cox 2000). Since cherry tomatoes often contain significantly higher amounts of lycopene (Kuti and Konuru 2005, George et al. 2004), it has been suggested that they may be useful varieties for improvement of nutritional and health benefits in tomato breeding programmes (Tigchelaar 1986). Recently, breeding programmes have paid close attention to the cultivars that are referred to as 'hp', since hybrids have shown a significant increment for lycopene content from 2.4- to 7-fold higher than those of commercial varieties (Fig. 1).

Vitamin C

Large variations in vitamin C levels have been reported among tomato cultivars and species. Abushita et al. (1997) found that the vitamin C content of tomatoes cultivated in Hungary ranged from 210 to 480 mg kg^{-1}, which is significantly higher than the values of 61.3 to 175.6 mg kg^{-1} reported by Bajaj et al. (1990) after analysing 35 varieties. George et al.

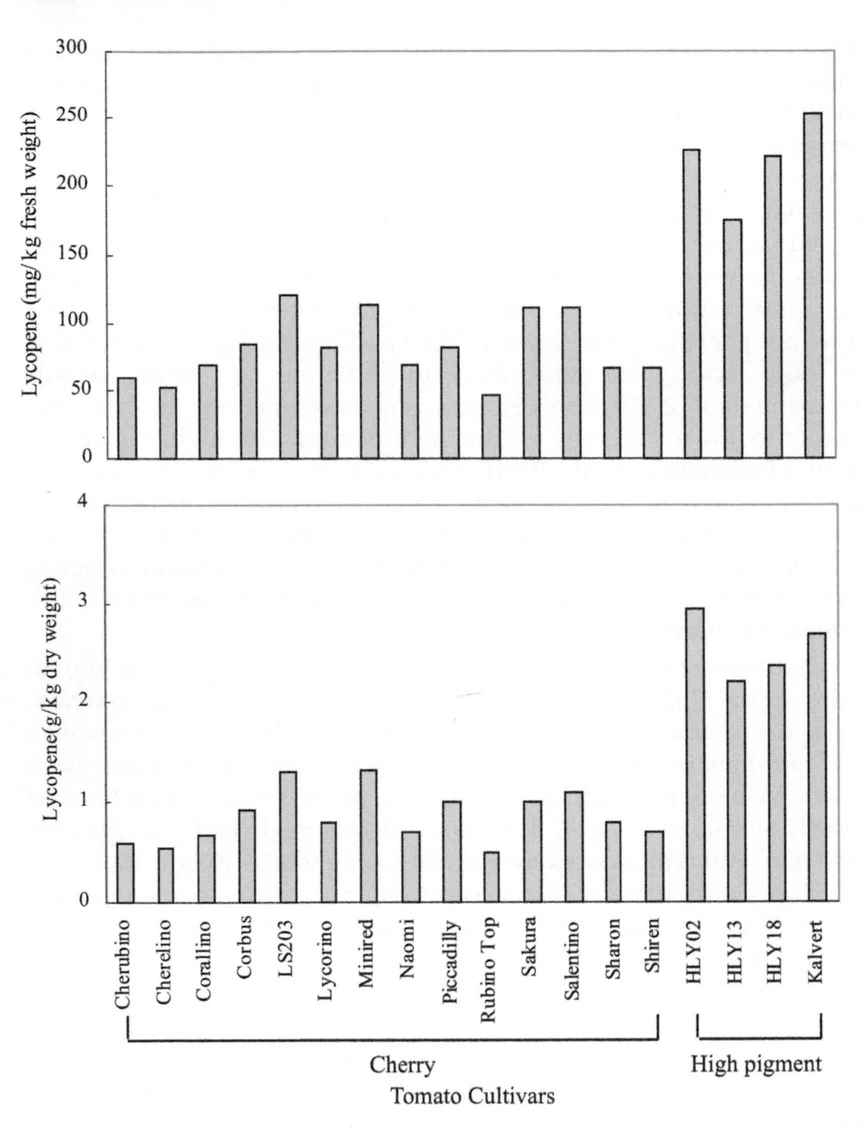

Fig. 1. Content of lycopene in various cultivars of cherry and high pigment tomatoes. Data are means ± standard deviation of three replicates (modified from Lenucci et al. 2006).

(2004) also observed significant differences between vitamin C contents among 13 cultivars, with values ranging from 84 to 324 mg kg^{-1} fresh weight in tomato pulp, and from 90 to 560 mg kg^{-1} fresh weight in tomato peel. Maclinn et al. (1937) found that vitamin C levels varied from 130 to 440 mg kg^{-1} among 98 cultivars. Particularly, Lincoln et al. (1943) found vitamin C concentrations ranging from 80 to 1190 mg kg^{-1} among species, cultivars, and strains. Wild species *Lycopersicon peruvianum* contains far higher level of vitamin C (1190 mg kg^{-1} fresh weight) than cultivated varieties (Stevens and Rick 1986). It has also been noted that cherry tomatoes often contain higher levels of vitamin C than other varieties (George et al. 2004, Causse et al. 2003), as shown in Table 3.

Table 3 Variations of vitamin C content in tomato as a function of genotype.

Varieties	*Vitamin C content (mg 100 g^{-1})*			
	Peel		*Pulp*	
	fwb	*dwb*	*fwb*	*dwb*
818 (cherry tomato)	56.0	454	32.4	383
BR-124 (cherry tomato)	40.6	362	29.0	455
T56 (cherry tomato)	50.8	462	28.6	413
DT-2	25.9	256	15.2	365
5656	20.1	196	13.4	270
7711	17.2	169	12.7	308
Rasmi	42.0	385	14.8	349
Pusa Gaurav	13.4	141	9.2	195
DTH-7	24.0	315	10.7	224
FA-180	45.5	415	9.4	171
FA-574	9.4	104	8.4	211
R-144	8.6	262	30.4	291

fwb: fresh weight basis; dwb: dry weight basis.
Source: George et al. 2004.

In contrast, Hamner et al. (1945) stated that the variations in the cultivated tomato fruit vitamin C content due to the variety are fairly small in comparison with those resulting from the growth conditions. Similarly, Abushita et al. (2000) found no great differences in vitamin C content between cultivars for fresh consumption (salad tomatoes) and those for processing. In fact, the variations in the vitamin C content of cultivated tomatoes during a long period of 40 years were remarkably small, considering the differing environmental factors and cultivars, though there was some indication that the vitamin C content of tomatoes has increased slightly since the 1950s (Davies and Hobson 1981).

Berger et al. (1966) indicated that the characteristic of high vitamin C level is an inherited one and can be transferred into cultivated tomato varieties. They noted that the level of vitamin C found in the fruits of the wild, green-fruited tomato species, *Lycopersicum minutum,* was much higher than that in *L. esculentum.* However, some lines derived from the interspecies cross *L. esculentum* × *L. minutum* showed vitamin C levels as high as 430 mg kg^{-1}, while standard tomato strains had 170 mg kg^{-1}. Gould (1992), in his recommendations for breeding varieties for processing purposes, suggested the need for developing varieties that have vitamin C in excess of 200 mg kg^{-1}. However, in spite of considerable effort to develop cultivars with higher vitamin C levels, few successful cultivars have been released because of an apparent relationship between high vitamin C levels and relatively poor yields (Stevens 1986, Dumas et al. 2003).

AGRICULTURAL MANAGEMENT

Nutrients

Lycopene

Nitrogen (N) availability is closely linked to carotenoid biosynthesis in higher plants (Weston and Barth 1997). However, the effects of N application on lycopene content in tomato fruits have been found to be inconsistent. Dumas et al. (2003) reported that lycopene content of tomato grown in nutrient solution tended to decrease with increasing N levels. Meanwhile, in tomatoes grown in pots filled with peaty loam soil receiving N at a rate up to 600 kg N ha^{-1} the fruit lycopene content increased with an increase in N rate by an average of 30% (Montagu and Goh 1990). Obviously, more studies are needed to investigate the effects of N supply on lycopene contents in tomatoes under various growing conditions.

As for phosphorus (P), Saito and Kano (1970) observed that, under hydroponic growth conditions, increasing the P supply from 0 to 100 mg P L^{-1} in the nutrient solution greatly increased the lycopene content of tomatoes. In contrast, Oke et al. (2005) observed that the lycopene levels did not differ greatly in response to phosphorus fertilization during different seasons. In fact, P application did not seem to affect the contents of many parameters of tomato quality such as organic acids, sugars, vitamin C, or volatile compounds (Oke et al. 2005).

Current evidence shows that potassium (K) application increases lycopene contents in tomato fruits. Trudel and Ozbun (1971) reported a 40% increase in lycopene concentration when the level of K in the nutrient solution was increased from 0 to 8 mM. Similarly, Fanasca et al. (2006) observed that a high proportion of K in the nutrient solution increased lycopene contents of tomatoes. Moreover, they noted a positive correlation between fruit tissue K and lycopene content, further confirming that K supply may enhance lycopene synthesis in tomatoes. Potassium may be involved in the process of lycopene biosynthesis through its influence on the activity of enzymes that regulate carbohydrate metabolism, such as pyruvate kinase and phosphofructokinase, as well as precursors of IPP (pyruvate and glyceraldehydes 3-phosphate).

The negative effects of calcium (Ca) addition on the lycopene contents of tomatoes were recorded. Fanasca et al. (2006) observed a decrease in lycopene content when the nutrient solution with a high proportion of Ca was used to grow tomatoes, and a negative correlation was recognized between fruit tissue Ca and lycopene content. It was also reported that the lycopene levels in the tomato fruits decreased with increasing Ca concentration from 0.2 to 20 mM in nutrient solution (Paiva et al. 1998). This is possibly due to the simultaneous decrease in K absorption that occurred because of cationic competition, or a possible Ca influence on ethylene biosynthesis (Fanasca et al. 2006). However, in another study, the fruit lycopene content significantly increased from a level of 85 mg kg^{-1} to levels of 263, 300, and 340 mg kg^{-1} on a fresh weight basis, in the case of sprays with 0.1, 0.2, and 0.3% Ca concentrations, respectively (Subbiah and Perumal, 1990).

Vitamin C

Numerous studies have shown that increasing the N rate tends to decrease the vitamin C content of the tomato fruits, though the influence of mineral nutrition on vitamin C content had first been thought to be relatively minor (Hamner et al. 1945). For tomatoes grown in the field, it was demonstrated that increasing N fertilization reduced the fruit vitamin C concentration (Kaniszewski et al. 1987). Giovanelli et al. (1998) found that the vitamin C accumulation under N deficiency tended to decrease linearly with N application (Montagu and Goh 1990). In fact, high N concentration in the nutrient solution favours plant leaf area development, thereby decreasing light penetration in the canopy and fruit vitamin C content (Locascio et al. 1984). As a result, it can be expected that the negative effects of N application on vitamin C contents exist in other vegetables such as potatoes (Weston and Barth 1997). The N source provided to the plants also plays a major role in determining the levels of lycopene in tomatoes. For example, N supply in the form of urea had almost no effect on vitamin C content in the tomato fruit compared to that in the NO_3 form (Kowalska and Sady 1996). Toor et al. (2006a) observed that vitamin C contents in tomatoes from organic or ammonium-fertilized treatments were higher than in tomatoes from nitrate-fertilized treatments (Table 4). Additionally, it was noted that the real effects of N application on the vitamin C of tomato fruits were possibly associated with other agricultural management practices, such as fertilization time and light

Table 4. Effects of N sources on vitamin C of tomato fruits.

Treatment	*Vitamin C (mg 100 g^{-1} fresh weight)*
NO_3^- dominant	97.5
NH_4^+ dominant + Cl^-	120.2
NH_4^+ dominant + SO_4^{2-}	128.8
Chicken manure	147.6
Grass-clover mulch	148.8

All treatments were supplied in the equivalent levels of total N, P and K; for the NO_3^- dominant treatments, N was supplied in the NO_3^-:NH_4^+ ratio 4:1; for the NH_4^+ dominant treatments, N was supplied in the NO_3^-:NH_4^+ ratio 1:4; both Cl^- and SO_4^{2-} were designed to balance the positive/negative ions of the N form.

Source: Toor et al. 2006a.

intensity. Somers et al. (1951) found that N-NO_3 supplied to tomatoes for one month prior to the onset of ripening reduced the fruit vitamin C content; however, NO_3-N supplied after the onset of ripening had no effect on the vitamin C content. Though Murneek et al. (1954) also observed a decrease in the tomato fruit vitamin C content with increasing rates of N application, the differences may be slight when the tomato crop is grown under a prevalently low light intensity but considerable in the presence of bright sunlight.

Some studies have been conducted to investigate the effects of other nutrients on vitamin C content of tomato fruits. Oke et al. (2005) did not observe a significant difference in the vitamin C levels of tomato fruits in response to P fertilization, which was in accordance with the finding that increasing the P supply in hydroponics did not significantly affect the vitamin C content of fruit (Saito and Kano 1970). However, increasing both P and N application (up to 140 kg P ha^{-1} and 150 kg N ha^{-1}, respectively) significantly increased the vitamin C content of tomato fruits (Dumas et al. 2003).

Calcium application markedly increased the concentration of vitamin C in tomato fruits. For example, Premuzic et al. (1998) noted a positive correlation between Ca and vitamin C levels in tomato fruits obtained in various growth environments. Subbiah and Perumal (1990) found a significant increase in the tomato fruit vitamin C content upon applying Ca sprays in a soil-pot experiment. Dong et al. (2004) suggested that spraying Ca solution on the leaves during anthesis greatly enhanced vitamin C synthesis in tomato fruits, increasing it by 50% compared with the control. However, Ca spray on the 3-wk-old fruit did not significantly affect vitamin C content. As for K, Weston and Barth (1997) reported that optimal levels of K elicit a favourable response in vitamin C content. However, little further information is available about the effects of K application on the vitamin C content of tomato fruits.

Electrical Conductivity (EC)

Lycopene

It is widely believed that fruits from tomato plants grown under moderately saline conditions are of higher quality. Most important, this higher EC growing condition resulted in significant increases in lycopene concentration. For example, Krauss et al. (2006) found that the lycopene content of tomato fruits reached 58 mg kg^{-1} for EC 10 dS m^{-1} from 43 mg kg^{-1} for EC 3 dS m^{-1} on a fresh weight basis. De Pascale et al. (2001) observed that the lycopene content of the tomato fruits expressed on both fresh- and dry-weight basis gradually increased from the non-sanitized control to the 4.4 dS m^{-1} treatment and they decreased at an EC of the irrigation water higher than 4.4 dS m^{-1} (Table 5). Similarly, Wu et al. (2004) showed that the lycopene concentration of tomato fruits was enhanced significantly without reducing the total yield by growing plants under a moderate salt stress provided by high salt concentration in the nutrient solution. Moreover, the increase in total lycopene in the tomato fruits was shown to be cultivar specific under high EC conditions, varying from 34 to 85% (Kubota et al. 2006). Increases in EC can be reached not only by adding NaCl to a nutritional solution, but also by applying water stress caused by limited irrigation, that is, by increasing the strength of the nutrient concentration, in order to enhance lycopene concentration of tomato fruits (Krauss et al. 2006, Matsuzoe et al. 1998, Zushi and Matsuzoe 1998).

The exact mechanisms underlying the enhancement of lycopene concentration under high EC stress are not clear, but evidence suggests that ethylene synthesis triggered by water and salt stress could be central to the increase in lycopene deposition within the flesh of the tomatoes (Kubota et al. 2006). Giuliano et al. (1993) indicated that the stress-induced up-regulation of the genes encoding for the enzymes involved in the key steps for lycopene biosynthesis may explain the increased lycopene level

Table 5. Effects of EC levels on lycopene of tomato fruits.

EC level (dS m^{-1})	*Lycopene*	
	(mg 100 g^{-1} fresh weight)	*(mg/fruit)*
0.5	5.89	6.11
2.3	8.31	8.15
4.4	10.22	9.15
8.5	7.88	6.55
15.7	6.71	4.22

Source: De Pascale et al. 2001.

detected in the salt-grown tomatoes. However, at excessive salinity, inhibition effects may take over, resulting in reduced lycopene. This was probably due to a high temperature-induced inhibition of lycopene biosynthesis in tomatoes exposed to high solar radiation arising from smaller leaf area and consequently more fruits directly exposed to sunlight in salt-stressed plants (De Pascale et al. 2001).

Vitamin C

Increased EC leads to higher contents of vitamin C in tomato fruits. Krauss et al. (2006) found that the vitamin C content in tomatoes grown at EC of 10 dS m^{-1} was 101 mg kg^{-1} fresh weight, whereas it was just 97 mg kg^{-1} fresh weight at EC of 3 dS m^{-1}. De Pascale et al. (2001) observed that the vitamin C content of the fruits increased with salinity and it was 60% higher in tomatoes grown at EC of 15.7 dS m^{-1}, compared with non-sanitized controls. Moreover, they noted that the rate of vitamin C increase with salinity was higher at low or moderate salinity and when ammonium was used as N source.

Temperature

Lycopene

Tomatoes exposed to direct sunlight in the field often develop a poor colour, mainly because fruits exposed to high temperatures have lower lycopene contents (Dumas et al. 2003). Davies and Hobson (1981) demonstrated that temperatures between 16 and 21°C favour lycopene synthesis, whereas temperatures above 30°C inhibit lycopene synthesis, and the fruits remained yellow rather than turning red. A similar phenomenon was found in other studies. Chang et al. (1977) reported that the concentration of lycopene was much lower in fruit ripened at 32°C than in fruit ripened at 21°C. Robertson et al. (1995) reported that a temperature between 18 and 26°C was optimum for lycopene production. On the other hand, in fruit cultured *in vitro* at 16°C, lycopene concentration was 1.7 times that of the 26°C treatment (Ishida, 1998, Ishida et al. 1998). They suggested that low temperature (16°C), which involves tomato *IAG1* gene activation, leads to an elevated lycopene production and may, in part, explain this high concentration.

When seasonal variations in lycopene contents of tomatoes were evaluated, Toor et al. (2006b) found that increase in temperature over the summer months had a negative effect on lycopene content in tomato fruits. Raffo et al. (2006) also observed that the hot temperatures of mid-summer in the Mediterranean basin may produce a significantly negative effect on lycopene accumulation in tomato fruits. Cherry tomatoes grown in the greenhouse have lower lycopene contents in fruits because of temperatures over 32°C in most cases (Kuti and Konuru 2005). As a result, it appears that some additional cooling of the greenhouse during the peak hours of solar radiation in summer months may help to increase the lycopene content of tomatoes.

Vitamin C

Raffo et al. (2006) investigated seasonal variations in antioxidant components of cherry tomatoes and found that high temperatures may contribute to the relatively high levels of vitamin C observed in fruits harvested in June and July. However, Murneek et al. (1954) suggested that fruit ripening occurring at relatively high temperatures, whether on or off the plant, along with low light intensity levels, probably leads to a decrease in the vitamin C content due to oxidation. Most interestingly, seasonal variations in vitamin C content observed in greenhouse-grown tomatoes at the mature-green stage may be a result of temperature variations under greenhouse conditions (Liptay et al. 1986). In general, limited information is available on the effects of temperature on vitamin C content of tomato fruits.

Light

Lycopene

Concentration of lycopene in tomatoes was reportedly altered by light quantity. Toor et al. (2006a,b) found that solar radiation had slightly positive effects. McCollum (1954) reported that, at lower temperatures, the rate of development of lycopene can be increased by greater illumination of the tomatoes. However, lycopene synthesis was inhibited when green fruits were exposed to 650 W m^{-2} for 1.5 to 4 hr, which indicated that lycopene synthesis is severely inhibited by exposure to intense solar radiation. It has been suggested that radiation injury to tomato fruit might be due to the general effects of overheating on irradiated tissues (Adegoroye and Jolliffe 1987, Dumas et al. 2003). Meanwhile, light quality also affects lycopene synthesis in tomato fruits. Red light and its intensity have a positive effect on carotenoid synthesis in detached mature-green fruit, and this effect was not temperature dependent (Thomas and Jen 1975). Moreover, Alba et al. (2000) found that brief red-light treatment of harvested mature-green fruit stimulated lycopene accumulation 2.3-fold during fruit development. This red-light-induced lycopene accumulation was reversed by subsequent treatment with far-red light. Establishing light-induced accumulation of lycopene in tomato is regulated by fruit-localized phytochromes.

Vitamin C

Light exposure is favourable to vitamin C accumulation (Dumas et al. 2003, Lee and Kader 2000). Hamner et al. (1945) observed that the very large variations in vitamin C content observed in tomatoes may be attributable to the light intensity a few days previous to harvest. In general, the lower the light intensity, the lower the content of vitamin C in plant tissues (Kader 1987). In addition, vitamin C concentration also generally increases with increased exposure to light, particularly in leafy greens (Nagy and Wardowski 1988, Shewfelt 1990). Thus, it can be expected that shading (by leaves or artificial covers) decreases the vitamin C content (El-gizawi et al. 1993, Venter 1977). Greenhouse-grown tomatoes were usually found to have lower vitamin C levels than those grown outdoors (Lopez-Andreu et al. 1986).

Water Availability

Lycopene

Irrigation is also believed to be an important pre-harvest factor influencing the biosynthesis of antioxidant compounds. However, the results of the effects of irrigation on lycopene are not conclusive. De Pascale et al. (2001) observed significant increases in lycopene contents when tomatoes were irrigated with moderately saline solution combined with different N fertilizer treatments. However, Naphade (1993) found that moisture stress reduced lycopene content in some tomato varieties, while increasing lycopene as well as β-carotene content in others. Irrigation methods also affect lycopene content in tomatoes. Kadam and Sahane (2002) found that tomatoes grown under drip irrigation recorded 3.2 mg kg^{-1} lycopene in fruits, a result significantly superior to that achieved with surface irrigation (2.6 mg kg^{-1}).

Vitamin C

Dumas et al. (2003) reviewed the effects of water availability on vitamin C content in tomato fruits and indicated that water shortages tend to increase the fruit's vitamin C contents. Lumpkin (2005) suggested that deficient irrigation often improved tomato fruit's nutritional values by a concentration effect when the fruit's water content was reduced, accompanying decreased water consumption by plants. In some cases, however, the effects of water deficits on vitamin C content on a fresh weight basis may be insignificant, depending on the particular cultivars (Dumas et al. 2003).

REFERENCES

Abushita, A.A. and E.A. Hebshi, H.G. Daood, and P.A. Biacs. 1997. Determination of antioxidant vitamins in tomatoes. Food Chem. 60: 207-212.

Abushita, A.A. and H.G. Daood, and P.A. Biacs. 2000. Changes in carotenoids and antioxidant vitamins in tomato as a function of varietal and technological factors. J. Agric. Food Chem. 48: 2075-2081.

Adegoroye, A.S. and P.A. Jolliffe. 1987. Some inhibitory effects of radiation stress on tomato fruit ripening. J. Sci. Food Agric. 39: 297-302.

Alba, R. and M.M. Cordonnier-Pratt, and L.H. Pratt. 2000. Fruit-localized phytochromes regulate lycopene accumulation independently of ethylene production in tomato. Plant Physiol. 123: 363-370.

American Diabetes Association. 2002. Nutrition & you: Trends 2002. Final report. http://www.eatright.org/cps/rde/xchg/ada/hs.xsl/media_1578_ENU_HTML.htm. Accessed 27 June 2007.

Bajaj, K.L. and R. Mahajan, P.P. Kaur, and D.S. Cheema. 1990. Chemical evaluation of some tomato varieties. J. Res. Punjab Agric. Univ. 27: 226-230.

Beecher, G.R. 1998. Nutrient content of tomatoes and tomato products. Proc. Soc. Exp. Biol. Med. 218: 98-100.

Berger, S. and T. Chmielewski, and A. Gronowska-Senger. 1966. Studies on the inheritance of high ascorbic acid level in tomatoes. Plant Foods Hum. Nutr. 13: 214-218.

Canene-Adams, K. and J.K. Campbell, S. Zaripheh, E.H. Jeffery, and J.W. Erdman, Jr. 2005. The tomato as a functional food. J. Nutr. 135: 1226-30.

Causse, M. and M. Buret, K. Robini, and P. Verschave. 2003. Inheritance of nutritional and sensory quality traits in fresh market tomato and relation to consumer preferences. J. Food Sci. 68: 2342-2350.

Chang, Y.H. and L.C. Raymundo, R.W. Glass, and K.L. Simpson. 1977. Effect of high temperature on CPTA-induced carotenoid biosynthesis in ripening tomato fruits. J. Agric. Food Chem. 25: 1249-1251.

Cox, S. 2000. I Say Tomayto, You Say Tomahto.
http://www.landscapeimagery.com/tomato.html

Davies, J.N. and G.E. Hobson. 1981. The constituents of tomato fruit—the influence of environment, nutrition, and genotype. CRC Crit. Rev. Food Sci. Nutr. 15: 205-280.

De Pascale, S. and A. Maggio, V. Fogliano, P. Ambrosino, and A. Ritieni. 2001. Irrigation with saline water improves carotenoids content and antioxidant activity of tomato. J. Hortic. Sci. Biotechnol. 76: 447-453.

Dong, C.X. and J.M. Zhou, X.H. Fan, H.Y. Wang, Z.Q. Duan, and C. Tang. 2004. Application methods of calcium supplements affect nutrient levels and calcium forms in mature tomato fruits. J. Plant Nutr. 27: 1443-1455.

Dorais, M. and A.P. Papadopoulos, and A. Gosselin. 2001. Greenhouse tomato fruit quality. Hortic. Rev. 26: 239-319.

Dumas, Y. and M. Dadomo, G. DiLucca, and P. Grolier. 2003. Review: Effects of environmental factors and agricultural techniques on antioxidant content of tomatoes. J. Sci. Food Agric. 83: 369-382.

El-Gizawi, A.M. and M.M.F. Abdallah, H.M. Gomaa, and S.S. Mohamed. 1993. Effect of different shading levels on tomato plants. 2. Yield and fruit quality. Acta Hortic. 323: 349-354.

Fanasca, S. and G. Colla, Y. Rouphael, and F. Saccardo. 2006. Evolution of nutritional value of two tomato genotypes grown in soilless cultures as affected by macrocation proportions. Hortscience 41: 1584-1588.

FAOSTAT. 2004. http://faostat.fao.org/site/408/DesktopDefault.aspx?PageID=408. Accessed 27 June 2007.

George, B. and C. Kaur, D.S. Khurdiya, and H.C. Kapoor. 2004. Antioxidants in tomato (*Lycopersicum esculentum*) as a function of genotype. Food Chem. 84: 45-51.

Giovanelli, G. and V. Lavelli, C. Peri, and L. Guidi. 1998. Variation of antioxidant content in tomato during ripening. Proc. Tomato and Health Seminar. Pamplona, Spain, 25-28 May, pp. 63-67.

Giovannucci, E. 1999. Tomatoes, tomato-based products, lycopene, and cancer: review of the epidemiologic literature. J. Nat. Cancer Inst. 91: 317-331.

Giuliano, G. and G.E. Bartley, and P.A. Scolnik. 1993. Regulation of carotenoids biosynthesis during tomato development. Plant Cell 5: 379-387.

Gould, W.A. 1992. Tomato Production, Processing and Technology. Baltimore: CTI Publ., Maryland, USA.

Hamner, K.C. and L. Bernstein, and L.A. Maynard. 1945. Effects of light intensity, day length, temperature, and other environment factors on the ascorbic acid content of tomatoes. J. Nutr. 29: 85-97.

Hart, D.J. and J. Scott. 1995. Development and evaluation of an HPLC method for the analysis of carotenoids in foods, and the measurement of the carotenoid content of vegetables and fruits commonly consumed in the UK. Food Chem. 54: 101-111.

Ishida, B.K. 1998. Activated lycopene biosynthesis in tomato fruit *in vitro*. Proc. Tomato and Health Seminar. Pamplona, Spain, 25-28 May, pp. 151-156.

Ishida, B.K. and S.M. Jenkins, and B. Say. 1998. Induction of *AGAMOUS* gene expression plays a key role in ripening of in vitro-grown tomato sepals. Plant Mol. Biol. 36: 733-739.

Kadam, J.R. and J.S. Sahane. 2002. Quality parameters and growth characters of tomato as influenced by NPK fertilizer briquette and irrigation methods. J. Maharashtra Agric. Univ. 27: 124-126.

Kader, A.A. 1987. Influence of preharvest and postharvest environment on nutritional composition of fruits and vegetables. Proc. 1st Int. Symp. Hort. & Human Health. ASHS Symp. Ser. 1: 18-32.

Kaniszewski, S. and K. Elkner, and J. Rumpel. 1987. Effect of nitrogen fertilization and irrigation on yield, nitrogen status in plants and quality of fruits of direct seeded tomatoes. Acta Hortic. 200: 195-202.

Kowalska, I., and W. Sady. 1996. Suitability of urea and nitrate forms of nitrogen for greenhouse tomato (*Lycopersicon esculentum* Mill.) grown using nutrient film technique (NFT). Folia Hortic. 8: 105-114.

Krauss, S. and W.H. Schnitzler, J. Grassmann, and M. Woitke. 2006. The influence of different electrical conductivity values in a simplified recirculating soilless system on inner and outer fruit quality characteristics of tomato. J. Agric. Food Chem. 54: 441-448.

Kubota, C. and C.A. Thomson, M. Wu, and J. Javanmardi. 2006. Controlled environments for production of value-added food crops with high phytochemical concentrations: lycopene in tomato as an example. Hortscience 41: 522-525.

Kuti, J.O. and H.B. Konuru. 2005. Effects of genotype and cultivation environment on lycopene content in red-ripe tomatoes. J. Sci. Food Agric. 85: 2021-2026.

Lee, S.K. and A.A. Kader. 2000. Preharvest and postharvest factors influencing vitamin C content of horticultural crops. Postharvest Biol. Technol. 20: 207-220.

Lenucci, M.S. and D. Cadinu, M. Taurino, G. Piro, and G. Dalessandro. 2006. Antioxidant composition in cherry and high-pigment tomato cultivars. J. Agric. Food Chem. 54: 2606-2613.

Leoni, C. The influence of processing techniques on the content and bioavailability of lycopene for humans. *In:* Role and Control of Antioxidants in the Tomato Processing Industry. Second Bulletin on the Advancement of Research. 1999. FAIR RTD Programme (FAIR CT 97-3233), pp. 13-18.

Lincoln, R.E. and F.P. Zscheile, J.W. Porter, G.W. Kohler, and R.M. Caldwell. 1943. Provitamin A and vitamin C in the genus *Lycopersicon*. Bot. Gaz. 105: 113-115.

Liptay, A. and A.P. Papadopouslos, H.H. Bryan, and D. Gull. 1986. Ascorbic acid levels in tomato (*Lycopersicon esculentum* Mill) at low temperatures. Agric. Biol. Chem. 50: 3185-3187.

Locascio, S.J. and W.J. Wiltbank, D.D. Gull, and D.N. Maynard. 1984. Fruit and vegetable quality as affected by nitrogen nutrition. pp. 617-626. *In*: Nitrogen in Crop Production. ASA-CSSA-SSSA, Madison, Wisconsin, USA.

Lopez-Andreu, F.J. and A. Lamela, R.M. Esteban, and J.G. Collado. 1986. Evolution of quality parameters in the maturation stage of tomato fruit. Acta Hortic. 191: 387-394.

Lumpkin, H. 2005. A comparison of lycopene and other phytochemicals in tomatoes grown under conventional and organic management systems. Tech. Bull. No. 34. AVRDC publication number 05-623. Shanhua, Taiwan; AVRDC—The World Vegetable Center.

Maclinn, W.A. and C.R. Fellers, and R.E. Buck. 1937. Tomato variety and strain differences in ascorbic acid (vitamin C) content. Proc. Am. Soc. Hortic. Sci. 34: 543-552.

Martinez-Valverde, I. and M.J. Periago, G. Provan, and A. Chesson. 2002. Phenolic compounds, lycopene and antioxidant activity in commercial varieties of tomato (*Lycopersicon esculentum*). J. Sci. Food Agric. 82: 323-330.

Matsuzoe, N. and K. Zushi and T. Johjima. 1998. Effect of soil water deficit on c oloring and carotene formation in fruits of red, pink and yellow type cherry tomatoes. J. Jpn. Soc. Hortic. Sci. 67: 600-606.

Maul, F. 1999. Flavor of fresh market tomato (*Lycopersicon esculentum* Mill) as influenced by harvest maturity and storage temperature. PhD dissertation, University of Florida.

McCollum, J.P. 1954. Effects of light on the formation of carotenoids in tomato fruits. Food Res. 19: 182-189.

Montagu, K.D. and K.M. Goh. 1990. Effects of forms and rates of organic and inorganic nitrogen fertilizers on the yield and some quality indices of tomatoes (*Lycopersicon esculentum* Mill.). NZ J. Crop Hortic. Sci. 18: 31-37.

Murneek, A.E. and L. Maharg and S.H. Wittwer. 1954. Ascorbic acid (vitamin C) content of tomatoes and apples. Univ. Missouri. Agric. Exp. Stn. Res. Bull. 568: 3-24.

Nagy, S. and W.F. Wardowski. Effects of agricultural practices, handling and storage on fruits. pp. 73-100. *In:* E. Karmas and R.S. Harris. [eds.] 1988. Nutritional Evaluation of Food Processing. Van Nostrand Reinhold, New York.

Naphade, A.S. 1993. Effect of water regime on the quality of tomato. Maharashtra J. Hortic. 7: 55-60.

Offord, E.A. 1998. Nutritional and health benefits of tomato products. Proc. Tomato and Health Seminar. Pamplona, Spain, 25-28 May, pp. 5-10.

Oke, M. and T. Ahn, A. Schofield, and G. Paliyath. 2005. Effects of phosphorus fertilizer supplementation on processing quality and functional food ingredients in tomato. J. Agric. Food Chem. 53: 1531-1538.

Paiva, E.A.S. and R.A. Sampaio, and H.E.P. Martinez. 1998. Composition and quality of tomato fruit cultivated in nutrient solutions containing different calcium concentrations. J. Plant Nutr. 21:2653-2661.

Petro-Turza, M. 1986. Flavor of tomato and tomato products. Food Rev. Int. 2: 309-351.

Premuzic, Z. and M. Bargiela, A. Garcia, A. Rendina, and A. Iorio. 1998. Calcium, iron, potassium, phosphorus, and vitamin C content of organic and hydroponic tomatoes. HortScience 33: 255-257.

Raffo, A. and G.L. Malfa, V. Fogliano, G. Maiani, and G. Quaglia. 2006. Seasonal variations in antioxidant components of cherry tomatoes (*Lycopersicon esculentum* cv. Naomi F1). J. Food Composition Anal. 19: 11-19.

Rao, V.S. and D.S. Yadav. 1988. Preliminary evaluation of some varieties for processing under Imphal valley conditions. Agric. Sci. Digest Karnal 8: 149-152.

Robertson, G.H. and N.E. Mahoney, N. Goodman, and A.E. Pavlath. 1995. Regulation of lycopene formation in cell suspension culture of VFNT tomato (*Lycopersicon esculentum*) by CPTA, growth regulators, sucrose, and temperature. J. Exp. Bot. 46: 667-673.

Saito, S. and F.J. Kano. 1970. Influence of nutrients on growth of solanaceous vegetable plants, quality and chemical composition in their fruit. (Part 1) On the effect of different phosphate levels on the lycopene content of tomatoes. J. Agric. Sci. Tokyo 14: 233-238.

Shewfelt, R.L. 1990. Sources of variation in the nutrient content of agricultural commodities from the farm to the consumer. J. Food Qual. 13: 37-54.

Shi, J. and M. Le Maguer. 2000. Lycopene from tomatoes: Chemical and physical properties affected by food processing. CRC Crit. Rev. Food Sci. Nutr. 110: 1-42.

Somers, G.F. and W.C. Kelly, and K.C. Hamner. 1951. Influence of nitrate supply upon the ascorbic content of tomatoes. Am. J. Bot. 38: 472-475.

Stevens, M.A. 1986. Inheritance of tomato fruit quality components. Plant Breeding Rev. 4: 273-311.

Stevens, M.A. and C.M. Rick. Genetics and breeding; fruit quality. pp. 84-96. *In*: J.G. Atherton and J. Rudich. [eds.] 1986. The Tomato Crop, A Scientific Basis for Improvement. Chapman and Hall, London.

Subbiah, K. and R. Perumal. 1990. Effect of calcium sources, concentrations, stages and number of sprays on physicochemical properties of tomato fruits. S. Indian Hortic. 38: 20-27.

Thakur, N.S. and B.B. Lal Kaushal. 1995. Study of quality characteristics of some commercial varieties and F1 hybrids of tomato grown in Himachal Pradesh in relation to processing. Indian Food Packer 25-31.

Thomas, R.L. and J.J. Jen. 1975. Red light intensity and carotenoid biosynthesis in ripening tomatoes. J. Food Sci. 40: 566-568.

Thompson, A.E. and M.L. Tomes, E.V. Wann, A.K. McCollum, and A.K. Stoner. 1964. Characterization of crimson tomato fruit color. Amer. Soc. Hortic. Sci. 86: 610-616.

Thompson, K.A. and M.R. Marshall, C.A. Sims, C.I. Wei, S.A. Sargent, and J.W. Scott. 2000. Cultivar, maturity and heat treatment on lycopene content in tomatoes. Food Chem. 65: 791-795.

Tigchelaar, E.C. Tomato breeding. pp. 135-171. *In*: M.J. Bassett. [eds.] 1986. Breeding Vegetable Crops. AVI Publ., Westport, Connecticut, USA.

Toor, R.K. and G.P. Savage, and A. Heeb. 2006a. Influence of different types of fertilisers on the major antioxidant components of tomatoes. J. Food Composition Anal. 19: 20-27.

Toor, R.K. and G.P. Savage, and C.E. Lister. 2006b. Seasonal variations in the antioxidant composition of greenhouse grown tomatoes. J. Food Composition Anal. 19: 1-10.

Trudel, M.J. and J.L. Ozbun. 1971. Influence of potassium on carotenoid content of tomato fruit. J. Am. Soc. Hortic. Sci. 96: 763-765.

Venter, F. 1977. Solar radiation and vitamin C content of tomato fruits. Acta Hortic. 58: 121-127.

Weston, L.A. and M.M. Barth. 1997. Preharvest factors affecting postharvest quality of vegetables. Hortscience 32: 812-816.

Wu, M. and J.S. Buck, and C. Kubota. 2004. Effects of nutrient solution EC, plant microclimate and cultivars on fruit quality and yield of hydroponic tomatoes (*Lycopersicon esculentum*). Acta Hortic. 659: 541-547.

Zushi, K. and N. Matsuzoe. 1998. Effects of soil water deficit on vitamin C, sugar, organic acid and carotene contents of large-fruited tomatoes. J. Jpn. Soc. Hortic. Sci. 67: 927-933.

3

Field Trials of Genetically Modified Tomato: Fruit Quality and Productivity

G.L. Rotino[2], T. Pandolfini[1], R. Lo Scalzo[3], E. Sabatini[2], M. Fibiani[3] and A. Spena[1]

[1]Faculty of Science, University of Verona, Strada Le Grazie 15, 37134 Verona, Italy

[2]CRA-ORL, Agricultural Research Council, Research Unit for Vegetable Crops, Via Paullese 28, 26836 Montanaso Lombardo (LO), Italy

[3]CRA-IAA, Agricultural Research Council, Research Unit for Agro Food Industry, Via Venezian, 26 20133, Milan, Italy

ABSTRACT

In the past twenty years, plant genetic engineering has been used to confer traits of both heuristic and applied interest to tomato. Genetic modifications have aimed to improve either the quality of the fruit and/or agronomical aspects of tomato cultivation. In open field trials, the novel genotype is tested under agronomical and environmental conditions similar to those used for production. Thus, field trials represent the most stringent test for any novel plant genotype, including genetically modified plants. In this Chapter, we discuss field trials performed with transgenic tomato plants that represent examples of different genetic strategies used in modern plant breeding. The examples chosen are genetic modifications related to three general aspects: (1) macro- and micro-nutrient content of the fruit, such as sugar content and beta-carotene content, (2) biological processes of either direct or indirect agronomical interest, such as parthenocarpy and resistance to disease, (3) features related to post-harvest and processing technology, such as fruit firmness and prolonged fruit vine-life. The discussion is focused on three main aspects: the genetic strategy used to confer the novel phenotypic trait(s), quality assessment of the genetically modified plant and fruit, and plant production. The results obtained confirm the value of plant genetic engineering as a means of improving tomato productivity and tomato products.

INTRODUCTION

Tomato was introduced into southern Europe at the beginning of the 16th century. By the end of the 18th century, tomato fruit consumption was very popular in this region, as indicated by its presence in several recipes (Mattioli 1568, Felici 1572, Corrado 1773). By the end of the 19th century, tomato fruit consumption and processing were common all over Europe and North America. When tomato was introduced in Europe, its fruit was often considered to be an aphrodisiac and unhealthy. Some of its names are reminiscent of this imaginary effect (love apple, pomme d'amour, pomo d'oro). A more rational basis for the initial resistance to tomato fruit is the presence, in plants belonging to the Solanaceae, of toxic glycoalkaloids, such as potato solanine and chaconine (Friedman 2002). However, the tomato glycoalkaloid tomatine not only is less toxic than solanine and chaconine, but also has several possible beneficial effects on human health (Friedman 2002).

Tomato breeding probably started soon after the widespread cultivation of tomato. Indeed, the cultivars imported from the New World had either yellow or red fruits, but varieties cultivated by the end of the 18th century had mostly red fruits. Thus, the breeding of tomato started with its consumption, and since then its evolution has been driven by human demand. At first, tomato breeding probably consisted of mere empirical selection and propagation of plants with the most interesting features. During the 20th century, tomato breeding has been based on genetic knowledge. In the last twenty years, new genetic tools have been developed to genetically modify plants and these have consequently been used in tomato breeding too (Gepts 2002).

In this chapter, we focus on field trials of tomato plants genetically modified with the aim of improving, either directly or indirectly, tomato fruit quality and productivity. Genetically modified tomato fruits can also be produced for purposes other than nutrition. Indeed, any crop plant can be designed to produce molecules of biotechnological interest not meant for human nutrition (Twyman et al. 2003).

In the past twenty years, tomato genetic engineering has implemented genetic modifications aimed at improving features related to fruit quality. Such modifications are broadly related to three general aspects: (1) macro- and micro-nutrient content of the fruit, (2) biological processes of either direct or indirect agronomical interest, such as parthenocarpy and resistance to disease, and (3) features related to post-harvest and processing technology.

The value of any new tomato variety is determined by its performance in farmers' fields. Activities concerned with genetic improvement and, particularly, genetic engineering are first carried out in a controlled environment. However, fruit quality parameters are influenced not only by the interaction between the introgressed (trans)gene and the genetic background, but also by cultivation practice and environmental conditions (Brandt et al. 2006, Collins et al. 2006). Thus, any new plant and its product must be assessed under the cultivation conditions used for crop production. In this regard, field trials are essential to provide quality and productivity information that cannot be otherwise gathered. First of all, productivity and quality data on genetically modified tomato fruits have to be analysed in comparison with their background genotype(s). Then, the new transgenic cultivar or line has to be evaluated by comparison with commercial cultivars different from the parental one. This is because centuries of tomato breeding, also exploiting its solanaceous wild relatives, have produced a large number of cultivars and lines. Such huge intra-species variability in tomato fruit quality and productivity has to be taken into consideration when evaluating genetically modified tomato fruit. Thus, a genetically modified fruit is defined as a "substantial equivalent" by comparison with its genetic background (i.e., the genotype in which the novel gene(s) has been introduced) and when its fruit quality parameters are within the range of variability of the species *Solanum lycopersicum*. This comparison should take into account the main macronutrients, liposoluble vitamins, vitamin C and tomatine content. With regard to the latter, it is interesting to note that for rather a long time, tomatine glycoalkaloids were considered to be as toxic as solanine and chaconine of potato (Keeler et al. 1991). However, recent findings have shown that tomatine has several properties that may have beneficial effects on human health and it is now estimated to be 20 times less toxic than solanine and chaconine (Friedman 2002).

Transgenic plants have been often tested under field conditions for trait(s) of interest. More rarely, fruit productivity and fruit quality parameters have been concomitantly analysed. In this article, we present examples of genetically modified tomatoes evaluated by field trials and discuss the genetic rationale and, when available, both the quality and productivity data (Table 1).

GENETIC ENGINEERING OF TOMATO

For the sake of clarity, we have divided the genetically modified plants discussed here into three groups: (1) tomato genetically modified in the macro- and micro-nutrient content of the fruit, (2) tomato genetically modified for biological processes of either direct or indirect agronomical interest, such as parthenocarpy and resistance to disease, and (3) tomato genetically modified to confer features related to post-harvest and processing technology.

Plant genetic engineering has so far used two general strategies. The fir st type aims to express novel genes in the chosen plant species. The trait(s) of interest are then achieved via the biochemical action of the product encoded by the introgressed gene. The gene product can be completely novel or already codified in the genome. In the latter case, the pattern and/or strength of expression of the gene are modified.

The second type of genetic strategy produces the trait(s) of interest by curtailing the function of the target gene(s). This strategy came about because of the discovery of RNA interference (Fire et al. 1998) and relies on the understanding of its pathways (Zamore and Haley 2005). RNA interference is a eukaryotic epigenetic mechanism controlling gene expression either transcriptionally or post-transcriptionally. Post-trascriptional mechanisms consist of either degradation of target mRNA and/or inhibition of translation of target mRNA. Small non-coding RNA molecules, either small interfering RNAs (siRNA) or microRNAs (miRNA), confer specificity to RNA interference. Specificity relies on sequence homology between the small RNAs and the target RNA. Degradation or translation inhibition is mediated by protein complexes such as RISC, the RNA-Induced Silencing Complex that contains SLICER (AGO protein), an endoribonuclease that degrades the target RNA annealed to the complementary siRNA (Hammond et al. 2000). RNA interference is easily triggered by expressing long (usually 100-500 bp long) double-stranded RNA molecules that are processed to siRNA by the activity of DICER ribonucleases (Bernstein et al. 2001). Thus, RNA interference allows us to silence any gene(s) of interest by using appropriate transgenic constructs (Horiguchi 2004). As a consequence, RNA interference has been used to silence the expression of genetic information either endogenous (i.e., present in the genome of the plant) or exogenous (e.g., present in the genome of either a plant virus or nematode). In this second type of genetic strategy, the trait(s) of interest are achieved by suppressing the expression of the target gene(s).

Both types of genetic strategies can be articulated and implemented in various ways. The novel transgene can have a plant origin. In such cases, it is derived from genetic information present in the genome of the same species or of other plant species. Alternatively, the novel gene can derive from genetic information present in the genome of species belonging to other kingdoms of life. The use of genes of plant origin is somewhat reminiscent of interspecies hybridization employed in classical plant breeding. The use of genetic information derived from genomes other than plants further enlarges the genetic repertoire amenable to plant

Table 1 Transgenic tomato field trials surveyed in this chapter. For each trial the author of the cited publication, the engineered phenotype, the transgene used, the location of the field trial, the genotypes employed and the presence of qualitative and/or productivity data are reported.

Authors	*Phenotype*	*Gene(s)*	*Location*	*Destination/Cultivars*	*Data measured*	*Yield quality*
Laporte et al. 1997, 2001	Improved productivity and sugar content	Sucrose-phosphate synthase	USA	Processing inbred lines	Y	Y
Giorio et al. 2007	Increased beta-carotene content	Lycopene beta-cyclase (tLcy-b)	Italy	Processing F1 hybrid	Y	Y
Rotino et al. 2005	Parthenocarpy	DefH9-RI-iaaM (auxin-synthesis)	Italy	Processing inbred cv UC 82	Y	Y
Accotto et al. 2005 Corpillo et al. 2004	Virus resistance (TMSV)	TSWV modified nucleoprotein gene	Italy	Fresh market F1 hybrid	Y	Y
FDA 1992 (Calgene)	Fruit firmness	Transgene (antisense) for silencing polygalacturonase	USA	Processing line CR3	N	Y
Porretta et al. 1998 (Zeneca)	Fruit firmness	Transgene for silencing polygalacturonase	Italy	Processing inbred cv	N	Y
Kalamaki et al. 2003 Powell et al. 2003	Fruit firmness/texture	Transgenes for silencing polygalacturonase and expansin 1	USA	Tomato cv Ailsa Craig	N	Y
Metha et al. 2002	Prolonged vine life Polyamine content Fruit ripening	ySAMdc gene (yeast S-adenosylmethionine decarboxylase)	USA	Processing commercial variety	N	Y

Y= data available; N= no data available.

breeding. The genetically modified plants chosen in this survey provide examples of the two main genetic strategies used to breed plants via genetic engineering.

Tomato Genetically Modified in Nutrient Content of Fruit

Sugars

Sugars are important macronutrients of the human diet. In plants, sucrose is the main transported form of carbohydrates. Moreover, sucrose regulates transcription and/or translation of target genes (Ciereszko et al. 2001, Wiese et al. 2005). Sucrose metabolism and transport might be altered by several genetic mechanisms. The chosen example shows that sucrose-phosphate synthase activity influences whole-plant carbon allocation affecting both tomato productivity and fruit sugar content (Laporte et al. 2001). This conclusion is based on field trials performed with tomato lines transgenic for the sucrose-phosphate synthase gene from maize (Laporte et al. 1997, 2001) driven either by the *CaMV35S* promoter or by the *rbcS* promoter. Interestingly, only plant lines with a moderate (i.e., two-fold) increase in sucrose-phosphate synthase activity showed an increased fruit production (on the average 20%; Laporte et al. 1997, 2001) and sometimes a slight increase of soluble solids in genetically modified fruits in comparison to wild-type control (Laporte et al. 1997). This study (Laporte et al. 2001) also shows that the choice of the proper gene (e.g., sucrose-phosphate synthase from maize) as well as an appropriate level of expression of the transgenic construct are crucial to achieve a valuable product. Indeed, only plants with a moderate increase in sucrose-phoshate synthase activity (i.e., activity increased twice in the leaves) display all the desired benefits, while a higher expression level of sucrose-phosphate synthase is not beneficial for plant growth and production.

Beta-carotene

Beta-carotene is a micronutrient component converted in our body to vitamin A. Beta-carotene is also used as natural food colouring additive. Among several strategies employed to increase beta-carotene in crop plants via genetic engineering, one strategy has been tested under open field cultivation conditions (Giorio et al. 2007). This strategy has genetically modified lycopene beta-cyclase gene expression in tomato (Giorio et al. 2007). Lycopene beta-cyclase converts lycopene to beta-carotene. Expression of lycopene beta-cyclase is down-regulated at the breaker stage of tomato fruit development, causing accumulation of lycopene (Pecker et al. 1996). Consequently, expression of lycopene beta-cyclase during late phases of fruit development increases beta-carotene content. The HighCaro tomato has been genetically modified in order to express the lycopene beta-cyclase gene under the control of the *CaMV35S* promoter, a promoter expressed also during late phases of fruit development. In HighCaro tomato fruits, lycopene beta-cyclase is expressed also after the breaker stage, causing the almost complete conversion of lycopene to beta-carotene (Giorio et al. 2007). HighCaro tomato fruit has an increased (more than 10-fold) beta-carotene content, while lycopene content is drastically reduced. Total carotenoids and lutein content was also lower. Other common fruit quality parameters, such as dry weight, soluble solids and titratable acidity, are increased in HighCaro fruits; however, their values are well within the range of intraspecies variability (Table 2). The HighCaro tomato line gave a marketable fruit yield statistically not different from its control. Total

Table 2 Tomato fruit genetically modified for high content of beta-carotene by lycopene beta-cyclase over-expression.

Fruit quality parameters	*Genotypes*		
	HP	*RS*	*HC*
Dry weight (%)	5.64 a	5.08 b	5.58 a
Soluble solids (°Brix)	4.13 b	3.96 c	4.36 a
Titratable acidity (mEq/100 g)	5.95 a	5.41 b	6.12 a
Total carotenoids (μg/g fw)	131.7 a	125.4 a	84.0 b
β-carotene (μg/g fw)	5.7 a	6.3 a	76.6 b
Lycopene (μg/g fw)	108.9 a	102.9 a	0.5 b
Lutein (μg/g fw)	11.4 a	10.9 a	2.3 b

HP, commercial hybrid; RS, background untransformed genotype; HC, transgenic line. Different letters mean significant differences in each row (modified after Giorio et al. 2007).

yield (marketable plus unmarketable fruit) was significantly lower than its control (Table 2).

Tomato Genetically Modified for Biological Processes of Agronomical Interest

Parthenocarpy

Parthenocarpy, or fruit set without pollination/fertilization, represents a useful trait in tomato. Parthenocarpic fruits are seedless, and the absence of seeds is advantageous for both the processing industry (e.g., tomato paste, tomato juice) and fresh consumption (Fig. 1). Other advantages of parthenocarpy are related to the increase of soluble solids content and increased plant productivity. The increased productivity, observed under greenhouse cultivation conditions (Acciarri et al. 2000), is mainly due to improved fruit set because parthenocarpic plants set fruits under environmental conditions adverse for pollination/fertilization. To be a valid product, a genetically modified parthenocarpic plant should have not only seedless fruits and an improved plant productivity under cultivation conditions adverse for fruit set, but also fruit quality at least equal to that of its genetic background. Parthenocarpic plants transgenic for an ovule-placenta-specific auxin synthesizing gene have also been tested under field trial conditions (Rotino et al. 2005). The parthenocarpic trait conferred on the tested tomato lines is facultative (i.e., when pollinated the genetically modified lines can develop seeds). Nevertheless, both the number of seeded fruits and the number of seeds per seeded fruit were decreased (Table 3).

Altogether, seed content was on the average approximately 90% lower (Table 3). Beta-carotene increased in both parthenocarpic genetically modified lines tested (Table 4).

Other fruit quality parameters were substantially equivalent to wild-type controls and the observed variability was well within the variability of the species (Tables 3 and 4). Fruit yield, measured as fruit fresh weight, was comparable in parthenocarpic and control plants (Rotino et al. 2005). However, the number of fruits per plant was significantly higher in transgenic parthenocarpic plants, while fruit weight was significantly lower. A cultivation practice adapted to the higher fruit setting capacity of the genetically modified parthenocarpic plants (e.g., watering, fertilization) might result in a less pronounced reduction in fruit weight and consequently in a higher fruit production. In conclusion, genetically modified parthenocarpic tomato fruit showed very little variation in comparison with untransformed control genotype except for seedlessness and beta-carotene content.

This case study also shows that a successful strategy must not only identify the proper gene to be used to confer the trait of interest, but also choose the appropriate tissue-specificity of expression (i.e., promoter). Moreover, the level of expression must be optimized in light of the genetic background (Fig. 1). In this case study, the *iaaM* gene was chosen to confer auxin synthesis, the *DefH9* promoter to confer ovule-placenta tissue specificity of expression and a DNA sequence derived from the *rolA* intron to optimize the level of expression in some tomato genotypes with a high auxin sensitivity (Pandolfini et al. 2002). Thus, gene action and organ/tissue specificity of expression are crucial elements for a successful genetic

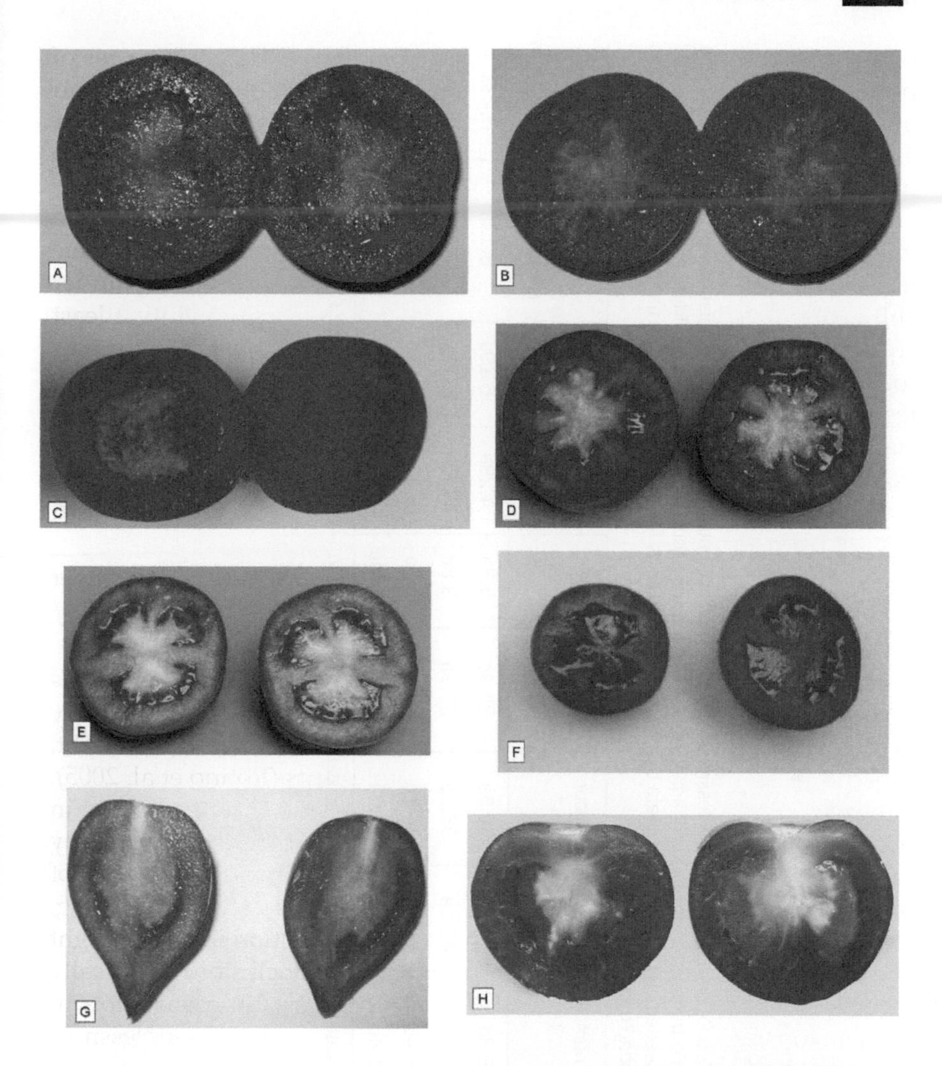

Fig. 1 Seedless parthenocarpic fruit from different tomato genotypes obtained by auxin-synthesis transgene expression. Cut fruits from different tomato genotypes engineered for the parthenocarpic trait. A, parental line L276 (Ficcadenti et al. 1999); B, parental line INB777 (our unpublished result); C hybrid "Giasone" (LH76PC × L276 iaaM 1-1) (Acciarri et al. 2000); D, an experimental hybrid using L276 iaaM 1-1 as male parent (Acciarri et al. 2000); E, cherry type line CM (Ficcadenti et al. 1999); F, cultivar UC 82 (Rotino et al. 2005); G, Italian tomato ecotype "Pizzutello" (our unpublished result); H, cultivar "Ailsa Craig" (our unpublished result). A, B, C, D, E and H tomato fruits transgenic for the auxin-synthesis gene *DefH9-iaaM*; F and G, tomato fruits transgenic for the *DefH9-RI-iaaM* gene (i.e., modified version of the *DefH9-iaaM* gene with reduced level of expression, see Pandolfini et al. 2002).

Table 3 Tomato fruit genetically modified for parthenocarpy: seedlessness and technological features. Allflesh (commercial hybrid); UC82 (background untransformed genotype); Ri4 (UC82 line transgenic for parthenocarpy); Ri5 (UC82 line transgenic for parthenocarpy). Mean values of percentage of fruits with seeds, number of seeds per seeded fruit, colour (coordinate L, a* e b*), Soluble Solids (S.S.), dry matter (DM), titratable acidity (tit. ac.), skin resistance (firmness) (modified after Rotino et al. 2005).

Genotypes	Fruit with seeds (%)	Seed number	Colour L*	Colour a*	Colour b*	S.S. °Brix	pH	DM (%)	Tit.ac. (mEq/100 ml NaOH 0.1 N)	Firmness (kg)
Allflesh	87a	68.8a	39.2b	36.5a	25.4c	5.3a	4.29a	6.19a	6.12a	0.43a
UC82	85a	36.5b	41.1a	35.8a	27.9a	3.8c	4.08bc	5.06bc	7.23a	0.40a
Ri4	27b	18.4c	39.9ab	34.1b	25.9bc	4.5b	3.96c	4.85c	6.12a	0.43a
Ri5	20b	11.4c	40.0ab	34.0b	27.1ab	4.2bc	4.11b	5.24b	6.70a	0.41a

For each trait at least one common letter indicates no significant difference according to the Duncan test ($\alpha = 0.05$). Colour coordinates: L, brightness; a*, red colour; b*, yellow colour.

Table 4 Tomato fruit genetically modified for parthenocarpy: biochemical analyses. Allflesh (commercial hybrid); UC82 (background untransformed genotype); Ri4 (UC82 line transgenic for parthenocarpy); Ri5 (UC82 line transgenic for parthenocarpy). Mean value of citric acid, tartaric acid, oxalic acid and vitamin C (mg/100 g dw), and β-carotene, lycopene and tomatine (μg/g dw), detected in the four genotypes tested (modified after Rotino et al. 2005).

Genotypes	Citric	Tartaric	Oxalic	Vit C	β-carotene	Lycopene	Tomatine
Allflesh	2194a	40.2a	37.3a	539.6a	491.1c	659.5a	189.4a
UC82	1434b	32.1a	36.0a	445.6a	566.6b	815.9a	248.2a
Ri4	2271a	30.4a	36.0a	436.5a	698.9a	928.5a	202.8a
Ri5	1837ab	40.6a	22.9b	374.3a	643.6a	811.1a	169.3a

For each trait at least one common letter indicates no significant difference according to the Duncan test ($\alpha = 0.05$).

strategy. However, to be employed in many different tomato genetic backgrounds the level of expression of the transgene might need to be optimized for the chosen genotypes.

Virus Resistance

Viral diseases represent one of the most limiting factors for tomato production. Chemicals are ineffective against plant virus diseases. Thus, virus protection is mainly based on genetic resistance. Genetic resistance to viral disease has been traditionally searched for in the genomes of various tomato wild relatives. Traditional plant breeding has been supported by the development of genetic engineering strategies to obtain virus-resistant transgenic tomato.

Tomato spotted wilt virus (TSWV), an extremely polyphagus virus, can have devastating effects on tomato. Consequently, a resistance gene (*Sw-5)* from *S. peruvianum* has been introgressed into modern F1 hybrids, but new isolates of TSWV able to overcome this resistance have been reported (Latham and Jones 1998). One of the first approaches to genetically engineer TSWV-resistant tomato plants consisted in the expression of a modified pathogen-derived nucleoprotein (*N*) gene (Glein et al. 1991). The original rationale envisaged that the expression of a dysfunctional protein would inhibit virus proliferation. Transgenic tomato hybrids resistant to TSWV proved, in two field trials performed in 1999, to have reliable TSWV protection (Accotto et al. 2005). In the first open field trial, the virus "pressure" was high. Under these conditions, the transgenic virus-resistant hybrids outperformed the corresponding untransformed controls, which were severely infected by the TSWV. In the second field trial, virus "pressure" was absent. Under these conditions, transgenic virus-resistant and control tomatoes showed a similar fruit production. The substantial equivalence of the transgenic TSWV-resistant hybrids was assessed by proteomic analysis. Transgenic tomato seedlings did not display any major change in their protein profile in comparison with their wild-type control genotypes (Corpillo et al. 2004).

The TSWV-resistant tomato plants produced aberrant RNA transcripts but not the *N* protein (Nervo et al. 2003). These data indicate that virus resistance was most likely caused by RNA interference. Thus, this and several other findings showing that expression of transgenic viral RNA by itself, and not the expression of viral protein(s), elicits resistance to many viruses have prompted the development of new genetic strategies based on the mechanisms of RNA interference (Lindbo and Dougherty 2005). In this regard, RNA interference represents a mechanism of defence against viruses that is presently exploited to confer virus resistance to plant species by eliciting the production either of siRNA or of newly designed miRNA (Lindbo and Dougherty 2005, Schwab et al. 2006).

It is worthwhile mentioning that not only was the development of plants resistant to virus diseases one of the first aims of plant biotechnology, but they also represent one of the first pieces of scientific evidence of the phenomenon of RNA interference. Later, the discovery that double-stranded RNA is the elicitor of RNAi (Fire et al. 1998) allowed the molecular understanding of this epigenetic mechanism of gene regulation.

Tomato Genetically Modified for Features Related to Post-harvest and Processing Technology

Fruit Firmness

Most tomato fruit is consumed after processing. Fruit firmness is valuable for both post-harvest and processing technology. Several strategies have been developed to improve fruit consistency and juice viscosity. During ripening, polygalacturonase (PG) hydrolyses pectins and causes fruit softening. Consequently, PG was one of the first targets chosen for modifying fruit firmness by down-regulation of PG gene expression. Down-regulation of PG inhibits cell wall degradation and consequently improves fruit firmness in tomato (Smith et al. 2002).

Down-regulation of PG gene expression and activity in tomato fruit has been achieved by several transgenes acting most likely through a common molecular mechanism: silencing of the PG gene. Thus, although not formally proven by showing the presence of siRNA homologous to the target gene, the case studies described here most likely represent genetic strategies based on eliciting *in planta* RNA interference. siRNA homologous to PG mRNA would guide the degradation of PG mRNA by the Slicer activity of the RISC complex. The selective degradation of PG mRNA would then cause a decreased PG mRNA steady-state level curtailing PG mRNA translation and consequently reducing PG content and activity.

The FLAV-SAVR^{TM} tomato has reduced PG gene expression and activity (Calgene 1992). The expression of an antisense PG gene construct silenced endogenous PG gene expression and consequently the production of the PG enzyme. The PG activity in FLAV-SAVR^{TM} fruit was lower than that found in control fruit (http://www.cfsan.fda.gov/~acrobat2/bnfMFLV.pdf). Polygalacturonase is associated with the breakdown of pectin, so its reduced expression and activity results in red fruit that remain firm for a longer time. FLAV-SAVR^{TM} tomatoes can consequently be left to ripen on the vine for a longer time, enabling fruit to develop an enhanced flavour. Field trials with FLAV-SAVR^{TM} tomatoes have evaluated some fruit quality parameters, such as vitamin C, vitamin A, and tomatine (http://www.cfsan.fda.gov/~acrobat2/bnfMFLV.pdf, http://vm.cfsan.fda.gov/~lrd/biopolcy.html#eval). Vitamin C and vitamin A content were within the range of commercial tomato varieties. Tomatine level in red ripe fruit was comparable to commercial fruits. No other glycoalkaloids were detected by HPLC analysis in comparison with parental tomato fruit.

Silencing of PG gene expression has also been achieved by transcription of only a fragment of the PG mRNA (http://www.cfsan.fda.gov/~rdb/bnfm003.html). Transgene transcription was under the control of the *CaMV35S* promoter and the PG gene fragment was transcribed either in the sense or in the antisense orientation. Both types of transgenic constructs silenced the expression of the endogenous PG gene, also in these cases most likely by eliciting RNA interference. The genetically modified tomato fruits had a decreased PG activity and a prolonged shelf-life. The only difference between the genetically modified and the parental variety of tomato is the molecular weight distribution of pectin molecules. No difference in tomatine, lycopene, soluble carbohydrates, fibres, minerals, malic and citric acid was found (http://www.cfsan.fda.gov/~rdb/bnfm003.html; VIB 2001).

Porretta et al. (1998) investigated the physicochemical and sensory properties of processed pulp from genetically modified PG silenced and control fruits. The PG-silenced fruit had improved viscosity, measured as Bostwick juice, syneresis, and sensory attributes. Another parameter showing a small difference was volatile acidity. Other parameters, reducing sugars, total acidity and pH, showed no significant differences (Table 5).

Table 5 Tomato fruit genetically modified for improved firmness by decreasing polygalacturonase gene expression. Physicochemical and sensory analyses between control and genetically modified tomato. Statistical differences, indicated by different letters, are evaluated between each genetically modified tomato sample and the corresponding control. (Modified from Porretta et al. 1998.)

Trait	*Control 1*	*Genetically modified 1*	*Control 2*	*Genetically modified 2*	*Control 3*	*Genetically modified 3*
Glucose (g/kg dw)	244 a	259 a	223 a	234 a	209 a	215 a
Fructose (g/kg dw)	288 a	298 a	270 a	277 a	252 a	250 a
Total acidity (% dw)	6.21 a	6.52 a	6.66 a	6.47 a	5.71 a	5.88 a
pH	4.55 a	4.48 a	4.51 a	4.43 a	4.55 a	4.53 a
Volatile acidity (g/kg dw)	0.024 a	0.035 b	0.011 a	0.026 b	0.027 a	0.018 b
Total pectin (% dw)	3.76 a	4.03 a	3.11 a	3.69 a	3.98 a	4.21 a
Bostwick juice (cm in 30 s)	9 a	6 b	10 a	5 b	11 a	5 b
Syneresis (%)	45 a	7 b	95 a	9 b	80 a	10 b
Viscosity (sensory result)	3.1 a	7.7 b	3.7 a	6.9 b	2.4 a	6.4 b

Under greenhouse cultivation conditions, transgenic tomato plants silenced in the expression of *LeExp1*, a ripening-regulated expansin, showed an improved fruit firmness only during early stages of fruit ripening (Brummel et al. 2002). Expansin and PG influence pectin and other cell wall polymers via different mechanisms of action. Consequently, they could act synergistically on fruit softening and texture. These considerations prompted Powell et al. (2003) to evaluate tomato plants suppressed in the expression of both PG and expansin (*LeExp1*) under open field trial conditions. Expression of PG was silenced by an antisense construct driven by the *CaMV35S* promoter, while expansin was silenced by a sense-suppression construct driven by the same type of promoter (Powell et al. 2003). Both genes were efficiently silenced. Polygalacturonase activity was reduced to less than 1% of control fruits, while expansin was barely detectable by western blot analysis. Under open field conditions and in the Ailsa Craig genotype, a fresh market tomato cultivar, silencing either the PG or the expansin gene alone was not sufficient to increase fruit firmness (Powell et al. 2003). Improvement of firmness during all ripening stages, i.e., from breaker to ripe red, was obtained by silencing the tomato plants for both PG and expansin. The double genetically modified plants were obtained by crossing two homozygous lines transgenic for either the PG antisense or the expansin-suppressing construct, respectively. In fully ripe tomatoes silenced for both genes fruit firmness was improved by 15%. The juice of the genetically modified fruits silenced in either PG or in both PG and expansin expression had higher viscosity than juice from parental fruits. Soluble solids and pH were measured during ripening and they were indistinguishable among all genotypes (Powell et al. 2003). The shelf life of fruits silenced for expansin was improved, and yet silencing the PG gene also did not further improve fruit shelf life. This study has used RNA interference to silence two target genes whose products are known to be involved in the cell wall disassembly that takes place during ripening. The concomitant suppression of both PG and expansin shows synergistic effects on fruit ripeness, while the prolonged shelf life and the increased juice viscosity appear to be due to silencing of only one of the two target genes: either the expansin or the PG gene, respectively (Powell et al. 2003). Ailsa Craig is a tomato cultivar for the fresh market. Kalamaki et al. (2003) have used industrial tomato hybrids, albeit cultivated under greenhouse conditions, to show that the combined suppression of *LeExp1* and *LePG* is a valid strategy to improve the quality of tomato fruit for the processing industry. Most likely, this genetic strategy will have to be optimized for the chosen genetic background and it might include the genetic modification of other genes involved in ripening-associated changes in firmness and texture.

Prolonged Vine Life

During tomato fruit ripening, polyamines content decreases (Galston and Sawhney 1990), while ethylene biosynthesis and action increases. Ethylene is a phytohormone that controls fruit ripening. Evidence has suggested that polyamines inhibit ethylene biosynthesis and that polyamines and ethylene interact during tomato fruit ripening. Thus, polyamines are considered to have an anti-ripening function (Cassol and Mattoo 2002).

The polyamines and ethylene biosynthetic pathways have a common substrate: S-adenosylmethionine (SAM). The enzyme SAM decarboxylase converts SAM into decarboxylated SAM, and thus it controls a rate-limiting step in polyamine biosynthesis. In the ethylene biosynthetic pathway, SAM is converted to 1-aminocyclopropane-carboxylic acid (ACC) by ACC synthase. The ACC is then converted to ethylene. These considerations prompted Metha and collaborators to genetically modify tomato with a transgene driving expression of the SAM decarboxylase from yeast under the control of the fruit-specific E8 promoter (Metha et al. 2002). The transgene was expressed only in the tomato fruit during ripening and it caused an increase in spermine and spermidine content of the fruit. Fruit ripening was significantly delayed, thus prolonging fruit vine life. Genetically modified fruits were of normal size. Lycopene content was more than doubled in genetically modified fruits. This increase is beneficial because of its antioxidant properties. Tomato fruits genetically modified for an increased polyamine content are also improved in quality parameters relevant for processing technologies: juice viscosity (on the average 40% increased) and precipitate weight ratio increment (on the average 50% increased) in comparison to controls. Despite a higher content of polyamines (spermine and spermidine), genetically modified fruits produced more ethylene than control fruits.

CONCLUSIONS

Most transgenic tomato plants have been raised for heuristic reasons. Thus, the main benefit obtained from transgenic tomato plants has been, so far, knowledge of plant biological phenomena. In this regard, crop improvement based on genetic knowledge is the consequent, and yet subordinate, product of our zest for knowledge. A genetic strategy to engineer a novel plant product, transgenic or not, must be tested under cultivation conditions used for production. This final assessment represents also the most severe test for our understanding of biological processes.

Fruit productivity and quality are both crucial for the successful assessment of any novel tomato cultivar or hybrid. Consequently, transgenic tomatoes must meet these criteria in order to represent an innovative product worthy of widespread cultivation. An innovative tomato plant, in order to be valuable for consumers, farmers and the processing industry, must be improved either in its quality or in its productivity, and preferably in both. In any case, an improved quality of the product should not curtail plant productivity.

Several studies on genetically modified plants have been performed within projects whose approval is required by public institutions of food safety (e.g., European Commission, US Food and Drug Administration), in order to ensure the safety of these products for human health and for the environment (EU 2001, FDA 1992, FAO/WHO 1996). An examination of the summary notifications that have been submitted under EU directives for the deliberate release into the environment of genetically modified organisms (http://genetically modifiedoinfo.jrc.it/genetically modifiedp_browse.aspx; http://biotech.jrc.it/deliberate/genetically modifiedo.asp) shows that the genetically modified tomato plants so far tested in field trials were mainly modified for virus resistance, delayed ripening, or parthenocarpic fruit development. In the United States (http://www.cfsan.fda.gov/~lrd/biocon.html), most of the tomato plants tested under field cultivation conditions were genetically modified either for disease resistances or for delayed ripening. The field trials so far performed have shown that the quality of genetically modified tomato is either equal or superior to their parental genotypes. In some cases, tomato plant productivity has been improved.

REFERENCES

Acciarri, N. and V. Ferrari, G. Vitelli, N. Ficcadenti, T. Pandolfini, A. Spena, and G.L. Rotino. 2000. Effetto della partenocarpia in ibridi di pomodoro geneticamente modificati. L'Informatore Agrario LVI(4): 117-121.

Accotto, G.P. and G. Nervo, N. Acciarri, L. Tavella, M. Vecchiati, M. Schiavi, G. Mason, and A.M. Vaira. 2005. Field evaluation of tomato hybrids engineered with Tomato spotted wilt virus sequences for virus resistance, agronomic performance and pollen-mediated transgene flow. Phytopathology 95(7): 800-807.

Bernstein, E. and A.A. Caudy, S.M. Hammond, and G.J. Hannon. 2001. Role for a bidentate ribonuclease in the initiation step of RNA interference. Nature 409: 363-366.

Brandt, S. and Z. Pék, E. Barna, A. Lugasi, and L. Helyes. 2006. Lycopene content and colour of ripening tomatoes as affected by environmental conditions. J. Sci. Food Agric. 86: 567-572.

Brummell, D.A. and W.J. Howie, C. Ma, and P. Dunsmuir. 2002. Postharvest fruit quality of transgenic tomatoes suppressed in expression of a ripening-related expansin. Postharvest Biol. Technol. 25: 209-220.

[CALGENE] 1992. Annual Report. Calgene, Inc., Davis, USA, 62 pp.

Cassol T. and A.K. Mattoo. 2002. Do polyamines and ethylene interact to regulate plant growth, development and senescence? pp. 121-132. *In:* P. Nath, A. Mattoo, S.R. Ranade and J.H. Weil. [eds.] Molecular Insights in Plant Biology. Science Publishers, Inc., Enfield, USA.

Collins, J.K. and P. Perkins-Veazie, and W. Roberts. 2006. Lycopene: from plants to humans. Hortic. Sci. 41: 1135-1144.

Ciereszko, I. and H. Johansson, and L. Kleczkowski. 2001. Sucrose and light regulation of a cold-inducible UDP-glucose pyrophosphorilase gene via exokinase-independent and abscisic acid-insentive pathway in *Arabidopsis*. Biochem. J. 354: 67-72.

Corpillo, D. and G. Gardini, A.M. Vaira, M. Basso, S. Aime, G.P. Accotto, and M. Fasano. 2004. Proteomics as a tool to improve investigation of substantial equivalence in genetically modified organism: the case of a virus-resistant tomato. Proteomics 4: 193-200.

Corrado V. 1773. Il cuoco galante Napoli, Stamperia Raimondiana.

[EU] European Union. 2001. Directive 2001/18/EC of the European Parliament and of the Council of 12 March 2001 on the deliberate release into the environment of genetically modified organisms and repealing Council Directive 90/220/EEC. Off. J. Eur. Commun. L 106: 1-39. http://europa.eu.int/smartapi/cgi/sga_doc?smartapi!celexapi!prod!CELEXnumdoc&lg=EN&numdoc=32001L0018&model=guichett.

[FAO/WHO] Food and Agriculture Organization of the United Nations. 1996. Biotechnology and Food Safety. Report of a Joint FAO/WHO Consultation, Rome, Italy, 30 September-4 October 1996. FAO Food and Nutrition Paper 61.

[FDA] Food and Drug Administration. 1992. Statement of policy: foods derived from new plant varieties. Federal Register 57: 22984-23005. http://www.cfsan.fda.gov/~acrobat/fr920529.pdf.

Felici, C. 1572. Scritti naturalistici, I, De l'insalata e piante che in qualunque modo vengono per cibo de l'homo. Manoscritti 1569-1572. Guido Arbizzoni editor, Quattroventi edizioni, Urbino.

Ficcadenti, N. and S. Sestili, T. Pandolfini, C. Cirillo, G.L. Rotino, and A. Spena. 1999. Genetic engineering of parthenocarpic fruit development in tomato. Mol. Breed. 5: 463-470.

Fire, A. and S. Xu, M.K. Montgomery, S.A. Kostas, and S.E. Driver, C.C. Mello. 1998. Potent and specific genetic interference by double-stranded RNA in *Caenorabditis elegans*. Nature 391: 806-811.

Friedman, M. 2002. Tomato glycoalkaloyds: role in the plant and in the diet. J. Agr. Food Chem. 50: 5751-5780.

Galston, A.W. and R.K. Sawhney. 1990. Polyamines in plant physiology. Plant Physiol. 94: 406-410.

Gepts, P. 2002. A comparison between crop domestication, classical plant breeding and genetic engineering. Crop Sci. 42: 1780–1790.

Giorio, G. and A.L. Stigliani, and C. D'Ambrosio. 2007. Agronomic performance and transcriptional analysis of carotenoid biosynthesis in fruits of transgenic HighCaro and control tomato lines under field conditions. Transgenic Res. 16: 15–28.

Glein, J.J.L. and P. de Haan, A.J. Kool, D. Peters, M.Q.J.M. van Grinsven, and R.W. Goldbach. 1991. Engineered resistance to tomato spotted wilt virus, a negative-strand RNA virus. Bio. Technology 9: 1363-1367.

Hammond, S.M. and E. Bernstein, D. Beach, and G.J. Hannon. 2000. An RNA-directed nuclease mediates post-transcriptional gene silencing in *Drosophila* cells. Nature 404: 293-296.

Horiguchi, G. 2004. RNA silencing in plants: a short cut to functional analysis. Differentiation 72: 65-73.

Kalamaki, M.S. and M.H. Harpster, J.M. Palys, J.M. Labavitch, D.S. Reid, and D.A. Brummell. 2003. Simultaneous transgenic suppression of LePG and LeExp1 influences rheological properties of juice and concentrates from a processing tomato variety. J. Agric. Food Chem. 51: 7456-7464.

Keeler, R.F. and D.C. Baker, and W. Gaffield. 1991. Teratogenic *Solanum* species and the responsible teratogens. pp. 83-99. *In:* R.F. Keeler and A.T. Tu. [eds.]

Handbook of Natural Toxins, Vol. 6 (Toxicology of Plant and Fungal Compounds). Marcel Dekker, New York, USA.

Laporte, M.M. and J.A. Galagan, J.A. Shapiro, M.R. Boersig, C.K. Shewmaker, and T.D. Sharkey. 1997. Sucrose-phosphate synthase activity and yield analysis of tomato plants transformed with maize sucrose-phosphate synthase. Planta 203: 253-259.

Laporte, M.M. and J.A. Galagan, A.L. Prash, P.J. Vanderveer, D.T. Hanson, C.K. Shewmaker, and T.D. Sharkey. 2001. Promoter strength and tissue specificity effects on growth of tomato plants transformed with maize sucrose-phosphate synthase. Planta 212: 817-822.

Latham L. and R.A.C. Jones. 1998. Selection of resistance breaking strains of tomato spotted wilt tospovirus. Ann. Appl. Biol. 133: 385-402.

Lindbo, J.A. and W.G. Dougherty. 2005. Plant pathology and RNAi: a brief history. Annu. Rev. Phytopathol. 43: 192-204.

Mattioli P.A. 1568. I discorsi nelli sei libri di Pedacio Dioscoride Anarzabeo della materia medicinale, Vincenzo Valgrisi, Venezia.

Metha, R.A. and T. Cassol, N. Li, N. Ali, A.K. Handa, and A.K. Mattoo. 2002. Engineered polyamine accumulation in tomato enhances phytonutrient content, juice quality, and vine life. Nature Biotechnol. 20: 613-618.

Nervo, G. and C. Cirillo, G.P. Accotto, and A.M. Vaira. 2003. Characterization of two tomato lines highly resistant to TSWV following transformation with the viral nucleoprotein gene. J. Plant Pathol. 85: 139-144.

Pecker, I. and R. Gabbay, F.X. Cunningham, and J. Hirschberg. 1996. Cloning and characterization of the cDNA for lycopene beta-cyclase from tomato reveals decrease in its expression during fruit ripening. Plant Mol. Biol. 30: 807-819.

Pandolfini, T. and G.L. Rotino, S. Camerini. R. Defez, and A. Spena. 2002. Optimisation of transgene action at the post-transcriptional level: high quality parthenocarpic fruits in industrial tomatoes. BMC Biotechnol. 2: 1.

Porretta, S. and G. Poli, and E. Minuti. 1998. Tomato pulp quality from transgenic fruits with reduced polygalacturonase (PG). Food Chem. 62: 283-290.

Powell, A.L.T. and M.S. Kalamaki, P.A. Kurien, S. Gurrieri, and A.B. Bennett. 2003. Simultaneous transgenic suppression of LePG and LeExp1 influences fruit texture and juice viscosity in a fresh market tomato variety. J. Agric. Food Chem. 51: 7450-7455.

Rotino, G.L. and N. Acciarri, E. Sabatini, G. Mennella, R. Lo Scalzo, A. Maestrelli, B. Molesini, T. Pandolfini, J. Scalzo, B. Mezzetti, and A. Spena. 2005. Open field trial of genetically modified parthenocarpic tomato: seedlessness and fruit quality. BMC Biotechnol. 5: 32.

Schwab, R. and S. Ossowski, M. Riester, N. Warthmann, and D. Weigel. 2006. Highly specific gene silencing by artificial microRNAs in Arabidopsis. Plant Cell 18: 1121-1133.

Smith, C.J.S. and C.F. Watson, J. Ray, C.R. Bird, P.C.Morris, W. Schuch, and D. Grierson. 2002. Antisense RNA inhibition of polygalacturonase gene expression in transgenic tomato. Nature 334: 724-726.

Twyman, R.M. and E. Stoger, S. Schillberg, P. Christou, and R. Fischer. 2003. Molecular farming in plants: host systems and expression technology. Trends Biotechnol. 21: 570-578.

VIB Flanders Interuniversity Institute for Biotechnology. 2001. Safety of genetically engineered crops. VIB publication, March 2001. HYPERLINK http://www.vib.be www.vib.be

Zamore P.D. and B. Haley. 2005. Ribognome: the big world of small RNAs. Science 309: 151-1524.

Wiese, A. and N. Elzinga, B. Wobbes, and S. Smeekens. 2005. Sucrose-induced translational repression of plant bZIP-type transcription factors. Biochem. Soc. Trans. 33: 272-275.

4

Post-harvest Ripening of Tomato

María Serrano[1], Pedro Javier Zapata[2], Fabián Guillén[2], Domingo Martínez-Romero[2], Salvador Castillo[2] and Daniel Valero[2]*

[1]Dept. Applied Biology, University Miguel Hernández, Ctra. Beniel km. 3.2, 03312, Orihuela Alicante, Spain

[2]Dept. Food Technology, University Miguel Hernández, Ctra. Beniel km. 3.2, 03312, Orihuela Alicante, Spain

ABSTRACT

The main changes occurring during the ripening process of tomato, either on plant or during post-harvest storage, are described in this chapter. These changes are associated with the increase in ethylene production and occur relatively early after harvesting; the fruits reach an over-ripe state considered unmarketable. Thus, most post-harvest storage technologies are focused on controlling the biosynthesis and action of ethylene in order to delay the evolution of these changes and in turn to extend the storage life with optimum quality before consumption. A revision of the researched post-harvest tools, including heat treatment, modified atmosphere packaging, edible coatings and 1-methylcyclopropene (1-MCP) treatment are addressed, the best results being obtained with the use of 1-MCP, although edible coatings seem to be a promising technology.

INTRODUCTION

Wild-type tomato species are thought to be native to western South America and specifically the dry coastal desert of Peru. Some authors have postulated that tomato originated from the Mexican region, where

**Corresponding author*

A list of abbreviations is given before the references.

the people called it "tomati". During the 16th century, Spanish explorers introduced tomato into Europe. Some believe the Spanish explorer Cortez might have been the first to carry the small yellow tomato to Europe after he captured the Aztec city of Tenochtítlan in 1521, now Mexico City, although others believe Christopher Columbus, also from Spain, discovered the tomato earlier in 1493 (Gould 1983). Initially, tomato was considered an ornamental plant from the New World and it became popular in Europe as an edible fruit from the 18th century.

Tomatoes (*Solanum lycopersicum*, formerly *Lycopersicon esculentum*) belong to the Solanaceae family. In the late 1880s, 171 cultivars were named, of which only 61 were truly different lines. Currently, the total production of tomatoes worldwide is over 123 million t (FAOSTAT 2006), with the main producers being China, the United States and Turkey. In Europe, the total production is more than 17 million t. The most important producer countries are those within the Mediterranean Sea (EUROSTAT 2006): Italy (42%), followed by Spain (21%) and Greece (11%).

Tomato fruit is characterized to follow a ripening-climacteric pattern, which is controlled by the plant hormone ethylene, involving a wide range of physical, chemical, biochemical and physiological changes, which start in the plant and follow after detachment from it. These changes occur relatively quickly after harvest and the fruits reach an over-ripe state considered unmarketable. Thus, most post-harvest storage technologies are focused on controlling the biosynthesis and action of ethylene in order to delay these changes and to extend the storage life with optimum quality before consumption.

In this chapter, the changes that occur during both development and ripening of tomato on plant and/or during post-harvest storage affecting the organoleptic, nutritive and functional properties are reviewed. In addition, conventional and recent innovative post-harvest tools to maintain tomato quality during storage and marketing are described.

CHANGES IN TOMATO FRUIT RIPENING ON PLANT AND DURING POST-HARVEST STORAGE

The growth cycle of tomato fruit follows a simple sigmoid curve with three distinct phases. Phase I is characterized by slow growth involving the cell division processes and reaching the maximum number of cells. It lasts 7-14 d after pollination, depending on cultivar, environmental factors and growth conditions. Phase II comprises cell expansion leading to rapid fruit growth and attainment of maximum fruit size, in 3-5 wk. The final phase involves ripening with low or null growth rate but with intense metabolic changes (Srivastava and Handa 2005).

Tomato fruit ripening is a genetically programmed complex of events that terminates with senescence. In tomato, as well as other climacteric fruits, the plant hormone ethylene is required for normal fruit ripening and thus considered a trigger of a wide range of physical, physiological, and biochemical changes, which make tomato fruit attractive for consumption. The conversion of tomato fruit from the mature green to the fully ripe stage involves the above changes affecting quality parameters such as colour, texture, flavour, and bioactive compounds, which can occur during development on plant and after harvest.

Colour Changes

The typical colour changes during tomato ripening from green to red are associated with chlorophyll breakdown and the synthesis of carotenoid pigments due to the transformation of chloroplasts to chromoplasts. The measurement of colour by reflexion using the Hunter Lab procedure shows that the parameter a* is a good indicator of the tomato colour changes, since it shows negative values for green colour and positive values for red colour. In fact, Fig. 1 shows the evolution of colour a* at six ripening stages selected according to the USDA colour chart (USDA 1991). At green stage, for example, the a* value was -14, which increased significantly along the ripening process, reaching values of +22 at the red stage. The increase of parameter a* was highly correlated ($r^2 = 0.978$) with the increase of lycopene concentration (Fig. 1). Lycopene is considered the predominant carotenoid of tomato fruit (80-90%), followed by β-carotene (5-10%) (Lenucci et al. 2006). Lycopene has been suggested as a good indicator of the level of ripening. At the breaker stage, and when coloration becomes evident, lycopene starts to accumulate and its concentration increases up to 500-fold in ripe fruits, although concentrations vary significantly among cultivars (Fraser et al. 1994, Kaur et al. 2006).

Texture

During the ripening of tomato fruit, a softening process occurs on account of the activity of several enzymes that alter the structural components of the cell wall and diminish cell adhesion. The main enzymes responsible for tomato softening are polygalacturonase, pectin-methyl-esterase, endo-β-mannase, α- and β-galactosidases and β-glucanases (Fischer and Bennett 1991, Brummel and Harpster 2001, Seymour et al. 2002, Carrari and Fernie 2006). The activity of these enzymes causes softening of the whole fruit by degrading some components of the cell wall and decreasing the adhesion of the mesocarpic cells. In addition, not only the adhesion of mesocarp

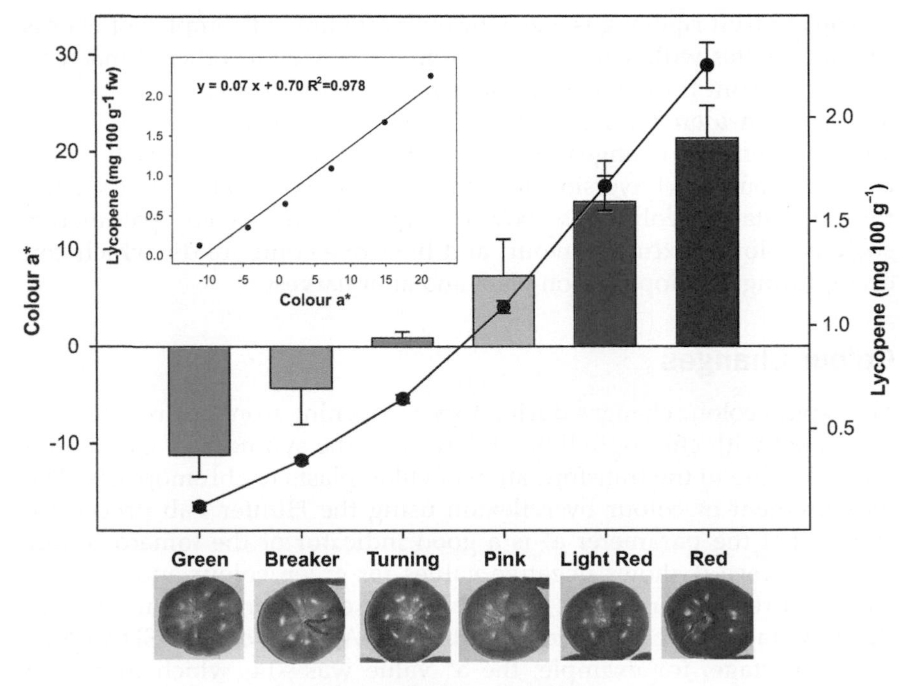

Fig. 1 Colour a* (bars) parameter and lycopene concentration (straight line) in tomato fruit from 'Raf' cultivar at different ripening stages. Data are the mean ± SE of 10 individual fruits. Inset plot shows the linear regression between the two parameters. Colour was measured using Minolta colorimeter as previously described (Serrano et al. 2005). Lycopene was extracted and determined by absorbance according to Fish et al. (2002).

parenchymatic cells but also the epidermis and cuticle are important structural components for the integrity of the tomato fruit, and their role in the softening process has been reported (Bargel and Neinhuis 2005). As can be seen in Fig. 2, there is a significant decrease in fruit firmness, expressed as deformation force, from the green to the red ripening stage, although the main firmness losses of about 70% of the initial levels occurred from green to pink stage.

Flavour

The typical taste of tomato is mainly attributed to soluble sugars, organic acids and volatile compounds. During tomato ripening there is an increase in total soluble solids (TSS), expressed as °Brix, and diminution of total acidity (TA) (Fig. 2). The main components of TSS are sugars, of which

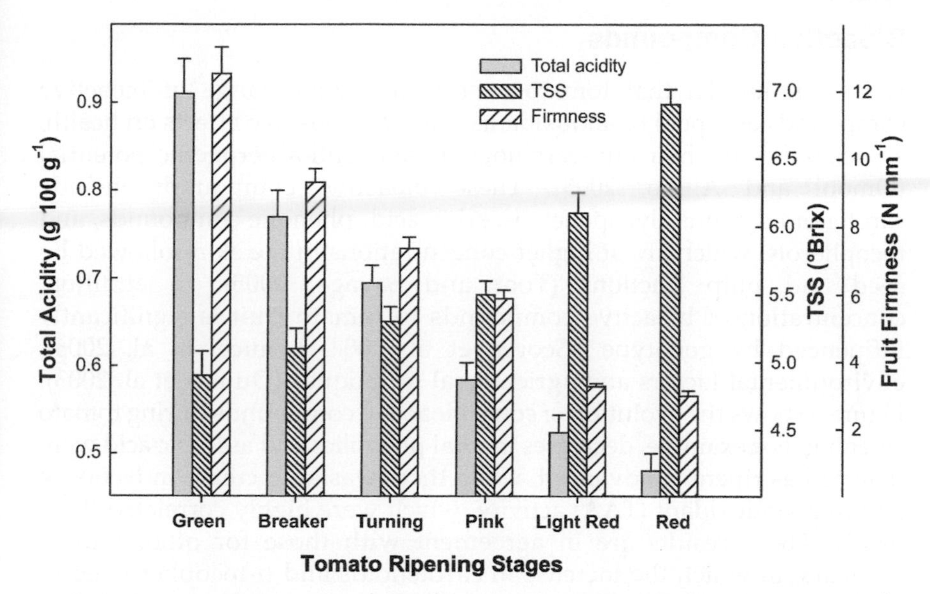

Fig. 2 Total acidity (g citric acid eq. 100 g^{-1}), total soluble solids (°Brix) and firmness (N mm^{-1}) at different ripening stage of 'Raf' tomato. Data are the mean ± SE of 10 individual fruits. Parameters were determined by refractometer (TSS), titration (TA) and texturometer (firmness) according to Serrano et al. (2005).

fructose and glucose are found to be predominant in domesticated tomatoes, while some wild tomato species accumulate sucrose. In relation to organic acids, citric acid is considered the main one responsible for acidity and it decreases from green to red ripening stage (Carrari and Fernie 2006). With respect to aroma, over 400 volatile compounds have been identified in fresh and processed tomato products (Petro-Turza 1987), although 20-30 have proved to be the most important compounds contributing to the aroma. Among these compounds, C6 compounds such as hexanal, *cis*-3-hexenal, *trans*-2-hexenal, *cis*-3-hexenol, and hexenol are considered predominant in macerated tomatoes (Buttery and Ling 1993, Yilmaz et al. 2001, Ruiz et al. 2005). The elevated presence of 2-methylbutanol, which derives from the amino acid leucine (Buttery and Ling 1993), has been described as one of the main components contributing to sweet/fresh ripe tomato flavour (Tandon et al. 2000, Maneerat et al. 2002, Baldwin et al. 2004). During post-harvest storage, there is an increase of most of the volatile compounds, and specifically hexanal correlated negatively to sourness and positively to sweetness (Krumbein et al. 2004).

Bioactive Compounds

It is well known that tomato possesses a wide range of bioactive compounds as a pool of antioxidants that have positive effects on health, associated with their anti-carcinogenic and anti-atherogenic potential (Omoni and Aluko, 2005). These bioactive compounds include carotenoids (mainly lycopene), ascorbic acid, phenolic compounds, and tocopherols, which are at higher concentrations in the skin followed by seed and pulp fractions (Toor and Savage, 2005). In addition, concentration of bioactive compounds in tomato fruit is significantly influenced by genotype (George et al. 2004, Lenucci et al. 2006), environmental factors and agricultural techniques (Dumas et al. 2003). Figure 3 shows the evolution of some bioactive compounds during tomato ripening. For example, decreases in total phenolics and ascorbic acid were detected as ripening advanced, while there was an increase in lycopene and total antioxidant (TAA) activity, which were highly correlated (R^2 = 0.972). These results are in agreement with those for other tomato cultivars, in which the increase in carotenoids and α-tocopherol led to increased TAA in the water-insoluble fraction (Raffo et al. 2002). However, these authors reported no significant variations in ascorbic acid or vitamin C as ripening stage advanced. During post-harvest storage, increases in lycopene concentration have been also reported, the enhancement being higher as temperature increased (Toor and Savage 2006), although TAA increased to higher extension at 5 than at 10°C, probably because of the stimulation of the biosynthesis of antioxidant active compounds by chilling stress (Javanmardi and Kubota 2006).

In relation to phenolic content, chlorogenic acid and rutin have been found to be the most important flavonoids in tomato. Butta and Spaulding (1997) reported high levels of phenolics at the earliest stage of tomato development but concentrations declined rapidly during fruit ripening, as well as during post-harvest ripening (Fig. 3), although other authors have shown that the content of total phenolics remained stable during ripening (Slimestad and Verheul 2005).

Physiological Changes

Tomato is considered a good model to study the biosynthesis and action of ethylene in climacteric fruits. Excellent and recent reviews are available about the ethylene biosynthesis, perception, signal transduction and its regulation at biochemical, genetic and biotechnological levels (Alexander and Grierson 2002, Giovannoni 2004, Carrari and Fernie 2006). Tomato

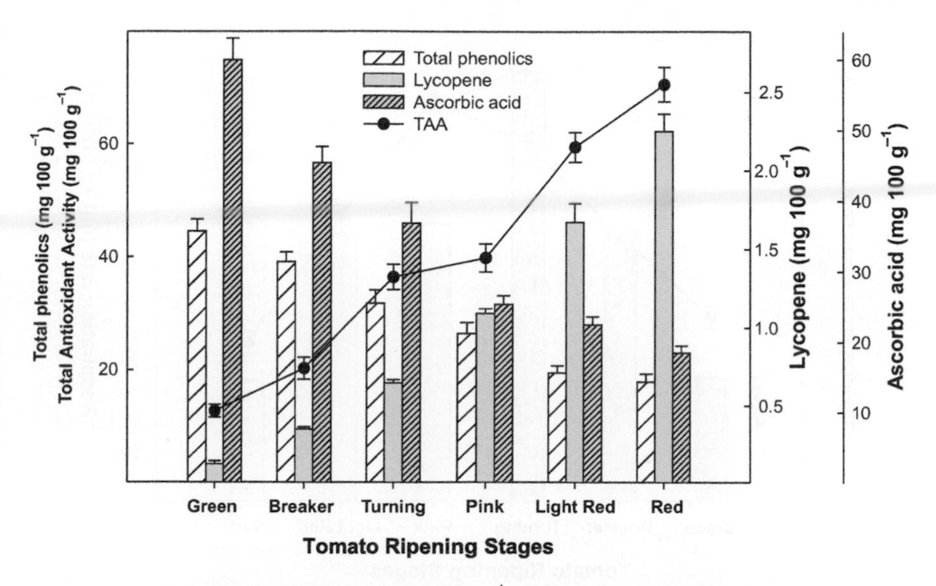

Fig. 3 Total phenolics (mg gallic acid eq. 100 g^{-1}), total antioxidant activity (mg ascorbic acid eq. 100 g^{-1}), lycopene (mg 100 g^{-1}) and ascorbic acid (mg 100 g^{-1}) in 'Raf' tomato at different ripening stages. Data are the mean ± SE of 10 individual fruits. Total phenolics and total antioxidant activity were measured by mass spectroscopy, while ascorbic acid was determined by HPLC (Serrano et al. 2005).

shows a climacteric physiological ripening pattern, with peaks in both ethylene production and respiration rates (Fig. 4). The maximum ethylene production is found at turning stage, while respiration rate peaked at pink stage. From green stage, at which the ethylene production began, the physicochemical parameters related to ripening started to change, mainly the diminution in total acidity and firmness, while the major changes in total soluble solids and lycopene accumulation occurred after the climacteric peak (turning stage), as can be seen in Figs. 1 and 2.

EFFECTS OF POST-HARVEST TECHNOLOGIES TO MAINTAIN TOMATO FRUIT QUALITY

Tomato fruit has a relatively reduced post-harvest life since many processes affecting quality loss take place after harvesting. The major limiting factors in the storage of tomato fruit are transpiration, fungal infection, acceleration of the ripening process and senescence. Under ambient temperature, tomato ripens rapidly and becomes unmarketable in a short period. The use of low temperature is effective in delaying and/

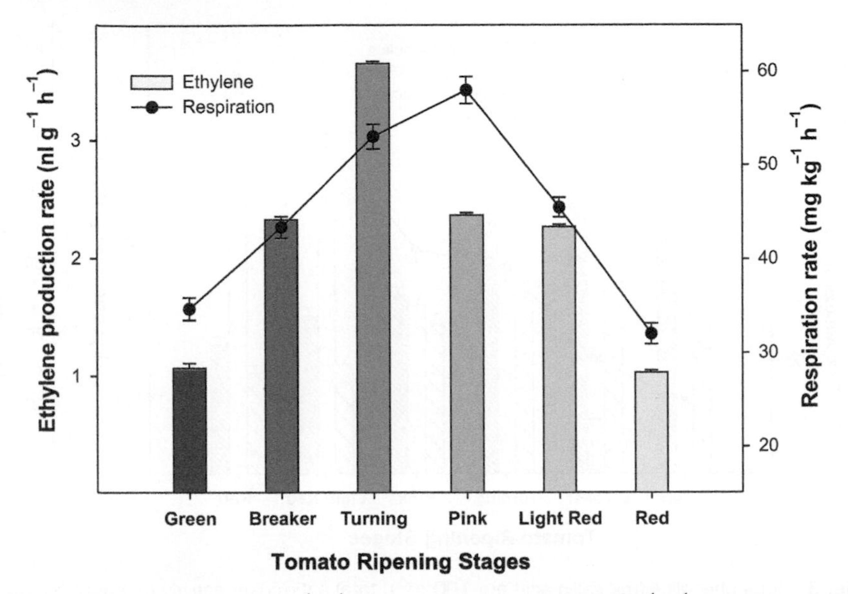

Fig. 4 Ethylene production (nL g^{-1} h^{-1}) and respiration rate (mg CO_2 kg^{-1} h^{-1}) of 'Raf' tomato at different ripening stages. Data are the mean ± SE of 10 individual fruits. Both parameters were analysed by GC-FID (ethylene) and GC-TCD (respiration rate) according to Guillén et al. (2007).

or reducing the ethylene production, but tomato fruits are sensitive to chilling injury (Cheng and Shewfelt 1988), and then temperatures over 11°C are advisable for post-harvest storage. Since the production of ethylene accelerates the above-mentioned biochemical and physiological changes that occur during ripening leading to senescence, any tool that prevents ethylene biosynthesis and/or action would delay the quality losses and in turn increase the post-harvest shelf life, with obvious commercial interest. Apart from low temperature storage, very few post-harvest technologies have been researched to maintain tomato quality parameters: they include heat treatments, modified atmosphere packaging, 1-methylcyclopropene (1-MCP) applications and the use of edible coatings.

Heat Treatment

The main objective of mild heat treatment prior to tomato storage has been to alleviate occurrence of chilling injury. For this purpose, hot water dips or hot air treatments have been used (McDonald et al. 1999, Ali et al. 2004, Saltveit 2005, Polenta et al. 2006), with several combinations of temperature and duration at ranges of 27-48°C and 15-90 min. Generally heat treatments can alleviate chilling injury symptoms in tomatoes by reducing decay, loss of firmness, and electrolyte leakage by protecting membrane components.

Modified Atmosphere Packaging (MAP)

The principles of fruit storage under modification of the atmosphere are based on changes in the internal atmosphere composition from refrigerated produce packaged in closed polymeric films. The gas composition is generated by the interaction between the respiration of the fruit or vegetable and the permeability of the film (Artés et al. 2006). In tomato, there is some evidence on the effect of MAP use on quality. Several modifications in gas levels have been reported with positive effects in different cultivars of tomatoes: 5kPa O_2 - 10 kPa CO_2 (Boukozba and Taylor 2002), 5 kPa O_2 - 8 kPa CO_2 (Suaparlan and Itoh 2003), 5 kPa O_2 - 3 kPa CO_2 (Artés et al. 2006), and 4 kPa O_2 - 10 kPa CO_2 (Bailén et al. 2006). The use of MAP led to reduction in weight losses and delay in both colour changes and softening, while the content of TSS and TA were unaffected. It is important to note that the use of MAP could not sufficiently control decay, which could be attributed to the high level of humidity inside the packages, which stimulates mould growth. In this sense, the addition of activated carbon impregnated with palladium (active packaging) showed diminutions in decay (60%) after one month of storage by reducing the ethylene accumulation with a net increase in shelf life compared to control MAP packages (Bailén et al. 2006). One of the disadvantages of using MAP to store tomatoes was the loss of volatile compounds due to the low oxygen conditions, although high losses were only found under low temperature storage (Boukozba and Taylor 2002). MAP has also been used on fresh-cut tomato slices, showing that shelf life could be extended by reducing microbial spoilage, decay and chilling injury symptoms when 1 kPa O_2 - 20 kPa CO_2 were reached (Hong and Gross 2001, Gil et al. 2002). In addition, lower colour changes and better appearance were detected with atmosphere modification of 2.5 kPa O_2 - 20 kPa CO_2, while TSS and TA were not significantly affected (Artés et al. 1999).

1-Methylcyclopropene

1-MCP was discovered by Sisler and Blankenship as a potent ethylene antagonist. It is able to bind the ethylene receptor with 10 times as much affinity as ethylene itself, being more active at much lower concentrations (Serek et al. 1995, Sisler and Serek 1999, Blankenship and Dole 2003). This compound has attracted great interest worldwide as a new, non-toxic agent for humans and environment (Environmental Protection Agency 2002) and safe post-harvest chemical in the agro-food industry capable of maintaining post-harvest quality of many fresh commodities, both climacteric and non-climacteric (Blankenship and Dole 2003). There are several reports that prove the efficacy of 1-MCP in inhibiting ethylene production in tomato and delaying the ripening process (Wills and Ku 2002, Mostofi et al. 2003, Opiyo and Ying 2005). In these papers, several combinations of 1-MCP dosage and duration of treatments were reported, but it has been recently proposed that the optimum treatment for tomato was 0.5 μl l^{-1} during 24 hr (Guillén et al. 2007a). These authors also showed that tomato cultivar and ripening stage at the time of 1-MCP treatment affect its efficacy in controlling post-harvest ripening (Guillén et al. 2006).

The efficacy of 1-MCP (at 0.5 μl l^{-1} during 24 hr) in seven tomato cultivars ('Boludo', 'Cherry', 'Daniela', 'Patrona', 'Raf', 'Roncardo' and 'Tyrade') was assayed during post-harvest storage. The results after 28 d of storage revealed that 1-MCP treatment was able to reduce weight loss (Fig. 5), delay the increase in colour parameter a* (Fig. 6), retard the

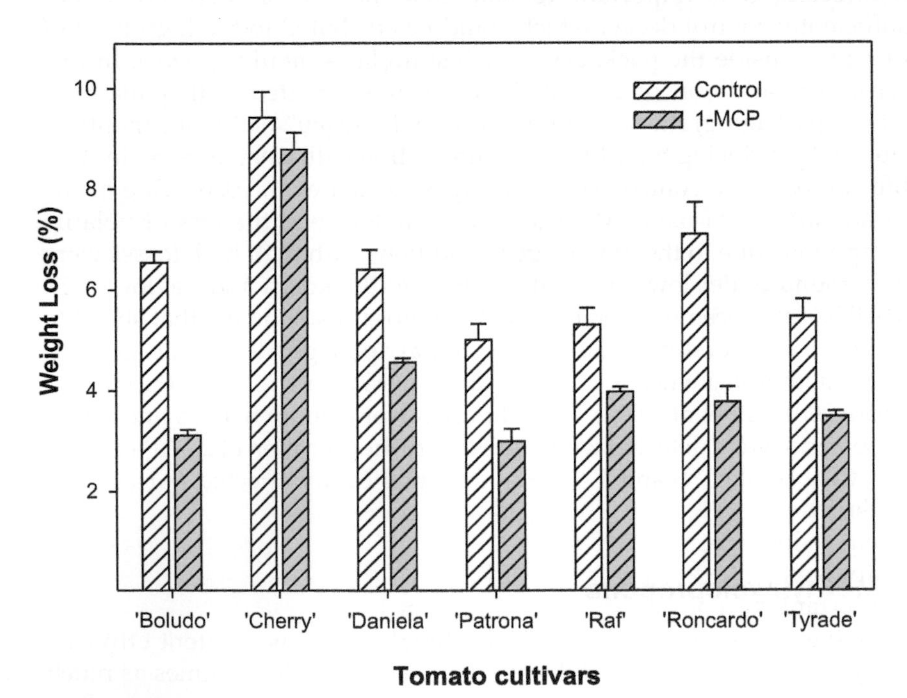

Fig. 5 Weight loss of different tomato cultivars after 28 d of storage at 10°C, treated or not (control) with 1-MCP at 0.5 μl l^{-1} during 24 hr after harvesting. Data are the mean ± SE of 3 replicates of 20 fruits.

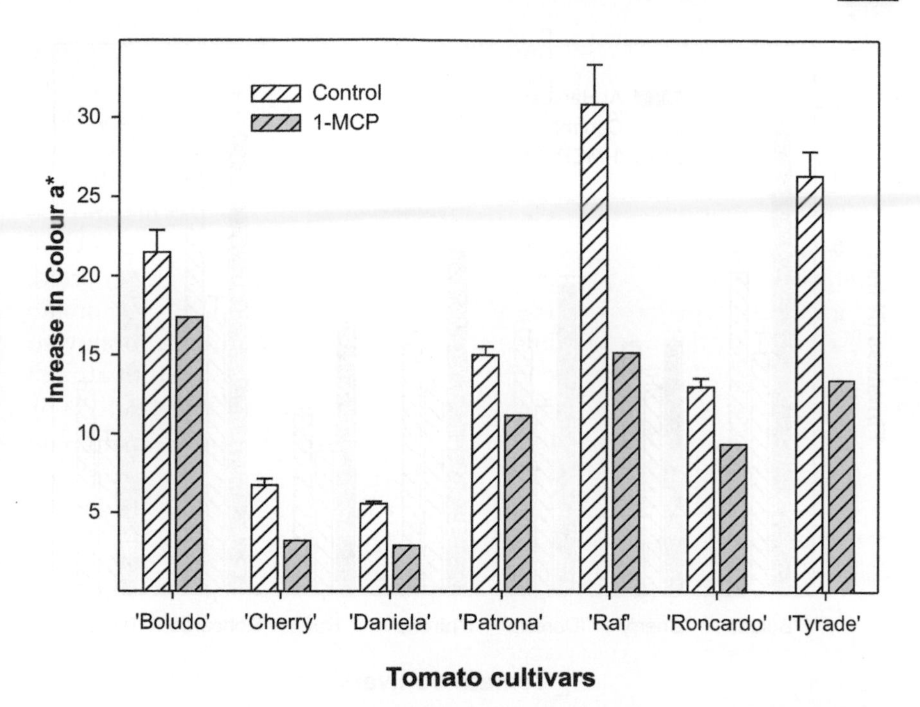

Fig. 6 Increase in colour a* parameter of different tomato cultivars after 28 d of storage at 10°C, treated or not (control) with 1-MCP at 0.5 μl l^{-1} during 24 hr after harvesting. Data are the mean ± SE of 3 replicates of 20 fruits.

softening process (Fig. 7) and reduce acidity loss (Fig. 8). No significant variation was obtained for TSS attributed to 1-MCP treatment for any cultivar (data not shown).

All these effects clearly indicated that tomato receptors were blocked by 1-MCP, the ethylene production was inhibited, and, in turn, the ripening process of tomato was delayed. Loss of weight is one of the most important causes of fruit quality deterioration, and very few reports describe the effect of 1-MCP on this parameter. On average, the observed reduction of weight loss was 10-50% depending on cultivar. This lower weight loss has an important economic repercussion during long shipment periods. In tomato, colour, firmness and acidity are very important for consumer acceptance, and their rapid evolution during post-harvest storage is responsible for the reduced shelf life of tomato. The magnitude of the inhibition of increase of colour a* parameter, and firmness and acidity retention by 1-MCP during storage was dependent on both cultivar and the value for each parameter at harvest.

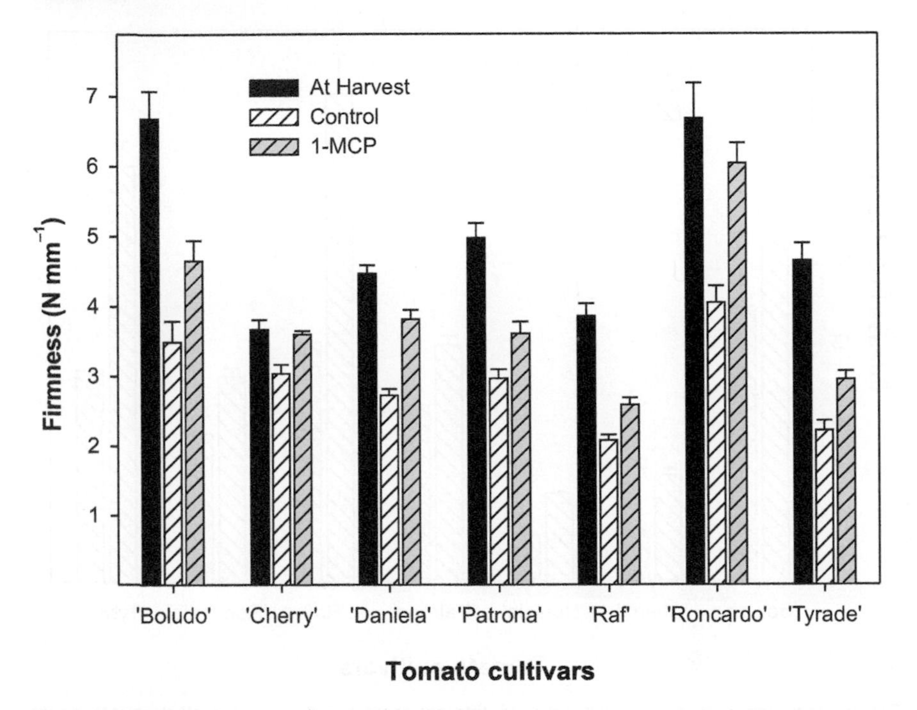

Fig. 7 Fruit firmness (N mm^{-1}) of different tomato cultivars at harvest and after 28 d of storage at 10°C, treated or not (control) with 1-MCP at 0.5 µl l^{-1} during 24 hr after harvesting. Data are the mean ± SE of 3 replicates of 20 fruits.

Edible Coatings

Edible coatings have been used to improve food appearance and preservation. They act as barriers during processing, handling and storage. They not only retard deterioration, enhancing food quality, but also improve food safety due to their natural biocide activity or the incorporation of antimicrobial compounds (Petersen et al. 1999, Cha and Chinnan 2004). Different compounds have been used as edible coatings to prevent commodity deterioration, including wax, milk proteins, celluloses, lipids, starch, zein, and alginate. In tomato particularly, there is little evidence about the effectiveness of edible coating as post-harvest treatment. We have found that coating of tomatoes with alginate or zein (at 5 or 10 % w/v) reduced the softening process and the increase in colour a* parameter after 9 d of storage at 20°C compared to control non-coated fruits (Fig. 9).

In fact, tomatoes coated with 10% zein retained more firmness than those with 5%, which could be related to the higher film thickness. Another explanation for the reduction of softening could be related to the

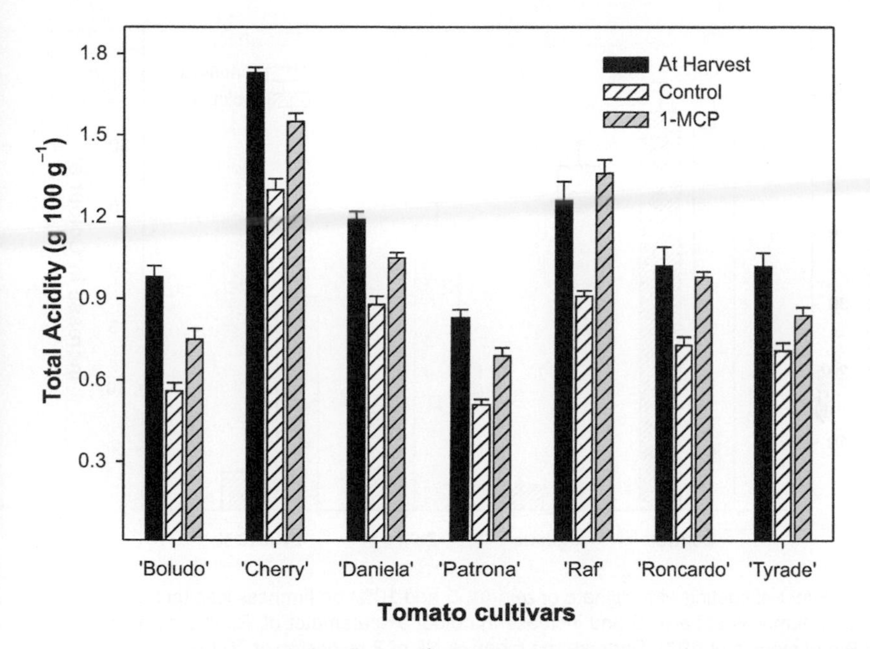

Fig. 8 Total acidity (g citric acid eq. 100 g^{-1}) of different tomato cultivars at harvest and after 28 d of storage at 10°C, treated or not (control) with 1-MCP at 0.5 μl l^{-1} during 24 hr after harvesting. Data are the mean ± SE of 3 replicates of 20 fruits.

lower weight loss observed (data not shown). The diminution of weight loss seems to be a general effect of coatings in fruits and might be based on the hygroscopic properties of the coatings, which enable formation of a barrier to water diffusion between fruit and environment, preventing its external transference (Morillon et al. 2002). The lower increase in colour a* parameter could be related to a delay of the ripening process probably due to a modification of the internal atmosphere (an increase in CO_2 level and a decrease in O_2), as was reported earlier (Park et al. 1994). In the case of colour, zein at 10% was also more effective than zein at 5% in delaying the colour evolution, although alginate was the most effective in controlling this parameter.

CONCLUDING REMARKS

In this chapter, the major changes occurring in tomato ripening on plant and during post-harvest storage have been described. Acceleration of the parameters related to ripening after harvest leads to a reduction in tomato quality affecting marketability and shelf life. Apart from low temperature

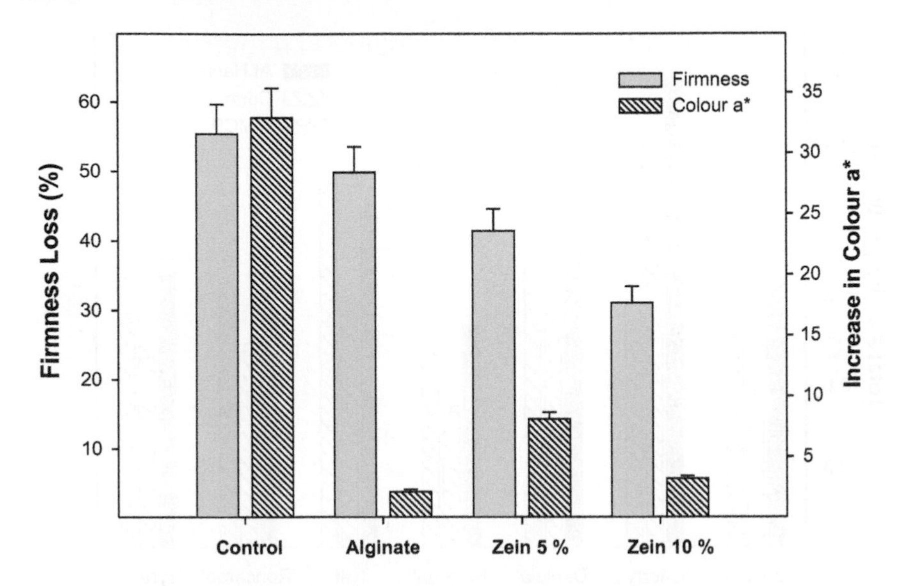

Fig. 9 Effect of coating with alginate or zein (at 5 and 10%) on firmness loss (percentage with respect to firmness at harvest) and increase in colour a* parameter of 'Rambo' tomato cultivar after 9 d of storage at 20°C. Data are the mean ± SE of 3 replicates of 20 fruits.

storage, there are few post-harvest technologies to retard the tomato ripening process and extend storage life. Some attempts have been carried out using heat treatments (to alleviate chilling injury), MAP technology, 1-MCP treatments and edible coatings. From the results reported herein, 1-MCP seems to have a promising future for commercial application. Currently, 1-MCP is being commercially used in cut flowers and fruits such as apples, bananas and melons. It is registered (under the trade name Smartfresh®, by Agrofresh, Inc, a subsidiary of Rohm & Haas, Spring House, Pennsylvania, USA) for use as post-harvest fruit treatment in 6 EU countries and 15 countries outside Europe, including the United States and Canada. In Spain, authorization and registration are pending for the use of 1-MCP in tomato. More research is necessary in the field of edible coatings, which could be considered safe for humans because of the use of natural products as ingredients. Finally, the combination of different tools to reduce and/or inhibit ethylene action in tomato deserves further research.

ACKNOWLEDGEMENTS

The authors thank Agrofresh (Rohm & Haas Italy S.R.L.) for 1-MCP support. Part of this work has been co-funded by the Spanish Ministry of

Science through Project AGL2006 04359/ALI and the European Commission with FEDER funds.

ABBREVIATIONS

1-MCP: 1-Methylcyclopropene; MAP: Modified atmosphere packaging; TAA: Total antioxidant activity; TSS: Total soluble solids; TA: Titratable acidity

REFERENCES

Alexander, L. and D. Grierson. 2002. Ethylene biosynthesis and action in tomato: A model for climacteric fruit ripening. J. Exp. Bot. 53: 2039-2055.

Ali, M.D. and K. Nakano, and S. Maezawa. 2004. Combined effect of heat treatment and modified atmosphere packaging on the color development of cherry tomato. Postharvest Biol. Technol. 34: 113-116.

Artés, F. and M.A. Conesa, S. Hernández, and M.I. Gil. 1999. Keeping quality of fresh-cut tomato. Postharvest Biol. Technol. 17: 153-162.

Artés, F. and P.A. Gómez, and F. Artés-Hernández.. 2006. Modified atmosphere packaging of fruits and vegetables. Stewart Postharvest Rev. 5:2.

Bailén, G. and F. Guillén, S. Castillo, M. Serrano, D. Valero, and D. Martínez-Romero. 2006. Use of activated carbon inside modified atmosphere packages to maintain tomato fruit quality during cold storage. J. Agric. Food Chem. 54: 2229-2235.

Baldwin, E.A. and K. Goodner, K. Pritchett, and M. Einstein. 2004. Effect of volatiles and their concentration on perception of tomato descriptors. J. Food Sci. 69: S310-S318.

Bargel, H. and C. Neinhuis. 2005. Tomato (*Lycopersicon esculentum* Mill.) fruit growth and ripening as related to the biomechanical properties of fruit skin and isolated cuticle. J. Exp. Bot. 56: 1049-1060.

Blankenship, S.M. and J.M. Dole. 2003. 1-Methylcyclopropene: a review. Postharvest Biol. Technol. 28: 1-25.

Boukozba, F. and A.J. Taylor. 2002. Effects of postharvest treatment on flavour volatiles of tomatoes. Postharvest Biol. Technol. 25: 321-331.

Brummel, D.A. and M.H. Harpster. 2001. Cell wall metabolism in fruit softening and quality and its manipulation in transgenic plants. Plant Mol. Biol. 41: 311-340.

Butta, J.G. and D.W. Spaulding. 1997. Endogenous levels of phenolics in tomato fruit during growth and maturation. J. Plant Growth Regul. 16: 43-46.

Buttery, R.G. and L. Ling. Enzymatic production of volatiles in tomatoes. pp. 137-146. *In:* P. Schreier and P. Winterhalter. [eds.] 1993. Progress in Flavour Precursor Studies. Allured Publishing Corporation, Carol Stream, USA.

Carrari, F. and A.R. Fernie. 2006. Metabolic regulation underlying tomato fruit development. J. Exp. Bot. 57: 1883-1897.

Cha, D.S. and M.Chinnan. 2004. Biopolymer-based antimicrobial packaging: A review. Crit. Rev. Food Sci. Nutr. 44: 223-237.

Cheng, T.S and R.L. Shewfelt. 1988. Effect of chilling exposure of tomatoes during subsequent ripening. J. Food Sci. 53: 1160-1162.

Dumas, Y. and M.Dadomo, G. Di Lucca, and P. Grolier. 2003. Review. Effects of environmental factors and agricultural techniques on antioxidant content of tomatoes. J. Sci. Food Agric. 83: 369-382.

Environmental Protection Agency. 2002. Federal Register, July 26, 67(144): 48796-48800.

EUROSTAT. http://ec.europa.eu/eurostat/. Accessed 2006.

FAOSTAT. http://faostat.fao.org/site/339/default.aspx. Accessed 2006.

Fischer, R.L. and A.B. Bennett. 1991. Role of cell wall hydrolases in fruit ripening. Annu. Rev. Plant Physiol. Plant Mol. Biol. 42: 675-703.

Fish, W.W. and P.Perkins-Veazie, and J.K. Collins. 2002. A quantitative assay for lycopene that utilizes reduced volumes of organic solvents. J. Food Compos. Anal. 15: 309-317.

Fraser, P.D. and M.R.Truesdale, C.R Bird, W. Schuch, and P.M. Bramley. 1994. Carotenoid biosynthesis during tomato fruit development. Plant Physiol. 105: 405-413.

George, B. and C. Kaur, D.S. Khurdiya, and H.C. Kapoor. 2004. Antioxidant in tomato (*Lycopersicum esculentum*) as a function of genotype. Food Chem. 84: 45-51.

Gil, M.I. and M.A. Consea, and F. Artés. 2002. Quality changes in fresh-cut tomato as affected by modified atmosphere packaging. Postharvest Biol. Technol. 25: 199-207.

Giovannoni, J.J. 2004. Genetic regulation of fruit development and ripening. Plant Cell 16: S170-S180.

Gould, W.A. 1983. Tomato Production, Processing and Quality Evaluation, 2nd ed. AVI Publishing Company, Inc. Westport, Connecticut, USA.

Guillén, F. and S. Castillo, P.J. Zapata, D. Martínez-Romero, M. Serrano, and D. Valero. 2006. Efficacy of 1-MCP treatment in tomato fruit. 2. Effect of cultivar and ripening stage at harvest. Postharvest Biol. Technol. 42: 235-242.

Guillén, F. and S. Castillo, P.J. Zapata, D. Martínez-Romero, M. Serrano, and D. Valero. 2007. Efficacy of 1-MCP treatment in tomato fruit. 1. Duration and concentration of 1-MCP treatment to gain an effective delay of postharvest ripening. Postharvest Biol. Technol. 43: 23-27.

Hong, J.H. and K.C. Gross. 2001. Maintaining quality of fresh-cut tomato slices through modified atmosphere packaging and low temperature storage. J. Food Sci. 66: 960-965.

Javanmardi, J. and C. Kubota. 2006. Variation of lycopene, antioxidant activity, total soluble solids and weight loss of tomato during postharvest storage. Postharvest Biol. Technol. 41: 151-155.

Kaur, D. and R. Sharma, A.A. Wani, B.S. Gill, and D.S. Sogi. 2006. Physicochemical changes in seven tomato (*Lycopersicon esculentum*) cultivars during ripening. Int. J. Food Protect. 9: 747-757.

Krumbein, A. and P. Peters, and B. Brückner. 2004. Flavour compounds and quantitative descriptive analysis of tomatoes (*Lycopersicon esculentum* Mill.) of different cultivars in short-term storage. Postharvest Biol. Technol. 32: 15-28.

Lenucci, M.S. and D. Cadinu, M. Taurino, G. Piro, and G. Dalessandro. 2006. Antioxidant composition in cherry and high-pigment tomato cultivars. J. Agric. Food Chem. 54: 2606-2613.

Maneerat, C. and Y. Hayata, H. Kozuka, K. Sakamoto, and Y. Osajima. 2002. Application of the Porapak Q column extraction method for tomato flavor volatile analysis. J. Agric. Food Chem. 50: 3401-3404.

McDonald, R.E. and T.G. McCollum, and E.A. Baldwin. 1999. Temperature of water heat treatments influences tomato fruit quality following low-temperature storage. Postharvest Biol. Technol. 16: 147-155.

Morillon, V. and F. Debeaufort, G. Blond, M. Capelle, and A. Voilley. 2002. Factors affecting the moisture permeability of lipid-based edible films: A review. Crit. Rev. Food Sci. Nutr. 42: 67-89.

Mostofi, Y. and P.M.A. Toivonen, H. Lessani, M. Babalar, and C. Lu. 2003. Effects of 1-methylcyclopropene on ripening of greenhouse tomatoes at three storage temperatures. Postharvest Biol. Technol. 27: 285-292.

Omoni, A.O. and R.E. Aluko. 2005. The anti-carcinogenic and anti-atherogenic effects of lycopene: a review. Trends Food Sci. Technol. 16: 344-350.

Opiyo, A.M. and T.J. Ying. 2005. The effects of 1-methylcyclopropene treatment on the shelf life and quality of tomato (*Lycopersicon esculentum*) fruit. Int. J. Food Sci. Technol. 40: 665-673.

Park, H.J. and M.S. Chinnan, and R.L. Shewfelt. 1994. Edible coating effects on storage life and quality of tomatoes. J. Food Sci. 59: 568-570.

Petersen, K. and P.V. Nielsen, M. Lawther, M.B. Olsen, N.H. Nilsson, and G. Mortensen. 1999. Potential of biobased materials for food packaging. Trends Food Sci. Technol. 10: 52-68.

Petro-Turza, M. 1987. Flavor of tomato and tomato products. Food Rev. Int. 2: 309-351.

Polenta, G. and C. Lucangeli, C. Budde, C.B. González and R.Murray. 2006. Heat and anaerobic treatments affected physiological and biochemical parameters in tomato fruits. LWT 39: 27-34.

Raffo, A. and C. Leonardi, V. Fogliano, P. Ambrosino, M. Salucci, L. Gennaro, R. Bugianesi, F. Giuffrida, and G. Quaglia. 2002. Nutritional value of cherry tomatoes (*Lycopersicon esculentum* Cv. Naomi F1) harvested at different ripening stages. J. Agric. Food Chem. 50: 6550-6556.

Ruiz, J.J. and A. Alonso, S. García-Martínez, M. Valero, P. Blasco, and F. Ruiz-Bevia. 2005. Quantitative analysis of flavour volatiles detects differences among closely related traditional cultivars of tomato. J. Sci. Food. Agric. 85: 54-60.

Saltveit M.E. 2005. Influence of heat shocks on the kinetics of chilling-induced ion leakage from tomato pericarp discs. Postharvest Biol. Technol. 36: 87-92.

Serek, M. and G. Tamari, E.C. Sisler, and A. Borochov. 1995. Inhibition of ethylene-induced cellular senescence symptoms by 1-methylcyclopropene, a new inhibitor of ethylene action. Physiol. Plant. 94: 229-231.

Serrano, M. and F. Guillén, D. Martínez-Romero, S. Castillo, and D. Valero. 2005. Chemical constituents and antioxidant activity of sweet cherry at different ripening stages. *J. Agric. Food Chem. 53:* 2741-2745.

Seymour, G.B. and K. Manning, E.M. Eriksson, A.H. Popovich, and G.J. King. 2002. Genetic identification and genomic organization of factors affecting fruit texture. J. Exp. Bot. 53: 2065-2071.

Sisler, E.C. and M. Serek. 1999. Compounds controlling the ethylene receptor. Bot. Bull. Acad. Sin. 40: 1-7.

Slimestad, R. and M. Verheul. 2005. Content of chalconaringenin and chlorogenic acid in cherry tomatoes is strongly reduced during postharvest ripening. J. Agric. Food Chem. 53: 7251-7256.

Srivastava, A. and A.K. Handa. 2005. Hormonal regulation of tomato fruit development: a molecular perspective. J. Plant Growth Regul. 24: 67-82.

Suaparlan, H.J. and K. Itoh. 2003. Combined effects of hot water treatment (HWT) and modified atmosphere packaging (MAP) on quality of tomatoes. Packaging Technol. Sci. 16: 171-178.

Tandon, K.S. and E.A. Baldwin, and R.L. Shewfelt. 2000. Aroma perception of individual volatile compounds in fresh tomatoes (*Lycopersicon esculentum*, Mill.) as affected by the medium of evaluation. Postharvest Biol. Technol. 20: 261-268.

Toor, R.K. and G.P. Savage. 2005. Antioxidant in different fractions of tomatoes. Food Res. Int. 38: 487-494.

Toor, R.K. and G.P. Savage. 2006. Changes in major antioxidant components of tomatoes during post-harvest storage. Food Chem. 99: 724-727.

USDA. 1991. United States Standard for Grades of Fresh Tomatoes. United States Department of Agriculture, Agricultural Marketing Service, p. 13.

Wills, R.B.H. and V.V.V. Ku. 2002. Use of 1-MCP to extend the time to ripen of green tomatoes and postharvest life of ripe tomatoes. Postharvest Biol. Technol. 26: 85-90.

Yilmaz, E. and K.S. Tandon, J.W. Scout, E. Baldwin, and R.L. Shewfelt. 2001. Absence of a clear relationship between lipid pathway enzymes and volatile compounds in fresh tomatoes. J. Plant. Physiol. 158: 1111-1116.

Tomato and Flavour

Christian Salles

UMR 1129 FLAVIC, ENESAD, INRA, University of Burgundy, 17 rue Sully, BP86510, F21065 Dijon, France

ABSTRACT

Tomato is one of the more popular fruits cultivated and consumed in a large number of countries throughout the world. Besides texture and colour, flavour characteristics are an important factor taken into account by the consumer. The characteristic flavour of tomato is mainly due to volatile and small, water-soluble non-volatile components. Their structure and origin are very different. Most volatile compounds responsible for tomato aroma derive mainly from fatty acids and amino acids but pigments such as lycopene and carotenoids can also be precursors of qualitatively important aroma compounds. Reducing sugars and free acids are at the origin of the sweet-sour taste of tomato but minerals and nitrogen compounds can also intervene as potential taste-active compounds. However, aroma and taste components quantitatively vary greatly according to many factors such as the tomato species, stage of ripeness, year of growth and conditions of cultivation. For the past few years, consumers seem increasingly to desire improved flavour of commercial tomato varieties; therefore, research on tomato flavour and selection of more "savoury" products are fully justified.

INTRODUCTION

Flavour is defined as the result of retronasal aroma, taste and trigeminal perceptions. In the formation of the characteristic flavour of tomato, most constituents may play a role. The effect is direct for aroma or taste compounds or indirect in the case of precursors or compounds that affect

A list of abbreviations is given before the references.

the availability of the stimulus. As with many foods, the characteristic flavour of tomato is due to the complex effect of volatile and non-volatile components and their interactions. The interactions can have a physico-chemical origin with the food matrix components, resulting in retention with pectin or proteins, or in release due to salting out effect with intrinsic mineral salts. Moreover, the very complex flavour compound composition of tomato can generate intra- and intermodal sensory interactions.

A wide range of sugars, acids, vitamins and volatile compounds was found in various cultivars of tomato (Davies and Hobson 1981, Buttery et al. 1988, Islam 1997). They are mainly implicated in tomato flavour. Concerning tomato aroma compounds, many studies have been carried out on the identification of volatile compounds (Petro-Turza 1986-1987). There are nearly 400 identified aroma compounds in tomato but only a few of them are aroma-active compounds and directly involved in the characteristic fresh tomato aroma. Some volatile compounds originate from non-volatile precursors by enzymatic reactions occurring very quickly after the fruit has been blended or chewed by consumers. The main consequence of this phenomenon is that the amounts of volatile compounds vary with time and new techniques were developed to get initial concentration of volatiles in the fruit and to follow the biosynthesis and aroma release kinetics of tomato aroma compounds. The main tomato volatile compound formation pathways and the main factors influencing the volatile compound composition in the fruit were reviewed by Boukobza (2001).

Substances affecting taste in tomato are small and water-soluble organic and mineral compounds that are the main constituents of the dry matter of tomato. It is well established that sugars, organic acids and minerals fully participate in tomato taste but the participation of other components such as nitrogen compounds is probable but still questionable. However, compared to tomato aroma compounds, even though many studies have been carried out on tomato composition, very few have been specifically dedicated to tomato taste.

This chapter starts with a presentation of the main methods used to extract and analyse volatile and non-volatile tomato components, then aroma and taste components of interest to fully explain tomato flavour are reviewed.

TOMATO AROMA COMPOUNDS

Techniques for Extraction and Analysis of Tomato Flavour

The extraction of volatile compounds, generally by steam distillation, dynamic headspace extraction under vacuum or more recently by solid phase micro extraction (SPME), is a key step in the analysis of aroma compounds of a particular food and the question of the representativity of the odour of the obtained extract compared to the odour of the crude product should be examined each time. Therefore, the different steps leading to the obtaining of tomato aroma extract must be optimized, as a universal extraction method does not exist. Etiévant and Langlois (1999) optimized each step such as fruit blending, enzyme deactivation and volatile compound extraction/distillation by comparing the quality of the aroma extracts with that of the crude product by sensory evaluation to get the most representative aroma extract.

The conventional method to identify volatile compounds from an aroma extract is the use of GC-MS (gas chromatography coupled with mass spectrometry) and the comparison of linear retention indexes of volatile compounds contained in the sample with reference compounds. Volatile compounds are then quantified by GC using an internal standard. When these techniques are not sufficient to identify volatiles, complementary techniques can be used. Infrared spectroscopy and particularly GC-FTIR (GC coupled with Fourier Transformation Infra Red) is an analytical tool for detection of functional groups by infrared absorption in molecules. Nuclear magnetic resonance (NMR) gives information on the structure of the molecules. Both techniques were used for the identification of tomato aroma compounds after extraction (Buttery et al. 1994). New strategies are now used, in the field of metabolomic, using full mass spectral alignment of GC-MS profiles for fast multivariate analyses (Tikunov et al. 2005).

The coupling of GC with olfactometry (GC-O) allows detection of odorant zones characterized by an olfactive index and by a description of the perceived odour from an aroma extract. The reliability of the results depends mainly on the representativity of the aroma extract. This is an interesting technique particularly because of the high sensitivity of the human nose, often higher than any electronic detector. Applying GC-O method to tomato aroma extracts, Langlois et al. (1996) were able to identify several aroma-active compounds of tomato.

With the advent of direct headspace or nose-space analyses linked to Atmospheric Pressure Ionization Mass Spectrometry (APCI-MS) or more recently to Proton Transfer Reaction (PTR) (Taylor and Linforth 2003), it is now possible to measure mixtures of volatiles in headspace and nose-space directly in real time. APCI-MS was first applied to tomatoes to measure volatile release following the enzymatic process during the eating of tomato fruits (Brauss et al. 1998). APCI-MS or PTR-MS does not need extraction steps, is capable of simultaneous detection and allows real time analysis but the identification of volatile compounds is not unequivocal. However, previous conventional analysis steps such as GC-O are sometimes necessary to determine what ion corresponding to active component this technique should follow.

The main active compounds involving taste are small non-volatile molecules, so, they can be easily analysed using conventional techniques. The main mineral cations are analysed by atomic absorption or flame ionization and mineral anions by colorimetric methods or high performance liquid chromatography (HPLC) using conductimetry as the detection method. Concerning organic compounds, sugars and acids can be analysed using specific enzymatic kit reagents or by HPLC using refractometry, UV/visible (Salles et al. 2003, Marconi et al. 2007), amperometric or conductimetric detection.

Biogenesis of Aroma Compounds

Tomato aroma compounds, like aroma compounds in numerous fruits and vegetables, mainly originate from proteins, sugars and lipids (Fig. 1).

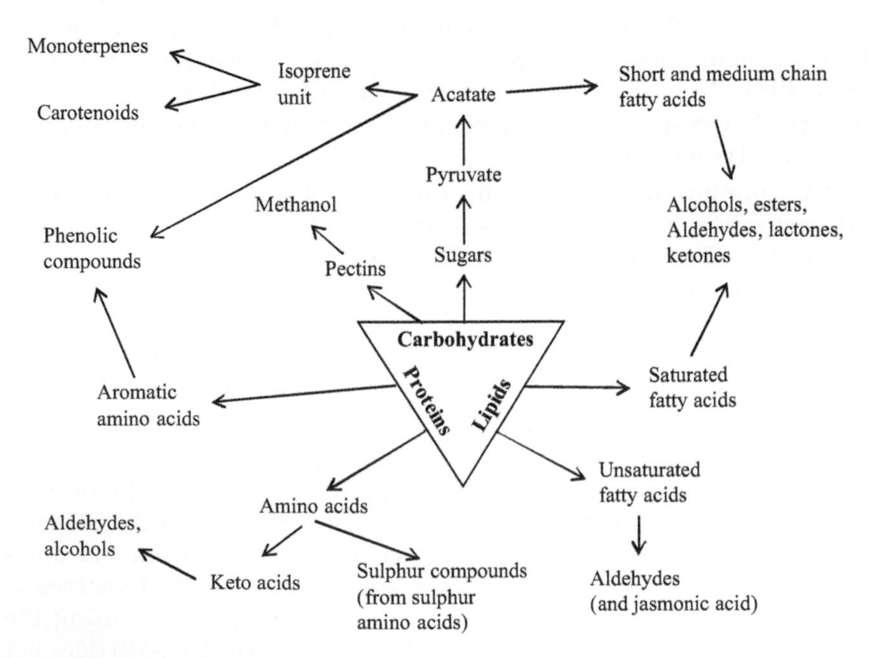

Fig. 1 Biosynthesis of volatile compounds in tomato.

Amino Acids Metabolism

Many volatiles are derived from amino acids. The absence of differences between crude and blended fruits for some volatiles suggested that they were formed during the ripening process. Moreover, concentrations of these compounds at different stages of ripening showed that the main formation of these compounds seems to occur between the breaker and the table ripe stage (Buttery and Ling 1993a). The use of C^{14} labelled amino acid precursors showed that it mainly concerns alcohols, carbonyls, acids and esters. Amino acid is first deaminated into the corresponding keto acid by an amino transferase, then, the α-keto acid is converted into an aldehyde by decarboxylation, as a result of which various volatile esters are formed (Petro-Turza 1986-1987). The main amino acids involved in this process are leucine, isoleucine, valine, alanine and phenylalanine (Buttery and Ling 1993a). For example, 3-methylbutanal originates from leucine (Fig. 2). Leucine is deaminated by a leucine aminotransferase to give a keto-acid that can be enzymatically both decarboxylated to give 3-methylbutanal and dehydrogenated to give an acyl CoA derivative. This acyl CoA intermediate can give by the action of alcohol acyl transferase several types of esters. 3-Methylbutanol derived from the corresponding aldehyde can also be esterified to give esters. The *in vitro* formation of 3-

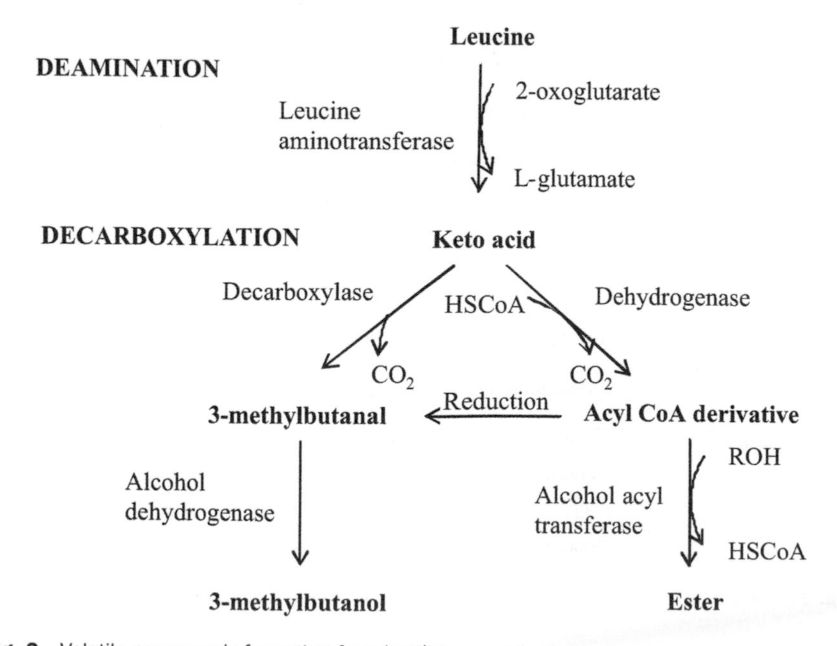

Fig. 2 Volatile compounds formation from leucine.

methylbutanol from leucine using a crude enzyme extract from tomato was reported by Yu et al. (1968a) and confirmed using labelled [^{14}C]-leucine (Yu et al. 1968b). Concerning aromatic compounds, Tieman et al. (2006a) reported that tomato aromatic amino acid decarboxylase participates in the synthesis of 2-phenylethanol and 2-phenylacetaldehyde from phenylalanine.

The presence of nitro-compounds has been reported in many studies (Petro-Turza 1986-1987). It could be formed by a Nef reaction from phenylalanine to 1-nitro-phenylethane and hence to phenylacetaldehyde (Buttery 1993). 2-isobutylthiazole was also reported in several studies (Petro-Turza 1986-1987). Hypotheses were proposed to explain its formation but need to be validated. Volatile sulphur and ammonium compounds were reported in fresh tomato by Petro-Turza (1986-1987).

Fatty Acid Metabolism

Some tomato volatiles are produced enzymatically from lipid oxidation after tissue damage (Kazeniac and Hall 1970, Galliard et al. 1977). A general scheme of the formation of volatile compounds from acyl lipids is presented in Fig. 3. Buttery et al. (1988) confirmed the formation of these volatiles during the disruption process of tomato fruit tissue. C6 aldehydes are produced in tomato in different parts of the plant (Buttery

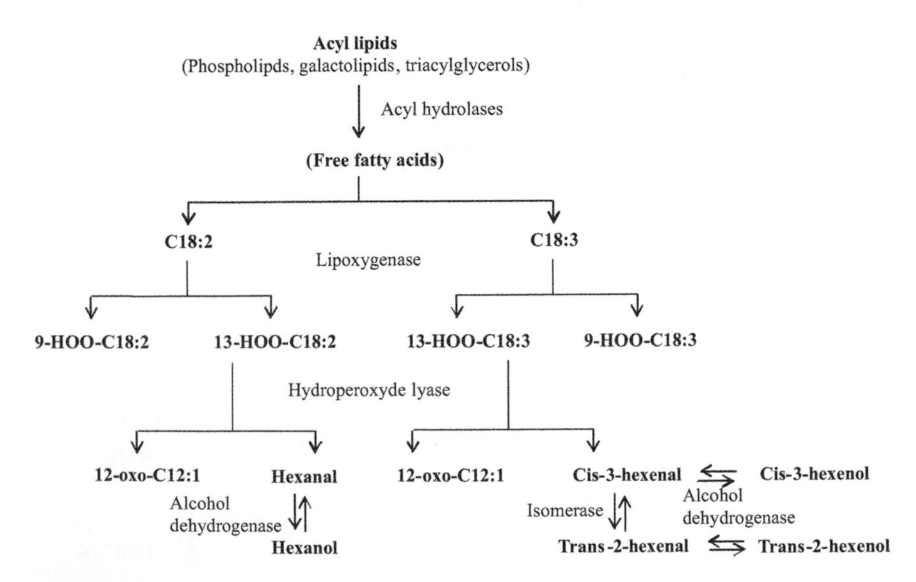

Fig. 3 Pathway of volatile compounds formation by enzymatic degradation of tomato acyl lipids.

and Ling 1993a). These aldehydes contribute to the "green note" of fresh tomato. They could be components of plant defence mechanism and it has been suggested that the lipoxygenase activity, in the case of pea, increases in response to drought stress (Oneill et al. 1996).

Tomato fruit tissue disruption results in the degradation of endogenous lipids by hydrolytic and oxidative enzymes resulting in the formation of free fatty acids (mainly linoleic C18:2 and linolenic C18:3), which are subsequently oxidized to their hydroperoxides by lipoxygenase action. The C18 hydroperoxides are cleaved to C6 aldehydes (hexanal and cis-3-hexenal) by a specific enzyme, hydroperoxide lyase, for 13-hydroperoxide isomers. The specific enzyme for 9-hydroperoxide isomers is not present in the tomato tissue, therefore 9-hydroperoxy-fatty acids accumulate in the cell, and their fate is not known. Cis-3-hexenal is rapidly converted into its isomer trans-2-hexenal by an isomerase and can readily occur chemically. Through the action of alcohol dehydrogenase, the aldehydes can be converted to their corresponding alcohols. The ratio of linolenic acid to linoleic acid and the ratio of hexanal to hexenal generated from macerated fruits were compared and found to be different according to tomato varieties (Gray et al. 1999).

Although lipoxygenase is widely distributed in plants, its role in cell physiology is not well defined. It catalyses oxidation of free fatty acids containing the cis,cis-1,4-pentadiene structure, mainly linoleic (18:2) and linolenic (18:3) acids, resulting in a hydroperoxide located at carbon 9 or carbon 13 depending on the isoenzyme (Riley et al. 1996). The lipoxygenase pathway is important for the production of flavour compounds as well as for plant defence, development and inter- and intra-plant communication (Bate et al. 1998).

Only a few data are related to subcellular distribution of the relevant enzymes (lipoxygenase, hydroperoxyde liase). A membrane-associated lipoxygenase was first reported by Todd et al. (1990) in green tomato, and appeared to be located in thylakoids of chloroplasts (Todd et al. 1990, Bowsher et al. 1992). Studies report data on the subcellular generation and distribution of lipid-derived volatiles during the ripening of tomato and in ripe tomato (Riley et al. 1996, Riley and Thompson 1997). Membrane-associated lipoxygenase activity increases during tomato development, peaking during the breaker stage, whereas hydroperoxide lyase exhibits constant activity from mature green through ripe stages of fruit development, and lipid particles and key flavour volatiles could be both derived from membranes (Riley et al. 1996). Lipidic particles could serve as a vehicle for moving flavour volatiles out of membranes and the volatiles subsequently partition out of the particles into the cytosol in accordance with partition coefficients (Riley and Thompson 1997).

Hydroperoxide lyase seems to exist in multiple locations in plant and has also been reported to be membrane bound in tomato (Galliard et al. 1977).

Cellular damage initiates a rapid chain of reactions that include the activation of the previously described enzymatic systems (lipase, lipoxygenase, hydroperoxide lyase, isomerase and alcohol dehydrogenase) utilizing O_2 as a substrate along with polyunsaturated fatty acids. Among all the enzymes, lipoxygenase and hydroperoxide lyase seem to be the most important since product specificity of lipoxygenase and substrate specificity of hydroperoxide lyase determine the pattern of volatile aldehydes formed (Hatanaka et al. 1986).

Other volatile compounds showing a relatively large increase in concentration when tomato fruits are blended are formed upon disruption of the tissue from the lipid oxidation pathway. Some compounds found at a negligible concentration in raw fruit but at measurable levels in blended fruits derive from the oxidation of free fatty acids by the action of lipoxygenase. However, degradation of hydroperoxides does not require hydroperoxide lyase; homolytic cleavage could also give rise to various types of aldehydes, ketones and alcohols.

Alcohols can be derived from their corresponding aldehydes by the action of alcohol dehydrogenase. 4,5-Epoxy-(E)-2-decenal is another important lipid oxidation product present in tomato (Buttery and Ling 1993a). Its pathway formation from linoleic acid has been proposed by Buettner and Schieberle (2001). Enzymatic cooxidation of carotenoid by tomato lipoxygenase could be responsible for the formation of certain volatiles. Purified tomato lipoxygenase oxidized β-carotene faster than α-carotene and lutein, whereas lycopene remained unaffected (Cabibel and Nicolas 1991). Geranyl acetone, β-ionone and 6-methyl-5-hepten-2-one are likely to be derived from the cooxidation of carotenoids (Buttery and Ling 1993b).

Fatty acid peroxyl radicals, formed by enzymatic oxidation of the free fatty acids, can be the oxidizing agent for carotenoids forming peroxy carotenoid radical. The peroxy radical reacts with the carotenoid, which becomes oxidized and is fragmented into some volatile molecules. Alternatively, as described by Wu et al. (1999) for β-carotene, the addition of peroxy radical can give fatty acid peroxyl radicals-carotenoid from which further derived products such as epoxides can be formed.

Isoprenoid Pathway

Volatile monoterpenes represent only a low fraction of the terpenoids present in tomato. Oxygenated monoterpene volatiles such as linalool, geranial and neral were identified in ripe tomato fruit, while α-Copaene was identified in green fruit (Buttery and Ling 1993b). The pathway of carotene biosynthesis from the isopentenyl diphosphate (isoprene unit) in tomato fruit has been established (Porter and Spurgeon 1979). This isoprenoid pathway becomes important during the ripening process. Camara et al. (1995) reported studies on changes and accumulation of carotenoid pigments in *Capsicum annuum* (same family as tomato). These studies showed, using isolated chromoplasts, that carotenoids were formed by the condensation of C5 units derived from mevalonate with several enzymatic steps and that phytoene is the first C_{40} precursor of carotenoids. An alternative to this pathway has been considered (Bach 1995). A biosynthetic pathway of isopentenyl pyrophosphate formation should be via the Rohmer pathway (Rohmer et al. 1996). Important correlations between specific aroma compounds and carotenoid precursors were shown (Stevens 1970). The conversion to volatile compounds appeared to be an oxidative process occurring at the first conjugated diene bonds. It was observed that ketones are more abundant than aldehydes, suggesting that oxidation preferentially occurs at methyl-substituted rather than unsubstituted double bonds or that the aldehydes formed are less stable. Phytoene, phytofluene, carotene, neurosporene and lycopene are respectively carotenoid precursors of farnesyl acetone, farnesyl acetone and geranyl acetone, geranyl acetone, 6-methyl-5-hepten-2-one and geranyl acetone, and 6-methyl-5-hepten-2-one. The production of these compounds might also be derived from other different chemical pathways such as thermal hydrolysis of tomato glycosides releasing high amounts of 6-methyl-5-hepten-2-one but no geranyl acetone. A large increase on maceration and absence of production in the glycoside hydrolysis for geranyl acetone demonstrate its origin from an oxidative process possibly coupled with the lipid oxidation (Buttery 1993). There is a good correlation between β-carotene content and β-ionone concentration. Crouzet and Seck (1982) described a pathway for the formation of β-ionone from β-carotene.

Other Pathways

Other pathways for the formation of minor volatile compounds found in tomato were also reported. Phenolic compounds such as eugenol, methyl salicylate and guaiacol were reported in highly variable amounts in tomato fruits. β-oxidation is also described as an important pathway for the formation of various aldehydes, ketones, esters from saturated and unsaturated fatty acids. The combination of the action of pyruvate decarboxylase and alcohol dehydrogenase on pyruvate can generate both ethanol and acetaldehyde. In lactic acid fermentation, lactate is formed by the action of lactate dehydrogenase and ethanol can accumulate in many plant species (Tadege et al. 1999).

Glycosides

Compared to grapes or wine, very few studies on glycoside aroma precursors were carried out on tomato fruits. Volatile aldehydes such as 3-methylbutanal and phenylacetaldehyde could be formed from enzymatic oxidation of the corresponding alcohols, released by enzymatic hydrolysis of glycosides during ripening. Isolation of the fresh tomato glycoside fraction yielded 3-methylbutyric acid and β-damascenone as major products upon hydrolysis (Buttery et al. 1990, Buttery 1993). Marlatt et al. (1991) isolated several glycosidically bound C13 norisoprenoids from tomato. Later, Marlatt et al. (1992) showed the presence of a larger variety of glycosides that can release volatile compounds using enzyme-mediated hydrolysis. The presence of furaneol glucoside and free furaneol were reported in intact tomatoes (Buttery et al. 1994). Krammer et al. (1994) reported the identification of 2,5-dimethyl-4-hydroxy-3(2H)-furanone glucoside in tomato. Measurement of volatiles after deactivation of enzymes using a saturated NaCl solution supports the view that furaneol was formed by slow enzymatic hydrolysis of the glucoside occurring during ripening (Buttery et al. 1995). Furaneol concentration was found above its threshold value in fresh fruit, suggesting that it contributes to the fresh aroma (Buttery et al. 2001). Krammer et al. (1995), using countercurrent chromatography system, collected highly purified fractions and characterized for the first time by NMR techniques 3-(4-hydroxyphenyl)propionic acid and benzyl-β-D-glucosides, and 2-phenylethyl-β-D -gentiobioside.

Aroma Compounds Formed during Processing

The maceration process and high temperature contribute to the formation of new volatiles in tomato. An important loss occurs for a number of compounds such as (Z)-3-hexenal, 1-penten-3-one, while other compounds such as dimethylsulfide, acetaldehyde and β-damascenone show a significant increase in concentration in processed tomato paste (Petro-Turza 1986-1987). Moreover, the quantity of some other components also changes upon heat treatment. Buttery (1993) formulated a synthetic tomato paste and its aroma, judged by a sensory panel, was found to be very similar to tomato paste. Glycosides are an important precursor of volatile aroma compounds in processed tomato (Buttery et al. 1990), as they are progressively hydrolysed in both hot and acid conditions.

Table 1 Volatile compounds quantified in three tomato varieties (unpublished results). The average of three replicates (μg/kg fruit) and the standard deviation (%) are reported.

Volatile compounds	*Cencara*		*Zucker*		*Lucy*	
3-methyl-butanal	24.6	*26.6*	92.0	*16.8*	179.2	*14.2*
2-methyl-butanal	10.9	*47.7*	65.8	*5.5*	108.0	*16.1*
pent-1-en-3-one	155.7	*17.8*	295.3	*24.1*	202.6	*4.5*
pentanal	47.8	*21.0*	52.4	*14.3*	38.1	*4.0*
pentan-3-one	33.9	*6.5*	48.0	*10.8*	33.8	*8.6*
2-methylbut-2-enal	44.9	*64.1*	246.1	*12.3*	349.8	*7.6*
2- and 3-methyl butan-1-ol	110.4	*58.2*	680.9	*4.8*	1043.8	*10.8*
pent-2-enal	35.3	*35.6*	93.8	*39.5*	57.9	*5.3*
hexanal	2100.0	*14.9*	2224.5	*15.5*	1001.0	*6.6*
(Z)-hex-3-enal	44.0	*28.7*	111.0	*18.7*	63.8	*9.5*
3-methylpentan-1-ol	5.1	*32.2*	1.5	*126.7*	6.0	*14.7*
(E)-hex-2-enal	1745.6	*22.1*	3979.9	*6.4*	1950.3	*9.9*
hex-3-en-1-ol	43.8	*34.8*	57.2	*23.3*	29.2	*6.9*
hexan-1-ol	7.9	*21.6*	7.6	*26.9*	4.4	*26.8*
2-methyl-thio-ethanol	9.5	*34.6*	41.2	*41.2*	8.7	*29.2*
3-methyl-thio-propanal	8.6	*68.8*	38.7	*25.8*	80.7	*8.0*
hept-2-enal	42.5	*21.1*	37.5	*26.1*	22.9	*7.5*
6-methylhept-5-ene-2-one	213.9	*24.9*	96.2	*15.5*	91.7	*9.1*
benzaldehyde	18.3	*4.6*	19.5	*21.5*	10.6	*7.2*
6 methylhept-5-ene-2-ol	23.8	*34.1*	4.6	*8.3*	4.3	*18.9*
isobutylthiazole	26.6	*27.3*	8.3	*38.0*	23.0	*9.7*
butanolactone + heptadienal	30.8	*12.2*	36.8	*26.1*	50.9	*13.8*
3-methyl-thio-propan-1-ol	9.0	*23.2*	22.0	*38.1*	20.5	*21.6*
2-methyl-butanolactone	12.4	*4.9*	6.2	*11.8*	6.7	*8.1*
2-phenyl-ethanal	141.5	*25.8*	63.8	*10.0*	94.3	*15.5*
nonanal	18.8	*68.0*	13.0	*19.3*	9.9	*9.2*
phenyl-methanol	88.9	*9.8*	31.7	*2.9*	52.0	*21.2*
2-methoxy-phenol	10.8	*80.1*	1.9	*62.4*	1.1	*105.6*
2-phenyl-ethanol	1043.2	*27.2*	846.8	*22.4*	853.9	*21.1*
decanal	21.3	*40.5*	12.9	*10.8*	9.2	*13.1*
methyl salicylate	50.3	*15.6*	1.9	*74.2*	48.0	*18.6*
2,3-epoxy-geranial	23.9	*22.4*	6.2	*13.2*	7.7	*11.3*
benzothiazole	17.6	*90.4*	21.4	*60.9*	21.9	*111.0*
neral	13.0	*19.5*	6.6	*7.9*	5.4	*17.3*
geranial	48.1	*18.7*	17.2	*14.9*	16.0	*14.0*
3(1,1-dimethyl-ethyl)phenol	2.7	*23.8*	3.9	*10.0*	2.6	*13.6*
geranylacetone	148.3	*10.0*	146.9	*26.2*	62.9	*5.1*
β-ionone	5.0	*2.1*	10.2	*5.3*	5.4	*3.9*
5,6-epoxy-β-ionone	6.1	*8.9*	7.2	*12.7*	3.9	*29.3*

Factors Influencing Fresh Tomato Flavour

Tomato aroma compound composition varies qualitatively and quantitatively according to the variety (Table 1). These compounds are mainly produced by the metabolism of lipids, amino acids, lycopene and carotenoids but large differences are observed between varieties (Table 2) that can influence flavour qualities of tomatoes. In order to evaluate the impact of these volatile compounds on tomato flavour and to evaluate to what extent aroma active components can characterize tomato varieties, Langlois et al. (1996) performed GC-O analysis by three judges with tomato varieties that were described by different flavour qualities. The results for one judge are reported in Table 3 as an example. Among the 40 volatile compounds identified in fresh tomato (Table 1), the three judges were able to detect a total of 32 odours but only 13 of them were clearly identified. Among these potent odorant compounds, 11 appeared particularly important in discriminating between varieties. Most of these compounds are related to the metabolism of amino acids (2- and 3-methylbutanol, 2- and 3-methylbutanal, 2-methyl-2-butenal, 2-phenyl-ethanol, phenylmethanol, benzaldehyde and 3-methylthiopropanal). Only geranylacetone is related to the metabolism of carotenes in this study, though important fruit aroma volatiles are derived from the degradation of carotenoid pigments and lycopene, the major pigment of red tomato, and give rise to several volatile compounds (Lewinsohon et al. 2005). Conversely, they found that the metabolism of unsaturated fatty acids

Table 2 Importance of volatile compounds originating from different metabolic ways in several tomato varieties (unpublished results).

Variety	*Fatty acids*	*Aliphatic amino acids*	*Aromatic amino acids*	*Lycopene*	*Carotenes and others*
Cencara	+		+++	+++	+++
Ibiza			+	++	++
Lucy		++++	++		
Melody		++++	Variable		
Zucker	++++	++	++		++++
Elena				Variable	Variable
Rondello	+	+++			
Daniela			Variable		
Lemon boy				—	—

+ to ++++: importance of the presence of volatile compounds originating from the same precursor family.
—: absence.

Variable: the importance of the presence of volatile compounds originating from the same precursor family varies according to the growing year.

Table 3 Gas chromatography–olfactometry analysis obtained for three tomato varieties by one subject (unpublished results).

Zucker				*Lucy*				*Cencara*			
LRI	*ND*	*Area*	*Odour description*	*LRI*	*ND*	*Area*	*Odour description*	*LRI*	*ND*	*Area*	*Odour description*
735	4	56	Unpleasant odour	735	5	108	Unpleasant odour	735	4	44	Unpleasant odour
763	7	328	Chemical strawberry	763	3	27	Chemical strawberry	763	4	65	Strawberry
773	1	15	Chemical odour	773	1	13	Plastic	773	2	23	Chemical, plastic
884	7	595	Green, grass	883	6	422	Grass	884	6	393	Grass
963	3	25	Apple	962	1	7	Apple	963	3	28	Apple
988	1	12	Green tomato	995	1	15	Green tomato	993	1	13	Green tomato
1012	7	834	Foot, socks	1011	5	367	Foot, socks	1011	5	264	Foot, socks
1045	9	5243	Potatoes	1045	7	1767	Potatoes	1046	7	1117	Potatoes
1083	6	493	Green bean	1083	3	72	Green bean	1083	4	194	Green bean
				1109	3	37	Neutral odour	1109	2	30	
1138	2	20	Green, green tomato and potatoes					1139	1	17	Chemical odour
1177	1	6	Sweat	1179	1	7	Sweat	1179	3	28	Sweat
1182	5	175	Floral	1182	5	160	Floral	1182	6	378	Floral
				1234	1	26	Unpleasant odour	1232	2	37	Chemical
1279	5	210	Floral	1278	4	60	Lip tip	1280	4	103	Floral
1302	7	1142	Wet wood	1302	4	203	Wet wood	1300	6	533	Wet wood
				1380	1	15		1381	1	26	
1455	8	1639	Green pepper								
1463	8	1701	Modelling clay	1463	5	425	Modelling clay	1455	8	3127	Modelling clay
				1509	1	26	Orange flower	1509	7	1104	Soap, banana
1564	7	1053	Green, vegetable	1564	5	305	Green	1564	7	779	Green

LRI: linear retention index; ND: number of dilutions

does not appear to produce more aroma-active compounds in particular varieties. Using GC-O, Krumbein and Auerswald (1998) identified (Z)-3-hexanal, hexanal, 1-octen-3-one, methional, 1-penten-3-one and 3-methylbutanal as the most odour-active volatile compounds in fresh tomato.

Harvesting conditions such as stage of maturity, handling procedure, and storage condition and external factors such as temperature, humidity, soil quality, and light have an important impact on fruit quality and should be taken into account for the development of flavour quality. For example, fruit size and soluble solid content can also be manipulated by controlling the electrical conductivity of the nutrient solution (Auerswald et al. 1999b, Krauss et al. 2006). Potassium supply can influence the acid content, colour and shape of the fruits (Wright and Harris 1985). Flavour quality at harvest is also influenced by the maturity of the fruit (Maul et al. 1998), the manner in which the fruits were ripened (Arias et al. 2000), and growing conditions (Moraru et al. 2004, Thybo et al. 2006). Temperature affects the rates of fruit development and carotenoid synthesis. When fruits are bruised, during harvesting or transportation, they release ethylene, which hastens the ripening process and reduces the shelf life of fruits. Tomato fruits are also sensitive to high temperature and reduced levels of oxygen, which can affect flavour quality.

The selection of new varieties includes as an important factor the increase of fruit firmness. Moreover, the shelf life of fruits has been increased with the development of modified atmosphere packaging and fruits can be stored several weeks according to their maturity stage. Temperature and atmosphere composition have a direct effect on fruit metabolism post-harvest and can lead to change in flavour compound metabolism.

Sugar and acid ratios of mature green or light pink fruits vary during storage at low temperature (Buescher 1975, Kader et al. 1978b). Changes in flavour attributes are also observed (Kader et al. 1978b, Auerswald et al. 1999a). Buttery et al. (1987) showed that storage of ripe fruit at cold temperature led to a decrease of the major volatiles. The flavour quality of the fruit is closely related to its maturity stage (Paull 1999). The final ripening temperature has also an important effect on flavour compound level (Stern et al. 1994). The reduction of oxygen level slows down the ripening process. O_2 level lower than 3% and CO_2 level higher than 15% can have negative effects on the flavour (Kader et al. 1989). Moreover, storage under low O_2 increases ethanol and acetaldehyde production (Ratanachinakorn et al. 1997). Boukobza and Taylor (2002) researched the effect of temperature, storage, and atmosphere on flavour change in tomato. Mimicking typical storage and transport scenarios, they were able to follow in real time and simultaneously nine volatile compounds released upon disruption that were subject to change following the effects of stress factors. Bailen et al. (2006) showed that the use of granular-activated carbon alone or impregnated with palladium as a catalyst inside modified atmosphere packages helps maintain tomato fruit qualities such as sweetness, firmness, juiciness, colour, odour and flavour during cold storage.

The main challenge of the breeders is now to improve fresh tomato quality. So far the focus has mainly been on production, colour, texture, and time of storage in taking into account flavour quality. Studies focused on the determination of molecular genetic factors responsible for each quality component were carried out to define quality criteria and their impact on consumer perception. The definition of genes and quantitative trait loci (QTLs) corresponding to quality components were reported (Causse et al. 2001) but their correlation with flavour compounds was rather weak. These authors observed clusters of QTLs for aroma, taste and texture attributes in fresh market tomato. They found that most of the favourable alleles came from the cherry tomato parent, showing the potential of this line for tomato quality improvement. Concerning taste compounds, Fulton et al. (2002) reported QTLs affecting sugars, organic acids and several other biochemicals that possibly contribute to flavour. Later, Tieman et al. (2006b) identified multiple loci affecting the composition of chemicals related to flavour in fresh tomatoes. In particular, they identified 25 loci significantly altered in one or more of 23 volatile compounds and 4 loci altered in citric acid content.

TASTE COMPOUNDS OF TOMATO

References for non-volatile taste compounds in the tomato are much less numerous than for volatile compounds. This fruit contains 5-7.5% dry matter, which is in particular constituted of fructose (25%), glucose (22%), citric acid (9%), malic acid (4%), dicarboxylic acids (2%), minerals (8%), peptidic and protein compounds (8%), and traces of ascorbic acid, pigments, vitamins, polyphenols, and monocarboxylic amino acids (Davies and Hobson 1981, Petro-Turza et al. 1989a, b). So, most of the dry matter content of tomato fruit is composed of sugars and organic acids. The sugars are primarily the reducing sugars fructose and glucose with a trace amount of sucrose, while the major organic acids in tomato are citric and malic, with citric predominant. Thus, the amount of reducing sugars in tomato generally correlates with soluble solids content (Stevens 1972, Stevens et al. 1977).

A large range of sugars, acids and vitamins were found in various cultivars of tomato (Davies and Hobson 1981, Islam 1997). Some of these compounds depend on several factors such as variety, soil composition, and nutrients (Davies and Hobson 1981, Heeb et al. 2005, 2006) and are indicative of the harvest stage of the fruit. They also mainly contribute to the total taste of tomato. Generally, an increase in organic acids and sugars resulted in a corresponding increase in overall flavour intensity (Buescher 1975, Stevens et al. 1977, Kader et al. 1978b), but the ratio of sugars to acids was also shown to be an important factor in defining flavour differences among tomato cultivars (Stevens 1972).

Reducing sugars and organic acids are significant components of sweet and sour tastes in tomato, respectively (Stevens et al. 1977). Their concentration level may significantly affect flavour acceptability. Malundo et al. (1995), demonstrated the impact of increasing the content of sugars and acids on flavour acceptability; maximum levels of acid were reached above which further increases negatively affected consumer acceptability. Studies have shown that the levels of sugars and acids in tomato affect not only tomato taste attributes of sweetness and sourness, but also the overall flavour as perceived by trained sensory judges (Stevens et al. 1977, Kader et al. 1978b, Hobson and Bedford 1989). Moreover, these authors pointed out some synergistic effects between citric acid and glucose. In particular, the concentration of citric acid seemed to enhance the importance of glucose compared to fructose in the sweet taste, which is also affected by the presence of salts (De Bryun et al. 1971). Stevens et al. (1977) reported that citric acid decreased the sweetness perception when the concentration of sugars was low in tomato and increased it when the concentration of sugar was high. This viewpoint is supported by a psychophysical study (Schifferstein 1994). Watada and Aulenbach (1979) and Jones and Scott (1984) found that although sweetness was probably highly dependent on sugar content, the perception of this taste was not explained simply by the presence of sugars. Concerning sourness, all authors agreed on the importance of organic acids, mainly citric and malic acids, in this taste characteristic (Petro-Turza 1986-1987). They reported a good correlation between sourness and titratable acidity. Though malic acid was sourer than citric acid (De Bryun et al. 1971), its influence was weaker than that of citric acid present in a higher concentration. At low citric acid and glucose concentrations, fructose reduced sourness (Stevens et al. 1977). Salles et al. (2003) confirmed this data (Table 4) using omission tests, which consisted in omitting one or several components in a model mixture mimicking the tomato juice and studying the effect of omission on perception. They found in particular that reducing sugars were totally responsible for

Table 4 Omission tests performed from a tomato model juice (adapted from Salles et al. 2003, copyright Elsevier).

Fraction	Mean taste intensity (/100)			
	Sweet	*Sour*	*Salty*	*Umami*
MJ	42.6[bcd]	45.9[defg]	25.3[cdefghi]	17.3[b]
MJ-Gluc-Fruc	8.3[hij]	66.2[abc]	52.8[a]	25.7[b]
MJ-Cit-Mal	64.5[a]	1.0[i]	3.2[j]	26.7[b]
MJ-Gluc	23.0[efgh]	49.0[cdef]	36.8[abcdef]	9.1[b]
MJ-Fruc	14.2[hij]	59.8[abcd]	41.3[abcd]	15,2[b]
MJ-Cit	40.4[cde]	25.5[h]	19.5[defghij]	16.9[b]
MJ-Mal	44.2[bcd]	33.3[fgh]	16.1[fghij]	17.1[b]
MJ-Gluc-Cit	21.2[fghi]	34.1[fgh]	18.6[efghij]	27.5[b]
MJ-Gluc-Mal	19.8[ghi]	28.5[gh]	24.2[cdefghi]	20.1[b]
MJ-Fruc-Cit	13.6[hij]	33.5[fgh]	35.7[abcdef]	18.5[b]
MJ-Fruc-Mal	8.9[hij]	40.1[efgh]	35.3[abcdef]	25.8[b]
MJ-Gluc-Fruc-Cit-Mal	3.1[ij]	2.25[i]	9.9[ghij]	47.8[a]
MJ-AA	43.5[bcd]	34.8[fgh]	17.5[efghij]	5.8[b]
MJ-Nu	34.7[defg]	46.7[defg]	29.7[bcdefg]	13.6[b]
MJ-Mn	42.2[bcd]	55.9[bcde]	30.5[bcdefg]	9.9[b]
MJ-AA-Mn	46.3[bcd]	52.3[cdef]	25.4[cdefghi]	8.3[b]
MJ-Gluc-Fruc-Mn	14.0[hij]	73.4[ab]	48.1[ab]	16.6[b]
MJ-Cit-Mal-Mn	55.9[abc]	12.0[i]	8.2[hij]	28.1[b]

For each attribute, the means with the same letter (a-j) are not significantly different at the level of 5% according to Newman-Keuls tests.

MJ, model juice; Gluc, glucose; Fruc, fructose; Cit, citric acid; Mal, malic acid; Nu, mononucleotides; Mn, minerals ($CaCl_2$, $MgCl_2$, Na_2HPO_4, K_2HPO_4); AA, amino acids.

sweetness, which is only slightly modulated by citric acid. They explained the contradiction of these results and those obtained by Jones and Scott (1984), who reported that only 50% of the perceived sweetness was due to reducing sugar, other fruit constituents contributing to part of the remaining 50%. The difference can be explained by the fact that only Salles et al. (2003) used nose-clips to suppress aroma perception. This difference in sweet perception intensity could be at least partly explained by sensorial interactions between aroma compounds and sweetness perception, as can occur in tomato (Baldwin et al. 1998).

Salles et al. (2003) observed that sugar had a masking effect on sourness (Table 3), as already reported by Stevens et al. (1977). Concerning the effect of acids on sweetness, neither omission of citric acid nor omission of malic acid has led to significant effects on the perception of sweetness, probably because of the concentration of sugars ranking in a range where the effect of citric acid was low. Salles et al. (2003) reported an effect on sweetness only when the two main acids were omitted together, showing a significant masking effect on sweetness only due to the effect of the sum of the two acids (Table 3). The opposite results reported by Stevens et al. (1977) are supported by significant interactions between acids and sweetness encountered in tomato only when the concentration of acids and sugars is high or when the ratio of acid to sugar is unbalanced.

Otherwise, Petro-Turza (1986-1987) and Petro Turza and Teleky-Vamosky (1989) showed the positive influence of the presence of salts on taste characteristics of tomato. In addition, minerals can have an indirect action on taste. Potassium can influence the free acid content, while phosphate can interact by its buffering capacity (Petro-Turza 1986-1987). Salles et al. (2003) showed that the association of organic acids and potassium ions was a determinant for the saltiness characteristic of tomato.

Regarding non-volatile nitrogen compounds (i.e., amino acids and peptides), very contradictory results have been reported. On the one hand, because of their low quantity, their contribution to taste was considered negligible (Kader et al. 1978a) and their buffer activities were of minor importance beside the main organic acids (Stevens 1972). On the other hand, Fuke and Shimizu (1993) considered that the predominant amino acids, glutamic and aspartic acids, have an effect on global tomato taste. Teleky-Vamossy and Szepes-Krell (1989) showed that the addition of amino acids to a model juice made of mineral salts, sugars and acids led to a significant improvement of its taste characteristics, but the presence of salts was important for self-assertion of their characteristic flavour. However, an excessive quantity of amino acids led to an unfavourable change in the taste of tomato (De Bryun et al. 1971). Using omission tests, Salles et al. (2003) showed that the neutral and alkaline amino acids had no effect on tomato taste. Acid amino acids such as glutamic, aspartic and gamma-aminobutyric acids were shown to develop umami taste in the model tomato juice only when organic acids and sugars were omitted. Indeed, though monosodiumglutamate, which is the main component responsible for umami, is present at a level higher than its perception threshold value, it is totally masked by the other taste components.

Bitterness is not characteristic of tomato and was rarely reported. However, Salles et al. (2003) reported a retentate of ultrafiltration of ripe tomato juice presenting mainly bitterness. Alpha-tomatine, which could be responsible for this taste in tomato (Davies and Hobson 1981), was not detected. This compound is generally more abundant in green tomato than in red tomato. Though peptides were a minor constituent of tomato juice, Salles et al. (2003) suggested that they could be at least partly responsible for the bitter taste of the fraction.

TOMATO FLAVOUR QUALITY

Relationships between sensory and instrumental data were frequently used in order to understand how a physicochemical property such as a particular flavour-active component, colour or texture characteristic acts to produce a more or less specific sensation or how a change of ingredient affects sensory properties between species, between ripening stages, or after a process. These relationships make an important contribution to our understanding of how physicochemical properties act to produce specific sensations. Many multivariate techniques were used to provide visualization tools that are helpful to observe relationships between variables.

The obtaining of such data sets requires many sensory evaluation sessions and the measurements of many physicochemical properties and the identification and quantification of many metabolites, which is expensive and time-consuming. Several works reported rapid new instrumental measurements to simplify and accelerate physicochemical analyses and the development of predictive models for quality assessment (Azodanlou et al. 2003).

The lack of flavour in fresh market tomatoes was noticed early in the development of post-harvest storage procedures (Kader et al. 1978b). That was confirmed later by analyses of volatiles carried out on fresh fruits stored at low temperature (Buttery et al. 1987). Improvements made to maintain a good quality of colour or texture in processed tomato products were not sufficient and consumers still complained about the lack of flavour of fresh fruits as well as the poor quality flavour of processed tomato products. Therefore, the improvement of flavour and the evaluation of consumer attitude became the aim of many researchers.

Several studies reported relationships between instrumental measurements and consumer preference. Malundo et al. (1995) showed that increasing total sugar and acid levels did not affect fresh tomato note but did significantly and positively affect flavour acceptability. However, too much acid in fresh tomato negatively affected consumer acceptability. Significant relations were found between the content of acid and quantitative descriptive analysis to smell, flavour, and aftertaste attributes such as sour, tomato-like, sweet, spoiled, sweetish and mouldy (Auerswald et al. 1999a). Auerswald et al. (1999a) found also significant relationships between fruit deformation and descriptive mouth-feel attributes. Concerning tomato quality during storage (4 to 7 d), they showed that tomatoes stored for a shorter time were preferred by consumers in flavour, aftertaste, mouth-feel, and appearance, with an equal or better score than fresh harvested tomatoes.

Fresh tomatoes are now marketed in various forms, sizes, colours and flavours. However, few studies have been published on consumer behaviour related to fresh tomato consumption. Simonne et al. (2006) showed through a survey that in the United States, consumers preferred low-price and high-lycopene-content fresh-market tomatoes. However, preferences were found to differ with age and sex of consumers. For example, younger participants (< 39 yr) gave greater importance to production method, lycopene content and tomato type, while older participants (39-57 yr) were the most price-sensitive, and females were less sensitive than males. Segmentation of consumers was also reported for preference for fresh tomatoes by Lê and Ledauphin (2006), Plaehn and Lundahl (2006), and Brueckner et al. (2007). Serrano-Megias and Lopez-Nicolas (2006) used agglomerative hierarchical clustering to identify consumer tomato preferences and showed the presence of four consumer clusters from physicochemical and sensory characteristics. However, the choice of the consumer seems more complex and is not based only on sensory or physicochemical criteria; nutrition (Jahns et al. 2001, Causse et al. 2003), growing conditions (Johansson et al. 1999) and gene technology use information also influence consumer attitude (Saba and Vassallo 2002).

ACKNOWLEDGEMENTS

Thanks to Dominique Langlois for providing data on tomato volatile compound studies and to Emma Butler for English language correction.

Abbreviations

APCI: Atmospheric pressure chemical ionisation; FTIR: Fourier transformation infra-red; GC: Gas chromatography; HPLC: High performance liquid chromatography; MS: Mass spectrometry; NMR: Nuclear magnetic resonance; O: Olfactometry; PTR: Proton reaction transfer; QTL: Quantitative Trait Locus; UV: Ultra-violet

REFERENCES

Arias, R. and T.C. Lee, D. Specca, and H. Janes. 2000. Quality comparison of hydroponic tomatoes (*Lycopersicon esculentum*) ripened on and off vine. J. Food Sci. 65: 545-548.

Auerswald, H. and P. Peters, B. Bruckner, A. Krumbein, and R. Kuchenbuch. 1999a. Sensory analysis and instrumental measurements of short-term stored tomatoes (*Lycopersicon esculentum Mill.*). Postharvest Biol. Technol. 15: 323-334.

Auerswald, H. and D. Schwarz, C. Kornelson, A. Krumbein, and B. Bruckner. 1999b. Sensory analysis, sugar and acid content of tomato at different EC values of the nutrient solution. Scientia. Hortic. 82: 227-242.

Azodanlou, R. and C. Darbellay, J.L. Luisier, J.C. Villettaz, and R. Amado. 2003. Development of a model for quality assessment of tomatoes and apricots. Lebensm.-Wiss. Technol. 36: 223-233.

Bach, T.J. 1995. Some new aspects of isoprenoid biosynthesis in plants: a review. Lipids 30: 191-202.

Bailen, G. and F. Guillen, S. Castillo, M. Serrano, D. Valero, and D. Martinez-Romero. 2006. Use of activated carbon inside modified atmosphere packages to maintain tomato fruit quality during cold storage. J. Agric. Food Chem. 54: 2229-2235.

Baldwin, E.A. and J.W. Scott, M.A. Einstein, T.M.M. Malundo, B.T. Carr, R.L. Shewfelt, and K.S. Tandon. 1998. Relationship between sensory and instrumental analysis for tomato flavour. J. Am. Soc. Hortic. Sci. 123: 906-915.

Bate, N.J. and S. Sivasankar, C. Moxon, J.M.C. Riley, J.E. Thompson, and S.J. Rothstein. 1998. Molecular characterization of an *Arabidopsis* gene encoding hydroperoxide lyase, a cytochrome P-450 that is wound inducible. Plant Physiol. 117: 1393-1400.

Boukobza, F. 2001. Effect of growing and storage conditions on tomato flavour. PhD thesis, University of Nottingham, UK.

Boukobza, F. and A.J. Taylor. 2002. Effect of postharvest treatment on flavour volatiles of tomatoes. Postharvest Biol. Technol. 25: 321-331.

Bowsher, C.G. and B.J.M. Ferrie, S. Ghosh, Todd, J., J.E. Thompson, and S.J. Rothstein. 1992. Purification and partial characterization of a membrane-associated lipoxygenase in tomato fruit. Plant Physiol. 100: 1802-1807.

Brauss, M.S. and R.S.T. Linforth, and A.J. Taylor. 1998. Effect of variety, time of eating, and fruit-to-fruit variation on volatile release during eating of tomato fruits (*Lycopersicon esculentum*). J. Agric. Food Chem. 46: 2287-2292.

Brueckner, B. and I. Schonhof, R. Schroedter, and C. Kornelson. 2007. Improved flavour acceptability of cherry tomatoes. Target group: Children. Food Qual. Pref. 18: 152-160.

Buescher, R.W. 1975. Organic acid and sugar levels in tomato pericarp as influenced by storage at low temperature. Hortscience 10: 158-159.

Buettner, A. and P. Schieberle. 2001. Aroma properties of a homologous series of 2,3-epoxyalkanals and trans-4,5-epoxyalk-2-enals. J. Agric. Food Chem. 49: 3881-3884.

Buttery, R.G. Quantitative and sensory aspects of flavour of tomato and other vegetables and fruits. pp. 259-285. *In:* T.E. Acree and R. Teranishi. [eds.] 1993. Flavour Science. American Chemical Society, Washington, DC.

Buttery, R.G. and L. Ling. Enzymatic production of volatiles in tomatoes. pp. 137-146. *In:* P. Schreier and P. Winterhalter. [eds.] 1993a. Progress in Flavour Precursor Studies. Allured Publishing Corporation, Carol Stream, USA.

Buttery, R.G. and L.C. Ling. Volatile components of tomato fruit and plant parts. pp. 23-34. *In:* R. Teranishi, R.G. Buttery and H. Sugisava. [eds.] 1993b. Bioactive Volatile Compounds from Plants. American Chemical Society, Washington, DC.

Buttery, R.G. and G. Takeoka, R. Teranishi, and L.C. Ling. 1990. Tomato aroma components: identification of glycoside hydrolysis volatiles. J. Agric. Food Chem. 38: 2050-2053.

Buttery, R.G. and G.R. Takeoka, G.E. Krammer, and L.C. Ling. 1994. Identification of 2,5-dimethyl-4-hydroxy-3(2h)-furanone (furaneol) and 5-methyl-4-hydroxy-3(2h)-furanone in fresh and processed tomato. Lebensm.-Wiss. Technol. 27: 592-594.

Buttery, R.G. and G.R. Takeoka, and L.C. Ling. 1995. Furaneol—odor threshold and importance to tomato aroma. J. Agric. Food Chem. 43: 1638-1640.

Buttery, R.G. and G.R. Takeoka, M. Naim, H. Rabinowitch, and Y. Nam. 2001. Analysis of furaneol in tomato using dynamic headspace sampling with sodium sulfate. J. Agric. Food Chem. 49: 4349-4351.

Buttery, R.G. and R. Teranishi, and L.C. Ling. 1987. Fresh tomato aroma volatiles: a quantitative study. J. Agric. Food Chem. 35: 540-544.

Buttery, R.G. and R. Teranishi, L.C. Ling, R.A. Flath, and D.J. Stern. 1988. Quantitative studies on origins of fresh tomato aroma volatiles. J. Agric. Food Chem. 36: 1247-1250.

Cabibel, M. and J. Nicolas. 1991. Lipoxygenase from tomato fruit (*Lycopersicon esculentum* L.)—Partial purification, some properties and invitro cooxidation of some carotenoid-pigments. Sci. Aliments 11: 277-290.

Camara, B. and P. Hugueney, F. Bouvier, M. Kuntz, and R. Moneger. 1995. Biochemistry and molecular biology of chromoplast development. Int. Rev. Cytol. 163: 175-247.

Causse, M. and M. Buret, K. Robini, and P. Verschave. 2003. Inheritance of nutritional and sensory quality traits in fresh market tomato and relation to consumer preferences. J. Food Sci. 68: 2342-2350.

Causse, M. and V. Saliba-Colombani, I. Lesschaeve, and M. Buret. 2001. Genetic analysis of organoleptic quality in fresh market tomato. 2. Mapping QTLs for sensory attributes. Theor. Appl. Genet. 102: 273-283.

Crouzet, J. and S. Seck. 1982. L'arome de la tomate, mecanismes de formation des constituants par voies biochimique et chimique. Parfums, Cosmétiques, arômes 44: 71-84.

Davies, J.N. and G.E. Hobson. 1981. The constituent of tomato fruit—the influence of environment, nutrition and genotype. Crit. Rev. Food Sci. Nutr. 15: 205-280.

De Bryun, J.W. and F. Garretsen, and E. Kooistra. 1971. Variation in taste and chemical composition of the tomato (*Lycopersicon esculentum*). Euphytica 20: 214-227.

Etiévant, P.X. and D. Langlois. 1999. L'amélioration de la représentativité olfactive des extraits aromatiques dans la recherche des composés clés de l'arôme des aliments. Rivista Italiana EPPOS 268-278.

Fuke, S. and T. Shimizu. 1993. Sensory and preference aspects of umami. Trends Food Sci. Technol. 4: 246-250.

Fulton, T.M. and P. Bucheli, E. Voirol, J. Lopez, V. Pétiard, and S.D. Tanksley. 2002. Quantitative trait loci (QTL) affecting sugars, organic acids and other biochemical properties possibly contributing to flavour, identified in four advanced backcross populations of tomato. Euphytica 127: 163-177.

Galliard, T. and J.A. Matthew, A.J. Wright, and M.J. Fishwick. 1977. The enzymatic breakdown of lipids to volatile and non volatile carbonyl fragments in disrupted tomato fruits. J. Sci. Food Agric. 28: 863-868.

Gray, D.A. and S. Prestage, R.S.T. Linforth, and A.J. Taylor. 1999. Fresh tomato specific fluctuations in the composition of lipoxygenase-generated C6 aldehydes. Food Chem. 64: 149-155.

Hatanaka, A. and T. Kajiwara, and J. Sekiya. Fatty acid hydroperoxide lyase in plant tissues. *In:* T.H. Parliment and R. Croteau. [eds.] 1986. Biogeneration of Aromas. American Chemical Society, Washington, DC.

Heeb, A. and B. Lundegardh, T. Ericsson, and G.P. Savage. 2005. Nitrogen form affects yield and taste of tomatoes. J. Sci. Food Agric. 85: 1405-1414.

Heeb, A. and B. Lundegardh, G.P. Savage, and T. Ericsson. 2006. Impact of organic and inorganic fertilizers on yield, taste, and nutritional quality of tomatoes. J. Plant Nutr. Soil Sci. 169: 535-541.

Hobson, G.E. and L. Bedford. 1989. The composition of cherry tomatoes and its relation to consumer acceptability. J. Hortic. Sci. 64: 321-329.

Islam, M.S. 1997. Variability in different physical and biochemical characteristics of six tomato genotypes according to stages of ripeness. Bangladesh J. Bot. 26: 137-147.

Jahns, G. and H.M. Nielsen, and W. Paul. 2001. Measuring image analysis attributes and modelling fuzzy consumer aspects for tomato quality grading. Comput. Electron. Agric. 31: 17-29.

Johansson, L. and Å. Haglund, L. Berglund, P. Lea, and E. Risvik. 1999. Preference for tomatoes, affected by sensory attributes and information about growth conditions. Food Qual. Pref. 10: 289-298.

Jones, R.A. and S.J. Scott. 1984. Genetic potential to improve tomato flavour in commercial F hybrids. J. Am. Soc. Hortic. Sci. 109: 318-321.

Kader, A.A. and L.L. Morris, M.A. Stevens, and M. Allbright-Holton. 1978a. Amino acid composition and flavour of fresh market tomatoes as influenced by fruit ripeness when harvested. J. Am. Soc. Hortic. Sci. 103: 541-544.

Kader, A.A. and L.L. Morris, M.A. Stevens, and M. Allbright-Holton. 1978b. Composition and flavour quality of fresh market tomatoes as influenced by some postharvest handling procedures. J. Am. Soc. Hortic. Sci. 103: 6-13.

Kader, A.A. and D. Zagory, and E.L. Kerbel. 1989. Modified atmosphere packaging of fruits and vegetables. Crit. Rev. Food Sci. Nut. 28: 1-30.

Kazeniac, S.J. and R.M. Hall. 1970. Flavour chemistry of tomato volatiles. J. Food Sci. 35: 519-530.

Krammer, G.E. and R.G. Buttery, and G.R. Takeoka. Studies on tomato glycosides. pp. 164-181. *In:* R.L. Rouseff and M.M. Leahy. [eds.] 1995. Fruit Flavours: Biogenesis Characterisation and Authentification. American Chemical Society, Washington, DC.

Krammer, G.E. and G.R. Takeoka, and R.G. Buttery 1994. Isolation and identification of 2,5-dimethyl-4-hydroxy-3(2H)-furanone glucoside from tomatoes, J. Agric. Food Chem. 42: 1595-1597.

Krauss, S. and W.H. Schnitzler, J. Grassmann, and M. Woitke. 2006. The influence of different electrical conductivity values in a simplified recirculating soilless system on inner and outer fruit quality characteristics of tomato. J. Agric. Food Chem. 54: 441-448.

Krumbein, A. and H. Auerswald. 1998. Characterization of aroma volatiles in tomatoes by sensory analyses. Nahrung 42: 395-399.

Langlois, D. and P.X. Etievant, P. Pierron, and A. Jorrot. 1996. Sensory and instrumental characterisation of commercial tomato varieties. Z. Lebensm. Unters. Forsch. 203: 534-540.

Lê, S. and S. Ledauphin. 2006. You like tomato, I like tomato: segmentation of consumers with missing values. Food Qual. Pref. 17: 228-233.

Lewinsohon, E. and Y. Sitrit, E. Bar, Y. Azulay, M. Ibdah, A. Meir, E. Yosef, D. Zamir, and Y. Tadmor. 2005. Not just colours – carotenoid degradation as a link between pigmentation and aroma in tomato and watermelon fruit. Trends Food Sci. Technol. 16: 407-415.

Malundo, T.M.M. and R.L. Shewfelt, and J.W. Scott. 1995. Flavour quality of fresh tomato (*Lycopersicon esculentum* Mill.) as affected by sugar and acid levels. Postharvest Biol. Technol. 6: 103-110.

Marconi, O. and S. Floridi, and L. Montanari. 2007. Organic acids profile in tomato juice by HPLC with UV detection. J. Food Qual. 30: 43-56.

Marlatt, C. and M. Chien, and C.T. Ho. 1991. C13 norisoprenoids bound as glycosides in tomato. J. Essent. Oil Res. 3: 27-31.

Marlatt, C. and C. Ho, and M. Chien. 1992. Studies of aroma constituents bound as glycosides in tomato. J. Agric. Food Chem. 40: 249-252.

Maul, F. and S.A. Sargent, M.O. Balaban, E.A. Baldwin, D.J. Huber, and C.A. Sims. 1998. Aroma volatile profiles from ripe tomatoes are influenced by physiological maturity at harvest: An application for electronic nose technology. J. Am. Soc. Hortic. Sci. 123: 1094-1101.

Moraru, C. and L. Logendra, T.C. Lee, and H. Janes. 2004. Characteristics of 10 processing tomato cultivars grown hydroponically for the NASA advanced life support (ALS) program. J. Food Compos. Anal. 17: 141-154.

Oneill, M. and R. Casey, C. Domoney, C. Forster, and S. Quarrie. 1996. A role for lipoxygenase in stress responses in pea. Plant Physiol. 111: 506.

Paull, R.E. 1999. Effect of temperature and relative humidity on fresh commodity quality. Postharvest Biol. Technol. 15: 263-277.

Petro Turza, M. 1986-1987. Flavour of tomato and tomato products. Food Rev. Int. 2: 309-351.

Petro Turza, M. and I. Szarfoeldi Szalma, K. Fuezesi Kardos, and A. Barath. 1989a. Study on taste substances of tomato. II. Analytical investigation into the free amino acid, peptide and protein contents. Acta Aliment. 18: 107-117.

Petro Turza, M. and K. Szarfoeldi Szalma, K. Fuezesi Kardos, and F. Hajdu. 1989b. Study on taste substances of tomato. I. Analytical investigations into the sugar, organic acid and mineral contents. Acta Aliment. 18: 97-106.

Petro Turza, M. and G. Teleky-Vamosky. 1989. Study on taste substances of tomato. III. Sensory evaluations. Nahrung 33: 387-394.

Plaehn, D. and D.S. Lundahl. 2006. A L-PLS preference cluster analysis on French consumer hedonics to fresh tomatoes. Food Qual. Pref. 17: 243-256.

Porter, J.W. and S.L. Spurgeon. 1979. Enzymatic synthesis of carotenes. Pure Appl. Chem. 51: 609-622.

Ratanachinakorn, B. and A. Klieber, and D.H. Simons. 1997. Effect of short-term controlled atmospheres and maturity on ripening and eating quality of tomatoes. Postharvest Biol. Technol. 11: 149-154.

Riley, J.C.M. and J.E. Thompson. 1997. Subcellular generation and distribution of lipid-derived volatiles in the ripe tomato. J. Plant Physiol. 150: 546-551.

Riley, J.C.M. and C. Willemot, and J.E. Thompson. 1996. Lipoxygenase and hydroperoxide lyase activities in ripening tomato fruit. Postharvest Biol. Technol. 7: 97-107.

Rohmer, M. and M. Seemann, S. Horbach, S. Bringer-Meyer, and H. Sahm. 1996. Glyceraldehyde 3-phosphate and pyruvate as precursors of isoprenic units in an alternative non-mevalonate pathway for terpenoid biosynthesis. J. Am. Chem. Soc. 118: 2564-2566.

Saba, A. and M. Vassallo. 2002. Consumer attitudes toward the use of gene technology in tomato production. Food Qual. Pref. 13: 13-21.

Salles, C. and S. Nicklaus, and C. Septier. 2003. Determination and gustatory properties of taste-active compounds in tomato juice. Food Chem. 81: 395-402.

Schifferstein, H.N.J. 1994. Sweetness suppression in fructose citric acid mixtures—A study of contextual effect. Percept. Psychophys. 56: 227-237.

Serrano-Megias, M. and J.M. Lopez-Nicolas. 2006. Application of agglomerative hierarchical clustering to identify consumer tomato preferences: influence of physicochemical and sensory characteristics on consumer response. J. Sci. Food Agric. 86: 493-499.

Simonne, A.H. and B.K. Behe, and M.M. Marshall. 2006. Consumers prefer low-priced and high-lycopene-content fresh-market tomatoes. Horttechnology 16: 674-681.

Stern, D.J. and R.G. Buttery, R. Teranishi, L. Ling, K. Scott, and M. Cantwell. 1994. Effect of storage and ripening on fresh tomato quality. 1. Food Chem. 49: 225-231.

Stevens, M.A. 1970. Relation between polyene-carotene content and volatile compound composition of tomatoes. J. Am. Soc. Hortic. Sci. 95: 461-464.

Stevens, M.A. 1972. Relationships between components contributing to quality variation among tomato lines. J. Am. Soc. Hortic. Sci. 97: 70-73.

Stevens, M.A. and A.A. Kader, M. Albright-Holton, and M. Algazi. 1977. Genotypic variation for flavour and composition in fresh market tomatoes. J. Am. Soc. Hortic. Sci. 102: 680-689.

Tadege, M. and I. Dupuis, and C. Kuhlemeier. 1999. Ethanolic fermentation: new functions for an old pathway. Trends Plant Sci. 4: 320-325.

Taylor, A.J. and R.S.T. Linforth. 2003. Direct mass spectrometry of complex volatile and non volatile flavour mixtures. Int. J. Mass Spectrom. 223-224: 179-191.

Teleky-Vamossy, G. and G. Szepes-Krell. 1989. Investigation on the sensory properties of some taste enhancing food additives. Nahrung 33: 395-404.

Thybo, A.K. and M. Edelenbos, L.P. Christensen, J.N. Sorensen, and K. Thrup-Kristensen. 2006. Effect of organic growing systems on sensory quality and chemical composition of tomatoes. Lebensm.-Wiss. Technol. 39: 835-843.

Tieman, D. and M. Taylor, N. Schauer, A.R. Fernie, A.D. Hanson, and J. Klee. 2006a. Tomato aromatic amino acid decarboxylase participate in synthesis of the flavour volatiles 2-phenylethanol and 2-phenylacetaldehyde. Proc. Natl. Acad. Sci. USA 103: 8287-8292.

Tieman, D.M. and M. Zeigler, E.A. Schmelz, M.G. Taylor, P. Bliss, M. Kirst, and H.J. Klee. 2006b. Identification of loci affecting flavour volatile emissions in tomato fruits. J. Exp. Bot. 57: 887-896.

Tikunov, Y. and A. Lommen, C.H. Ric de Vos, H.A. Verhoeven, R.J. Bino, R.D. Hall, and A.G. Bovy. 2005. Novel approach for nontargeted data analysis for metabolomics. Large-scale profiling of tomato fruit volatiles. Plant Physiol. 139: 125-1137.

Todd, J.F. and G. Paliyath, and J.E. Thompson. 1990. Characteristics of a membrane-associated lipoxygenase in tomato fruit. Plant Physiol. 94: 1225-1232.

Watada, A.E. and B.B. Aulenbach. 1979. Chemical and sensory qualities of fresh market tomatoes. J. Food Sci. 44: 1013-1016.

Wright, D.H. and N.D. Harris. 1985. Effect of nitrogen and potassium fertilization on tomato flavour. J. Agric. Food Chem. 33: 355-358.

Wu, Z. and D.S. Robinson, R.K. Hughes, R. Casey, D. Hardy, and S.I. West. 1999. Co-oxidation of beta-carotene catalyzed by soybean and recombinant pea lipoxygenases. J. Agric. Food Chem. 47: 4899-4906.

Yu, M.H. and D.K. Salunkhe, and L.E. Olson. 1968a. Production of 3-methylbutanal from L-leucine by tomato extract. Plant Cell Physiol. 9: 633.

Yu, M.H. and L.E. Olson, and D.K. Salunkhe. 1968b. Precursors of volatile components in tomato fruit. III. Enzymic reaction products. Phytochemistry 7: 561-565.

6

Antioxidant Activity in Tomato: A Function of Genotype

Charanjit Kaur[1*] and H.C. Kapoor[2]

[1]Division of Post Harvest Technology, Indian Agricultural Research Institute, New Delhi 110012, India

[2]Division of Biochemistry, Indian Agricultural Research Institute, New Delhi 110012, India

ABSTRACT

A continuous understanding of the relationship between food genotype and diet-related diseases has initiated an exciting era for nutritional and medicinal science. The proven beneficial effects of the "Mediterranean diet", a phenomenon partly attributed to lycopene, have fuelled great interest in exploring antioxidants in tomatoes (*Lycopersicon esculentum* Mill.). Antioxidant composition in tomatoes is a strong environment-regulated genetic response and strategies to enhance antioxidant levels involve mediation of the cultivars, temperature, light, mineral content and post-harvest ripening. Huge genetic variability among commercial and wild genotypes offers considerable opportunity to improve functional quality in tomatoes. Cherry tomatoes and Mediterranean cultivars seem to be valuable genotypes with indisputable high antioxidant composition. *Lycopersicon pennelli* var. *puberulum* and accession LA 1996 seem to be promising wild genotypes for producing tomatoes rich in flavonoid and anthocyanin through genetic manipulations. Antioxidant labeling for defining functional quality in tomatoes relies on selection of a standardized assay system including both hydrophilic and lipophilic components.

**Corresponding author*

A list of abbreviations is given before the references.

INTRODUCTION

Discovery of antioxidant phytochemicals and their promising health-promoting effects has paved the way to a food revolution, promising an age of good health (Rodriguez et al. 2006). Health concerns are driving plant breeders and food producers to develop food products with functional health ingredients comprising of both nutritive and non-nutritive antioxidants. Epidemiological studies and successive scientific investigations have transformed fruits and vegetables into functional foods and neutraceuticals for promoting health and reducing the need for health care tools (Kaur and Kapoor 2001). In this context, it is customary to find the health-promoting compounds present in raw materials, their concentrations and total antioxidant effects (Van der Sluis et al. 2001). Genetics seems to have a prime role in the ability of cultivars to respond to biotic and abiotic stresses, which ultimately can impact their antioxidant capacity. Evaluation of different varieties and cultivars and selection and identification of the superior genotype for high antioxidant capacity have been a subject of active research in the last decade (Brovelli 2006). It is important to characterize crops on the basis of total antioxidant capacity, which depicts a total interaction between plant genotype and environment rather than a single specific antioxidant compound (Scalzo et al. 2005). This offers greater opportunity for enhancing the levels of antioxidants and their anti-proliferative and chemo-protective properties through both conventional and transgenic genetic manipulations. Exploration of genetic variability for the production of healthy foods rich in antioxidants or neutraceuticals is a recent trend in plant science.

Tomato (*Lycopersicon esculentum* Mill.), among antioxidant-rich commodities, has achieved a spectacular status because of its rich composition and widespread consumption. It is one of the major vegetable crops, grown in almost every country of the world. Studies indicate that regular intake of cooked tomato as a part of the vegetable regimen appears to be the major nutritional factor accounting for lower risk of prostrate cancer, digestive tract cancer and coronary heart diseases in the Mediterranean region, particularly in southern Italy (Giovannucci et al. 1995, La Vecchia 1997). Supported by overwhelming and convincing epidemiological studies, tomato has been called a functional food and an effective preventive strategy against major lifestyle diseases such as cardiovascular diseases and cancer, and it is said to protect cells from DNA damage (Weisberger 2002, Sesso et al. 2003, Willcox et al. 2003, Canene-Adams et al. 2005).

ANTIOXIDANT CONSTITUENTS AND FUNCTIONAL QUALITY

Tomatoes have been ranked first as a source of lycopene (71.6%), second as a source of vitamin C (12.0%), pro-vitamin A carotenoids (14.6%) and other carotenoids (17.2%), and third as a source of vitamin E (6.0%) (Garcia-Closas et al. 2004). In the Italian diet, tomatoes have been assessed as the second most important source of vitamin C after orange juice (La Vecchia 1997). The major antioxidant compounds in tomatoes are characterized as lipid-soluble and water-soluble. Lycopene represents the predominant lipid-soluble compound and constitutes more than 80% of total tomato carotenoids in fully red-ripe fruits. Although devoid of pro-vitamin A activity, it is a well-known bio-molecule of interest in epidemiological studies (Rao and Agarwal 1999). Lycopene principally occurs in all-*trans* form with small amounts of *cis* isomers (Nguyen and Schwartz 1999) and shows strong antioxidant activity (AOX) both *in vitro* and *in vivo*, having a physical quenching rate constant with singlet oxygen almost twice that of β-carotene. Generally, processing, lycopene degradation and agronomic and physiochemical parameters of crop cycle in the field (temperature, sun exposure, fertilizer and irrigation regimes) result in irreversible isomerization of all-*trans* lycopene to *cis* isomers. Most of the lycopene is attached to the insoluble and fibrous parts of the tomato and the skin may contain about five times as much lycopene as the pulp. Other carotenoids in tomato constitute β-carotene and small amounts of phytoene, phytofluene, α-carotene, γ-carotene, neurosporene, and lutein (Khachik et al. 2002). β-carotene is of special interest for its pro-vitamin A activity and constitutes nearly 7-10% of total tomato carotenoid synthesis (Nguyen and Schwartz 1999).

Tomatoes also contain moderate amounts of water-soluble phenolics, flavonoids (quercetin, kaempferol and naringenin) and the hydrocinnamic acids (caffeic, chlorogenic, ferulic and *p*-coumaric acids), mainly concentrated in skin (Martinez-Valverde et al. 2002, Minoggio et al. 2003). Rutin, caffeoyl, quinic and chalconaringenin have also been identified in tomato fruit (Proteggente et al. 2002). With the exception of tomato containing *Aft* gene, tomato fruit does not contain significant levels of anthocyanins (Jones et al. 2003). Vitamin C content in tomato is moderate (84 to 590 mg/kg), but its contribution to diet is significant because of its high consumption. One hundred grams of tomato can supply about 20% and 40% of adult US recommended daily allowance of vitamins A and C respectively. Tomatoes are also a good source of potassium, folate and vitamin E, soluble and insoluble dietary fibres (pectins, hemicelluloses and celluloses, with a concentration ranging from 0.8 to 1.3 g/100 g fresh weight) (USDA National Nutrient Data Base for standard reference, http: www.nal.usda.gov/fnic/foodcomp).

ANTIOXIDANT ACTIVITY: METHODS AND SAMPLE PREPARATION

Methods

The health benefits of fruits and vegetables are a function of additive and synergistic combinations of various antioxidants and phytochemicals; thus, it becomes essential to measure total AOX apart from evaluating antioxidants (Liu et al. 2004). Antioxidant activity is a unique parameter that quantifies the ability of a complex biological sample to reduce free radicals and scavenge reactive oxygen species. Carotenoids, polyphenols, flavonoids, anthocyanins and vitamin C are the major contributors to AOX. In the last decade, there has been an unprecedented rise in research activities focusing on and stressing the need to estimate AOX as an essential biomarker tool to establish the functional quality of raw material and processed products. A large number of methods have been developed, but there has been no consensus on an approved and preferred method (Prior et al. 2005). Lack of standardized assays has led to use of multiple assays to generate an antioxidant profile encompassing reactivity towards both aqueous and lipid/organic radicals directly via radical quenching and radical reducing mechanisms and indirectly via metal complexing. No single assay can elucidate a full profile of antioxidants and a multi-functional approach using multiple methods is probably the only answer to end the present chaos in methodologies (Frankel and Meyer 2000). Evaluation of AOX in tomato is a challenge, as there are few methods that permit evaluation in both water- and lipid-soluble component in food samples. The most popular and widely used assays are primarily designed to evaluate hydrophilic components and the lipophilic component is often neglected. Most commonly employed assay systems used for studying AOX in tomatoes are TEAC, ORAC, FRAP and Folin-Ciocalteu assay (Table 1). The First International Congress on Antioxidant Methods held in Orlando, Florida, in 2004 recommends all of them as standard methods for evaluation of AOX, barring FRAP (Prior et al. 2005).

The TEAC method employs $ABTS^{+}$ radical cation to evaluate AOX in both hydrophilic and lipophilic constituents in the same sample. The method is operationally easier, accurate, and quickly applied (Arnao et al. 2001) and recent modifications have further improved its practicability (Ozgen et al. 2006). The ORAC developed by Cao et al. (1996) and recently modified by Ou et al. (2002) and Wu et al. (2004) is closely related to biological functions of chain-breaking antioxidants and evaluates *in vivo* responses to dietary antioxidant manipulation. ORAC measures antioxidant inhibition of peroxyl radical-induced oxidations and thus

Table 1 Antioxidant activity in tomato genotypes.

Research groups	*Methods*	*Antioxidant activity*	*Phenolic content*	*Country*[b]	*Solvents used*
1. Wang et al. 1996	ORAC	1.89 [a]	–	USA	Ethanol
2. Paganaga et al. 1996	ABTS	1.60 [a]	–	–	–
3. Scalfi et al. 2000	ABTS	Hydrophilic 0.30-0.51[c] Lipophilic 0.73-1.55[d]	–	Italy	Aqueous and petroleum ether
4. Martinez-Valverde et al. 2002	ABTS DPPH	0.58-3.50 mmol/l1.27 $\times 10^{-6}$ 13.46 $\times 10^{-6}$	7.19-43.59	Spain	80% aqueous methanol (1%-HCl) mg/kg (Quercetin)
5. Ou et al. 2002	FRAP ORAC	2.35–4.20[a] 1.65-3.35[a]	–	USA	Ethanol
6. Cano et al. 2003	ABTS	Hydrophilic 1.92-2.18[a] Lipophilic 0.33-0.88 [a]	–	Spain	Sodium phosphate buffer (50 mM)
7. George et al. 2004	FRAP	0.6-2.2[a]	–	India	80% ethanol
8. Wu et al. 2004	ORAC	Hydrophilic 0.24[a] Lipophilic 3.26 [a] Total 3.4[a]	–	USA	Acetone : water : acetic acid (10:29.5:0.5)
9. Toor and Savage 2005	ABTS	1.16-1.63[a]	–	New Zealand	Acetone : water : acetic acid (70:29.5:0.5)
10. Rotino et al. 2005	ABTS	Hydrophilic 2.11-2.60 [a] Lipophilic 0.39-0.42 [a]	–	Italy	Ethanol : water (80:20)
11. Chun et al. 2005	ABTS	29.44 mg*VCE /100 g	23.69 mg GAE/100 g	USA	
12. Zhou and Yu 2006	ABTS	1.27-4.8 [a]	–	USA	50% Acetone
13. Toor et al. 2006	ABTS	2.4-2.6 [a]	287-328 mg GAE/100 g dry matter	New Zealand	Acetone : water : acetic acid (70:29.5:0.5)
14. Raffo et al. 2006	ABTS	Hydrophilic 1.70-4.20 [a] Lipophilic 0.26-0.34 [a] Total 1.90-4.50 [a]	8.66-17.38 mg/100 g	Italy	Phosphate buffer (pH 7.0)
15. Lenucci et al. 2006	FRAP	2.16-4.53 [a]	970-1330 mg GAE/kg	Italy	Phosphate buffer

[a]Antioxidant activity expressed as µmol/g fresh weight. [b]Countries in which varieties were grown. [c]mg ascorbic acid/g. [d]mg BHT/g.

ABTS, 2,2'-azino-bis-3-ethylbenzthiazoline-6-sulfonic acid diammonium salt; DPPH, 2,2 diphenyl picryl hydrazyl; FRAP, ferric reducing antioxidant power; ORAC, oxygen radical absorbance capacity; *VCE, vitamin C equivalent activity.

reflects classical radical chain-breaking AOX by hydrogen atom transfer. Inclusion of β-cyclodextrin in the ORAC assay improves correlation between ORAC and lycopene concentration, thus expanding the scope of the ORAC assay to include an additional fat-soluble antioxidant (Bangalore et al. 2005). The FRAP method developed by Benzie and Strain (1996) measures the ability of compound to reduce Fe^{3+} (ferric ion) to Fe^{2+} (ferrous ion) in the presence of antioxidants. It is popular among researchers as it is simple and consistent and requires no sophisticated equipments. However, FRAP is not suitable for lipophilic components because of the inability of the carotenoids to reduce Fe^{3+} to Fe^{2+} (Pulido et al. 2000, George et al. 2004). Basically, the FRAP and $ABTS^{+}$ assay are the same as they both take advantage of electron transfer using a ferric salt with a redox potential similar to that of $ABTS^{+}$. Folin-Ciocalteu assay routinely used for measuring the total phenolic content in natural products is based on the basic mechanism of oxidation and reduction reaction and is yet another assay used for AOX measurement.

Sample Preparation

Measurement of AOX encompasses a series of steps in sample preparation, involving extraction methods, appropriate temperature, sequential solvent processing and concentration steps, which dramatically influence the final results. The task of recovery of desired constituents is complicated as fruits constitute a natural matrix with a high enzyme activity and, hence, extreme care must be taken to ensure correct extraction devoid of chemical modification, which will invariably result in artefacts. The ultimate goal is the preparation of sample extract uniformly enriched in all compounds of interest and free from interfering matrix compounds. Considering the lipophilic and hydrophilic nature of tomato constituents, two different solvent extractions have been considered that are as mild as possible to avoid oxidation and thermal degradation of lycopene and phenolics. For hydrophilic fractions, mild extractants such as methanol, ethanol, acetone or sodium phosphate buffer (pH 7.5) have been mostly used and found to favour the recovery of most of the phenols in tomatoes. Methanol has been considered to be the most suitable solvent for extraction of both flavanones and flavanone glycosides. However, for lipophilic fractions, acetone or hexane have been used as most suitable solvents to completely extract the pigmented antioxidants such as carotenoids followed by concentration under nitrogen to avoid any possible oxidation or thermal degradation of carotenoids. Often, measurement of both hydrophilic and lipophilic AOX in the same medium is the major limiting factor. Wu et al. (2004), using $ORAC_{FL}$, have stressed the need for evaluating the antioxidant properties of lipophilic and hydrophilic antioxidant in different solvents separately and clubbing the values together to prevent the possibility of any lipophilic component being extracted along with the hydrophilic components. Sequential solvent extraction with water, acetone extraction is found to contribute the most towards AOX (Pellegrini et al. 2007).

ANTIOXIDANTS IN RESPONSE TO ENVIRONMENT AND GENOTYPES

Antioxidants and phytochemicals accumulate in plants in an organ-specific manner and their distribution and accumulation is governed by a variety of environmental factors, pre- and post-harvest conditions and genotypes or cultivars (Brovelli 2006). A close look at the antioxidant contents in tomato reveals that it is a strong environment-regulated genetic response and absence of the requisite genetic or environmental conditions may result in poor quality. Since antioxidant content also governs the technological parameters such as taste, flavour and colour of tomatoes it is essential to assess their genetic performance of cultivars at several locations over years.

Environmental Factors

The functional quality and antioxidant constituents are significantly affected by environmental factors (temperature, light, water availability, nutrients) and their interaction with agronomic practices and genotypes (Dumas et al. 2003, Yahia et al. 2005). Depending on the prevailing weather conditions in any region, tomatoes are grown either in the field or under greenhouse conditions. Environmental factors have less impact in the greenhouse than in the field. However, changes in solar radiation and variable temperature during different seasons may affect the antioxidant components of greenhouse-grown tomatoes as well (Dumas et al. 2003, Toor et al. 2006). Greenhouse-grown tomatoes receive less ultraviolet radiation than field-grown tomatoes, which might affect the levels of antioxidants (Davey et al. 2000, Stewart et al. 2000). The available literature on the effects of environmental factors in tomato has been mainly concentrated on lycopene and a few reports exist for ascorbic acid and phenolics.

Lycopene

Tomato is a moderate salinity-tolerant crop that requires relatively high irradiance and temperature for optimum yield and quality. Lycopene formation is strictly dependent on the temperature and is severely affected by exposure to intense solar radiation. The favourable temperature range is 12 to 32°C with an optimum range of 26°C at pink-ripe stage. Excessive sunlight inhibits lycopene and best conditions are sufficiently high temperature along with dense foliage (Andgoroyes and Jolliffe 1987, Leoni 1992, Dumas et al. 2003). At favourable temperature (22-25°C), the rates of synthesis during ripening can be increased by illumination. Martinez-Valverde et al. (2002) and Raffo et al. (2006) clearly demonstrated that the hot temperature of mild summer in the Mediterranean basins has significant negative effect on lycopene accumulation because of stimulation in the conversion of lycopene to β-carotene (Leoni 1992, Robertson et al. 1995, Hamauza et al. 1998). In an earlier study (Cabibel and Ferry 1980), field-grown tomatoes showed higher lycopene followed by those grown under plastic tunnel and glass. The light-promoting effects have been attributed to light-induced lycopene accumulation regulated by fruit-localized phytochrome, with red light having a positive effect (Thomas and Jen 1975).

Considerable variations have been observed in total lycopene content in tomato cultivars grown under field and greenhouse conditions. Cherry tomatoes responded more favourably to field conditions, whereas cluster and round tomatoes responded favourably to greenhouse conditions (Kuti and Konuru 2005). The genetic accumulation of lycopene appears to be regulated by over-expression of phytoene desaturase (lycopene synthesis) or repression of lycopene cyclase (β-carotene synthesis) with *hp* locus, shown to regulate response to light (Yen et al. 1997).

Water management is a critical factor influencing lycopene accumulation. Controlled level of salinity in irrigation water has proven to increase the concentration of total carotenoids, lycopene and AOX of tomato fruit (De Pascale et al. 2001). Sgherri et al. (2007) reported that tomatoes can grow well in diluted sea water and produce more natural antioxidants, representing a valid approach to tackle the problem of stress under drought conditions. A recent attempt to engineer drought-resistant tomatoes resulted in over-expression of the gene AVP1 responsible for stronger and larger root systems (Park et al. 2005). Since sensitivity to salt treatment is genotype-dependent, it will be worthwhile to explore its effects on the antioxidant composition of modern cultivars and identify salt-tolerant genotypes. The studies are extremely relevant in the present scenario of changing global climatic patterns of water scarcity and increased salinity.

There is a distinct effect of mineral nutrients and growth regulators on lycopene content; potassium results in a sharp increase, while there are contradictory reports for phosphorus and calcium that need confirmation (Dumas et al. 2003). A high proportion of K in the nutrient solution increased the lycopene content of tomato fruit, whereas a high proportion of Ca improved tomato fruit yield and reduced the incidence of blossom-end rot. The highest AOX was observed in the treatment with a high proportion of Mg in the 'Lunarossa' cultivar (Fanasca et al. 2006). Growth regulator CPTA markedly increased the â-carotene content. The ability of CPTA to bring about carotenogenesis when tested on tomato cell suspension cultures brought about an increase of around 60-fold in total carotenes and around 70-sfold in lycopene (Dumas et al. 2003).

Ascorbic Acid

Ascorbic acid has been shown to increase from green to red-ripe stage during post-harvest ripening; increase of 2.14- and 1.4-fold was observed in cultivars 'Vermone' and 'Heinz 9478' during ripening (Venter 1977, Shi et al. 1999). Light exposure and water stress have been found to be favourable to ascorbic acid accumulation in tomato fruit (Lee and Kader 2000, Dumas et al. 2003). An increase of 60% in the ascorbic acid content of ripe fruit resulted when plants were transferred from the shade to sunshine at mature green stage (Liptay et al. 1986, Vanderslice et al. 1990). Raffo et al. (2006) reported a direct correlation between ascorbic acid content and temperature and attributed this to cultivar, salinity and sunny climate. Excessive nitrogen fertilization and greenhouse conditions were usually found to yield tomato fruits with lower vitamin C levels (Lopez-Andreu et al. 1986, Mozafar 1993).

Phenolics

Genetic control of phenolics represents the main factor in determining their accumulation in vegetables, although external factors may also have a significant effect (Macheix et al. 1990). The phenolics of tomatoes are found to occur in the skin and 98% of flavonols detected in tomatoes occur as conjugates (quercetin and kaempferol). The content in current commercial cultivars in Spain, characterized as flavonoids and hydroxycinnamic acid, ranged from 7.2 to 43.6 mg/kg. Additionally, chalconaringenin, naringenin, naringenin-7-glucoside, and *m*- and *p*-coumaric acids were characterized in tomato cultivars ('Alisa Craig', 'Alicante' and 'Grower's Pride'). In cherry tomatoes, plants grown in greenhouse under high light accumulated approximately twice as much soluble phenol content as low-light plants (Wilkens et al. 1996). However, low light and red light were found to favour formation of chalconaringenin. Raffo et al. (2006) observed great variability among phenolic compounds during different times of the year, with narningenin content ranging from 1.84 to 9.04 mg/100 g and rutin content from 1.79 to 6.61 mg/100 g in cherry tomatoes (*Lycopersicon esculentum* cv. Naomi F1).

Marked variations, with no definite seasonal trend, were also observed in quercetin levels in cherry tomatoes grown at different times of the year (Crozier et al. 1997, Stewart et al. 2000). Phenolic content of tomato fruits has been reported to be significantly affected by the cultivar (Giuntini 2005) and spectral quality of solar UV radiation (Luthria et al. 2006).

Genotypes

Variety manipulations in tomato have been recognized as a powerful tool for modifying antioxidant fruit patterns and contents. A large number of genotypes and cultivars have been screened for their compositional profile, in order to identify genotypes with enhanced levels of antioxidants for health (Lenucci et al. 2006, Martinez-Valverde et al. 2002, George et al. 2004). The major emphasis has been on the cultivars from the Mediterranean region (Italy, Spain, Hungry), the United States and New Zealand. Identification of lycopene-rich tomato lines or germplasm has been the major focus of research workers because of its importance in imparting a rich colour to processed products such as sauce, puree and paste. In recent years, however, the focus has shifted to phenolics and flavonoids because of their significant health-promoting effects.

Total AOX of average red tomatoes weighing 123 g has been reported to be in the range of 116-517 μmol in ORAC assay, 290-423 μmol in FRAP assay and 232-314 μmol in TEAC assay (Wang et al. 1996, Ou et al. 2002, Proteggente et al. 2002, Halvorsen et al. 2002, Wu et al. 2004, Beer et al. 2004). When converted on unit basis, the values are 1.16-5.17 μmol Trolox/g in ORAC, 2.90-4.23 μmol Trolox/g in FRAP and 2.32-3.14 μmol Trolox/g in TEAC. The reported Folin's assay values range from 11 to 30 mg GAE/100 g (George et al. 2004, Chun et al. 2005, Toor and Savage 2005, Lenucci et al. 2006). Cherry tomatoes (*Lycopersicon esculentum* var. *cerasiforme*), among tomato genotypes, have universally been shown to contain high lycopene content and high total AOX (Kuti and Konuru 2005, Lenucci et al. 2006). Hydrophilic AOX in tomato is significantly affected by genotype and processing and is almost independent of the ripening stage; on the other hand, lipophilic AOX is influenced more by ripening stage than by genotype (Raffo et al. 2006). The hydrophilic components account for 83-92% of total AOX in comparison to lipohilic components.

Information on AOX in tomato genotypes is scarce. In most cases, market samples have been used and specific cultivars have not been mentioned. According to the most elaborate studies, exploring genetic variability in antioxidant components and their activity in tomatoes (Martinez-Valverde et al. 2002, Abushita et al. 2000, George et al. 2004, Kuti and Konuru 2005, Lenucci et al. 2006, Toor and Savage 2005, Toor et al. 2006, Raffo et al. 2006, Frusciante et al. 2007), Italian varieties seem to dominate the tomato scene. The high-pigmented cultivars from Italy have been reported to contain lycopene content as high as > 200 mg/kg fresh weight. The highest content of 253 mg/kg fresh weight (2.7 g/kg dry weight) was observed in the variety 'Kalvert' among the pigmented cultivars and 129 mg/kg fresh weight (1.3 g/kg dry weight) in LS-203 among cherry cultivars (Lenucci et al. 2006). Critical evaluation of these results based on dry matter shows that these cultivars are also high in dry matter ranging from 7 to 9% in comparison to an average dry matter content of 5-6% observed in common tomato cultivars and this could be the reason for high lycopene content observed in these cultivars, as most of the lycopene is associated with dry matter. These cultivars were also found to be rich in total phenolics, the pigmented cultivar 'HLY13' showing 1330 mg of GAE/kg and cherry cultivar 'Lycorino' showing 1370 mg GAE/kg. The AOX as measured by FRAP assay was found to be in the range of 2.16 μmol FRAP/g in 'Sharon' and 'Saentino' to 4.53 μmol FRAP/g in 'Corbus'. Italian cherry varieties 'LS 203' and 'corbus' appear to be cultivars with the highest content of lipophilic and hydrophilic antioxidants among cherry tomatoes. Considering their promising traits, researchers feel that they are ideal germplasm for breeding programmes as well as for processing (George et al. 2004).

Cherry tomatoes ('Pomodoro di Pachino' cv. Naomi F1) grown in Italy showed nearly 2.36-fold variation in AOX depending on the season of cultivation. Tomato samples harvested in March showed markedly higher content of 4.5 μmol/g than those harvested in April, which showed 1.9 μmol/g. These higher values are attributed to higher rutin content in the hydrophilic fractions (Raffo et al. 2006). In another report, Martinez-Valverde et al. (2002), using DPPH and ABTS assay, identified Spanish cultivars 'Pera' and 'Durina' as most promising with high lycopene, phenolic and AOX. Total AOX among nine cultivars ranged from 0.58 to 3.5 mmole/L in ABTS and 1.90×10^{-6} to 13.46×10^{-6} in DPPH assay. Total extractable phenolics ranged from 25.2 to 50.0 mg/100 g and 'Daniella' was found to contain high levels of flavonoids. The quercetin content ranged from 7.19 in 'Rambo' to 43.59 mg/100 g in 'Daniella'. In a comprehensive study on New Zealand tomato cultivars ('Excell', 'Tradiro', 'Flavourine', and 'Camapri'), AOX ranged from 11.96 to 19.60 μmol/g (Toor and Savage 2005). Cultivar 'Campari', among the cultivars studied, had 1.4 times the AOX of normal-sized cultivars. High AOX was correlated with its high total flavonoid (254 mg rutin/100 g dry matter), phenolics and ascorbic acid. The hydrophilic extract contributed to more than 92% of total observed AOX, with skin containing higher levels than flesh. In another study, on Indian cultivars, AOX ranged from 0.63 to 2.3

ìmol FRAP/g, the highest being in cherry cultivar 818; moderate values ranging from 1 to 1.19 μmol FRAP/g were observed in 'BR-124', 'Rasmi', 'DTH' and 'Pusa Gaurav' (George et al. 2004).

Ecotypes of *Corbarini* small tomatoes, grown in Italy (Scalfi et al. 2000), and Campari', a small New Zealand cultivar (Toor and Savage 2005), were correlated with shape and surface ratio between surface (skin) and total fruit weight. The AOX as estimated with the DMPD (*N,N'*-dimethyl-*p*-phenylenediamine) method in the water-soluble fractions and with ABTS in the water-insoluble fractions in 16 ecotypes of *Corbarini* small tomatoes were found to be strongly related to each other and varied significantly among the ecotypes (Scalfi et al. 2000). The round tomatoes were richer in both water-soluble and water-insoluble antioxidants than the long tomatoes.

Abushita et al. (2000) critically evaluated the antioxidant constituents of Hungarian salad and processing tomato varieties: 'Jovanna', with highest lycopene content, was identified as the most promising one for industrial use. Other lycopene-rich varieties were 'Sixtina' followed by 'Simeone', 'K-541', and 'Soprano', while 'Amico', 'Gobe', and 'Zaphyr' had high β-carotene content. Processing tomatoes contained almost the same amounts of ascorbic acid but had higher α-tocopherol content than salad tomatoes. α-tocopherol ranged from 122 to 326 μg/100 g, with 'Monika' containing the highest levels (612 μg/100 g).

Frusciante et al. (2007) have proposed an index for defining antioxidant nutritional quality (I_{QUAN}) in tomatoes. The index highlights the importance of all the antioxidant components including lycopene, β-carotene, polyphenols, and flavonoids apart from technological parameters of dry matter and dietary fibre. The index seems to be useful for plant breeders and processors who may like to consider each individual component that contributes to yield and quality of tomatoes for better selection for commercial purposes. Tomatoes to be analysed were divided into three categories according to their I_{QUAN} values: high antioxidant quality ($I_{QUAN} > 90$), intermediate antioxidant quality ($I_{QUAN} =$ 70-90), and low antioxidant quality ($I_{QUAN} < 70$). Lycopene showed the major significant variations among all genotypes analysed followed by β-carotene and other carotenoids. The lines 'Motelle' and 'Poly 20' showed the highest average lycopene content (16.9 and 16.0 mg/100 g, respectively). The average dry matter content was observed to be 5.5%. Vitamin C showed significant variations among lines. The ascorbic acid in cultivars ranged from 11.4 to 16.3 mg/100 g, in the following decreasing order: Heline > Motelle > 981 > Momor > Poly 56. Vitamin E content ranged from 0.17 to 0.62 mg/100 g in the following decreasing order: Poly 56 > Momor > Motelle. The content of phenolic acids was also significantly different among lines ranging from 2.04 (Cayambe) to 3.75 mg/100 g (Motelle).

GENETIC MANIPULATIONS FOR ENHANCED ANTIOXIDANTS

There is a worldwide surge in genetic improvement of tomato for improving health-promoting effects either by traditional breeding (Ronen et al. 1999, Zhiang and Stommel 2000) or by transgenic incorporation (Giuliano et al. 2000). Continuous domestication and modern breeding have significantly depleted the natural variation of tomato and flavonoids are nearly absent from fruits of cultivated tomato (*Lycopersicon esculentum* Mill.). Undue domestication through inadvertent counter-selection has resulted in selection against nutritional traits in favour of agronomic ones. Although lycopene has been the main target compound (Mehta et al. 2002), restoration of functional pathway for tomatoes of high flavonoid content has also been targeted for enhancing the health-promoting compounds (Schijlen et al. 2006). Wild-type tomato and flavonoid-enriched tomato were found to significantly reduce basal human cardiovascular risk protein concentrations by 43 and 56%, respectively (Rein et al. 2006). A HighCaro, a transgenic tomato with high content of pro-vitamin A, sufficient to cover three times the daily recommended allowance for an adult (6 mg pro-vitamin A per day), has been identified (http://www.invenia.es/tech:04_it_susi_0bi6).

Flavonol biosynthesis has been well characterized through the use of a commercial processing tomato variety, FM6203, to develop transgenic tomatoes with increased levels of flavonols (Verhoeyen et al. 2002). Tomato transformed with a sequence encoding the enzyme chalcone isomerase (which utilizes naringenin chalcone as a substrate to produce dihydrofavonols) from *P. hybrida* showed an overall (~78-fold) increase in total fruit flavonols (Bovy et al. 2002). In an alternative approach, *P. hybrida* sequences encoding key enzymes chalcone synthase, chalcone isomerase, flavanone-3-hydroxylase and flavonol synthase were expressed simultaneously to get transgenics with very high levels of quercetin glycosides and significant increased levels of kaempferol and naringenin-glycosides in the peel (Colliver et al. 2002). Since quercetin is a more potent antioxidant it may be desirable to enhance its levels in tomato flesh by over-expressing flavanone-3-hydroxylase, the enzyme that converts dihydrokaempferol to dihydroquercetin, which in turn is converted to quercetin.

Metabolite profiling of wild-type tomato genotype 'Alisa Craig' and mutant high pigment (hp-1) and 'LA 3771' has confirmed that perturbations in carotenoid biosynthesis generally do not alter phenolic or flavonoid content and these lines can act as the hosts for further genetic manipulation for increased AOX content (Long et al. 2006).

High AOX activity associated with anthocyanins and their role in preventing the risk of diseases is prompting plant breeders to incorporate the anthocyanin fruit (*Aft*) (anthocyanin) into carotene-rich tomatoes for producing purple tomatoes (Jones et al. 2003). Normal tomato genotypes routinely have anthocyanin in vegetative tissues, and petunidin 3-(p-coumaryl rutinoside)-5-glucoside and malvidin 3-(p-coumaryl rutinoside)-5-glucoside as the principal anthocyanins have been identified in vegetative tissues of tomato (Bovy et al. 2002). The simple introgression of *Aft* gene in the existing germplasm provides the opportunity to develop new cultivars rich in water- and lipid-soluble antioxidants. Transgenic approach for inducing parthenocarpy in tomato is yet another promising area for improving technological properties of tomato in the ketchup industry (Rotino et al. 2005).

Use of wild genotypes in modern breeding programmes offers significant promise to improve the hydrophilic component of tomato. Recently, Willits et al. (2005) developed a high-flavonoid tomato using a wild genotype *Lycopersicon* accession LA 1926 via a non-transgenic metabolic approach. The strategy, using an RNA-based approach to screen potential germplasm, is a viable method to trace the genetic potential of wild germplasm resources vis-á-vis a non-transgenic manipulation of metabolic pathways and to produce healthier tomatoes rich in flavonoids. In yet another innovative attempt, introgression of wild germplasm of green-fruited species *L. pennelli* improved the nutritional quality of domesticated tomato, *L. esculentum* (Rousseaux et al. 2005). A total of 20 quantitative trait loci were identified, including five for total antioxidant capacity in water-soluble fraction, six for ascorbic acid, and nine for total phenolics. Some of the identified quantitative trait loci showed increased levels of ascorbic acid and phenolics as compared to the parental line *L. esculentum.* Apart from improving antioxidant composition, wild genotypes possess higher regenerative ability, and transfer of this desired trait into cultivated tomato via somatic embryogenesis is a promising technique for mass production of genetically improved tomato cultivars (Bhatia et al. 2004).

FUTURE PROSPECTS

Lycopene and flavonoids represent interesting bio-molecules for increasing the overall antioxidant capacity and health-promoting capacity of tomatoes. Strategy for improvement involves selection of genotypes showing favourable response to environmental variations and biotic stresses. Changing global climatic patterns, water stress, salinity and UV radiation are challenges for plant breeders to develop adaptable cultivars with desirable traits through conventional and transgenic means. There is an urgent need to tap the unlocked potential of wild genotypes of tomato for improving quality. Integrating knowledge on the effects of environmental factors on the antioxidant quality and incorporation of wild genotypes will yield functionally laced and healthier tomatoes. Encouraging results from transgenic tomatoes with enhanced flavonoids represent a trend for production of neutraceutical tomatoes with cardio-protective and anticancerous effects. However, the issue of yield and disease susceptibility of high-antioxidant cultivars needs to be addressed to systematically capitalize on the consumer market. Genotypic labelling for antioxidant quality tomatoes requires consistent use of standardized assays and indexes incorporating both antioxidant quality and technological parameters.

ABBREVIATIONS

ABTS: 2,2′-azino-bis-3-ethylbenzthiazoline-6-sulphonic acid diammonium salt; AOX: Antioxidant activity; CPTA: 2-4–chloro phenyl thiotriethylamine hydrochloride; DMPD: *N,N′*-dimethyl-*p*-phenylenediamine; DPPH: 2,2 diphenyl picryl hydrazyl; FRAP: Ferric reducing antioxidant power; I_{QUAN}: Index for antioxidant nutritional quality; ORAC: Oxygen radical absorbance capacity; TEAC: Trolox equivalent antioxidant capacity

REFERENCES

Abushita, A.A. and H.G. Daood, and P.A. Biacs. 2000. Change in carotenoids and antioxidant vitamins in tomato as a function of varietal and technological factors. J. Agric. Food Chem. 48: 2075-2081.

Andgoroyes, A.S. and P.A. Jolliffe. 1987. Some inhibitory effects of radiation stress on tomato fruit ripening. J. Sci. Food Agric. 39: 297-302.

Arnao, M.B. and A. Cano, and M. Acosta. 2001. The hydrophilic and lipophilic contribution to total antioxidant activity. Food Chem. 73: 239-244.

Bangalore, D.V. and W. McGlynn, and D.D. Scott. 2005. Effect of β-cyclodextrin in improving the correlation between lycopene concentration and ORAC values. J. Agric. Food Chem. 53: 878 -1883.

Beer, C. and R.A. Myers, J.H. Sorenson, and L.R. Bucci. 2004. Comprehensive comparison of the antioxidant activity of fruits and vegetables based on typical servings sizes from common methods. Curr. Topics Nutr. Res. 2: 227-250.

Benzie, I.E.F. and J.J. Strain. 1996. The ferric reducing ability of plasma (FRAP) as a measure of antioxidant power: the FRAP assay. Anal. Biochem. 239: 70-76.

Bhatia, P. and N. Ashwath, T. Senaratna, and D. Midmore. 2004. Tissue culture studies of tomato (*Lycopersicon esculentum*). Plant Cell Tissue Organ Cult. 78: 1-21.

Bovy, A. and R. de Vos, M. Kemper, E. Schijlen, M.A. Pertezo, S. Muir, G. Collins, S. Robinson, M. Verhoeyen, S. Hughes, C. Santos-Buelga, and A. van Tunen. 2002. High flavonol tomatoes resulting from the hetrologous expression of the maize transcription factor genes *LC* and *C1*. Plant Cell. 14: 2509-2526.

Brovelli, E.A. 2006. Pre- and postharvest factors affecting nutraceutical properties of horticultural products. Stewart Postharvest Rev. 2: 1-6.

Cabibel, M. and P. Ferry. 1980. Evolution de la teneur en carotenoides de la tomate en function de la maturation et des conditions culturales. Annales de Technol. Agricole (Paris). 29: 27-46.

Canene-Adams, K. and J.K. Campbell, S. Zaripheh, E.H. Jeffery, and J.W. Erdman, Jr. 2005. The tomato as a functional food. J. Nutr. 135: 1226-1230.

Cano, A. and M. Acosta, and M.B. Arnao. 2003. Hydrophilic and lipophilic antioxidant activity changes during on-vine ripening of tomatoes (*Lycopersicon esculentum* Mill.). Post-harvest Biol. Technol. 28: 59-65.

Cao, G. and E. Sofie, and R.L. Prior. 1996. Antioxidant capacity of tea and common vegetables. J. Agric. Food Chem. 44: 3426-3431.

Chun, O.K. and D-Ok Kim, N. Smith, D. Schroeder, J.T. Han, and C.Y. Lee. 2005. Daily consumption of phenolics and total antioxidant capacity from fruit and vegetables in the American diet. J. Sci. Food Agric. 85: 1715-1724.

Colliver, S. and A. Bovy, G. Collins, S. Muir, S. Robinson, C.H.R. de Vos, and M.E. Vorhoeyen. 2002. Improving the nutritional content of tomatoes through reprogramming their flavonoid biosynthetic pathway. Phytochem. Rev. 1: 113-123.

Crozier, A. and M.E.J. Lean, M.S. McDonald, and C. Black. 1997. Quantitative analysis of the flavonoid content of commercial tomatoes, onions, lettuce and celery. J. Agric. Food Chem. 45: 590-595.

Davey, M.W. and M. van Montagu, D. Inze, M. Sanmartin, A. Kanellis, N. Smirnoff, I.J.J. Benzie, J.J. Strain, D. Favell, and J. Fletcher. 2000. Plant L-ascorbic acid chemistry, function, metabolism, bioavailability and effects of processing. J. Sci. Food Agric. 80: 825-860.

De Pascale, S. and A. Maggio, V. Fogliano, P. Ambrosino, and A. Ritieni. 2001. Irrigation with saline water improves carotenoids contents and antioxidant activity of tomato. J. Hortic. Sci. Biotechnol. 76: 447-453.

Dumas, Y. and M. Dadomo, G. Di Lucca, and P. Grolier. 2003. Effects of environmental factors and agricultural techniques on antioxidant content of tomatoes. J. Sci. Food Agric. 83: 369-382.

Fanasca, S. and G. Colla, G. Maiani, E. Venneria, Y. Rouphael, E. Azzini, and F. Saccardo. 2006. Changes in antioxidant content of tomato fruits in response to cultivar and nutrient solution composition. J. Agric. Food Chem. 54: 4319-4325.

Frankel, E.N. and A.S. Meyer. 2000. The problems of using one-dimensional methods to evaluate multifunctional food and biological antioxidants. J. Sci. Food Agric. 80: 1925-1941.

Frusciante, L. and P. Carli, M.R. Ercolano, R. Pernice, A. Di Matteo, V. Fogliano, and N. Pellegrini. 2007. Antioxidant nutritional quality of tomato. Mol. Nutr. Food Res. 51: 609-617.

Garcia-Closas, R. and A. Berenguer, M.J. Tormo, M.J. Sanchez, and J.R. Quiors. 2004. Dietary sources of vitamin C, vitamin E and specific carotenoids in Spain. Brit. J. Nutr. 91: 1005-1011.

George, B. and C. Kaur, D.S. Khurdia, and H.C. Kapoor. 2004. Antioxidants in tomato (*Lycopersicon esculentum*) as a function of genotype. Food Chem. 84: 45-51.

Giovannucci, E. and A. Ascherio, E.B. Rimm, M.J. Stampfer, G.A. Colditz, and W.C. Willet. 1995. Intake of carotenoids and retinol in relation to risk of prostate cancer. J. Natl. Cancer Inst. 87: 1767-1776.

Giuntini, D. and G. Graziani, B. Lercari, V. Fogliano, G.F. Soldatini, and A. Ranieri. 2005. Changes in carotenoid and ascorbic acid contents in fruits of different tomato genotypes related to the depletion of UV-B radiation. J. Agric. Food Chem. 53: 3174-3181.

Giuliano, G. and R. Aquilani, and S. Dharmapuri. 2000. Metabolic engineering of plant carotenoids. Trends Plant Sci. 5: 406–409.

Halvorsen, B.L. and K. Holte, M.C.W. Myhrstad, I. Barikmo, E. Hvattum, S.F. Remberg, A.B. Wold, K. Haffner, H. Baugerod, L.F. Andersen, J.O. Moskaug, D.R. Jacobs, and R. Blomhoff. 2002. Systematic screening of total antioxidants in dietary plants. Am. Soc. Nutr. Sci. J. Nut. 132: 461-471.

Hamauza, Y. and K. Chachin, and Y. Ueda. 1998. Effect of postharvest temperature on the conversion of 14C-mevalonic acid to carotene in tomato fruit. J. Jpn. Soc. Hortic. Sci. 67: 549-555.

Huang, D. and B. Ou, M. Hampsch-Woodill, J.A. Flanagan, and E.K. Deemar. 2002. Development and validation of oxygen radical absorbance capacity for lipophillic antioxidants using randomly methylated β-cyclodextrin as the solubility enhancer. J. Agric. Food Chem. 50: 1815-1821.

Jones, C.M. and P. Mes, and J.R. Myers. 2003. Characterization and inheritance of the anthocyanin fruit (*Aft*) tomato. J. Hered. 94: 449-456.

Kaur, C. and H.C. Kapoor. 2001. Antioxidants in fruits and vegetables—the millennium's health. Int. J. Food Sci. Technol. 36: 703-725.

Khachik, F. and L. Carvalho, P.S. Bernstein, G.J. Muir, D.Y. Zhao, and N.B. Katz. 2002. Chemistry, distribution, and metabolism of tomato carotenoids and their impact on human health. Exp. Biol. Med. 227: 845-851.

Kuti, J.O. and H.B. Konuru. 2005. Effects of genotype and cultivation environment on lycopene content in red-ripe tomatoes. J. Sci. Food Agric. 85: 2021-2026.

La Vecchia, C. 1997. Mediterranean epidemiological evidence on tomatoes and the prevention of digestive tract cancers. Proc. Soc. Exp. Biol. Med. 218: 125-128.

Lee, S.K. and A.A. Kader. 2000. Preharvest and postharvest factors influencing vitamin C content of horticultural crops. Postharvest Biol. Technol. 20: 207-220.

Lenucci, M.S. and D. Cadinu, M. Taurino, G. Piro, and G. Dalessandro. 2006. Antioxidant composition in cherry and high-pigment tomato cultivars. J. Agric. Food Chem. 54: 2606-2613.

Leoni, C. 1992. Industrial quality as influenced by crop management. Acta Hortic. 301: 177-184.

Liptay, A. and A.P. Papadopoulos, H.H. Bryan, and D. Gull. 1986. Ascorbic acid levels in tomato (*Lycopersicon esculentum* Mill.) at low temperature. Agric. Biol. Chem. 50: 3185-3187.

Liu, Y.S. and S. Roof, Z.B. Ye, C. Barry, A. van Tuinen, J.C. Vrebalov, C. Bowler, and J. Giovannoni. 2004. Manipulation of light signal transduction as means of modifying fruit nutritional quality in tomato. Proc. Natl. Acad. Sci. USA 101: 9897-9902.

Long, M. and D.J. Millar, Y. Kimura, G. Donovan, J. Rees, P.D. Frasler, P.M. Bramley, and G.P. Bolwell. 2006. Metabolite profiling of carotenoid and phenolic pathways in mutant and transgenic lines of tomato: Identification of a high antioxidant fruit line. Phytochemistry 67: 1750-1757.

Lopez-Andreu, F.J. and A. Lamela, R.M. Esteban, and J.G. Collado. 1986. Evaluation of quality parameters in the maturation stage of tomato fruit. Acta Hortic. 191: 387-394.

Luthria, D. and S. Mukhopadhyay, and D.T. Krizek. 2006. Content of total phenolics and phenolic acids in tomato (*Lycopersicon esculentum* Mill.) fruits as influenced by cultivar and solar UV radiation. J. Food Comp. Anal. 19: 771-777.

Macheix, J.J. and A. Fleuriet, and J. Billot. 1990. Fruit Phenolics. CRC Press, Boca Raton, Florida, USA, pp. 68-71, 113.

Martínez-Valverde, I. and M.J. Periago, G. Provan, and A. Chesson. 2002. Phenolic compounds, lycopene and antioxidant activity in commercial varieties of tomato (*Lycopersicon esculentum*). J. Sci. Food Agric. 82: 323-330.

Mehta, R.A. and T. Cassol, N. Li, A.K. Handa, and A.K. Mattoo. 2002. Engineered polyamine accumulation in tomato enhance phytonutrient content, juice quality and vine life. Nat. Biotechnol. 20: 613-618.

Minoggio, M. and L. Bramati, P. Simmonetti, C. Gardana, L. Iemoli, E. Santangelo, P.L. Mauri, P, Siigno, G.P. Soressi, and P.G. Pietta. 2003. Polyphenol pattern and antioxidant activity of different tomato lines and cultivars. Ann. Nutr. Metab. 47: 64-69.

Mozafar, A. 1993. Nitrogen fertilization and the amount of vitamin in plant; a review. J. Plant Nutr. 16: 2479-2506.

Nguyen, M.L. and S.J. Schwartz. 1999. Lycopene: chemical and biological properties. Food Technol. 53: 38-45.

Ou, B. and D. Huang, M. Hampschwoodwill, T.A. Flanagan, and E.K. Deemer. 2002. Analysis of antioxidant activities of common vegetables employing oxygen radical absorbance capacity and FRAP assays: A comparative study. J. Agric. Food Chem. 50: 3122-3128.

Ozgen, M. and R.N. Reese, A.Z. Tulio Jr., J.C. Scheerens, and A.R. Miller. 2006. Modified 2,2-azino-bis-3-ethylbenzothiazoline-6-sulfonic acid (ABTS) method to measure antioxidant capacity of selected small fruits and comparison to ferric reducing antioxidant power (FRAP) and 2,2'-diphenyl-1-picrylhydrazyl (DPPH) methods. J. Agric. Food Chem. 54: 1151-1157.

Paganga, G. and H. Al-Hashim, H. Khodr, B.C. Scott, O. Aruoma, R.C. Hider, B. Halliwell, and C.A. Rice-Evans. 1996. Mechanism of antioxidant activities of quercetin and catechin. Redox Report. 359-364.

Park, S. and J. Li, J.K. Pittman, G.A. Berkowitz, H. Yang, S. Undurraga, J. Morris, K.D. Hirschi, and R.A. Gaxiola. 2005. Up-regulation of a H^+-pyrophosphatase (H^+-PPase) as a strategy to engineer drought-resistant crop plants. PNAS 102: 18830-18835.

Pellegrini, N. and B. Colombi, S. Salvatore, O.V. Brenna, G. Galaverna, D.D. Rio, M. Bianchi, R.N. Bennett, and F. Brighenti. 2007. Evaluation of antioxidant capacity of some fruit and vegetable foods: efficiency of extraction of a sequence of solvents. J. Sci. Food Agric. 87: 103-111.

Prior, R.L. and X. Wu, and K. Schaich. 2005. Standardized methods for the determination of antioxidant capacity and phenolics in food and dietary supplements. J. Agric. Food Chem. 53: 4290-4302.

Pulido, R. and L. Bravo, and F. Saura-Calixto. 2000. Antioxidant activity of dietary polyphenols as determined by a modified ferric reducing/antioxidant power assay. J. Agric. Food Chem. 48: 3396-3402.

Proteggente, A.R. and A.S. Pannala, G. Paganga, and L. Van Buren. 2002. The antioxidant activity of regularly consumed fruits and vegetables reflect their phenolic and vitamin C composition. Free Radic. Res. 36: 217-233.

Raffo, A. and G. La Malfa, V. Fogliano, G. Maiania, and G. Quaglia. 2006. Seasonal variation in antioxidant components of cherry tomatoes (*Lycopersicon esculentum* cv. Naomi F1). J. Food Comp. Anal. 19: 11-19.

Rao, A.V. and S. Agarwal. 1999. Role of lycopene as antioxidant carotenoid in the prevention of chronic diseases: A review. Nutr. Res. 19: 305-323.

Rein, D. and E. Schijlen, T. Kooistra, K. Herbers, L. Verschuren, R. Hall, U. Sonnewald, A. Bovy, and R. Kleemann. 2006. Transgenic flavonoid tomato intake reduces c-reactive protein in human c-reactive protein transgenic mice more than wild-type tomato. J. Nutr. 136: 2331-2337.

Robertson, G.H. and N.E. Mahoney, N. Goodman, and A.E. Pavlath. 1995. Regulation of lycopene formation in cell suspension culture of VFNT tomato (*Lycopersicon esculentum*) by CPTA, growth regulators, sucrose and temperature. J. Exp. Bot. 46: 667-673.

Rodriguez, E.B. and M.E. Flavier, D.B. Rodriguez-Amaya, and J. Amaya-Farfan. 2006. Phytochemicals and functional foods. Current situation and prospect for developing countries. Segurança Alimentar e Nutricional Campinas 13: 1-22.

Ronen, G. and M. Cohen, D. Zamir, and J. Hireschberg. 1999. Regulation of carotenoid biosynthesis during tomato fruit development. Expression of gene for lycopene epilon-cyclase down regulated during ripening and is elevated in the mutant Delta. Plant J. 17: 341-351.

Rotino, G.L. and N. Acclarri, E. Dabatini, G. Mennellia, R.L. Scalzo, A. Maestralli, B. Molesini, T. Pandolfini, J. Scalzo, B. Mezzetti, and A. Spena. 2005. Open field trial of genetically modified parthenocarpic tomato: seedlessness and fruit quality. BMC Biotechnol. 5: 32-39.

Rousseaux, M.C. and C.M. Jones, and D. Adams. 2005. QTL analysis of fruit antioxidants in tomato using *Lycopersicon pennelli* introgression lines. Theor. Appl. Genet. 111: 1396-1408.

Scalfi, L. and V. Fogliano, A. Pentangelo, G. Graziani, I. Giordano, and A. Ritieni. 2000. Antioxidant activity and general fruit characteristics in different ecotypes of corbarini small tomatoes J. Agric. Food Chem. 48: 1363-1366.

Scalzo, J. and A. Politi, N. Pellegrini, B. Mezzetti, and M. Battino. 2005. Plant genotype affects total antioxidant capacity and phenolic contents in fruits. Nutrition 21: 207-213.

Sesso, H.D. and S. Liu, J.M. Gaziano, and J.E. Buring. 2003. Dietary lycopene, tomato-based food products and cardiovascular disease in women. J. Nutr. 133: 2336-2341.

Schijlen, E. and C.H. Riedevos, H. Jonkeer, H. Van den Broeck, J. Molthoff, A. vanTunen, S. Martens, and A. Bovy. 2006. Pathway engineering of healthy phytochemicals leading to production of novel flavonoid in tomato fruit. Plant Biotechnol. 4: 433-444.

Sgherri, C. and F. Navari-Izzo, A. Pardossi, G.P. Soressi, and R. Izzo. 2007. The influence of diluted sea water and ripening stage on the content of antioxidants in fruits of different tomato genotypes. J. Agric. Food Chem. 55: 2452-2458.

Shi, J. and M. LeMaguer, Y. Kakuda, A. Liptay, and F. Niekamp. 1999. Lycopene degradation and isomerisation in tomato dehydration. Food. Res. Int. 32: 15-21.

Stewart, A.J. and S. Bozonner, W. Mullen, G.I. Jenkins, M.E.J. Lean, and A. Crozier. 2000. Occurrence of flavonols in tomatoes and tomato based products. J. Agric. Food Chem. 48: 2663-2669.

Thomas, R.L. and J.J. Jen. 1975. Red light intensity and carotenoid biosynthesis in ripening tomatoes. J. Food Sci. 40: 566-568.

Toor, R.K. and G.P. Savage. 2005. Antioxidant activity in different fractions of tomato. Food Res. Int. 38: 487-494.

Toor, R.K. and G.P. Savage, and C.E. Lister. 2006. Seasonal variations in antioxidant composition of greenhouse grown tomatoes. J. Food Comp. Anal. 19: 1-10.

Vanderslice, J.T. and D.J. Higgs, J.M. Hayes, and G. Block. 1990. Ascorbic acid and dehydroascorbic acid content of foods-as-eaten. J. Food Comp. Anal. 3: 105-118.

Van der Sluis, A.A. and M. Dekker, A. De Jager, and W.M.F. Jongen. 2001. Activity and concentration of polyphenolic antioxidants in apple: Effect of cultivar, harvest year, and storage conditions. J. Agric. Food Chem. 49: 3606-3613.

Venter, F. 1977. Solar radiation and vitamin C content of tomato fruit. Acta Hortic. 58: 121-127.

Verhoeyen, M.E. and A. Bovy, G. Collins, S. Muir, S. Robinson, C.H.R. de Vos, and S. Colliver. 2002. Increasing antioxidant levels in tomatoes through modification of the flavonoid biosynthetic pathway. J. Exp. Bot. 53: 2099-2106.

Wang, H. and G. Cao, and R.L. Prior. 1996. Total antioxidant capacity of fruits. J. Agric. Food Chem. 44: 701-705.

Weisberger, J.H. 2002. Lycopene and tomato products in health promotion. Exp. Biol. Med. 227: 924-927.

Wilkens, R.T. and J.M. Spoerke, and N.E. Stamp. 1996. Differential response of growth and two soluble phenolics of tomato to resource availability. Ecology 77: 247-258.

Willcox, J. and G. Catignani, and S. Lazarus. 2003. Tomatoes and cardiovascular health. Crit. Rev. Food Sci. Nutr. 43: 1-18.

Willits, M.G. and C.M. Crammer, R.T.N. Prata, V. De Luca, B.G. Potter, J.C. Steffens, and G. Graser. 2005. Utilization of the genetic resources of wild species to create a nontransgenic high flavonoid tomato. 53: 1231-1236.

Wu, X. and G.R. Beecher, J.M. Holden, D.B. Haytowitz, S.E. Gebhardt and R.L. Prior. 2004. Lipophilic and hydrophilic antioxidant capacities of common foods in the United States. J. Agric. Food Chem. 52: 4026-4037.

Yahia, E.M. and X. Hao, and A.P. Papadopoulos. Influence of crop management decisions on postharvest quality of greenhouse tomatoes. pp. 379-405. *In:* Ramdane Dris [Ed.] 2005. Crops: Quality, Growth and Biotechnology. Ramdane Dris, WFL Publishers, Helsinki, Finland.

Yen, H.C. and B.A. Shelton, L.R. Howard, S. Lee, and J. Vrebalov. 1997. The tomato high pigmented (hp) locus maps to chromosome 2 and influences plastome copy number and fruit quality. Theor. Appl. Genet. 95: 1069-1079.

Zhiang, Y. and J.R. Stommel. 2000. RAPD and AFLP tagging and mapping of Beta (B) and Beta modifier (Mo[B]) two genes which influence β-carotene accumulation in fruit of tomato (*Lycopersicon esculentum* Mill.). Theor. Appl. Genet. 100: 368-375.

Zhou, K. and L. Yu. 2006. Total phenolic contents and antioxidant properties of commonly consumed vegetables grown in Clorado. LWT-Food Sci. Technol. 39: 1155-1162.

7

Carotenoids in Tomato Plants

B. Stephen Inbaraj and B.H. Chen*

Department of Food Science, Fu Jen University, Taipei 242, Taiwan

ABSTRACT

Tomato (*Lycopersicon esculentum*) and tomato-based products are important sources of many traditional nutrients and a predominant source of several carotenoids. As numerous studies suggest an association of tomato carotenoids, especially lycopene, to reduced risk of various chronic diseases such as cardiovascular disease and cancer, its presence in the diet is of utmost importance. This chapter reviews the composition, variety, content and distribution of carotenoids in tomato plants and tomato products, various growth factors that influence the synthesis of carotenoids in tomato plants, methods of analysis, and stability during processing and storage. The structural and biochemical properties as well as bioavailability and biological activity of tomato carotenoids are also emphasized.

INTRODUCTION

Carotenoids, a class of micronutrients found predominantly in tomatoes (*Lycopersicon esculentum*), are frequently studied for their established association with prevention of chronic diseases such as cardiovascular disease and prostate cancer (Giovannucci 1999, Giovannucci et al. 2002). Among foods typically consumed by humans, tomatoes are a particularly rich source of several carotenoids, especially lycopene. Of the 14 carotenoids found in human serum, tomato and tomato-based products

---

*Corresponding author

A list of abbreviations is given before the references.

contribute to 9 and are the principal source of about half of the carotenoids (Khachik et al. 1995). The red color and quality of tomatoes can be attributed mainly to the presence of the major carotenoid, lycopene, representing a key factor for their commercial value. In plants, carotenoids are localized in cellular plastids and are associated with light-harvesting complexes in the thylakoid membranes or present as semicrystalline structures derived from the plastids (Kopsell and Kopsell 2006). Carotenoids assist in harvesting light energy, mostly in the blue-green wavelength range, which is transferred to the photosynthetic reaction centers of plants. This distinctive feature is due to their extensive conjugated double bond system that serves as the light-absorbing chromophore, imparting yellow, orange or red color to tomatoes.

The composition of carotenoids synthesized by tomato plants can be affected by various factors such as cultivar and genotype variety, maturity stage, climate or geographic site of production, seasonal variation, nutrient composition, harvest and post-harvest handling, processing and storage (Gross 1991). This unfolds the scope of carotenoid research in tomatoes regarding their variety, content and degradation pattern under the variation in aforesaid conditions, besides evaluating their biological activities and developing methods for analysis. This chapter aims to cover all these issues incorporating recent developments, in addition to presenting some of the important basic aspects of tomato carotenoids.

CAROTENOID CHEMISTRY AND PLANT BIOSYNTHESIS—AN OVERVIEW

Structure, Nomenclature and Classification

Carotenoids are isoprenoid polyenes formed by the joining of eight C_5 isoprenoid units so that the arrangement of the units is reversed at the centre of the molecule, leaving the two central methyl groups in a 1,6 position relative to each other and non-terminal methyl groups in a 1,5 relationship. Food carotenoids are usually C_{40} tetraterpenoids derived from the acyclic $C_{40}H_{56}$ conjugated polyene lycopene by various reactions.

The numbering of carotenoid molecule is 1 to 15 from the left end to the center, 16 to 20 for the additional methyl groups and 1′ to 15′ from the right end to the center. Semi-systematic names that provide the structural information of carotenoids are based on their parent stem name "carotene" prefixed by any two end group designations (ψ, β, ε, κ, φ or χ) of carotenes as detailed elsewhere (Britton 1995, Gross 1991). The numbering of carbon position and semisystematic names are illustrated for the two biologically important tomato carotenoids, lycopene and β-carotene (Fig. 1).

Carotenoids are basically classified into hydrocarbons collectively

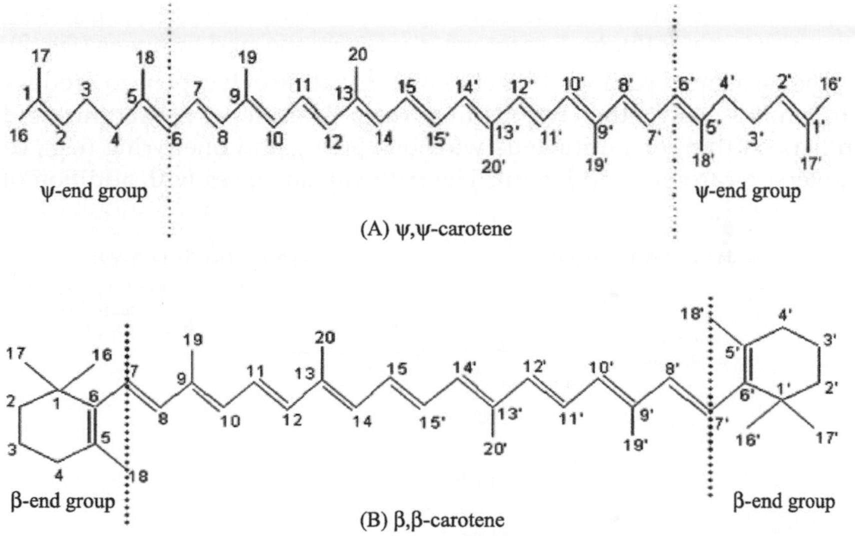

Fig. 1 Illustration of carbon position, numbering and semisystematic names for (A) lycopene and (B) β-carotene.
Dotted lines identify the two end-group moieties in the carotenoid molecule; (A) semisystematic names of lycopene; (B) semisystematic name of β-carotene.

called *carotenes* and their oxygenated derivatives termed *xanthophylls.* A hydroxy substituent occurs chiefly at C-3 in the ε- or β-ring. The double bonds in the 5,6 and 5′,6′ of β-ring of cyclic carotenoids are susceptible to epoxidation, while unconjugated double bond in the ε-ring is not.

Biosynthesis of Carotenoids in Plants

Carotenoids are biosynthesized in plant plastids using the precursor isopentenyl pyrophosphate (IPP), which is formed from mevalonic acid (mevalonate pathway) after successive phosphorylation by kinase enzymes and adenosine triphosphate followed by decarboxylation (Gross 1991, Kopsell and Kopsell 2006). A recently identified 1-deoxy-D-xylulose-5-phosphate (DOXP) pathway produced IPP by condensation of a C_2 unit, derived from pyruvate decarboxylation, with glyceraldehyde 3-phosphate and a transposition yielding the branched C_5 skeleton of isoprenic units (Bramley 2002). The IPP on isomerization yields dimethylallyl diphosphate (DMAPP), which in turn condenses with successive addition of IPP forming a C_{20} geranylgeranyl pyrophosphate (GGPP) mediated by the enzyme GGPP synthase. The condensation of two molecules of GGPP yields the first C_{40} carotenoid, phytoene. The sequential introduction of double bonds at alternate sides of phytoene gives rise to phytofluene, ζ-carotene, neurosporene, and lycopene (Fig. 2).

The carotenoid pathway branches by cyclization of lycopene to produce carotenoids with either two β-rings (e.g., β-carotene, zeaxanthin and antheraxanthin) or carotenoids with one β-ring and one ε-ring (e.g., α-carotene, γ-carotene and lutein). The pathway advances with addition of

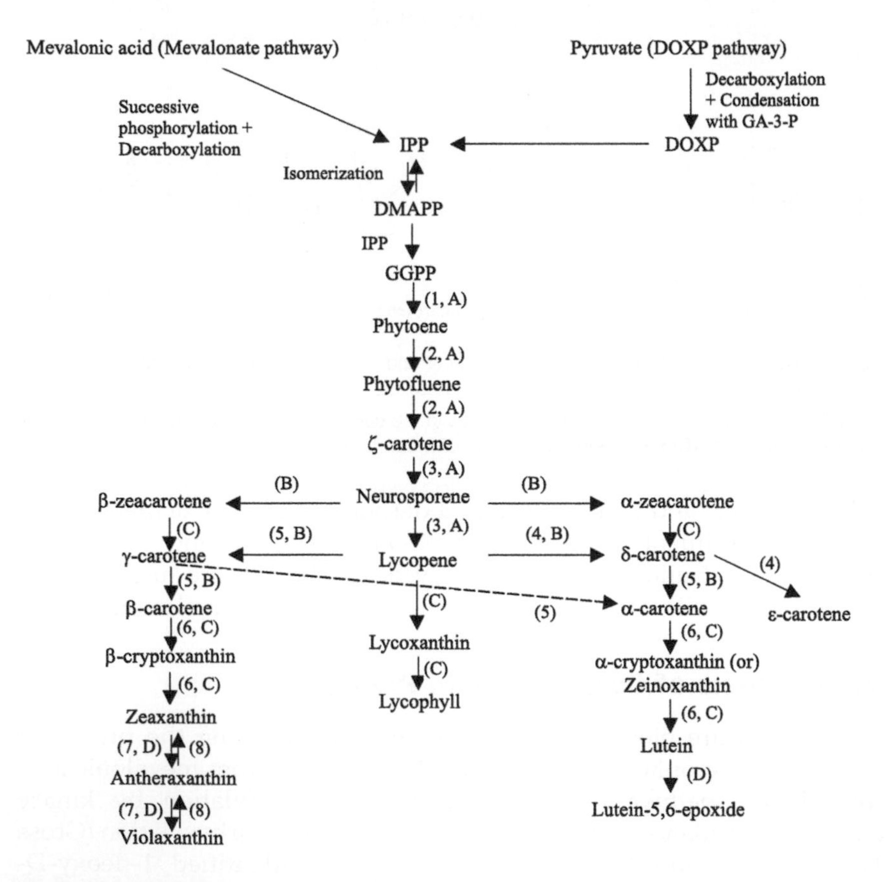

Fig. 2 The biosynthetic pathway of carotenoids in plants. DOXP, 1-deoxy-D-xylulose; GA-3-P, glyceraldehyde 3-phosphate; IPP, isopentyl diphosphate; DMAPP, dimethylallyl diphosphate; GGPP, geranylgeranyl pyrophosphate; 1, phytoene synthase; 2, phytoene desaturase; 3, ζ-carotene desaturase; 4, lycopene ε-cyclase; 5, lycopene β-cyclase; 6, β-carotene hydroxylase; 7, zeaxanthin epoxidase; 8, violaxanthin de-epoxidase; A, desaturation; B, cyclization; C, hydroxylation; D, epoxidation.

oxygen moieties, which convert the hydrocarbon α- and β-carotenes into xanthophylls (Fig. 2). A number of epoxy carotenoids are synthesized from xanthophylls. The red-ripe tomatoes rich in lycopene lack sufficient cyclase activity to convert lycopene to γ-carotene and β-carotene efficiently.

CAROTENOIDS IN TOMATO PLANT

Variety, Content and Distribution in Tomato Fruit

The visible red color of the tomato fruit is due to its major carotenoid, lycopene, making up 80 to 90% of the total carotenoids (Shi and Le Maguer 2000). It also contains colorless carotenoid precursors such as phytoene and phytofluene (15-30%), xanthophylls (free and esterified, 6%) and minor tomato hydrocarbon carotenes such as β-carotene, γ-carotene, ζ-carotene, α-carotene and δ-carotene (Gross 1991). The carotenoid composition varies in some tomato strains. The yellow and orange tomato strains either do not synthesize lycopene or other carotenes may predominate. β-Carotene and δ-carotene are the major carotenoids in beta and delta strains, respectively. Tangerine-type tomatoes synthesize a poly cis-isomer of lycopene (prolycopene) instead of all-trans-lycopene (Gross 1991).

The total carotenoid content of tomato varies between 7 and 19 mg/100 g fresh weight (Gross 1991). Lycopene content ranges from 0.5 to 7 mg/100 g for various genotypes of tomato fruits (Kuti and Konuru 2005), though a higher amount is also shown for some other varieties. The lycopene content in whole tomato fruit is about 9.27 mg/100 g (Tonucci et al. 1995), while some deep-red and yellow varieties contain 15 and 0.5 mg/100 g, respectively (Hart and Scott 1995). The amount of β-carotene in tomato is approximately one-tenth of lycopene.

The distribution of carotenoids in tomato fruit is not uniform. The outer pericarp constitutes the largest amount of total carotenoids and lycopene, while the locule contains a high proportion of carotene (Shi and Le Maguer 2000). About 12 mg of lycopene per 100 g fresh weight was found in tomato skin, while the whole tomato fruit contained only 3.4 mg/100 g fresh weight (Al-Wandawi et al. 1985). According to Sharma and Le Maguer (1996b), the concentration of lycopene in outer pericarp (53.9 mg/100 g) was about 3- to 6-fold higher than in whole tomato pulp (11 mg/100 g). Thus, tomato skin is a rich source of lycopene, indicating that lycopene is attached to the insoluble fiber portion of tomatoes, whereas β-carotene is equally distributed in tomatoes.

Carotenoids in Other Parts of Tomato Plant

Unlike in pericarp, tomato seeds of two red and one tangerine-type varieties contain lutein (64.4 and 90.2%) as the major carotenoid, followed by β-carotene (21.5 and 5.3%) and lycopene (14.1 and 4.5%), respectively (Rymal and Nakayama, 1974). On the contrary, β-carotene (54.4%) is the dominant carotenoid in cherry tomato seeds, followed by lycopene (34.6%), γ-carotene (3.4%), α-carotene (2.5%), phytofluene (1.7%) and β-zeacarotene (1.3%) (Rodriguez et al. 1975). In flowers and leaves, the major pigments are xanthophylls making up 90 and 70% of total carotenoids, respectively, with lutein representing the predominant xanthophyll in both (Bramley et al. 1992). A low level of neoxanthin (14%), neurosporene (10%), and α- and β-carotenes (10%) are also present in both flowers and leaves.

Carotenoid Composition in Tomato Products

Over 80% of tomatoes produced are consumed in the form of various processed products such as juice, paste, sauce, soup, puree, catsup and spaghetti sauce. Tonucci et al. (1995) determined lycopene to be present in highest concentration, followed by γ-carotene, phytoene, neurosporene, phytofluene, β-carotene, ζ-carotene, lycopene-5,6-diol and lutein, with the level of individual carotenoids in tomato products exceeding their corresponding concentrations in whole tomatoes (Table 1). Mechanical and/or thermal treatments carried out during processing should account for the enhanced release of carotenoids from the tomato matrix.

GROWTH, ENVIRONMENTAL AND TECHNOLOGICAL FACTORS AFFECTING CAROTENOID CONTENT

Effect of Temperature

Carotenoids are among the antioxidants severely affected by temperature in tomato fruits. Generally, high temperature due to excessive direct sunlight and intense solar radiation reduces the bioaccumulation of carotenoids except β-carotene. The rate of lycopene synthesis was inhibited at both high (> 30°) and low temperatures (< 12°) (Dumas et al. 2003). Thus, the optimal temperature for lycopene formation was reported to range between 12 and 32°, but within that the critical optimum levels vary with variety, cultivar and other environmental and growth conditions of tomato plants (Dumas et al. 2003). For instance, in tomatoes harvested for processing, the lycopene levels rose by 5 and 33% for

Table 1 Carotenoid content (mg/100 g) in whole tomatoes and tomato products (Tonucci et al. 1995).

Sample	*β-carotene*	*γ-carotene*	*ζ-carotene*	*Lutein*	*Lycopene*	*Neurosporene*	*Phytoene*	*Phytofluene*	*Lycopene-5,6-diol*
Whole tomatoes	0.23 ± 0.04[a]	1.50 ± 0.28	0.21 ± 0.05	0.08 ± 0.02	9.27 ± 1.02	1.11 ± 0.13	1.86 ± 0.29	0.82 ± 0.14	0.11 ± 0.01
Paste	1.27 ± 0.24	9.98 ± 1.15	0.84 ± 0.08	0.34 ± 0.11	55.45 ± 0.90	6.95 ± 0.84	8.36 ± 0.80	3.63 ± 0.38	0.44 ± 0.08
Puree	0.41	2.94	0.25	0.09	16.67	2.11	2.40	1.08	0.17
Sauce	0.45 ± 0.13	3.17 ± 0.64	0.29 ± 0.03	tr[b]	17.98 ± 1.47	2.48 ± 0.54	2.95 ± 0.43	1.27 ± 0.20	0.16 ± 0.02
Juice	0.27 ± 0.04	1.74 ± 0.20	0.18 ± 0.03	0.06 ± 0.02	10.77 ± 1.07	1.23 ± 0.27	1.90 ± 0.19	0.83 ± 0.14	0.11 ± 0.03
Soup	0.23 ± 0.05	1.95 ± 0.41	0.17 ± 0.02	0.09 ± 0.02	10.92 ± 2.92	1.53 ± 0.20	1.72 ± 0.17	0.72 ± 0.18	0.11 ± 0.03
Catsup	0.59 ± 0.12	3.03 ± 1.25	0.33 ± 0.05	nd[c]	17.23 ± 2.18	2.63 ± 0.73	3.39 ± 0.38	1.54 ± 0.18	0.18 ± 0.05
Spaghetti sauce	0.44 ± 0.07	3.02 ± 0.42	0.34 ± 0.07	0.16 ± 0.04	15.99 ± 0.90	3.15 ± 0.90	2.77 ± 0.66	1.56 ± 0.53	0.17 ± 0.03

[a]mean ± standard deviation (number of observations: 4 to 9); [b]traces (below 0.005 mg/100 g); [c]not detected.

temperatures at 30-34° and 37°, respectively, while it decreased at temperatures above 30° in salad varieties. The most appropriate conditions in enhancing lycopene synthesis are sufficiently high temperature (22-25°) along with dense foliage to protect fruits from direct exposure to the sun.

Effect of Light

Under favorable temperatures, the carotenoid biosynthesis can be increased by illuminating tomato plants during ripening. Red light had a more pronounced effect on synthesis of carotenoids than white or green light, while far-red light inhibited it in ripening tomatoes, indicating a phytochrome-mediated response. Harvested mature-green fruit exposed to red light raised lycopene accumulation by 2.3-fold from 3.7 (control in dark) to 8.7 mg/100 g (at red ripe stage) and reversed back to 5.2 mg/100 g by subsequent treatment with far-red light (Alba et al. 2000). Tomatoes ripened in the dark or under light showed the same carotenoid pattern (Gross 1991), suggesting that carotenogenesis in tomato plants is independent of light. Light may augment carotenoid synthesis but is not required for its induction.

Effect of Water Availability and Salinity

Water deficit in terms of soil moisture content increased both total carotenoid and lycopene contents (Matsuzoe et al. 1998). In other words, moisture stress must be avoided to obtain maximum carotenoid levels. Water with varying salinity (0.5 to 15.7 dS/m) showed a gradual rise in lycopene until 4.4 dS/m treatment (approximately 0.25% NaCl, w/v) and decreased thereafter, indicating increased susceptibility of lycopene loss at electrical conductivity of water >4.4 dS/m (De Pascale et al. 2001). Further studies are required for a more comprehensive understanding of this effect.

Effect of Mineral Nutrients

Nutrient solutions containing low levels of nitrogen generally increased the lycopene content in tomatoes. Tomatoes grown in nutrient solution with three different levels of nitrate-nitrogen showed the highest lycopene content at a low nitrogen dosage, yielding 6.8, 4.4 and 3.8 mg/100 g fresh weight of lycopene for nitrogen levels at 1.0, 12.9 and 15.8 meq/L, respectively (Dumas et al. 2003). On the contrary, raising phosphorus or potassium supply in nutrient solution greatly increased the lycopene content (Dumas et al. 2003, Fanasca et al. 2006). However, high calcium and chloride as well as low sulfur levels in fertilizers were found to reduce lycopene content in tomatoes (Paiva et al. 1998, Toor et al. 2006). A comparative study of tomatoes (Corfù and Lunarossa cultivars) grown in nutrient solutions containing different potassium, calcium and magnesium proportions revealed tomatoes fed with a high potassium nutrient solution to produce a larger level of lycopene and β-carotene, followed by magnesium- and calcium-enriched nutrient solutions, with lutein showing a reverse trend (Fanasca et al. 2006). Furthermore, tomatoes grown by the conventional and organic agricultural practices did not show any significant difference in the carotenoid content (Caris-Veyrat et al. 2004).

Effect of Growth Regulators

Carotenoid synthesis can also be influenced by several growth regulators such as 2-(4-chlorophenylthio)triethylamine hydrochloride (CPTA), which exhibited a positive influence on carotenoid formation in tomato fruits except for β-carotene, with the effect being more pronounced at 20° than 30° (Dumas et al. 2003). The seedlings grown in greenhouse at 24/18° day/night temperature after treatment with different doses of 2-(3, 4-dichlorophenoxy)triethylamine (DCPTA) with 0.1% Tween 80 at 24° for 6 hr produced a high amount of lycopene (5.8 to 11.8 mg/100 g) and β-carotene (0.22 to 0.57 mg/100 g) for the DCPTA level from 0-5 mg/100 g (Keithly et al. 1990). Gibberellic acid and cycocel (2-chloroethyl trimethyl ammonium 3-chloride) increased the β-carotene content in field-grown tomatoes (Dumas et al. 2003). Several other growth regulators found to elevate carotenoid level in tomatoes include chlormequat, alar (succinic acid 2,2-dimethylhydrazide) (Gabr et al. 2006), brassinosteroids (Vardhini and Rao 2002), ethephon (Dumas et al. 2003), auxins (Cohen 1996) and osmotic solutions such as sucrose (Télef et al. 2006).

Effect of Ripening and Post-harvest Storage

Carotenoids are largely accumulated in tomato during fruit ripening by the disappearance of chlorophylls and transformation of chloroplast into chromoplast during the lag phase that precedes maturation. This results in a change in tomato fruit color from green to yellow, yellow to orange, orange to pink and pink to red during ripening. Changes in the content of individual carotenoids during different ripening stages of cherry tomatoes as investigated by Raffo et al. (2002) are shown in Table 2. It can be seen that lycopene is synthesized rapidly over the entire ripening period, while the levels of phytoene, phytofluene, lycopene 1,2-epoxide and 5,6-dihydroxy-5,6-dihydrolycopene gradually increased. However, γ-carotene occurred only at a later ripening stage, whereas ζ-carotene

Table 2 Carotenoid content (mg/100 g) at different ripening stages of cherry tomatoes (Raffo et al. 2002).

Carotenoid compound	*Ripening stage*				
	G-Y	*G-O*	*O-R*	*L-R*	*R*
Lycopene	0.45[a]	2.23[b]	4.51[c]	6.92[d]	10.44[e]
β-Carotene	0.34[a]	0.71[b]	0.90[d]	0.84[c]	1.07[e]
Phytoene	0.05[a]	0.19[b]	0.40[c]	0.46[c]	0.58[d]
Phytofluene	0.02[a]	0.12[b]	0.26[c]	0.27[cd]	0.30[d]
ζ-Carotene	0.19[c]	0.11[b]	0.09[b]	0[a]	0[a]
γ-Carotene	0[a]	0[b]	0.02[b]	0.03[c]	0.05[d]
5,6-dihydroxy-5,6-dihydrolycopene	0[a]	0.01[b]	0.02[c]	0.02[d]	0.04[e]
Lycopene 1,2-epoxide	0[a]	0.02[b]	0.07[c]	0.09[d]	0.18[e]
Lycopene/β-carotene	1.3	3.1[b]	5.0	8.2	9.7
Total carotenoids	1.08[a]	3.39[b]	6.28[c]	8.66[d]	12.71[e]

G-Y, green-yellow (~ 30% yellow skin); G-O, green-orange (50% orange skin); O-R, orange-red (90% orange or red skin); L-R, light red (fully orange or red skin); R, red (fully red skin); a-e, values with same letter in a row are not significantly different ($p > 0.01$).

markedly decreased during ripening. Though β-carotene followed an increased trend during ripening, conflicting data on its accumulation rate have been shown in the literature. Some studies reported β-carotene to reach a maximum prior to full ripeness (Davies and Hobson 1981), while others showed a gradual decrease during ripening (Gross 1991). As lycopene concentration varies significantly from one maturity stage to another, it is commonly used as a maturity index.

Carotenogenesis in tomato fruits continues even after harvest and vine-ripened tomatoes were reported to produce more carotenoids than those ripened in storage (Gross 1991). Nevertheless, maintenance of appropriate post-harvest conditions was shown to enhance the level of carotenoids in tomatoes (Giovanelli et al. 1999). Controlled-atmosphere storage of relatively high CO_2 and low O_2 levels for optimized exposure time has been proved beneficial in delaying ripening to extend the post-harvest life span by suppressing chlorophyll degradation and ethylene synthesis, which would prevent carotenoid accumulation (Sozzi et al. 1999, Batu and Thompson 1998).

Effect of Genotype and Cultivar Variety

The carotenoid content in tomato can also vary with genotype and cultivar variety. The lycopene and its cis-isomer content determined in 39 tomato genotype varieties ranged from 0.6 to 6.4 mg/100 g and 0.4 to 11.7 mg/100 g for greenhouse and field-grown tomatoes, respectively (Kuti and Konuru 2005). Similarly, different cultivar varieties have been shown to possess varied lycopene concentrations (Abushita et al. 2000, Fanasca et al. 2006, Hart and Scott 1995, Thompson et al. 2000, Toor and Savage 2005).

STRUCTURAL AND BIOCHEMICAL PROPERTIES OF TOMATO CAROTENOIDS

The color, photochemical and antioxidant properties of carotenoids are due to their extended conjugated double bonds. Most carotenoids have maximum absorption at three wavelengths and the greater the number of conjugated double bonds, the higher the λ_{max} values. Thus, ζ-carotene having 7 conjugated double bonds absorbs at a lower wavelength than lycopene, while phytoene and phytofluene with less than 7 are colorless (Davies 1976). Some structural characteristics of tomato carotenoids are summarized in Table 3.

Carotenoids can undergo isomerization around each C = C bond in the polyene chain. Though theoretically a large number of stereoisomers are possible, in reality only a few stereoisomers exist because of steric hindrance effect. The formation of stereoisomers from various carotenoids by light absorption was related to their molecular structure (Zechmeister 1962). The presence of a cis double bond creates greater steric hindrance between nearby hydrogen atoms and/or methyl groups and hence cis-isomers are thermodynamically less stable than the trans form. During processing, the predominant all-trans-lycopene isomer in tomato undergoes isomerization to various cis-lycopene isomers such as 5-cis, 9-cis 13-cis and 15-cis-lycopene (Fig. 3), which were identified in tomato products and human serum as well (Lin and Chen 2005a, Rajendran et al. 2005, Schierle et al. 1997, Wang and Chen 2006). Change in isomeric forms has been reported to cause variation in biological activity (Boileau et al. 1999). The size and shape of a carotenoid molecule also affects its properties and functions. Cis-isomers because of their kinked structure have a lesser tendency to aggregate or crystallize and therefore may be more readily solubilized in lipids, absorbed and transported than their all-trans counterparts (Boileau et al. 1999, Britton 1995). Bohm et al. (2002) reported that cis-isomers of lycopene, α-carotene and β-carotene possessed a higher Trolox equivalent antioxidant capacity than their corresponding trans forms.

Carotenoids react with free radicals by three different mechanisms: (1) radical addition (adduct formation), (2) electron transfer to the radical, or (3) allylic hydrogen abstraction (Krinsky and Yeum 2003). The antioxidant

Table 3 Tomato carotenoids and their structural characteristics.

Carotenoid	*Semisystematic name*	*Conjugated double bonds*	*Absorption spectrum* (λ_{max})	*Extinction coefficient* ($\varepsilon^{1\%}_{1cm}$)
Lycopene	ψ,ψ-carotene	11	444, 470, 502	3450[a]
β-Carotene	β,β-carotene	11 (2 in ring)	425, 450, 477	2592[b]
α-Carotene	β,ε-carotene	10 (1 in ring)	422, 445, 473	2800[a]
γ-Carotene	β,γ-carotene	11 (1 in ring)	437, 462, 494	2760[b]
ζ-Carotene	7,8, 7′8′-tetrahydro-ψ, ψ-carotene	7	378, 400, 425	2555[b]
δ-Carotene	εψ-carotene	10	431, 456, 489	3290[a]
Phytoene	7,8,11,12,7′,8′,11′,12′-octahydro-ψ,ψ-carotene	3	276, 286, 297	1250[b]
Phytofluene	7,8,11,12,7′,8′-hexahydro-ψ,ψ-carotene	5	331, 348, 367	1577[b]
Lutein	β,ε-carotene-3,3′-diol	10 (1 in ring)	422, 445, 474	2550[c]
β-cryptoxanthin	β,β-caroten-3-ol	11 (2 in ring)	425, 449, 476	2386[a]

λ_{max}, absorption wavelength maximum; $\varepsilon^{1\%}_{1cm}$, specific extinction coefficient of a 1% carotenoid solution measured in a 1 cm light-path spectrophotometer cuvette; a-c, $\varepsilon^{1\%}_{1cm}$ measured with carotenoid dissolved in [a]petroleum ether, [b]hexane or [c]ethanol solvents.

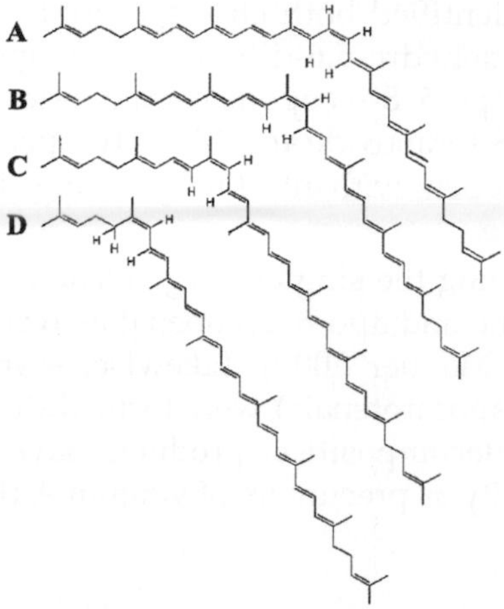

Fig. 3 Major cis-isomers of lycopene in tomato products and human serum as well as tissues. (A) 15-cis-lycopene; (B) 13-cis-lycopene; (C) 9-cis-lycopene; (D) 5-cis-lycopene.

activity in terms of singlet oxygen and peroxy radical quenching were shown to be higher for lycopene than for other carotenoids (Miller et al. 1996, Stahl and Sies 1996). Apparently, the antioxidant activity of carotenoids is mainly due to their conjugated double bonds and is less influenced by its cyclic or acyclic end groups or nature of substituents on the end groups (Stahl and Sies 1996). Nevertheless, the cyclic carotenoids derive antioxidant activity partly also from the 5,6 and 5′,6′ double bonds in their cyclic end groups (Mortensen et al. 2001). Since lycopene lacks a β-ionone ring, it does not exhibit pro-vitamin A activity. However, β-carotene, α-carotene and γ-carotene, all of which contain one or two β-ionone rings, do show pro-vitamin A activity. The vitamin A activity of β-carotene, α-carotene, γ-carotene, 9-cis-β-carotene, 13-cis-β-carotene, 9-cis-α-carotene and 13-cis-α-carotene was reported to be 100, 53, 42-50, 38, 53, 13 and 16%, respectively (Gross 1991). The vitamin A value of commercial tomatoes is 200 mg retinol equivalents/kg (Gross 1991).

Oxidative and Cleavage Products of Carotenoids

Hydroxylated derivatives of lycopene have been detected in human serum and tomato as well as tomato-based products. It was proposed that lycopene may undergo oxidation during tomato processing or *in vivo* metabolism to form lycopene 5,6-epoxide, which may be rearranged to 2,6-cyclolycopene-1,5-epoxides, followed by hydrolysis to their corresponding diols in both acidic tomatoes catalyzed by enzymes and human stomach in the presence of acids (Khachik et al. 2002).

Ferreira et al. (2003) identified both cleavage products (3-keto-apo-13-lycopenone and 3,4-dehydro-5,6-dihydro-15,15′-apo-lycopenal) and oxidative products (2-apo-5, 8-lycopenal-furanoxide, lycopene-5,6,5′,6′-diepoxide, lycopene-5, 8-furanoxide and 3-keto-lycopene-5′,8′-furanoxide) during incubation of lycopene with intestinal post-mitochondrial fraction of rat and soy lipoxygenase.

Lycopene on quenching the singlet oxygen leads to the formation of 2-methyl-2-hepten-6-one and apo-6′-lycopenal as well as other reaction products (Shi and Le Maguer 2000). Likewise, several epoxides and carbonyl compounds (apocarotenals) were formed during radical attack on β-carotene. These decomposition products have been reported to possess biological activity as precursors of vitamin A (Krinsky and Yeum 2003).

BIOAVAILABILITY OF TOMATO CAROTENOIDS

Bioavailability of carotenoids from consumption of tomato or tomato products is influenced by many factors, such as quantity of consumption, degree of release from the tomato matrix, co-ingestion with dietary lipids and fibers, absorption in the intestinal tract, transportation within the lipoprotein fractions, biochemical conversions and tissue-specific depositions as well as the nutritional status of the human subject (Kopsell and Kopsell 2006). The bioavailability is usually determined using three complementary models: (1) an *in vitro* digestion model to evaluate bioaccessibility, (2) the human Caco-2 (TC-7 clone) intestinal cell line and (3) postprandial chylomicron responses in humans (Reboul et al. 2005). Likewise, gerbils and ferrets, which absorb carotenoids similarly to humans, are the animals most commonly subjected to bioavailability studies (Cohen 2002).

Carotenoids are more bioavailable from processed tomatoes than from raw tomatoes (Gärtner et al. 1997). Mechanical or heat treatment facilitates disruption of cell matrix in tomato leading to a greater release of tomato carotenoids. A combination of homogenization and heat treatment processes was reported to enhance carotenoid bioavailability (van het Hof et al. 2000). Cis-isomers of lycopene are more bioavailable than its trans-isomer (Boileau et al. 2002). Though lycopene in tomato and tomato products is predominantly in all-trans form, its cis-isomers make up > 50% of the total lycopene in human serum and tissues (Clinton et al. 1996). Probably, cis-isomers are more soluble in bile acid micelles and preferentially incorporated into chylomicrons, which may be caused by their lower tendency to aggregate and form crystals (Boileau et al. 1999). In addition, the level of cis-isomers of lycopene increased during incubation of tomato puree with a model gastric juice, revealing that acidic conditions in stomach could result in lycopene isomerization (Re et al. 2001). However, only a slight rise of cis-isomers in stomach (6.2%) and intestine (17.5%) was observed in an *in vivo* study of ferrets, which did not account for a high percentage of cis-isomers found in intestinal mucosa (58.8%) (Boileau et al. 2002). Thus, the mechanism of selective incorporation of cis-isomers into micelles remains poorly understood. A recent report showed that the bioavailability of β-carotene from tomato puree was higher than lycopene, which may be attributed to a better solubility of the former in micelles or to a different localization of carotenoids in tomato cells (Reboul et al. 2005). Carotenoids are more bioavailable when supplemented with dietary lipids, while dietary fibers (especially pectin) decreased the carotenoid absorption (Gärtner et al. 1997, Shi and Le Maguer 2000). Also, the bioavailability can be affected by interaction of lycopene with other carotenoids (Shi and Le Maguer 2000). Lycopene and β-carotene levels in blood and human tissues are summarized in the literature (Boileau et al. 2002, Rajendran et al. 2005, Stahl and Sies 1996).

BIOLOGICAL ACTIVITY OF TOMATO CAROTENOIDS

Several epidemiological studies have shown a positive correlation between increased intake of tomato or tomato-based products and reduced risk of various diseases including cardiovascular disease and cancer (Giovannucci 1999, Giovannucci et al. 2002, Petr and Erdman, 2005). The important underlying mechanism often cited is the potential of tomato to modulate radical-mediated oxidative damage, which may contribute to the initiation and progression of many chronic diseases (Hadley et al. 2003, Lehucher-Michel et al. 2001). However, the presence of some other functional components in tomato and/or other potential mechanisms may also play a vital role in preventing chronic disease (Boileau et al. 2001, Clinton et al. 1996). Prevention of cardiovascular disease could possibly be due to reduction in inflammation and low-density lipoprotein oxidation, inhibition of cholesterol synthesis and enhancement of immune function (Heber and Qing-Yi 2002, Petr and Erdman 2005). Prevention of cancer risk can be attributed to the ability of tomato carotenoids to inhibit tumor cell proliferation and cell transformation, increase gap junction communication through stabilization of connexin 43 (tumor suppressor gene) and modulate the expression of gene determinants and cell-cycle progression as well as increase pro-differentiation activity (Campbell et al. 2004, Heber and Qing-Yi 2002, Kucuk et al. 2001, Wang et al. 2003). It also prevents endogenous levels of DNA strand-breaks in human lymphocytes by resisting the oxidative stress (Heber and Qing-Yi 2002).

Among 72 epidemiological studies reviewed, only 35 reported inverse associations between tomato intake or blood lycopene level and risk of cancer with statistical significance (Giovannucci 1999). Evidence of strong association was shown for lung, stomach and prostate cancers, while evidence was only suggestive for cancers of the cervix, breast, oral cavity, pancreas, colorectum and esophagus. Though the consumption of tomato and its products reduced the risk of a variety of cancers, the direct benefit of its major carotenoid lycopene has not been proved, suggesting a possible synergistic effect of lycopene with various tomato micronutrients (Giovannucci 1999, Giovannucci et al. 2002).

ANALYTICAL METHODS FOR DETERMINATION OF TOMATO CAROTENOIDS

The analysis of carotenoids, especially lycopene, is quite complicated as they are sensitive to light, heat, oxygen and acids and therefore extreme care should be taken to perform all the analytical operations under dim light and inert conditions (under N_2 or vacuum); solvents used must be pure and free of oxidizing compounds to avoid isomerization or oxidation.

Color Index Method

The red color of tomatoes has been shown to be directly proportional to its quality and concentration of carotenoids (particularly lycopene), and thus evaluation of color offers a less tedious, non-destructive assessment of tomato ripening and change in carotenoid concentration during processing than the other chemical methods (Barreiro et al. 1997, Thompson et al. 2000). Color measurement is based on the Hunter's *L*, *a* and *b* tristimulus values and color parameters such as ΔE (total color difference), SI (saturation index), D (hue angle) and *a/b* ratio derived from *L*, *a* and *b* values have been commonly used for quality specification in tomato and tomato products. After calibrating the tristimulus colorimeter with an appropriate color standard, the parameters ΔE, SI and hue angle were calculated from the *L*, *a* and *b* values of tomato sample directly read from the instrument using the following equations:

$$\Delta E = \sqrt{(L_0 - L)^2 + (a_0 - a)^2 + (b_0 - b)^2}$$

$$SI = \sqrt{a^2 + b^2}$$

$$\text{Hue angle} = \tan^{-1}(b/a)$$

An *a/b* ratio of 2.0 and above is indicative of an excellent color, while a value below 1.8 is considered unacceptable. The hue angle value of 0 or 360° represents red hue and angles of 90, 180 and 270° denote yellow, green and blue hues, respectively (Barreiro et al. 1997). The inability of the method to distinguish clearly between isomerization and oxidation responsible for decrease in color is the major limitation. In addition, the off-shades resulting from Maillard browning reactions may be misleading. Nevertheless, this method is extensively used for determining tomato quality during ripening and online quality monitoring of tomato products during processing.

Extraction of Carotenoids

Carotenoids are routinely extracted with organic solvents because of their fat-soluble nature. Solvents such as ethanol, acetone, petroleum ether, hexane, benzene, and chloroform can be used for carotenoid extraction, but solvent mixtures including hexane, acetone and methanol or ethanol are used most often (Lin and Chen 2003, Periago et al. 2004). Initially, the tomato sample to be analyzed must be made into a paste by blending or mechanical grinding to achieve maximum extraction. Fresh samples are preferred and dehydrated material requires moistening with a water-miscible solvent prior to extraction with a water-immiscible solvent (diethyl ether) to achieve greater extraction efficiency (Gross 1991). A saponification procedure is normally adopted in carotenoid analysis in order to remove unwanted lipids, chlorophylls and other water-soluble impurities to minimize interference during subsequent separation and identification. It is often carried out by adding about 1 or 2 mL of 40-60% potassium hydroxide to the extract and allowing the mixture to stand in the dark under N_2 gas for 12-16 hr (Gross 1991).

Some rapid extraction methods such as microwave solvent extraction and pressurized accelerated solvent extraction can achieve a high lycopene recovery ranging from 98.0 to 98.6% (Sadler et al. 1990, Benthin et al. 1999). Extraction using supercritical CO_2 is emerging as a non-toxic (solvent-free) and eco-friendly alternative, which reduces the organic solvent consumption and time required for extraction and concentration (Baysal et al. 2000, Gómez-Prieto et al. 2003, Rozzi et al. 2002). Supercritical CO_2 possesses solvent strength approaching that of non-polar liquid solvents. For optimizing conditions to achieve maximum extraction, the solvent strength of supercritical CO_2, which is directly related to its density, can be adjusted by changing the extraction pressure and temperature. Wang and Chen (2006) obtained a high yield of lycopene from tomato pulp by using the optimized conditions of pressure at 350 bar and temperature at 70°, with cis-isomers making up 41.4% of total lycopene.

Spectrophotometric Method

Spectrophotometric method allows quantitative determination of carotenoids by measuring the absorbance (or optical density) of an isolated pure carotenoid compound (dissolved in an organic solvent) at maximum absorption wavelength and the amount of carotenoid in micrograms per gram material can be determined using the expression

$$\mu g\ carotenoid/g = \frac{ABS \times V \times 10^6}{\varepsilon_{1cm}^{1\%} \times 100 \times W}$$

where V is the total volume (mL) containing W g of sample, and $\varepsilon_{1cm}^{1\%}$ is the extinction coefficient of the carotenoid analyzed. The extinction coefficients for various tomato carotenoids are given in Table 3 (Davies 1976).

HPLC Method

Open-column and thin-layer chromatographic techniques have been extensively discussed in earlier reviews (Davies 1976, Gross 1991). High performance liquid chromatography (HPLC) is a rapid, reproducible and highly sensitive technique used with wide varieties of stationary phases, mobile phase systems and detectors for determination of carotenoids in tomato and tomato products. The standard method for determination of carotenoids fails to resolve cis-isomers (AOAC 1995). Both C_{18} and C_{30} stationary phase columns are used, but the former allows only partial separation of cis-isomers, whereas the latter provides a higher resolution of cis-isomers. Research (2000-2006) on HPLC methods with their conditions used in the determination of carotenoids and/or lycopene in tomato and tomato products is summarized in Table 4, while the earlier HPLC methods have been reviewed by Shi and Le Maguer (2000).

Identification

Carotenoids are usually identified by their chromatographic and spectrophotometric behavior. The absorption spectrum of each chromatographically pure pigment is recorded and compared with values from the literature. For identifying the geometric isomers, photoisomerization is induced by illuminating authentic standards of

Table 4 HPLC[a] methods for determination of carotenoids and/or lycopene in raw tomatoes and tomato products.

Extraction	*HPLC[a] column*	*Mobile phase/flow rate/detector*	*References*
0.2 g spray-dried tomato powder with 40 mL tetrahydro-furan/methanol (1:1, v/v)	C_{18} (250 × 4.6 mm i.d., 5 μm) Vydac column	Acetonitrile/methanol/2-propanol (44:54:2, v/v/v); flow rate 1 mL/min; PDA[b]	Anguelova and Warthesen (2000)
2 g tomato paste extracted with hexane/ acetone/absolute alcohol/ toluene mixture (or) 53 g tomato paste extracted with supercritical fluid CO_2 at 55°, pressure 300 bar, flow rate 4 kg/h	C_{18} (150 × 3.9 mm i.d., 5 μm) Waters Symmetry column	Methanol/tetrahydrofuran/water (67:27:6, v/v/v); flow rate 1.5 mL/min; PDA[b]	Baysal et al. (2000)
5 g tomato pulp extracted with 50 mL methanol and filter cake with 50 mL acetone/hexane (1:1, v/v) after homogenizing with 1 g $CaCO_3$, 4 g Celite and 50 mL methanol for 1 min and filtered	C_{30} (250 × 4.6 mm i.d., 3 μm) YMC column with C_{30} guard column	Gradient elution using methanol/methyl tert-butyl ether (95:5, v/v for initial 5 min and changed to 30:70, v/v for 55 min); flow rate 1 mL/min; PDA[b]	Tiziani et al. (2006)
150 g tomato/tomato juice/50 g tomato paste homogenized with 10% Celite, 10% $MgCO_3$ and 250 mL tetrahydrofuran, followed by extraction in 250 mL tetrahydrofuran	C_{18} (250 × 4.6 mm i.d., 5 μm) Rainin Dynamax column with a guard column	Linear gradient using acetonitrile/methanol/ dichloromethane/hexane (85:10:2.5:2.5 to 45:10:22.5:22.5% at 40 min); flow rate 0.8 mL/min; PDA[b]	Takeoka et al. (2001)
10 g fresh/processed tomato homogenized with 50 mL methanol, 1 g $CaCO_3$ and 3 g Celite followed by extraction with acetone/hexane (1:1, v/v)	C_{30} (250 × 4.6 mm i.d., 3 μm) polymeric column with a C_{30} guard column	multi-step linear gradient using 15 to 50% methyl-tert-butyl ether in methanol for 55 min; flow rate 1 mL/min; PDA[b]	Nguyen et al. (2001)
1-10 g tomato pericarp extracted with 20 mL dichloromethane after	C_{30} (250 × 4.6 mm i.d., 3 μm) YMC column	Methyl-tert-butyl ether/methanol/ethyl acetate (40:50:10, v/v/v); flow rate 1 mL/min; PDA[b]	Ishida et al. (2001)

Contd.

homogenizing with 10-20 mL methanol, 0.01% ethylenediamine tetraacetic acid and butylated hydroxyanisole (1-10 mg)			
2 g tomato seeds and skins extracted by sonication with 20 mL chloroform for 30 min (or) Supercritial fluid CO_2 at 32 to 86°, pressure 138 to 483 bar, flow rate 2.5 mL/min	C_{18}(2) (150 × 4.6 mm i.d., 3 µm) Phenomenex Luna column	Gradient elution using A: methanol/0.2 M ammonium acetate (90:10, v/v) and B: methanol/ 1-propanol/1.0 M ammonium acetate (78:20:2, v/v/v); electrochemical detector	Rozzi et al. (2002)
10 g tomato slurry extracted by shaking with a mixture of 5 mL chloroform 3 mL acetone and 15 mL hexane for 5 min	C_{30} (250 × 4.6 mm i.d., 3 µm) YMC column	Methyl tert-butyl ether/methanol (7:3, v/v); flow rate 1 mL/min; UV-VIS[c] at 471 nm	Dewanto et al. (2002)
0.5 g freeze-dried tomato extracted with supercritical fluid CO_2 at 0.25–0.90 g/mL density, temperature 40°, flow rate 4 mL/min	C_{30} (250 × 4.6 mm i.d., 5 µm) Develosil UG column at 20°	Linear gradient using A: methanol-water (96:4 v/v) B: methyl tert-butyl ether from 83(A):17(B) to 33(A):67(B) at 60 min; flow rate 1 mL/min; PDA[b]	Gómez-Prieto et al. (2003)
8 g tomato juice agitated with 0.2 g $MgCO_3$ and 40 mL ethanol/hexane (4:3, v/v) for 30 min	C_{30} (250 × 4.6 mm i.d., 5 µm) YMC column	Gradient elution with A: 1-butanol/ acetonitrile (30:70, v/v) and B: methylene chloride; flow rate 2 mL/min; PDA[b]	Lin and Chen (2003)
2 g tomato juice/sauce/soup/ baked tomato slices homogenized with 400 mg MgO, and 500 µL echinenone followed by extraction with 35 mL methanol/ tetrahydrofuran (1:1, v/v) in ice for 5 min	C_{30} (250 × 4.6 mm i.d., 5 µm) YMC column at 23°	Methanol/methyl-tert-butyl ether; flow rate 1.3 mL/min; PDA[b]	Seybold et al. (2004)

Contd.

6 kg of tomato puree with 250 mL acetone/hexane (50:50 v/v) after homogenizing with 500 mL ethanol	C_{30} (250 × 4.6 mm i.d., 5 µm) YMC column with a C_{18} guard column at 25°	Variable gradient condition with methyl-tert-butyl ether/methanol for oleoresins from different tomato varieties; flow rate 1 mL/min; PDA[b]	Hackett et al. (2004)
5 g tomato puree with 120 mL hexane/acetone/ethanol (2:1:1, v/v/v)	C_{30} (250 × 4.6 mm i.d.) YMC column) at 24°	Methanol/methyl-tert-butyl ether/ethyl acetate (50:40:10, v/v/v); flow rate 1.5 mL/min; UV-VIS[c]	Qiu et al. (2006)
3 g of tomato extracted by stirring with 100 mL hexane/acetone/ethanol (50:25:25, v/v/v) for 30 min	C_{18} (300 × 2 mm i.d., 10 µm) µBondapack column with a C_{18} precolumn (20 × 3.9 mm i.d., 10 µm) at 30°	Methanol/acetonitrile (90:10 v/v) plus 9 µM triethylamine; flow rate 0.9 mL/min; UV-VIS[c] at 475 nm	Barba et al. (2006)
250 mg raw/cooked tomato extracted in acetone/ethanol/hexane mixture of 1 mL each (vortexed for 30 s)	C_{18} (250 × 4.6 mm i.d., 5 µm) Discovery column	Methanol/acetone (90:10, v/v); flow rate 0.8 mL/min; PDA[b]	Mayeaux et al. (2006)

[a]HPLC, high performance liquid chromatography; [b]PDA, photodiode array; [c]UV-VIS, ultraviolet-visible.

anticipated carotenoids and subjecting to HPLC analysis under the same condition as that of unknown carotenoid extract. By comparing the retention time and absorption spectrum, the geometric isomer can be tentatively determined. A cis-isomer differs from the all-trans-isomer in its adsorption affinity and absorption spectrum. The absorption maximum of a cis-isomer shifts to lower wavelength compared to its all-trans counterpart. The di-cis-isomers may shift to a lower wavelength than mono-cis-isomers. The identification of carotenoids based on spectral characteristics has been well described in the literature (Davies 1976, Goodwin 1981). Further confirmation of identified carotenoids can be made by performing cochromatography test using isomerized and non-isomerized carotenoid standards.

A photodiode array detector (PDA) coupled with HPLC provides online monitoring of absorption spectrum of carotenoid peaks at multiple wavelengths. However, it fails to provide complete structural and stereochemical information of carotenoids. Mass spectrometer enables the quantification and elucidation of carotenoid structure on the basis of molecular mass and characteristic fragmentation pattern. Figure 4A shows the negative-ion atmospheric pressure chemical ionization (APCI) product ion tandem mass spectrum of lycopene molecular ion radical (m/z 536.43) (van Breemen 2005). Though lycopene, α-carotene and β-carotene show similar molecular ion peaks, lycopene can be selectively monitored using its characteristic fragment ion of m/z 467 obtained by elimination of terminal isoprene group, while α-carotene and β-carotene fail to form this fragment ion because of their terminal ring groups. β-Carotene gives a characteristic fragment ion m/z 444 because of loss of toluene, as illustrated in a positive-ion APCI spectrum (Fig. 4B) (Lacker et al. 1999). Structurally significant fragment ions of α-carotene are m/z 480 and 388, which correspond to the retro-Diels-Alder fragment ion, $[M-56]^{+\cdot}$, and a loss due to both toluene and retro-Diels-Alder fragmentation $[M-92-56]^{+\cdot}$, respectively (van Breemen et al. 1996). The collision-induced dissociation mass spectrum of lutein was found to show fragment ions of m/z 551, 459 and 429 due to the losses of water [MH-18], both water and toluene [MH-18-92] and the terminal ring [MH-140], respectively (van Breemen et al. 1996). However, mass spectrometer fails to distinguish among stereoisomers.

High resolution nuclear magnetic resonance (NMR) spectroscopy with higher magnetic field strength (800-900 MHz) and the application of two-dimensional pulse sequences assist in elucidating the double-bond stereochemistry, which is the key area for identification of most carotenoids and their isomers. Tiziani et al. (2006) recently profiled carotenoids in a lipophilic mixture of tomato juice by a combination of

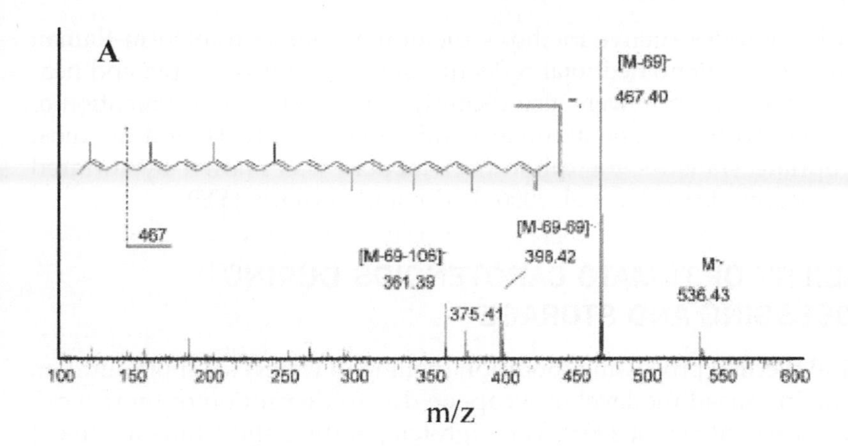

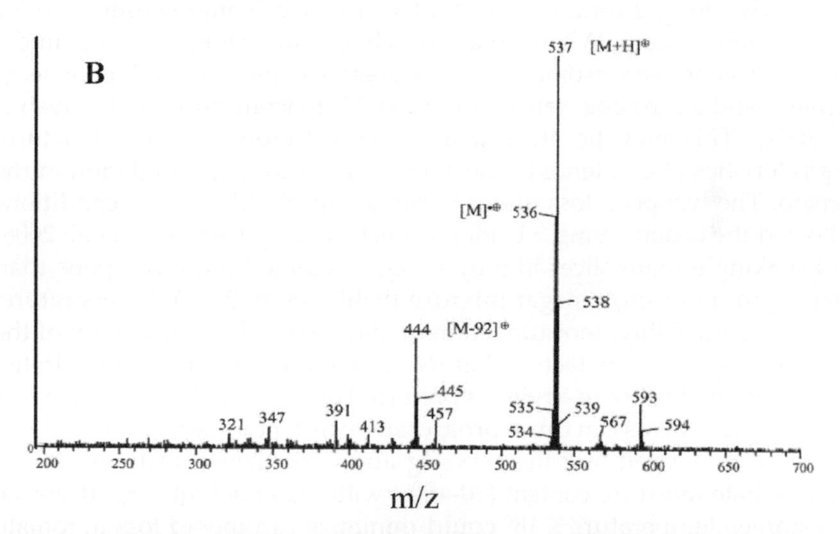

Fig. 4 Mass spectrum of (A) lycopene and (B) β-carotene (van Breeman 2005, Lacker et al. 1999). (A) negative-ion APCI (atmospheric pressure chemical ionization) product ion tandem mass spectrum of the lycopene molecular ion radical of m/z 536.43 (note the m/z 467 fragment ion due to elimination of an isoprene group in lycopene does not occur for α- and β-carotenes); (B) positive-ion APCI mass spectrum of β-carotene.

homonuclear and heteronuclear 2D NMR techniques, and ^1H and ^{13}C NMR assignments for all-trans-lycopene, 5-cis-lycopene, 9-cis-lycopene, 13-cis-lycopene, all-trans-β-carotene and 15-cis-phytoene have been reported.

Other non-destructive methods including Fourier transform-Raman spectroscopy, attenuated total reflection infrared spectroscopy, and near infrared spectroscopy were also recently employed for determination of lycopene and β-carotene in tomato and tomato products, and the most appropriate statistics for calibration models was shown by infrared spectroscopy (Baranska et al. 2006, Pedro and Ferreira 2005).

STABILITY OF TOMATO CAROTENOIDS DURING PROCESSING AND STORAGE

Typical thermal processing condition (boiling in water) for short duration of time increased the level of lycopene due to disruption of tomato cell matrix (Dewanto et al. 2002). The high-temperature-short-time treatment (121° for 4 s) of tomato juice enhanced the leaching of lycopene with minimum degradation (Lin and Chen 2005a). However, high temperature along with long-duration treatment can cause isomerization and/or degradation to occur. When tomato was boiled in water for 30 min, lutein and β-carotene were isomerized to a greater extent, while lycopene, γ-carotene and δ-carotene were relatively stable to isomerization (Nguyen et al. 2001). This may be attributed to the difference in the structural characteristics of carotenoids and their localization/accumulation in the tomato. The lycopene loss under various household cooking conditions followed the order: frying > baking > microwaving (Mayeaux et al. 2006) and soaking tomato slices in only vinegar retained more lycopene than soaking in oil or oil/vinegar mixture (Sahlin et al. 2004). Temperature, oxygen permeability, moisture content and texture (fine or coarse) of the tomato sample are the factors that determine the stability of dehydrated tomato products. Freeze-drying is preferred over air- and spray-drying, as the latter treatments can cause progressive carotenoid degradation due to exposure to oxygen and heat (Wang and Chen 2006). Dehydration to intermediate moisture content (20-40%) with water activity (a_w) 0.69-0.86 at a storage temperature ≤ 18° could minimize carotenoid loss in tomato products (Zanoni et al. 2000).

Complete inactivation of natural enzymes in tomato by blanching can prevent carotenoid degradation in tomato during frozen storage (Fellows 2000). Several studies have demonstrated that the higher the storage temperature, the faster the degradation rate (Lin and Chen 2005b, Sharma and Le Maguer 1996a). Lycopene degradation under different storage conditions followed the order: air and light > air and dark > vacuum and dark (Sharma and Le Maguer 1996a). Canned tomato juice stored either in dark or under light showed minimum carotenoid degradation due to less exposure to oxygen or light (Lin and Chen, 2005b).

Carotenoid loss in tomato can be minimized by treating tomato products with osmotic solution (sucrose or high dextrose equivalent maltodextrin) under optimized conditions prior to processing or storage (Dermesonlouoglou et al. 2007, Shi et al. 1999). In addition, subjecting the tomato sample to high hydrostatic pressure before storage could offer a sterilization effect to inactivate enzymes and reduce carotenoid degradation (Qiu et al. 2006).

FUTURE CONSIDERATIONS

1. More studies need to be explored on tomato carotenoids other than lycopene and β-carotene.
2. More frequent investigation should be carried out on the impact of various growth, environmental and technological factors.
3. New pretreatment techniques must be developed to prevent carotenoid loss during processing and storage of tomato and tomato products.
4. The mechanism of selective incorporation of cis-isomers into bile acid micelles should be explored, and various factors influencing the bioavailability of cis-isomers and their importance in human health should be further investigated.
5. In addition to antioxidant effect, other potential mechanisms of biological action by tomato and tomato products need to be elucidated. In addition, the biological activity of oxidative and cleavage metabolites of tomato carotenoids and its mechanism require further investigation.
6. The importance and mechanism of lycopene interaction with other carotenoids and micronutrients in tomato requires to be characterized for evaluating the possible synergistic effect in biological action. Also, the pharmacokinetic studies on lycopene remain uncertain and epidemiologic studies focusing upon the relationship between biomarkers of tomato product intake, such as serum lycopene or other phytochemicals, and biomarkers of health risk should be critically examined.

ABBREVIATIONS

ABS: Absorbance; AOAC: Association of Official Analytical Chemists; APCI: Atmospheric pressure chemical ionization; CPTA: 2-(4-Chlorophenylthio)triethylamine hydrochloride; DCPTA: 2-(3,4-dichlorophenoxy)triethylamine; DMAPP: Dimethylallyl diphosphate;

DOXP: 1-Deoxy-D-xylulose-5-phosphate; GGPP: Geranylgeranyl pyrophosphate; HPLC: High performance liquid chromatography; IPP: Isopentyl pyrophosphate; NMR: Nuclear magnetic resonance; PDA: Photodiode array; SI: Saturation index

REFERENCES

Abushita, A.A. and H.G. Daood, and P.A. Biacs. 2000. Change in carotenoids and antioxidant vitamins in tomato as a function of varietal and technological factors. J. Agric. Food Chem. 48: 2075-2081.

Alba, R. and M.M. Cordonnier-Pratt, and L.H. Pratt. 2000. Fruit-localized phytochromes regulated lycopene accumulation independently of ethylene production in tomato. Plant Physiol. 123: 363-370.

Al-Wandawi, H. and M. Abdul-Rahman, and K. Al-Shaikhly. 1985. Tomato processing waste an essential raw materials source. J. Agric. Food Chem. 33: 804-807.

Anguelova, T. and J. Warthesen. 2000. Lycopene stability in tomato powders. J. Food Sci.—Food Chem. Toxicol. 65: 67-70.

AOAC. 1995. Official Methods of Analysis. Association of Official Analytical Chemists, 16th ed. Arlington, Virginia, USA.

Baranska, M. and W. Schütz, and H. Schulz. 2006. Determination of lycopene and β-carotene content in tomato fruits and related products: Comparison of FT-Raman, ATR-IR, and NIR spectroscopy. Anal. Chem. 78: 8456-8461.

Barba, A.I.O. and M.C. Hurtado, M.C.S. Mata and V.F. Ruiz, and M.L. Sáenz de Tejada. 2006. Application of a UV-vis detection-HPLC method for a rapid determination of lycopene and β-carotene in vegetables. Food Chem. 95: 328-336.

Barreiro, J.A. and M. Milano, and A.J. Sandoval. 1997. Kinetics of colour change of double concentrated tomato paste during thermal treatment. J. Food Eng. 33: 359-371.

Batu, A. and A.K. Thompson. 1998. Effect of short term high carbon dioxide treatment on tomato ripening. Tr. J. Agric. Forest. 22: 405-410.

Baysal, T. and S. Ersus, and D.A.J. Starmans. 2000. Supercritical CO_2 extraction of β-carotene and lycopene from tomato paste waste. J. Agric. Food Chem. 48: 5507-5511.

Benthin, P. and H. Danz, and M. Hamburger. 1999. Pressurized liquid extraction of medicinal plants. J. Chromatogr. A. 837: 211-219.

Bohm, V. and N.L. Puspitasari-Nienaber, M.G. Ferruzzi, and S.J. Schwartz. 2002. Trolox equivalent antioxidant activity of different geometrical isomers of α-carotene, β-carotene, lycopene, and zeaxanthin. J. Agric. Food Chem. 50: 221-226.

Boileau, A.C. and N.R. Merchen, K. Wasson, C.A. Atkinson, and J.W. Erdman. 1999. Cis-lycopene is more bioavailable than trans-lycopene *in vitro* and *in vivo* in lymph-cannulated ferrets. J. Nutr. 129: 1176-1181.

Boileau, T.W. and S.K. Clinton, Z. Liao, M.H. Monaco, S.M. Donovan, and J.W. Erdman. 2001. Lycopene, tomato powder and dietary restriction influence survival of rats with prostate cancer induced by NMU and testosterone. J. Nutr. 131: 191S-199S.

Boileau, T.W.M. and A.C. Boileau, and J.W. Erdman. 2002. Bioavailability of all-trans and cis-isomers of lycopene. Exp. Biol. Med. 227: 914-919.

Bramley, P.M. 2002. Regulation of carotenoid formation during tomato fruit ripening and development. J. Exp. Bot. 53: 2107-2113.

Bramley, P. and C. Teulieres, I. Blain, C. Bird, and W. Schuch. 1992. Biochemical characterization of transgenic tomato plants in which carotenoid synthesis has been inhibited through the expression of antisense RNA to pTOM5. Plant J. 2: 343-349.

Britton, G. 1995. Structure and properties of carotenoids in relation to function. FASEB J. 9: 1551-1558.

Campbell, J.K. and K. Canene-Adams, B.L. Lindshield, T.W.M. Boileau, S.K. Clinton, and J.W. Erdman. 2004. Tomato phytochemicals and prostate cancer risk. J. Nutr. 134: 3486S-3492S.

Caris-Veyrat, C. and M.J. Amiot, V. Tyssandier, D. Grasselly, M. Buret, M. Mikolajczak, J.C. Guilland, C. Bouteloup-Demange, and P. Borel. 2004. Influence of organic versus conventional agricultural practice on the antioxidant microconstituent content of tomatoes and derived purees; consequences on antioxidant plasma status in humans. J. Agric. Food Chem. 52: 6503-6509.

Clinton, S.K. and C. Emenhiser, S.J. Schwartz, D.G. Bostwick, A.W. Williams, B.J. Moore, and J.W. Erdman. 1996. Cis-trans lycopene isomers, carotenoids, and retinol in the human prostrate. Cancer Epidemiol. Biomarkers Prev. 5: 823-833.

Cohen, J.D. 1996. *In vitro* tomato fruit cultures demonstrate a role for indole-3-acetic acid in regulating fruit ripening. J. Am. Soc. Hortic. Sci. 121: 520-524.

Cohen, L.A. 2002. A review of animal model studies of tomato carotenoids, lycopene, and cancer prevention. Exp. Biol. Med. 227: 864-868.

Davies, B.H. Carotenoids. pp. 67-84. *In*: T.W. Goodwin. [ed.] 1976. Chemistry and Biochemistry of Plant Pigments, Vol. II. Academic Press, New York, USA.

Davies, J. and J.M. Hobson. 1981. The constituents of tomato fruit. The influence of environment, nutrition, and genotype. Crit. Rev. Food Sci. Nutr. 15: 205-280.

De Pascale, S. and A. Maggio, V. Fogliano, P. Ambrosino, and A. Ritieni. 2001. Irrigation with saline water improves carotenoids content and antioxidant activity of tomato. J. Hortic. Sci. Biotechnol. 76: 447-453.

Dermesonlouoglou, E.K. and M.C. Giannakourou, and P.S. Taoukis. 2007. Kinetic modeling of the degradation of quality of osmo-dehydrofrozen tomatoes during storage. Food Chem. 103: 985-993

Dewanto, V. and X. Wu, K.K. Adom, and R.H. Liu. 2002. Thermal processing enhances the nutritional value of tomatoes by increasing total antioxidant activity. J. Agric. Food Chem. 50: 3010-3014.

Dumas, Y. and M. Dadomo, G.D. Lucca, and P. Grolier. 2003. Effects of environmental factors and agricultural techniques on antioxidant content of tomatoes. J. Sci. Food Agric. 83: 369-382.

Fanasca, S. and G. Colla, G. Maiani, E. Venneria, Y. Rouphael, E. Azzini, and F. Saccardo. 2006. Changes in antioxidant content of tomato fruits in response to cultivar and nutrient solution composition. J. Agric. Food Chem. 54: 4319-4325.

Fellows, P.J. Blanching. pp. 233-240. *In*: P.J. Fellows. [ed.] 2000. Food Processing Technology—Principles and Practice. Woodhead Publishing Limited, Cambridge.

Ferreira, A.L.D. and K.J. Yeum, R.M. Russell, N.I. Krinsky, and G. Tang. 2003. Enzymatic and oxidative metabolites of lycopene. J. Nutr. Biochem. 14: 531-540.

Gabr, S. and A. Sharaf, and S. El-Saadany. 2006. Effect of chlormequat and alar on some biochemical constituents in tomato plants and fruits. Food/Nahrung 29: 219-228.

Gärtner, C. and W. Stahl, and H. Sies. 1997. Lycopene is more bioavailable from tomato paste than from fresh tomatoes. Am. J. Clin. Nutr. 66: 116-122.

Giovanelli, G. and V. Lavelli, C. Peri, and S. Nobili. 1999. Variation in antioxidant components of tomato during vine and post-harvest ripening. J. Sci. Food Agric. 79: 1583-1588.

Giovannucci, E. 1999. Tomatoes, tomato-based products, lycopene, and cancer: review of the epidemiologic literature. J. Natl. Can. Inst. 91: 317-331.

Giovannucci, E. and E.B. Rimm, Y. Liu, M.J. Stampfer, and W.C. Willett. 2002. A prospective study of tomato products, lycopene, and prostate cancer risk. J. Natl. Can. Inst. 94: 391-398.

Gómez-Prieto, M.S. and M.M. Caja, M. Herraiz, and G. Santa-María. 2003. Supercritical fluid extraction of all-trans-lycopene from tomato. J. Agric. Food Chem. 51: 3-7.

Goodwin, T.W. 1981. The Biochemistry of Carotenoids Vol. I. Chapman and Hall, New York, USA.

Gross, J. 1991. Pigments in Vegetables. van Nordstrand Reinhold, New York, USA.

Hackett, M.M. and J.H. Lee, D. Francis, and S.J. Schwartz. 2004. Thermal stability and isomerization of lycopene in tomato oleoresins from different varieties. J. Food Sci.—Food Chem. Toxicol. 69: 536-541.

Hadley, C.W. and S.K. Clinton, and S.J. Schwartz. 2003. The consumption of processed tomato products enhances plasma lycopene concentrations in association with a reduced lipoprotein sensitivity to oxidative damage. J. Nutr. 133: 727-732.

Hart, D.J. and K.J. Scott. 1995. Development and evaluation of an HPLC method for the analysis of carotenoids in foods, and the measurement of the carotenoid content of vegetables and fruits commonly in the UK. Food Chem. 54: 101-111.

Heber, D. and L. Qing-Yi. 2002. Overview of mechanisms of action of lycopene. Exp. Biol. Med. 227: 920-923.

Ishida, B.K. and J. Ma, and B. Chan. 2001. A simple, rapid method for HPLC analysis of lycopene isomers. Phytochem. Anal. 12: 194-198.

Keithly, J.H. and H. Yokohama, and H. Gausman. 1990. Enhanced yield of tomato in response to 2-(3,4-dichloriphenoxy)triethylamine (DCPTA). Plant Growth Regul. 9: 127-136.

Khachik, F. and G. Beecher, and J.C. Smith. 1995. Lutein, lycopene, their oxidative metabolites in chemoprevention of cancer. J. Cell Biochem. Suppl. 22: 236-246.

Khachik, F. and L. Carvalho, P.S. Bernstein, G.J. Muir, D.Y. Zhao, and N.B. Katz. 2002. Chemistry, distribution, and metabolism of tomato carotenoids and their impact on human health. Exp. Biol. Med. 227: 845-851.

Kopsell, D.A. and D.E. Kopsell. 2006. Accumulation and bioavailability of dietary carotenoids in vegetable crops. Tr. Plant Sci. 11: 499-507.

Krinsky, N.I. and K.J. Yeum. 2003. Carotenoid-radical interactions. Biochem. Biphys. Res. Comm. 305: 754-760.

Kucuk, O. and F.H. Sarkar, W. Sakr, Z. Djuric, M.N. Pollak, F. Khachik, Y.W. Li, M. Banerjee, D. Grignon, J.S. Bertram, J.D. Crissman, E.J. Pontes, and D.P. Wood. 2001. Phase II randomized clinical trial of lycopene supplementation before radical prostatechomy. Cancer Epidemiol. Biomarkers Prev. 10: 861-868.

Kuti, J.O. and H.B. Konuru. 2005. Effects of genotype and cultivation environment on lycopene content in red-ripe tomatoes. J. Sci. Food Agric. 85: 2021-2026.

Lacker, T. and S. Strohschein, and K. Albert. 1999. Separation and identification of various carotenoids by C_{30} reversed-phase high-performance liquid chromatography coupled to UV and atmospheric pressure chemical ionization mass spectrometric detection. J. Chromatogr. A 854: 37-44.

Lehucher-Michel, M.P. and J.F. Lesgards, O. Delubac, P. Stocker, P. Durand, and M. Prost. 2001. Oxidative stress and human disease. Current knowledge and perspectives for prevention. Presse Med. 30: 1076-1081.

Lin, C.H. and B.H. Chen. 2003. Determination of carotenoids in tomato juice by liquid chromatography. J. Chromatogr. A 1012: 103-109.

Lin, C.H. and B.H. Chen. 2005a. Stability of carotenoids in tomato juice during processing. Eur. Food Res. Technol. 221: 274-280.

Lin, C.H. and B.H. Chen. 2005b. Stability of carotenoids in tomato juice during storage. Food Chem. 90: 837-846.

Matsuzoe, N. and K. Zushi, and T. Johjima. 1998. Effect of soil water deficit on coloring and carotene formation in fruits of red, pink and yellow type cherry tomatoes. J. Jpn. Soc. Hortic. Sci. 67: 600-606.

Mayeaux, M. and Z. Xu, J.M. King, and W. Prinyawiwatkul. 2006. Effects of cooking conditions on the lycopene content in tomatoes. J. Food Sci. 71: 461-464.

Miller, N.J. and J. Sampson, L.P. Candeias, P.M. Bramley, and C.A. Rice-Evans. 1996. Antioxidant activities of carotenes and xanthophylls. FEBS Lett. 384: 240-242.

Mortensen, A. and L.H. Skibsted, and T.G. Truscott. 2001. The interaction of dietary carotenoids with radical species. Arch. Biochem. Biophys. 385: 13-19.

Nguyen, M.L. and D. Francis, and S.J. Schwartz. 2001. Thermal isomerization susceptibility of carotenoids in different tomato varieties. J. Sci. Food Agric. 81: 910-917.

Paiva, E.A.S. and R.A. Sampaio, and H.E.P. Martinez. 1998. Composition and quality of tomato fruit cultivated in nutrient solutions containing different calcium concentration. J. Plant Nutr. 21: 2653-2661.

Pedro, A.M.K. and M.M.C. Ferreira. 2005. Nondestructive determination of solids and carotenoids in tomato products by near-infrared spectroscopy and multivariate calibration. Anal. Chem. 77: 2505-2511.

Periago, M.J. and F. Rincón, M.D. Agüera, and G. Ros. 2004. Mixture approach for optimizing lycopene extraction from tomato and tomato products. J. Agric. Food Chem. 52: 5796-5802.

Petr, L. and J.W. Erdman. Lycopene and risk of carodiovascular disease. pp. 204-217. *In*: L. Packer, U. Obermueller-Jevic, K. Kramer and H. Sies. [eds.] 2005. Carotenoids and Retinoids: Biological Actions and Human Health. AOCS Press, Champaign, Illinois, USA.

Qiu, W. and H. Jiang, H. Wang, and Y. Gao. 2006. Effect of high hydrostatic pressure on lycopene stability. Food Chem. 97: 516-523.

Raffo, A. and C. Leonardi, V. Fogliano, P. Ambrosino, M. Salucci, L. Gennaro, R. Bugianesi, F. Giuffrida, and G. Quaglia. 2002. Nutritional value of cherry tomatoes (*Lycopersicon esculentum* Cv. Naomi F1) harvested at different ripening stages. J. Agric. Food. Chem. 50: 6550-6556.

Rajendran, V. and Y.S. Pu, and B.H. Chen. 2005. An improved HPLC method for determination of carotenoids in human serum. J. Chromatogr. B 824: 99-106.

Re, R. and P.D. Fraser, M. Long, P.M. Bramley, and C. Rice-Evans. 2001. Isomerization of lycopene in the gastric milieu. Biochem. Biophys. Res. Comm. 281: 576-581.

Reboul, E. and P. Borel, C. Mikail, L. Abou, M. Charbonnier, C. Caris-Veyrat, P. Goupy, H. Portugal, D. Lairon, and M.J. Amiot. 2005. Enrichment of tomato paste with 6% tomato peel increases lycopene and β-carotene bioavailability in men. J. Nutr. 135: 790-794.

Rodriguez, D.B. and T.C. Lee, and C.O. Chichester. 1975. Comparative study of the carotenoid composition of the seeds of ripening Momardica charantia and tomatoes. Plant Physiol. 56: 626-629.

Rozzi, N.L. and R.K. Singh, R.A. Vierling, and B.A. Watkins. 2002. Supercritical fluid extraction of lycopene from tomato processing byproducts. J. Agric. Food Chem. 50: 2638-2643.

Rymal, K.S. and T.O.M. Nakayama. 1974. Major carotenoids of the seeds of three cultivars of the tomato. J. Agric. Food Chem. 22: 715-717.

Sadler, G. and J. Davis, and D. Dezman. 1990. Rapid extraction of lycopene and β-carotene from reconstituted tomato paste and pink grapefruit homogenates. J. Food Sci. 55: 1460-1461.

Sahlin, E. and G.P. Savage, and C.E. Lister. 2004. Investigation of the antioxidant properties of tomatoes after processing. J. Food Comp. Anal. 17: 635-647.

Schierle, J. and W. Bretzel, I. Bühler, N. Faccin, D. Hess, K. Steiner, and W. Schüep. 1997. Content and isomeric ratio of lycopene in food and human blood plasma. Food Chem. 59: 459-465.

Seybold, C. and K. Fröhlich, R. Bitsch, K. Otto, and V. Böhm. 2004. Changes in contents of carotenoids and vitamin E during tomato processing. J. Agric. Food Chem. 52: 7005-7010.

Sharma, S.K. and M. Le Maguer. 1996a. Kinetics of lycopene degradation in tomato pulp solids under different processing and storage conditions. Food Res. Int. 29: 309-315.

Sharma, S.K. and M. Le Maguer. 1996b. Lycopene in tomatoes and tomato pulp fractions. Ital. J. Food Sci. 2: 107-113.

Shi, J. and M. Le Maguer. 2000. Lycopene in tomatoes: chemical and physical properties affected by food processing. Crit. Rev. Food Sci. Nutr. 40: 1-42.

Shi, J. and M. Le Maguer, Y. Kakuda, A. Liptay, and F. Niekamp. 1999. Lycopene degradation and isomerization in tomato dehydration. Food. Res. Int. 32: 15-21.

Sozzi, G.O. and G.D. Trinchero, and A.A. Fraschina. 1999. Controlled-atmosphere storage of tomato fruit: low oxygen or elevated carbon dioxide levels alter galactosidase activity and inhibit exogenous ethylene action. 79: 1065-1070.

Stahl, W. and H. Sies. 1996. Perspectives in biochemistry and biphysics—Lycopene: A biologically important carotenoid for humans. Arch. Biochem. Biophys. 336: 1-9.

Takeoka, G. R. and L. Dao, S. Flessa, D.M. Gillespie, W.T. Jewell, B. Huebner, D. Bertow, and S.E. Ebeler. 2001. Processing effects on lycopene content and antioxidant activity of tomatoes. J. Agric. Food Chem. 49: 3713-3717.

Télef, N. and L. Stammitti-Bert, A. Mortain-Bertrand, M. Maucourt, J.P. Carde, D. Rolin, and P. Gallusci. 2006. Sucrose deficiency delays lycopene accumulation in tomato fruit pericarp discs. Plant Mol. Biol. 62: 453-469.

Thompson, K.A. and M.R. Marshall, C.A. Sims, C.I. Wei, S.A. Sargent, and J.W. Scott. 2000. Cultivar, maturity and heat treatment on lycopene content in tomatoes. J. Food Sci.—Food Chem. Toxicol. 65: 791-795.

Tiziani, S. and S.J. Schwartz, and Y. Vodovotz. 2006. Profiling of carotenoids in tomato juice by one- and two-dimensional NMR. J. Agric. Food Chem. 54: 6094-6100.

Tonucci, L.H. and J.M. Holden, G.R. Beecher, F. Khachik, C.S. Davis, and G. Mulokozi. 1995. Carotenoid content of thermally processed tomato-based food products. J. Agric. Food Chem. 43: 579-586.

Toor, R.K. and G.P. Savage. 2005. Antioxidant activity in different fractions of tomatoes. Food Res. Int. 38: 487-494.

Toor, R.K. and G.P. Savage, and A. Heeb. 2006. Influence of different types of fertilizers on the major antioxidant components of tomatoes. J. Food Comp. Anal. 19: 20-27.

van Breemen, R.B. 2005. How do intermediate endpoint markers respond to lycopene in men with prostrate cancer or benign prostrate hyperplasia? J. Nutr. 135: 2062S-2064S.

van Breemen, R.B. and C.R. Huang, Y. Tan, L.C. Sander, and B. Alexander. 1996. Liquid chromatography/mass spectrometry of carotenoids using atmospheric pressure chemical ionization. J. Mass Spectrometry 31: 975-981.

van het Hof, K.H. and B.C.J. de Boer, L.B.M. Tijburg, B.R.H.M. Lucius, I. Zijp, C.E. West, J.G.A.J. Hautvast, and J.A. Weststrate. 2000. Carotenoid bioavailability in humans from tomatoes processed in different ways determined from the carotenoid response in the triglyceride-rich lipoprotein fraction of plasma after a single consumption and in plasma after four days of consumption. J. Nutr. 130: 1189-1196.

Vardhini, B.V. and S.S. Rao. 2002. Acceleration of ripening of tomato pericarp discs by brassinosteroids. Phytochemistry 6: 843-847.

Wang, C.Y. and B.H. Chen. 2006. Tomato pulp as source for the production of lycopene powder containing high proportion of cis-isomer. Eur. Food Res. Technol. 222: 347-353.

Wang, S. and V.L. DeGroff, and S.K. Clinton. 2003. Tomato and soy polyphenols reduce insulin-like growth factor-1-stimulated rat prostate cancer cell proliferation and apoptotic resistance in vitro via inhibition of intracellular signaling pathways involving tyrosine kinase. J. Nutr. 133: 2367-2376.

Zanoni, B. and E. Pagliarini, and R. Foschino. 2000. Study of the stability of dried tomato halves during shelf-life to minimize oxidative damage. J. Sci. Food Agric. 80: 2203-2208.

Zechmeister, L. 1962. Cis-trans Isomeric Carotenoids, Vitamin A and Arylpolyenes. Academic Press, New York, USA.

Tomato Lectin

Willy J. Peumans[1] and Els J.M. Van Damme[2]
[1]Nievelveldweg 9, 9310 Aalst, Belgium
[2]Laboratory of Biochemistry and Glycobiology, Dept. of Molecular Biotechnology, Ghent University, Coupure Links 653, 9000 Gent, Belgium

ABSTRACT

Tomato is a classic example of a food plant that contains a potent hemagglutinating protein in its edible part. Though the tomato lectin (also called *Lycopersicum esculentum* lectin or LEA) was already isolated and characterized in some detail in 1980, the exact molecular structure has not yet been elucidated. A reinvestigation of the relevant literature combined with a detailed analysis of the publicly accessible genome and transcriptome sequence data revealed that tomato fruits express a mixture of three structurally and evolutionarily related lectins: (1) a previously cloned lectin-related protein called Lycesca, (2) a previously isolated lectin-related 42 kDa chitin-binding protein and (3) the genuine lectin (LEA). Complete sequences are available for Lycesca and the 42 kDa protein but not for the lectin itself. However, a fairly accurate model could be elaborated for LEA based on partial sequences. The lectin consists of two sugar-binding modules (comprising two in tandem arrayed hevein domains) separated by a Ser/Pro-rich linker. All four hevein domains are active, rendering LEA a potent tetravalent hemagglutinating lectin. In contrast, both Lycesca and the 42 kDa protein possess a single active hevein domain and are monovalent lectins devoid of hemagglutinating activity. Recent specificity studies revealed that LEA is not a chitin-binding lectin *sensu strictu* because it interacts equally well with complex and high mannose N-glycans. It is predominantly but not exclusively expressed in fruits, where it accumulates at an exponential rate during the first 10 days of

A list of abbreviations is given before the references.

fruit development. Striking varietal differences indicate that the tomato fruit lectin content is genetically determined. At present, the physiological role of the tomato fruit lectin is poorly understood. There are no indications for acute or chronic toxicity of LEA in mammals. However, the documented biological activities of LEA and the closely related potato lectin suggest that they might affect the immune system on oral uptake. Moreover, LEA might cause latex-fruit syndrome in sensitive individuals. This raises the question whether low-lectin tomatoes are possibly safer for the consumer.

BRIEF INTRODUCTION TO PLANT LECTINS

Many plants contain proteins that are commonly known as lectins, agglutinins or phytohemagglutinins (for recent reviews see Van Damme et al. 1998, 2004a, 2007b). This terminology has a historical background and refers to the capability of lectins to cause a macroscopically visible clumping of human and animal red blood cells. The mechanism underlying the erythrocyte agglutination activity remained unclear until it was eventually demonstrated that it relies on a specific sugar-binding activity of lectins. Once this crucial discovery had been made lectins were no longer simply considered clumping agents but acquired their status of carbohydrate-binding proteins. For a long time all plant lectins were classified into a single group of proteins on the basis of one common but rather aspecific biological activity. However, detailed biochemical analyses in the late 1970s and early 1980s provided the first firm evidence for major differences between lectins from different plant species. Subsequent structural and molecular studies confirmed the existence of different types of plant lectins. Moreover, sequence data generated by molecular cloning of lectin genes and transcriptome analyses revealed a fairly widespread occurrence of chimeric proteins comprising a carbohydrate-binding domain fused to an unrelated domain with a totally different structure and activity/function. The latter finding put the notion "lectin" in a new perspective and urged a rethinking of both the definition and classification of (plant) lectins in terms of carbohydrate-binding domain(s). In principle, the presence of at least one non-catalytic domain that binds reversibly to a specific carbohydrate can be used as a simple criterion for a protein to be considered a lectin. Accordingly, plant lectins were defined as "all plant proteins possessing at least one non-catalytic domain, which binds reversibly to a specific mono- or oligosaccharide". Hitherto, nine different sugar-binding domains/motifs have been identified with certainty in plants. Using these domains as basic structural units, a relatively simple system was elaborated that allows classification of virtually all known plant lectins in less than 10 families of structurally and evolutionarily related proteins. Most plant lectins belong to one of the seven "widespread" families, which are (in alphabetic order): the amaranthins, the Cucurbitaceae phloem lectins, the *Galanthus nivalis* agglutinin family, the lectins with hevein domain(s), the jacalin-related lectins, the legume lectins and the ricin-B family (Van Damme et al. 1998, 2007b). In addition, two families with a narrow taxonomic distribution were identified only recently, namely orthologues of the fungal *Agaricus bisporus* agglutinin (which are found in the liverwort *Marchantia polymorpha* and the moss *Tortula ruralis*) (Peumans et al. 2007) and catalytically inactive homologues of class V chitinases (which are apparently confined to the family Fabaceae) (Van Damme et al. 2007a).

HISTORICAL OVERVIEW OF THE DISCOVERY, ISOLATION AND CHARACTERIZATION OF THE TOMATO LECTIN

As early as 1926, Marcusson-Begun mentioned in a paper describing the results of his research on a hemagglutinin in potato tubers that Schiff previously discovered hemagglutinating activity in the sap of tomatoes. A first report on the extraction and preliminary characterization of a tomato (*Lycopersicum esculentum* Mill. now called *Solanum esculentum* L., Solanaceae) lectin (abbreviated LEA, *Lycopersicum esculentum* agglutinin) appeared in 1968 when Hossaini demonstrated that a saline extract of tomato seeds was able to agglutinate human type A, B and O erythrocytes (Hossaini 1968). Isolation of the fruit lectin was achieved independently in 1980 by two different groups using affinity chromatography on immobilized A+H blood group substance (Nachbar et al. 1980) and adsorption to trypsin-treated human type B erythrocytes (Kilpatrick 1980). Using these affinity-purified preparations the biochemical properties, sugar-binding specificity and some biological activities of the tomato fruit lectin could be corroborated in some detail (Nachbar et al. 1980). In addition, it became evident that the tomato lectin closely resembled the potato tuber lectin that was isolated and characterized a few years before (Allen and Neuberger 1973, Allen et al. 1978). During the next two decades rapid progress was made in the study of the sugar-binding specificity, biological activities and applications of the tomato lectin in biological and biomedical research. However, the exact molecular structure of the tomato lectin as well as that of all other Solanaceae lectins remained unclear. Only recently, molecular cloning of the potato lectin allowed unambiguous determination of the molecular structure of a typical Solanaceae lectin (Van Damme et al. 2004b). A structurally similar homologue of the potato lectin was also found in tomato (Peumans et al. 2004). However, the cloned tomato homologue does not correspond to the genuine tomato fruit lectin.

DEFINITION AND NOMENCLATURE OF TOMATO LECTINS AND RELATED PROTEINS

As is discussed below in detail tomatoes contain a fairly complex mixture of a genuine lectin and at least two lectin-related proteins. In the absence of a uniform nomenclature these proteins are still referred to by the names introduced by the authors who described them for the first time. In this contribution the following nomenclature is used:

(1) the genuine hemagglutinating tomato lectin is referred to by the standard acronym "LEA", which stands for *Lycopersicum esculentum* agglutinin.

(2) the non-hemagglutinating 42 kDa lectin-related protein described by Naito et al. (2001) is referred to as "42 kDa lectin-related protein".

(3) the expressed and cloned but still unidentified lectin-related protein (Peumans et al. 2004) is referred to by the name "Lycesca". This name (which is not considered an abbreviation) was introduced in the original report to distinguish it from the genuine lectin.

CUMBERSOME ELUCIDATION OF THE MOLECULAR STRUCTURE OF TOMATO LECTIN

Early Work on Isolation and Biochemical Characterization of Tomato Fruit Lectin(s)

The first detailed description of the purification and biochemical characterization of the tomato lectin dates back to 1980 (Nachbar et al. 1980). Nachbar and colleagues isolated the lectin in a single step from a cleared fruit extract using affinity chromatography on a column of immobilized hog A + H blood group substance and analysed the agglutinin by the biochemical techniques available at that time (including polyacrylamide gel electrophoresis, isoelectric focusing, amino acid and carbohydrate analysis, N-terminal amino acid sequencing and analytical ultracentrifugation). From their results, Nachbar et al. concluded that the tomato lectin is a monomeric 71 kDa protein that contains approximately 50% covalently bound carbohydrate (85% arabinose and 15% galactose). The lectin is especially rich in cysteine (Cys) and hydroxyproline (Hyp) and contains a methionine (Met) residue at its N-terminus. Isoelectric focusing further revealed that the lectin is a natural mixture of two isoforms with an isoelectric point of 8.8 and 10.0, respectively. Hapten inhibition assays demonstrated that the lectin specifically interacts with oligomers of N-acetylglucosamine (GlcNAc) and suggested that the binding site can accommodate at least four GlcNAc-residues.

It should be mentioned here that Nachbar et al. (1980) were not the first to report the isolation of the tomato lectin. A few months earlier, Kilpatrick (1980) had already described the purification of the same lectin by affinity adsorption on trypsin-treated erythrocytes and the inhibition of its agglutinating activity by oligomers of GlcNAc. Moreover, Kilpatrick demonstrated that the tomato lectin is a glycoprotein that cross-reacts immunologically with the lectin from *Datura stramonium* (thorn-apple).

After the pioneering work by Kilpatrick (1980) and Nachbar et al. (1980), several other papers were published on the isolation and characterization of the tomato lectin. Kilpatrick et al. (1983) introduced chromatofocusing to isolate the tomato lectin. In addition, novel purification schemes based on affinity chromatography on immobilized ovomucoid (Merkle and Cummings 1987a) and erythroglycan-Sepharose (Zhu and Laine 1989) were developed. However, none of these papers provided better insight into the molecular structure of the tomato lectin. The same applies to a paper describing the purification and characterization of the lectin from "cherry" tomatoes (Saito et al. 1996). In that paper, the authors made a clear distinction between "normal" and cherry tomatoes, which is rather unfortunate because they give the wrong impression that the corresponding lectins are different (which evidently is not correct because all tomato varieties belong to the very same species).

Development of Models of a Multi-domain Structure of the Tomato Lectin

In contrast to the tomato lectin, substantial progress was made in the elucidation of the molecular structure of the potato lectin during the 1990s. Early studies of especially the potato lectin led to the hypothesis that the Solanaceae lectins are chimeric proteins comprising a Cys-rich chitin-binding domain equivalent to wheat germ agglutinin and a domain resembling the cell wall glycoprotein extensin (Allen 1983). Partial amino acid sequencing confirmed the modular structure of the potato lectin and indicated that the Cys-rich lectin domain might be located at the N-terminus of the protein (Kieliszewski et al. 1994). Later, more detailed analyses prompted Allen and collaborators to the conclusion that the potato lectin is a three-domain glycoprotein built up of an N-terminal domain of approximately 100 residues that is rich in proline (Pro) but poor in Hyp, a middle extensin-like domain of at least 104 residues that is extremely rich in glycosylated Hyp-residues and a C-terminal domain comprising two tandemly arrayed hevein domains (Allen et al. 1996).

Taking into account the presumed close relationship between Solanaceae lectins from different species one could reasonably expect to resolve the structure of the tomato lectin by a similar approach. However, when applied to the fruit lectin from cherry tomatoes, a combination of amino acid analysis and protein/peptide sequencing of native and deglycosylated lectin yielded a model quite different from those proposed for the potato lectin (Naito et al. 2001). It was concluded that the genuine tomato lectin is a three-domain protein consisting of an N-terminal extensin-like domain, a middle Cys-rich domain resembling wheat germ agglutinin, and a glutamine (Gln)-rich C-terminal domain equivalent to the large subunit of the tomato seed 2S albumin. According to this model the tomato lectin comprises a C-terminal domain that has not yet been identified in any other Solanaceae lectin. While studying the genuine cherry tomato lectin, Naito et al. (2001) identified an additional chitin-binding protein that is devoid of agglutinating activity but cross-reacts with antibodies against the lectin. Following the same approach as for the lectin, the authors concluded that the newly found "42 kDa" chitin-binding protein corresponds to a homologue of the lectin that lacks the N-terminal extensin-like domain.

Taking into account the very similar biochemical properties (in terms of size, amino acid composition, carbohydrate content, serological relationship), the striking differences in overall domain structure of the proposed models of the tomato and potato lectins are difficult to explain. Since models, which are based on a few partial sequences, always suffer an inherent uncertainty, one has to conclude that the long-lasting enigma of molecular structure of the Solanaceae lectins can be resolved only by determining the complete amino sequence of the proteins or cloning of the corresponding genes.

Molecular Cloning of Tomato Lectin(s)

The identification and sequencing of a complete cDNA encoding a potato tuber lectin eventually made it possible to determine the molecular structure of a typical Solanaceae lectin (Van Damme et al. 2004b). Analysis of the sequence demonstrated that the potato lectin consists of two nearly identical chitin-binding modules (each comprising two tandemly arrayed hevein domains) that are interconnected by a serine/proline (Ser/Pro)-rich linker of approximately 60 amino acid residues, and in addition contains two short unrelated domains located at its N- and C-terminus, respectively. Molecular cloning of the potato lectin confirmed the "canonical" chimeric nature of the Solanaceae lectins but at the same time demonstrated that none of the previously proposed models was correct.

Using the information generated by the cloning of the potato lectin we were able to identify an (incomplete) expressed sequence tag (EST) encoding a putative homologue in tomato leaves. Further analysis revealed that this EST (clone cLET21N20) encodes a putative lectin (called Lycesca) that possesses the same overall modular domain structure as the potato lectin but contains a shorter (only 11 residues) Ser/Pro-rich linker between the two chitin-binding modules (Peumans et al. 2004) (see Fig. 1A). However, since the precursor of Lycesca does not comprise internal sequences that match the determined N-terminal and internal sequences of the genuine lectin (Naito et al. 2001) and, in addition, cannot yield a mature (glycosylated) polypeptide of 71 kDa, we had to conclude that clone cLET21N20 does not correspond to the genuine high molecular mass tomato lectin. Moreover, since the precursor of Lycesca does not contain

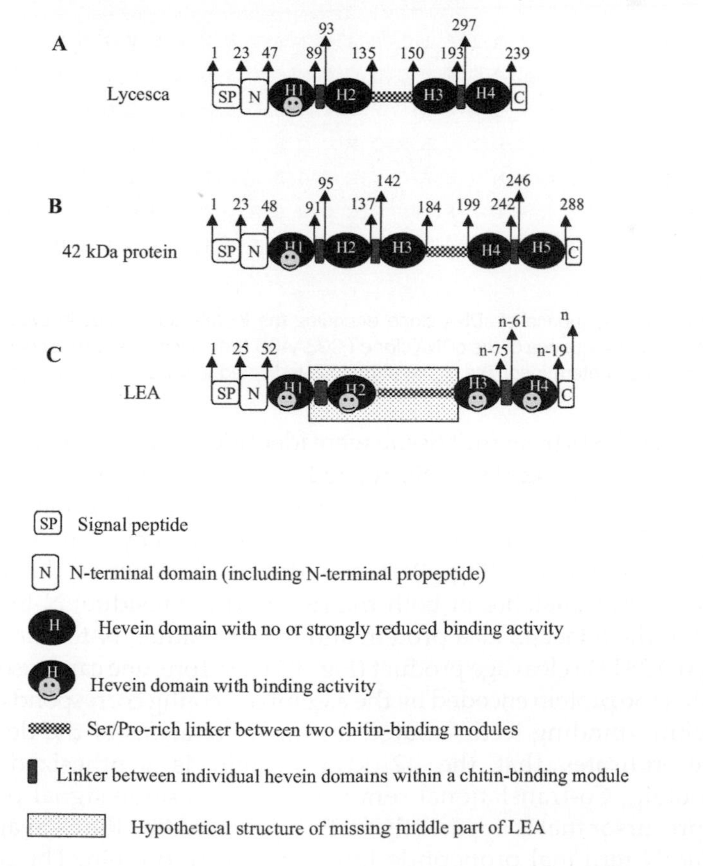

Fig. 1 Schematic representation of the tomato fruit lectin and related chitin-binding proteins. Modular organization of (A) the expressed lectin-related protein Lycesca, (B) the 42 kDa lectin-related protein, and (C) the genuine lectin LEA (*Lycopersicum esculentum* agglutinin).

an internal sequence matching the N-terminus of the 42 kDa lectin-related protein, this protein also can be excluded as the product of clone cLET21N20.

Fortunately, the recent release of a large number of novel ESTs (especially from tomato fruit development and ripening transcriptome analysis programs) as well as a set of genomic clones allowed researchers to readdress the issue of the molecular structure of the tomato lectins. A comprehensive analysis of the accessible sequence data revealed that at least three different lectins are expressed in tomato fruits.

First, 16 new EST clones, all from fruit tissues, could be identified that are identical to cLET21N20 (Fig. 2; Table 1). This implies that Lycesca is also expressed in fruits.

```
M K E M R I S I L A L L A L F L L E A V L A N E V F D V P M N E T I G V
E S I N A S V G G Y P R C G A Q G D G G N C P S G M C C S V W G W C G K
T Y G Y C A P Q N C Q K Q C P A P Y P E G R C G W Q A D G K S C P N G K
C C S Y G G W C G T T S D Y C A R Q N C Q K Q C I L P S P P P P P P P P
P G F P R P E C G L Q K N G E R C T K P G E C C S I W G L C G A T Y K Y
C D P Q H C Q K Q C S A P F P P G R C G W Q A D G R P C P T G Q C C S F
S G W C G T T S A H C T Y P Q C V S Q C N D P R F P S S L N N R I Q S F
M L
```

Fig. 2 Amino acid sequence of cDNA clone encoding the lectin-related protein Lycesca. The deduced amino acid sequence of the cDNA clone FC03AA06 (from *Lycopersicum esculentum* cv. Micro-Tom fruit) was determined. The signal peptide is indicated in italic.

Second, two ESTs from fruit tissue were identified that apparently match the N-terminus of the 42 kDa lectin-related protein described by Naito et al. (2001) (Table 1). Both ESTs are short and comprise only the N-terminus of the 42 kDa protein. However, a contig could be assembled by combining EST and genomic sequences (Fig. 3). The deduced sequence of this contig comprises perfect matches of both the reported (30 residue) N-terminal sequence of the intact 42 kDa protein and the (9 residue) N-terminus of a (proteolytic) 28 kDa cleavage product (Fig. 3). Therefore, one can reasonably assume that the protein encoded by the assembled contig corresponds to the 42 kDa chitin-binding lectin-related lectin. An analysis of the deduced sequence indicates that the 42 kDa protein is synthesized as a preproprotein. Co-translational removal of a 22 residue signal peptide yields a precursor that is apparently further processed by the cleavage of a 22 residue N-terminal propeptide into the mature protein. The overall domain structure of the mature 42 kDa protein resembles that of Lycesca in that it comprises two chitin-binding modules interconnected by a 13

```
M K E T A I S V L G L L S L F L L E V V S A N E I I N G T F V L E G I N
K N A S G G V F A N G D E C G M Q A N H R S K C P S G M C C S I W G W C
G T T S E Y C G S G F C Q N Q C T G P S P H G S C G M Q G G G T K C P S
G Q C C S L L G W C G T G S D F C K P E I C Q S Q C S G P P F P N G R C
G W Q A D G R L C P R G Q C C S V D G W C G T T T D Y C A S G L C Q S Q
C P F T P P P S P P P P S P P P S Q Y Q C G M Q N G G T K C N R T G E C
C G I S G M C G N T Y E Y C F P G Y C Q M Q C P G P Y P E G R C G W Q A
D G K S C P T G Q C C G N A G W C G I G P G F C D P I F C Q S Q C S G A
P I S T A K R D G G I R S
```

Alignment of deduced and experimental sequences

```
Native_42_kDa_protein      ANGDECGMQANHRSKCPSGMCCSIWGWCG
                           ***** *************** * * **
Deduced_sequence           ANGDEMGMQANHRSKCPSGMCCXIXGXCG
                                          ******* *
28 kDa fragment            ---------------CPSGMCCXI-----
```

Fig. 3 Deduced amino acid sequence of the 42 kDa chitin-binding fruit protein. The sequence encoding the 42 kDa chitin-binding fruit protein described by Naito et al. (2001) is translated from a contig assembled from EST LeTFR44 and genomic clones LE_HBa0222N02 and LE_HBa0163K05. The signal peptide is indicated in italic and the N-terminal propeptide is boxed black. Sequences corresponding to the reported N-terminal sequences of the native 42 kDa protein and a 28 kDa proteolytic fragment are boxed grey. Stars represent conserved positions in the amino acid sequence.

residue Ser/Pro-rich linker (Fig. 1B). However, there is one major difference because the N-terminal chitin-binding module consists of three tandemly arrayed hevein domains (instead of two). Taking into account that the native 42 kDa protein contains 10% carbohydrate (Naito et al. 2001), glycosylation of the 257 residue polypeptide chain (calculated Mr: 26631.8 Da) should in principle yield a mature protein of approximately 30 kDa. This calculated value is considerably lower than the molecular mass estimated by sodium dodecyl sulphate polyacrylamide gel electrophoresis (SDS-PAGE). Most probably this discrepancy has to be ascribed to the anomalous migration of the protein in the gel. Because of the presence of the densely glycosylated Ser/Pro-rich linker sequence, Solanaceae lectins migrate considerably more slowly than "normal" proteins upon SDS-PAGE (Van Damme et al., 2004b). It is worth mentioning in this context that the genuine tomato lectin, which has a molecular mass of approximately 70 kDa (as estimated by ultracentrifugation), also migrates with a much higher apparent molecular mass (> 100 kDa) in an SDS-polyacrylamide gel.

A major conclusion to be drawn from the sequence data is that the model of the 42 kDa protein proposed by Naito et al. (2001) can no longer

be maintained. It can be ruled out, indeed, that the protein comprises at its C-terminus a Gln-rich domain resembling the large subunit of the tomato seed 2S albumin.

Third, four other EST sequences (all from fruits) were deposited that encode a protein comprising sequences that almost perfectly match the N-terminus and/or two proteolytic fragments of the genuine tomato fruit lectin (LEA) (Table 1). At present, no complete contig could be assembled (Fig. 4). However, the available sequence data provide some interesting

Table 1 Summary of expressed sequence tags encoding the tomato lectin and related proteins.

cDNA library	*ESTs[1] corresponding to:*		
	Lycesca	*42 kDa protein*	*LEA*
L. esculentum cv. Micro-Tom fruit	FC03AA06 FC14DD09 LEFL2007I12 LEFL2037G23 LEFL2022B01 LEFL2036N09 LEFL2046015 LEFL2027I12		FB03DB07 FA03CH05, FB12BG06
L. esculentum maturing fruit (cv. Micro-Tom)	FB05BE09 FA18DC05 FA18CG05 FB19BC12 FA24CF07 FB17BF04		
L. esculentum var. cerasiforme fruit 15 d post-anthesis	LE15BG11		
L. esculentum var. cerasiforme fruit 8 d post-anthesis			LE08BA02
Tomato fruit related cDNA library-yeast signal sequence trap 1 (cultivar Ailsa Craig)	LeTFR9	LeTFR44 LeTFR49	
Tomato mixed elicitor, BTI (*S. lycopersicum* cultivar Rio Grande PtoR)	Clone cLET21N20		

The NCBI database was searched for expressed sequence tags corresponding to the genuine tomato lectin (LEA, *Lycopersicum esculentum* agglutinin), the 42 kDa lectin-related protein and the expressed lectin-related protein called Lycesca.

[1]Numbers refer to NCBI database.

information about LEA. cDNA clone FA03CH05 encodes a 91 residue polypeptide in which sequences corresponding to the N-terminus (residues 36-45) and two fragments of the lectin (residues 67-80 and 81-91, respectively) can be recognized. Since the sequence corresponding to the N-terminus of the native lectin starts with Met^{36}, one can reasonably assume that next to the cotranslational removal of the 24 residue signal peptide an 11 residue N-terminal propeptide is cleaved. A comparison of the deduced and experimentally determined N-terminal sequence indicates that the Pro residues at positions 41, 42 and 43 are hydroxylated (these residues are indicated by "O") and possibly substituted with O-linked galactose or (oligo)arabinoside. It should also be mentioned here that as early as 1980 Nachbar et al. reported that N-terminal analysis of their lectin preparation yielded only methionine. Moreover, the lectin from

N-terminus

M K E T L I I S V L C V V T L Q Y L F L V S A D R L S L P H N E T F G M
P L S S P P P H E P S P P P P Y P R C G M G G G D G K C K S N E C C S I
W S W C G T T E S Y C A P Q N C Q S Q

Middle part is still missing: Xn

C-terminus

G G K C K S N E C C S I W S W C G T T E S Y C A P Q N C Q S Q C P H T P
S P S P P P T P P P P Y P R C G M G G G G G K C K S N E C C S I W S W C
G T T E S Y C A P Q N C Q S Q C K K N I I S S V M N P M N V T Y G I E S
F

Alignment of deduced and experimental N-terminal sequences

N-terminus (Naito et al. 2001)	^{1}MPLSSOOOO
Deduced sequence	^{36}MPLSSPPPHE^{44}
N-terminus (Wang and Ng 2007)	^{1}MPLSHEHPHE^{10}
	****

Alignment of deduced and experimental internal sequences (Naito et al. 2001)

Peptide B4	NECCSIWSWCGTTE
	**************
Deduced sequence:	NECCSIWSWCGTTE

Peptide A2	SFCAPQNCQSQCPWT
	*:*********** *
Deduced sequence:	SYCAPQNCQSQCPHT

Fig. 4 Partial sequence of the genuine tomato fruit lectin. The sequence of the genuine tomato fruit lectin (LEA, *Lycopersicum esculentum* agglutinin) is translated from cDNA clones FA03CH05 (N-terminus) and FB03DB07 (C-terminus) from *L. esculentum* maturing fruit. The signal peptide is indicated in italic and the N-terminal propeptide is boxed black. Sequences corresponding to the reported N-terminal and internal sequences of LEA are boxed grey. Stars represent conserved positions in the amino acid sequence.

"samta tomatoes" (which is just another variety of the species *L. esculentum*) also yielded a sequence that closely resembles the deduced sequence except for residues 5-7 (Wang and Ng 2006). Most probably the latter discrepancy has to be ascribed to problems inherent in the sequencing of Hyp containing proteins. Taking into account that the deduced sequence almost perfectly matches three different experimentally determined sequences, one can reasonably assume that cDNA clone FA03CH05 encodes the N-terminal part of LEA.

Besides a cDNA clone encoding the N-terminus of LEA, another incomplete cDNA clone (FB03DB07) that apparently corresponds to the 109 residue C-terminal part of the genuine lectin could be retrieved. FB03DB07s comprises two sequences that are identical to peptide B4 and one sequence that differs at only two positions from peptide A2 (Fig. 4). It seems likely, therefore, that clone FB03DB07 corresponds to the C-terminal part of LEA.

Combining cDNA clones FA03CH05 and FB03DB07 yields an incomplete sequence covering 56 and 109 amino acid residues of the N- and C-terminus of LEA, respectively (Fig. 4). The length and sequence of the missing middle part are still unknown. However, taking into account that LEA closely resembles the potato lectin in its amino acid composition and carbohydrate content but is definitely larger (approximately 70 kDa versus 55 kDa for potato lectin), one can reasonably assume that the LEA polypeptide is at least 300 residues long. This implies that roughly half of the sequence remains to be determined and that accordingly no definitive model can be given of the structure of the tomato lectin. However, it seems likely that LEA has the same domain structure as the potato lectin except that the Ser/Pro-rich domain is 15-25 residues longer (Fig. 1C).

Though no definitive model can be elaborated yet it is evident that the "domain construction" of the tomato lectin described by Naito et al. (2001) is not correct. First, there is no hydroproline-rich glycoprotein domain at the N-terminus of the lectin but just a short (14 amino acid residue) Ser/Pro-rich peptide. Second, the lectin definitely does not comprise a Gln-rich C-terminal domain that resembles the large subunit of the 2S albumin. As we pointed out earlier (Peumans et al. 2004), the presence of a polypeptide resembling the large subunit of the 2S albumin has to be ascribed most probably to contamination of the lectin preparation isolated by Naito et al. (2001) by non-covalently bound fruit 2S albumin. The latter protein is strongly amphiphilic and hence tends to interact through non-specific interactions with other proteins.

Active and Inactive Hevein Domains in Tomato Fruit Lectins: Predictions Based on Analysis of Deduced Sequences

Lycopersicum esculentum agglutinin has for a long time been known as a potent hemagglutinin and accordingly must be a multivalent carbohydrate-binding protein. The 42 kDa lectin-related protein also binds chitin but possesses no agglutination activity, indicating that it might be monovalent. Since Lycesca has not been isolated yet its possible sugar-binding and agglutinating activity are not supported by experimental evidence. Taking into account that the three tomato lectins comprise at least four hevein domains, they are in principle all multivalent and hence should behave as agglutinins. Since this is clearly not the case for the 42 kDa protein and perhaps also for Lycesca, the question arises whether all hevein domains are functional. To address this issue it was checked whether the residues that are involved in the binding of GlcNAc residues (of chitooligosaccharides) to the saccharide-binding site of hevein (Ser19, Trp21, Trp23 and Tyr30) (Anderson et al. 1993) are conserved (or alternatively replaced by functionally homologous residues) in the individual hevein domains of the respective tomato lectins. As shown in Fig. 5, both in the 42 kDa protein and Lycesca only the first hevein domain possesses a fully active binding site. Owing to substitution(s) of one or more essential residues, all other hevein domains have a binding site with a strongly reduced or completely abolished activity. This implies that both proteins are monovalent and hence are devoid of agglutinating activity, which is in perfect agreement with the data reported by Naito et al. (2001). In contrast, all three identified hevein domains of LEA possess a fully active site. Evidently, this obvious multivalency explains why the genuine tomato lectin behaves as a potent hemagglutinin.

In summary, it can be concluded that tomato fruits contain three different proteins with multiple hevein domains. Complete sequences are available for Lycesca and a 42 kDa lectin-related protein. It appears that both proteins are monovalent lectins and hence are devoid of agglutinating activity. No complete sequences were reported of the genuine lectin. However, there is sufficient sequence information to confirm the multivalency of LEA and build a fairly accurate model. All evidence suggests that LEA closely resembles the previously described potato lectin, the main difference being that the Ser/Pro-rich linker between the N- and C-terminal chitin-binding modules is substantially longer in LEA.

A. Lycesca

```
HEV    EQCGRQAGGKLCP-NNLCCSQWGWCGSTDEYCSPDHNCQSNCKD    Active
H1     PRCGAQGDGGNCP-SGMCCSVWGWCGKTYGYCAPQN-CQKQCPA    Active
H2     GRCGWQADGKSCP-NGKCCSYCGWCGTTSDYCARQN-CQKQCIL    Inactive
H3     PECGLQKNGERCTKPGECCSIWGLCGATYKYCDPQH-CQKQCSA    Inactive
H4     -RCGWQADGRPCP-TGQCCSFSGWCGTTSAHCTYPQ-CVSQCND    Inactive
        .** * .* *. . *** * ** * :* : * .:*
```

Conclusion: LycescA is a monovalent lectin devoid of agglutinating activity

B. 42 kDa protein

```
HEV    EQCGRQAGG-KLCPNNLCCSQWGWCGSTDEYCSPDHNCQSNCKD    Active
H1     DECGMQANHRSKCPSGMCCSIWGWCGTTSEYCGSGF-CQNQCTG    Active
H2     GSCGMQGGG-TKCPSGQCCSLLGWCGTGSDFCKPEI-CQSQCSG    Inactive
H3     GRCGWQADG-RLCPRGQCCSVDGWCGTTTDYCASGL-CQSQCPF    Inactive
H4     YQCGMQNGGTKCNRTGECCGISGMCGNTYEYCFPGY-CQMQCPG    Inactive
H5     GRCGWQADG-KSCPTGQCCGNAGWCGIGPGFCDPIF-CQSQCSG    Inactive
         ** * . . **. * ** :* . ** :*
```

Conclusion: the 42 kDa protein is a monovalent lectin devoid of agglutinating activity

a is a monovalent lectin devoid of agglutinating activity

C. LEA

```
HEV    EQCGRQAGGKLCPNNLCCSQWGWCGSTDEYCSPDHNCQSNCKD    Active
H1     PRCGMGGGDGKCKSNECCSIWSWCGTTESYCAPQN-CQSQXXX    Active
H2
H3     XXXXXXXXGGKCKSNECCSIWSWCGTTESYCAPQN-CQSQCPH    Active
H4     PRCGMGGGGGKCKSNECCSIWSWCGTTESYCAPQN-CQSQCKK    Active
          . * .* *** *.***:*:.**:*:: ***:
```

Conclusion: LEA is a multivalent lectin with agglutinating activity

Fig. 5 Alignment of the deduced amino acid sequences of the hevein domains of tomato lectin and related proteins compared to the sequence of hevein. The deduced sequences of the individual hevein domains of the 42 kDa lectin-related protein, the expressed lectin-related protein Lycesca and the genuine tomato lectin (LEA, *Lycopersicum esculentum* agglutinin) are aligned and compared with the sequence of mature hevein (NCBI: gi|42543425|). Residues identical or functionally identical to the amino acids forming the binding site of hevein are boxed grey. Residues in white and boxed black indicate functionally not conserved residues. Key to symbols: *, identity; :, conservative replacement; ., non-conservative replacement.

CARBOHYDRATE-BINDING SPECIFICITY OF THE TOMATO LECTIN

Though it was demonstrated that the 42 kDa protein is retained on chitin and can be eluted with GlcNAc, no further details are available for what concerns the sugar-binding properties of this lectin-related protein. In contrast, the specificity of the genuine lectin LEA has been studied in great detail.

Initially LEA was described as a lectin that specifically binds GlcNAc-oligomers. According to hemagglutination inhibition studies, the binding site was most complementary to three or possibly four β-(1-4)-linked GlcNAc units (Nachbar et al. 1980). Therefore, LEA was—like all other

Solanaceae lectins—for a long time considered a "chitin-binding" lectin. However, in a later study Merkle and Cummings (1987b) demonstrated that immobilized LEA interacts with a high affinity with glycopeptides containing long poly-N-acetyllactosamine chains (which are built up of the repeating disaccharide [3Galβ1-4GlcNAcβ1]). This observation was of paramount importance because it demonstrated that LEA does not exclusively bind chitin but also interacts with complex type N-linked glycans substituted with long poly-N-acetyllactosamine chains (and accordingly made clear that LEA reacts not only with a typical fungal cell wall polysaccharide but also with typical animal glycoproteins). A more recent (qualitative) study of the carbohydrate-binding specificity using lectin blot analysis not only confirmed that LEA reacted with glycoproteins containing tri- and tetra-antennary complex type N-glycans but also revealed that the lectin bound to animal glycoproteins containing high mannose type N-glycans as well as to a horseradish peroxidase (Oguri 2005). Moreover, using a combination of LEA blot analysis and exoglycosidase treatment of the glycoproteins it could be demonstrated that LEA interacted with the N-acetyllactosaminyl side chains of the complex type N-glycans and the proximal chitobiose core of the high mannose type N-glycans, respectively. Surprisingly, however, the reactive glycoproteins were not capable of inhibiting the hemagglutinating activity of LEA. At present, there is no simple explanation for the apparent discrepancy between the blotting and hemagglutination inhibition experiments. One possible explanation is that the complex type and high mannose type N-glycans are recognized by different hevein domains (or even a different set of hevein domains). It should be mentioned in this context that the sites that accommodate the N-glycans might be more extended than the "canonical" sugar-binding site found in hevein.

Further analyses using the high performance glycan array systems developed by the Consortium for Functional Glycomics made it possible to determine the specificity of LEA in more detail and—what is more important—yielded fairly accurate indications for what concerns the relative affinity of the lectin for different carbohydrates. For example, the relative fluorescence units observed for (GlcNAcβ1-4)6, N-acetyllactosamine and the high mannose N-glycan [Manα1-3(Manα1-6)Manβ1-4GlcNAcβ1-4GlcNAcβ-] were 49045, 47289 and 42301, respectively. Though these data have to be interpreted with care they indicate that LEA interacts with a comparable affinity with all three types of glycans. Full details of the specificity analyses can be found in the publicly accessible data of the Consortium for Functional Glycomics (http://www.functionalglycomics.org/glycomics/publicdata/primaryscreen.jsp) (select plant lectins; select LEA).

In a recent report a totally different specificity was attributed to the fruit lectin from "samta" tomatoes (Wang and Ng 2006). According to the authors the lectin exhibits specificity towards rhamnose and O-nitrophenyl-β-D-galactopyranoside. However, the results of these studies are useless because none of the relevant sugars/glycans (GlcNAc-oligomers, complex and high mannose type N-glycans) were included in the inhibition assays.

In summary, it can be concluded that LEA is a polyspecific lectin that reacts with several structurally different glycans such as GlcNAc-oligomers and complex type and high mannose type N-glycans. Therefore, LEA can no longer be considered a lectin with an exclusive specificity towards the fungal cell wall polysaccharide chitin but should be regarded as a protein that is also capable of interacting with animal and plant glycoproteins. These novel insights in the specificity are highly relevant because they put the possible role of LEA in a totally different perspective (see below).

BIOLOGICAL ACTIVITIES OF THE TOMATO LECTIN

Effects on Isolated Cells

Lycopersicum esculentum agglutinin strongly agglutinates human and animal erythrocytes. Minimal concentrations required for agglutination are in the μg/ml concentration range depending on the species and preparation (e.g. protease or sialidase treatment) of the red blood cells. With regard to human erythrocytes LEA exhibits no preference for any blood group within the ABO system. Though hemagglutination is a very sensitive and versatile technique for the detection of lectin activity the observed effect (i.e., clumping of red blood cells) on itself is of little biological relevance because no further attention is given to possible reactions within the cells.

At the time of the discovery and purification of LEA, plant lectins were primarily known as "mitogens" (agents capable of inducing or stimulating division and proliferation of lymphocytes). Especially the lectins from *Canavalia ensiformis* (concanavalin A or ConA), *Phaseolus vulgaris* (phytohemagglutinin or PHA) and *Phytolacca americana* (pokeweed mitogen) were very popular because they were the only readily available mitogens for biological and biomedical research. Owing to the "canonical" association between lectins and mitogens, virtually all newly discovered plant lectins in the 1970s and 1980s were extensively tested for possible mitogenic activity. Therefore, it is not surprising that even the first full paper on the tomato lectin described in some detail the effects of LEA on lymphocytes. It was non-mitogenic for mouse and chicken lymphocytes. On the contrary, the tomato lectin exhibited a dose-dependent inhibition of the mitogenic effects brought about by ConA, PHA and pokeweed mitogen in chicken lymphocytes (Nachbar et al. 1980). A similar antagonistic effect of LEA was observed on PHA and pokeweed mitogen-induced transformation of human peripheral blood lymphocytes (Kilpatrick 1983, Kilpatrick et al. 1986). Moreover, in this test system the inhibition was more potent with antigen (e.g., tuberculin) mediated transformation.

Effects on Whole Organisms

Hitherto there have been no reports of acute toxic effects of LEA on whole organisms. Feeding trials with artificial diets indicated that the tomato lectin is non-toxic to the insects *Ostrinia nubilalis* and *Diabrotica undecimpunctata* (Czapla and Lang 1990). There are also no indications for acute toxicity to mammals or other vertebrates.

Though it is unlikely that LEA exerts direct toxic effect(s), the lectin as well as Lycesca and the 42 kDa protein are potential allergens that indirectly affect human health. The potential allergenicity of the tomato lectins is due to the presence of hevein domains. Hevein is a major IgE-binding allergen for patients allergic to natural rubber latex (Wagner and Breiteneder 2002). It is estimated that 30-50% of the individuals who are allergic to natural rubber latex show an associated hypersensitivity to some plant-derived foods, and especially to freshly consumed fruits. This association of latex allergy and allergy to plant-derived foods has been called "latex-fruit syndrome". Besides avocado, banana, chestnut, kiwi and peach, tomato, potato and bell pepper are also associated with this syndrome. It is believed that allergen cross-reactivity is mediated by IgE antibodies that recognize structurally similar epitopes on evolutionarily conserved proteins from different plant species.

Severe allergic reactions to tomatoes are rather rare. Though the responsible allergens have not been identified yet with certainty, two independent reports associate allergic reactions to tomato with a 43-44 kDa protein (Reche et al. 2001, Zacharisen et al. 2002). Taking into consideration that the 42 kDa lectin-related protein is one of the most abundant proteins in at least cherry tomatoes, it might well correspond to the 43-44 kDa allergen.

APPLICATIONS OF TOMATO LECTIN IN BIOLOGICAL AND BIOMEDICAL RESEARCH

Many plant lectins are widely used in fundamental and applied biological and biomedical research. Some applications directly exploit the sugar-binding activity/specificity of lectins. For example, lectins are used as molecular probes for the detection and/or localization of specific glycans or as highly specific affinity sorbents for the isolation of specific glycans or glycoproteins. Other applications are based on effects provoked in cells (e.g., mitogenic lectins) or organisms (e.g., insecticidal lectins). In addition, there are some specialized applications whereby lectins are used as specific carriers for other bioactive agents (e.g., immunotoxins based on ricin, lectin-mediated drug delivery).

The tomato lectin was (virtually) not used as a tool until Merkle and Cummings (1987b) observed that LEA exhibits a marked specificity towards poly-N-acetyllactosamine chains. After this crucial observation LEA was rapidly introduced as a favourite tool for the detection/localization of poly-N-acetyllactosamine structures on proteins and cells, and for the purification of poly-N-acetyllactosamine-containing glycoproteins by affinity chromatography on the immobilized lectin. The first documented application of LEA was the purification of murine cell surface T200 glycoprotein (carrying poly-N-acetyllactosamine) by affinity chromatography on tomato lectin-Sepharose 4B (Gilbert et al. 1988). Soon thereafter, LEA was successfully used as a molecular probe for the localization of poly-N-acetyllactosamine-containing membrane proteins in gastric parietal cells and for the subsequent identification of a tomato lectin binding 60-90 kDa membrane glycoprotein of tubulovesicles (Callaghan et al. 1990). In the meantime, only a few more glycoproteins were isolated (e.g., pig gastric H+/K(+)-ATPase complex, Callaghan et al. 1992; *Trypanosoma brucei* flagellar pocket proteins, Nolan et al. 1999). Besides proteins, there is also a report of the isolation of bovine choriocapillary endothelial cells using LEA-coated Dynabeads (Hoffman et al. 1998). Most probably, the limited availability hampers a large-scale use of immobilized LEA for the purification of proteins and cells. Since applications of LEA as a molecular probe require only very small quantities there were far fewer restrictions on use of tomato lectin in histochemistry and lectin blotting experiments. As a result, applications of LEA as a histochemical tool are fairly well documented. Early work described the use of LEA for differential and specific labeling of epithelial and vascular endothelial cells of the rat lung (Bankston et al. 1991) and for staining of poly-N-acetyllactosamine-containing lysosomal membrane glycoproteins in human leukaemia cells (Wang et al. 1991). After studies with rat brain tissue revealed that LEA specifically stains macrophages and microglia (Acarin et al. 1994), the tomato lectin became an important tool in the histochemistry of brain and nerve tissue of various animal species including fish (Velasco et al. 1995), mouse (Vela Hernandez et al. 1997), human (Andjelkovic et al. 1998), rabbit (Bass et al. 1998), chicken (Miskevich, 1999) and pig (Salazar et al. 2004). Recent studies with rat, mouse and guinea pig at different developmental stages also revealed that the tomato lectin can be used as an effective and versatile endothelial marker of normal and tumoral blood vessels in the central nervous system (Mazzetti et al. 2004). *Lycopersicum esculentum* agglutinin has also been used to visualize the pattern of tomato lectin binding sites in the basement membranes of the developing chick embryo (Ojeda and Icardo 2006). Thereby, a marked heterogeneity was observed of the chick embryo basement membranes during development. Moreover, LEA appeared to constitute an excellent marker for the primordial germ cells.

Besides applications as a tool, LEA was also used as a model lectin for the development of oral drug delivery systems. The underlying idea is to develop a bioadhesive drug delivery system on the basis of molecules that selectively bind to the small intestinal epithelium by specific, receptor-mediated mechanisms (Naisbett and Woodley 1990, Lehr et al. 1992). Because LEA resists digestion in the mammalian alimentary canal and binds to intestinal villi without deleterious effects (Kilpatrick et al. 1985), it is a suitable carrier molecule. *Ex vivo* studies with different structures of rat intestinal mucosa demonstrated that coating with the tomato lectin enhanced the interaction of latex particles with the mucus gel layer (Irache et al. 1996). Experiments with rats confirmed that coating with tomato lectin enhanced subsequent intestinal transcytosis of orally administered colloidal particulates (*in casu* polystyrene nanoparticles) to which it is bound (Hussain et al. 1997). Further studies confirmed that LEA might be a suitable candidate for the development of an oral drug delivery system. However, even though there is a strong belief that the concept of bioadhesion via lectins can be applied to the gastrointestinal tract as well as to other biological barriers (like the nasal mucosa, the lung, the buccal cavity, the eye and the blood-brain barrier), it is evident that much work remains to be done and that before advanced drug delivery systems using lectins can be realized, rigorous evaluation of their toxicity and immunogenicity is required (Woodley 2000, Bies et al. 2004).

THE BIOLOGY OF TOMATO LECTIN

Tomato Lectin is Preferentially but Not Exclusively Expressed in Fruits

Until 2002, all tomato lectin preparations were isolated from cleared juice. Accordingly, LEA is usually considered a fruit-specific protein. However, recent experiments with the cultivar 'Moneymaker' demonstrated that leaves contain a lectin that in its agglutination properties and molecular structure is indistinguishable from the fruit lectin. Moreover, the overall yield of the fruit lectin (2 mg/kg) was approximately 4-fold that of the leaf lectin (0.5 mg/kg) (Peumans et al. 2004). It should be mentioned here that despite the relatively high lectin content, extracts from leaves exhibited no clearly visible agglutination activity. However, the failure to agglutinate the red blood cells was not due to a lack of lectin activity but to the rapid lysis of the erythrocytes (by saponins that are present in leaves but absent from fruits).

Within the fruits the lectin is not uniformly distributed. Semi-quantitative agglutination assays with extracts from different tissues revealed that 64% of the lectin is located in the locular fluid and 20% in the locular gel (Merkle and Cummings 1987a). The remaining 16% was found in the placenta. No lectin activity was detected in seeds, outer and inner wall pericarp, radial wall and skin.

Developmental Control of LEA Accumulation in Fruits

For historical and practical reasons, LEA was always isolated from ripe tomatoes. Though it is evident that LEA is synthesized during fruit development, the timing and kinetics of its accumulation were only recently studied. Using semi-quantitative agglutination assays, Naito et al. (2001) made a detailed analysis of the accumulation of lectin activity in developing cherry tomato fruits. The total lectin content increased logarithmically until 10 d after pollination, after which it remained constant till complete maturity. Since only LEA is a genuine agglutinin, the observed changes in agglutination activity reflect the accumulation of LEA. Parallel determinations of the fresh weight indicated that the fruit size also increased logarithmically to reach a plateau 15 d post-anthesis. This implies that the accumulation of LEA is completed approximately 5 d before the fruits reach their final size.

It is not clear whether the results obtained with the cherry tomatoes can be extrapolated to other varieties. However, one can reasonably expect that the overall shape of the lectin accumulation curve will be similar but that there might be substantial differences in the exact timing (depending on the total length of the fruit development period).

Lectin Content of Tomato Fruits is Genetically Determined

An overview of the reported data reveals that the overall yield of purified LEA varies from 2.5 to 25 mg per kg fresh fruits (Nachbar et al. 1980,

Merkle and Cummings 1987a, Naito et al. 2001, Wang and Ng, 2006). Though these values are not necessarily a correct measure of the lectin concentration of the starting material, they indicate that there might be major differences in LEA content between individual tomato varieties. Moreover, there is a reasonable chance that lectinologists searched tomatoes with a high lectin content, which implies that the lectin content of many varieties is even (much) lower than the reported values. To corroborate this issue we compared the agglutinating activity of the locular fluid from a series of tomato varieties. As shown in Table 2, there are marked varietal differences in lectin content. The locular fluid of some varieties contains up to 1 mg lectin/ml, whereas in others no agglutination activity could be detected. This apparent lack of lectin activity does not necessarily imply that these varieties are lectin-deficient. It can only be concluded that their lectin content falls below the level of detection (approximately 5 μg/ml) by the agglutination assay.

The data summarized in Table 2 clearly demonstrate that there are striking differences in lectin content between tomato varieties. Since all varieties were grown under identical conditions and assayed in the same

Table 2 Comparison of the lectin content in the locular fluid of different tomato varieties checked for agglutination activity.

Variety	*Titre*[1]	*Variety*	*Titre*	*Variety*	*Titre*
Ambiance	160	Drw 49-90	320	Rz74-52	< 5
Blitz	5	Drw 51-80	160	S&g 61-26	80
Bst 44-28	40	Drw 51-53	160	S&g 61-97	< 5
Bst 29-14	480	Durinta	160	S&g 62-63	< 5
Bst 26-37	160	E 20-29-972	60	Tradiro	40
Bst 26-38	160	Fausto	80	Ws 86-29-74	80
Cabrion	480	F 17-204	240	Ws 84-44-13	640
Drw 50-16	160	F 61-97	5	Ws 87-94-58	1280
Drw 51-33	160	Gc 62-63	10	Ws 87-92-18	160
Drw 51-49	480	Mississippi	< 5[2]	Ws 85-22-26	10
Drw 50-07	10	Rapsodie	15		
Drw 46-28	40	Rz 74-21	160	LEA (1 mg/ml)	960

[1]The agglutination titre is defined as the highest dilution of the locular fluid that still yields a visible agglutination of a 1% suspension of trypsin-treated rabbit red blood cells. Parallel assays with purified LEA indicated that a titre of 960 corresponds to a lectin content of 1 mg/ml.

[2]A titre < 5 means that no lectin activity could be detected. This does not imply that there is no lectin but just that the lectin content falls below the level of detection (approximately 5 μg/ml).

experiment, the observed differences strongly argue that the lectin content of tomato fruits is genetically determined.

PHYSIOLOGICAL ROLE OF TOMATO LECTIN

At present, the physiological role of the tomato lectin(s) and related Solanaceae lectins is still far from understood. In contrast to most other highly expressed plant lectins it is difficult to attribute a storage and/or defence role (Peumans and Van Damme 1995) to the group of Solanaceae lectins. For example, the expression levels and spatio-temporal regulation of the tomato and potato lectins can hardly be reconciled with a storage role. Moreover, there are no indications that these lectins contribute to the plant's defence against herbivores or phytophagous invertebrates. Furthermore, it is difficult to attribute a specific role within the plant cell (analogous to that of the inducible cytoplasmic/nuclear plant lectins) (Van Damme et al. 2004a) to the Solanaceae lectins.

Most probably the Solanaceae lectins do not fulfil an essential role in or outside the plant. It should be emphasized, indeed, that the Solanaceae lectins are a small, highly specialized subfamily of lectins that are confined to a narrow taxonomic group covering just a few genera of a single family. If these lectins were really indispensable they would certainly be more widespread.

As was pointed out in a previous paper (Van Damme et al. 2004b), the final elucidation of the domain structure of the potato lectin urged revision of the physiological role of this chimeric lectin. It was suggested that the presence of a long rigid linker between the two chitin-binding modules is a crucial feature because it allows cross-linking of two distant carbohydrate receptors. Furthermore, it was speculated that the potato lectin might be a structural analogue of some animal collectins (collagen-like lectins), which play an important role in the recognition and binding of microorganisms (Lu et al. 2002). Though still speculative, the presumed (distant) structural and functional similarity between the potato lectin and the collectins might explain why the potato lectin specifically interacts with some strains of *Pseudomonas solanacearum* (Sequeira and Graham 1977) and the *D. stramonium* lectin interferes with bacterial motility (Broekaert and Peumans 1986). Possibly the Solanaceae lectins are a group of specialized defence proteins that were developed in a common ancestor of the genera *Solanum* and *Datura* to specifically act against bacteria and/or other microorganisms.

TOMATO LECTIN AND FOOD SAFETY

Lectins, especially those found in food and feed plants, are an important issue for food safety because many plant lectins are notorious toxins or antinutrients (Peumans and Van Damme 1996). Most knowledge about toxic or harmful lectins comes from feeding trials with experimental animals. However, a few rare cases of accidental food poisoning (e.g., insufficiently cooked beans) illustrate how devastating the effects of a food-borne plant lectin can be. In general, potentially dangerous or harmful lectins are inactivated during food processing. For example, proper heating completely abolishes the toxic lectins in beans and soybeans. Evidently, this does not apply to plants or plant products that are eaten raw (like some vegetables and fruits). Though there are no vegetables or fruits that contain toxic lectins, (supposedly) harmless lectins are quite common in fruits (e.g., banana, tomato, jack fruit) and vegetables (e.g., onions, leek, cucumber, melon) (Peumans and Van Damme 1996). None of these lectins exerts a known acute or long-term toxic effect. However, taking into account that most of these dietary lectins survive in the gut and recognize and bind animal and human glycans, they almost certainly interact with glycoconjugates exposed on the surface of the gastrointestinal tract. Such an interaction does not necessarily provoke adverse or deleterious effect(s) but it cannot be ruled out that especially upon repeated or prolonged exposure some of these lectins eventually cause more or less severe damage. The latter considerations also apply to the tomato lectin. *Lycopersicum esculentum* agglutinin survives passage through the gut (Kilpatrick et al. 1985) and is capable of interacting with complex and high mannose type N-glycans (Oguri 2005) exposed along the gastrointestinal tract. At present, there are no reports of noxious effect of LEA on animals or humans. However, it has been shown that the tomato lectin suppresses mitogen-induced proliferation of animal and human lymphocytes (Kilpatrick et al. 1986). If the lectin exerts a similar effect on the immunocompetent cells in the gastrointestinal tract there is always a possibility that it affects the immune system, especially in individuals who consume tomatoes on a daily basis. Moreover, there is another reason to worry about dietary LEA. In a recent report, it was demonstrated that potato lectin activates basophils and mast cells of atopic subjects by its interaction with core chitobiose of cell-bound non-specific immunoglobulin E (Pramod et al. 2007). On the basis of these observations, it was suggested that dietary potato lectin might increase the severity of the clinical symptoms of non-allergic food hypersensitivity in atopic subjects. Most probably, LEA is capable of having similar effects, in light of its very similar carbohydrate-binding specificity. Even in the absence of evidence for a possible harmful effect of the tomato lectin it may be worth considering promoting the production of tomatoes with a strongly reduced or no lectin content. According to the results summarized in Table 2, such varieties are already available. An additional advantage of lectin-less or low-lectin varieties is that they contain fewer antigens that can cause the latex-fruit syndrome in sensitive individuals.

OTHER TOMATO LECTINS

Hitherto, only LEA and related chitin-binding proteins have been identified in tomatoes. However, it is very likely that tomato fruits contain minute amounts of other lectins. Screening of the tomato transcriptome databases revealed that developing fruits express two different types of lectin: (1) an orthologue of the tobacco leaf lectin (called Nictaba, encoded by cDNA clone FC17CB09), which is classified in the family of Cucurbitaceae phloem lectins (Chen et al. 2002), and (2) a mannose-binding jacalin homologue (encoded by cDNA clone FB02CE01).

ABBREVIATIONS

cDNA: copy DNA; ConA: concanavalin A; Cys: cysteine; EST: expressed sequence tag; GlcNAc: N-acetylglucosamine; Gln: glutamine; Hyp: hydroxyproline; LEA: *Lycopersicum esculentum* agglutinin; Man: mannose; Met: methionine; Pro: proline; PHA: *Phaseolus vulgaris* agglutinin; SDS-PAGE: Sodium dodecyl sulphate polyacrylamide gel electrophoresis.

REFERENCES

Acarin, L. and J.M. Vela, B. Gonzalez, and B. Castellano. 1994. Demonstration of poly-N-acetyl lactosamine residues in ameboid and ramified microglial cells in rat brain by tomato lectin binding. J. Histochem. Cytochem. 42: 1033-1041.

Allen, A.K. 1983. Potato lectin—a glycoprotein with two domains. Prog. Clin. Biol. Res. 138: 71-85.

Allen, A.K. and J.A. Neuberger. 1973. The purification and properties of the lectin from potato tubers—a hydroxyproline-containing glycoprotein. Biochem. J. 135: 307-314.

Allen, A.K. and N.N. Desai, A. Neuberger, and J.M. Creeth. 1978. Properties of potato lectin and the nature of its glycoprotein linkages. Biochem. J. 171: 665-674.

Allen, A.K. and G.P. Bolwell, D.S. Brown, C. Sidebottom, and A.R. Slabas. 1996. Potato lectin: a three-domain glycoprotein with novel hydroxyproline-containing sequences and sequence similarities to wheat-germ agglutinin. Int. J. Biochem. Cell Biol. 28: 1285-1291.

Anderson, N.H. and B. Cao, A. Rodriguez-Romero, and B. Arreguin. 1993. Hevein: NMR assignment and assessment of solution-state folding for the agglutinin-toxin motif. Biochemistry 32: 1407-1422.

Andjelkovic, A.V. and B. Nikolic, J.S. Pachter, and N. Zecevic. 1998. Macrophages/microglial cells in human central nervous system during development: an immunohistochemical study. Brain Res. 814: 13-25.

Bankston, P.W. and G.A. Porter, A.J. Milici, and G.E. Palade. 1991. Differential and specific labeling of epithelial and vascular endothelial cells of the rat lung by *Lycopersicon esculentum* and *Griffonia simplicifolia* I lectins. Eur. J. Cell Biol. 54: 187-195.

Bass, W.T. and G.A. Singer, and F.J. Liuzzi. 1998. Transient lectin binding by white matter tract border zone microglia in the foetal rabbit brain. Histochem. J. 30: 657-666.

Bies, C. and C.M. Lehr, and J.F. Woodley. 2004. Lectin-mediated drug targeting: history and applications. Adv. Drug Deliv. Rev. 56: 425-435.

Broekaert, W.F. and W.J. Peumans. Lectin release from seeds of *Datura stramonium* and interference of the *Datura stramonium* lectin with bacterial motility. pp. 57-66. *In:* T.C. Bog-Hansen and E. Van Driessche. [eds.] 1986. Lectins: Biology, Biochemistry, Clinical Biochemistry, Vol. 5, W. De Gruyter, Berlin, Germany.

Callaghan, J.M. and B.H. Toh, J.M. Pettitt, D.C. Humphris, and P.A. Gleeson. 1990. Poly-N-acetyllactosamine-specific tomato lectin interacts with gastric parietal cells. Identification of a tomato-lectin binding 60-90 × 10(3) Mr membrane glycoprotein of tubulovesicles. J. Cell Sci. 95: 563-576.

Callaghan, J.M. and B.H. Toh, R.J. Simpson, G.S. Baldwin, and P.A. Gleeson. 1992. Rapid purification of the gastric H+/K(+)-ATPase complex by tomato-lectin affinity chromatography. Biochem. J. 283: 63-68.

Chen, Y. and W.J. Peumans, B. Hause, J. Bras, M. Kumar, P. Proost, A. Barre, P. Rougé, and E.J.M. Van Damme. 2002. Jasmonic acid methyl ester induces the synthesis of a cytoplasmic/nuclear chitooligosaccharide-binding lectin in tobacco leaves. FASEB J. 16: 905-907.

Czapla, T.H. and B.A. Lang. 1990. Effect of plant lectins on the larval development of European corn borer (Lepidoptera:Pyralidae) and southern corn rootworm (Coleoptera:Chrysomelidae). J. Econ. Entomol. 83: 2480-2485.

Gilbert, C.W. and M.H. Zaroukian, and W.J. Esselman. 1988. Poly-N-acetyllactosamine structures on murine cell surface T200 glycoprotein participate in natural killer cell binding to YAC-1 targets. J. Immunol. 140: 2821-2828.

Hoffmann, S. and C. Spee, T. Murata, J.Z. Cui, S.J. Ryan, and D.R. Hinton. 1998. Rapid isolation of choriocapillary endothelial cells by *Lycopersicon esculentum*-coated Dynabeads. Graefes Arch. Clin. Exp. Ophthalmol. 236: 779-784.

Hossaini, A.A. 1968. Hemolytic and hemagglutinating activities of 222 plants. Vox Sang. 15: 410-417.

Hussain, N. and P.U. Jani, and A.T. Florence. 1997. Enhanced oral uptake of tomato lectin-conjugated nanoparticles in the rat. Pharm. Res. 14: 613-618.

Irache, J.M. and C. Durrer, D. Duchene, and G. Ponchel. 1996. Bioadhesion of lectin-latex conjugates to rat intestinal mucosa. Pharm. Res. 13: 1716-1719.

Kieliszewski, M.J. and A.M. Showalter, and J.F. Leykam 1994. Potato lectin: a modular protein sharing sequence similarities with the extensin family, the hevein lectin family, and snake venom disintegrins (platelet aggregation inhibitors). Plant J. 5: 849-861.

Kilpatrick, D.C. 1980. Purification and some properties of a lectin from the fruit juice of the tomato (*Lycopersicon esculentum*). Biochem J. 185: 269-272.

Kilpatrick, D.C. 1983. Tomato (*Lycopersicon esculentum*) lectin and serologicaly related molecules. Prog. Clin. Biol. Res. 138: 63-70.

Kilpatrick, D.C. and J. Weston, and S.J. Urbaniak. 1983. Purification and separation of tomato isolectins by chromatofocusing. Anal. Biochem. 134: 205-209.

Kilpatrick, D.C. and A. Pusztai, G. Grant, C. Graham, and S.W. Ewen. 1985. Tomato lectin resists digestion in the mammalian alimentary canal and binds to intestinal villi without deleterious effects. FEBS Lett. 185: 299-305.

Kilpatrick, D.C. and C. Graham, and S.J. Urbaniak. 1986. Inhibition of human lymphocyte transformation by tomato lectin. Scand. J. Immunol. 24: 11-9.

Lehr, C.M. and J.A. Bouwstra, W. Kok, A.B. Noach, A.G. de Boer, and H.E. Junginger. 1992. Bioadhesion by means of specific binding of tomato lectin. Pharm. Res. 9: 547-553.

Lu, J. and C. Teh, U. Kishore, and K.B. Reid. 2002. Collectins and ficolins: sugar pattern recognition molecules of the mammalian innate immune system. Biochim. Biophys. Acta 1572: 387-400.

Marcusson-Begun, H. 1926. Untersuchungen über das Hämagglutinin der kartoffelknolle. Z. Immunitätsforsch. Exp. Ther. 45: 49-73.

Mazzetti, S., and S. Frigerio, M. Gelati, A. Salmaggi, and L. Vitellaro-Zuccarello. 2004. *Lycopersicon esculentum* lectin: an effective and versatile endothelial marker of normal and tumoral blood vessels in the central nervous system. Eur. J. Histochem. 48: 423-428.

Merkle, R.K. and R.D. Cummings. 1987a. Tomato lectin is located predominantly in the locular fluid of ripe tomatoes. Plant Sci. 48: 71-78.

Merkle, R.K. and R.D. Cummings. 1987b. Relationship of the terminal sequences to the length of poly-N-acetyllactosamine chains in asparagine-linked oligosaccharides from the mouse lymphoma cell line BW5147. Immobilized tomato lectin interacts with high affinity with glycopeptides containing long poly-N-acetyllactosamine chains. J. Biol. Chem. 262: 8179-8189.

Miskevich, F. 1999. Laminar redistribution of a glial subtype in the chick optic tectum. Brain Res. Dev. Brain Res. 115: 103-109.

Nachbar, M.S. and J.D. Oppenheim, and J.O. Thomas. 1980. Lectins in the U.S. Diet. Isolation and characterization of a lectin from the tomato (*Lycopersicon esculentum*). J. Biol. Chem. 255: 2056-2061.

Naisbett, B. and J. Woodley. 1990. Binding of tomato lectin to the intestinal mucosa and its potential for oral drug delivery. Biochem. Soc. Trans. 18: 879-980.

Naito, Y., and T. Minamihara, A. Ando, T. Marutani, S. Oguri, and Y. Nagata. 2001. Domain construction of cherry-tomato lectin: relation to newly found 42-kDa protein. Biosci. Biotechnol. Biochem. 65: 86-93.

Oguri, S. 2005. Analysis of sugar chain-binding specificity of tomato lectin using lectin blot: recognition of high mannose-type N-glycans produced by plants and yeast. Glycoconj. J. 22: 453-461.

Ojeda, J.L. and J.M. Icardo. 2006. Basement membrane heterogeneity during chick development as shown by tomato (*Lycopersicon esculentum*) lectin binding. Histol. Histopathol. 21: 237-248.

Peumans, W.J. and E.J.M. Van Damme. 1995. Lectins as plant defence proteins. Plant Physiol. 109: 347-352.

Peumans, W.J. and E.J.M. Van Damme. 1996. Prevalence, biological activity and genetic manipulation of lectins in foods. Trends Food Sci. Technol. 7: 132-138.

Peumans, W.J. and P. Rougé, and E.J.M. Van Damme. 2004. The tomato lectin consists of two homologous chitin-binding modules separated by an extension-like linker. Biochem. J. 376: 717-724.

Peumans, W.J. and E. Fouquaert, A. Jauneau, P. Rougé, N. Lannoo, H. Hamada, R. Alvarez, B. Devreese, and E.J.M. Van Damme. 2007. The liverwort *Marchantia polymorpha* expresses orthologs of the fungal *Agaricus bisporus* agglutinin family. Plant Physiol., in press.

Pramod, S.N. and Y.P. Venkatesh, and P.A. Mahesh. 2007. Potato lectin activates basophils and mast cells of atopic subjects by its interaction with core chitobiose of cell-bound non-specific immunoglobulin E. Clin. Exp. Immunol., 148: 391-401.

Reche, M. and C.Y. Pascual, J. Vicente, T. Caballero, F. Martin-Munoz, S. Sanchez, and M. Martin-Esteban. 2001. Tomato allergy in children and young adults: cross-reactivity with latex and potato. Allergy 56: 1197-1201.

Saito, K. and H. Yagi, K. Baba, I.J. Goldstein, and A. Misaki. 1996. Purification, properties and carbohydrate-binding specificity of cherry tomato (*Lycopersicon esculentum* var. cherry) lectin. J. Appl. Glycosc. 43: 331-345.

Salazar, I., and P. Sanchez Quinteiro, M. Lombardero, N. Aleman, and P. Fernandez de Troconiz. 2004. The prenatal maturity of the accessory olfactory bulb in pigs. Chem. Senses 29: 3-11.

Sequeira, L. and T.L. Graham. 1977. Agglutination of avirulent strains of *Pseudomonas solanacearum* by potato lectin. Physiol. Plant Pathol. 11: 43-54.

Van Damme, E.J.M. and W.J. Peumans, A. Barre, and P. Rougé. 1998. Plant lectins: a composite of several distinct families of structurally and evolutionary related proteins with diverse biological roles. Crit. Rev. Plant Sci. 17: 575-692.

Van Damme, E.J.M. and A. Barre, P. Rougé, and W.J. Peumans. 2004a. Cytoplasmic/nuclear plant lectins: a new story. Trends Plant Sci. 9: 484-489.

Van Damme, E.J.M. and A. Barre, P. Rougé, and W.J. Peumans. 2004b. Potato lectin: an updated model of a unique chimeric plant protein. Plant J. 37: 34-45.

Van Damme, E.J.M. and R. Culerrier, A. Barre, R. Alvarez, P. Rougé, and W.J. Peumans. 2007a. A novel family of lectins evolutionarily related to class V chitinases: an example of neofunctionalization in legumes. Plant Physiol., 144: 662-672.

Van Damme, E.J.M. and P. Rougé, and W.J. Peumans. Carbohydrate-protein interactions: Plant lectins. *In*: J.P. Kamerling, G.J. Boons, Y.C. Lee, A. Suzuki, N. Taniguchi and A.G.J. Voragen. [eds.] 2007b. Comprehensive Glycoscience—From Chemistry to Systems Biology. Elsevier, New York, USA. pp. 563-599.

Velasco, A. and E. Caminos, E. Vecino, J.M. Lara, and J. Aijon. 1995. Microglia in normal and regenerating visual pathways of the tench (*Tinca tinca* L., 1758; Teleost): a study with tomato lectin. Brain Res. 705: 315-324.

Vela Hernandez, J.M. and I. Dalmau, B. Gonzalez, and B. Castellano. 1997. Abnormal expression of the proliferating cell nuclear antigen (PCNA) in the spinal cord of the hypomyelinated Jimpy mutant mice. Brain Res. 747: 130-139.

Wagner, S. and H. Breiteneder. 2002. The latex-fruit syndrome. Biochem. Soc. Trans. 30: 935-940.

Wang, W.C. and N. Lee, D. Aoki, M.N. Fukuda, and M. Fukuda. 1991. The poly-N-acetyllactosamines attached to lysosomal membrane glycoproteins are increased by the prolonged association with the Golgi complex. J. Biol. Chem. 266: 23185-23190.

Wang, H. and T.B. Ng. 2006. A lectin with some unique characteristics from the samta tomato. Plant Physiol. Biochem. 44: 181-185.

Woodley, J.F. 2000. Lectins for gastrointestinal targeting—15 years on. J. Drug Target. 7: 325-333.

Zacharisen, M.C. and N.P. Elms, and V.P. Kurup. 2002. Severe tomato allergy (*Lycopersicon esculentum*). Allergy Asthma Proc. 23: 149-152.

Zhu, B.C. and R.A. Laine. 1989. Purification of acetyllactosamine-specific tomato lectin by erythroglycan-sepharose affinity chromatography. Prep. Biochem. 19: 341-350.

9

Presence of Zingiberene and Curcumene in Wild Tomato Leaves

George F. Antonious

Department of Plant and Soil Science, Water Quality/Environmental Toxicology, Land Grant Program, 218 Atwood Research Center, Kentucky State University, Frankfort, Kentucky 40601, USA

ABSTRACT

Two monocyclic sesquiterpene hydrocarbons, zingiberene and curcumene, were found by gas chromatography and mass spectrometry (GC/MS) analysis to be the most abundant sesquiterpenes in the leaf oil of the wild tomato *Lycopersicon hirsutum* f. *typicum* (Solanaceae). Volatile metabolites of *Lycopersicon* species such as zingiberene and curcumene have a role in tomato flavor, human health-related properties, and host plant defense against arthropod herbivores. Many of these compounds are known for their medicinal and anti-oxidative properties. Zingiberene [5-(1,5-dimethyl-4-hexenyl)-2-methyl-1,3-cyclohexadiene] is known as the main constituent of ginger oil from the rhizomes of *Zingiber officinale*, Roscoe (Zingiberaceae). An authentic specimen of zingiberene isolated from commercial ginger oil and an authentic specimen isolated from wild tomato leaves of *Lycopersicon hirsutum* f. *typicum* (PI-127826 and PI-127827) have identical zingiberene and curcumene GC/MS spectrum. The main sesquiterpene hydrocarbons identified in the leaves of wild tomato species were α-zingiberene, α-curcumene, and trans-caryophyllene. Leaves of six wild tomato accessions of *L. hirsutum* f. *glabratum* (Mull), three accessions of *L. hirsutum* f. *typicum* (Humb & Bonpl.), two accessions of *L. pennellii* Corr. (D' Arcy), one accession of *L. pimpinellifolium*, and leaves of cultivated tomato *L. esculentum* cv. Fabulous (Solanaceae) were analyzed for

An abbreviation is given before the references.

chemical composition. Crude extracts of wild tomato leaves were purified using an open glass chromatographic column containing alumina and the contents were separated for GC/MS identification and quantification. Analysis of *L. hirsutum* f. *typicum* accessions (PI-127826, PI-127827) indicated the presence of zingiberene, curcumene, and other lipophilic secondary metabolites in their leaves. Zingiberene and curcumene were found mainly in type VI glandular trichomes of typicum accessions. Sesquiterpene content of tomato leaf oil varied considerably among species. Seasonal variation affected the amount of zingiberene and curcumene per unit area of leaves by controlling concentration of zingiberene and curcumene in the glandular tips of type VI trichomes present on the leaves of wild tomato species. Total leaves collected from each plant of accessions PI-127826 and PI-127827 provided 10.7 and 9.8 g of zingiberene, respectively. Wild tomato accessions (PI-127826 and PI-127827) of *L. hirsutum* f. *typicum* can be explored as a biorational source of zingiberene.

INTRODUCTION

Ginger, *Zingiber officinale* Roscoe (Zingiberaceae), is a widely used spice, flavoring agent, and herbal medicine and is also employed in the perfume industry. It was well known in England as early as the 11th century and became a major item of the spice trade in the 13th and 14th centuries (Wohlmuth et al. 2006). Ginger is a perennial plant that grows in India, China, Mexico, and several other countries (Bartley and Foley 1994, Menut et al. 1994, Onyenekwe and Hashimoto 1999). It is now cultivated in many tropical and subtropical areas, the main producers being India, China, Indonesia, and Nigeria (Sutarno et al. 1999). The characteristic pleasant aroma of ginger oil from fresh ginger rhizomes is due to its contents of monoterpene hydrocarbons, zingiberene and curcumene. In food and pharmaceutical industries, the efficacy of ginger rhizomes for the prevention of nausea, dizziness, and vomiting as symptoms of motion sickness (kinetosis), as well as postoperative vomiting and vomiting of pregnancy, has been well documented in numerous clinical studies (Langner et al. 1998). Overwhelming evidence from various studies indicated that the foliage of the wild tomato *Lycopersicon hirsutum* f. *glabratum* (Mull), *L. hirsutum* f. *typicum* (Humb & Bonpl.), and *L. pennellii* Corr. (D'Arcy) is covered with a dense vesture of type IV and type VI glandular trichomes (plant hairs) (Luckwill 1943, Lin et al. 1987, Snyder et al. 1993, Antonious and Snyder 1993, Eigenbrode et al. 1996, Antonious 2001a, b). Structure of type IV and type VI glandular trichomes and their surface areas were recently described (Antonious 2001a). Several major chemicals in the glandular tips of type VI trichomes of *Lycopersicon* species were identified and quantified by gas chromatography and mass spectrometry (GC/MS). Many monoterpenes, including limonene, have been found in the leaves of the domestic tomato *Lycopersicon esculentum*, and most are also found in wild tomato species. Research has shown that the sesquiterpene content of tomato leaf oil varies considerably among species. During the course of evolution, plants have synthesized thousands of secondary compounds that are not essential to a plant's primary metabolic processes but serve other adaptive roles (Antonious et al. 1999). Several active compounds such as sesquiterpene hydrocarbons, sesquiterpene acids, methylketones, and glucolipids in glandular trichomes on the leaves of wild tomato accessions can be extracted and quantified.

Terpenoids in ginger may be regarded as important protectants against gastric lesions, thus supporting the use of ginger as a natural stomatic medicine. Ginger contains several active ingredients such as α-zingiberene and α-curcumene (Fig. 1). The sesquiterpene hydrocarbon zingiberene [5-(1,5- dimethyl-4-hexenyl)-2-methyl-1,3-cyclohexadiene] has been shown to have a considerable spectrum of biological activity such as antiviral, antiulcer (Yamahara et al. 1988) and antifertility (Ni et al. 1989) effects. From an entomological point of view, zingiberene is also associated with resistance to the Colorado potato beetle (*Leptinotarsa decemlineata* Say) (Carter et al. 1989) and beet armyworm (*Spodoptera exigua*) (Eigenbrode and Trumble 1993, Eigenbrode et al. 1996). Curcumene, a C_{15} sesquiterpene hydrocarbon in ginger rhizomes, also has antiulcer (Yamahara et al. 1992) and insecticidal properties (Agarwal et al. 2001).

GROWING WILD TOMATO PLANTS FOR MASS PRODUCTION OF LEAVES

Lycopersicon hirsutum is a non-cultivated, green-fruited relative of domestic tomato (*L. esculentum*). Glandular trichomes and their secretions are very different between *L. esculentum* and *L. hirsutum* (Antonious et al. 2005). Generally, trichomes on the wild tomato *L. hirsutum* are associated with an abundance of oil-like compounds, but these compounds are largely absent on *L. esculentum*.

Wild tomato seeds of six accessions of *L. hirsutum* f. *glabratum* Mull (PI-126449, PI-134417, PI-134418, PI-251305, PI-251304, and LA-407), three accessions of *L. hirsutum* f. *typicum* Humb & Bonpl. (PI-127826, PI-127827, and PI-308182), two accessions of *L. pennellii* Corr. (D' Arcy) (PI-246502 and PI-414773), and one accession of *L. pimpinellifolium* (PI-1335) were obtained from the USDA/ARS, Plant Genetic Resources Unit, Cornell University Geneva, New York (USA). Seeds of the cultivated tomato

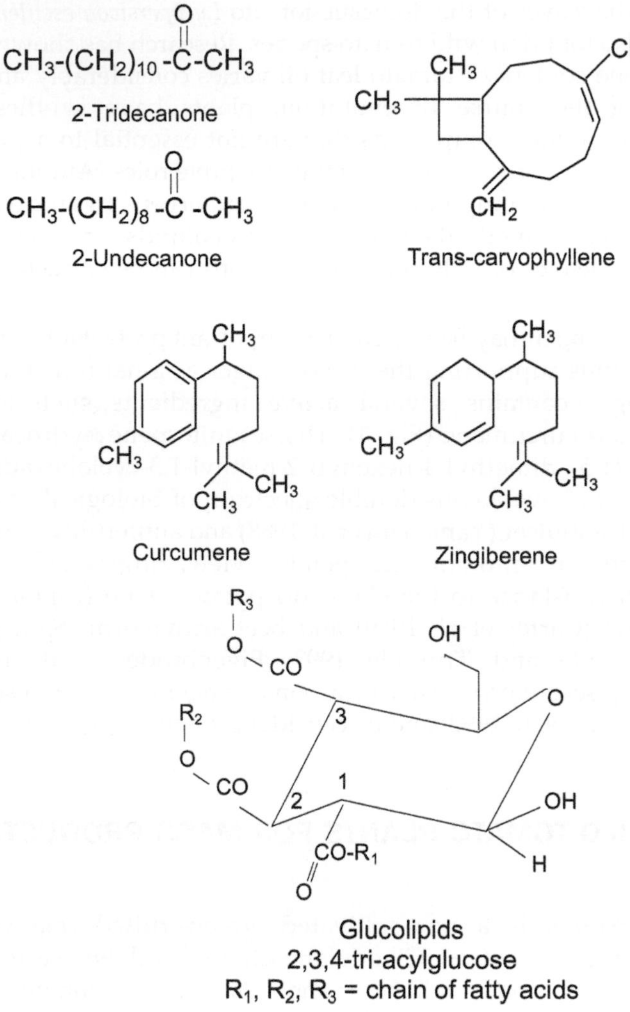

$CH_3\text{-}(CH_2)_{10}\text{-}C(=O)\text{-}CH_3$

2-Tridecanone

$CH_3\text{-}(CH_2)_8\text{-}C(=O)\text{-}CH_3$

2-Undecanone

Trans-caryophyllene

Curcumene

Zingiberene

Glucolipids
2,3,4-tri-acylglucose
R_1, R_2, R_3 = chain of fatty acids

Fig. 1 Chemical structures of two methylketones (2-tridecanone and 2-undecanone), three sesquiterpene hydrocarbons (trans-caryophyllene, α-zingiberene and α-curcumene), and glucolipids (sugar-esters) detected in glandular trichomes of *Lycopersicon hirsutum* f. *glabratum*, *L. hirsutum* f. *typicum*, and *L. penellii*, respectively.

(*L. esculentum* cv. Fabulous-F1), obtained from Holmes Seed Co., Canton, Ohio (USA), were included as a control. All seeds were germinated in the laboratory on moistened filter paper in Petri dishes kept in the dark. After germination, all seedlings were transplanted into plastic pots containing commercial Pro Mix (Premier Horticulture Inc., Red Hill, Pennsylvania, USA) and maintained under fluorescent light in the laboratory. At the six-leaf stage, plants from each accession were transported into the greenhouse and transplanted into plastic pots containing Pro-Mix and grown under natural daylight conditions supplemented with sodium lamps providing additional photosynthetic photon flux of 110 µmol s^{-1} m^{-1}. Pots were distributed on the greenhouse benches in a randomized complete block design. Plants were irrigated daily and fertilized twice a month with water containing 200 ppm of Peters (Scotts Co., Marysville, Ohio) as a general purpose fertilizer of N, P, and K (20:20:20). Average greenhouse temperature and relative humidity were 30 ± 3.9°C and 49.5 ± 11.8%, respectively. No insects were observed in the greenhouse on both the wild tomato foliage and the cultivated tomato (Fabulous-F1) plants and no insecticides were applied.

When plants were 90 d old, three leaves free of visible defects from each accession were sampled from equivalent positions below the plant apex (designated as node number below the apical meristem). The second, fourth and sixth pairs of leaves were considered upper, middle and lower leaves of each plant, respectively. One leaflet of each pair of leaves was used to obtain adaxial trichome counts and the corresponding opposite leaflet was used for abaxial counts. Only Type VI trichomes were counted using a light microscope (Antonious 2001a) at magnification of 100 × (10 × ocular and 10 × objective). Three counts were made for each leaflet surface (within interveinal areas) at the top, near the center, and at the bottom of each leaflet. Leaflet lengths and number of Type VI trichomes per mm^2 and on each leaflet were recorded. Leaflet surface area in mm^2 was calculated using a laser leaf-area meter (CID, Inc., Vancouver, Washington, USA). A regression line for each accession was established on the basis of the relationship between leaflet surface area (cm^2) and leaflet weight (g). Number of type VI trichomes per leaflet was determined by multiplying trichome density per mm^2 by leaflet surface area. At harvest, leaves were weighed and their weight and surface area were determined per plant using regression lines.

OCCURRENCE OF TYPE VI GLANDULAR TRICHOMES ON WILD TOMATO LEAVES

The morphology of Type VI glandular trichome tip differed between species. When magnified, type VI tips on *L. esculentum* (cultivated tomato) leaves appeared to have four lobes, due to marked divisions between the four lobes. On the leaves of the wild tomato relative, *L. hirsutum* f. *typicum* (Fig. 2), these divisions are less apparent, resulting in a globular appearance of the tip (Fig. 3). Type IV and Type VI glandular trichomes were the most prevalent on leaflets of the wild tomato species examined.

Fig. 2 Two wild tomato accessions of *Lycopersicon hirsutum* f. *typicum* (Family: Solanaceae), PI-127826 (upper photo) and PI-127827 (lower photo) grown under greenhouse conditions at KSU Research Farm, Franklin County, Frankfort, Kentucky.

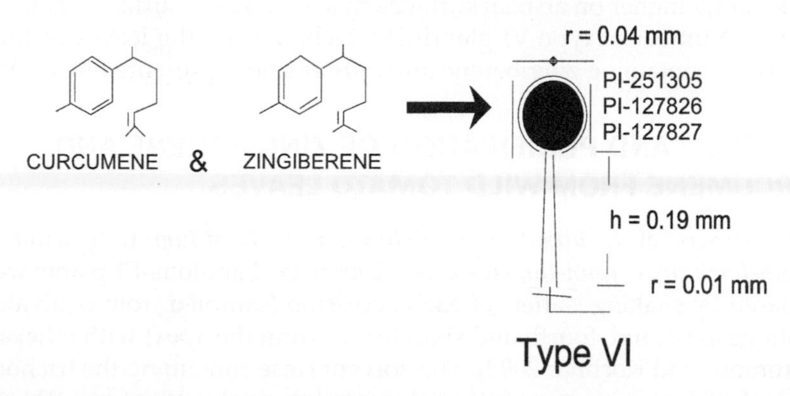

Fig. 3 Structure of type VI glandular trichomes (containing α-zingiberene and α-curcumene) on the leaves of wild tomato accessions PI-251305 of *Lycopersicon hirsutum* f. *glabratum*, PI-127826 and PI-127827 of *L. hirsutum* f. *typicum*. Note that type VI glandular trichomes on the leaflets of these wild tomato species are about 0.2 mm long, topped by a nearly round cap of 0.04 mm radius, and a base of 0.01 mm radius.

Type IV trichomes as described by Luckwill (1943) were absent on *L. esculentum*. The hairs of type VI trichomes on leaflets of *L. hirsutum* (*hirsutum* f. *glabratum* and *hirsutum* f. *typicum*) are about 0.2 mm long, consisting of a stalk cell and a neck cell, and are topped by a nearly round cap, resulting in a globular appearance of the tip, that makes up about one third of the length of the hair. The morphology of the type VI tip differs between tomato species; the rounded morphology of *hirsutum* type VI tip is different from the four-lobed structure of the *esculentum* type VI tip. Leaflets of the cultivated tomato *L. esculentum* cv. Fabulous contain only one type of glandular trichome (type VI), which consists of a stalk cell (about 0.1 mm long) and a neck cell, topped by a cap that makes up about one third of the length of the hair; the cap contains four lobe cells. Microscopic examination of the two types of trichomes (type IV and type VI) in *hirsutum* leaflets indicated that the surface area of type VI trichomes is 3.4 times the surface area of type IV trichomes, considering that the surface area of a trichome tip is similar to the surface area of a sphere ($4\,\pi r^2$) and the surface area of the trichome stalk is similar to the surface area of a cylinder ($2\,\pi r \times h$), where r is the radius of the sphere and h is the height of the cylinder (Antonious 2001a).

Type IV trichomes were significantly more abundant on adaxial than abaxial surfaces, averaging 56,851 trichomes $\cdot$ g^{-1} leaflets on the former, and 39,527 trichomes · g^{-1} leaflets on the latter. Type IV trichomes occurred at much higher densities than type VI trichomes over all accessions and both abaxial and adaxial leaflet surfaces. Type VI trichome densities were significantly higher on abaxial surfaces than on adaxial surfaces (data not shown). Number of type VI glandular trichomes on the leaves of three accessions containing zingiberene and curcumene is presented in Table 1.

ISOLATION AND PURIFICATION OF ZINGIBERENE AND CURCUMENE FROM WILD TOMATO LEAVES

Leaf extracts of *L. hirsutum* f. *glabratum, L. hirsutum* f. *typicum, L. pimpinellifolium, L. pennellii,* and *L. esculentum* cv. Fabulous-F1 plants were prepared by shaking leaflets of each accession (sampled from equivalent positions as second, fourth and sixth leaves from the apex) with n hexane (Antonious and Kochhar 2003). The solvent rinse containing the trichome and leaf surface contents was then decanted through a Whatman 934-AH glass microfiber filter and the filtrate was evaporated to dryness under vacuum using a rotary vacuum evaporator (Buchi Rotovapor Model 461, Switzerland) at 35°C followed by a gentle stream of nitrogen gas (N_2). The concentrated extract was then redissolved in 10 mL of n-hexane and applied to the top of a glass chromatographic column (20 × 1.1 cm) containing 10 g alumina grade-II that was pre-wetted with n-hexane. The column was eluted with 75 mL of n-hexane to obtain a fraction (Fraction-1) rich in α-zingiberene and α-curcumene (Fig. 4) as identified by their mass spectral data. This alumina column also can be used to separate methylketones from sesquiterpene hydrocarbons and from glucolipids (glucose esters) present in some wild tomato species as shown in Fig. 4. Glucolipids are non-ionic surfactants containing sucrose as hydrophilic group and fatty acids as lipophilic group (Juvik et al. 1994), while methylketones (such as 2-tridecanone and 2-undecanone) are major constituents of type VI glandular trichomes of the wild tomato relative, *L. hirsutum* f. *glabratum* (Antonious 2001a). Leaves of three wild tomato accessions, PI-127826, PI-127827, and PI-251305, which have shown considerable levels of α-zingiberene and α-curcumene (Table 2) were fortified with zingiberene and curcumene (prepared from ginger rhizomes) to test the efficiency of the extraction procedure to recover the two sesquiterpenes. The actual amount of zingiberene and curcumene recovered from each sample was obtained from the difference between the amount detected in the sample before and after fortification. Recovery values of added compounds averaged 93 and 89% for zingiberene and curcumene, respectively. Mass production of zingiberene and curcumene from whole plant leaves of *L. hirsutum* f. *typicum* accessions containing

Table 1 Average number of type VI glandular trichomes (containing zingiberene and curcumene) on the leaves of three wild tomato accessions grown in two greenhouses at Kentucky State University Research Farm, Franklin County, Kentucky (January to December, 2002).

Tomato accession	*Taxon*	*No. of type VI/g (two surfaces)*	*No. of type VI/cm^2 (two surfaces)*	*No. of type VI leaflet^{-1}*	*No. of type VI plant^{-1} (thousands)*
PI 251305	*L. hirsutum* f. *glabratum*	60,505 ± 22,359	774 ± 312	8,069 ± 2,172	12,113 ± 1,158
PI 127826	*L. hirsutum* f. *typicum*	72,722 ± 23,835	803 ± 163	10,868 ± 2,675	23,059 ± 1,893
PI 127827	*L. hirsutum* f. *typicum*	64,024 ± 20,397	683 ± 117	9,720 ± 1,633	20,708 ± 1,544

Each value in the table is an average ± SE obtained from five samples collected monthly during the study period (January to December, 2002).

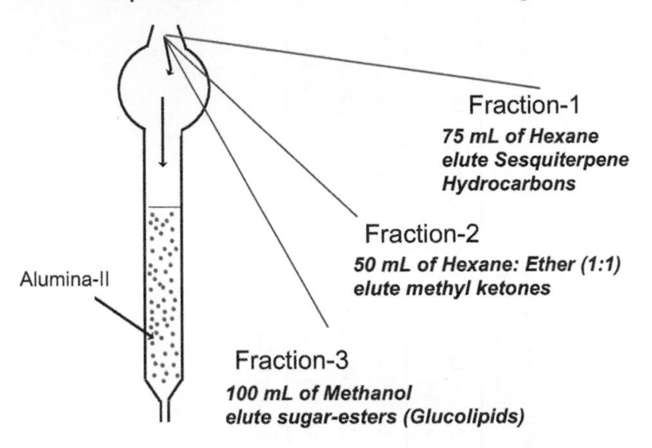

Fig. 4 Glass chromatographic column (1.2 × 22 cm) containing alumina grade-II used for separation of sesquiterpene hydrocarbons (Fraction-1), methylketones (Fraction-2), and glucolipids (Fraction-3) present in glandular trichomes on the leaves of wild tomato species.

considerable concentration of zingiberene and curcumene was achieved by steeping the leaves in a 20 L tank. A bulk crude extract of zingiberene and curcumene was prepared by soaking 5 kg wild tomato leaves collected from the greenhouse-grown wild tomato plants in 10 L of water containing 15 mL of 2% Sur-Ten (sodium dioctyl sulfosuccinate), obtained from Aldrich Chemical Company, Milwaukee, Wisconsin (USA) (Antonious 2004). After continuous manual shaking, the mixture was filtered through cheesecloth and each liter of the filtrate was partitioned with 200 mL of n-hexane in a separatory funnel. The hexane layers were combined and the procedure was completed as described above.

OCCURRENCE OF ZINGIBERENE AND CURCUMENE IN TYPE VI TRICHOMES OF WILD TOMATO LEAVES

Leaflets of all accessions investigated were collected from the greenhouse-grown plants and wiped with a cotton swab soaked with methanol to remove trichome exudates. Analysis of the leaflets of PI-127826 and PI-127827 wiped with a cotton swab soaked with methanol indicated that methanol removed 90% of zingiberene and curcumene from the surface of the wiped leaves, confirming that these exudates are present on the leaf surface rather than in the leaf interior matrix. The strong correlation (r = + 0.96, P = 0.0001) between trichome density per mm^2 and per g fresh leaflets indicated that the selection of accessions based on fresh weight of leaflets would be a suitable criterion in selection of accessions for exudates extraction.

Table 2 Chemical composition of hexane[†] and methanol[‡] extracts prepared from the leaves of 12 wild tomato accessions and a cultivated tomato (*L. esculentum* cv. Fabulous) grown under greenhouse conditions at KSU Research Farm, Franklin County, Kentucky.

Tomato accession	*Taxon*	*Sesquiterpene hydrocarbons*[†], *µg/g fresh leaves*			*Methylketones*[†], *µg/g fresh leaves*		*Glucolipids*[‡], *mg g*$^{-1}$ *Fresh Leaves*
		α-curcumene	*α-zingiberene*	*trans-caryophyllene*	*2-undecanone*	*2-tridecanone*	
PI 251304	*glabratum*	ND	ND	8.64 ± 3.22	113.22 ± 24.80	60.35 ± 20.45	90.7 ± 15.2
PI 126449	*glabratum*	ND	ND	5.03 ± 2.54	98.56 ± 18.42	52.35 ± 12.8	110.8 ± 20.0
PI 251305	*glabratum*	0.04 ± 0.01	170.41 ± 34.67	ND	4.77 ± 0.92	ND	123.2 ± 18.6
PI 134417	*glabratum*	ND	ND	8.20 ± 3.15	98.24 ± 19.33	53.27 ± 17.90	25.5 ± 3.8
PI 134418	*glabratum*	ND	ND	5.72 ± 2.77	99.98 ± 22.82	59.65 ± 12.50	245.5 ± 44.5
LA 407	*glabratum*	ND	ND	5.38 ± 2.44	93.15 ± 17.60	60.90 ± 22.65	39.1 ± 5.2
PI 1335	*pimpinellifolium*	ND	ND	0.58 ± 0.13	ND	ND	93.5 ± 8.8
PI 308182	*typicum*	ND	ND	ND	ND	ND	164.4 ± 19.4
PI 127826	*typicum*	0.06 ± 0.01	560.65 ± 122.80	ND	ND	ND	132.5 ± 13.6
PI 127827	*typicum*	0.35 ± 0.12	440.33 ± 98.56	ND	ND	ND	124.4 ± 18.5
PI414773	*pennellii*	ND	ND	ND	ND	ND	1,678.0 ± 121.4
PI-246502	pennellii	ND	ND	ND	ND	ND	2,029.6 ± 153.9
Fabulous	*esculentum*	ND	ND	ND	ND	0.73 ± 0.22	128.2 ± 23.5

ND indicates not detectable at the detectability limits of 0.001 µg/g fresh leaves. Each value in the table is an average ± SE of three replicates of three locations within each plant.

† Indicates the composition of hexane extract (Sesquiterpene) hydrocarbons and methyl ketones).

‡ Indicates the composition of methanol extract (glucolipids).

Methanol was used rather than another solvent mainly to minimize possible damage to the leaves caused by removal of epicuticular leaf waxes along with trichome exudates. Leaflets were then extracted with n-hexane and 1 μL of the purified extract was injected into a GC/MS (Hewlett Packard (HP) gas chromatograph (GC), model 5890 equipped with mass selective detector (MSD) and HP 7673 automatic injector) for identification and quantification. The instrument was auto-tuned with perfluorotributylamine (PFTBA) at m/z 69, 219, and 502. Electron impact mass spectra were obtained using an ionization potential of 70 eV. The operating parameters of the GC were as follows: injector and detector temperatures 210 and 275°C, respectively. Oven temperature was programmed from 70 to 230°C at a rate of 10°C/min (2 min initial hold). Injections onto the GC column were made in splitless mode using a 4 mm ID single taper liner with deactivated wool. A 25 m × 0.2 mm ID capillary column containing 5% diphenyl and 95% dimethyl-polysiloxane (HP-5 column) with 0.33 μm film thickness using helium as a carrier gas was used at a flow rate of 5.2 mL/min.

Identification of α-zingiberene, m/z 204 (7), 161 (2), 119 (88), 93 (100), 77 (30), 69 (44), 56 (28), 27 (19) and α-curcumene, m/z 202 (28), 145 (26), 132 (72), 119 (100), 105 (44), 95 (7), 83 (16) was confirmed by comparing the retention time and electron impact spectrum of α-zingiberene and α-curcumene with those from the oil of ginger rhizomes (*Zingiber officinale* Rosco) (Antonious and Kochhar 2003, Antonious and Snyder 2006). Trans-caryophyllene, m/z 204 (14), 189 (16), 161 (33), 147 (28), 133 (74), 121 (37), 107 (42), 93 (100), 79 (56), 69 (98) was confirmed by comparing the retention time and mass spectrum of analytical-grade trans-caryophyllene. 2-undecanone, m/z 170 (5), 155 (2), 127 (2), 112 (5), 85 (7), 71 (30), 58 (98), 43 (100), 27 (19) and 2-tridecanone, m/z 198 (2), 140 (2), 96 (2), 85 (5), 71 (30), 58 (93), 43 (100), 27 (16) were similarly confirmed.

Identification of the chemical contents in glandular heads of type VI trichomes was based on several samples of 100 intact mature glands collected from 4-mon-old plants of each wild tomato accession. Using dissecting microscope, a glass probe was used to rupture the glands and transfer their contents directly into n-hexane. The probe was rinsed between sample collections to prevent contamination of subsequent samples. After about 3,000 glands were collected from the greenhouse plants, the contents of type VI glandular trichome tips were dissolved in 250 μL of n-hexane and 1 μL was injected into the GC. α-Zingiberene detected in *L. hirsutum* f. *typicum* Humb. & Bonpl. occurred in the tips of type VI trichomes of accessions PI-127826 and PI-127827 of *L. hirsutum* f. *typicum* and was undetectable in other trichomes on the leaflet matrix. These findings are in agreement with those of Carter et al. (1989), who reported that zingiberene is present in type VI glandular trichomes and apparently does not occur at substantial levels in any other trichomes or in the leaf matrix.

ZINGIBERENE AND CURCUMENE IN GINGER RHIZOMES

Fresh ginger rhizomes (three replicates of 50 g each) were obtained from the local market (Meijer's Super Market, Lexington, Kentucky) and chopped into small pieces. The pieces were ground in a mortar and stirred five times with n-hexane for 6 h to furnish a golden-colored ginger oil. Ginger oil extracts were passed through an alumina glass chromatographic column (as described in Fig. 4). Diluted solutions of pure ginger oil (95% purity) purchased from Spectrum Bulk Chemicals (SBC, 14422 South San Pedro St, Gardena, California, USA) and fresh ginger oil prepared from fresh ginger rhizomes were injected separately into the GC for comparison and confirmation purposes. Quantification of sesquiterpene hydrocarbons in wild tomato leaves and fresh ginger rhizome extracts was carried out using 1 μL aliquot injections of diluted extracts. Detection of sesquiterpenes in the leaves of *L. hirsutum* f. *typicum* plants and ginger fresh rhizomes was confirmed by GC/MSD analysis of the hexane extracts. Under the GC conditions described above, retention times were 14.46, 14.52, 14.68, and 14.82 min for α-curcumene, α-zingiberene, β-bisabolene, and β-sesquiphellandrene, respectively. The retention time and mass spectra of zingiberene and curcumene isolated from the leaf samples (PI-127826, PI-127827, and PI-251305) matched those from ginger rhizomes. The two sesquiterpenes were also identified as α-zingiberene and α-curcumene by comparing their mass spectra with those reported in the literature (EPA/NIH 1998, Agarwal et al. 2001). Three sesquiterpenes were detected in purified ginger root hexane extract and identified by GC-MS analysis as α-zingiberene, β-bisabolene, and α-curcumene. Other constituents (beyond the scope of this chapter) detected in ginger rhizomes included 1,8-cineole, β-citronellol, z-citral, trans-geraniol, citral, α-pinene, β-farnesene, and farnesol (Antonious and Kochhar 2003).

Chemical composition of wild tomato leaves was compared to solutions of zingiberene and curcumene prepared from ginger rhizomes and purified using alumina grade-II glass chromatographic column as described above. Linearity over the range of concentrations was determined using regression analysis. Concentrations of sesquiterpenes were determined per leaflet area in cm^2 and per g fresh leaves of each wild tomato accession. Means were analyzed and separated using Duncan's LSD test (SAS Institute 2001). Correlation coefficients between mean estimates of glandular trichome densities and sesquiterpene concentrations were also determined.

QUANTIFICATION OF ZINGIBERENE AND CURCUMENE BY GC/MS

Analysis of purified hexane extracts prepared from the leaves of two accessions of *L. hirsutum* f. *typicum* (PI-127826 and PI-127827) and one accession of *L. hirsutum* f. *glabratum* (PI-251305) revealed the presence of α-zingiberene and α-curcumene. No zingiberene or curcumene was detected in other *L. hirsutum* f. *glabratum, L. pennellii, L. pimpinellifolium* accessions, *L. esculentum,*or PI-308182 of *L. hirsutum* f. *typicum* (Table 2). Mass spectrometric analysis of leaf extracts prepared from the three wild tomato accessions (PI-127826, PI-127827 and PI-251305) and from fresh ginger rhizomes showed fragments with identical molecular ions at m/z 202 (Fig. 5) and at m/z 204 (Fig. 6), along with other characteristic fragment ions that are consistent with the assignment of the molecular formula of curcumene ($C_{15}H_{22}$) and zingiberene ($C_{15}H_{24}$), respectively, known as major constituents of ginger rhizomes. Zingiberene spectral data, which showed a molecular ion peak (M+) at m/z 204, were in agreement with those previously reported in ginger rhizomes (Agarwal et al. 2001). Curcumene was also assigned on the basis of its mass fragmentation pattern, which gave a molecular ion peak (M+) at m/z 202, along with other characteristic fragment ion peaks.

The use of dried rhizomes of the ginger plant as a medicinal agent and biorational alternative source for antiviral and anticancer activity is well established. It is an important medicinal and culinary herb, known worldwide for its health-promoting properties (Ma and Gang 2006). Owing to the complexity of the active ingredients in ginger extracts, commercial synthesis is likely to be impractical. In addition, since ginger does not reproduce by seed but is clonally propagated via rhizome division and replanting, it is susceptible to accumulation and transmission of pathogens from generation to generation. Using standard agricultural practices under greenhouse conditions, two *L. hirsutum* f. *typicum* accessions produced considerable amounts of zingiberene and curcumene (Table 2). Because these amounts do not include exudates from type VI glandular trichomes on the plant stem, which are also found in large numbers, the amount of sesquiterpene hydrocarbons from plant foliage when including the stem is expected to be more than what is obtained from the leaves only.

Monthly variations in number of type VI trichomes per g fresh leaves and per leaflet surface area (Fig. 7, upper graph) and total concentrations

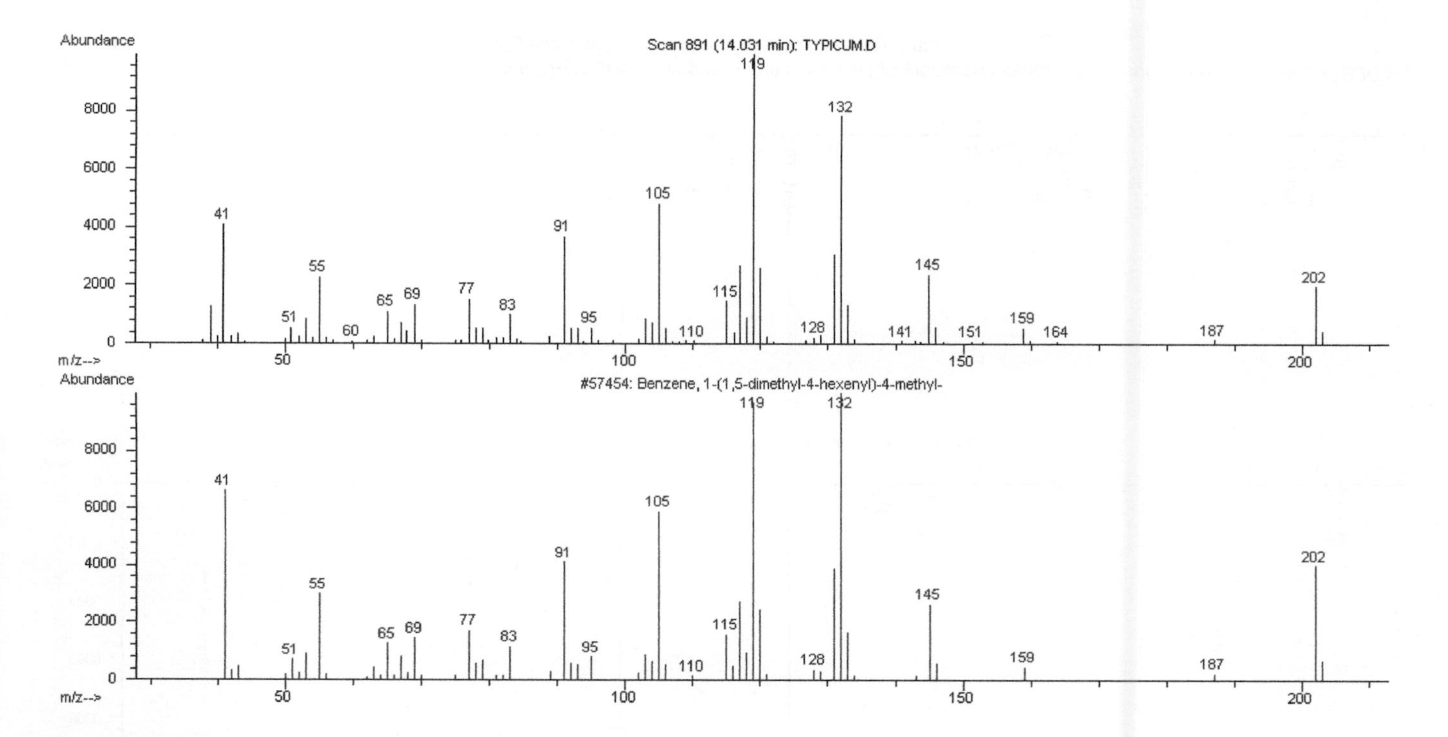

Fig. 5 Electron impact mass spectrum of curcumene ($C_{15}H_{22}$) extracted from the leaves of *Lycopersicon hirsutum* f. *typicum* accessions (PI-127826 and PI-127827) indicating the molecular ion of m/z 202, along with characteristic fragment ions.

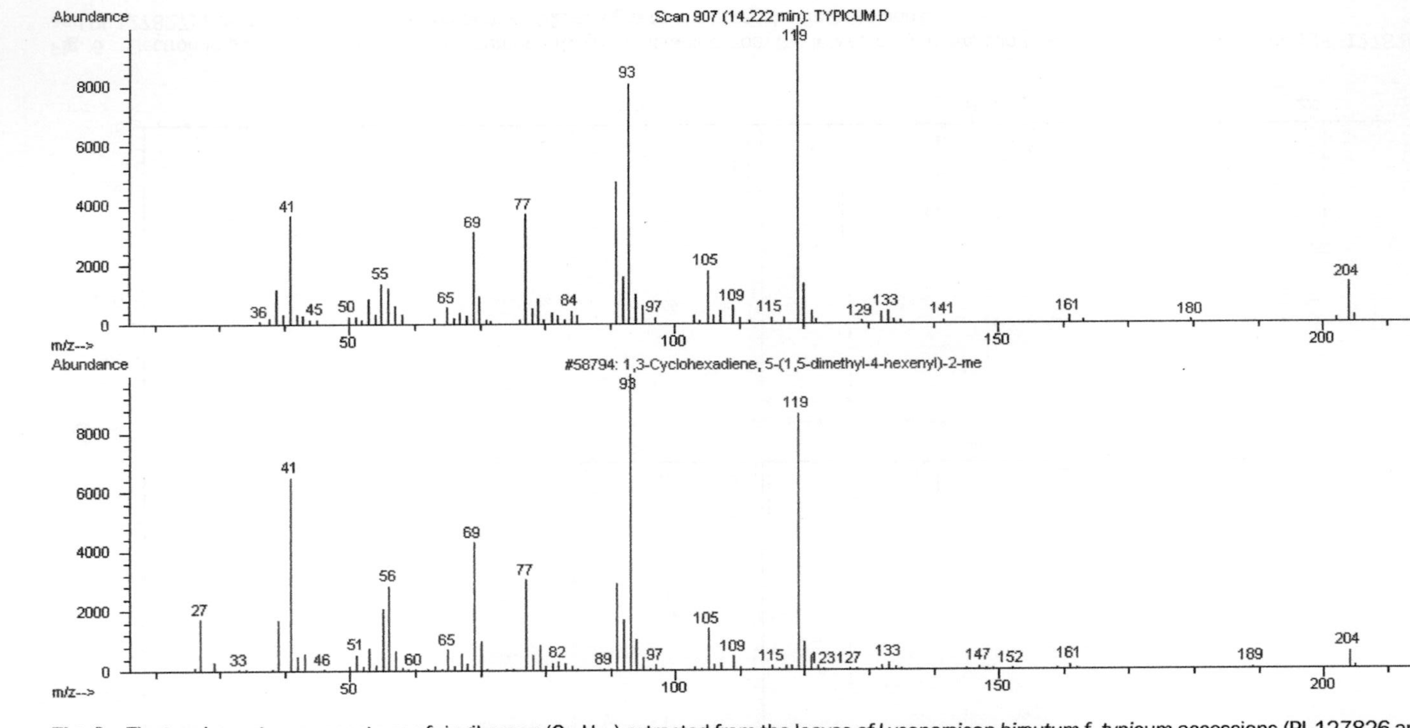

Fig. 6 Electron impact mass spectrum of zingiberene ($C_{15}H_{24}$) extracted from the leaves of *Lycopersicon hirsutum* f. *typicum* accessions (PI-127826 and PI-127827) indicating the molecular ion of m/z 204, along with other characteristic fragment ions.

of zingiberene and curcumene in the leaves of *L. hirsutum* f. *glabratum* accession PI-251305, an accession with relatively low levels of zingiberene and curcumene (Fig. 7, lower graph), indicated variation in trichome density during sampling periods. Total concentrations of zingiberene and curcumene per accession per year were obtained by adding amounts obtained from each harvest time per plant. These total concentrations were 2,305 and 13 mg per plant of zingiberene and curcumene, respectively (Fig. 7, lower graph). In PI-127826 the corresponding concentrations were 10,725 and 24 mg per plant, respectively (Fig. 8, lower graph) and in PI-271827 they were 9,730 and 52 mg per plant, respectively (Fig. 9, lower graph). Photoperiod affects the amount of zingiberene per unit surface area and per unit weight of leaves as reported by Gianfagna et al. (1992), who found that photoperiod affects concentration of zingiberene by controlling zingiberene content per trichome. Working with subspecies of *L. hirsutum* f. *glabratum*, Kennedy et al. (1981) found that type VI trichome density was greater in plants grown under long day than in short day conditions, as was 2-tridecanone (a methylketone) content in trichomes. Snyder and Hyatt (1984) also found higher trichome

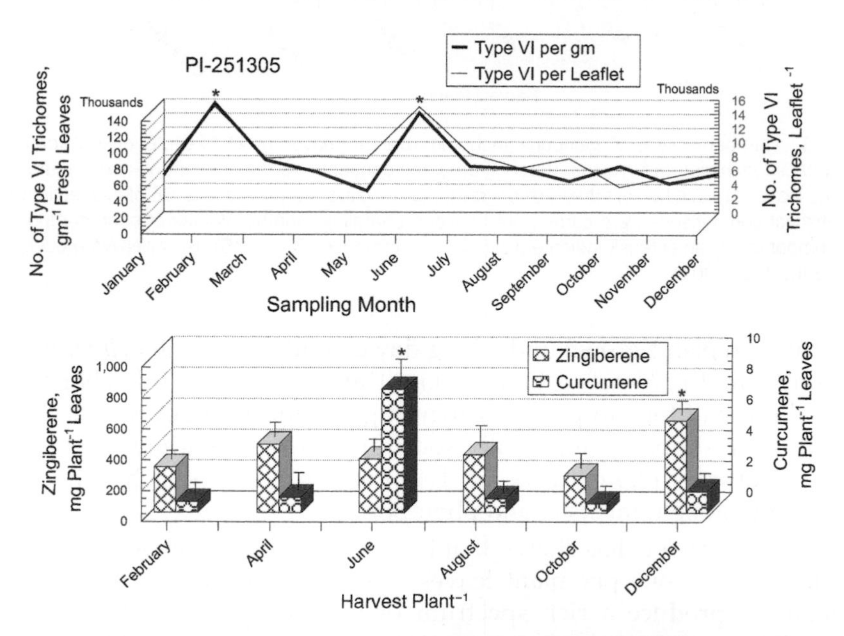

Fig. 7 Number of type VI glandular trichomes on the leaves of PI-251305 of wild tomato, *Lycopersicum hirsutum* f. *glabratum* (upper graph) and concentration of zingiberene and curcumene in the leaves at each harvest time (lower graph). Each value is an average ± SE of five replicates. Statistical comparisons were carried out between sampling months. Values of each compound accompanied by an asterisk indicate a significant difference ($P < 0.05$) using ANOVA procedure (SAS Institute 2001).

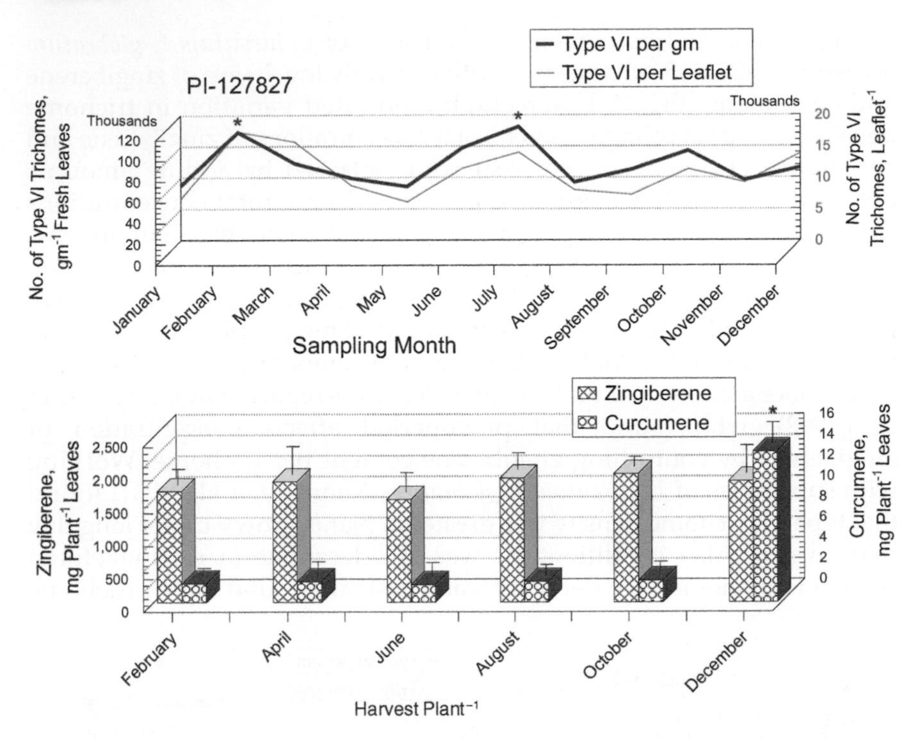

Fig. 8 Number of type VI glandular trichomes on the leaves of PI-126826 of wild tomato, *Lycopersicum hirsutum* f. *typicum* (upper graph) and concentration of zingiberene and curcumene in the leaves at each harvest time (lower graph). Each value is an average ± SE of five replicates. Statistical comparisons were carried out between sampling months. Values of each compound accompanied by an asterisk indicate a significant difference ($P < 0.05$) using ANOVA procedure (SAS Institute 2001).

densities in plants grown under long day conditions in both *L. hirsutum* f. *hirsutum* and *L. hirsutum* f. *glabratum*. Data reported in this chapter is based on wild tomato plants grown in greenhouse under natural light conditions supplemented with sodium lamps providing consistent photosynthetic photon flux of 110 μmol s^{-1} m^{-1}, which minimize variability in light intensity and photoperiod.

Research review has shown that type VI glandular trichomes of some *typicum* accessions per plant leaves are abundant (Table 1). Type VI trichomes produce a rich spectrum of sesquiterpenes (Table 2). The sesquiterpene content of tomato leaf oil varies considerably among species, with caryophyllene being widespread (Cloby et al. 1998). Curcumene occurs in *L. hirsutum* f. *typicum* at lower concentrations than zingiberene. Although the concentration of zingiberene and curcumene on the leaves of wild tomato is low compared to the concentration of these

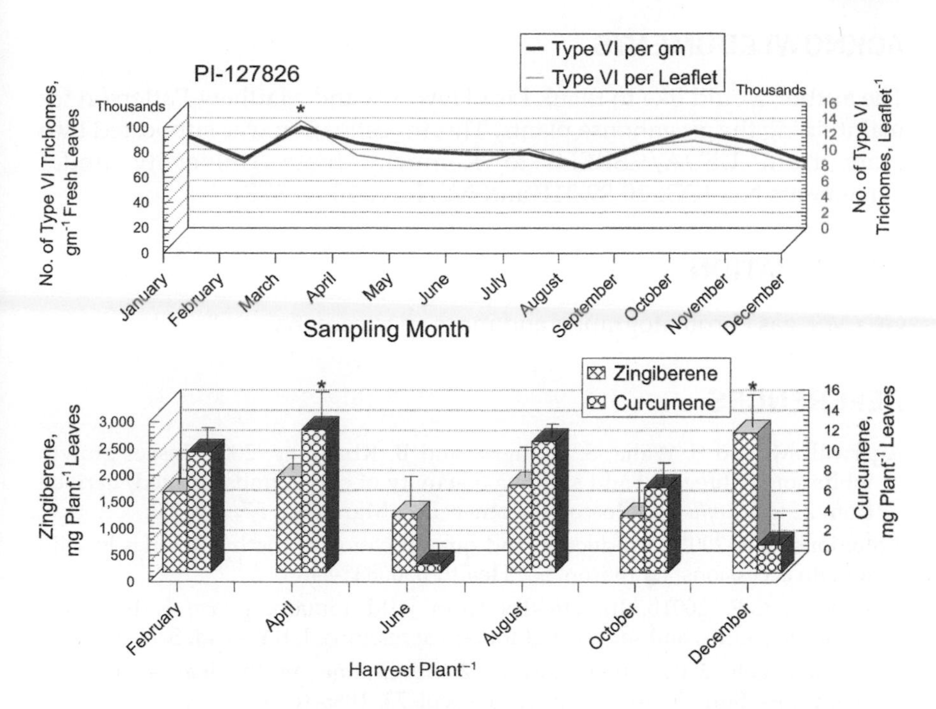

Fig. 9 Number of type VI glandular trichomes on the leaves of PI-127827 of wild tomato, *Lycopersicum hirsutum* f. typicum (upper graph) and concentration of zingiberene and curcumene in the leaves at each harvest time (lower graph). Each value is an average ± SE of five replicates. Statistical comparisons were carried out between sampling months. Values of each compound accompanied by an asterisk indicate a significant difference ($P < 0.05$) using ANOVA procedure (SAS Institute 2001).

sequiterpenes in ginger rhizomes, wild tomato foliage of *L. hirsutum* f. *typicum* (PI-271826 and PI-271827) (Fig. 2) can be considered a biorational source of zingiberene and curcumene, and this chapter provides background information on the level of these two sesquiterpene hydrocarbons in wild tomato species that may provide an opportunity for many farmers and organic growers who may be able to grow wild tomato foliage for use in crop production as medicinal agents for industrial use. The extraction of a few milligrams of zingiberene from wild tomato foliage is a simple procedure. However, more studies are needed in growing wild tomato plants with nutrient and soil management practices to identify optimum growing conditions and harvest periods through time course measurement of foliage yield and constituent concentration.

ACKNOWLEDGMENTS

The author would like to thank Lisa Hawkins and Matthew Patterson for maintaining the greenhouse plants. This investigation was supported by a grant from USDA/CSREES to Kentucky State University under agreements No. KYX-10-99-31P and No. KYX-10-03-37P

ABBREVIATION

GC/MS: gas chromatography and mass spectrometry

REFERENCES

Agarwal, M. and S. Walia, S. Dhingra, and B. Khambay. 2001. Insect growth inhibition, antifeedant and antifungal activity of compounds isolated/derived from *Zingiber officinale* Roscoe rhizomes. Pest Mgmt. Sci. 57: 289-300.

Antonious, G.F. 2001a. Production and quantification of methyl ketones in wild tomato accessions. J. Environ. Sci. Health B36: 835-848.

Antonious, G.F. 2001b. Insecticides from wild tomato: phase I—breeding, trichome counts, and selection of tomato accessions. J. KY Acad. Sci. 62: 79.

Antonious, G.F. 2004. Persistence of 2-tridecanone on the leaves of seven vegetables. Bull. Environ. Contam. Toxicol. 73: 1086-1093.

Antonious, GF and T.S. Kochhar. 2003. Zingiberene and curcumene in wild tomato. J. Environ. Sci. Health B38: 489-500.

Antonious, G.F. and J.C. Snyder. 1993. Trichome density and pesticide retention and half-life. J. Environ. Sci. Health B28: 205-219.

Antonious, G.F. and J.C. Snyder. 2006. Natural products: Repellency and toxicity of wild tomato leaf extracts to the two spotted spider mite, *Tetranychus urticae* Koch. J. Environ. Sci. Health B41: 43-55.

Antonious, G.F. and J.C. Snyder, and D.L. Dahlman. 1999. Tomato cultivar susceptibility to Egyptian cotton leafworm (Lepidoptera: Noctuidae) and Colorado Potato Beetle (Coleoptera: Chrysomelidae). J. Entomol. Sci. 34: 171-182.

Antonious, G.F. and T.S. Kochhar, and A.M. Simmons. 2005. Natural products: Seasonal variation in trichome counts and contents in *Lycopersicum hirsutum* f. *glabratum*. J. Environ. Sci. Health B40: 619-631.

Bartley, J.P. and P. Foley. 1994. Supercritical fluid extraction of Australian-grown ginger (*Zingiber officinale*). J. Sci. Food Agric. 66: 365-371.

Carter, C.D. and T.J. Gianfagna, and J.N. Sacalis. 1989. Sesquiterpenes in glandular trichomes of wild tomato species and toxicity to Colorado potato beetle. J. Agric. Food Chem. 37: 1425-1428.

Cloby, S.M. and J. Crock, B. Dowdle-Rizzo, P.G. Lemaux, and R. Croteau. 1998. Germacrene C synthase from *Lycopersicon esculentum* cv. VFNT Cherry tomato. Proc. Natl. Acad. Sci. 95: 2216-2221.

Eigenbrode, S.D. and J.T. Trumble. 1993. Antibiosis to beet armyworm (*Spodoptera exigua*) in *Lycopersicon* accessions. HortScience 28: 932-934.

Eigenbrode, S.D. and J.T. Trumble, and K.K. White. 1996. Trichome exudates and resistance to beet armyworm (Lepidoptera: Noctuidae) in *Lycoperiscon hirsutum* f. *typicum* accessions. Environ. Entomol. 25: 90-95.

EPA/NIH Mass Spectral Database National Institute of Standards and Technology. 1998. Gaithersburg, Maryland, USA.

Gianfagna, T.J. and C.D. Carter, and T.N. Sacalis. 1992. Temperature and photoperiod influence trichome density and sesquiterpene content of *Lycopersicon hirsutum* f. *hirsutum*. Plant Physiol. 100: 1403-1405.

Juvik, J.A. and J.A. Shapiro, T.E. Young, and Mutschler M.A.. 1994. Acylblucoses from wild tomato alter behavior and reduce growth survival of *Helicoverpa zea* and *Spodoptera exigua* (Lepidoptera: Noctuidae). J. Econ. Entomol. 87: 482- 492.

Kennedy, G.G. and R.T. Yamamoto, M.B. Dimock, W.G. Williams, and J. Bordner. 1981. Effect of day length and light intensity on 2-tridecanone levels and resistance in *Lycopersicon hirsutum* f. *glabratum*. J. Chem. Ecol. 7: 707-716.

Langner, E. and S. Greifenberg, and J. Gruenwald. 1998. Ginger: history and use. Adv. Ther. 15: 25-44.

Lin, S.Y.H. and J.T. Trumble, and J. Kumamoto. 1987. Activity of volatile compounds in glandular trichomes of *Lycopersicon* species against two insect herbivores. J. Chem. Ecol. 13: 837-850.

Luckwill, L.C. 1943. The genus *Lycopersicon*. An historical, biological, and taxonomic survey of wild and cultivated tomatoes. Aberdeen University Studies No. 120, Aberdeen University Press, Aberdeen, UK, 44 pp.

Ma, X. and D.R. Gang. 2006. Metabolic profiling of *in vitro* micropropagated and conventionally greenhouse grown ginger (*Zingiber officinale*). Phytochemistry 67: 2239-2255.

Menut, C. and G. Lamaty, J.M. Bessiere, and J. Koudou. 1994. Aromatic plants of tropical central Africa. XIII. Rhizome volatile components of two zingiberales from the Central African Republic. J. Essential Oil Res. 6: 161-164.

Ni, M. and Z. Chen, B. Yan, and H.X. Huadong. 1989. Synthesis of optically active sesquiterpenes and correlation of their antifertility effect. Chem. Abs. 111: 195149w.

Onyenekwe, P.C. and S. Hashimoto. 1999. The composition of the essential oil of dried Nigerian ginger (*Zingiber officinale* Rosocoe). Eur. Food Res. Technol. 209: 407-410.

SAS Institute SAS/STAT Guide, SAS Inc. 2001, SAS Campus Drive, Cary, North Carolina, USA.

Snyder, J.C. and J.P. Hyatt. 1984. Influence of daylength on trichome densities and leaf volatiles of *Lycopersicon* species. Plant Sci. Lett. 37: 177-181.

Snyder, J.C. and Z. Guo, R. Thacker, J.P. Goodman, and J.S. Pyrek. 1993. 2,3-Dihydrofarnesoic acid a unique terpene from trichomes of *Lycopersicon hirsutum,* repels spider mites. J. Chem. Ecol. 19: 2981-2997.

Sutarno, H. and E.A. Hadad, and M. Brink. *Zingiber officinale* Rosocoe. pp. 238-244. *In:* C.C. Guzman and J. S. Siemonsoma. [eds.] 1999. Spices. Backhuys, Leiden.

Wohlmuth, H. and M.K. Smith, L.O. Brooks, S.P. Myers and, D.N. Leach. 2006. Essential oil composition of diploid and tetraploid clones of ginger (*Zingiber officinale* Roscoe) grown in Australia. J. Agric. Food Chem. 54: 1414-1419.

Yamahara, J. and S. Hatakeyama, K. Taniguchi, M. Kawamura, and M. Yoshikawa. 1992. Stomatic principles of ginger. Yakugaku-Zasshi 112: 645-655.

Yamahara, J. and M. Mochizuki, H.Q. Rong, H. Matsuda, and H. Fujimura. 1988. The anti-ulcer effects in rats of ginger constituents. J. Ethnopharmacol. 23: 299-304.

10

The Role of Intracellular and Secreted Purple Acid Phosphatases in Tomato Phosphate Nutrition

Gale G. Bozzo[1] **and William C. Plaxton**[2*]

[1]Horticultural Sciences Dept., Univ. of Florida, P.O. Box 110690, Gainesville, FL 32611-690, USA

[2]Dept. of Biology, Queen's Univ., Kingston, Ontario, Canada K7L 3N6

ABSTRACT

Phosphate (Pi) is a crucial but limiting macronutrient for plant growth and metabolism. Agricultural Pi deficiency is alleviated by the massive application of Pi fertilizers. However, Pi assimilation by fertilized crops is quite inefficient and unsustainable, as a large proportion of the applied Pi becomes immobile in the soil or may run off into, and thereby pollute, nearby surface waters. The projected depletion of global rock-Pi reserves has prompted plant scientists to develop strategies and molecular tools for engineering Pi-efficient transgenic crops. However, for this to succeed we need to achieve a thorough understanding of the intricate molecular and biochemical adaptations of Pi-deprived (–Pi) plants, which include the *de novo* synthesis of purple acid phosphatases (PAPs). Purple acid phosphatases likely function in –Pi plants to recycle and scavenge Pi from intra- and extracellular organic Pi-esters. Three Pi-starvation inducible (PSI) PAP isozymes demonstrating distinctive physical and kinetic characteristics have been recently purified and characterized from –Pi tomato suspension cell cultures. Two are secreted monomeric PAPs having M_rs of 84 and 57 kDa, whereas the third is a 142 kDa heterodimeric intracellular

*Corresponding author

A list of abbreviations is given before the references.

PAP composed of 63 and 57 kDa subunits. The three PSI tomato PAPs efficiently hydrolyzed Pi from a wide variety of Pi-esters under acidic conditions and are multifunctional proteins exhibiting significant alkaline peroxidase activity, indicating a potential additional function in the metabolism of reactive oxygen species. Time-course immunoblot studies of tomato cells and seedlings undergoing a transition from Pi sufficiency to Pi deficiency revealed a close relationship between total acid phosphatase activity and relative amounts of the three antigenic PSI PAPs. These results corroborated recent transcript profiling studies suggesting that PSI proteins are subject to both temporal and tissue-specific syntheses in –Pi plants. The discovery of intracellular and secreted PSI tomato PAP isozymes may lead to biotechnological strategies to increase Pi efficiency in tomato plants grown on Pi-limited soils. The aim of this chapter is to present an overview of the properties and functions of PSI PAPs in tomato Pi nutrition.

INTRODUCTION

Phosphorus (P) is a key macronutrient required for metabolism and energy transduction processes in all living cells. It is a structural component of pivotal biomolecules such as ATP, NADPH, nucleic acids, phospholipids, and sugar phosphates. In humans, P is an important nutrient for maintaining regular bone growth and formation, as well as serving as a blood buffering component (Yates et al. 1998). The daily P requirement for humans ranges from 275 to 1250 mg/d, and the richest P sources for the human diet are milk, meats, nuts and legumes. Absorption of orthophosphate ($H_2PO_4^-$, Pi) by the kidney helps to regulate Pi homeostasis in humans. P homeostasis in plants is also important for their optimal growth and metabolism. Unlike animals, plant cells are dependent on soluble soil Pi as the only form of P that they can directly assimilate from the environment. Pi plays a central role in virtually all major metabolic processes in plants, including photosynthesis, respiration, and C and N allocation. Despite its importance, Pi is one of the least available macronutrients in most terrestrial and aquatic ecosystems. Although plentiful in the earth's crust, soil Pi forms strong ionic interactions with Al^{3+}, Ca^{3+}, or Fe^{3+} cations in acidic soils, resulting in the formation of insoluble mineral Pi complexes (Plaxton and Carswell 1999, Vance et al. 2003, Plaxton 2004). This is particularly relevant to tomato plants, which are typically cultivated in acidic soils, with optimal growth in the pH range 5.5-6.5. Similarly, soil organic Pi-esters originating from decaying biomatter cannot be directly absorbed by roots but must be first enzymatically dephosphorylated to liberate soluble Pi for its subsequent uptake by low or high affinity plasmalemma Pi transporters of root epidermal cells (Raghothama 1999).

The typical concentration of soluble Pi in unfertilized soils is often 10^3- to 10^4-fold lower than intracellular Pi concentrations (5-20 mM) present in nutrient-sufficient plants (Bieleski 1973). Pi deficiency in plants including tomato and other crops is associated with a marked reduction in biomass accumulation. In comparison to many other crops, tomato plants require significantly more soil Pi for their optimal growth and fruit development. The maximal deleterious influence of Pi deficiency on tomato seedling growth and development was evident in 3- to 5-wk-old plants (Besford 1978). Indeed, tomato seedlings cultivated for 24 d under -Pi conditions accumulated only 65% of the biomass relative to Pi-sufficient (+Pi) control plants (Bozzo et al. 2006). Similarly, a 60% decrease in the fresh weight of heterotrophic tomato suspension cells was observed in -Pi vs +Pi cultures following their 14 d growth period (Bozzo et al. 2006). This marked inhibition in the growth of -Pi tomato suspension cells was correlated with a ~50-fold reduction in their intracellular free Pi levels (Fig. 1).

The massive use of Pi fertilizers in agriculture demonstrates that the soluble Pi levels of most soils are suboptimal for plant growth. Of the approximately 40 million t of Pi fertilizers currently applied annually, less than 20% is absorbed by crops during their first growth season (Vance et al. 2003). Pi runoff from fertilized fields into nearby surface waters is associated with environmentally damaging processes, particularly aquatic eutrophication. Unlike the global N cycle, Pi fertilizers rely on extraction from non-renewable rock-Pi reserves, which are mainly derived from fossilized bone deposits. The application of Pi fertilizers is quite expensive, particularly in developing countries, and the problem is exacerbated in tropical and sub-tropical regions containing acidic soils. The projected depletion of global rock-Pi reserves within the next 60-80 years raises an interesting dilemma in the face of the world's population explosion (Plaxton 2004). In order to ensure agricultural sustainability and a reduction in Pi fertilizer use, plant scientists need to bioengineer Pi-efficient transgenic crops. The design of biotechnological strategies to enhance Pi acquisition in crops should follow detailed and precise analyses of the complex adaptive mechanisms of -Pi plants as a means of identifying suitable engineering targets. The aim of this article is to provide a brief overview of the remarkable adaptations of -Pi plants, with an emphasis on the properties and roles of intracellular and secreted purple acid phosphatase (PAP) isozymes in facilitating tomato acclimation to nutritional Pi deprivation.

THE PLANT PHOSPHATE-STARVATION RESPONSE

Plants have evolved the ability to acclimate, within species-dependent limits, to extended periods of Pi deprivation. Plant Pi deprivation elicits a

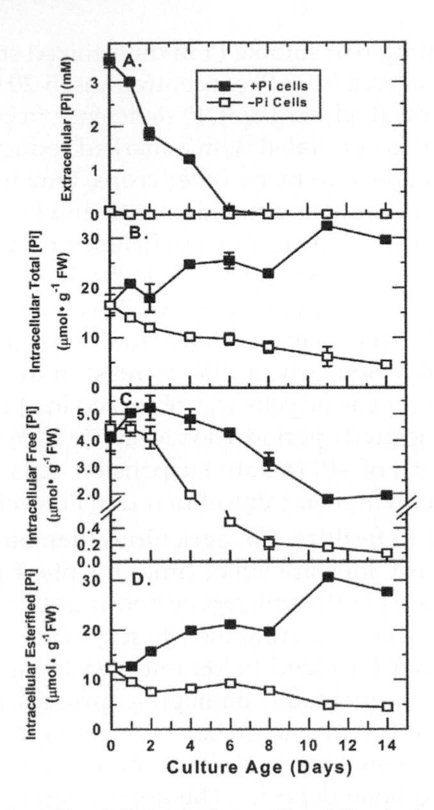

Fig. 1. Influence of nutritional Pi deprivation on Pi and Pi-ester levels of heterotrophic tomato suspension cells. Time-course of (A) extracellular (clarified culture filtrate or 'CCF') Pi levels, and intracellular (B) total, (C) free, and (D) esterified Pi levels of tomato suspension cells cultured in MS media containing 3.5 mM Pi (+Pi; ■) or 0 mM Pi (–Pi; □). "Esterified [Pi]" represents the difference between the respective levels of total *versus* free Pi. All values represent means ±SEM of n = 3 separate flasks (from Bozzo et al. 2006; reproduced with permission of Blackwell Publishing Ltd.).

complex array of morphological, physiological and biochemical adaptations, collectively known as the "Pi starvation response". This response arises in part from the coordinated induction of hundreds of Pi-starvation inducible (PSI) genes in order to reprioritize internal Pi use and maximize external Pi acquisition (Raghothama 1999, Hammond et al. 2004, Ticconi and Abel 2004).

Morphological Adaptations of –Pi Plants

Morphological adaptations that help to increase plant Pi acquisition from soils containing suboptimal Pi levels include (1) alterations in root architecture and diameter, (2) a shift from primary to lateral root growth as a means of promoting topsoil exploration, (3) increased root hair growth and density, and (4) an increase in the root:shoot growth ratio (Plaxton and Carswell 1999, Vance et al. 2003). For example, 35-d-old -Pi tomato seedlings displayed a 4-fold increase in their root:shoot fresh weight ratio relative to +Pi plants (Bozzo et al. 2006). All of these morphological adaptations have evolved as a strategy to significantly increase the root surface area for absorption of the limiting Pi nutrient. An analogous adaptive strategy of most -Pi plants, including tomato, is the formation of symbiotic associations between their roots and beneficial mycorrhizal fungi.

Mycotrophic vs Non-mycotrophic Plants

Pi acquisition by so-called mycotrophic plants is significantly enhanced by the presence of arbuscular mycorrhizae formed between soil-inhabiting fungi of the order *Glomales* and roots of about 90% of known plant species (excluding Chenopodiaceae, Cruciferae, Cyperaceae, Junaceae, and Proteaceae families) (Bolan 1991). The symbiosis develops in the plant roots where the fungus colonizes the apoplast and cells of the cortex to access photosynthate (sucrose) supplied by the host plant. Root colonization of non-fertilized tomato plants by mycorrhizal fungi increases the plant's shoot and root growth and Pi uptake capacity (Kim et al. 1998). However, disruption of beneficial mycorrhizal associations due to Pi fertilization and soil tilling has been an undesirable consequence of modern agriculture. It is notable that many non-mycotrophs, such as buckwheat, white lupin, and harsh hakea plants, are notorious for their ability to thrive on infertile soils. This reflects the view that relative to mycotrophic plants, the non-mycotrophs have evolved to allow more efficient acclimation to low Pi conditions (Murley et al. 1998).

Biochemical Adaptations of Pi-starved Plants

The plant Pi starvation response is associated with numerous biochemical adaptations (Plaxton and Carswell 1999, Plaxton 2004, Plaxton and Podestá 2006). These include the accumulation of the protective pigment anthocyanin in leaves and the induction of high affinity plasmalemma Pi transporters, as well as metabolic Pi recycling and alternative bypass enzymes to Pi- and adenylate-dependent glycolytic and respiratory electron transport reactions (Fig. 2). Anthocyanin accumulation has been observed in leaves of many -Pi plants including tomato and functions to protect chloroplasts against photoinhibition (Plaxton and Carswell 1999, Bozzo et al. 2006, Fig. 4A). Other secondary metabolites such as flavonols (i.e., quercitin) accumulate in the skins of mature green fruit of -Pi tomato

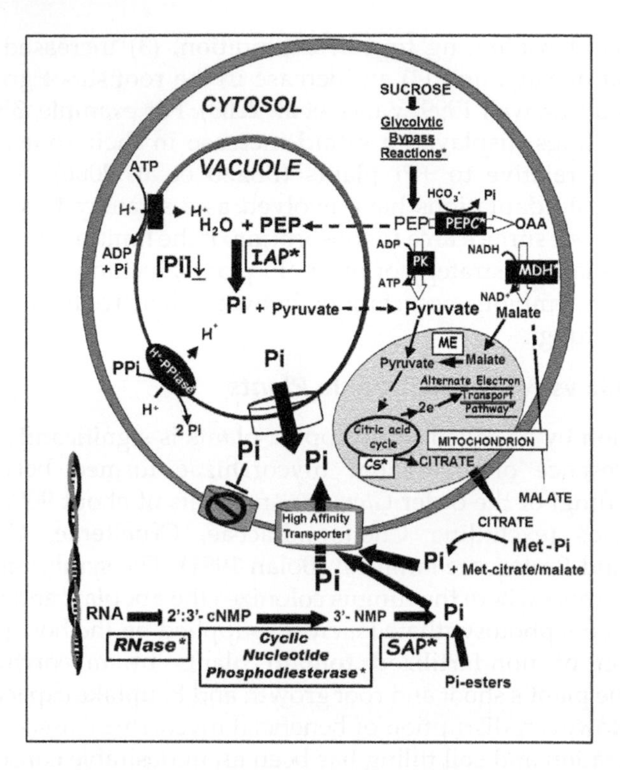

Fig. 2. A model suggesting various adaptive metabolic processes (indicated by asterisks) that are believed to help plants to acclimate to nutritional Pi deficiency. Alternative pathways of cytosolic glycolysis, mitochondrial electron transport, and tonoplast H^+-pumping facilitate respiration and vacuolar pH maintenance by –Pi plant cells because they negate the dependence on adenylates and Pi, the levels of which become markedly depressed during severe Pi starvation. Large quantities of organic acids produced by PEP carboxylase (PEPC), malate dehydrogenase (MDH), and citrate synthase (CS) may also be excreted by roots to increase the availability of mineral-bound Pi (by solubilizing Ca-, Fe- and Al-phosphates = "Met-Pi"). A key component of this model is the critical secondary role played by metabolic Pi recycling systems such as PEPC during Pi deprivation. Apart from a role as a possible PEP phosphatase glycolytic "bypass" to ADP-limited pyruvate kinase, the vacuolar IAP may also recycle Pi from other non-essential Pi-esters and nucleic acid fragments. Likewise, SAP is proposed to scavenge Pi from extracellular Pi-monoester and nucleic acid fragment pools for its eventual uptake by PSI high-affinity Pi transporters of the plasma membrane.

plants, but their precise roles are largely unknown (Stewart et al. 2001). Several PSI glycolytic bypass enzymes such as pyrophosphate-dependent phosphofructokinase, phosphoenolpyruvate (PEP) phosphatase, and PEP carboxylase promote intracellular Pi recycling in –Pi plants, as Pi is a by-product of the reactions catalyzed by each of these enzymes (Fig. 2) (Plaxton and Podestá 2006). Their reactions may also facilitate respiration and/or organic acid excretion, while generating free Pi for its reassimilation into the metabolism of the -Pi cells. -Pi plants also scavenge and conserve Pi by replacing their membrane phospholipids with non-P-containing amphipathic galacto- and sulfonyl lipids. In addition, the induction of secreted ribonuclease, phosphodiesterase, and acid phosphatases functions in the systematic catabolism of soil localized nucleic acids and their degradation products to mobilize Pi, which is then made available for uptake by root-localized high affinity Pi transporter(s) (Fig. 2) (Abel et al. 2000). Scavenging of soil Pi by -Pi plants also arises from the enhanced excretion of photosynthate-derived organic acids. Roots and suspension cell cultures of -Pi plants markedly up-regulate PEP carboxylase (Duff et al. 1989b, Vance et al. 2003, Plaxton and Podestá 2006). PEP carboxylase induction during Pi stress has been correlated with the excretion of significant quantities of malate and citrate. This causes acidification of the rhizosphere, which contributes to the solubilization and assimilation of mineral Pi from the environment.

PLANT ACID PHOSPHATASES

Acid phosphatases (APases) catalyze the hydrolysis of Pi from a broad and overlapping range of Pi-monoesters with an acidic pH optimum (Duff et al. 1994). Eukaryotic APases exist as a wide variety of tissue- and/or cellular compartment-specific isozymes that display variation in their physical and kinetic properties. The induction of intracellular and secreted APases (IAP and SAP, respectively) is thought to be pivotal to the ability of -Pi plants to remobilize and scavenge Pi from non-essential Pi-monoester pools (Fig. 2). The probable function of IAPs is to scavenge and recycle Pi from expendable intracellular Pi-monoester pools (Fig. 2). This is accompanied by a marked reduction in cytoplasmic P-metabolites during extended Pi stress (Duff et al. 1989b, 1994, Plaxton 2004, Bozzo et al. 2006). Secreted APases belong to a group of PSI phosphohydrolases secreted by roots and cell cultures of -Pi plants to hydrolyze Pi from external organophosphates, known to comprise up to 80% of total soil P (Ticconi and Abel 2004). For example, the combined action of secreted PSI ribonucleases, phosphoesterases, and SAPs allows -Pi tomato cell cultures and seedlings to efficiently scavenge extracellular nucleic acids as their sole source of nutritional Pi (Abel et al. 2000, Bosse and Köck 1998, Ticconi and Abel 2004). High-affinity plasmalemma Pi transporters are also induced in -Pi plants, including tomato, to facilitate extracellular Pi uptake against a steep (up to 10,000-fold) concentration gradient (Raghothama 1999).

Purple Acid Phosphatases

Purple APases are distinguished biochemically from other APases by their characteristic pink or purple color in solution, binuclear transition metal active site center, and insensitivity to L-tartrate inhibition (Olczak et al. 2003). They were initially purified and characterized from several microorganisms and mammalian tissues. Over the past 15-20 years, however, a number of plant PAPs have been characterized (Table 1). For many years the main distinction between animal and plant PAPs was thought to be that the former exist as 35 kDa monomers containing a binuclear Fe(III)-Fe(II) center and the latter as larger 110 kDa homodimers containing either an Fe(III)-Zn(II) or Fe(III)-Mn(II) center (Table 1) (Olczak et al. 2003). However, reports of a number of low M_r PAPs in *Arabidopsis* and a plant-like 55 kDa PAP in humans and other animals suggest that there are some similarities between plant and animal PAP proteins (del Pozo et al. 1999, Li et al. 2002, Flanagan et al. 2006). The biochemical characterization of three PSI tomato PAPs added further complexity to the molecular characteristics of plant PAPs, as the two secreted tomato PAPs exist as monomers of varying M_r, whereas the corresponding intracellular PAP exists as a heterodimer (Bozzo et al. 2002, 2004a). A family of 29 putative *Arabidopsis* PAPs has been grouped into subfamilies of high and low M_r isozymes (Li et al. 2002). Structurally distinct plant PAP isozymes may have specific functions in plant Pi metabolism.

BIOCHEMICAL PROPERTIES OF PSI TOMATO PAP ISOZYMES

Tomato suspension cell cultures were generated from seedling cotyledons of *Moneymaker* tomato, a common greenhouse variety. These heterotrophic suspension cells have been widely used to examine tomato Pi starvation responses as their biochemical and molecular responses to Pi stress closely parallel many of those that occur *in planta* (Köck et al. 1998, Raghothama 1999, Bozzo et al. 2006). Suspension cell cultures represent an undifferentiated tissue in which a homogeneous cell population is equally exposed to the environmental parameters prevalent in the liquid culture. Moreover, relatively large quantities of secreted and intracellular proteins can be generated from –Pi cultures, which greatly facilitates the eventual purification and detailed analyses of PSI secreted and intracellular enzymes, including PSI PAPs.

Eight days following subculture of tomato suspension cells into –Pi growth media, secreted (culture media) APase activity increased from undetectable levels to a maximum of 7.5 units/mg protein (Bozzo et al. 2002), whereas intracellular APase activity increased 4-fold from 0.4 to 1.6 units/mg protein (Bozzo et al. 2004a). Two monomeric SAPs (84 kDa

Table 1. Properties and proposed functions of purified plant purple acid phosphatases.

Source	Subcellular localization	Preferred substrate	Physical properties				Proposed function	Reference
			Native M_r (kDa)	Subunit M_r(s) (kDa)	A_{max} (nm)	Metal center		
Yellow lupin seeds	n.d.[a]	Phenyl-P	94	50, 44	n.d.	n.d.	Pi-mobilization	Olczak et al. 2003
Red kidney beans	Secreted	ATP	130	55	560	Fe, Zn	Pi-mobilization	Beck et al. 1986
Soybean cotyledon	Secreted	Phytate	130	70	n.d.	n.d.	Phytate degradation	Hegeman and Grabau 2001
Soybean suspension cells	Secreted	n.d.	130	58	556	Mn(II)	Pi-metabolism	LeBansky et al. 1992
Spinach leaf	n.d.	*p*NPP[b]	92	50	530	Mn	Pi-metabolism	Fujimoto et al. 1977
Duckweed	Cell wall	*p*NPP	100	57	556	Fe,Mn	Pi-recycling	Nakazato et al. 1998
Sweet potato tuber	n.d.	*p*NPP	110	55	515	Mn(III)	Pi-metabolism	Sugiura et al. 1981
Arabidopsis seedlings (AtPAP17)	n.d.	n.d.	34	34	n.d.	n.d.	Pi-recycling/ROS[c] metabolism	Del Pozo et al. 1999
Arabidopsis suspension cells (AtPAP26)	Vacuolar	PEP	100	55	520	n.d.	Pi-recycling/ROS metabolism	Veljanovski et al. 2006
Tomato suspension cells								
(SAP1)	Secreted	PEP	84	84	518	n.d.	Pi-scavenging/ROS metabolism	Bozzo et al. 200
(SAP2)	Secreted	*p*NPP	57	57	538	n.d.	Pi-scavenging/ROS metabolism	Bozzo et al. 2002
(IAP)	Vacuolar	PEP	142	63,57	546	n.d.	Pi-recycling/ROS metabolism	Bozzo et al. 2004a
Rice suspension cells	Cell wall	n.d.	155	65	555	Mn	Cell wall regeneration	Igaue et al. 1976

[a]n.d., not determined.
[b]pNPP, *p*-nitrophenylphosphate.
[c]ROS, reactive oxygen species.

SAP1 and 57 kDa SAP2) were resolved during cation-exchange chromatography and fully purified from the media harvested from the 8-d-old -Pi tomato cultures (Bozzo et al. 2002), whereas a single IAP was purified from clarified extracts of the -Pi tomato cells. They were designated as PAPs because of their (1) violet color in solution, (2) amino acid sequence similarity to putative or previously characterized plant PAPs, and (3) insensitivity to tartrate. Glycosylation is a post-translational modification common to plant APases (Duff et al. 1994). Lectin-affinity chromatography (Con-A Sepharose) is commonly used to bind the glycan side chains of many plant APs, including the 3 PSI tomato PAPs, during their purification. The three tomato PAPs were confirmed to be glycoproteins (Bozzo et al. 2002, 2004a).

Several lines of evidence indicated that the three PSI tomato PAPs are structurally distinct isozymes that differ from the majority of the 100-130 kDa homodimeric plant PAPs examined to date (Table 1). Electrophoresis and analytical gel filtration indicated that SAP1 and SAP2 exist as monomeric glycoproteins of 84 and 57 kDa, respectively. Conversely, the purified IAP is a 142 kDa heterodimer, composed of an equivalent ratio of 63 and 57 kDa subunits (α- and β-subunit, respectively). N-terminal and tryptic peptide sequences of all three tomato PAPs showed similarity with deduced portions of different members of the Arabidopsis PAP family. CNBr peptide maps of SAP1 and SAP2 were quite dissimilar, whereas CNBr cleavage patterns indicated that IAP's α- and β-subunits were structurally distinct. Immunoblots probed with rabbit antibodies raised against the purified SAP1 or IAP indicated the immunological distinctiveness of the 84 kDa SAP1 relative to IAP and SAP2 (Bozzo et al. 2006). SAP1, SAP2 and IAP also displayed different A_{max} values of 518, 538 and 546 nm, respectively (Table 1). Both tomato SAPs were relatively heat-stable, as 100% activity was retained following their incubation at 70°C for 5 min, whereas the IAP was relatively heat-labile, losing 80% of its activity when incubated under identical conditions.

Interrogation of the *Arabidopsis* genome with PAP motif sequences (i.e., the domains containing the conserved metal ligating residues) identified 29 PAPs (Li et al. 2002). According to transcript profiling studies, two of these putative PAPs (AtPAP11 and AtPAP12) were up-regulated by Pi stress. Biochemical analysis of PSI APases from -Pi *Arabidopsis* suspension cells and seedlings identified AtPAP26, a 100 kDa homodimeric vacuole-localized PAP, as the predominant PSI intracellular PAP of *Arabidopsis* (Veljanovski et al. 2006). Although a maximal 5-fold increase in AtPAP26 polypeptide levels and activity occurred following Pi deprivation, no corresponding change in AtPAP26 mRNA expression level was observed, since AtPAP26 appears to be constitutively transcribed in all *Arabidopsis* tissues examined to date. Thus, transcriptional controls exert little influence on AtPAP26 levels, relative to translational and/or proteolytic controls. It is interesting to note that the deduced amino acid sequence of AtPAP26 was highly similar (76% identity) to the deduced sequence for the 57-kDa β-subunit of the heterodimeric PSI IAP previously isolated from -Pi tomato suspension cells (Bozzo et al. 2002, Veljanovski et al. 2006), suggesting that this PAP ortholog plays a key role in mediating intracellular Pi scavenging and recycling by -Pi plant cells. Moreover, comparison of the N-terminal sequence of the 57 kDa β-subunit of the PSI tomato IAP with its deduced full-length sequence revealed that its 31 amino acid signal peptide is cleaved at an identical position as AtPAP26 (Veljanovski et al. 2006).

Plant and mammalian PAPs characterized to date display non-specific substrate selectivity, as they are generally capable of catalyzing Pi hydrolysis from a broad spectrum of Pi-monoesters (Table 1). There are a few reports of vacuolar PAPs exhibiting a substrate preference for the glycolytic intermediate PEP (Duff et al. 1989a), whereas a PAP purified from germinating soybeans displayed preferential utilization of phytate (Hegeman and Grabau, 2001). Many of the plant PAPs examined to date have been proposed to function in the hydrolysis of Pi-monoesters during Pi starvation (Table 1). In fact, all three PSI tomato PAP isozymes were classified as non-specific APases (Table 1), capable of hydrolyzing Pi from a wide range of physiologically relevant Pi-esters (Bozzo et al. 2002, 2004a).

Differential Synthesis of PSI Tomato PAP Isozymes is Correlated with Intracellular Pi Levels

Prolonged Pi deprivation of tomato suspension cells and seedlings led to decreased biomass accumulation, and in the case of seedlings an increased root:shoot ratio. This altered growth pattern was associated with a marked (up to 50-fold) decrease in intracellular Pi concentration (Fig. 1) (Bosse and Köck 1998, Bozzo et al. 2006). This magnitude of intracellular Pi decline has been well documented in various plants (Duff et al. 1989b, Theodorou and Plaxton 1994, Plaxton and Carswell 1999, Veljanovski et al. 2006). The biochemical Pi starvation response of plants can be divided into an early and late stage. In the early Pi starvation response, cytoplasmic Pi is maintained at the expense of vacuolar Pi. Prolonged Pi starvation results in depletion of vacuolar Pi pools and subsequent marked reductions in cytoplasmic Pi and P-metabolite levels (Plaxton 2004, Bozzo et al. 2006). Tomato suspension cells undergoing a transition to Pi deficiency experienced a 10- and 50-fold reduction in intracellular free Pi within 6 d (early phase) and 14 d (late phase) of their transfer into -Pi media (Fig. 1).

The shift of tomato suspension cells and seedlings from +Pi to –Pi conditions resulted in increased SAP1, SAP2, and IAP polypeptide synthesis and activity (Bozzo et al. 2006). The induction of PSI PAP isozyme synthesis and activity in –Pi tomato suspension cells occurred 5 d following the depletion of Pi in the growth media (Fig. 3) (Bozzo et al. 2006), and in maize roots this phenomenon was correlated with the depletion of vacuolar Pi stores (Plaxton and Carswell 1999). This points to an intracellular Pi-sensor that may trigger the Pi starvation response, as has been suggested for the induction of tomato PSI ribonucleases and APases (Köck et al. 1998). The marked delay in IAP induction in leaves and stems relative to roots of the –Pi tomato seedlings (Bozzo et al. 2006) agrees with reports that –Pi plants preferentially redistribute limiting Pi to the leaves at the expense of roots (Theodorou and Plaxton 1994, Mimura et al. 1996). This has been hypothesized to represent an adaptive strategy to sequester limiting Pi to the leaves so as to maintain optimal rates of photosynthesis for as long as possible in –Pi plants (Theodorou and Plaxton 1994) and is in accord with our observation that free Pi levels of leaves of the 30-d-old –Pi tomato seedlings were at least 3-fold those of the corresponding roots (Bozzo et al. 2006).

That enhanced IAP synthesis plays an important *in vivo* role recycling Pi from intracellular P-metabolites in the 6 to 11 d –Pi tomato cells was supported by the concomitant 2-fold reduction in levels of intracellular esterified Pi over the same time period (Figs. 1 and 3). In addition, purified PSI IAP from –Pi tomato suspension cells displayed potent allosteric inhibition by Pi (K_i = 1.6 mM) (Bozzo et al. 2004a). The intracellular Pi concentration of the 0 to 4 d –Pi cells (~2.0-4.5 mM, assuming 1 g FW ~1 ml) (Fig. 1C) should exert significant IAP inhibition *in vivo*. Conversely, the subsequent > 20-fold reductions in intracellular Pi levels of the 6-11 d –Pi cells is expected to relieve the PSI tomato IAP from Pi inhibition. Between day 11 and 14, the IAP activity and concentration of the –Pi suspension cells decreased to that of +Pi cells (Fig. 3A and 3C). This could be due to IAP-mediated depletion of intracellular P-metabolites to a minimal level necessary to sustain cellular viability.

It is notable that the large reduction in intracellular Pi content of the –Pi tomato suspension cells was associated with the temporal induction of IAP (day 6) and SAP (day 8) activities and corresponding antigenic IAP and SAP1/SAP2 polypeptides respectively present in clarified cell extracts and cell culture filtrate (Fig. 3). Similarly, SAP1 polypeptide accumulation began at least 7 d following the up-regulation of IAP protein in roots of the –Pi seedlings (Bozzo et al. 2006). Likewise, genome-wide expression studies of –Pi *Arabidopsis* seedlings implicated a variety of PSI genes that display short-term expression, whereas prolonged

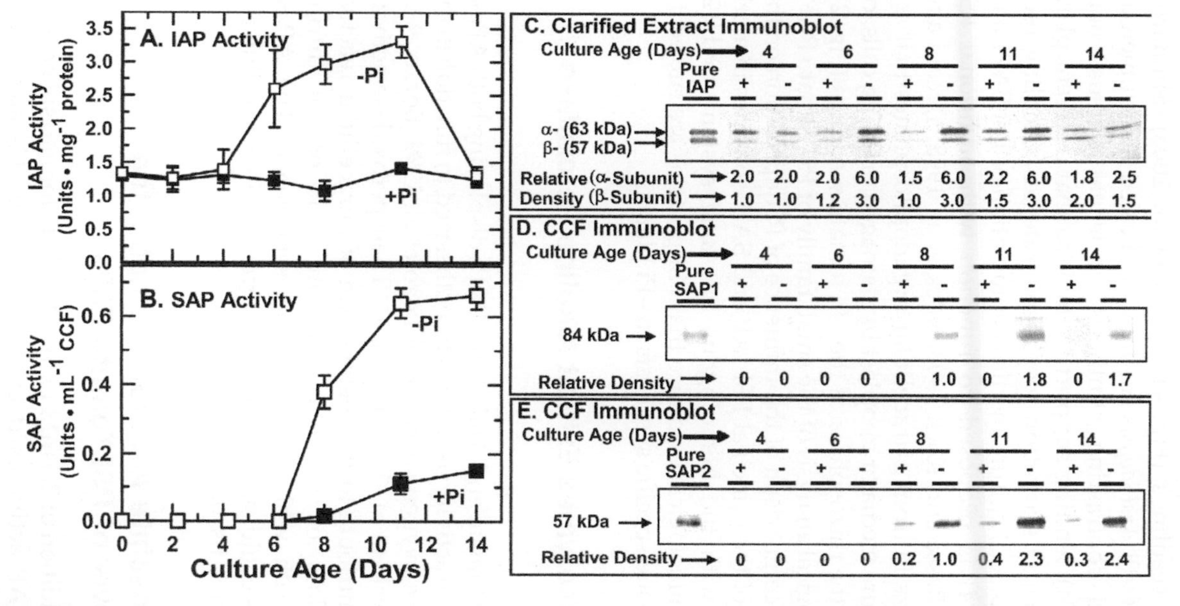

Fig. 3. Acid phosphatase activities and immunological detection of intracellular and secreted Pi-starvation inducible purple acid phosphatase isozymes of tomato suspension cells. The cells were cultured in Murashige-Skoog media containing 3.5 mM Pi (+Pi; ■) or 0 mM Pi (–Pi; □). (A and B) Time-course of intracellular and secreted APase activity (IAP and SAP) respectively present in clarified cell extracts (A) and CCF (B). All values represent the means ± S.E. of $n = 3$ separate flasks. (C, D, and E) Immunoblot analyses of PSI tomato PAP isozymes. Purified IAP (C) (50 ng), SAP1 (D) (20 ng), SAP2 (E) (20 ng), and clarified cell extract proteins (C) (5 μg/lane) or CCF proteins (D and E) (1.4 μg/lane) were resolved by SDS-PAGE and blotted as described in Bozzo et al. (2006). Immunoblots were probed with a 1:50 dilution of affinity-purified anti-(tomato IAP)-IgG (C and E) or a 1:2000 dilution of anti-(tomato SAP1) immune serum (D). Relative amounts of antigenic polypeptides were estimated via laser densitometry. Lanes labeled "+" and "–" denote clarified cell extracts or CCF from +Pi and –Pi cultures, respectively (from Bozzo et al. 2006; reproduced with permission of Blackwell Publishing Ltd).

Pi-deficiency elicited a switch to the expression of alternate PSI genes (Hammond et al. 2004).

Of further interest was the apparent tissue-specific synthesis of SAP1 in roots of –Pi tomato seedlings (Fig. 4) (Bozzo et al. 2006). Similarly, transcript profiling of –Pi *Arabidopsis* seedlings indicated that numerous PSI genes exhibited contrasting synthesis in leaves versus roots (Hammond et al. 2004), lending support to the idea that different plant organs exhibit distinct Pi starvation response strategies. Seedling root surface and suspension cell culture media accumulation of PSI SAP1 and SAP2 (Figs. 3 and 4) is proposed to function with additional PSI phosphohydrolases (Bosse and Köck 1998, Köck et al. 1998, Ticconi and Abel 2004) to mobilize Pi from extracellular organophosphates that are released into the culture media or apoplast by damaged or dying cells, or otherwise present in the rhizosphere of the –Pi tomato seedlings (Fig. 2). The recent immunological analysis of the small family of PSI tomato PAPs provided the first (1) comparison of the influence of Pi deprivation on the coordinate synthesis of the principal PSI IAP and SAP isozymes in a cell suspension culture with the homologous *in planta* system, and (2) biochemical and immunological evidence that PSI proteins are subject to both temporal and tissue-specific synthesis in –Pi plants.

PSI Tomato PAP Isozymes Exhibit Alkaline Peroxidase Activity

Pi-starvation induced tomato and *Arabidopsis* PAPs displaying both APase and alkaline peroxidase activity have been hypothesized to function in the metabolism of reactive oxygen species (del Pozo et al. 1999, Bozzo et al. 2002, 2004a, Veljanovski et al. 2006) (Table 1). Their peroxidase activity was suggested to contribute to the metabolism of intracellular reactive oxygen following loss of vacuolar integrity that occurs during programmed cell death which accompanies senescence. The APase activity of these PAPs could simultaneously participate in intracellular Pi recycling from senescing tissues.

Phosphate or Phosphite Addition Promotes the Proteolytic Turnover of PSI Tomato PAPs

Within 48 hr of the addition of 2.5 mM Pi to 8-d-old –Pi tomato suspension cells, (1) SAP and IAP activities decreased by about 12- and 6-fold, respectively, and (2) immunoreactive PAP polypeptides either disappeared (SAP1 and SAP2) or were substantially reduced (IAP) (Bozzo et al. 2004b). In-gel protease assays correlated the disappearance of both

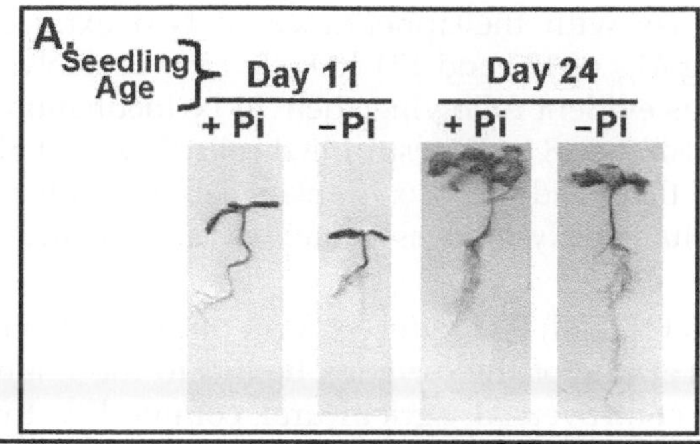

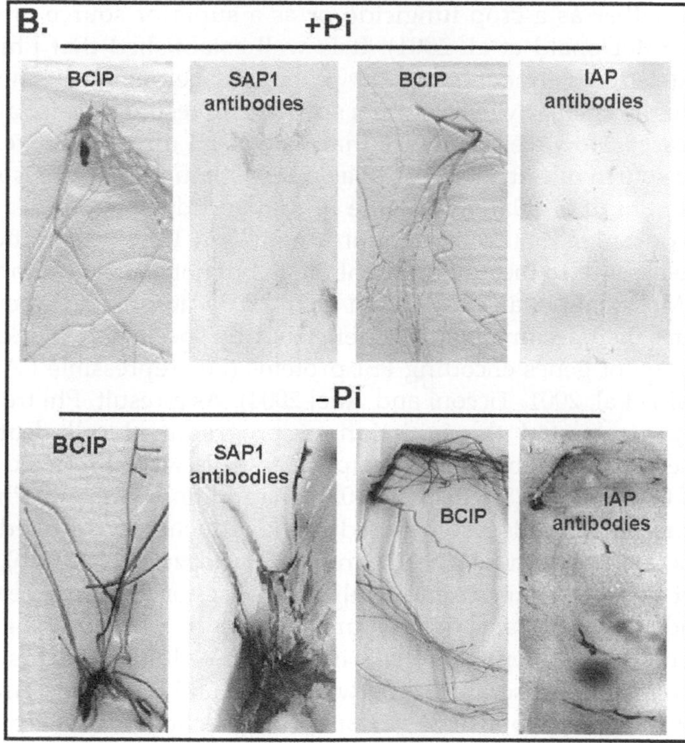

Fig. 4. Influence of Pi starvation on tomato seedling growth and root secretion of extracellular purple acid phosphatase isozymes. (A) Tomato seedlings cultivated on Murashige-Skoog-solidified agar supplemented with 0 mM (–Pi) or 2.5 mM Pi (+Pi) were harvested at 11 d (shown on left) and 24 d (shown on right) post-imbibition. (B) Localization of root surface SAP activity and protein. +Pi and –Pi tomato seedlings (35-d-old) were harvested, and excised root material incubated with 5-Br-4-Cl-3-indoyl-P (BCIP) (colorless APase substrate). Root surface SAP activity is visualized as a dark color, indicating BCIP hydrolysis. Following *in situ* SAP activity staining, immunoblots of the intact root preparations were probed with a 1:2000 dilution of rabbit anti-(tomato SAP1) immune serum or a 1:50 dilution of affinity-purified anti-(tomato IAP)-IgG. Antigenic polypeptides were visualized using an alkaline phosphatase-tagged secondary antibody (from Bozzo et al. 2006; reproduced with permission of Blackwell Publishing Ltd).

PSI SAP isozymes with the appearance of two extracellular serine proteases having M_rs of 127 and 121 kDa. *In vitro* proteolysis of purified tomato SAPs was evident following their 24 hr incubation with culture media filtrate from the 48 hr Pi-resupplied cells (Bozzo et al. 2004b). The results indicated that Pi addition to -Pi plant cells not only represses PAP genes, but simultaneously induces proteases that specifically target PSI PAPs.

Phosphite (HPO_3^-, Phi) is a reduced form of Pi in which a non-acidic hydrogen replaces one of the oxygens bound to the P atom. Phi is an important, but controversial agricultural commodity that is widely marketed either as a crop fungicide or as a superior source of crop P nutrition (McDonald et al. 2001). It is well established that Phi-based fungicides effectively control plant pathogens belonging to the order Oomycetes (particularly *Phytophthora* sp.) that are responsible for a host of crop diseases. However, evidence that plants can directly utilize Phi as their sole source of nutritional P is lacking. Although plant cells readily assimilate Phi, it is relatively stable *in vivo* and does not appear to be readily oxidized or metabolized. Interestingly, low Phi concentrations are highly deleterious to the development of -Pi but not +Pi plants, including tomato (McDonald et al. 2001, Singh et al. 2003, Ticconi and Abel 2004). Phi negates their acclimation to Pi deficiency by specifically blocking the derepression of genes encoding PSI proteins (i.e., repressible PAP, etc.) (McDonald et al. 2001, Ticconi and Abel 2004). As a result, Phi treatment markedly accelerated the initiation of programmed cell death that otherwise occurs when *B. napus* suspension cells were subjected to one month of Pi starvation (Singh et al. 2003). The addition of 2.5 mM Phi to -Pi tomato suspension cells also resulted in the induction of secreted serine proteases that target the PSI SAP1 and SAP2 (Bozzo et al. 2004b). While Phi is not a substrate in enzyme-catalyzed phosphoryl transfer reactions, plant and yeast Pi transporters participating in Pi uptake or signal transduction components involved in sensing cellular Pi status do not appear to efficiently discriminate between the Pi and Phi anions. Although Phi is pivotal in the control of *Phytophthora* infection of numerous crop plants, it is clearly phytotoxic to -Pi plants. Thus, its widespread marketing in agriculture as a supposed superior source of crop P nutrition is very questionable, since -Pi crops treated with Phi will be at a distinct metabolic and physiological disadvantage relative to similar plants receiving Pi fertilizers (McDonald et al. 2001, Singh et al. 2003).

CONCLUDING REMARKS

The design of biotechnological strategies to enhance crop Pi acquisition from organic and inorganic pools of soil Pi is of great interest to plant scientists and agronomists wishing to reduce the overuse of non-renewable and polluting Pi fertilizers in agriculture. Characterization of intra- and extracellular PSI tomato PAP isozymes (1) indicated their probable physiological role in Pi scavenging and Pi recycling by –Pi tomato, (2) corroborated recent transcriptomic studies suggesting that PSI proteins are subject to both temporal and tissue-specific synthesis in –Pi plants, and (3) led to the discovery of Pi and Phi inducible secreted serine proteases that target PSI tomato SAPs. It will also be of interest to assess whether these proteases are also involved in the degradation of additional secreted PSI hydrolytic enzymes of –Pi tomato cell cultures, including PSI ribonucleases and phosphodiesterases. Any gains in plant Pi nutrition that might arise from the successful overexpression of intracellular or secreted PSI tomato PAP isozymes could be lost if crops were cultivated in Pi-rich soils or Pi-deficient soils where Phi is applied either as a fungicide or as a supposed Pi-fertilizer substitute. Therefore, biotechnological strategies for engineering Pi-efficient crops should bear in mind that PSI protein overexpression in transgenic plants may be enhanced by modified protease expression and/or the design of protease-resistant PSI proteins.

ACKNOWLEDGMENTS

WCP is indebted to past and present members of his laboratory who have examined the biochemical adaptations of vascular plants, green algae, and yeast to Pi scarcity. We thank Ms. Vicki Knowles for providing critical comments on the manuscript. Research and Equipment Grants funding by the Natural Sciences and Engineering Research Council of Canada (NSERC) is also gratefully acknowledged.

ABBREVIATIONS

APase: acid phosphatase; CCF: cell culture filtrate; IAP: intracellular acid phosphatase; PAP: purple acid phosphatase; PEP: phosphoenolpyruvate; Phi: phosphite; Pi: phosphate; +Pi: phosphate-sufficient; –Pi: phosphate-deprived; PSI: phosphate-starvation inducible; SAP: secreted acid phosphatase

REFERENCES

Abel, S. and T. Nürnberger, V. Ahnert, G.-J. Krauss, and K. Glund. 2000. Induction of an extracellular cyclic nucleotide phosphodiesterases as an accessory ribonucleolytic activity during phosphate starvation of cultured tomato cells. Plant Physiol. 122: 543-552.

Beck, J.L. and L.A. McConachie, A.C. Summors, W.N. Arnold, J. de Jersey, and B. Zerner. 1986. Properties of a purple phosphatase from red kidney bean: a zinc-iron metalloenzyme. Biochim. Biophys. Acta 869: 61-68.

Besford, R.T. 1978. Phosphorus nutrition and acid phosphatase activity in the leaves of seven plant species. J. Sci. Food Agric. 30: 281-285.

Bieleski, R.L. 1973. Phosphate pools, phosphate transport, and phosphate availability. Ann. Rev. Plant Physiol. 24: 225-252.

Bolan, N.S. 1991. A critical review on the role of mycorrhizal fungi in the uptake of phosphorus by plants. Plant Soil 134: 189-207.

Bosse, D. and M. Köck. 1998. Influence of phosphate starvation on phosphohydrolases during development of tomato seedlings. Plant Cell Environ. 21: 325-332.

Bozzo, G.G. and K.G. Raghothama, and W.C. Plaxton. 2002. Purification and characterization of two secreted purple acid phosphatase isozymes from phosphate-starved tomato (*Lycopersicon esculentum*) cell cultures. Eur. J. Biochem. 269: 6278-6286.

Bozzo, G.G. and K.G. Raghothama, and W.C. Plaxton. 2004a. Structural and kinetic properties of a novel purple acid phosphatase from phosphate-starved tomato (*Lycopersicon esculentum*) cell cultures. Biochem J. 377: 419-428.

Bozzo, G.G. and V.K. Singh, and W.C. Plaxton. 2004b. Phosphate or phosphite addition promotes the proteolytic turnover of phosphate-starvation inducible tomato purple acid phosphatase isozymes. FEBS Lett. 573: 51-54.

Bozzo, G.G. and E.L. Dunn, and W.C. Plaxton. 2006. Differential synthesis of phosphate-starvation inducible purple acid phosphatase isozymes in tomato (*Lycopersicon esculentum*) suspension cells and seedlings. Plant Cell Environ. 29: 303-313.

Davis R.L. and H. Zhang, J.L. Schroder, J.J. Wang, M.E. Payton, and A. Zazulak. 2005. Soil characteristics and phosphorus level effect on phosphorus loss in runoff. J. Environ. Qual. 34: 1640-1650.

Del Pozo, J.C. and I. Allona, V. Rubio, A. Leyva, A. de la Peña, C. Aragoncillo, and J. Paz-Ares. 1999. A type 5 acid phosphatase gene from *Arabidopsis thaliana* is induced by phosphate starvation and by some other types of phosphate mobilising/oxidative stress conditions. Plant J. 19: 579-589.

Duff, S.M.G. and D.D. Lefebvre, and W.C. Plaxton. 1989a. Purification and characterization of a phospho*enol*pyruvate phosphatase from *Brassica nigra* suspension cells. Plant Physiol. 90: 724-741.

Duff, S.M.G. and G.B.G. Moorhead, D.D. Lefebvre, and W.C. Plaxton. 1989b. Phosphate starvation inducible 'bypasses' of adenylate and phosphate dependent glycolytic enzymes in *Brassica nigra* suspension cells. Plant Physiol. 90: 1275-1278.

Duff, S.M.G. and G. Sarath, and W.C. Plaxton. 1994. The role of acid phosphatases in plant phosphorus metabolism. Physiol. Plant. 90: 791-800.

Flanagan, J.U. and A.I. Cassady, G. Schenk, L.W. Guddat, and D.A. Hume. 2006. Identification and molecular modeling of a novel, plant-like, human purple acid phosphatase. Gene 377: 12-20.

Fujimoto, S. and T. Nakagawa, and A. Ohara. 1977. Purification and some properties of violet-colored acid phosphatase from spinach leaves. Chem. Pharm. Bull. 25: 1459-1462.

Hammond, J.P. and M.R. Broadley, and P.J. White. 2004. Genetic responses to phosphorus deficiency. Ann. Bot. 94: 323-332.

Hegeman, C.E. and E.A. Grabau. 2001. A novel phytase with sequence similarity to purple acid phosphatases is expressed in cotyledons of germinating soybean seedlings. Plant Physiol. 126: 1598-1608.

Igaue, I. and H. Watabe, K. Takahashi, M. Takekoshi, and A. Morota. 1976. Violet-colored acid phosphatase isozymes associated with cell wall preparations from rice plant cultured cells. Agr. Biol. Chem. 40: 823-825.

Kim, K.Y. and D. Jordan, and G.A. McDonald. 1998. Effect of phosphate-solubilizing bacteria and vesicular-arbuscular mycorrhizae on tomato growth and soil microbial activity. Biol. Fertil. Soils 26: 79-87.

Köck, M. and K. Theierl, I. Stenzel, and K. Glund. 1998. Extracellular administration of phosphate-sequestering metabolites induces ribonucleases in cultured tomato cells. Planta 204: 404-407.

LeBansky, B.R. and T.D. McKnight, and L.R. Griffing. 1992. Purification and characterization of a secreted purple phosphatase from soybean suspension cultures. Plant Physiol. 99: 391-395.

Li, D. and H. Zhu, K. Liu, X. Liu, G. Leggewie, M. Udvardi, and D. Wang. 2002. Purple acid phosphatases of *Arabidopsis thaliana*. Comparative analysis and differential regulation by phosphate deprivation. J. Biol. Chem. 277: 27772-27781.

McDonald, A.E. and B.R. Grant, and W.C. Plaxton. 2001. Phosphite: Its relevance in the environment and agriculture, and influence on the plant Pi starvation response. J. Plant Nutr. 24: 1505-1519.

Mimura, T. and K. Sakano, and T. Shimmen. 1996. Studies on the distribution, re-translocation and homeostasis of inorganic phosphate in barley leaves. Plant Cell Environ. 19: 311-320.

Murley, V.R. and M.E. Theodorou, and W.C. Plaxton. 1998. Phosphate-starvation inducible pyrophosphate-dependent phosphofructokinase occurs in plants whose roots do not form symbiotic associations with mycorrhizal fungi. Physiol. Plant. 103: 405-414.

Nakazato, H. and T. Okamoto, M. Nishikoori, K. Washio, N. Morita, K. Haraguchi, G.A. Thompson, and H. Okuyama. 1998. The glycosylphosphatidylinositol-anchored phosphatase from *Spirodela oligorrhiza* is a purple acid phosphatase. Plant Physiol. 118: 1015-1020.

Oehl, F. and E. Sieverding, P. Mader, D. Dubois, K. Ineichen, T. Boller, and A. Wiemken. 2004. Impact of long-term conventional and organic farming on the diversity of arbuscular mycorrhizal fungi. Oecologia 138: 574-583.

Olczak, M. and M. Bronislawa, and W. Watorek. 2003. Plant purple acid phosphatases – genes, structures and biological function. Acta Biochim. Pol. 50: 1245-1256.

Plaxton, W.C. Plant response to stress: biochemical adaptations to phosphate deficiency. pp. 976-980. *In:* R. Goodman. [ed.] 2004. Encyclopedia of Plant and Crop Science. Marcel Dekker, New York, USA.

Plaxton, W.C. and M.C. Carswell. Metabolic aspects of the phosphate starvation response in plants. pp. 349-372. *In:* H.R. Lerner. [ed.] 1999. Plant Responses to Environmental Stresses: From Phytohormones to Genome Reorganization. Marcel Dekker, New York, USA.

Plaxton, W.C. and F.E. Podestá. 2006. The functional organization and control of plant respiration. CRC Crit. Rev. Plant Sci. 25: 159-198.

Raghothama, K.G. 1999. Phosphate acquisition. Annu. Rev. Plant Physiol. Plant Mol. Biol. 50: 665-693.

Singh, V. and S.M. Wood, V.L. Knowles, and W.C. Plaxton 2003. Phosphite-accelerated programmed cell death in phosphate-starved oilseed rape (*Brassica napus*) suspension cell cultures. Planta 218: 233-239.

Stewart, A.J. and W. Chapman, G.I. Jenkins, I. Graham, T. Martin, and A. Crozier. 2001. The effect of nitrogen and phosphorus deficiency on flavonol accumulation in plant tissues. Plant Cell Environ. 24: 1189-1197.

Sugiura, Y. and H. Kawabe, H. Tanaka, S. Fujimoto, and A. Ohara. 1981. Purification, enzymatic properties and active site determination of a novel manganese(III)-containing acid phosphatase. J. Biol. Chem. 256: 10664-10670.

Theodorou, M.E. and W.C. Plaxton. 1994. Induction of PPi-dependent phosphofructokinase by phosphate starvation in seedlings of *Brassica nigra*. Plant Cell Environ. 17: 287–294.

Ticconi, C.A. and S. Abel. 2004. Short on phosphate: plant surveillance and countermeasures. Trends Plant Sci. 9: 548-555.

Vance, C.P. and C. Udhe-Stone, and D.L. Allan. 2003. Phosphorus acquisition and use: critical adaptations by plants for securing a nonrenewable resource. New Phytol. 157: 423-447.

Veljanovski, V. and B. Vanderbeld, V.L. Knowles, W.A. Snedden, and W.C. Plaxton. 2006. Biochemical and molecular characterization of AtPAP26, a vacuolar purple acid phosphatase up-regulated in phosphate-deprived Arabidopsis suspension cells and seedlings. Plant Physiol. 142: 1282-1293.

Yates, A.A. and S.A. Schlicker, and C.W. Suitor. 1998. Dietary reference intakes: the new basis for recommendations for calcium and related nutrients, B vitamins, and choline. J. Am. Diet. Assoc. 98: 699-706.

11

Nutritional Characterization of Tomato Juices

Concepción Sánchez-Moreno, Begoña de Ancos, Lucía Plaza, Pedro Elez-Martínez and M. Pilar Cano

Department of Plant Foods Science and Technology
Instituto del Frío, Consejo Superior de Investigaciones Científicas (CSIC)
José Antonio Novais 10
Ciudad Universitaria, E-28040 Madrid, Spain

ABSTRACT

Tomato (*Lycopersicon esculentum* Mill.) is the second most important vegetable crop. Tomato and tomato products such as tomato juice have remarkably high concentrations of micronutrients, such as vitamin C, vitamin E, folate, and minerals. In addition to their micronutrient content, tomatoes and in consequence tomato juice also contain valuable phytochemicals or bioactive components: phenolic compounds and carotenoids such as lycopene and the provitamin A β-carotene. There are several factors affecting the nutritional value of tomato products, among them tomato variety and maturity, and processing and storage conditions. The bioactive components in tomato are thought to contribute to the reduced risk of cardiovascular disease and prostate cancer in humans by reducing oxidant status and inflammation, inhibiting cholesterol synthesis, improving immune function, or acting as chemopreventive agent. *In vitro* studies, epidemiological studies, cellular, animal and human intervention studies provide support for significant health effects of tomato juice consumption, although the molecular pathways of tomato components need further investigation.

A list of abbreviations is given before the references.

INTRODUCTION

Tomato (*Lycopersicon esculentum* Mill.) is the second most important vegetable crop next to potato. Present world production is about 100 million t fresh fruit from 3.7 million ha (FAOSTAT 2001). Tomatoes are the second highest produced and consumed vegetable in the United States today. Apart from its macronutrient composition (Table 1), tomato has remarkably high concentrations of folate, vitamin C, vitamin E, and carotenoids, such as lycopene and the provitamin A β-carotene (Beecher 1998). Consequently, tomatoes and tomato-based foods may provide a convenient matrix by which nutrients and other health-related food components can be supplied to humans. Accordingly, protective activity of tomato products on *in vivo* markers of lipid oxidation (low density lipoprotein (LDL) oxidizability, and F_2-isoprostanes) has been reported (Visioli et al. 2003). Moreover, the consumption of tomato products has been associated with a lower risk of developing digestive tract and prostate cancers (Giovannucci et al. 2002). These protective effects may be due to the ability of lycopene and other antioxidant components to prevent cell damage through synergistic interactions (Friedman 2002, George et al. 2004).

More than 80% of tomatoes produced are consumed in the form of processed products such as juice, paste, puree, ketchup, sauces and soups (Willcox et al. 2003). There is some confusion in the literature concerning the definitions of concentrated tomato juice, pulp, puree, and paste, but the following definitions are found (Hayes et al. 1998): Tomato pulp refers to crushed tomatoes either before or after the removal of skins and seeds. Tomato juice refers to juice from whole crushed tomatoes from which the skins and seeds have been removed and which has been subject to fine screening, and is intended for consumption without dilution or concentration. Tomato paste is the product resulting from the concentration of tomato pulp, after the removal of skins and seeds, and contains 24% or more natural tomato soluble solids. Tomato paste that is

Table 1 Nutritional composition of tomato juice (macronutrients and energy).

Macronutrients	*Units*	*Values per 100 g*	*Values per 243 g*
Water	g	93.90	228.18
Carbohydrate	g	4.24	10.30
Total dietary fiber	g	0.4	1.0
Protein	g	0.76	1.85
Total lipid (fat)	g	0.05	0.12
Energy	kcal	73	177

Source: USDA 2006a.

marketed to the consumer in small packs and sold as a condiment may also be described as tomato puree. Tomato puree is the term applied to lower concentrations of tomato paste (containing 8-24% natural tomato soluble solids). Rather unfortunately, tomato puree in the United States can also be called "tomato pulp" or, if it satisfies certain legislative criteria, "concentrated tomato juice".

In any case, the nutritional value of tomato products depends on processing and storage conditions (Shi and Le Maguer 2000, Anese et al. 2002). Homogenization, heat treatment, and the incorporation of oil in processed tomato products leads to increased lycopene bioavailability, while some of the same processes cause significant loss of other nutrients. Nutrient content is also affected by variety and maturity (Willcox et al. 2003). Special concerns arise in the case of juice or other tomato beverage products because of the losses of vitamin C (Hayes et al. 1998).

MICRONUTRIENTS

The main micronutrients in tomato juice are vitamin C, total folate, vitamin A, vitamin E, and potassium (Table 2). The primary contributors to daily vitamin intake are fruit juices.

Vitamin C

A serving of about 243 g of tomato juice provides 44.5 mg of vitamin C (Table 2, USDA 2006a). Vitamin C is particularly susceptible to destruction by oxidation during processing and storage of tomato juice. For a long time, vitamin C has been studied as a quality parameter in the processing of tomato juice (Sidhu et al. 1975). It has been shown that the retention of ascorbic acid by processed juices was 90.3% for juice made by cold extraction and 93.7% for juice made by hot extraction, both using a screw-type equipment, and 83.2% for juice made by cold extraction and 85.5% for juice made by hot extraction, both using a superfine pulper with stainless steel paddles and strainer. In this context, the effects of processing on the quality attributes of tomato juice were studied, showing that ascorbic acid was destroyed to a high extent during mechanical extraction for tomato juice elaboration. In addition, there was a remarkable variation among heat treatments in their effects on the quality components of tomato juice (Daood et al. 1990). In another study, the vitamin C content of different tomato products (juice, baked tomatoes, sauce, and soup) decreased during the thermal processing of tomatoes (Gahler at al. 2003). Sánchez-Moreno et al. (2006a) found a mean value for ascorbic acid and vitamin C of 17.9 ± 4.31 mg/100 mL and 20.0 ± 4.19 mg/100 mL, respectively, in different commercial tomato juices analyzed. The authors showed that the

Table 2 Nutritional composition of tomato juice (micronutrients—vitamins and minerals).

Micronutrients	*Units*	*Values per 100 g*	*Values per 243 g*
Vitamins			
Vitamin C, total ascorbic acid	mg	18.3	44.5
Thiamin	mg	0.047	0.114
Riboflavin	mg	0.031	0.075
Niacin	mg	0.673	1.635
Pantothenic acid	mg	0.250	0.608
Vitamin B6	mg	0.111	0.270
Folate, total	mcg	20	49
Vitamin A, IU	IU	450	1094
Vitamin A, RAE	mcg RAE	23	56
Vitamin E (alpha-tocopherol)	mg	0.32	0.78
Vitamin K (phylloquinone)	mcg	2.3	5.6
Minerals			
Calcium	mg	10	24
Iron	mg	0.43	1.04
Magnesium	mg	11	27
Phosphorus	mg	18	44
Potassium	mg	229	556
Sodium	mg	10	24
Zinc	mg	0.15	0.36
Copper	mg	0.061	0.148
Manganese	mg	0.070	0.170
Selenium	mcg	0.3	0.7

Source: USDA 2006a.

stability of vitamin C was affected by the type of container (Table 3). Podsêdek et al. (2003) showed variable contents of ascorbic acid in tomato juices, varying from 0.69 mg/100 g to 26.87 mg/100 g of tomato juice.

Traditionally, foods have been preserved by using heat (commercial sterilization, pasteurization and blanching), but the industry is developing alternatives to the use of heat preservation. Recently, interest in non-thermal food preservation technologies, such as high-pressure and pulsed electric fields technologies, has increased appreciably in order to affect the nutritional quality of the food as little as possible. In this context, vitamin C was measured in tomato purees subjected to high-pressure (as emerging, non-convectional, non-thermal or even new technology) and traditional thermal technologies. In general, high-pressure and thermal treatments showed a decrease in total vitamin C content (about 29%), with no statistically significant differences among them. Although thermal treatments were relatively short (30 s and 1 min), the temperature use for

Table 3 Vitamin C content (mg/100 mL) of freshly made (A) and commercial traditional pasteurized (B-G) tomato juices.[a]

Sample	*Ascorbic acid*	*Total vitamin C*
A	13.3 ± 0.13b	16.5 ± 0.81b
B	23.6 ± 1.03f	25.4 ± 1.74d
C	7.65 ± 0.15a	9.19 ± 0.52a
D	14.4 ± 0.47c	15.7 ± 0.36b
E	20.8 ± 0.22e	23.1 ± 0.65d
F	17.3 ± 0.40d	19.3 ± 0.60c
G	59.4 ± 0.71g	67.6 ± 4.55e

[a] Values are means ± standard deviation, $n = 6$. Means within a column with different letters are significantly different at $P < 0.05$ (Sánchez-Moreno et al. 2006a).

the pasteurization (70°C and 90°C) could be responsible for the depletion in vitamin C. This was in accordance with the large body of evidence supporting vitamin C destruction in tomato products within this temperature range. The authors also found depletion in vitamin C content in the high-pressurized tomato puree, which could be due to the long treatment time (15 min). No studies dealing with the effect on vitamin C in tomato products preserved by high-pressure treatment were found in the literature (Sánchez-Moreno et al. 2006b). Min et al. (2003) showed that tomato juice processed by pulsed electric fields retained more ascorbic acid than thermally processed (92°C for 90 s) tomato juice stored at 4°C for 42 d.

Vitamin A

According to the USDA National Database, 100 g of tomato juice has 450 IU of vitamin A. Vitamin A is provided from plant foods as carotenoids that can be biologically transformed to active vitamin A (Rodríguez-Amaya 1996). In a study in which commercial tomato juices were nutritionally evaluated, the provitamin A β-carotene, and the vitamin A value, ranged from 182 to 321 μg/100 mL and from 17.6 to 31.5 retinol activity equivalents/100 mL, respectively. In freshly made tomato juice these values were slightly higher (369 μg/100 mL and 32.8 RAE/100 mL, respectively) (Sánchez-Moreno et al. 2006a).

Vitamin E

Vitamin E refers to a family of eight molecules having a chromanol ring, four tocopherols (α-, β-, γ-, and δ-) and four tocotrienols (α-, β-, γ-, and δ-). Vitamin E in tomatoes is predominantly represented by α-tocopherol, the content of which is 0.32 mg per 100 g in the case of tomato juice. Homogenization and sterilization of tomatoes during production of tomato juice resulted in significant losses of contents of α-tocopherol on wet as well as dry weight bases (Seybold at al. 2004). The α-tocopherol content for tomato juices from a local market in Poland was 1.60 mg/100 g of tomato juice (Podsędek et al. 2003).

B Vitamins

Cataldi et al. (2003) showed the content of flavins obtained for tomato juice. They obtained a concentration for flavin mononucleotide (290.8 μg/L) relatively high compared with the mean values of flavin adenine dinucleotide (15 μg/L) and riboflavin (137.3 μg/L).

Minerals

Tomato juice is a rich source of potassium. Glucose, fructose, citric and malic acids and their potassium salts were shown to be the main taste-active components of tomato juice (Salles et al. 2003).

BIOACTIVE CONSTITUENTS IN TOMATO

Tomatoes contain many bioactive components including carotenoids and phenolic compounds (Table 4).

Table 4 Nutritional composition of tomato juice (bioactive compounds).

Bioactive compounds	*Units*	*Values per 100 g*	*Values per 243 g*
Carotenoids			
Carotene, beta	mcg	270	656
Carotene, alpha	mcg	0	0
Cryptoxanthin, beta	mcg	0	0
Lycopene	mcg	9037	21960
Lutein + zeaxanthin	mcg	60	146
Phytoene	mcg	1900	4617
Phytofluene	mcg	830	2017
Polyphenols			
Apigenin	mg	0.00	0.00
Luteolin	mg	0.00	0.00
Kaempferol	mg	0.06	0.144
Myricetin	mg	0.05	0.12
Quercetin	mg	1.46	3.504

Source: USDA 2006a, b, Tonucci et al. 1995.

Carotenoids

Tomatoes and tomato-based food products such as juice are the major sources of lycopene (10.77 mg/100 g) and a number of other carotenoids, such as phytoene (1.90 mg/100 g), phytofluene (0.83 mg/100 g), ζ-carotene (0.18 mg/100 g), γ-carotene (1.74 mg/100 g), β-carotene (0.27 mg/100 g), neurosporene (1.23 mg/100 g), and lutein (0.06 mg/100 g) (Khachik et al. 2002) (Fig. 1).

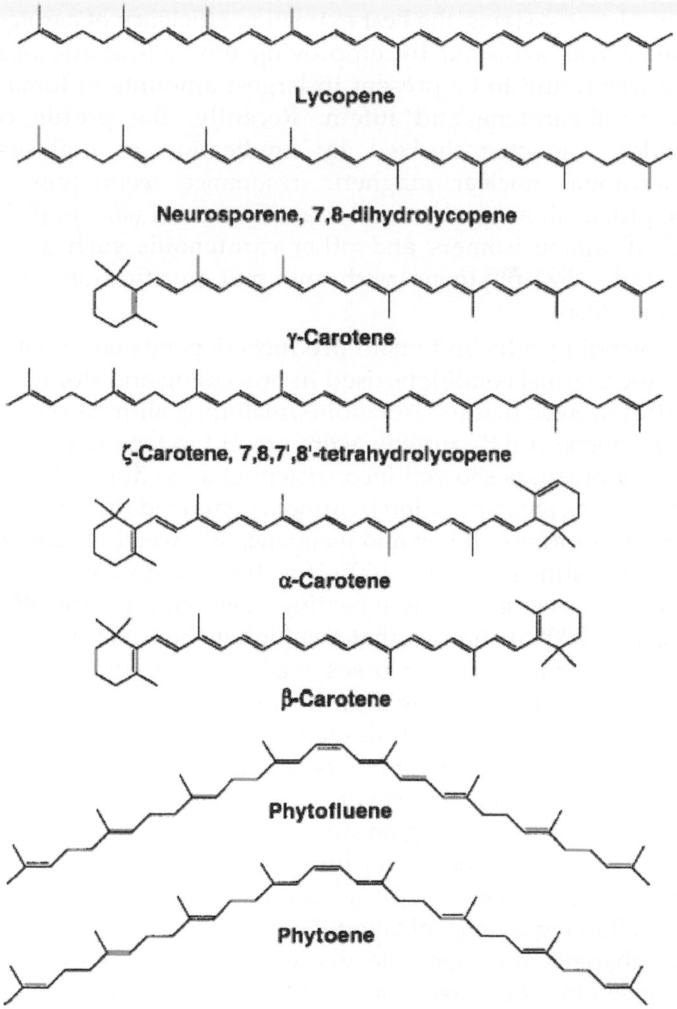

Lycopene

Neurosporene, 7,8-dihydrolycopene

γ-Carotene

ζ-Carotene, 7,8,7',8'-tetrahydrolycopene

α-Carotene

β-Carotene

Phytofluene

Phytoene

Fig. 1 Chemical structures of carotenoids in tomatoes and tomato-based food products (Khachik et al. 2002).

Podsędek et al. (2003) found a mean carotenoid content in tomato juices from Poland of 5.67 mg/100 g; Markovic et al. (2006) reported a lycopene content of 20.10 ± 13.83 mg/100 g wet weight for tomato juice commonly consumed in Croatia. The distribution of lycopene and related carotenoids in tomatoes and tomato-based food products has been determined by extraction and high-performance liquid chromatography-UV/visible photodiode array detection. In this context, Lin and Chen (2003) identified and quantified 16 carotenoids, including all-*trans*-lutein, all-*trans*-β-carotene, all-*trans*-lycopene and their 13 *cis* isomers. Of the various extraction solvent systems, the best extraction efficiency of carotenoids in tomato juice was achieved by employing ethanol-hexane (4:3, v/v). Lycopene was found to be present in largest amounts in tomato juice, followed by β-carotene and lutein. Recently, the profile of these biomolecules was characterized by application of high-resolution multidimensional nuclear magnetic resonance techniques using a cryogenic probe, allowing the rapid identification of (*all-E*)-, (5Z)-, (9Z)-, and (13Z)-lycopene isomers and other carotenoids such as (*all-E*)-β-carotene and (15Z)-phytoene with minimal purification procedures (Tiziani et al. 2006).

The carotenoid profile in tomato products depends on tomato variety as well as the thermal conditions used in processing and storage. Lin and Chen (2005a) showed that 16 carotenoids, including all-*trans* plus *cis* forms of lutein, lycopene and β-carotene, were present in tomato juice. Most *cis* isomers of carotenoids showed inconsistent change during heating. The high-temperature, short-duration treatment generated the highest yield of all-*trans* plus *cis* forms of lutein and lycopene, followed by heating at 90°C for 5 min and heating in water at 100°C for 30 min. Only a minor change in β-carotene was observed for these heating treatments. On the other hand, Lin and Chen (2005b) showed that the higher the tomato juice storage temperature, the greater are the losses of all-trans plus cis forms of lutein, β-carotene and lycopene during illumination. All-*trans*-lycopene showed the highest degradation loss, followed by all-*trans*-β-carotene and all-*trans*-lutein. More *cis*-isomers of lycopene than lutein or β-carotene were generated during storage. However, the type of major isomers formed may be inconsistent, depending on storage conditions. In a study in which tomato juice was produced under industrial-like conditions, on wet as well as dry bases, contents of (*E*)-lycopene and (*E*)-β-carotene significantly decreased following homogenization (Seybold et al. 2004). In contrast, no consistent changes in lycopene levels were observed as fresh tomatoes were processed into hot break juice, but statistically significant decreases in lycopene levels of 9-28% occurred as the tomatoes were processed into paste (Takeoka et al. 2001). Moreover, the effect of microwave processing on the nutritional characteristics of tomato juice was investigated and compared with conventional extraction methods such as hot break and cold break extraction. The authors showed that microwave-processed juice had low electrolytes, high viscosity and high retention of ascorbic acid, total carotenoids and lycopene contents compared to conventionally processed tomato juice (Kaur et al. 1999). Generally, studies on carotenoid content of tomato juices have not systematically followed changes throughout processing. In some cases it is not known whether observed changes are due to improved extraction during the processing or to the effects of the heat treatment.

In another study, commercial conventional thermal pasteurized tomato juices were nutritionally evaluated, at the same point in their commercial shelf life, for their carotenoid contents. Higher lycopene epoxide, lycopene, γ-carotene, β-carotene contents, and vitamin A values were found in those juices obtained from a concentrated tomato source, suggesting structural changes in the tomato tissue due to the concentration process (Sánchez-Moreno et al. 2006a) (Table 5).

The same authors compared the impact of high-pressure (emerging technology) and low pasteurization, high pasteurization, freezing, and high pasteurization plus freezing (traditional thermal technologies) on carotenoids of tomato puree. Individual and total carotenoids, and provitamin A carotenoids, were significantly higher in high-pressurized tomato puree than in the untreated and other treated tomato purees (Sánchez-Moreno et al. 2006b). It has been reported that high-pressure treatment can affect the membranes in vegetable cells (Smelt 1998). In addition, carotenoids are tightly bound to macromolecules, in particular to protein and membrane lipids, and high-pressure processing is known to affect macromolecular structures such as proteins and polymer carbohydrates (Butz and Tauscher 2002). In this context, in a study designed to investigate the effect of combined treatments of high-pressure and natural additives (citric acid and sodium chloride) on carotenoid extractability and antioxidant activities of tomato puree, the carotenoid extractability was at a maximum when tomato puree was subjected to pressures from 200 to 400 MPa, and without additives (Sánchez-Moreno et al. 2004). In contrast, Fernández-García et al. (2001) did not observe effect of pressurization in the concentration of lycopene and β-carotene, extracted with tetrahydrofuran, but lower recovery of carotenoids with petroleum ether suggested structural changes in the tomato pulp tissue due to processing.

Table 5 Carotenoid content and vitamin A value (RAE/100 mL) of freshly made (A) and commercial traditional pasteurized (B-G) tomato juices.[a]

Sample	*Lutein*	*Lycopene epoxide*	*Lycopene*	*γ-Carotene*	*β-Carotene*	*Total carotenoids*	*Vitamin A*
A	29.4 ± 4.31a	53.2 ± 5.02a	1024 ± 98.3a	48.3 ± 4.23a	369 ± 20.0d	1524 ± 160.3a	32.8 ± 3.02d
B	33.6 ± 1.96a	78.7 ± 4.30a	2129 ± 270b	56.4 ± 5.95a	182 ± 8.79a	2480 ± 272b	17.6 ± 0.72a
C	70.3 ± 5.29cd	136 ± 19.4c	2449 ± 284bc	103 ± 16.1cd	265 ± 40.0b	3023 ± 221c	26.4 ± 3.94bc
D	51.2 ± 7.98b	183 ± 23.0d	2930 ± 297d	109 ± 16.0d	314 ± 38.4bc	3587 ± 309d	30.7 ± 3.19cd
E	61.3 ± 4.17c	133 ± 22.4bc	2544 ± 260c	101 ± 9.00cd	290 ± 40.2bc	3129 ± 187c	28.4 ± 3.69bcd
F	70.6 ± 10.09d	108 ± 21.9b	2369 ± 189bc	85.6 ± 7.49bc	265 ± 25.5b	2899 ± 249c	25.7 ± 2.39b
G	64.2 ± 4.49cd	155 ± 12.8c	3435 ± 226e	115 ± 5.47d	321 ± 14.4c	4090 ± 232e	31.5 ± 1.42d

[a] Values are means ± standard deviation, $n = 6$. Means within a column with different letters are significantly different at $P < 0.05$ (Sánchez-Moreno et al. 2006a).

Polyphenols

Among phenolic compounds, tomatoes contain, primarily as conjugates, the flavonoids quercetin and kaempferol. Tomato juice was found to be a rich source of flavonols with total flavonol contents of 15.2-16.9 μg/mL (Stewart et al. 2000). In contrast to tomato fruit, which contains almost exclusively conjugated quercetin, up to 30% of the quercetin in processed produce was in the free form. These authors did not observed the hydrolysis of flavonol conjugates during cooking of tomatoes, so the accumulation of quercetin in juices, puree, and paste may be a consequence of enzymatic hydrolysis of rutin and other quercetin conjugates during pasteurization and processing (Stewart et al. 2000). The authors suggest that the concentration of flavonols in tomato juice is likely to depend on the extraction of flavonols from the skin into the juice during initial processing, which often involves heating, and also on the amount of skin remaining in the tomato juice following filtration. Tokuşoğlu et al. (2003) found a content of 17.6 mg/L of quercetin in tomato juice. Volikakis and Efstathiou (2005) found 17 mg/L of quercetin in tomato juice quantified as quercetin equivalents (quercetin + 1.06[kaempferol] + 1.37 [myricetin]). Hertog et al. (1993) reported that tomato juice contains 13 mg/L quercetin. Stewart et al. (2000) reported that two different types of tomato juice (commercial composite) contained 14.4 and 16.2 μg/mL quercetin. Tokuşoğlu et al. (2003) found kaempferol levels in fresh tomatoes (0.2-0.6 mg/kg) similar to that in tomato juice (1.7 mg/L) and tomato salsa (0.9-1.1 mg/kg). In this context, Stewart et al. (2000) found that tomato juice contained 0.7-0.8 μg/mL kaempferol after hydrolysis. Regarding myricetin content, both Tokuşoğlu et al. (2003) and Hertog et al. (1993) found less than 0.5 mg/L in tomato juice. Podsędek et al. (2003) reported a mean total polyphenol content of 36.77 mg/100 g of tomato juice.

Other Bioactive Compounds

Plant sterols occur both as free sterols and as bound conjugates, i.e., fatty acid esters (mainly C16 and C18 fatty acids), esters of phenolic acids, glycosides (most commonly with *b*-D-glucose) and acylated glycosides (esterified at the 6-hydroxy group of the sugar moiety). The plant sterol content in a given plant may depend on many factors, such as genetic background, growing conditions, tissue maturity and post-harvest changes. Piironen et al. (2003) showed that the total plant sterol content in cucumber, potted lettuce, onion, potato and tomato was < 100 mg/kg, but no data have been found regarding sterol content in tomato juice.

HEALTH-RELATED PROPERTIES

In vitro Antioxidant Aspects

Antioxidant can neutralize free radicals, which cause oxidative damage to biological molecules, especially to lipids, proteins, and nucleic acids. Vitamins C and E, carotenoids, and polyphenols are the major antioxidant compounds of fruits and vegetables. Tomato juice is rich in these micronutrients and bioactive compounds with antioxidant properties. There are several studies describing the antioxidative activity of tomato juice. The trolox equivalent antioxidant capacity (TEAC) was evaluated in tomato juice, along with other lycopene-containing foods, showing that the foods with the highest antioxidant capacity per serving did not have the highest lycopene levels (Djuric and Powell 2001). Podsędek et al. (2003) reported that the average antioxidative capacities of hydrophilic fractions by the 2,2′-azinobis(3-ethylbenzothiazoline-6-sulfonic acid) (ABTS) method were 1.40 μmol TEAC/g for tomato juices. Mean values for lipophilic fractions were 0.31 μmol TEAC/g for tomato juices. The authors concluded that the higher activity of hydrophilic fraction could be due to high antioxidant content and/or high antioxidative activity of polyphenols. In addition, Sánchez-Moreno et al. (2006a) reported that radical-scavenging capacity (by the 2,2-diphenyl-1-picrylhydrazyl (DPPH) method) was higher in the aqueous fractions than in the organic fractions of tomato juices. Vitamin C was mainly responsible for the radical-scavenging capacity of the aqueous tomato juice fractions, whereas lutein and lycopene were the carotenoids responsible, not only for the EC_{50} (amount of antioxidant necessary to decrease by 50% the initial DPPH concentration) but also for the kinetic of the DPPH changes of the organic tomato juice fractions. The antioxidant capacity of tomato juice had been also analyzed by the oxygen radical absorbance capacity (ORAC) assay. In this study, among the commercial fruit juices, grape juice had the highest ORAC activity, followed by grapefruit juice, tomato juice, orange juice, and apple juice (Wang et al. 1996). Common European fruit and vegetable juices were also evaluated according to their scavenging capacity against three reactive oxygen species, peroxyl and hydroxyl radicals and peroxynitrite. Nearly all juices (apple, beetroot, blueberry, carrot, elderberry, lemon, lingonberry, orange, pink grapefruit, sauerkraut, and tomato) were most efficient against peroxyl radicals; a much smaller number were efficient against peroxynitrite, and even fewer were efficient against hydroxyl radicals (Lichtenthaler and Marx 2005).

Naturally occurring antioxidants could be significantly lost as a consequence of processing and storage. In particular, thermal treatments are believed to be the main cause of the depletion in natural antioxidants.

Therefore, considering the role of these compounds as health-promoting factors, the original antioxidant properties of raw foods should be maintained through the use of optimized food processing conditions. In this context, the changes in the antioxidant properties (chain breaking and oxygen scavenging activities) of tomato juices as a consequence of heat treatments were studied. The results showed that heating caused an increase in the overall antioxidant potential of the tomato juice, as a consequence of the formation of melanoidins during the advanced steps of the Maillard reaction. However, short heat treatment promoted an initial reduction in the original antioxidant potential, suggesting that compounds with pro-oxidant properties were formed during the early stages of the Maillard reaction (Anese et al. 1999). Gahler et al. (2003) studied changes in the antioxidant capacity by processing tomatoes into different products. The hydrophilic antioxidant capacity was measured by three different methods: TEAC assay, ferric reducing antioxidant power (FRAP) test, and the photochemiluminescence assay (PCL). A significant increase (FRAP and PCL) resulted only after the homogenization step. Sterilization and bottling led to decreases in antioxidant capacity. The authors suggested that this decrease might be based on the interaction of the juice with oxygen during the processing (Gahlet et al. 2003).

Fernández-García et al. (2001) found that the high-pressure processing of tomato homogenate did not lead to changes in the antioxidant activity, by means of the radical cation assay, within the water-soluble fraction; however, during cold storage it exceeded that of the controls, which could be due to pressure-induced changes in enzyme activity. When the impact of high pressure (emerging technology) was compared with that of low pasteurization, high pasteurization, freezing, and high pasteurization plus freezing (traditional thermal technologies) on radical scavenging activity of tomato puree, no significant differences were found among high-pressurized, low pasteurized, high pasteurized, and high pasteurized plus frozen tomato purees. In the study, an inverse significant correlation between the ascorbic acid content and the total carotenoid content was found in the different processed tomato purees, suggesting a protecting effect of the hydrophilic component ascorbic acid over the lipophilic components during processing (Sánchez-Moreno et al. 2006b).

It is worthy to mention some studies in which the tomato juice was functionalized by adding vegetable byproduct extracts (Larrosa et al. 2002) or synthesized hydroxytyrosol (Larrosa et al. 2003). In these two studies, the antioxidant activity of functional tomato juice significantly increased over control juice according to a set of *in vitro* antioxidant assays (i.e., inhibition of lipid peroxidation determined by the ferric thiocyanate method and scavenging of both ABTS and DPPH free radicals). More recently, tomato juice had been functionalized with soy with the idea that a soy-enhanced tomato juice could potentially deliver several health benefits as a result of a synergistic effect arising from combining soy isoflavones and tomato lycopene (Goerlitz and Delwiche 2004, Tiziani and Vodovotz 2005).

Cardiovascular Disease and Cancer Prevention

Among bioactive components in tomatoes, lycopene (the main carotenoid) is often assumed to be responsible for the positive health effects seen with increased intake of tomato and tomato products. The antioxidant effect of lycopene is potentially beneficial in prevention of cardiovascular disease (Arab and Steck 2000). In addition, it has been suggested that increased plasma concentrations of lycopene are associated with a decreased risk of prostate cancer (Giovannucci 2002). In this regard, lycopene and tomato products such as juice could possibly reduce the disease development by reducing oxidant status and inflammation, inhibiting cholesterol synthesis, improving immune function, or acting as chemopreventive agent.

Epidemiological Evidence

Some epidemiological studies have reported an inverse association between consumption of tomato juice or tomato products and prevention of cardiovascular disease. A prospective cohort of 39,876 middle-aged and older women initially free of cardiovascular disease and cancer were enrolled in a study designed to determine whether the intake of lycopene or tomato-based products was associated with the risk of cardiovascular disease. Although the authors found no association between total dietary lycopene intake and the risk of cardiovascular disease in women, there was evidence that overall intake of tomato-based products may be associated with a reduced risk of cardiovascular disease (Sesso et al. 2003). Lycopene accounts for about ~50% of the carotenoids found in human blood. It is believed that the major dietary sources of lycopene are tomatoes and tomato products (Mangels et al. 1993, Scott et al. 1996). However, tomatoes and tomato products vary in their lycopene bioavailability, depending on whether they are processed, consumed raw, or cooked (Stahl and Sies 1992). In the over 521,000 subjects who volunteered to participate in a study by the European Prospective Investigation into Cancer and Nutrition (EPIC), the consumption of tomatoes (raw and cooked) and tomato products (sauces, pastes, ketchup) was measured with country-specific dietary questionnaires across 10 countries (Denmark, France, Germany, Greece, Norway, Italy, Spain,

Sweden, the Netherlands, and the United Kingdom). The authors found that the Greek EPIC cohort had one of the highest intakes of tomato and tomato products in this study. However, the correlation of plasma lycopene with the Mediterranean diet score in the EPIC Cross-Sectional Study was determined to be weak in most EPIC regions (Jenab et al. 2005). In the Health Professionals Follow-Up Study, an intake of two or more servings a week of tomato products (including tomatoes, tomato sauce, tomato juice, pizza, salsa, taco sauce, and ketchup) resulted in a lower risk of prostate cancer (Giovannucci et al. 2002). In general, future epidemiologic studies should examine populations with relatively high intakes of tomato products, be sufficiently large to evaluate moderate relative risks, and examine a wide range of age populations.

Experimental Evidence—Intervention Studies

There are several human intervention studies dealing with the plasma carotenoid response to the intake of selected foods, such as tomato juice (Micozzi et al. 1992, Stahl and Sies 1992, Paetau et al. 1998, 1999, Muller et al. 1999, Hadley et al. 2003, Allen et al. 2003, Cohn et al. 2004, Porrini et al. 2005, Frohlich et al. 2006). In all these studies, healthy subjects consumed realistic amounts of tomato juice during intervention periods ranging from 15 d to 6 wk. In general, the subjects involved in these studies showed increased concentrations of plasma carotenoids after the intervention periods, and changes in plasma carotenoid concentrations occurred rapidly with variation in dietary intake.

Carotenoids such as lycopene are highly lipophilic and are commonly found within cell membranes and other lipid components. It is therefore thought that the ability of carotenoids to scavenge free radicals may be greatest in a lipophilic environment. Studies that support the *in vivo* oxidation of LDL contribute to the hypothesis that dietary components may be important for LDL oxidative modification. In this context, Oshima et al. (1996) found that the LDLs from a healthy volunteer following a long-term supplementation with tomato juice accumulated cholesteryl eter hydroperoxides more slowly than the LDL prepared before supplementation when the suspensions containing these LDLs were subjected to a singlet oxygen-generating system. A study was undertaken to investigate the effect of dietary supplementation of lycopene on LDL oxidation in 19 healthy human subjects. Dietary lycopene was provided using tomato juice, spaghetti sauce, and tomato oleoresin for 1 wk each. Dietary supplementation of lycopene significantly increased serum lycopene levels by at least 2-fold. Although there was no change in serum cholesterol levels (total, LDL, or high-density lipoprotein), serum lipid peroxidation and LDL oxidation were significantly decreased (Agarwal and Rao 1998). Another study concluded that the fact that the 3- to 6-fold enrichments of LDL with β-carotene achieved by dietary supplementation of healthy humans with tomato juice were more effective in inhibiting oxidation than the 11- to 12-fold enrichments achieved by an *in vitro* method suggested that dietary supplementation is a more appropriate procedure for studies involving the enrichment of lipoprotein with carotenoids (Dugas et al. 1999). There are also studies measuring the LDL oxidation in response to dietary intervention with tomato juice in different patients. Sutherland et al. (1999) found that increased oxidative stress and susceptibility of LDL to oxidation may not be reduced by increasing plasma lycopene levels with regular consumption of tomato juice for 4 wk in 15 patients with a kidney graft. This same research group found that consumption of commercial tomato juice for 4 wk increased plasma lycopene levels and the intrinsic resistance of LDL to oxidation almost as effectively as supplementation with high dose of vitamin E, which also decreased plasma levels of C-reactive protein, a risk factor for myocardial infarction, in 57 patients with diabetes (Upritchard et al. 2000). The effects of tomato juice supplementation on the carotenoid concentration in lipoprotein fractions and the oxidative susceptibility of LDL were also investigated in 31 healthy Japanese female students. The authors conclude that α-tocopherol was the major determinant in protecting LDL from oxidation, while lycopene from tomato juice supplementation may contribute to protect phospholipids in LDL from oxidation (Maruyama et al. 2001). A more recent study designed to evaluate in 17 healthy subjects the effects of a long-term tomato-rich diet, consisting of various processed tomato products, on bioavailability and antioxidant properties of lycopene concluded that at the end of the treatment (4 wk), serum lycopene level and total antioxidant potential increased significantly, and lipid and protein oxidation was reduced significantly (Rao 2004).

As a conclusion from these studies it could be stated that the reduced susceptibility of lipoproteins from subjects fed a carotenoid-rich tomato product diet to *ex vivo* oxidative stress suggests a protective effect against oxidative stress *in vivo*.

In addition to the oxidative modification of LDL, considered to play an important role in the pathogenesis of atherosclerosis, lipid oxidation products are involved in the formation of mutagenic DNA adducts, which may contribute to carcinogenesis. Moreover, it has been suggested that the antioxidant and anti-inflammatory properties of some phytochemicals, such as carotenoids and tocopherols present in tomato juice, may be responsible for the anticancer effects of diets rich in fruits and vegetables. In consequence, there are cellular and animal studies (Okajima et al. 1998, Balestrieri et al. 2004, Cetin et al. 2006, Kasagi et al. 2006) and several human intervention studies evaluating the effect of tomato juice on total antioxidant capacity of plasma, markers of inflammation, immunomodulation, oxidative stress, and DNA damage (PoolZobel et al. 1997, Rao and Agarwal 1998, Steinberg and Chait 1998, Bohm and Bitsch 1999, Watzl et al. 1999, 2000, 2003, Bub et al. 2000, 2002, Durak et al. 2003, Briviba et al. 2004, Tyssandier et al. 2004, Aust et al. 2005, Porrini et al. 2005, Madrid et al. 2006, Riso et al. 2006a, b).

These studies suggest that tomato juice can significantly modulate carotenoid levels in the bloodstream. In addition, they imply that consumption of tomato juice may provide protection from *in vivo* oxidative damage.

In conclusion, tomatoes and tomato products such as tomato juice are an important source of micronutrients and some phytochemicals. Certain carotenoids found in tomatoes and tomato products appear to be more highly associated with potential health benefits. However, synergistic or additive effects of several tomato micronutrients and phytochemicals may mediate these findings. Therefore, future studies of tomato juice micronutrients and bioactive components on bioavailability, metabolism, and bioactivity will provide more insight into these potential health benefits suggested by the epidemiological and experimental evidence.

ACKNOWLEDGEMENTS

The authors acknowledge the financial support of Ministerio de Educación y Ciencia (AGL2005-03849, AGL2006-12758-C02-01/ALI and 200670I081), and European Union (FP6-FOOD-023140).

ABBREVIATIONS

ABTS: 2,2′-azinobis(3-ethylbenzothiazoline-6-sulfonic acid); FRAP: ferric reducing antioxidant power; DPPH: 2,2-diphenyl-1-picrylhydrazyl; LDL: low density lipoprotein; ORAC: oxygen radical absorbance capacity; PCL: photochemiluminescence; RAE: retinol activity equivalents; TEAC: trolox equivalent antioxidant capacity

REFERENCES

Agarwal, S. and A.V. Rao. 1998. Tomato lycopene and low density lipoprotein oxidation: a human dietary intervention study. Lipids 33: 981-984.

Allen, C.M. and S.J. Schwartz, N.E. Craft, E.L. Giovannucci, V.L. de Groff, and S.K. Clinton. 2003. Changes in plasma and oral mucosal lycopene isomer concentrations in healthy adults consuming standard servings of processed tomato products. Nutr. Cancer 47: 48-56.

Anese, M. and L. Manzocco, M.C. Nicoli, and C.R. Lerici. 1999. Antioxidant properties of tomato juice as affected by heating. J. Sci. Food Agric. 79: 750-754.

Anese, M. and P. Falcone, V.Fogliano, M.C. Nicoli, and R. Massini. 2002. Effect of equivalent thermal treatments on the color and the antioxidant activity of tomato purées. J. Food Sci. 67: 3442-3446.

Arab, L and S. Steck. 2000. Lycopene and cardiovascular disease. Am. J. Clin. Nutr. 71: 1691S-1695S.

Aust, O. and W. Stahl, H. Sies, H. Tronnier, and U. Heinrich. 2005. Supplementation with tomato-based products increases lycopene, phytofluene, and phytoene levels in human serum and protects against UV-light-induced erythema. Int. J. Vitam. Nutr. Res. 75: 54-60.

Balestrieri, M.L. and R. de Prisco, B. Nicolaus, P. Pari, V.S. Moriello, G. Strazzullo, E.L. Iorio, L. Servilllo, and C. Balestrieri. 2004. Lycopene in association with alpha-tocopherol or tomato lipophilic extracts enhances acyl-platelet-activating factor biosynthesis in endothelial cells during oxidative stress. Free Radic. Biol. Med. 36: 1058-1067.

Beecher, G.R. 1998. Nutrient content of tomatoes and tomato products. Proc. Soc. Exp. Biol. Med. 218: 98-100.

Bohm, V. and R. Bitsch. 1999. Intestinal absorption of lycopene from different matrices and interactions to other carotenoids, the lipid status, and the antioxidant capacity of human plasma. Eur. J. Nutr. 38: 118-125.

Briviba, K. and K. Schnabele, G. Rechkemmer, and A. Bub. 2004. Supplementation of a diet low in carotenoids with tomato or carrot juice does not affect lipid peroxidation in plasma and feces of healthy men. J. Nutr. 134: 1081-1083.

Bub, A. and B. Watzl, L. Abrahamse, H. Delincee, S. Adam, J. Wever, H. Muller, and G. Rechkemmer. 2000. Moderate intervention with carotenoid-rich vegetable products reduces lipid peroxidation in men. J. Nutr. 130: 2200-2206.

Bub, A. and S. Barth, B. Watzl, K. Brivika, B.M. Herbert, P.M. Luhrmann, M. Neuhauser-Berthold, and G. Rechkemmer. 2002. Paraoxonase 1 Q192R (PON1-192) polymorphism is associated with reduced lipid peroxidation in R-allele-carrier but not in QQ homozygous elderly subjects on a tomato-rich diet. Eur. J. Nutr. 41: 237-243.

Butz, P. and B. Tauscher. 2002. Emerging technologies: chemical aspects. Food Res. Int. 35: 279-284.

Cataldi, T.R.I. and D. Nardiello, V. Carrara, R. Ciriello, and G.E. De Benedetto. 2003. Assessment of riboflavin and flavin content in common food samples by capillary electrophoresis with laser-induced fluorescence detection. Food Chem. 82: 309-314.

Cetin, R. and E. Devrim, B. Kilicoglu, A. Avci, O. Candir, and I. Durak. 2006. Cisplatin impairs antioxidant system and causes oxidation in rat kidney tissues: possible protective roles of natural antioxidant foods. J. Appl. Toxicol. 26: 42-46.

Cohn, W. and P. Thurmann, U. Tenter, C. Aebischer, J. Schierle, and W. Schalch. 2004. Comparative multiple dose plasma kinetics of lycopene administered in tomato juice, tomato soup or lycopene tablets. Eur. J. Nutr. 43: 304-312.

Daood, H.G. and M.A. Alqitt, K.A. Bshenah, and M. Bouragba. 1990. Varietal and chemical aspect of tomato processing. Acta Aliment. 19: 347-357.

Djuric, Z. and L.C. Powell. 2001. Antioxidant capacity of lycopene-containing foods. Int. J. Food Sci. Nutr. 52: 143-149.

Dugas, T.R. and D.W. Morel, and E.H. Harrison. 1999. Dietary supplementation with beta-carotene, but not with lycopene, inhibits endothelial cell-mediated oxidation of low-density lipoprotein. Free Radic. Biol. Med. 26: 1238-1244.

Durak, I. and H. Biri, A. Avci, S. Sozen, and E. Devrim. 2003. Tomato juice inhibits adenosine deaminase activity in human prostate tissue from patient with prostate cancer. Nutr. Res. 23: 1183-1188.

FAOSTAT. 2001. http://faostat.fao.org/

Fernández-García, A. and P. Butz, and B. Tauscher. 2001. Effects of high-pressure processing on carotenoid extractability, antioxidant activity, glucose diffusion, and water binding of tomato puree (*Lycopersicon esculentum* Mill.). J. Food Sci. 66: 1033-1038.

Friedman, M. 2002. Tomato glycoalkaloids: Role in the plant and in the diet. J. Agric. Food Chem. 50: 5751-5780.

Frohlich, K. and K. Kaufmann, R. Bitsch, and V. Bohm. 2006. Effects of ingestion of tomatoes, tomato juice and tomato puree on contents of lycopene isomers, tocopherols and ascorbic acid in human plasma as well as on lycopene isomer pattern. Br. J. Nutr. 95: 734-741.

Gahler, S. and K. Otto, and V. Bohm. 2003. Alterations of vitamin C, total phenolics, and antioxidant capacity as affected by processing tomatoes to different products. J. Agric. Food Chem. 51: 7962-7968.

George, B. and C. Kaur, D.S. Khurdiya, and H.C. Kapoor. 2004. Antioxidants in tomato (*Lycopersicum esculentum*) as a function of genotype. Food Chem. 84: 45-51.

Giovannucci, E. 2002. A review of epidemiologic studies of tomatoes, lycopene, and prostate cancer. Exp. Biol. Med. 227: 852-859.

Giovannucci, E. and E.B. Rimm, Y. Liu, M. Stampfer, and W.C. Willett. 2002. A prospective study of tomato products, lycopene, and prostate cancer risk. J. Natl. Cancer Inst. 94: 391-398.

Goerlitz, C.D. and J.F. Delwiche. 2004. Impact of label information on consumer assessment of soy-enhanced tomato juice. J. Food Sci. 69: S376-S379.

Hadley, C.W. and S.K. Clinton, and S.J. Schwartz. 2003. The consumption of processed tomato products enhances plasma lycopene concentrations in association with a reduced lipoprotein sensitivity to oxidative damage. J. Nutr. 133: 727-732.

Hayes, W.A. and P.G. Smith, and A.E.J. Morris. 1998. The production and quality of tomato concentrates. Crit. Rev. Food Sci. Nutr. 38: 537-564.

Hertog, M.G.L. and P.C.H. Hollman, and B. Vandeputte. 1993. Content of potentially anticarcinogenic flavonoids of tea infusions, wines, and fruit juices. J. Agric. Food Chem. 41: 1242-1246.

Jenab, M. and P. Ferrari, M. Mazuir, A. Tjonneland, F. Clavel-Chapelon, J. Linseisen, A. Trichopoulou, R. Tumino, H.B. Bueno-de-Mesquita, E. Lund, C.A. Gonzalez, G. Johansson, T.J. Key, and E. Riboli. 2005. Variations in lycopene blood levels and tomato consumption across European countries based on the European Prospective Investigation into Cancer and Nutrition (EPIC) Study. J. Nutr. 135: 2032S-2036S.

Kasagi, S. and K. Seyama, H. Mori, S. Souma, T. Sato, T. Akiyoshi, H. Suganuma, and Y. Fukuchi. 2006. Tomato juice prevents senescence-accelerated mouse P1 strain from developing emphysema induced by chronic exposure to tobacco smoke. Am. J. Physiol.-Lung Cell. Mol. Physiol. 290: L396-L404.

Kaur, C. and D.S. Khurdiya, R.K. Pal, and H.C. Kapoor. 1999. Effect of microwave heating and conventional processing on the nutritional qualities of tomato juice. J. Food Sci. Technol.-Mysore 36: 331-333.

Khachik, F. and L. Carvalho, P.S. Bernstein, G.J. Muir, D.-Y. Zhao, and N.B. Katz. 2002. Chemistry, distribution, and metabolism of tomato carotenoids and their impact on human health. Exp. Biol. Med. 227: 845-851.

Larrosa, M. and J.C. Espín, and F.A. Tomás-Barberán. 2003. Antioxidant capacity of tomato juice functionalised with enzymatically synthesized hydroxytyrosol. J. Sci. Food Agric. 83: 658-666.

Larrosa, M. and R. Llorach, J.C. Espín, and F.A. Tomás-Barberán. 2002. Increase of antioxidant activity of tomato juice upon functionalisation with vegetable byproduct extracts. LWT-Food Sci. Technol. 35: 532-542.

Lichtenthaler, R. and F. Marx. 2005. Total oxidant scavenging capacities of common European fruit and vegetable juices. J. Agric. Food Chem. 53: 103-110.

Lin, C.H. and B.H. Chen. 2003. Determination of carotenoids in tomato juice by liquid chromatography. J. Chromatogr. A 112: 103-109.

Lin, C.H. and B.H. Chen. 2005a. Stability of carotenoids in tomato juice during processing. Eur. Food Res. Technol. 221: 274-280.

Lin, C.H. and B.H. Chen. 2005b. Stability of carotenoids in tomato juice during storage. Food Chem. 90: 837-846.

Madrid, E. and D. Vasquez, F. Leyton, C. Mandiola, and J.A. Escobar. 2006. Short-term *Lycopersicum esculentum* consumption may increase plasma high density lipoproteins and decrease oxidative stress. Rev. Medica Chile 134: 855-862.

Mangels, A.R. and J.M. Holden, G.R. Beecher, M.R. Forman, and E. Lanza. 1993. Carotenoid content of fruits and vegetables: an evaluation of analytic data. J. Am. Diet. Assoc. 93: 284-296.

Markovic, K. and M. Hruskar, and N. Vahcic. 2006. Lycopene content of tomato products and their contribution to the lycopene intake of Croatians. Nutr. Res. 26: 556-560.

Maruyama, C. and K. Imamura, S. Oshima, M. Suzukawa, S. Egami, M. Tonomoto, N. Baba, M. Harada, M. Ayaori, T. Inakuma, and T. Ishikawa. 2001. Effects of tomato juice consumption on plasma and lipoprotein carotenoid concentrations and the susceptibility of low density lipoprotein to oxidative modification. J. Nutr. Sci. Vitaminol. 47: 213-221.

Micozzi, M.S. and E.D. Brown, B.K. Edwards, J.G. Bieri, P.R. Taylor, F. Khachik, G.R. Beecher, and J.C. Smith. 1992. Plasma carotenoid response to chronic intake of selected foods and beta-carotene supplements in men. Am. J. Clin. Nutr. 55: 1120-1125.

Min, S. and Z.T. Jin, and Q.H. Zhang. 2003. Commercial scale pulsed electric field processing of tomato juice. J. Agric. Food Chem. 51: 3338-3344.

Muller, H. and A. Bub, B. Watzl, and G. Rechkemmer. 1999. Plasma concentrations of carotenoids in healthy volunteers after intervention with carotenoid-rich foods. Eur. J. Nutr. 38: 35-44.

Okajima, E. and M. Tsutsumi, S. Ozono, H. Akai, A. Denda, H. Nishino, S. Oshima, H. Sakamoto, and Y. Konishi. 1998. Inhibitory effect of tomato juice on rat urinary bladder carcinogenesis after N-butyl-N-(4-hydroxybutyl)nitrosamine initiation. Jpn. J. Cancer Res. 89: 22-26.

Oshima, S. and F. Ojima, H. Sakamoto, Y. Ishiguro, and J. Terao. 1996. Supplementation with carotenoids inhibits singlet oxygen-mediated oxidation of human plasma low-density lipoprotein. J. Agric. Food Chem. 44: 2306-2309.

Paetau, I. and D. Rao, E.R. Wiley, E.D. Brown, and B.A. Clevidence. 1999. Carotenoids in human buccal mucosa cells after 4 wk of supplementation with tomato juice or lycopene supplements. Am. J. Clin. Nutr. 70: 490-494.

Paetau, I. and F. Khachik, E.D. Brown, G.R. Beecher, T.R. Kramer, J. Chittams, and B.A. Clevidence. 1998. Chronic ingestion of lycopene-rich tomato juice or lycopene supplements significantly increases plasma concentrations of lycopene and related tomato carotenoids in humans. Am. J. Clin. Nutr. 68: 1187-1195.

Piironen, V. and J. Toivo, R. Puupponen-Pimiä, and A.-M. Lampi. 2003. Plant sterols in vegetables, fruits, and berries. J. Sci. Food Agric. 83: 330-337.

Podsędek, A. and D. Sosnowska, and B. Anders. 2003. Antioxidative capacity of tomato products. Eur. Food Res. Technol. 217: 296-300.

PoolZobel, B.L. and A. Bub, H. Muller, I. Wollowski, and G. Rechkemmer. 1997. Consumption of vegetables reduces genetic damage in humans: first results of a human intervention trial with carotenoid-rich foods. Carcinogenesis 18: 1847-1850.

Porrini, M. and P. Riso, A. Brusamolino, C. Berti, S. Guarnieri, and F. Visioli. 2005. Daily intake of a formulated tomato drink affects carotenoid plasma and lymphocyte concentrations and improves cellular antioxidant protection. Br. J. Nutr. 93: 93-99.

Rao, A.V. 2004. Processed tomato products as a source of dietary lycopene: bioavailability and antioxidant properties. Can. J. Diet. Pract. Res. 65: 161-165.

Rao, A.V. and S. Agarwal. 1998. Bioavailability and in vivo antioxidant properties of lycopene from tomato products and their possible role in the prevention of cancer. Nutr. Cancer 31: 199-203.

Riso, P. and A. Brusamolino, A. Martinetti, and M. Porrini. 2006a. Effect of a tomato drink intervention on insulin-like growth factor (IGF)-1 serum levels in healthy subjects. Nutr. Cancer 55: 157-162.

Riso, P. and F. Visioli, S. Grande, S. Guarnieri, C. Gardana, P. Simonetti, and M. Porrini. 2006b. Effect of a tomato-based drink on markers of inflammation, immunomodulation, and oxidative stress. J. Agric. Food Chem. 54: 2563-2566.

Rodríguez-Amaya, D.B. 1996. Assessment of provitamin A contents of foods—the Brazilian experience. J. Food Compos. Anal. **9:** 196-230.

Salles, C. and S. Nicklaus, and C. Septier. 2003. Determination and gustatory properties of taste-active compounds in tomato juice. Food Chem. 81: 395-402.

Sánchez-Moreno, C. and L. Plaza, B. de Ancos, and M.P. Cano. 2004. Effect of combined treatments of high-pressure and natural additives on carotenoid extractability and antioxidant activity of tomato puree (*Lycopersicum esculentum* Mill.). Eur. Food Res. Technol. 219: 151-160.

Sánchez-Moreno, C. and L. Plaza, B. de Ancos, and M.P. Cano. 2006a. Nutritional characterisation of commercial traditional pasteurised tomato juices: carotenoids, vitamin C and radical-scavenging capacity. Food Chem. 98: 749-756.

Sánchez-Moreno, C. and L. Plaza, B. de Ancos, and M.P. Cano. 2006b. Impact of high-pressure and traditional thermal processing of tomato purée on carotenoids, vitamin C and antioxidant activity. J. Sci. Food Agric. 86: 171-179.

Scott, K. J. and D.I. Thurnham, D.J. Hart, S.A. Bingham, and K. Day. 1996. The correlation between the intake of lutein, lycopene and beta-carotene from vegetables and fruits, and blood plasma concentrations in a group of women aged 50–65 years in the UK. Br. J. Nutr. 75: 409-418.

Sesso, H.D. and S.M. Liu, J.M. Gaziano, and J.E. Buring. 2003. Dietary lycopene, tomato-based food products and cardiovascular disease in women. J. Nutr. 133: 2336-2341.

Seybold, C. and K. Frohlich, R. Bitsch, K. Tto, and V. Bohm. 2004. Changes in contents of carotenoids and vitamin E during tomato processing. J. Agric. Food Chem. 52: 7005-7010.

Shi, J. and M. Le Maguer. 2000. Lycopene in tomatoes: chemical and physical properties affected by food processing. Crit. Rev. Food Sci. Nutr. 40: 1-42.

Sidhu, J.S. and V.K. Jain, and G.S. Bains. 1975. Effect of processing on tomato juice quality and ascorbic acid content. Indian J. Nutr. Diet. 12: 139-141.

Smelt, J.P.P.M. 1998. Recent advances in the microbiology of high pressure processing. Trends Food Sci. Technol. 9: 152-158.

Stahl, W. and H. Sies. 1992. Uptake of lycopene and its geometrical isomers is greater from heat-processed than from unprocessed tomato juice in humans. J. Nutr. 122: 2161-2166.

Steinberg, F.M. and A. Chait. 1998. Antioxidant vitamin supplementation and lipid peroxidation in smokers. Am. J. Clin. Nutr. 68: 319-327.

Stewart, A.J. and S. Bozonnet, W. Mullen, G.I. Jenkins, M.E.J. Lean, and A. Crozier. 2000. Occurrence of flavonols in tomatoes and tomato-based products. J. Agric. Food Chem. 48: 2663-2669.

Sutherland, W.H.F. and R.J. Walker, S.A. de Jong, and J.E. Upritchard. 1999. Supplementation with tomato juice increases plasma lycopene but does not alter susceptibility to oxidation of low-density lipoproteins from renal transplant recipients. Clin. Nephrol. 52: 30-36.

Takeoka, G.R. and L. Dao, S. Flessa, D.M. Gillespie, W.T. Jewell, B. Huebner, D. Bertow, and S.E. Ebeler. 2001. Processing effects on lycopene content and antioxidant activity of tomatoes. J. Agric. Food Chem. 49: 3713-3717.

Tiziani, S. and Y. Vodovotz. 2005. Rheological characterization of a novel functional food: tomato juice with soy germ. J. Agric. Food Chem. 53: 7267-7273.

Tiziani, S. and S.J. Schwartz, and Y. Vodovotz. 2006. Profiling of carotenoids in tomato juice by one- and two-dimensional NMR. J. Agric. Food Chem. 54: 6094-6100.

Tokuşoğlu, Ö. and M.K. Ünal, and Z. Yildirim. 2003. HPLC-UV and GC-MS characterization of the flavonol aglycons quercetin, kaempferol, and myricetin in tomato pastes and other tomato-based products. Acta Chromatogr. 13: 196-207.

Tonucci, L. and J. Holden, G. Beecher, F. Khackik, C. Davis, and G. Mulokozi. 1995. Carotenoid content of thermally processed tomato-based food products. J. Agric. Food Chem. 43: 579-586.

Tyssandier, V. and C. Feillet-Coudray, C. Caris-Veyrat, J.C. Guilland, C. Coudray, S. Bureau, M. Reich, M.J. Amiot-Carlin, C. Bouteloup-Demange, Y. Boirie, and P. Borel. 2004. Effect of tomato product consumption on the plasma status of antioxidant microconstituents and on the plasma total antioxidant capacity in healthy subjects. J. Am. Coll. Nutr. 23: 148-156.

Upritchard, J.E. and W.H.F Sutherland, and J.I. Mann. 2000. Effect of supplementation with tomato juice, vitamin E, and vitamin C on LDL oxidation and products of inflammatory activity in type 2 diabetes. Diabetes Care 23: 733-738.

USDA. 2006a. National Nutrient Database for Standard Reference, Release 19.

USDA. 2006b. USDA Database for the Flavonoid Content of Selected Foods, Release 2.

Visioli, F. and P. Riso, S. Grande, C. Galli, and M. Porrini. 2003. Protective activity of tomato products on in vivo markers of lipid oxidation. Eur. J. Nutr. 42: 201-206.

Volikakis, G.J. and C.E. Efstathiou. 2005. Fast screening of total flavonols in wines, tea-infusions and tomato juice by flow injection/adsorptive stripping voltammetry. Anal. Chim. Acta 551: 124-131.

Wang, H. and G.H. Cao, and R.L. Prior. 1996. Total antioxidant capacity of fruits. J. Agric. Food Chem. 44: 701-705.

Watzl, B. and A. Bub, B.R. Brandstetter, and G. Rechkemmer. 1999. Modulation of human T-lymphocyte functions by the consumption of carotenoid-rich vegetables. Br. J. Nutr. 82: 383-389.

Watzl, B. and A. Bub, K. Brivika, and G. Rechkemmer. 2003. Supplementation of a low-carotenoid diet with tomato or carrot juice modulates immune functions in healthy men. Ann. Nutr. Metab. 47: 255-261.

Watzl, B. and A. Bub, M. Blockhaus, B.M. Herbert, P.M. Luhrmann, M. Neuhauser-Berthold, and G. Rechkemmer. 2000. Prolonged tomato juice consumption has no effect on cell-mediated immunity of well-nourished elderly men and women. J. Nutr. 130: 1719-1723.

Willcox J.K. and G.L. Catignani, and S. Lazarus. 2003. Tomatoes and cardiovascular health. Crit. Rev. Food Sci. Nutr. 43: 1-18.

12

Acrylamide in Tomato Products

Fernando Tateo* **and Monica Bononi**
Analytical Research Laboratories - Food and Env - Di.Pro.Ve., Faculty of Agriculture
University of Milan, 2, Via Celoria, 20133 Milan, Italy

ABSTRACT

In 2004, initial research from the Swedish National Food Administration and Stockholm University indicated that acrylamide formation is associated with traditional high-temperature cooking processes for certain carbohydrate-rich foods. Among the breadth of products consumed in the Mediterranean diet, the tomato occupies a position of primary importance. Acrylamide levels in several tomato sauces common in the Italian market were evaluated. The results indicated that acrylamide values were < 50 μg/kg for simple tomato sauces and ≤ 103 μg/kg for tomato sauces containing seasoning ingredients: in practice the previous published data were confirmed. These results confirm the safe use of processed tomato derivatives. The data reported should not be considered indicative of food product choices by consumers.

INTRODUCTION

Our previous studies have investigated the content of acrylamide in tomato sauces, which were not mentioned in other studies, including one by the US Food and Drug Administration (FDA 2006) examining acrylamide contents in various foodstuffs. The aims of the FDA study were to evaluate the health risks associated with acrylamide consumption through food and to inform the public of progress in this area of research.

**Corresponding author*

A list of abbreviations is given before the references.

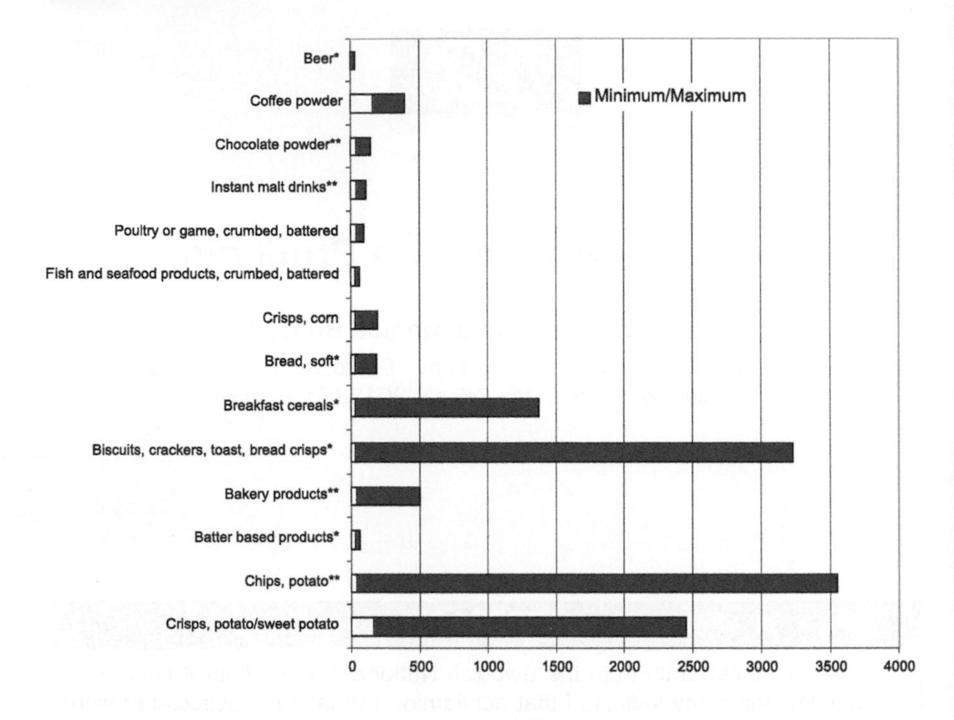

(*) and (**): less than 30 and 50 μg/kg

Fig. 1 Acrylamide levels in different foods and food products from Norway, Sweden, Switzerland, the United Kingdom, and the United States.

Preliminary data concerning acrylamide levels in different foods and food products groups from Norway, Sweden, Switzerland, the United Kingdom and the United States were reported in the Opinion of the Scientific Committee on Food (European Commission 2002) (Fig. 1).

Table 1 reports minimum and maximum levels of acrylamide in various product samples, deduced from data published by the US Food and Drug Administration. Data corresponding to tomato sauces are not mentioned in the list: gravies and seasonings reported cannot be identified as tomato sauces.

The first data on acrylamide in tomato sauces from the Italian market were reported by Tateo et al. (2007). In Table 2 are assembled results derived from two papers by Tateo and Bononi (2003) and Tateo et al. (2007): nine tomato sauces most commonly found on the Italian market were considered. Tomato sauces containing various ingredients were

Table 1 Acrylamide values in food product samples (minimum and maximum values deduced from data collected by US Food and Drug Administration, Department of Health and Human Services, from December 2002 to July 2006)

	Min (μg/kg)	*Max (μg/kg)*
Baby food	< 10	130
Breads and bakery products	< 10	364
Canned fruits and vegetables	< 10	83
Cereals	< 10	1057
Cookies	< 20	955
Crackers	37	1540
Dairy	< 10	43
Dried foods	< 10	1184
French fries	20	1250
Frozen vegetables	< 10	
Fruit and vegetables products	< 10	239
Gravies and seasonings	< 10	151
Olives	< 20	1925
Potato chips	117	2762
Snack foods (other than potato chips)	< 20	1340

Table 2 Acrylamide levels in various food samples from the Italian market. Minimum and maximum values deduced from data published by F. Tateo, M. Bononi and G. Andreoli, University of Milano, Italy (Tateo and Bononi 2003, Tateo et al. 2007).

	Min (μg/kg)	*Max (μg/kg)*
Boiled rice	< 50	
Seasoned risotto	< 50	113
Tomato sauces	< 50	
Various fast food (Italian market)	< 50	150
Biscuits	< 50	403
Homogenized meat	< 50	
Oven-baked foods	126	385
Snack foods	< 50	148

analysed and the increase of acrylamide level was found to depend on the use of other ingredients (such as olive and capers).

This study considered a large number of tomato sauce samples available on the Italian market. From a chemistry standpoint, Friedman (2003) reported that acrylamide (Fig. 2) in food is largely derived from heat-induced reactions between the amino group of the free amino acid asparagine and the carbonyl group of reducing sugars such as glucose.

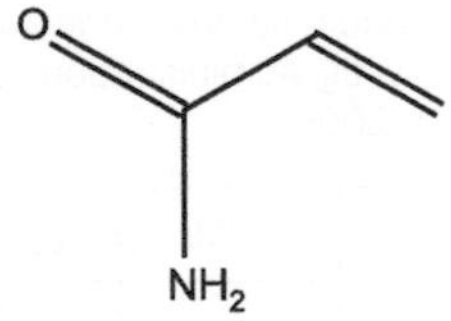

O
NH2

Fig. 2 Molecular structure of acrylamide.

We examined the possibility of acrylamide formation during tomato processing by applying an original method (Tateo and Bononi 2003) and a modified method (Bononi et al. 2005). The aim of this research was to determine whether transformation of the tomato by cooking or concentration results in the production of detectable quantities of acrylamide.

STUDY DESIGN

Chemicals and Equipment

Acrylamide (99+%) (catalogue number 14.866-0) was obtained from Sigma Aldrich (Milan, Italy); *n*-hexane (98+%), 2-propanol (99.8%), and anhydrous sodium sulphate were obtained from Merck (Darmstadt, Germany).

The quantification of acrylamide levels in food was performed on a Shimadzu 2010 GC (Shimadzu, Milan, Italy) coupled to a Shimadzu QP2010 quadrupole MS (Shimadzu, Milan, Italy). The GC column was a Supelcowax™-10 fused silica capillary column (30 m × 0.25 mm i.d., 0.25 μm film thickness) (Supelco, Milan, Italy).

Analysis of Acrylamide by GC/MS

Approximately 10 g of homogenized sample was weighed and dehydrated by adding 10-50 g of anhydrous sodium sulphate, depending on the water content of the sample. Each sample was defatted with 80 mL of hexane at room temperature, stirring for 30 min. Most of the solvent was removed by Pasteur pipette, and the residual solvent was removed by vacuum. A total of 100 mL 2-propanol was added to the defatted sample in a sealed flask, and the mixture was stirred for 15 min and shaken for 1 min in an ultrasonic bath. The 2-propanol phase was recovered by filtration (~70 mL) and concentrated in a rotary evaporator to < 2 mL. The residue was carefully transferred to a graduated vial, then diluted to 2 mL and fast filtered. Using an autosampler, 1 μL of the sample was injected in splitless mode (15 s), and then a splitting ratio of 1:20 was used.

The temperature programme for the GC was as follows: isothermal for 1 min at 60°C, temperature increase at a rate of 5°C/min to 240°C, then isothermal for 10 min. Analysis was performed using EI (70 eV) and selected ion monitoring. The ions monitored for identification of the analyte were *m/z* 55, 71, using *m/z* 71 for quantification.

Quantification was performed by comparison with a calibration curve (150-1000 µg/L of standard acrylamide in 2-propanol) corresponding to 40 and 285 µg/kg if 70 mL of 2-propanol extract were concentrated. Corrections for percentage recovery were required. Recovery tests were repeatedly performed by quantification of acrylamide in fresh tomato before and after the addition of acrylamide. Samples containing more than 400 µg/kg acrylamide were diluted up to a factor of two in the first extraction step.

Samples

Twenty tomato sauces from the Italian market were analysed. Special ingredients declared in labels for tomato sauces are listed in Table 3.

Table 3 Acrylamide levels in 20 tomato sauces from the Italian market.

Sample	*Product**	*Acrylamide level (µg/kg)*
1.	Peeled tomato GS	< 50
2.	Peeled tomato "Pomidori Pelati" Cirio	< 50
3.	Sauce "Verace" Cirio	< 50
4.	Sauce "Classica" Santa rosa Bertolli	< 50
5.	Sauce "Pummarò" Star	< 50
6.	Sauce "Ortolina" Rodolfi Mansueto	< 50
7.	Sauce Knorr with Bolognese meat spaghetti sauce	72
8.	Sauce "Scarpariello" Voiello with sweet peppers	55
9.	Sauce with basil "1" Althea	< 50
10.	Sauce GS with Arrabbiata hot pepper sauce	< 50
11.	Sauce Star with pesto and ricotta	< 50
12.	Sauce Star with Amatriciana pork bacon sauce	82
13.	Sauce Bertolli Puttanesca with olives and marrows	103
14.	Sauce Barilla with Bolognese meat spaghetti sauce	66
15.	Sauce Barilla with Arrabbiata hot pepper sauce	< 50
16.	Sauce Coop	< 50
17.	Sauce Cordoro	< 50
18.	Tomato pulp Scelgobio	< 50
19.	Tomato pulp "PolpaPiù" Cirio	< 50
20.	Tomato pulp "1" Franzese	< 50

* Tomato sauces most commonly found on the Italian market

ACRYLAMIDE CONTENT IN TOMATO PRODUCTS

The acrylamide contents of the 20 tomato sauces most commonly found on the Italian market are presented in Table 3. Acrylamide values were < 50 μg/kg for simple tomato sauces and ≤ 103 μg/kg for tomato sauces containing seasoning ingredients. The use of seasoning ingredients (olives, capers, meat, olive oil, onion, dye, mushrooms) may increase the acrylamide content as a consequence of cooking operations.

The highest acrylamide value reported for bread, the most common product in the diet, is ~49 μg/kg (Tateo and Bononi 2003). The consumption of tomato sauces in the diet is lower than that of bread, and for this reason, quantification values < 50 μg/kg were not considered significant.

The recovery data of the adopted analytical method and CV values were reported in a previous study (Tateo et al. 2007).

The data reported here cover only a limited number of brands because the aim of this work was to obtain exploratory data and not to allow absolute comparisons or to generate statistical data representative of the standard products. The same criterion was adopted by the FDA in its study on acrylamide in food (FDA 2006).

It was clear from previous works that substantial acrylamide formation did not result from heating tomatoes at moderate temperatures in the presence of water for the purpose of concentration to produce tomato sauce.

Thus, the data reported in this note confirm the safe use of processed tomato derivatives.

ABBREVIATIONS

FDA: Food and Drug Administration; GC: gas chromatograph; MS: mass spectrometer; CV: coefficient of variation; i.d.: internal diameter; EI: electronic impact

REFERENCES

Bononi, M., and F. Tateo, G. Andreoli, and A.Varani. 2005. Determinazione dell'acrilammide in matrici alimentari contenenti suoi precursori. Ind. Alim. 443: 33-37.

European Commission. 2002. Opinion of the Scientific Committee on Food on new findings regarding the presence of acrylamide in food. Health & Consumer Protection Directorate-General. http://ec.europa.eu/food/fs/sc/scf/out131_en.pdf

FDA (US Food and Drug Administration). 2006. Survey Data on Acrylamide in Food: Individual Food Products. US Dept. of Health and Human Services, Center for Food Safety and Applied Nutrition. http://www.cfsan.fda.gov/~dms/acrydata.html

Friedman, M. 2003. Chemistry, biochemistry, and safety of acrylamide. A review. J. Agric. Food Chem. 51: 4504-4526.

Tateo, F. and M. Bononi. 2003. Preliminary study on acrylamide in baby foods on the Italian market. Ital. J. Food Sci. 15: 593-599.

Tateo, F., and M. Bononi, and G. Andreoli. 2007. Acrylamide level in cooked rice, tomato sauces and some fast food on the Italian market. J. Food Comp. Anal. 20: 232-235.

PART 2

Cellular and Metabolic Effects of Tomato and Related Products or Components

13

Tomato Leaf Crude Extracts for Insects and Spider Mite Control

George F. Antonious[1] and John C. Snyder[2]

[1]Department of Plant and Soil Science, Water Quality/Environmental Toxicology, Land Grant Program, 218 Atwood Research Center, Kentucky State University, Frankfort, Kentucky 40601, USA

[2]Department of Horticulture, University of Kentucky, Lexington, Kentucky 40546, USA

ABSTRACT

Phytochemicals present in wild *Lycopersicon* taxa have been associated with pest resistance. Secondary compounds from the leaves of the wild tomato relatives *L. hirsutum* f. *glabratum*, *L. hirsutum* f. *typicum*, and *L. pennellii* (Solanaceae) have been identified as methylketones (2-tridecanone, 2-undecanone, 2-dodecanone, 2-pentadecanone), sesquiterpene hydrocarbons (zingiberene and curcumene), and glucolipids (sugar esters), respectively. The main constituents in crude extracts of tomato leaves were separated, purified, and screened for their insecticidal/acaricidal activity. Methylketones provided a strong insecticidal and acaricidal efficacy against the green peach aphid, tobacco hornworm, tobacco budworm, Colorado potato beetle, whitefly, and two-spotted spider mite. Two spider mite bioassays, one a measure of antibiosis and the other a measure of repellency, were used to determine the acaricidal performance of wild tomato leaf extracts. Wild tomato leaves possess pest-resistance mechanisms associated with their glandular trichomes and the exudates they produce. Type IV and type VI glandular trichomes on the leaves of six wild tomato accessions of *L. hirsutum* f. *glabratum* (PI 126449, PI 134417, PI 134418, PI 251304, PI 251305, and LA 407) were counted. Major chemical compounds from glandular leaf trichomes of the accessions tested were quantified using gas chromatography (GC) and mass spectrometry (GC/MS). The toxicity of two methylketones, 2-undecanone and 2-tridecanone—the major constituents of trichome secretions from the *L. glabratum* accessions tested—to adults of the sweet potato whitefly, *Bemisia tabaci* Gennadius, and fourth instar larvae of the Colorado potato beetle, *Leptinotarsa decemlineata* Say, was determined using no-choice bioassays. 2-undecanone caused 80% mortality of the fourth instar larvae of Colorado potato beetle at the highest concentration tested (100 mg 2-undecanone mL^{-1} of acetone), while 2-tridecanone caused 72% mortality of whiteflies at 20 mg 2-tridecanone mL^{-1} of ethanol. The concentration of 2-undecanone was greatest on the leaves of LA 407, and that of 2-tridecanone was greatest on PI 134417 compared to other accessions tested. Density of type VI trichomes varied among accessions and among sampling seasons. PI 134417 and LA 407 produced the highest number of type VI trichomes during the month of June. The two accessions of *L. pennellii* (PI 246502 and PI 414773) produced considerable concentrations of glucolipids. The two wild tomato accessions of *L. hirsutum* f. *glabratum* (PI 134417 and LA 407) are promising sources of 2-tridecanone and 2-undecanone that may be used for control of many insects and spider mites.

INTRODUCTION

Tomato, *Lycopersicon esculentum* Mill., is a well-known vegetable used around the world. This cultivated species is highly variable, with fruit ranging in diameter from 1.5 cm to more than 10 cm. Fruit color, when ripe, is usually red, but some varieties have yellow or green fruit when ripe. There are several undomesticated species of *Lycopersicon,* including *L. pimpinellifolium, L. hirsutum, L. pennellii, L. cheesmanii,* and *L. peruvianum. Lycopersicon pimpinellifolium* is the only undomesticated species that has red fruit when ripe, but the fruit are very small, usually less than 1.5 cm in diameter. All of the other undomesticated species also have small fruit. However, the undomesticated species of *Lycopersicon* harbor a great deal of genetic variability and have been studied extensively as sources of genes for genetic improvement of tomato. The undomesticated (wild) species have provided genes that confer resistance to disease, insects and nematodes. Two of the wild species, *L. hirsutum* and *L. pennellii,* are notably resistant to insects. In fact, resistance to more than 20 species of arthropods has been reported to occur in the genus *Lycopersicon* (Table 1). When mechanisms for arthropod resistance have been established, often leaf trichomes have been implicated as causes of resistance. Seven types of trichomes (leaf hairs) are generally recognized in *Lycopersicon* (Luckwill 1943). Of these seven types, the glanded Type IV and Type VI trichomes have most often been associated with arthropod resistance. In fact, the composition of trichome secretion can be very important with regard to

Table 1 Economically important insect and spider mite pests of tomato to which tomato breeders have successfully bred host-plant resistance.

Arthropod pest	*Common name*	*Reference*
Aculops lycopersici	Tomato russet mite	Kamau et al. 1992
Aphis gossypii	Cotton aphid	Williams et al. 1980
Bemisia argentifollii	Silverleaf whitefly	Heinz and Zalom 1995
Bemisia tabaci	Tobacco whitefly	Muigai et al. 2003, Channarayappa et al. 1992
Epitrix hirtipennis	Tobacco flea beetle	Gentile and Stoner 1968
Frankliniella occidentalis	Western flower thrips	Krishna Kumar et al. 1995
Heliothis armigera	Tomato fruitworm	Kashyap and Verma 1986
Heliothis zea	Tomato fruitworm	Fery and Cuthbert 1974, Cosenza and Green 1979
Keiferia lycopersicella	Tomato pinworm	Schuster 1977, Lin and Trumble 1986
Leptinotarsa decemlineata	Colorado potato beetle	Schalk and Stoner 1976
Liriomyza sativae	Vegetable leafminer	Webb et al. 1971
Liriomyza trifolii	American serpentine leafminer	Bethke et al. 1987
Macrosiphum euphorbiae	Potato aphid	Stoner et al. 1968, Quiros et al. 1977
Manduca sexta	Tomato hornworm	Kennedy and Henderson 1978
Neoleucinodes elegantalis	Tomato fruit borer	Salinas et al. 1993
Phthorimaea operculella	Potato moth	Juvik et al. 1982
Plusia chalcites	Tomato looper	Juvik et al. 1982
Spodoptera eridania	Southern armyworm	Schuster 1977
Spodoptera exigua	Beet armyworm	Juvik et al. 1982, Eigenbrode and Trumble 1993
Spodoptera littoralis	Egyptian cotton leaf worm	Juvik et al. 1982
Tetranychus cinnabarinus	Carmine spider mite	Stoner and Stringfellow 1967
Tetranychus marianae	Mariana mite	Wolfenbarger 1965
Tetranychus urticae	Two-spotted spider mite	Gilbert et al. 1966, Aina et al. 1972
Trialeurodes vaporariorum	Greenhouse whitefly	Ponti et al. 1980
Tuta absoluta	Tomato leafminer	Azevedo et al. 2003

arthropod resistance. Some secretions from wild tomato are toxic to arthropods, and others are repellent. Chemical composition can vary widely, from straight chain methylketones in certain lines of *L. hirsutum* f. *glabratum* to sesquiterpene hydrocarbons in lines of *L. hirsutum* f. *typicum*. The sesquiterpenes, depending on the particular line or accession of *L. hirsutum*, are structurally diverse. Methylketones and sesquiterpenoids are not the only components of trichome secretions in the wild species of

Lycopersicon. Trichome secretions on *L. pennellii* are often composed of acylated sugars (glucolipids), usually glucose and/or sucrose. Depending on the particular accession, however, the composition of the acyl sugar can vary with regard to the chain length and branching of the acyl group, as well as the sugar. While both *L. hirsutum* and *L. pennellii* are known for insect resistance, they are also known to produce a wide array of compounds in their trichome secretions that may be useful as botanical pesticides.

Insects and spider mites have developed resistance to many, in some cases all, of the synthetic insecticides used for their control. Insecticides and acaricides from wild tomato leaves may offer a partial solution as substitutes for synthetic pesticides, particularly when two or more active components are combined. Combination of more than one active ingredient in one formulation of botanical insecticides has the advantage of providing novel modes of action against a wide variety of insects. The risk of cross resistance will be reduced because insects will have difficulty adapting to a diverse group of bioactive compounds. Fewer pesticide applications will be required, with a significant saving for growers and farmers with limited resources.

This investigation is an approach to develop a new insecticide/acaricide from wild tomato leaves. While significant research effort is currently directed toward biological and cultural control strategies against agricultural pests, the application of synthetic pesticides remains an essential activity in many production systems. Pesticide resistance is increasing and the development and registration rate of new pest control chemicals on vegetable "minor crops" is low compared to pesticides used for large-acreage crops. New approaches to these problems are needed.

One way to protect crops is host-plant resistance. Plant resistance to insects enables a plant to avoid, tolerate, or recover from the injurious effects of insect feeding. Most commercial tomato varieties are susceptible to infestation by several insects (Antonious et al. 1999) and spider mites. As long as yields are comparable, the three advantages of controlling insects by resistant varieties are (1) little or no expense to the grower, (2) reduced chance for adverse environmental impacts from pesticide residue, and (3) reduced probability of toxic residues on the crop reaching the consumer. The elimination of even one spray application per season can mean a significant saving to the grower.

Because of the inherent toxicity of most existing synthetic pesticides to non-target organisms and because of their persistence in the environment (Antonious 2003, 2004a), there is increasing pressure on the agricultural industry to find more acceptable pest control alternatives to synthetic pesticides. In addition, some of the commonly used insecticides on vegetable crops either break down to more toxic metabolites (Antonious and Snyder 1994, Antonious 1995) or have other serious environmental effects (Antonious and Byers 1997, Antonious 1999, 2003, Antonious and Patterson 2005). Concerns about pesticide safety usually involve two sides, the environment and the end-user. To protect the environment, the general trend is to use reduced levels of active ingredients. This trend creates a need for pesticide formulations with improved efficacy at low application rates. To protect the end-user, environmentally safe formulations that eliminate organic solvent-based formulations are needed. The focus of this investigation is the development of an efficient natural product with low mammalian toxicity and little or no impact on environmental quality for use against vegetable pests that have gained resistance against many classes of insecticides.

The two-spotted spider mite, *Tetranychus urticae* Koch, is a well-known herbivorous pest of domestic tomato and many other crops. Often, mite-susceptible crops must be protected with synthetic acaricides during the hot and dry seasons that favor severe outbreaks of spider mites. Many studies have indicated the potential ecological damage due to the widespread use of synthetic pesticides (Sances et al. 1992, Antonious and Snyder 1994, Antonious et al. 1998, Strang 1998). The US Food Quality Protection Act in 1996 initiated a systematic effort to identify and reduce potential risks posed by synthetic pesticides to safeguard public health. Among the provisions of the Act is a requirement for the US Environmental Protection Agency to reassess all synthetic pesticide tolerances (9,700+) within 10 years of passage of the Act. Among those that are significant to varying degrees to many vegetable growers are azinphos-methyl (Guthion), chlopyrifos (Lorsban), phosmet (Imidan), diazinon and malathion (Cythion) (Strang 1998). Some agrochemical companies are already dropping organophosphorus insecticides and shifting to pyrethroids (synthetic insecticides) and natural products. However, in some specific situations replacements for pesticides that fall into the range specified in the Act are not currently available.

The use of natural plant products for insect control (Xia and Johnson 1997, Pillmoor 1998, Rice et al. 1998, Antonious 2001a, b, Antonious et al. 2007) may impart a selective advantage to plants by inhibiting, repulsing, and even killing non-adapted organisms that feed on or compete with the plant. Production of toxic chemical compounds against insects is one method by which wild tomato trichomes (leaf hairs) can impart resistance against insects. Research on the wild tomato *L. hirsutum* f. *glabratum* has demonstrated that glandular trichomes and the exudates they produce contribute to host resistance to insects (Xia and Johnson 1997, Eigenbrode et al. 1996, Guo et al. 1993). However, introgression of host-plant resistance from *L. hirsutum* into commercially acceptable cultivars has not been successful (Hartmann and St. Clair 1998). Although breeding has not been successful, the observation that *L. hirsutum* plants are resistant to herbivorous organisms makes glandular secreting trichomes of wild tomato an attractive system for study of biopesticides that may have activity against vegetable insects that have become resistant to all major classes of modern synthetic insecticides. There appear to be no environmental studies conducted with specific reference to the use of methylketones, especially 2-tridecanone, on vegetables as an organic insecticide. Four methylketones (2-undecanone, 2-dodecanone, 2-tridecanone, and 2-pentadecanone) were detected in *L. hirsutum* f. *glabratum* accessions (PI 251304, PI 126449, PI 134417, PI 134418, and LA 407) (Antonious 2001a, Antonious et al. 2003, 2005).

Recent research on *L. hirsutum* f. *glabratum* (Fig. 1) has demonstrated that their type VI glandular trichomes (Fig. 2) and the exudates they produce contribute to insect resistance (Antonious et al. 2003). In addition to methylketones, the sesquiterpene hydrocarbon zingiberene [5-(1,5-dimethyl-4-hexenyl)-2-methyl-1,3-cyclohexadiene] has been associated with resistance to the Colorado potato beetle (*Leptinotarsa decemlineata* Say) (Carter et al. 1989) and the beet armyworm (*Spodoptera exigua* Hübner) (Eigenbrode and Trumble 1993, Eigenbrode et al. 1996). Curcumene, another sesquiterpene hydrocarbon found in some accessions of *L. hirsutum* f. *typicum* Humb & Bonpl., has also shown insecticidal efficacy (Agarwal et al. 2001). *Lycopersicon hirsutum* f. *typicum* plants were highly resistant to the beet armyworm, as indicated by reduced survival and growth on excised leaflets (Eigenbrode et al. 1996). Resistance to a number of pests, including the carmine spider mite (*T. cinnabarinus* Boisduval), two-spotted spider mite (*T. urticae* Koch), greenhouse whitefly (*Trialeurodes vaporariorum* Westwood), cotton bollworm (*Heliothis armigera* Hubner), cotton leaf worm (*S. littoralis* Boisduval), and potato aphid (*Macrosiphum euphorbiae* Thomas), has been reported in *L. pennellii* (Fig. 3), a wild relative of the cultivated tomato (Rodriguez et al. 1993), due to the presence of viscous exudates (glucolipids) produced by their type IV glandular trichomes (Figs. 4 and 5). Characterization of insecticidal glucolipids has been reported (Chortyk et al. 1996, 1997, Xia et al. 1997).

Trichome-borne chemicals, such as the antibiotic 2-tridecanone and repellent 2,3-dihydrofarnesoic acid in *L. hirsutum* plants, have been associated with resistance of *L. hirsutum* plants to spider mites, *T. urticae.* Resistance of *L. hirsutum* to the whitefly *Bemisia argentifolii* was reported (Heinz and Zalom 1995, Muigai et al. 2003). Some accessions of *L. hirsutum* f. *typicum* were highly resistant to the beet armyworm (*S. exigua*).

Fig. 1 Leaves of the wild tomato *Lycopersicon hirsutum* f. *glabratum* grown at Kentucky State University Research Farm (Franklin County, Kentucky, USA). PI 134417 is resistant to most arthropods (insects and spider mites) because of the presence of type IV and type VI glandular trichomes on the leaf surface.

Resistance to a number of pests, including the carmine spider mite (*T. cinnabarinus* Boisduval), two-spotted spider mite (*T. urticae* Koch), greenhouse whitefly (*T. vaporariorum* Westwood), cotton bollworm (*H. armigera* Hubner), cotton leaf worm (*S. littoralis* Boisduval), and potato aphid (*M. euphorbiae* Thomas), has been reported in *L. pennellii* (Corr.) D'Arcy (Rodriguez et al. 1993), a wild relative of the cultivated tomato (*L. esculentum* Mill.). Several active compounds such as sesquiterpene

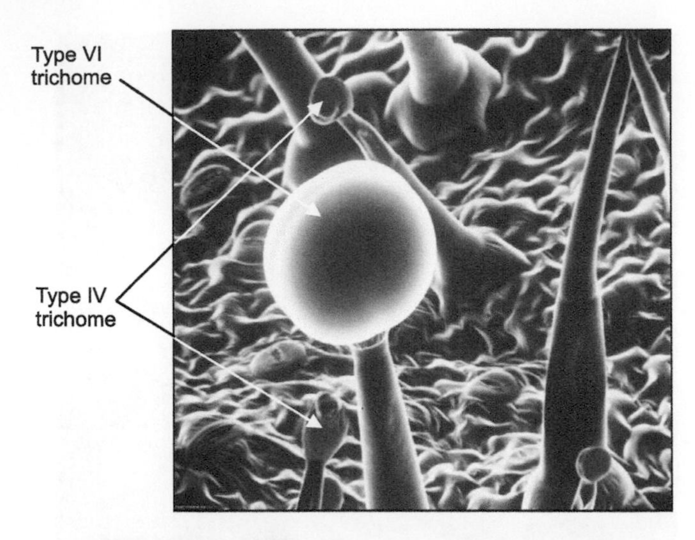

Fig. 2 Photomicrographs (400×) of type IV and type VI glandular trichomes on the leaves of *Lycopersicon hirsutum* f. *glabratum* obtained by scanning electron microscopy.

Fig. 3 Leaves of the wild tomato *Lycopersicon pennellii* grown at Kentucky State University Research Farm. PI 246502 is resistant to most arthropods (insects and spider mites) because of the presence of type IV and type VI glandular trichomes on the leaf surface.

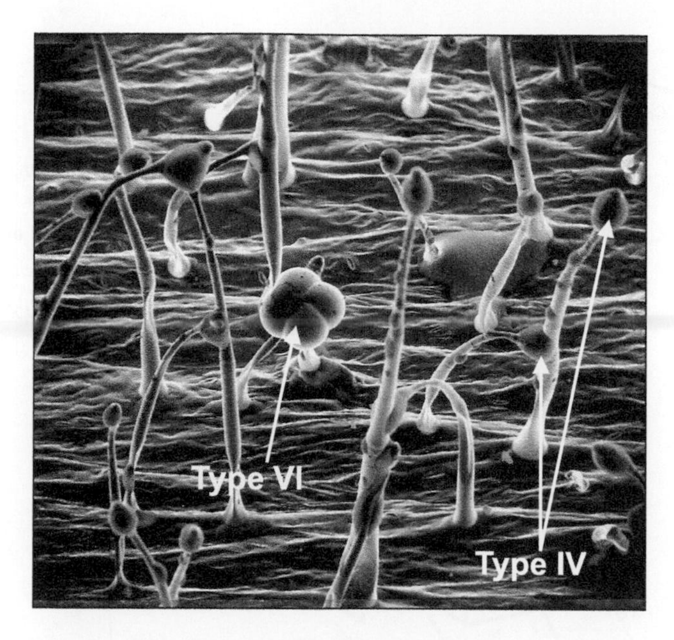

Fig. 4 Photomicrographs (165×) of type IV and type VI glandular trichomes on the leaves of *Lycopersicon pennellii* obtained by scanning electron microscopy. Resistance is due to entrapment of the insects and spider mites in the viscous exudates of type IV glandular trichomes on the leaf surface.

hydrocarbons, sesquiterpene acids, methylketones, and glucolipids (sugar esters) in glandular trichomes of wild tomato contribute to resistance by a variety of mechanisms including toxicity (Walters et al. 1990, Chatzivasileiadis and Sabelis 1997) and repellency (Walters et al. 1990, Snyder et al. 1993, Chatzivasileiadis and Sabelis 1997, Antonious and Snyder 2006). The purpose of the following is to explore the possibility of developing biopesticides from wild tomatoes. Bioactivity and production of extracts and development of formulations are all aspects of this development.

CRUDE EXTRACTS FROM WILD TOMATO LEAVES—METHODS

Seeds of wild tomato accessions were obtained from the USDA/ARS, Plant Genetic Resources Unit, Cornell University Geneva, New York (USA). Seeds of *L. esculentum* cv. Fabulous (included as control) were obtained from Holmes Seed Co. (Canton, Ohio, USA). Wild tomato plants studied included six accessions of *L. hirsutum* f. *glabratum* (PI 126449, PI 134417, PI 134418, PI 251304, PI 251305, and LA 407), three accessions of *L.*

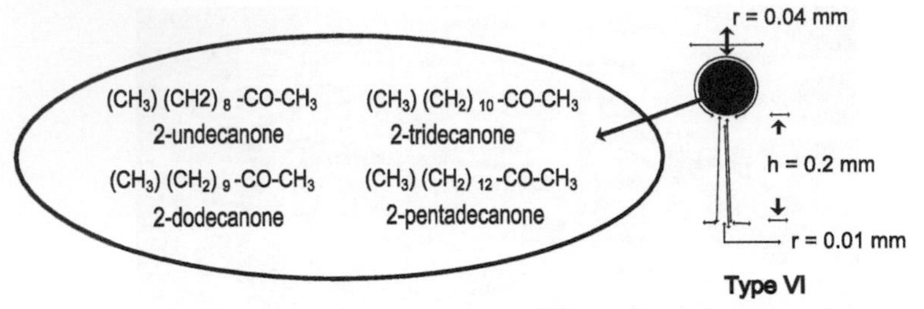

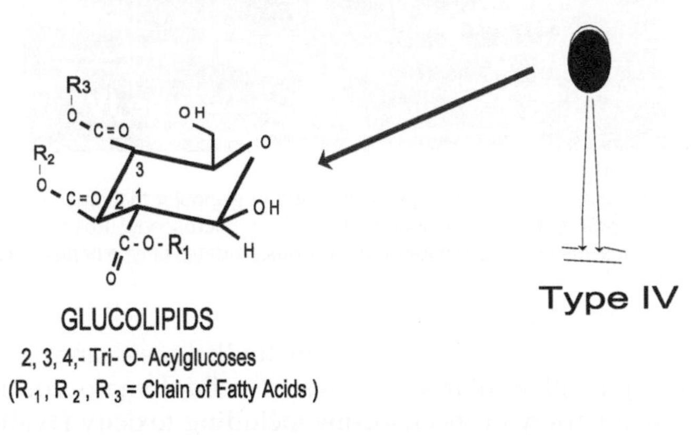

Fig. 5 Structure of type VI glandular trichomes (containing 2-undecanone, 2-tridecanone, 2-dodecanone, and 2-pentadecanone) on the leaves of wild tomato accessions PI-126449, PI-134417, PI-134418, PI-251304, and LA 407) of *Lycopersicon hirsutum* f. *glabratum* (upper structure) and structure of type IV glandular trichomes (containing glucolipids) on the leaves of *L. pennellii* (lower structure). Note that type VI glandular trichomes are about 0.2 mm long, topped by a nearly round cap of 0.04 mm radius, and a base of 0.01 mm radius.

hirsutum f. *typicum* (PI 127826, PI 127827, and PI 308182), two accessions of *L. pennellii* (PI 246502 and PI 414773), and one domestic tomato, *L. esculentum* cv. Fabulous. Seeds were germinated in the laboratory on moistened filter paper in Petri dishes kept in the dark. After potted seedlings attained six leaves, plants were transported into the greenhouse and transplanted into 20 cm diameter plastic pots containing Pro-Mix (Premier Horticultural Inc., Red Hill, Pennsylvania, USA) and grown under natural day-length conditions with sodium lamps providing additional photosynthetic photon flux of 110 μMoles s^{-1} m^{-1}. Plants were irrigated daily and fertilized twice a month with water containing 200 ppm of general purpose fertilizer with the elements N, P, and K (20:20:20).

Average greenhouse temperature and relative humidity were 30 ± 4°C and 50 ± 12% respectively. To obtain trichome counts newly mature leaves free of visible defects were sampled. One leaflet from a leaf was used to obtain adaxial trichome counts and the corresponding opposite leaflet was used for abaxial counts. Only type VI trichomes were counted using a light microscope (Antonious 2001a) at magnification of 100 × (10 × ocular and 10 × objective). Three counts were made for each leaflet surface (within interveinal areas) at the top, near the center, and at the bottom of each leaflet.

Crude extracts of leaflets were prepared by shaking 5 g of leaflets of each accession with chloroform for 10 min. Similar extracts were prepared in n-hexane. The extracts were filtered through a Whatman 934-AH glass microfiber filter of 55 mm diameter (Fisher Scientific, Pittsburg, Pennsylvania). One μL of each extract was injected into a GC/MS for quantification of methylketones. In addition, an aliquot was taken from each extract for quantification of glucolipids (sugar esters). Quantification of glucolipids was carried out by a colorimetric sugar assay (Moore and Antonious 2003) to determine the amount of glucose released from the sugar ester in each accession tested after hydrolyzing the ester using 0.4N NaOH.

For bulk extraction of trichome contents, all leaves of five plants from each accession of *L. hirsutum* f. *glabratum* were harvested and weighed. A bulk crude extract of 2-undecanone and 2-tridecanone was prepared by soaking leaves at the rate of 50 g of leaves per liter of water containing 3 mL of 2% Sur-Ten (sodium dioctyl sulfosuccinate), obtained from Aldrich Chemical Company, Milwaukee, Wisconsin (USA). After continuous manual shaking, portions of the mixture were filtered through a Whatman 934-AH glass microfiber filter and portions of the filtrate were partitioned with 200 mL of n-hexane in a separatory funnel. The hexane layers were combined and the solvent was evaporated to dryness using a rotary vacuum evaporator (Buchi Rotavapor Model 461, Switzerland) at 35°C followed by a gentle stream of nitrogen gas (N_2) and a concentrated bulk crude extract of wild tomato leaves was prepared. 2-undecanone and 2-tridecanone in the wild tomato crude extracts were identified and quantified on a Hewlett-Packard (HP) gas chromatograph (GC), model 5890 equipped with mass selective detector (HP 5971) and a HP 7673 automatic injector. The instrument was auto-tuned with perfluorotributylamine at m/s 69, 219, and 502. The instrument detection limit (the lowest concentration of 2-undecanone and 2-tridecanone that the instrument can detect) was 12.5 ng of each compound using the total ion mode. Electron impact mass spectra were obtained using an ionization potential of 70 eV. The operating parameters of the GC were as follows:

injector and detector temperatures 210 and 275°C, respectively. Oven temperature was programmed from 70 to 230°C at the rate of 10°C min^{-1} (2 min initial hold). Injections on to the GC column were made in splitless mode using a 4-mm ID single taper liner with deactivated wool. A 25 m × 0.2 mm ID capillary column containing 5% diphenyl and 95% dimethyl-polysiloxane (HP-5 column) with 0.33 μm film thickness was used. Carrier gas (He) flow rate was 5.2 mL min^{-1}. Quantification was based on average peak areas from two consecutive injections. Retention times of 2-undecanone and 2-tridecanone under these conditions were 11.39 and 14.57 min, respectively. Peak areas were determined on a HP Model 3396 Series II integrator. Area units were compared to an external standard solution of 2-undecanone, 99% purity, and 2-tridecanone, 99% purity (Aldrich Chemical Company). Linearity over the range of concentrations was determined using regression analysis. The retention time and mass of each methylketone isolated from *L. hirsutum* f. *glabratum* leaf samples (PI 251304, PI 126449, PI 134417, PI 134418, and LA 407) matched those of the authentic standards. Detectability limits (minimum detectable quantity (μg) divided by sample weight (g)) were similar for 2-undecanone and 2-tridecanone and averaged 0.013 μg g^{-1} fresh leaves using the conditions described above.

TOXICITY OF WILD TOMATO LEAF EXTRACTS TO VEGETABLE INSECTS

Our previous work on natural products indicated that wild tomato leaf extracts could be explored as an alternative to synthetic pesticides. Three methylketones (2-tridecanone, 2-dodecanone, and 2-undecanone) were effective against the tobacco hornworm, *Manduca sexta* L. 2-tridecanone was the most effective methylketone against tobacco hornworm and tobacco budworm, *H. virescens* (LC_{50} of 0.015 μMoles cm^{-2}) (Antonious et al. 2003).

Two methyl ketones, 2-tridecanone and 2-dodecanone, were more effective against the tobacco hornworm, *M. sexta* (LC_{50} of 0.015 and 0.028 μM·cm^{-2}, respectively) than 2-undecanone (LC_{50} of 0.096) and 2-pentadecanone. 2-pentadecanone at the highest concentration tested (5 μMoles/cm^2) killed only 27% of the tobacco budworms, 15% of the tobacco hornworms, 15% of the two-spotted spider mites, and none of the green peach aphids (data not shown). It was found that 2-tridecanone was the most effective methyl ketone against tobacco hornworm and budworm (LC_{50} of 0.015 μM·cm^{-2} of filter paper surface area) (Antonious et al. 2003).

2-tridecanone also was effective against adults of the green peach aphid, *Myzus persicae* (LC_{50} of 0.07 μ Moles cm^{-2}), at a significantly lower dose than 2-undecanone. Crude leaf extracts of *L. hirsutum* f. *glabratum* Mull (accessions PI 134417, PI 134418, and PI 126449) were the most effective against larvae of the two *Lepidoptera* species, tobacco hornworm and tobacco budworm. The toxicity of two methylketones (2-undecanone and 2-tridecanone), the major constituents of *L. hirsutum* f. *glabratum* accessions tested, to adults of the sweet potato whitefly and fourth instar larvae of the Colorado potato beetle (CPB) was determined using no-choice bioassays. 2-undecanone caused 80% mortality of the fourth instar larvae of the CPB at the highest concentration tested (100 mg 2-undecanone mL^{-1} of acetone), while 2-tridecanone caused 72% mortality of whiteflies at 20 mg 2-tridecanone $mL^{-1}$of ethanol (Antonious et al. 2005). 2-tridecanone, which has an herbaceous, spicy odor (Vernin et al. 1998), was found toxic to a number of insect species by contact, ingestion, or vapor action (Farrar et al. 1994, Muigai et al. 2002).

The CPB, *L. decemlineata* (Say), and the sweet potato whitefly, *B. tabaci* (Gennadius), are two important global vegetable insects that have developed resistance to many synthetic pesticides. The CPB can quickly destroy a plant and has become the most destructive pest of potatoes world-wide. It has become resistant to esfenvalerate, carbofuran, phosmet, and endosulfan (Wyman et al. 1995, Sikinyi et al. 1997). Costs of CPB control will continue to increase because most of the current insecticides are no longer effective and newly developed products are up to five times as costly (Grafius 1997).

Likewise, the sweet potato whitefly is a pervasive pest and vector of plant viruses affecting food and industrial crops and has become a major constraint to development in tropical and subtropical agriculture (Morales 2001) and greenhouse production systems (Oliveira et al. 2001). Historically, it has been difficult to control using conventional insecticides in agronomic and horticultural production systems (Palumbo et al. 2001) and has developed a high degree of resistance to several chemical classes of insecticides, including organophosphates, carbamates, pyrethroids, insect growth regulators and chlorinated hydrocarbons (Elbert and Nauen 2000).

Fourth instar larvae of the CPB were collected from unsprayed potato plants (*Solanum tuberosum* L.) grown at Kentucky State University Research Farm for bioassays. Beetles were subsequently stored at 10°C to slow their development until needed. At testing, larvae were removed from cold storage and warmed to room temperature. Fresh potato leaves were provided daily to the larvae for 10 d before testing (Antonious et al. 1999). A series of glass Petri dishes (9 cm diameter by 2.2 cm depth) were prepared. A small piece of cotton soaked in distilled water was placed in each plate to provide moisture. After 5 hr starvation, larvae (3 replicates of 10 larvae each) were placed in the dishes and offered 10 potato leaf disks (2.5 cm diameter) treated by topical application with 60 μL of each of the two methylketones (2-undecanone or 2-tridecanone) prepared in acetone. The solvent was allowed to evaporate before the leaf disks were introduced to the starved beetles. Disks treated with acetone only were used as controls. The glass lid was then placed on the plate and the plates were held for 8 hr at 20°C. After the exposure period, numbers of dead and live larvae were recorded. Larvae were considered alive if they moved at least one leg when probed. The test was replicated and four concentrations (10, 25, 50, and 100 g mL^{-1} of acetone) of each of the two methylketones were tested. Statistical comparisons were carried out using ProProbit analysis (Sakuma 1998).

Adults of the sweet potato whitefly were obtained from a greenhouse colony reared on a mixture of several vegetables (Simmons 1994). Plastic Petri dishes (9 cm diameter by 2.2 cm depth) were used as test arenas. A hole (about 0.5 cm^2) was drilled into the cover of the dish to allow insertion of the test insects. The bottom of the Petri dish was lined with two filter papers (Whatman No. 3), which were moistened with deionized water. 2-undecanone and 2-tridecanone solutions were prepared in 65% ethanol at 10, 15, and 20 mg mL^{-1} of ethanol. Leaves of commercial tomato plants, *L. esculentum* (Mill) cv. Homestead, free of defects were obtained from the greenhouse and examined to ensure the absence of whitefly eggs before the assays were run. Both surfaces of each leaf were sprayed with one of two methylketones applied at a distance of 20 cm using a Preval sprayer (Precision Valve Corporation, Yonkers, New York) so that the spray deposits on each leaf were uniform. Tomato leaves sprayed with 65% ethanol only were used as controls. The leaves were held vertically with forceps during spraying. They were allowed to air-dry and each was then placed in a Petri dish. The dish was covered with the lid. One hundred adult female whiteflies were placed into each dish via the entrance hole in the lid, and the hole was plugged with a cotton pad. The dishes were held for 8 hr in an environmental chamber with constant light at 26°C and 70 ± 5% relative humidity. After the exposure period, the number of dead whiteflies and number of eggs laid on both leaf surfaces were counted using a dissecting microscope. Whiteflies were considered dead if they exhibited no movement. The experiment was replicated five times for each concentration of each of the two methylketones tested.

Results of bioassays conducted to determine the toxicity of 2-undecanone and 2-tridecanone to adults of the sweet potato whitefly and the fourth instar larvae of the CPB indicated a concentration gradient effect. 2-undecanone caused the highest mortality (80%) of the fourth instar larvae of the CPB at the highest concentration tested (100 mg of 2-undecanone mL^{-1} of acetone), while 2-tridecanone caused 72% mortality of whiteflies at 20 mg of 2-tridecanone mL^{-1} of ethanol. The two natural products, 2-undecanone and 2-tridecanone, significantly reduced the numbers of whitefly eggs compared to the control (Antonious et al. 2005). The fact that the two insects (whitefly and CPB) differed greatly in their response to each of the two methylketones suggests that insect resistance in an accession may result from the synergistic or additive effects involving more than one chemical compound.

In other assays, 2-dodecanone and 2-tridecanone were about equal in toxicity against adult aphids *M. persicae* and required a significantly lower dose than 2-undecanone. Spider mite (*T. urticae*) was more sensitive to 2-undecanone and 2-dodecanone than 2-tridecanone and was also insensitive to 2-pentadecanone (15 % mortality at 5 μM cm^{-2} of treated surface area: Antonious et al. 2003). Clearly, there are differences in sensitivity among arthropods in their sensitivity to components of potential biopesticides.

PRODUCTION OF BIOPESTICIDES

Using standard agricultural practices under greenhouse conditions, some wild tomato accessions produced high levels of methylketones (Fig. 6). Concentrations of 2-tridecanone were greatest in PI 134417, while 2-undecanone concentration was greatest in LA 407. Antonious and Snyder (2003) reported that an average 3-mo-old wild tomato accession of *L. hirsutum* f. *glabratum* (PI 134417) that has 227.5 g leaves averaging about 49,872 cm^2 exposed leaflet surface area (not including the surface area of the plant stem, in which trichomes are also found in large numbers) would produce 411.3 mg 2-tridecanone, 0.44 mg 2-dodecanone, 98.9 mg 2-undecanone, and 33.1 mg 2-pentadecanone. Accordingly, it is possible to obtain about 2.4 mg of methylketones from one gram of wild tomato leaflets. Because these amounts do not include exudates from trichomes present on the plant stem, the amount of methylketones from 1 kg of plant foliage that includes plant stems is expected to be greater than what is obtained only from the leaves. Type VI trichome counts on the leaves of PI 134417 and LA 407 were significantly higher during the month of June (Fig. 7, upper graph). The greater weight of fresh leaves was obtained from accession PI 134417 compared to LA 407 (Fig. 7, lower graph). Differences in toxicity among accessions tested can be arranged as: PI 126449 > PI 251304 > LA 407 (Fig. 8). *Heliothis* is generally less sensitive to toxicity of crude extracts compared to *Manduca*. This contrasts with results obtained when using pure compound of each methylketone alone as reported by Antonious et al. (2003). In these bioassays pure 2-tridecanone

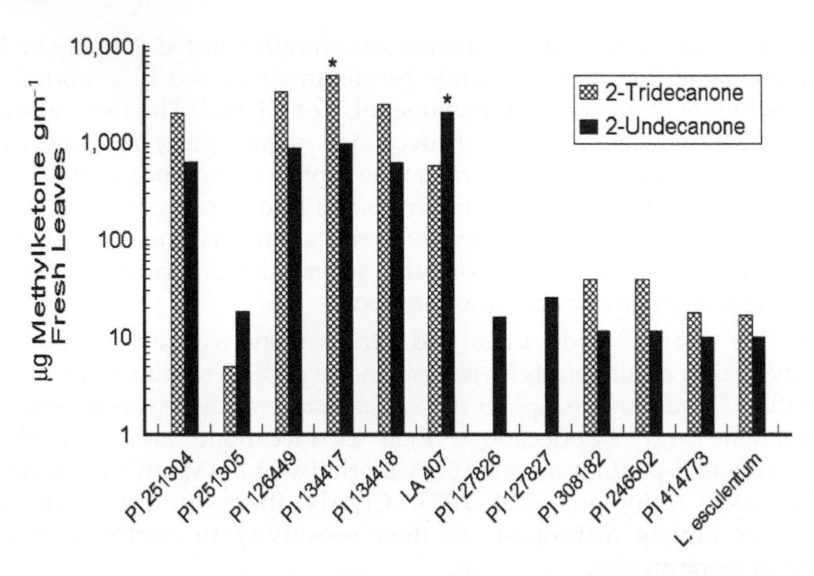

Fig. 6 Concentrations of two methylketones (2-tridecanone and 2-undecanone) in chloroform leaf extracts of 11 wild tomato accessions and one domestic tomato, *Lycopersicon esculentum* cv. Fabulous. Statistical comparisons were carried out for each compound among all accessions. Each value is an average of three replicates. Values accompanied by an asterisk indicate a significant difference between the concentration on that accession and all other accessions ($P < 0.05$) based on results from the ANOVA procedure (SAS Institute 2001).

and 2-undecanone were equally toxic to *Manduca* and *Heliothis*. 2-undecanone was somewhat more toxic to *Heliothis* than *Manduca*. Apparently, compounds present in trichome secretions but absent in bioassays of pure compounds are responsible for the observed differences. This indicates that toxicity of trichome exudates is due to the presence of more than one active compound in complex mixtures of crude extracts.

Lycopersicon pennellii, like *L. hirsutum*, is resistant to a number of arthropod pests (Rodriguez et al. 1993). Host resistance to the carmine spider mite (*T. cinnabarinus*), two-spotted spider mite (*T. urticae*), greenhouse whitefly (*T. vaporariorum*), cotton bollworm (*H. armigera*), cotton leaf worm (*S. littoralis*), and potato aphid (*M. euphorbiae*) has been reported. Often, the sugar esters produced by type IV trichomes on leaves of *L. pennellii* have been implicated in arthropod resistance. Thus, examination of the production of these sugar esters is one aspect of their development as biopesticides.

Sugar-esters are nonionic surfactants consisting of glucose as hydrophilic moiety and fatty acids as lipophilic moiety. A colorimetric sugar assay was used to determine the amount of glucose released from the sugar-ester after hydrolyzing the ester using 0.4 N NaOH (Moore and

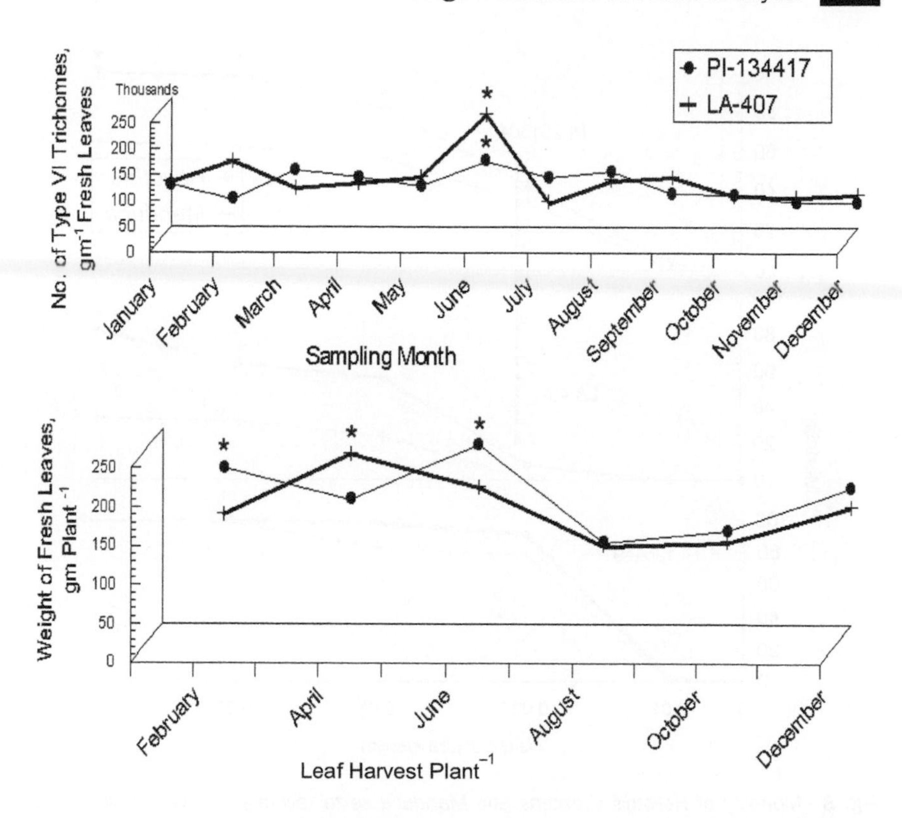

Fig. 7 Number of type VI glandular trichomes per gram of fresh leaves of two wild tomato accessions (PI 134417 and LA 407) of *L. hirsutum* f. *glabratum* (upper graph) and weight (g) of leaves collected at each harvest time (lower graph). Statistical comparisons were carried out between sampling months. Each value is an average of three replicates. Values accompanied by an asterisk indicate a significant difference ($P < 0.05$) between the two accessions for each sampling time using ANOVA procedure (SAS Institute 2001).

Antonious 2003). As shown in Fig. 9, only the two accessions of *L. pennellii* (PI 246502 and PI 414773) contained the high concentrations of sugar-esters (glucolipids) in their leaves compared to *L. hirsutum* f. *glabratum* accessions and *L. esculentum* (domestic tomato). Furthermore, concentrations tended to be higher in the summer than in the winter.

REPELLENCY AND TOXICITY OF TOMATO LEAF EXTRACTS TO SPIDER MITES

Using repellent chemicals for crop protection is a unique way to prevent insects and spider mites from laying eggs on target plants, and to prevent leaf and fruit damage. Defense of tomato against spider mites is

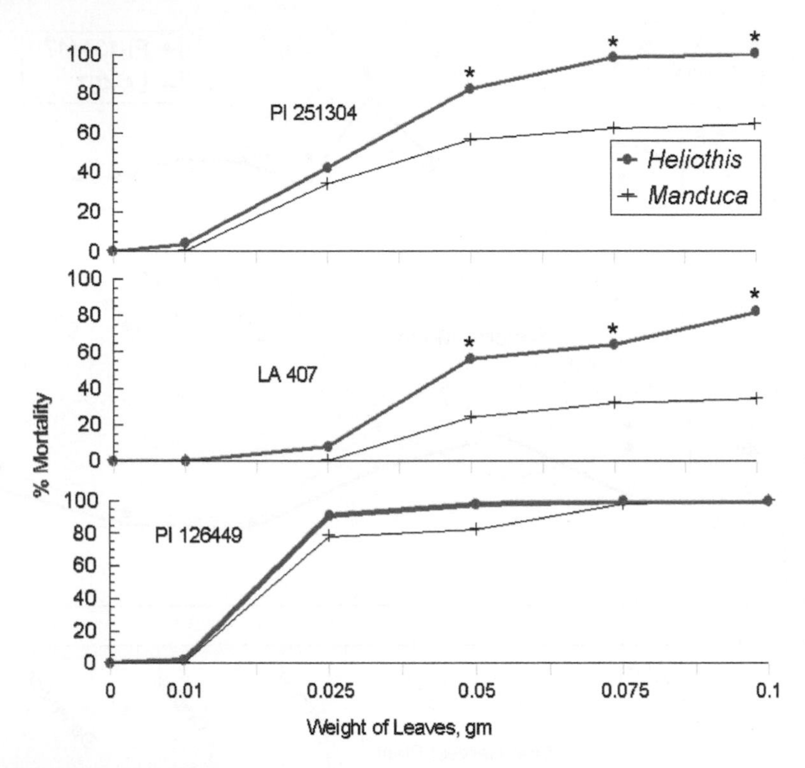

Fig. 8 Mortality of *Heliothis virescens* and *Manduca sexta* (average of 10 replicates of 10 neonate larvae) exposed to ethanol leaf extracts from three wild tomato accessions (PI 251304, LA 407, and PI 126449 from *Lycopersicon hirsutum* f. *glabratum*). Values accompanied by an asterisk indicate a significant difference ($P < 0.05$) between the two insects for each weight of leaves (SAS Institute 2001).

considered to be based mainly on glandular trichomes (Farrar and Kennedy 1991). Cultivated tomato (*L. esculentum*) is more susceptible to attack by the two-spotted spider mite than wild tomato species (*L. hirsutum*). Both physical and chemical effects of glandular trichomes on spider mites have been reported. Mites have been shown to become entrapped in the sticky exudates of trichomes on the leaves of wild tomato. Volatile exudates of wild tomato include 2-tridecanone and 2-undecanone or the terpenoids zingiberene and curcumene, depending on the tomato species. Wild tomato leaves from which the trichomes have been removed were still more resistant to two-spotted spider mites than cultivated tomato leaves without trichomes, suggesting that factors other than trichomes may also be involved in resistance in wild tomato, i.e., the glycoalkaloid α-tomatine, chlorogenic acid and rutin present in tomato leaves (Hoffland et al. 2000).

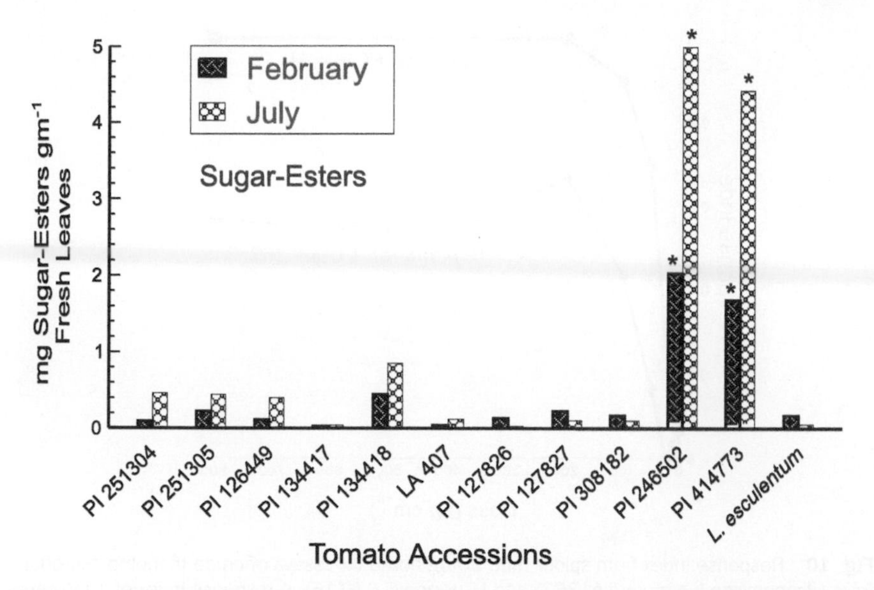

Fig. 9 Concentrations of sugar esters (glucolipids) in the leaves of 11 wild tomato accessions and one domestic tomato (*Lycopersicon esculentum* cv. Fabulous). Statistical comparisons were carried out between accessions. Bars compound accompanied by an asterisk indicate a significant difference ($P < 0.05$) between accessions for each sampling month (SAS Institute 2001).

Bioassays for repellency of crude extracts to spider mites were carried out using the diving board bioassay assembly that was developed by Guo et al. (1993). One gravid female mite (n=30) was placed in the center of a filter paper strip (bridge) and given the freedom to exit over the treatment or control strip. Exits over treatment or control strips were recorded and exit ratios were calculated (treatment to control) for each of 30 mites per accession or concentration tested. Exit ratios of diving board bioassays of crude trichome extracts of *L. hirsutum* (LA 1363) and *L. pennellii* (LA 716) are presented in Fig. 10. Extracts were tested in 0–100 μg cm^{-2}. A high exit ratio indicates repellency. Therefore, the extract from LA 1363 was more repellent than that from LA716.

To measure and test spider mite repellency, 2,3-dihydrofarnesoic acid (DFA), a known spider mite repellent, and 2-pentadecanone, a methylketone that has lesser degree of spider mite repellency (Antonious and Snyder 2006), were tested over a wide range of concentrations. Results have shown that the higher the concentration of DFA on filter paper rings, the longer mites stay inside the ring, indicating greater repellency of the compound. This trend was less pronounced when 2-pentadecanone was tested for repellency. Compared to the ethanol and chloroform crude leaf

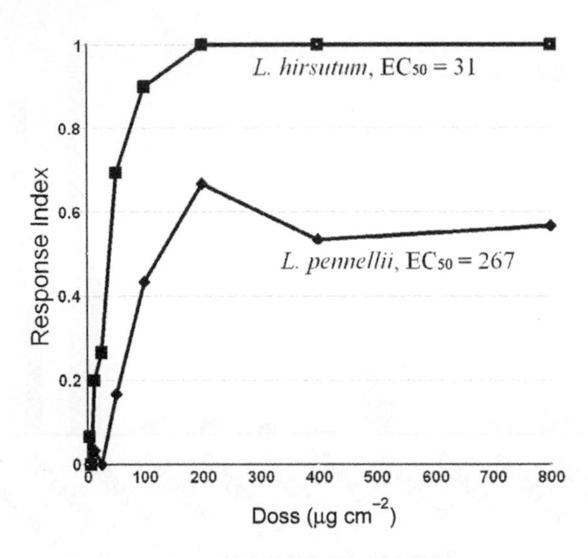

Fig. 10 Response index from spider mite diving board bioassays of crude trichome secretions from *Lycopersicon hirsutum* (LA1363) and *L. pennellii* (LA716). A response index of 1 indicates that all mites exited over the control, and a response index of 0 indicates that 50% of the mites exited over the control, and the other 50% exited over the trichome secretions. Trichome secretions from *L. hirsutum* were more repellent than those from *L. pennellii*.

extracts, the hexane leaf extracts exhibited greatest spider mite repellency from *L. hirsutum* f. *glabratum* accessions (PI 251304, PI 134417, PI 134418, and PI 126449).

Solvent used for extraction also can lead to differences in mortality. The chloroform extracts were considerably more toxic to spider mites than other solvents tested. The extracts from four *L. hirsutum* f. *glabratum* accessions (PI 251304, PI 126449, PI 134417, and PI 134418) were considerably more toxic than those from the domestic tomato cultivar, *L. esculentum* cv. Fabulous and all other accessions tested. Toxicity of the hexane extract from PI 126449 also was significantly greater than hexane extracts from the other accessions. Spider mite mortality was highest when exposed to the chloroform extracts of PI 251304, PI 134417, PI 134418, and PI 126449, suggesting that toxic action of certain phytochemicals in or on their leaves may contribute to this mortality. On the other hand, the fact that the repellency and response of spider mites differ greatly among the accessions tested indicates that repellency may result from the synergistic effects involving more than one chemical compound in wild tomato crude extracts. Performance of crude extracts from wild tomato leaves having a potentially acaricidal performance can be explored for developing a natural product for use as a biodegradable alternative to many synthetic acaricides especially in small-acreage, high-value crops.

CHEMICAL DIFFERENTIATION AMONG TOMATO ACCESSIONS

We investigated differences in chemical composition of the extracts that may explain the observed differences in toxicity and repellency among extracts from the 11 wild tomato accessions. Mass chromatographic analysis indicated the presence of 2-undecanone, trans-caryophyllene, α-curcumene, 2-tridecanone, and α-zingiberene. Generally, the hexane leaf extracts of *L. hirsutum* f. *glabratum* were dominated by two methylketones, 2-undecanone and 2-tridecanone, while only low concentrations of 2-tridecanone were detected in the domestic commercial tomato cultivar *L. esculentum* cv. Fabulous. This finding is consistent with previous data (Antonious and Snyder 2006) obtained from the analysis of *L. hirsutum* plants grown under similar greenhouse conditions. Chatzivasileiadis and Sabelis (1997) tested the effects of 2-tridecanone and 2-undecanone in the ratio found in the wild accession PI 134417 of *L. hirsutum* f. *glabratum* on two strains of *T. urticae,* from tomato and cucumber in greenhouses. Their results indicated that the LC_{50}s (1.44-1.84 mg mL^{-1}) were comparable to that of amitraz (a non-systemic synthetic acaricide/insecticide).

AN APPROACH TO DEVELOP A NEW INSECTICIDE FROM WILD TOMATO LEAVES

A bulk crude extract of 2-tridecanone was prepared by soaking 5 kg wild tomato leaves collected from greenhouse wild tomato plants in 10 L of water containing 15 mL of 2% Sur-Ten (sodium dioctyl sulfosuccinate), obtained from Aldrich Chemical Company. After continuous manual shaking, the mixture was filtered through cheesecloth and each liter of the filtrate was partitioned with 200 mL of n-hexane in a separatory funnel. The hexane layers were combined and the solvent was evaporated to dryness using a rotary vacuum evaporator (Buchi Rotavapor Model 461, Switzerland) at 35°C followed by a gentle stream of nitrogen gas (N_2). To the concentrated wild tomato leaf extract, Triton X-100, an inert nonionic surfactant (t-octylphenoxypolyethoxy ethanol) was added at 0.1% (v/v) as an emulsifier and the contents were mixed and used for spraying 45-d-old pepper (*Capsicum annum* cv. Aristotle X3R), squash (*Cucurbita maxima* cv. Blue Hubbard), radish (*Raphanus sativus* cv. Brio), domestic tomato (*L. esculentum* cv. Ponderasa), broccoli (*Brassica oleracea* cv. Italian Green), Swiss chard (*Beta vulgaris* cv. Lucullus), and watermelon (*Citrullus lanatus* cv. Stars & Stripes). To purify and determine the concentration of 2-tridecanone in the crude extract, 1 mL of the extract was re-dissolved in 10 mL of n-hexane and applied to the top of a glass chromatographic column (20 × 1.1 cm) containing 10 g alumina grade-II that had been pre-wetted and eluted with n-hexane (Fobes et al. 1985). The eluent was evaporated to dryness and reconstituted in n-hexane for GC/MSD quantification. 2-tridecanone in the prepared insecticide formulation was used at the rate of 0.44 g active ingredient (AI) L^{-1} of water in such a manner that the plants were drenched to the point of runoff (Shafer and Bukovac 1989). Spraying was carried out using a microsprayer (Plant Care Sprayer, Delta Industries, Pennsylvania) of 475 mL capacity capable of producing monosize droplets of 200 µm diameter. Spraying resulted in 20-30 drops cm^{-2} of treated leaf surface as determined by use of water-sensitive paper (Syngenta Crop Protection AG, CH-4002 Basle, Switzerland). Leaves were randomly collected at intervals of 1, 3, 5, 24, and 48 hr following spraying from both sprayed and unsprayed plants. Twenty grams of leaves were blended with 100 mL of chloroform for 2 min at high speed. After homogenization, the mixture was vacuum filtered through a Buchner funnel containing a glass microfiber filter containing 10 g anhydrous Na_2SO_4 and the extract was concentrated by rotary vacuum evaporator as described above. Clean-up of the concentrated extracts was achieved using the procedure described by Fobes et al. (1985) and Antonious and Snyder (1993). The retention time and mass of 2-tridecanone isolated from *L. hirsutum* f. *glabratum* leaf samples (PI 251304, PI 126449, PI 134417, PI 134418, and LA 407) matched those of the authentic standard. Minimum detectable levels averaged 0.02 to 0.005 µg g^{-1} leaves. Results indicated that 2-tridecanone spectral data, which showed a molecular ion peak (M^+) at m/z 198, along with other characteristic fragment ion peaks are consistent with the assignment of the molecular formula of tridecanone ($C_{13}H_{26}O$).

PERSISTENCE OF 2-TRIDECANONE ON PLANT TISSUES

Decline of 2-tridecanone residues on the greenhouse sprayed vegetable leaves as a function of time indicated that half-life ($T_{1/2}$) values (calculated as described by Anderson 1986) of 2-tridecanone were 1.3 hr on squash leaves and 4.0 hr on broccoli leaves. Maximum residue limits of 2-tridecanone on vegetables have not been established. The short persistence of 2-tridecanone on the leaves of the greenhouse vegetables tested so far can be recognized as a desirable chemical characteristic. Ideally, safe pesticides remain in the target area long enough to control the specific pest, then degrade into harmless compounds. The minimum intervals for worker reentry into treated areas are based upon pesticide deposits following spraying, repeated exposure, and residues remaining on the treated foliage and fruits (Antonious 1995). Foliage may accept and retain much greater pesticide deposits longer than fruits. This suggests that the plant foliage is important when considering worker reentry into pesticide-treated vegetables, since workers may be exposed to much greater surface areas of foliage than of fruits (Antonious 2002, 2004b). No phytotoxicity was observed on the leaves following spraying with 2-tridecanone formulation at 0.44 g L^{-1} of water. Many leaf surfaces represent the most unwettable of known surfaces. This is due to the hydrophobic nature of the leaf surface, which is actually covered with crystalline wax of straight chain paraffin alcohols. An important factor that can affect the biological efficacy of foliar spray application of pesticide is the extent to which the liquid wets and covers the foliage surface. The persistence of 2-tridecanone on vegetables and fruits can be tailor made by using appropriate surface active agents (Stevens 1993, Liu and Stansly 2000) and/or photo stabilizers for use under field conditions as described by Johnson et al. (2003). Most formulations of pesticides contain surfactants as emulsifiers and wetting agents to provide greater coverage and retention. In plant protection systems, the search for new efficient and safe compounds is highly desirable. The medium chain length methylketones, in particular 2-tridecanone, have been shown to be potent agents against a variety of insects and spider mites. With more research on the persistence and environmental toxicology of 2-tridecanone on vegetables and fruits, this compound could be a substitute for many synthetic pesticides used in plant protection. The genus *Lycopersicon* is characterized by great diversity within and among its species. Host-plant resistance to arthropods has been studied most intensively in *L. esculentum*, *L. hirsutum*, and *L. pennellii*. However, use of crude extracts and preparation of a new formulation from wild tomato leaf constituents having insecticidal and acaricidal performance for use in crop protection have received little attention.

FUTURE PLANS

Tomato is often considered a high-risk crop, requiring intensive use of pesticides. Current levels of resistance to arthropods in domestic tomato cultivars are not high enough to permit a significant decrease in amount of pesticides applied. This investigation provides evidence of the toxicity of methylketones (2-tridecanone and 2-undecanone) against the green peach aphid (*M. persicae*), tobacco hornworm (*M. sexta*), tobacco budworm (*H. virescens*), and two-spotted spider mite (*T. urticae*), which are important worldwide tomato pests. Future trends in agriculture and crop protection chemical technology will necessitate the discovery of agrochemicals with high selectivity, high mammalian and environmental safety, low rates of application, and low costs. This investigation provides background information on the level of methylketones in wild tomato species and may provide an opportunity for many farmers and organic growers who may be able to grow wild tomato foliage for use as botanical insecticides in their farms, which could be an effective tool in pest management programs against a wide variety of arthropod pests. The extraction of methylketones, glucolipids and sesquiterpenes from wild tomato foliage is a simple procedure. Many farmers may be able to grow wild tomato foliage from different species for use as multi-purpose insecticide on cultivated tomato and other crops in their farms. Such alternatives, which have few or no side-effects on the environment, low toxicity to warm-blooded animals and humans, high efficacy against insects and spider mites, and lower potential for insect resistance development, are in great demand.

For agrochemical companies, performance of methylketones as potential insecticidal and acaricidal products from wild tomato leaves can be explored for developing natural products for use as biodegradable alternative to synthetic pesticides. A formulation prepared from the leaves of selected accessions may therefore create a mixture with the desired level of constituents. However, if methylketones and other constituents in wild tomato leaves are to be used within the conventional and organic production systems to control vegetable insects and spider mites, further work is needed to investigate their performance under field conditions and the impact of methylketones on natural enemies.

ACKNOWLEDGMENTS

This investigation was supported by two grants from USDA/CSREES to Kentucky State University under agreements No. KYX-10-99-31P and No. KYX-10-03-37P.

REFERENCES

Agarwal, M. and S. Walia, S. Dhingra, and B. Khambay. 2001. Insect growth inhibition, antifeedant and antifungal activity of compounds isolated/derived from *Zingiber officinale* Roscoe rhizomes. Pest Manag. Sci. 57: 289-300.

Aina, O.J. and J.G. Rodriguez, and D.E. Knavel. 1972. Characterizing resistance to *Tetranychus urticae* in tomato. J. Econ. Entomol. 65: 641-643.

Anderson, A.C. 1986. Calculating biodegradation rates. J. Environ. Sci. Health B21: 41-56.

Antonious, G.F. 1995. Analysis and fate of acephate and its metabolite, methamidophos in pepper and cucumber. J. Environ. Sci. Health B30: 377-399.

Antonious, G.F. 1999. Efficiency of grass buffer strips and cropping system on off-site dacthal movement. Bull. Environ. Contam. Toxicol. 63: 25-32.

Antonious, G.F. 2001a. Production and quantification of methyl ketones in wild tomato accessions. J. Environ. Sci. Health B36: 835-848.

Antonious, G.F. 2001b. Insecticides from wild tomato: phase I-breeding, trichome counts, and selection of tomato accessions. J. Kentucky Acad. Sci. 62-79.

Antonious, G.F. 2002. Persistence and performance of esfenvalerate residues on broccoli. J. Pest Manag. Sci. 58: 85-91.

Antonious, G.F. Soil infiltration by pesticides. pp. 1-4. *In:* D. Pimentel. [ed.] 2003. Encyclopedia of Pest Management, Vol. 3. Marcel Dekker, Inc., New York.

Antonious, G.F. 2004a. Trifluralin residues in runoff and infiltration water from tomato production. Bull. Environ. Contam. Toxicol. 72: 962-969.

Antonious, G.F. 2004b. Residues and half-lives of pyrethrins on field-grown pepper and tomato. J. Environ. Sci. Health B39: 1-13.

Antonious, G.F. and M.E. Byers. 1997. Fate and movement of endosulfan under field conditions. J. Environ. Toxicol. Chem. 64: 644-649.

Antonious, G.F. and A.M. Patterson. 2005. Napropamide residues in runoff and infiltration water from pepper production. J. Environ. Sci. Health B40: 385-396.

Antonious, G.F. and J.C. Snyder. 1993. Trichome density and pesticide retention and half-life. J. Environ. Sci. Health B28: 205-219.

Antonious, G.F. and J.C. Snyder. 1994. Residues and half-lives of acephate, methamidophos, and pirimiphos-methyl in leaves and fruit of greenhouse-grown tomatoes. Bull. Environ. Contam. Toxicol. 52: 141-148.

Antonious, G.F. and J.C. Snyder. 2003. Insecticides from wild tomato leaves. Fruit and Vegetable Crops Research Report, University of Kentucky, Agricultural Experiment Station, USA, PR-488.

Antonious, G.F. and J.C. Snyder. 2006. Natural products: Repellency and toxicity of wild tomato leaf extracts to the two spotted spider mite, *Tetranychus urticae* Koch. J. Envir. Sci. Health B41: 43-55.

Antonious, G.F. and M.E. Byers, and J.C. Snyder. 1998. Residues and fate of endosulfan on field-grown pepper and tomato. Pesticide Sci. 54: 61-67.

Antonious, G.F. and J.C. Snyder, and D.L. Dahlman. 1999. Tomato cultivar susceptibility to Egyptian cotton leafworm (Lepidoptera: Noctuidae) and Colorado potato beetle (Coleoptera: Chrysomelidae). J. Entomol. Sci. 34: 171-182.

Antonious, G.F. and D.L. Dahlman, and L.M. Hawkins. 2003. Insecticidal and acaricidal performance of methylketones in wild tomato leaves. Bull. Environ. Contam. Toxicol. 71: 400-407.

Antonious, G.F. and T.S. Kochhar, and A.M. Simmons. 2005. Natural products: Seasonal variation in trichome counts and contents in *Lycopersicon esculentum* f. *glabratum*. J. Environ. Sci. Health B40: 619-631.

Antonious, G.F. and J.E. Meyer, J.A. Rogers, and Y.H. Hu. 2007. Growing hot pepper for cabbage looper, *Trichopulsia ni* (Hübner) and spider mite, *Tetranychus urticae* (Koch) control. J. Environ. Sci. Health B42: 559-567.

Azevedo, S.M. and M.V. Faria, W.R. Maluf, A.C. Oliveira, and J.D. Freitas. 2003. Zingiberene-mediated resistance to the South American tomato pinworm derived from *Lycopersicon hirsutum* var. *hirsutum*. Euphytica 134: 347-351.

Bethke, J.A. and M.P. Parrella, J.T. Trumble, and N.C. Toscano. 1987. Effect of tomato cultivar and fertilizer regime on the survival of *Liriomyza trifolii* (Diptera: Agromyzidae). J. Econ. Entomol. 80: 200-203.

Carter, C.D. and T.J. Gianfagna, and J.N. Sacalis. 1989. Sesquiterpenes in glandular trichomes of wild tomato species and toxicity to Colorado potato beetle. J. Agric. Food Chem. 37: 1425-1428.

Channarayappa, G.S. and V. Muniyappa, and R.H. Frist. 1992. Resistance of *Lycopersicon* species to *Bemisia tabaci*, a tomato leaf curl virus vector. Canad. J. Bot. 70: 2184-2192.

Chatzivasileiadis, E.A. and M. Sabelis. 1997. Toxicity of methyl ketones from tomato trichomes to *Tetranychus urticae* Koch. Exp. Appl. Acarol. 21(6-7): 473-484.

Chortyk, O.T. and J.C. Pomonis, and A.W. Johnson. 1996. Synthesis and characterization of insecticidal sucrose esters. J. Agric. Food Chem. 44: 51-57.

Chortyk, O.T. and S.J. Kays, and Q. Teng. 1997. Characterization of insecticidal sugar esters from petunia. J. Agric. Food Chem. 45: 270-275.

Cosenza, G.W. and H.B. Green. 1979. Behaviour of the tomato fruitworm, *Heliothis zea* (Boddie), on susceptible and resistant lines of processing tomatoes. HortScience 14: 171-173.

Eigenbrode, S.D. and J.T. Trumble. 1993. Antibiosis to beet armyworm (*Spodoptera exigua*) in *Lycopersicon* accessions. HortScience 28: 932-934.

Eigenbrode, S.D. and J.T. Trumble, and K.K. White. 1996. Trichome exudates and resistance to beet armyworm (Lepidoptera: Noctuidae) in *Lycoperiscon hirsutum* f. *typicum* accessions. Environ. Entomol. 25: 90-95.

Elbert, A. and R. Nauen. 2000. Resistance of *Bemisia tabaci* to insecticides in southern Spain with special reference to neonicotinoids. Pest Manag. Sci. 56: 60-64.

Farrar, R.R. and G.G. Kennedy. Insect and mite resistance in tomato. pp. 122-142. *In:* G. Kalloo. [ed.] 1991. Genetic Improvement of Tomato. Monographs on Theoretical and Applied Genetics. Springer-Verlag, Berlin, Heidelberg.

Farrar, R.R. and J.D. Barbour, and G.G. Kennedy. 1994. Field evaluation of insect resistance in a wild tomato and its effects on insect panasitoids. Entomol. Exp. Appl. 71: 211-226.

Fery, R.L. and F.P. Cuthbert. 1974. Effect of plant density on fruitworm damage in the tomato. HortScience 9: 140-141.

Fobes, J.F. and J.B. Mudd, and M.F. Marsden. 1985. Epicuticular lipid accumulation on the leaves of *L. pennellii.* Plant Physiol. 77: 567-570.

Gentile, A.G. and A.K. Stoner. 1968. Resistance in *Lycopersicon* spp. to the tobacco flea beetle. J. Econ. Entomol. 61: 1347-1349.

Gilbert, J.C. and J.T. Chinn, and J.S. Tanaka. 1966. Spider mite tolerance in multiple disease resistant tomatoes. Proc. Amer. Soc. Hort. Sci. 89: 559-562.

Grafius, E. 1997. Economic impact of insecticide resistance in the Colorado potato beetle (Coleoptera: Chrysomelidae) on the Michigan potato industry. J. Econ. Entomol. 90: 1144-1151.

Guo, Z. and P.A. Weston, and J.C. Snyder. 1993. Repellency to two-spotted spider mite, *Tetranychus urticae* Koch, as related to leaf surface chemistry of *Lycopersicon esculentum* accessions. J. Chem. Ecol. 19: 2965-2979.

Hartmann, J.B. and D.A. StClair. 1998. Variation for insect and horticultural traits in tomato inbred backcross populations derived from *Lycopersicon pennellii*. Crop Sci. 38: 1501-1508.

Heinz, K.M. and F.G. Zalom. 1995. Variation in trichome-based resistance to *Bemisia argentifolii* (Homoptera: Aleyrodidae) oviposition on tomato. J. Econ. Entomol. 88: 1494-1502.

Hoffland, E. and M. Dicke, W.V. Tintelen, H. Dijkman, and M.V. Beusichem. 2000. Nitrogen availability and defense of tomato against two-spotted spider mite. J. Chem. Ecol. 26: 2697-2711.

Johnson, S. and P. Dureja, and S. Dhingras. 2003. Photostabilizers of azadirachtin-A. J. Environ. Sci. Health B38: 451-462.

Juvik, J.A. and M.J. Berlinger, T. Ben David and J. Rudich. 1982. Resistance among accessions of the genera *Lycopersicon* and *Solanum* to four of the main insect pests of tomato in Israel *Spodoptera littoralis, Plusia chalcites, Heliothis armigera, Phthorimaea operculella*. Phytoparasitica 10: 145-156.

Kamau, A.W. and J.M. Mueke, and B.M. Khaemba. 1992. Resistance of tomato varieties to the tomato russet mite, *Aculops lycopersici* (Massee) (Acarina: Eriophyidae). Insect Sci. Appl. 13: 351-356.

Kashyap, R.K. and A.N. Verma. 1986. Screening of tomato germplasm for susceptibility to the fruitborer, *Heliothis armigera* Hubner. Indian J. Entomol. 48: 46-53.

Kennedy, G.G. and W.R. Henderson. 1978. A laboratory assay for resistance to the tobacco hornworm in *Lycopersicon* and *Solanum* spp. J. Amer. Soc. Hort. Sci. 103: 334-336.

Krishna Kumar, N.K. and D.E. Ullman, and J.J. Cho. 1995. Resistance among *Lycopersicon* species to *Frankliniella occidentalis* (Thysanoptera: Thripidae). J. Econ. Entomol. 88: 1057-1065.

Lin, S.Y.H. and J.T. Trumble. 1986. Resistance in wild tomatoes to larvae of a specialist herbivore, *Keiferia lycopersicella*. Entomol. Exp. et Appl. 41: 53-60.

Liu, T.X. and P.A. Stansly. 2000. Insecticidal activity of surfactants and oils against whitefly (*Bemisia argentifolii*). Pest Manag. Sci. 56: 861-866.

Luckwill, L.C. 1943. The genus *Lycopersicon*. An historical, biological, and taxonomic survey of wild and cultivated tomatoes. Aberdeen University Studies No. 120, Aberdeen University Press, Aberdeen, UK.

Moore, K. and G.F. Antonious. 2003. Screening wild tomato accessions for sugar-ester contents. Posters-at-the-Capitol Annual Meeting, Governor's State of the Commonwealth Building, Frankfort, Kentucky, USA, February 6, 2003. Published Abstract, 16 pp.

Morales, F.J. 2001. Conventional breeding for resistance to *Bemisia tabaci*-transmitted geminiviruses. Crop Prot. 20: 825-834.

Muigai, S.G. and D.J. Schuster, J.C. Snyder, J.W. Scott. M.J. Bassett, and H.J. McAuslane. 2002. Mechanisms of resistance in *Lycopersicon* germplasm to the whitefly *Bemisia argentifolii*. Phytoparasitica 30: 347-360.

Muigai, S.G. and M. J. Bassett, D.J. Schuster, and J.W. Scott. 2003. Greenhouse and field screening of wild *Lycopersicon* germplasm for resistance to the whitefly *Bemisia argentifolii*. Phytoparasitica 31: 27-38.

Oliveira, M.R.V. and T.J. Henneberry, and Anderson, P. 2001. History, current status, and collaborative research projects for *Bemisia tabaci*. Crop Prot. 20: 709-723.

Palumbo, J.C. and A.R. Horowitz, and N. Prabhaker. 2001. Insecticidal control and resistance management for *Bemisia tabaci*. Crop Prot. 20: 739-765.

Pillmoor, J.B. 1998. Carbocyclic coformycin: a case study of the opportunities and pitfalls in the industrial search for new agrochemicals from nature. Pest Sci. 52: 75-80.

Ponti, O.M.B. and N.G. Hogenboom. Breeding tomato (*Lycopersicon esculentum*) for resistance to the greenhouse whitefly (*Trialeurodes vaporariorum*). *In:* A.K. Minks and P. Gruys. [eds.] 1980. Integrated Control of Insect Pests in the Netherlands. Center for Agricultural Publishing and Documentation, Wageningen. pp. 187-190.

Quiros, C.F. and M.A. Stevens, C.M. Rick, and M.L. Kok-Yokomi. 1977. Resistance in tomato to the pink form of the potato aphid (*Macrosiphum euphorbiae* Thomas): the role of anatomy, epidermal hairs, and foliage composition. J. Amer. Soc. Hort. Sci. 102: 166-171.

Rice, M.J. and M. Legg, and K.A. Powell. 1998. Natural products in agriculture – a view from the industry. Pest. Sci. 52: 184-188.

Rodriguez, A.E. and W.M. Tingey, and M.A. Mutschler. 1993. Acylsugars of *Lycopersicon pennellii* deter settling and feeding of the green peach aphid (Homoptera: Aphididae). J. Econ. Entomol. 86: 34-39.

Sakuma, M. 1998. ProProbit and ProProbit NM. Division of Applied Biosciences, Graduate School of Agriculture, Kyoto University, Kyoto, Japan

Salinas, H. and F.A. Vallejo Cabrera, and S.E. Estrada. 1993. Evaluation of resistance to the tomato fruit borer *Neoleucinodes elegantalis* (Guenee) in material of *Lycopersicon hirsutum* Humb. & Bonpl. and *L. pimpinellifolium* (Just.) Mill. and its transfer to cultivated material of tomato, *L. esculentum* Mill. Acta Agronomica, Universidad Nacional de Colombia 43: 44-56.

Sances, F.V. and N.S. Toscano, and L.K. Gaston. 1992. Minimization of pesticide residues on head lettuce. J. Econ. Entomol. 85: 202-207.

SAS Institute. 2001. SAS/STAT Guide, Release 0.03 Edition, SAS Inc., SAS Campus Drive, Cary, North Carolina, USA.

Schalk, J.M. and A.K. Stoner. 1976. A bioassay differentiates resistance to the Colorado potato beetle on tomatoes. J. Amer. Soc. Hort. Sci. 101: 74-76.

Schuster, D.J. 1977. Effect of tomato cultivars on insect damage and chemical control. Florida Entomologist 60: 27-232.

Shafer, W.E. and M.J. Bukovac. 1989. Effects of triton X-100 on sorption of 2-(1-naphthyl) acetic acid by tomato fruit cuticles. J. Agric. Food Chem. 37: 486-492.

Sikinyi, E. and D.J. Hannapel, P.M. Imerman, and H.M. Stahr. 1997. Novel mechanism for resistance to Colorado potato beetle (Coleoptera: Chrysomelidae) in wild *Solanum* species. J. Econ. Entomol. 90: 689-696.

Simmons, A.M. 1994. Oviposition on vegetables by *Bemisia tabaci* (Homoptera: Aleyrodidae). Temporal and leaf surface factors. Environ. Entomol. 23: 381-389.

Snyder, J.C. and Z. Guo, R. Thacker, J.P. Goodman, and J. Pyrek. 1993. 2,3-Dihydrofarnesoic acid, a unique terpene from trichomes of *Lycopersicon hirsutum*, repels spider mites. J. Chem. Ecol. 19: 2981-2997.

Stevens, P.G. 1993. Organosilicone surfactants as adjuvants for agrochemicals. Pest. Sci. 38: 103-122.

Stoner, A.K. and T. Stringfellow. 1967. Resistance of tomato varieties to spider mites. Proc. Amer. Soc. Hort. Sci. 90: 324-329.

Stoner, A.K. and R.E. Webb, and A.G. Gentile. 1968. Reaction of tomato varieties and breeding lines to aphids. HortScience 3: 77.

Strang, J. 1998. Kentucky Fruit Facts. Cooperative Extension Service, US Department of Agriculture, University of Kentucky, College of Agriculture, Lexington, Kentucky.

Vernin, G. and C. Vernin, J.C. Pieribattesti, and C. Roque. 1998. Analysis of volatile compounds of *Psidium Cattleianum* Sabine fruit from Reunion island. J. Essential Oil Res. 10: 353-362.

Walters, D.S. and R. Graig, and R.O. Mumma. 1990. Effects of tall glandular trichome exudate of geraniums on the mortality and behavior of foxglove aphid. J. Chem. Ecol. 16: 877-886.

Webb, R.E. and A.K. Stoner, and A.G. Gentile. 1971. Resistance to leaf miners in *Lycopersicon* accessions. J. Amer. Soc. Hort. Sci. 96: 65-67.

Williams, W.G. and G.G. Kennedy, R.T. Yamamoto, J.D. Thacker, and J. Bordner. 1980. 2-Tridecanone: a naturally occurring insecticide from the wild tomato *Lycopersicon hirsutum* f. *glabratum*. Science 207: 888-889.

Wolfenbarger, D.A. 1965. Tomato, *Lycopersicon esculentum*, and *Lycopersicon* species and genetic markers in relation to mite, *Tetranychus marianae*, infestations. J. Econ. Entomol. 58: 891-893.

Wyman, J.A. and K.J. Kung, and D. Sexson. 1995. Developing best management practices to prevent pesticide leaching. pp. 179-182. Conf. Proc. Clean-water Clean-environment, 21st Century. Kansas City, Missouri, USA.

Xia, Y. and A.W. Johnson, and O.T. Chortyk. 1997. Enhanced toxicity of sugar esters to tobacco aphid (Homoptera: Aphididae) using humectants. J. Econ. Entomol. 90: 1015-1021.

Xia, Y. and A.W. Johnson. 1997. Effects of leaf surface moisture and relative humidity on the efficacy of sugar esters from *Nicotiana gossei* against the tobacco aphid. J. Econ. Entomol. 90: 1010-1014.

14

Bioactive Polysaccharides from Tomato

Giuseppina Tommonaro, Alfonso De Giulio, Giuseppe Strazzullo#, Salvatore De Rosa, Barbara Nicolaus and Annarita Poli

Istituto di Chimica Biomolecolare, Consiglio Nazionale delle Ricerche (C.N.R.)
Via Campi Flegrei, 34
80078 Pozzuoli, Napoli, Italy

ABSTRACT

Recent studies concerning the isolation and purification of exopolysaccharides from suspension-cultured cells of tomato (*Lycopersicon esculentum* L. var. San Marzano) and the description of a simple and rapid method with low environmental impact to obtain polysaccharides from solid wastes of tomato-processing industry (*Lycopersicon esculentum* variety "Hybrid Rome") are reported. Their chemical composition, rheological properties and partial primary structure were determined on the basis of spectroscopic analyses (ultraviolet, infra red, gas chromatography-mass spectroscopy, ^{1}H, ^{13}C nuclear magnetic resonance spectroscopy).

Moreover, the anticytotoxic activities of exopolysaccharides from culture cells of tomato were tested in brine shrimp bioassay and the achievement of biodegradable film by using chemical processes from polysaccharide of solid waste tomato industry was also reported.

INTRODUCTION

Polysaccharides from natural sources raise remarkable interest among several biotechnological products and have a large range of industrial and commercial applications. Some of them, for example, showing strong anti-

#Present address: ISPA-CNR, Sezione di Torino, Via Leonardo Da Vinci, 44
10095 Grugliasco - Torino

A list of abbreviations is given before the references.

genic and pathogenic activities, are employed successfully by the pharmaceutical industry for the formulation of vaccines, and others are used as industrial food additives because of their chemical-physical properties (e.g., emulsifying, viscoelasticity, polyelectrolyte, adherence, bio-compatible, stabilizer) (Gross 1986, Tong and Gross 1990, MacDougall et al. 1996, Maugeri et al. 2002, Arena et al. 2006).

Polysaccharides, because of their unusual multiplicity and structural complexity, are able to contain many biological messages and accordingly perform several functions. Moreover, these biopolymers have the ability to interact with other polymers, such as proteins and lipids, as well as polysaccharides (Koch and Nevis 1989, Pazur et al. 1994, Steele et al. 1997).

Vegetables are the most important sources of polysaccharides (cellular wall or stock products). More recently it has been seen that many microorganisms (bacteria and cyanobacteria) are also able to produce polysaccharides. Microbial polysaccharides are located in the cell wall, attached to the cells forming capsules or secreted into the extracellular environment in the form of slime (exopolysaccharide, EPS) (Nicolaus et al. 1999, 2002, Maugeri et al. 2002, Poli et al. 2004), but few data are available in literature regarding polysaccharides from the cell cultures of Solanaceae (Walker-Simmons and Ryan 1986).

In this chapter we report the isolation and chemical characterization of water-soluble bioactive polysaccharides from suspension-cultured cells of tomato (*Lycopersicon esculentum* L. var. San Marzano) and their anticytotoxic activities tested in a brine shrimp bioassay (Poli et al. 2006).

The chapter is also focused on the illustration of a rapid method conceived to recover high-grade polysaccharide in high yield from solid waste tomato processing industry (*Lycopersicon esculentum* variety "Hybrid Rome"). The tomato polysaccharide obtained from this natural and renewable source was characterized and used to realize useful biodegradable film (Strazzullo et al. 2003) and to obtain cheaper bacterial biomasses (Romano et al. 2004). The management of wastes represents the main issue of the food industry, which is an important sector of the world economy. Besides the manipulation of fresh products, new biotechnologies allow the recycling of wastes in order to obtain bio-products with high added value (Rosales et al. 2002). Several tons of tomatoes are processed annually worldwide and the desiccation of solid waste tomato produced represents the main approach to produce fertilizers; the goal to recover biopolymers from such a solid waste (harmful for the environment and economically disadvantageous for the industry) represents an excellent alternative for their exploitation according to the new philosophies concerning sustainable industrial development (Leoni 1997). Here, we report a rapid polysaccharide extraction method suitable to manipulate the solid waste and to transform it into an excellent source of products of high added value (Strazzullo et al. 2003).

EXOPOLYSACCHARIDES FROM CELL CULTURE OF *LYCOPERSICON ESCULENTUM* L. VAR. SAN MARZANO

The cell culture of *Lycopersicon esculentum* L. var. San Marzano was able to produce two main water-soluble EPSs and the presence of extracellular polysaccharides was observed from the high viscosity of culture medium. The callus was inducted from a sterile explant of *L. esculentum* cultured on MS basal medium supplemented with (mg L^{-1}): myoinositol (100), nicotinic acid (0.5), pyridoxine hydrochloride (0.5), thiamine hydrochloride (0.1), glycine (2), sucrose (30000), and agar (9000) (De Rosa et al. 1996). After 4 wk, the initial callus was transferred to fresh medium and cultured and maintained as above. Suspension cultures were obtained from the fourth generation callus by transferring about 3 g callus into 100 mL liquid medium. The suspension cultures were further maintained in 250 mL flasks by transferring about 1 g of fresh weight tissue (about 10 mL) into 100 mL fresh medium every 21 days. Cultures were maintained at 24°C, 150 rpm, in continuous light. Polysaccharide fraction was collected from the culture medium of tomato suspension cells (1 L) after 4 weeks of growth. The cell suspension was filtered and EPS fraction (260 mg) was obtained by EtOH precipitation of free cells culture broth (Fig. 1). The raw material, tested for sugar content (70%), protein content (10%) and nucleic acid content (1%), was purified by gel chromatography (Sepharose CL-6B DEAE) with a yield of 89%, and the resulting compounds comprised three different fractions, EPS(1) 9%, EPS(2) 60%, and EPS(3) 31%, all containing less than trace amounts of protein and nucleic acids (Fig. 1). EPS(1) was eluted in H_2O, representing the neutral fraction, while EPS(2) and EPS(3) were eluted at different salt concentrations (0.3 and 0.4 M NaCl, respectively) representing the acidic fractions.

Structural Characterization

Sugar mixtures of each fraction, native and carboxyl reduced, were identified by high pressure anion exchange-pulsed amperometric detection (HPAE-PAD) of hydrolysed polysaccharide and by gas-liquid

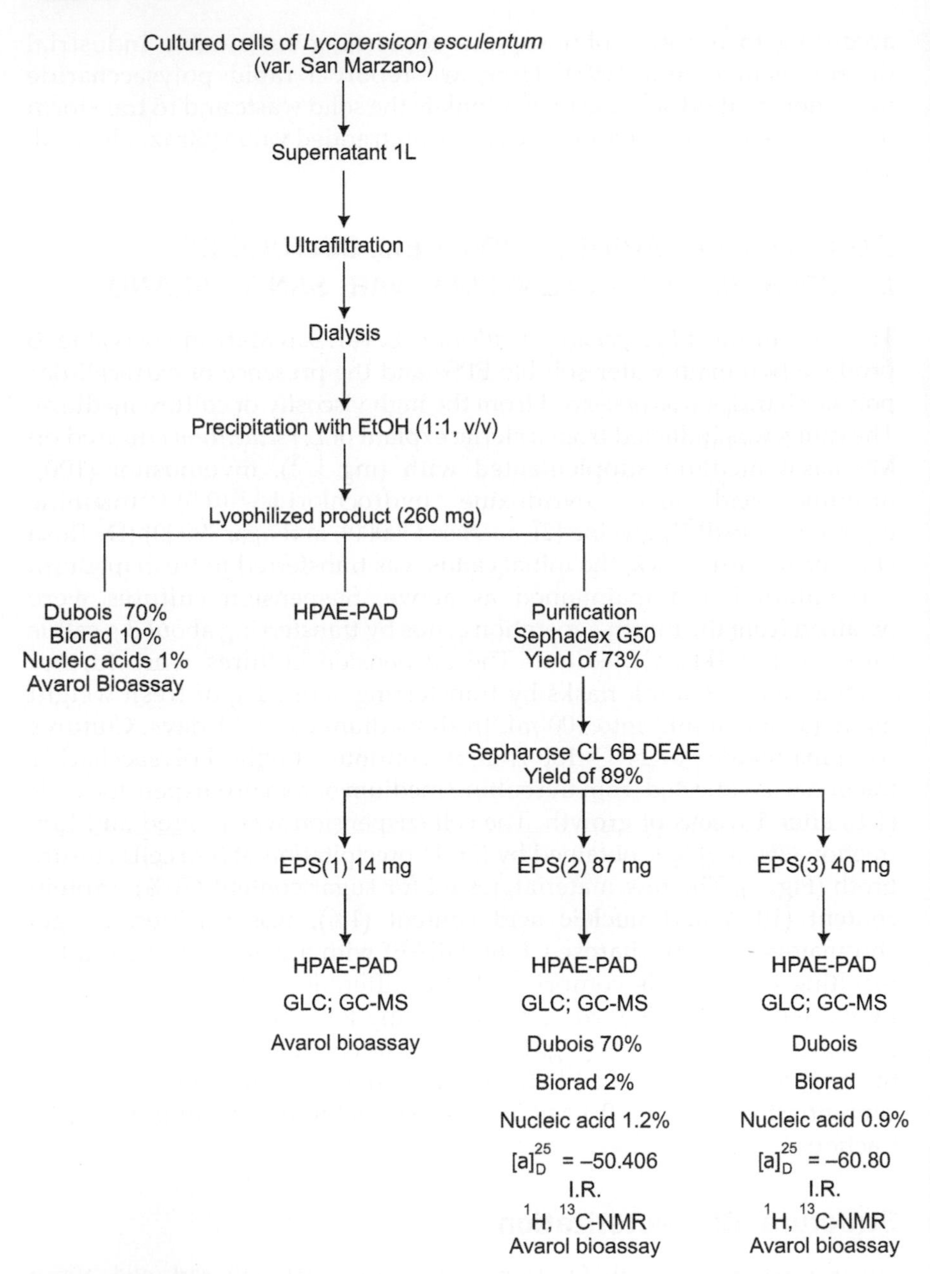

Fig. 1 Flow chart of polysaccharides isolation from suspension-cultured cells of tomato (*Lycopersicon esculentum* var. San Marzano).

chromatography (GLC) and gas chromatography-mass spectroscopy (GC-MS) of alditol acetates and methyl glycoside acetates. The sugar analysis of native EPSs indicated that EPS(1) was composed of Arabinose: Galactose: Glucose: Mannose in a relative ratio of 0.7:1.0:0.4:0.9, EPS(2) was composed of Arabinose: Galactose in a relative ratio of 0.3:1.0, and EPS(3) was composed of Arabinose: Mannose in a relative ratio of 1.0:0.5. The results of sugar analyses of methyl glycoside acetates indicated that EPS(2) was constituted of L-Arabinose: D-Galactose: L-Arabinuronic acid (0.5:1.0:0.2) and EPS(3) of L-Arabinose: D-Mannose: L-Arabinuronic acid (0.5:0.3:1.0).

Only EPS(2) and EPS(3) were further analysed because EPS(1) was less pure and recovered in a low yield. EPS(2) and EPS(3) were analysed by chemical and spectroscopic analysis. The quantities of uronic acid (Jansson et al. 1994) varied with different preparations reaching 170 μg/mg EPS(2) and 240 μg/mg EPS(3). The specific rotations of EPS(2) and EPS(3) were $[\alpha]^{25}_{D}$-50.40 and -60.80 (concentration of 5 mg/mL H_2O) respectively.

The absolute configuration of carbohydrates, performed as described by Leontein et al. (1978), using optically active (+)-2-butanol by GLC of their acetylated (+)-2-butyl glycosides, was shown to be D-Galactose and L-Arabinose for EPS(2) and D-Mannose and L-Arabinose for EPS(3). It was evident that the sugar residues in both EPSs were pyranosidic, while arabinose was furanosidic.

The molecular weight of EPSs was estimated from calibration curve of standard dextrans obtained by gel filtration on Sepharose CL-6B and also by density gradient centrifugation. In both methods, the molecular weights were approximately 8.0×10^5 Da for EPS(2) and 9.0×10^5 Da for EPS(3).

The infrared (IR) spectra of EPSs were similar to those of bacterial polysaccharides (Manca et al. 1996). A broad absorption band attributable to OH was observable at 3400 cm^{-1}. The absence of sulfate groups in EPS(2) as well as in EPS(3) was confirmed by IR spectra and also by a negative colour reaction with sodium rhodizonate (Fig. 2).

Analysis of the partially methylated alditol acetates, obtained from the permethylated EPSs after acid hydrolysis, showed that in both polysaccharides there is the presence of hexose chains linked on C1-C2, C1-C3 and C1-C6, and the presence of side chains on C1-C2-C6 (Table 1).

The ^{1}H and ^{13}C nuclear magnetic resonance spectroscopy (NMR) spectra recorded in H_2O at 70°C were quite complex (δ chemical shifts are expressed in parts per million, ppm) (Fig. 2). In the non-anomeric proton

Table 1 GC-MS of EPS(2) and EPS(3) from tomato cells of *Lycopersicon esculentum* var. San Marzano.

EPS(2)

t_r (min)	Sugar	% area	Link position
9.8	2,3,4,6-tetra-OMe exose	27.96	Terminal exose binds in C1
11.98	3,4,6-tri-OMe exose	17.43	Internal exose binds in C1 and C2
12.44	2,4,6-tri-OMe exose	7.08	Internal exose binds in C1 and C3
12.90	2,3,4-tri-OMe exose	17.37	Internal exose binds in C1 and C6
15.4	3,4-di-OMe exose	30.16	Branching exose binds in C1,C2 and C6

EPS(3)

t_r (min)	Sugar	% area	Link position
9.8	2,3,4,6-tetra-OMe exose	39.23	Terminal exose binds in C1
12.0	3,4,6-tri-OMe exose	18.91	Internal exose binds in C1 and C2
12.4	2,4,6-tri-OMe exose	7.36	Internal exose binds in C1 and C3
12.90	2,3,4-tri-OMe exose	5.70	Internal exose binds in C1 and C6
15.4	3,4-di-OMe exose	28.8	Branching exose binds in C1,C2 and C6

Methylation of the polysaccharides was carried out according to Manca et al. (1996).
The methylated material (0.5 mg) was hydrolysed with 2 M Trifluoro-acetic acid (TFA) at 120°C for 2 hr and then transformed in partially methylated alditol acetates by reduction with $NaBH_4$, followed by acetylation with Ac_2O-pyridine (1:1) at 120°C for 3 hr. Unambiguous identification of sugars was obtained by Gas-Chromatography Mass Spectroscopy (GC-MS) using sugar standards. GC-MS was performed on a Hewlett-Packard 5890-5970 instrument equipped with an HP-5-MS column and with an N_2 flow of 50 mL min^{-1}; the programme temperature was: 170°C (1 min), from 170° to 250°C at 3°C min^{-1}.
t_r retention time.

region, several overlapping spin systems were evident. The ^{1}H NMR spectrum of EPS(2) showed, in the anomeric region, four major signals at δ 5.26 (1H, d, J = 3.0 Hz), 4.71 (1H, d, J = 1.5 Hz), 4.52 (1H, d, J = 8.0 Hz), 4.49 (1H, d, J = 8.3 Hz) (Table 2). The ^{1}H NMR spectrum of EPS(3) showed five major anomeric signals at δ 5.47 (1H, d, J = 2.9 Hz), 5.44 (1H, d, J = 1.5 Hz), 5.39 (1H, d, J = 2.0 Hz), 5.32 (1H, d, J = 3.9 Hz), 4.55 (1H, d, J = 7.8 Hz) (Table 2). The ^{13}C NMR spectrum of EPS(2) showed four signals at δ 111.9, 106.0, 106.3 and 106.4 in the anomeric region, confirming the presence of four residues in the repeating unit, and a small signal at d 178.4 (COOH) due to the presence of uronic acid in minute quantities. The ^{13}C NMR spectrum of EPS(3) showed the presence of five signals at δ 111.5, 110.7, 105.1, 105.0 and 100.9, confirming the presence of five residues in the repeating unit, and an intense signal at δ 176.9 (COOH) indicative of the presence of uronic acid. The ^{1}H and ^{13}C chemical shifts and the C-H coupling constants of each anomeric carbon were assigned by

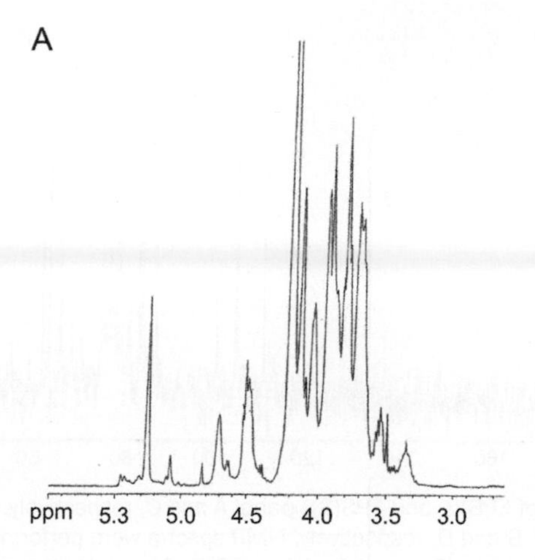
A
ppm 5.3 5.0 4.5 4.0 3.5 3.0

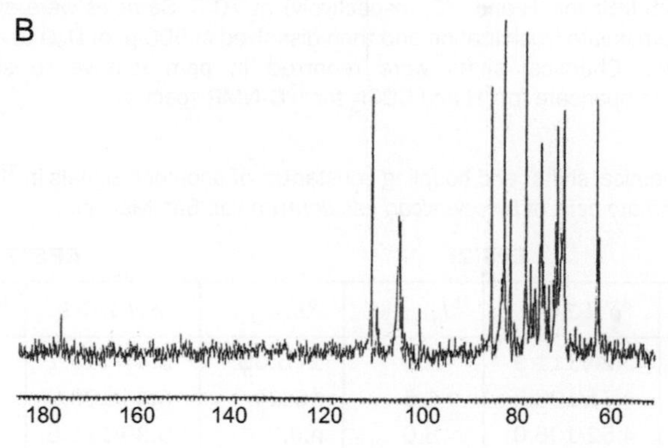
B
180 160 140 120 100 80 60

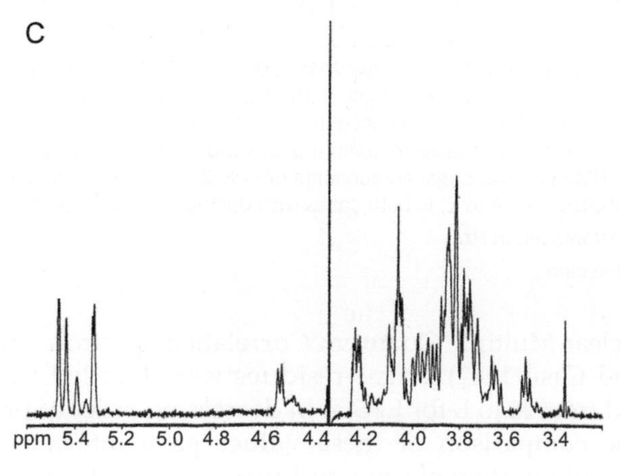
C
ppm 5.4 5.2 5.0 4.8 4.6 4.4 4.2 4.0 3.8 3.6 3.4

Fig. 2 (Contd.)

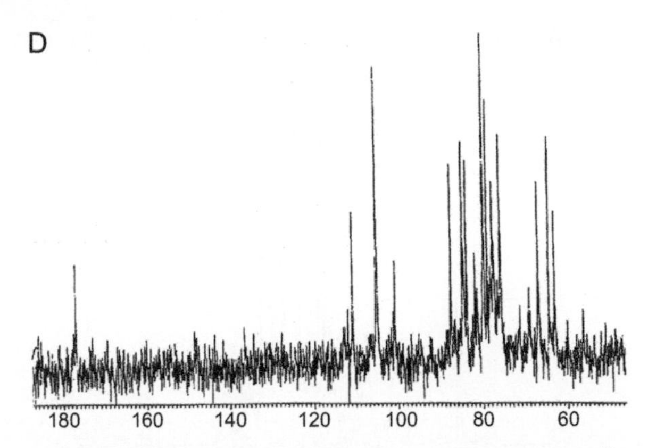

Fig. 2 ^{1}H NMR spectra of EPS(2) and EPS(3), panel A and C, respectively. ^{13}C NMR spectra of EPS(2) and EPS(3), panel B and D, respectively. NMR spectra were performed on a Bruker AMX (500 and 125 MHz for ^{1}H and ^{13}C, respectively) at 70°C. Samples were exchanged twice with D_2O with intermediate lyophilization and then dissolved in 500 μl of D_2O to a final concentration of 40 mg/ml. Chemical shifts were reported in ppm relative to sodium 2,2,3,3-d_4-(trimethylsilyl)propanoate for ^{1}H and $CDCl_3$ for ^{13}C-NMR spectra.

Table 2 Chemical shifts[a] and coupling constants[b] of anomeric signals in ^{1}H and ^{13}C spectra of EPSs from tomato cells of *Lycopersicon esculentum* var. San Marzano.

	EPS(2)			**EPS(3)**		
Residue	$^{a}\delta$ *H-1/C-1*	$^{b}J_{H-1,H-2}$	$^{b}J_{H-1,C-2}$	$^{a}\delta$ *H-1/C-1*	$^{b}J_{H-1,H-2}$	$^{b}J_{H-1,C-2}$
A	5.26/111.9	3.0	175.05	5.47/110.7	2.9	172.5
B	4.71/106.3	1.5	161.4	5.44/100.9	1.5	163.4
C	4.52/106.0	8.0	n.d.	5.39/111.5	2.0	176.0
D	4.49/106.4	8.3	n.d.	5.32/105.1	3.9	168.3
E				4.55/105	7.8	170.6

NMR spectra were performed on a Bruker AMX (500 and 125 MHz for ^{1}H and ^{13}C, respectively) at 70°C. Samples were exchanged twice with D_2O with intermediate lyophilization and then dissolved in 500 μl of D_2O to a final concentration of 40 mg/ml. [a]Chemical shifts were reported in parts per million (ppm) relative to sodium 2,2,3,3-d_4-(trimethylsilyl) propanoate for ^{1}H and $CDCl_3$ for ^{13}C-NMR spectra. Sugar components of EPS(2) are labelled from A to D, and those of EPS(3) are labelled from A to E, in both cases with decreasing chemical shifts.

[b]Coupling constants are in Hz.

n.d. = not detected

Heteronuclear Multiple Quantum Correlation spectroscopy experiments (Perlin and Casu 1982). Sugar residues were labelled from A to D for EPS(2) and from A to E for EPS(3) in decreasing order of proton chemical shifts. The comparison of these values gave information about the anomeric configuration of some residues (Table 2). In ^{1}H NMR spectrum

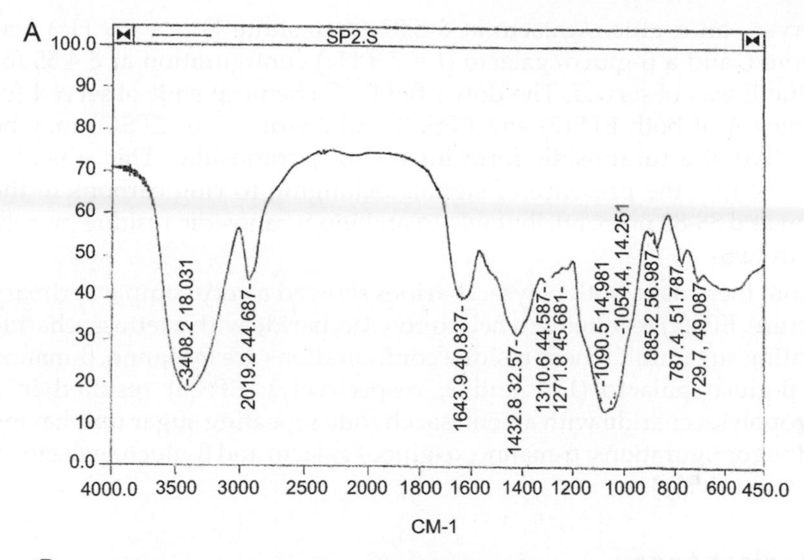

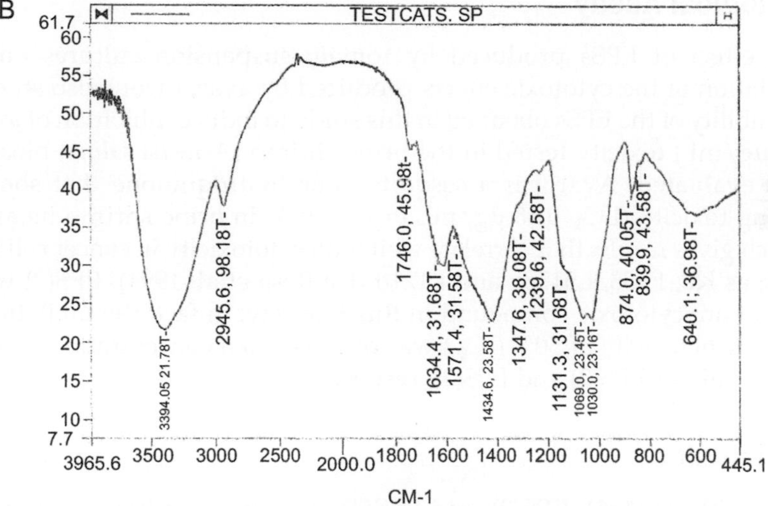

Fig. 3 Infra red spectra of EPS(2) and EPS(3), panels A and B, respectively. Fourier Transform Infra Red spectrum was recorded by using a Perkin-Elmer Paragon 500 single-beam spectrophotometer. The sample as powder was grounded with KBr and put under beam as diskette and the spectrum was collected after 16 scans under nitrogen.

of EPS(2) the signal of the residue B at δ 4.71 was a β-manno (*J* = 1.5 Hz), while the signals at δ 4.52 and 4.49 of the residues C and D were typical of a β-gluco/galacto (*J* = 8.0-8.3 Hz) configuration. In the ^{1}H NMR spectrum of EPS(3) at δ 5.44 an α-manno (*J* = 1.5 Hz) configuration for residue B was observed, an α-gluco/galacto at δ 5.32 for residue D (J = 3.9 Hz) was observed, and a β-gluco/galacto (J = 7.8 Hz) configuration at δ 4.55 for residue E was observed. The down field ^{13}C chemical shift observed for residues A of both EPS(2) and EPS(3) and residue C of EPS(3) may be indicative of a furanosidic form instead of pyranosidic. This was also confirmed by the presence of signals belonging to ring carbons in the region at d 88-80 ppm, attributable to arabino furanosidic residue present in both EPSs.

From these data, both polysaccharides showed a very complex primary structure. EPS(2) resulted in a heteropolysaccharide with a tetrasaccharide repeating sugar unit whose residue configurations are α-manno, β-manno and β-gluco/galacto (1, 1 and 2, respectively); EPS(3) resulted in a heteropolysaccharide with a pentasaccharide repeating sugar unit having residue configurations: α-manno, α-gluco/galacto and β-gluco/galacto (3, 1, 1, respectively).

Biological Assay

The effect of EPSs produced by tomato suspension cultures on the inhibition of the cytotoxic effects produced by avarol were also studied. The ability of the EPSs obtained in this study to induce inhibition of avarol (10 μg/mL) toxicity tested in the brine shrimp (*Artemia salina*) bioassay was evaluated. Avarol is a sesquiterpene hydroquinone that showed strong toxicity (LC_{50} 0.18 μg/mL or 0.57 nM) in brine shrimp bioassay, which gives results that correlate well with cytotoxicity in cancer cell lines such as KB, P388, L5178y and L1210 (De Rosa et al. 1994). EPS(2) was a potent anticytotoxic compound in this bioassay; in fact, the inhibition of the avarol toxicity of 50% (IC_{50}) was observed at a concentration of 3 and 11 μg/mL for EPS(2) and EPS(3), respectively.

Conclusion

Three EPSs, EPS(1), EPS(2), and EPS(3), were isolated from suspension-cultured *L. esculentum* (var. San Marzano) cells. The partial primary structures, hypothesized on the basis of spectroscopic analyses, resulted in a peculiar complex primary structure for all EPSs. In particular, EPS(2) was a heteropolysaccharide characterized by a tetrasaccharide repeating unit and EPS(3) was a heteropolysaccharide characterized by a pentasaccharide repeating unit. The anticytotoxic activities of EPSs, tested in brine shrimp bioassay, showed a potential role in host defence mechanisms and further studies are necessary to test other biological activities.

POLYSACCHARIDES FROM TOMATO WASTE (*LYCOPERSICON ESCULENTUM* VARIETY "HYBRID ROME")

The extraction method of tomato polysaccharide from solid wastes of tomato processing industry, produced by mechanical tomato pressing for the production of pulp and puree, is outlined in Fig. 4.

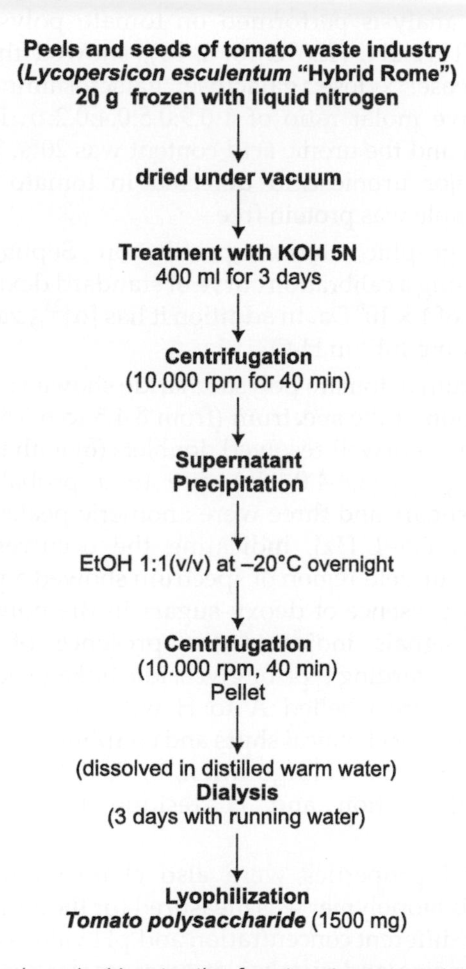

Fig. 4 Flow chart of polysaccharide extraction from tomato waste material.

The lyophilized biomass (20 g, peels and seeds, rotten and unripe tomatoes) was treated with 5 N KOH under stirring for 3 d and then centrifuged at 10.000 g for 40 min. After centrifugation the supernatant was precipitated with cold ethanol (v/v) at –20°C for a night. The pellet, collected after centrifugation, was dissolved in hot water, dialysed against running water for 3 d and dried under vacuum obtaining 1.5 g of tomato polysaccharide.

Structural Characterization

The HPAE-PAD analysis performed on tomato polysaccharide, after hydrolysis with TFA 2N, (120°C for 2 hr), showed the neutral sugar composition glucose: xylose: galactose: galactosamine: glucosamine: fucose in a relative molar ratio of 1:0.9:0.5:0.4:0.2:*tr*. Its carbohydrate content was 100% and the uronic acid content was 20%. The galacturonic acid was the major uronic acid detected in tomato polysaccharide. Moreover, the sample was protein free.

The chromatographic elution profile on Sepharose CL-6B of polysaccharide, using a calibration curve of standard dextrans, indicated a molecular weight of 1×10^6 Da. In addition it has $[\alpha]^{25}_D$ values of – 0,186 at concentration of 1 mg ml^{-1} in H_2O.

^1H NMR spectrum of tomato polysaccharide showed a complex profile. The anomeric region of the spectrum (from δ 4.5 to δ 5.5) exhibited eight peaks; five of them were well-resolved doublets (δ) with the same value of constant coupling $J_{1\text{-}2}$ (3.8-4.0 Hz) due to a probably *gluco-galacto* configuration of sugars and three were anomeric peaks, almost singlets with a small $J_{1\text{-}2}$ (0.5–1 Hz), indicating the occurrence of a *manno* configuration. The upfield region of spectrum showed a peak doublet at δ 1.20 indicative of presence of deoxy-sugars in the polysaccharide. The eight anomeric signals indicated the presence of eight different monosaccharides, regarding type or glycosidic linkage position. The eight monosaccharides were labelled A to H with respect to increasing δ (Table 3). On the base of chemical shifts and coupling constant data A, D, F residues have an α *manno* configuration, B, C, E, G residues have an α *gluco-galacto* configuration and H residue has a β *gluco-galacto* configuration.

The rheological properties were also characterized. The specific viscosity (η) of this biopolymer was measured for the aqueous solutions of polysaccharide at different concentration and pH values and proved to be influenced by the size and number of macromolecules in solution. As concentration increases coils start to overlap and become entangled, with viscosity showing a more marked dependence on concentration reaching η = 1.7 at 4% of concentration (Fig. 5). The viscosity does not change drastically respect to the increase of pH and its maximum value was obtained at pH 3.0 for a 1% polysaccharide solution in 50 mM citrate buffer (η = 3.29).

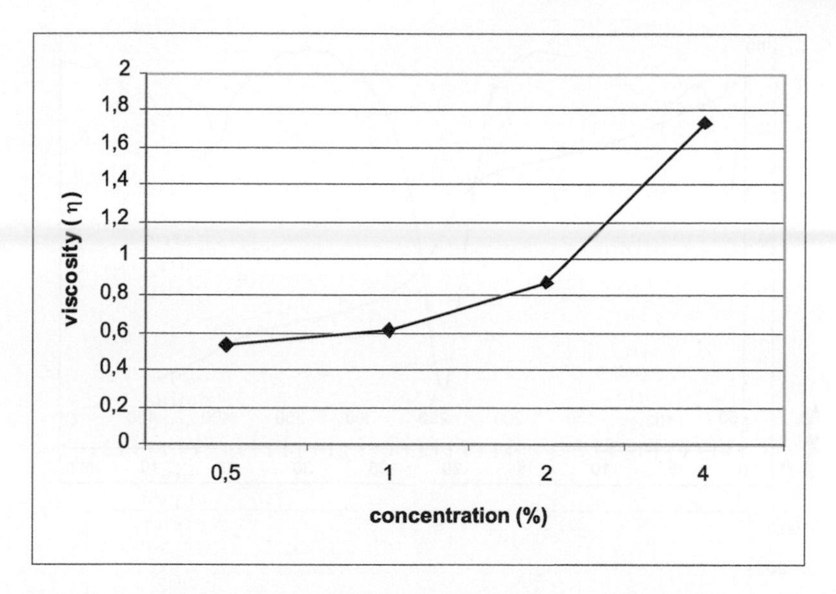

Fig. 5 The viscosity of aqueous solution of waste tomato polysaccharide at different concentration values was determined using Cannon-Ubbelohde 75 suspended level viscometer at 25°C.

Table 3 ^{1}H-NMR data[a] for anomeric region of spectrum of waste tomato polysaccharide.

Type[b]	δ^1H	*Molteplicity of signals*	$J_{1\text{-}2}$[c]	*Configuration*
A	5.30	pseudo s	0.5-1 Hz	*manno*
B	5.27	D	3.8-4.0 Hz	*gluco-galacto*
C	5.26	D	3.8-4.0 Hz	*gluco-galacto*
D	5.18	pseudo s	0.5-1 Hz	*manno*
E	5.09	D	3.8-4.0 Hz	*gluco-galacto*
F	5.07	pseudo s	0.5-1 Hz	*manno*
G	5.06	D	3.8-4.0 Hz	*gluco-galacto*
H	4.94	D	3.8-4.0 Hz	*gluco-galacto*

[a]Bruker AVANCE 400MHz; sample was exchanged twice with D_2O with intermediate lyophilization and then dissolved in 500 μl of D_2O to a final concentration of 30 mg/ml. δ values (ppm) referred to sodium 2,2,3,3-d_4-(trimethylsilyl) propanoate.

[b]Labels referred to different monosaccharides, regarding type of glycosidic linkage position.

[c]Coupling constants are in Hz.

The temperature of degradation of tomato polysaccharide (10 mg) was 250°C in 20 min with a residue of about 5 mg (Fig. 6A). Moreover, the infrared spectrum of this biopolymer showed the characteristic peak signals of polysaccharides: OH stretching at 3.400 cm^{-1}, CH and C = O

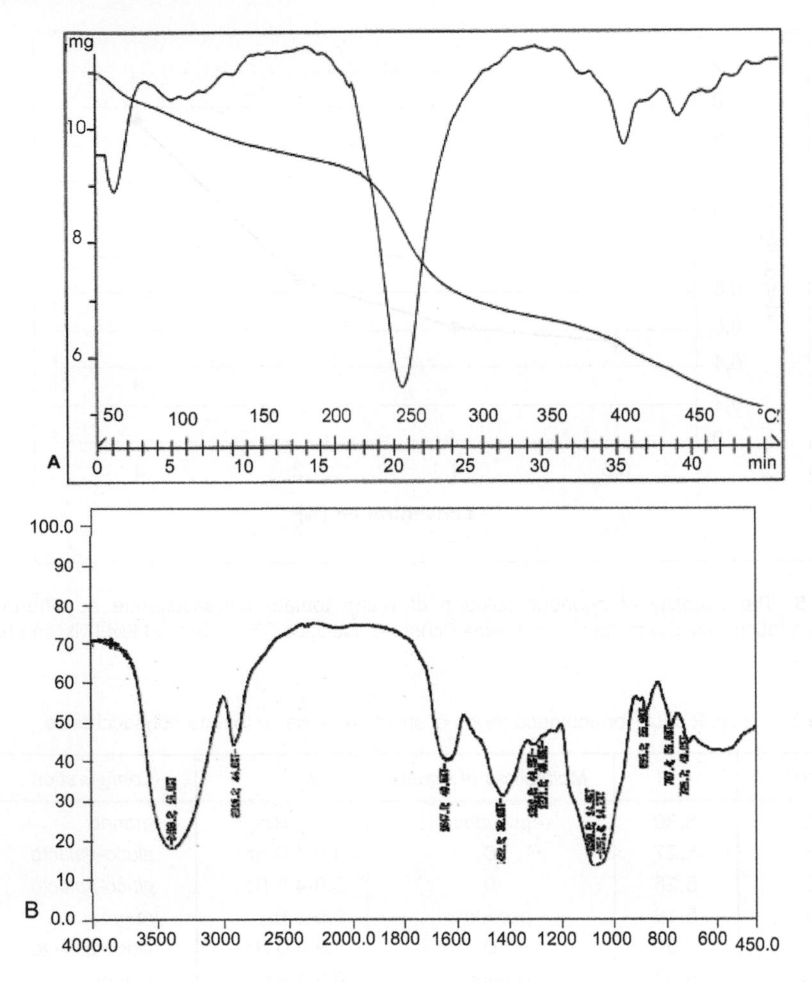

Fig. 6 (A) Thermogravimetric analysis of tomato polysaccharide was performed using Mettler Toledo Star System equipped with thermo analytical balance in a temperature range of 50° to 450°C with a temperature programme of 10 min at 50°C following by 10°C/min. (B) Fourier Transform Infra Red spectrum of tomato polysaccharide was recorded using a Perkin-Elmer Paragon 500 single-beam spectrophotometer. The sample as powder was grounded with KBr and put under beam as diskette and the spectrum was collected after 16 scans under nitrogen.

stretching at 2929 cm^{-1} and 1730–1660 cm^{-1} respectively. SO group was absent; in fact no signals were detected at 1240 cm^{-1} (Fig. 6B).

Biodegradability Test and Biotechnological Features

In order to verify the biodegradability of tomato polysaccharide, growth test was performed using a new thermohalophilic *Thermus thermophilus* strain (Samu-Sa1 strain) isolated from hot springs of Mount Grillo (Baia,

Naples, Italy) (Romano et al. 2004). This strain was grown on M162 medium (Degryse et al. 1978) modified with 0.2% NaCl at pH 7.2 and using as sole carbon source 0.1% of tomato polysaccharide. After 30 hr of batch incubation at 75°C (the optimal temperature of Samu-Sa1strain) the tomato polysaccharide was completely hydrolysed as resulted from phenol/sulphuric acid method tested on cell-free cultural broth (Dubois et al. 1956) (Fig. 7). Moreover, the growth curves of strain on polysaccharide medium was 2.6 times that obtained in TH medium (standard medium) with a yield of 1.2 g of dry cells/L (Fig. 7).

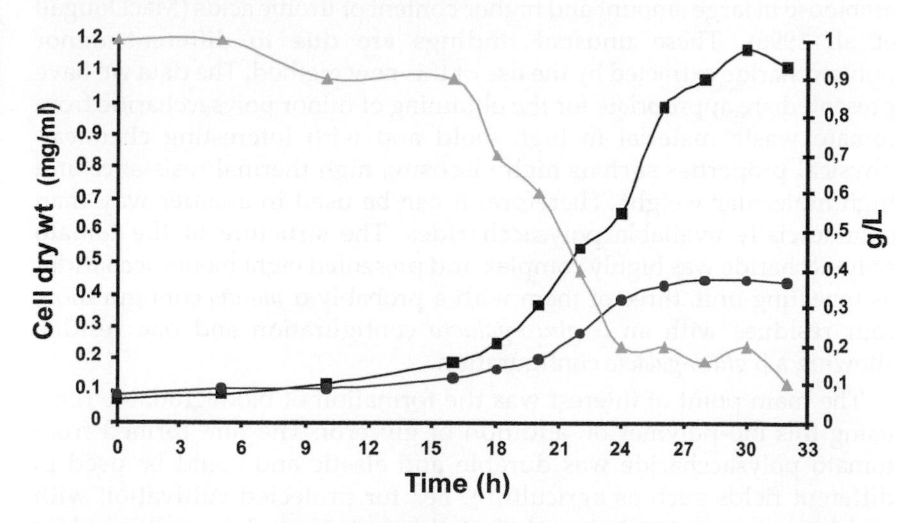

Fig. 7 Strain Samu-SA1 cell growth (expressed as dry cell weight for ml of culture broth, mg/ml) on polysaccharide medium (■-■-■) and on TH medium (●-●-●) at 75°C. The growth was followed by measuring absorbance at 540 nm and converted to the cell dry weight by an appropriate calibration. The depletion of polysaccharide reported as g/L followed using Dubois method (▲-▲-▲).

The tomato polysaccharide was used for the preparation of biodegradable films by solubilizing 50 mg of polysaccharide in 5 ml of distilled water at room temperature and adding 5 mg of glycerol as plasticizer (Strazzullo et al. 2003). The biofilms were clear, elastic, solid and durable when recovered from small static deformations produced by the applied tensile stress.

Conclusion

Food canning industries represent an important area of the Italian economy, in particular the industrial conversion of tomato into tomato purée, pulp and minced tomato from seed producer firms. One of the main problems of the food industry is the management of waste and its conversion into products of higher added value. Contemporary eco-compatible technologies promote the use of food waste to obtain biopolymers that can be re-used in the same sector as the raw materials.

The method here described is a rapid procedure with high yield cell-wall polysaccharide production (7.5%) with very low environmental impact. The sugar analysis of this polymer revealed the presence of glucose and xylose as major components, and a low level of uronic acids, in contrast to that of the cell-wall pectic polysaccharides that contain arabinose in large amount and higher content of uronic acids (MacDougall et al. 1996). These unusual findings are due to different minor polysaccharide extracted by the use of this new method. The data we have presented are appropriate for the obtaining of minor polysaccharide from tomato waste material in high yield and with interesting chemical-physical properties such as high viscosity, high thermal resistance and high molecular weight. Therefore, it can be used in a better way than commercially available polysaccharides. The structure of the tomato polysaccharide was highly complex and presented eight monosaccharides as repeating unit, three of them with a probably α *manno* configuration, four residues with an α *gluco-galacto* configuration and one residue showing a β *gluco-galacto* configuration.

The main point of interest was the formation of biodegradable films using this bio-polymer on addition of glycerol. The film formed from tomato polysaccharide was durable and elastic and could be used in different fields such as agriculture, i.e., for protected cultivation with mulching operation technique. In fact, the plastic material usually used for mulching and solarization has optimum mechanical characteristics and low cost but it is not biodegradable or reusable, and used film is discarded as a special waste.

An additional biotechnological use of the tomato polysaccharide could be as a cheaper substrate to obtained microbial biomasses. *Thermus thermophilus* strain Samu-SA1 possesses many hydrolytic enzymes with potential biotechnological applications and was able to grow on very cheap medium. In fact, 1.2 g/L of dry cells were obtained when tomato waste polysaccharide, extracted from discarded industrial tomato processing, was used as sole carbon source with a biomass yield 2.6 times that obtained with standard medium.

ACKNOWLEDGEMENT

This work was financed by Regione Campania.

ABBREVIATIONS

EPS: Exopolysaccharide; GLC: gas-liquid chromatography; GC-MS: gas chromatography-mass spectroscopy; HPAE-PAD: high pressure anion exchange-pulsed amperometric detection; IR: infra red; LC_{50}: lethal concentration at which 50% mortality is observed; NMR: nuclear magnetic resonance spectroscopy

REFERENCES

Arena, A. and T.L. Maugeri, B. Pavone, D. Iannello, C. Gugliandolo, and G. Bisognano. 2006. Antiviral and immunoregulatory effect of a novel exopolysaccharide from a marine thermo tolerant *Bacillus licheniformis*. Int. Immunopharmacol. 6: 8-13.

Degryse, E. and N. Glansdorff, and A. Pierard. 1978. A comparative analysis of extreme thermophilic bacteria belonging to the genus *Thermus*. Arch. Microbiol. 117: 189-196.

De Rosa, S. and A. De Giulio, and C. Iodice. 1994. Biological effects of prenylated hydroquinones: structure-activity relationship studies in antimicrobial, brine shrimp, and fish lethality assays. J. Nat. Prod. 57: 1711-1716.

De Rosa, S. and A. De Giulio, and G. Tommonaro. 1996. Aliphatic and aromatic glycosides from the cell cultures of *Lycopersicon esculentum*. Phytochemistry 42: 1031-1034.

Dubois, M. and K.A. Gilles, J.K. Hamilton, P.A. Rebers, and F. Smith. 1956. Colorimetric methods for determination of sugars and related substances. Anal. Chem. 28: 350-356.

Gross, K.C. 1986. Composition of ethanol-insoluble polysaccharides in water extracts of ripening tomatoes. Phytochemistry 25: 373-376.

Jansson, P.E. and B. Lindberg, M.C. Manca, W. Nimmich, and G. Widmalm. 1994. Structural studies of the capsular polysaccharide from *Klebsiella* type 38: a reinvestigation. Carbohyd. Res. 261: 111-118.

Koch, J.L. and D.J. Nevins. 1989. Tomato fruit cell wall. Plant Physiol. 91: 816-822.

Leoni, C. 1997. "Scarti" in the tomato processing industry: a contribution to disentanglement among culls rejected tomatoes. Production rejected and processing waste. Industria Conserve. 73: 278-290.

Leontein, K. and B. Lindberg, and J. Lonngren. 1978. Assignment of absolute configuration of sugars by g.l.c. of their acctylated glycosides formed from chiral alcohols. Carbohyd. Res. 62: 359-362.

MacDougall, A.J. and P.W. Needs, N.M. Rigby, and S.G. Ring. 1996. Calcium gelation of pectic polysaccharides isolated from unripe tomato fruit. Carbohydr. Res. 923: 235-249.

Manca, M.C. and L. Lama, R. Improta, E. Esposito, A. Gambacorta, and B. Nicolaus. 1996. Chemical composition of two exopolysaccharides from *Bacillus thermoantarcticus*. Appl. Environ. Microb. 62: 3265-3269.

Maugeri, T.L. and C. Gugliandolo, D. Caccamo, A. Panico, L. Lama, A. Gambacorta, and B. Nicolaus. 2002. A halophilic thermotolerant *Bacillus* isolated from a marine hot spring able to produce a new exopolysaccharide. Biotechnol. Lett. 24: 515-519.

Nicolaus, B. and A. Panico, L. Lama, I. Romano, M.C. Manca, A. De Giulio, and A. Gambacorta. 1999. Chemical composition and production of exopolysaccharides from representative members of heterocystous and non-heterocystous cyanobacteria. Phytochemistry 52: 639-647.

Nicolaus, B. and L. Lama, A. Panico, V. Schiano Moriello, I. Romano, and A. Gambacorta. 2002. Production and characterization of exopolysaccharides excreted by thermophilic bacteria from shallow, marine hydrothermal vents of flegrean areas (Italy). Syst. Appl. Microbiol. 25: 319-325.

Pazur, J.H. Neutral polysaccarides. pp. 73-124. *In:* M.F. Chaplin and J.F. Kennedy. [eds.] 1994. Carbohydrate Analysis. The Practical Approach Series, Irl Press, Oxford.

Perlin, A.S. and B. Casu. Spectroscopic methods. pp. 133-193. *In:* G.O. Aspinal. [ed.] 1982. The Polysaccharides, Vol. 1. Academic Press Ltd, London.

Poli, A. and V. Schiano Moriello, E. Esposito, L. Lama, A. Gambacorta, and B. Nicolaus. 2004. Exopolysaccharide production by a new *Halomonas* strain CRSS isolated from saline lake Cape Russell in Antarctica growing on complex and defined media. Biotechnol. Lett. 26: 1635-1638.

Poli, A. and M.C. Manca, A. De Giulio, G. Strazzullo, S. De Rosa, and B. Nicolaus. 2006. Bioactive exopolysaccharides from the cultured cells of tomato *Lycopersicon esculentum* L. var. San Marzano. J. Nat. Prod. 69: 658-661.

Romano, I. and L. Lama, V. Schiano Moriello, A. Poli, A. Gambacorta, and B. Nicolaus. 2004. Isolation of a new thermohalophilic *Thermus thermophilus* strain from hot spring, able to grow on a renewable source of polysaccharides. Biotechnol. Lett. 26: 45-49.

Rosales, E. and S. Rodriguez Couto, and A. Sanroman. 2002. New uses of food waste: application to laccase production by *Trametes hirsuta*. Biotechnol. Lett. 24: 701-704.

Steele, N.M. and M.C. McCann, and K. Roberts. 1997. Pectin modification in cell walls of ripening tomatoes occurs in distinct domains. Plant Physiol. 114: 373-381.

Strazzullo, G. and V. Schiano Moriello, A. Poli, B. Immirzi, P. Amazio, and B. Nicolaus. 2003. Solid wastes of tomato-processing industry (*Lycopersicon esculentum* "Hybride Rome") as renewable sources of polysaccharides. J. Food Technol. 1: 102-105.

Tong, C.B.S. and K.C. Gross. 1990. Stimulation of ethylene production by a cell wall component from mature green tomato fruit. Physiol. Plant. 80: 500-506.

Walker-Simmons, M. and C.A. Ryan. 1986. Proteinase Inhibitor I Accumulation in Tomato Suspension Cultures: Induction by Plant and Fungal Cell Wall Fragments and an Extracellular Polysaccharide Secreted into the Medium. Plant Physiol. 80: 68-71.

15

Ingestion of Tomato Products and Lycopene Isomers in Plasma

Volker Böhm and Kati Fröhlich

Institute of Nutrition, Friedrich Schiller University Jena, Dornburger Str. 25-29, 07743 Jena, Germany

ABSTRACT

Several epidemiologic studies indicated a beneficial effect of tomato consumption in the prevention of some major chronic diseases. One of the main phytochemicals in tomato products is lycopene, an aliphatic hydrocarbon with 13 double bonds, 11 of them being conjugated. The molecule exists in trans-*(all-E)*-form as well as in several cis-*(Z)*-forms. Around 85% of the lycopene ingested comes from tomatoes and tomato products. Lycopene is absorbed together with lipids and lipid-soluble vitamins. Food processing as well as the type of food matrix affect the absorption of lycopene and other carotenoids. The main isomer in all food samples investigated was *(all-E)*-lycopene. In contrast, in plasma the main lycopene isomers were *(all-E)*-lycopene and *(5Z)*-lycopene. In several human intervention studies, the *(all-E)*:*(Z)*-ratio of lycopene changed from 40:60 to 25-30:70-75 during the depletion period and returned approximately to the baseline values with intervention. Several hypotheses exist in literature to explain why lycopene is found in the human body as more than 50% *(Z)*-isomers in contrast to more than 85% *(all-E)*-lycopene in foods.

INTRODUCTION

Several epidemiological studies indicated a beneficial effect of tomato consumption in the prevention of some major chronic diseases, such as some types of cancer (Giovannucci et al. 2002) and cardiovascular diseases (Klipstein-Grobusch et al. 2000). One of the major phytochemicals in tomato products contributing to the prevention of cancer is lycopene. Reports from epidemiological studies, studies in animals and cell culture experiments suggested that lycopene has anticarcinogenic properties (Rao and Agarwal 1999, Etminan et al. 2004). In addition to its antioxidant properties (DiMascio et al. 1989, Böhm et al. 2002), lycopene has been shown to induce cell-cell communication (Zhang et al. 1991, Stahl et al. 2000), activate phase II enzymes (Breinholt et al. 2000), inhibit tumour cell proliferation (Levy et al. 1995), repress insulin-like growth factor receptor activation (Karas et al. 2000), and improve antitumour immune responses (Clinton 1998). The mechanisms by which lycopene might exert its biological activities are still unknown.

The general structure of lycopene is an aliphatic hydrocarbon with 11 conjugated carbon-carbon double bounds, making it soluble in lipids and red in colour. Being acyclic, lycopene has no vitamin A activity. Recent investigations showed that lycopene derivatives could activate retinoid receptors. However, the physiological significance has to be shown in future studies (Sharoni et al. 2004). Lycopene from tomatoes and tomato-based foods exists predominantly in *(all-E)*-configuration, the thermodynamically most stable form (Porrini et al. 1998). In contrast, various *(Z)*-isomers account for over 50% of blood lycopene and for over 75% of tissue lycopene (Clinton et al. 1996, Ferruzzi et al. 2001). The processes that influence isomer patterns and the mechanisms of interconversion are still an essentially unexplored area of research. Isomerization of lycopene may have significant consequences since the large three-dimensional differences between these geometric isomers may influence their pharmacological properties (Holloway et al. 2000).

GENERAL ASPECTS

Lycopene is an aliphatic hydrocarbon ($C_{40}H_{56}$) with 13 double bonds, 11 of them being conjugated. The molecule exists in trans-*(all-E)*-form as well as in several cis-*(Z)*-forms due to rotation of the substituents. *(Z)*-double bonds lead to sterical hindrance between hydrogen atoms and methyl groups. Thus, the *(all-E)*-isomer is generally thermodynamically more stable than the *(Z)*-isomers. Theoretically, lycopene can exist as 2048 (2^{11}) different geometrical isomers. However, only 7 of the 11 conjugated double bonds are "stereochemically active" (Fig. 1). So, lycopene can be determined as 72 different isomers (Zechmeister 1944). Up to now, the mechanism of formation of lycopene *(Z)*-isomers is nearly unknown. In plants, isomerization takes place partly by enzymes. *(7Z,9Z,7′Z,9′Z)*-lycopene in tangerine tomatoes, for example, isomerizes to *(all-E)*-

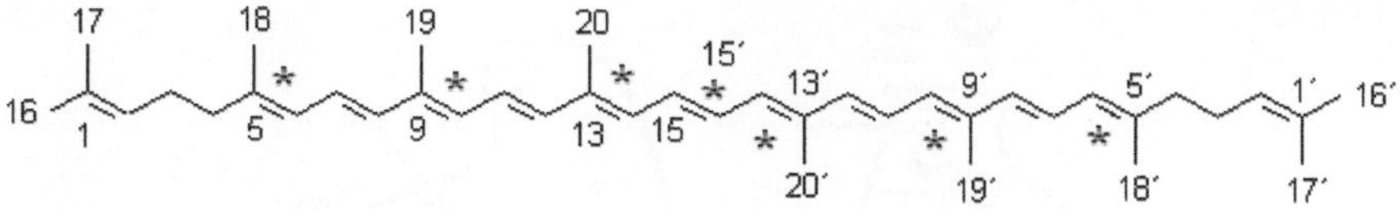

Fig. 1 Molecular structure of *(all-E)*-lycopene, stereochemically effective double bonds are marked with *.

lycopene by catalysis of a specific prolycopene isomerase (CrtlISO). In contrast, *(5Z)*-, *(9Z)*- and *(13Z)*-lycopene are isomerized non-enzymatically (Breitenbach et al. 1999, Giuliano et al. 2002, Breitenbach and Sandmann 2005). Lee and Chen (2001) postulated the formation of *mono-(Z)*-isomers from *(all-E)*-lycopene and the further formation of *di-(Z)*-isomers from the *mono-(Z)*-isomers. In plant products, lycopene normally exists as *(all-E)*-lycopene, the thermodynamically most stable isomer. In tangerine tomatoes, with yellow to orange fruits, the *tetra-(Z)*-isomer *(7Z,9Z,7′Z,9′Z)*-lycopene is the main isomer (Isaacson et al. 2002, Lewinsohn et al. 2005). Schierle et al. (1997) determined *(7Z, 9Z)*-lycopene as the main isomer in a special species of apricots (*Prunus armeniaca* L.) with only 8% of the lycopene as *(all-E)*-lycopene. In human blood, lycopene was determined with more than 50% as *(Z)*-isomers, in tissues over 75% of the lycopene was analysed as *(Z)*-isomers (Krinsky et al. 1990, Stahl et al. 1992, Clinton et al. 1996, Fröhlich et al. 2006). In human blood as well as in human tissues, 20-30 different lycopene isomers were detected (Clinton et al. 1996, Clinton 1998). Only a few of them were structurally unambiguously characterized.

ABSORPTION OF LYCOPENE

Around 85% of the lycopene ingested comes from tomatoes and tomato products. Watermelon, papaya and pink grapefruit also deliver lycopene (Grünwald et al. 2002). Lycopene is absorbed together with lipids and lipid-soluble vitamins. After liberation of the carotenoids from the food matrix and breakage of carotenoid protein complexes they are solubilized in lipid droplets (Fig. 2). Within the duodenum, mixed micelles are formed from gallic acids, free fatty acids, monoglycerides and phospholipids. Lycopene as unpolar compound is located mainly in the inner part of the micelles, while the more polar xanthophylls are located at the outer surface. The micelles are mainly absorbed by passive diffusion into the enterocytes of the mucosa. However, the involvement of special receptors in absorption of carotenoids was also discussed (Erdman et al. 1993, Gustin et al. 2004, Zaripheh and Erdman 2005). Carotenoids are incorporated into chylomicrons and transported to the liver and to other

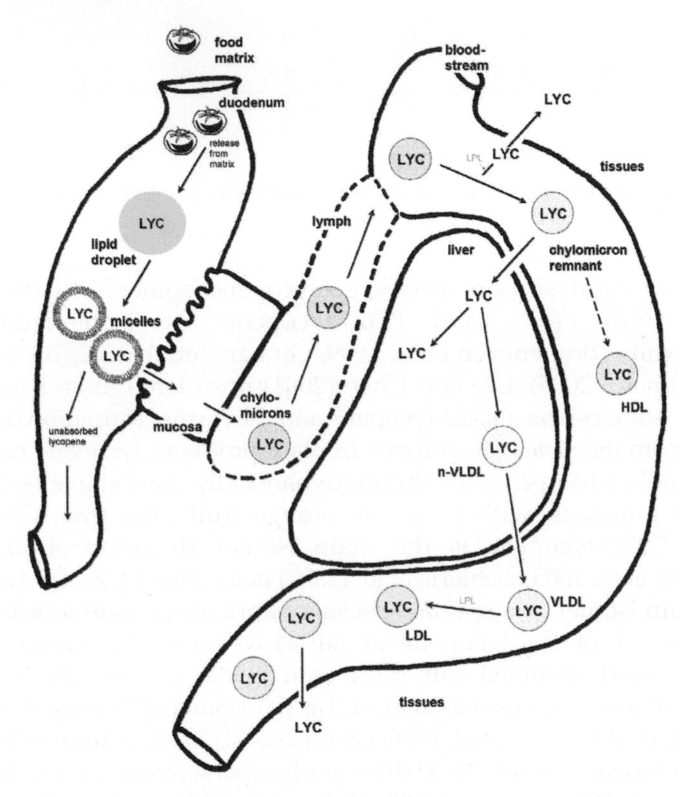

Fig. 2 Schematical overview on pathway of absorption of lycopene within the human body: LYC, lycopene; LPL, lipoprotein lipase; VLDL, very low density lipoprotein; LDL, low density lipoprotein; HDL, high density lipoprotein.

tissues. After partial storage in the liver, carotenoids are transported via lipoproteins to other places in the organism, carotenes, as lycopene, mainly in low density lipoproteins (LDL) and xanthophylls in LDL particles as well as in high density lipoproteins (HDL) (Romanchik et al. 1995). Lycopene is mainly located in tissues with a high number of LDL receptors, for example, liver and prostate (Kaplan et al. 1990, Schmitz et al. 1991). Some scientists postulated the existence of specific lycopene-binding proteins responsible for the tissue-specific uptake (Zaripheh and Erdman 2005). Up to now, specific proteins binding or transporting lycopene were not identified.

BIOAVAILABILITY OF LYCOPENE

Often bioavailability is defined as the fraction of diet ingested that is available within the body for metabolism and for different activities (Shi and Le Maguer 2000). The absorption of carotenoids is affected at different steps after ingestion of foods containing these compounds. Food processing as well as the type of food matrix and the type and content of lipids are some important aspects affecting the liberation of carotenoids and their absorption. In contrast to other food ingredients the bioavailability of carotenoids is very low. While only 5% of the carotenoids were absorbed from raw fruits and vegetables the absorption rate was > 50% from micellar solutions (Olson and Krinsky 1995). After uptake of synthetic lycopene or tomato lycopene only 1.2% or 0.2-0.3% of the lycopene were determined in human blood (Tang et al. 2005). Other studies found 0.2% lycopene circulating in the blood system after uptake of tomato paste (Gärtner et al. 1997, Richelle et al. 2002). Several studies showed a lower intestinal absorption for lycopene compared to other carotenoids (Micozzi et al. 1992, Carughi and Hooper 1994, Clark et al. 1998, 2000, Sugawara et al. 2001). *(Z)*-isomers of lycopene showed a better solubility in mixed micelles compared to *(all-E)*-lycopene as well as a better incorporation in chylomicrons due to a shorter structure and a slightly higher polarity than *(all-E)*-lycopene (Parker 1997, Boileau et al. 2002).

Uptake of lycopene-containing food led to lycopene concentrations in plasma/serum of up to 1 µmol/L. Dietary supplements as well as ingestion of lycopene-rich food products were able to increase the content of lycopene in plasma to values of 1-1.5 µmol/L (Fröhlich 2007). Dietary interventions with 5 mg/d (Böhm and Bitsch 1999) as well as with 70-75 mg/d (Paetau et al. 1998) led to a similar plateau of the concentration of lycopene in plasma (0.2-0.3 µmol/L). Investigations with lycopene doses between 10 and 120 mg/d showed comparable amounts absorbed for all doses. Calculation of lycopene mass in the plasma compartment suggested a saturation of absorption with doses of around 5 mg/d (Diwadkar-Navsariwala et al. 2003).

LYCOPENE ISOMERS IN FOOD PRODUCTS

Different commercial tomatoes and tomato products were analysed for their carotenoid contents. Besides nine samples of tomatoes, nine samples (per category) of tomato juices, tomato ketchups, pizza tomatoes (cans), and tomato pastes (tetrapack) and eight samples of tomato sauces were investigated. The samples were homogenized, extracted and analysed with C_{30}-HPLC-DAD (Seybold et al. 2004). Lycopene was the main carotenoid (89-99%) in all samples investigated. Tomatoes had the lowest content (Fig. 3) of total lycopene (9.4 ± 2.9 mg/100 g), while tomato juices contained the highest concentrations (19.8 ± 2.2 mg/100 g). The main

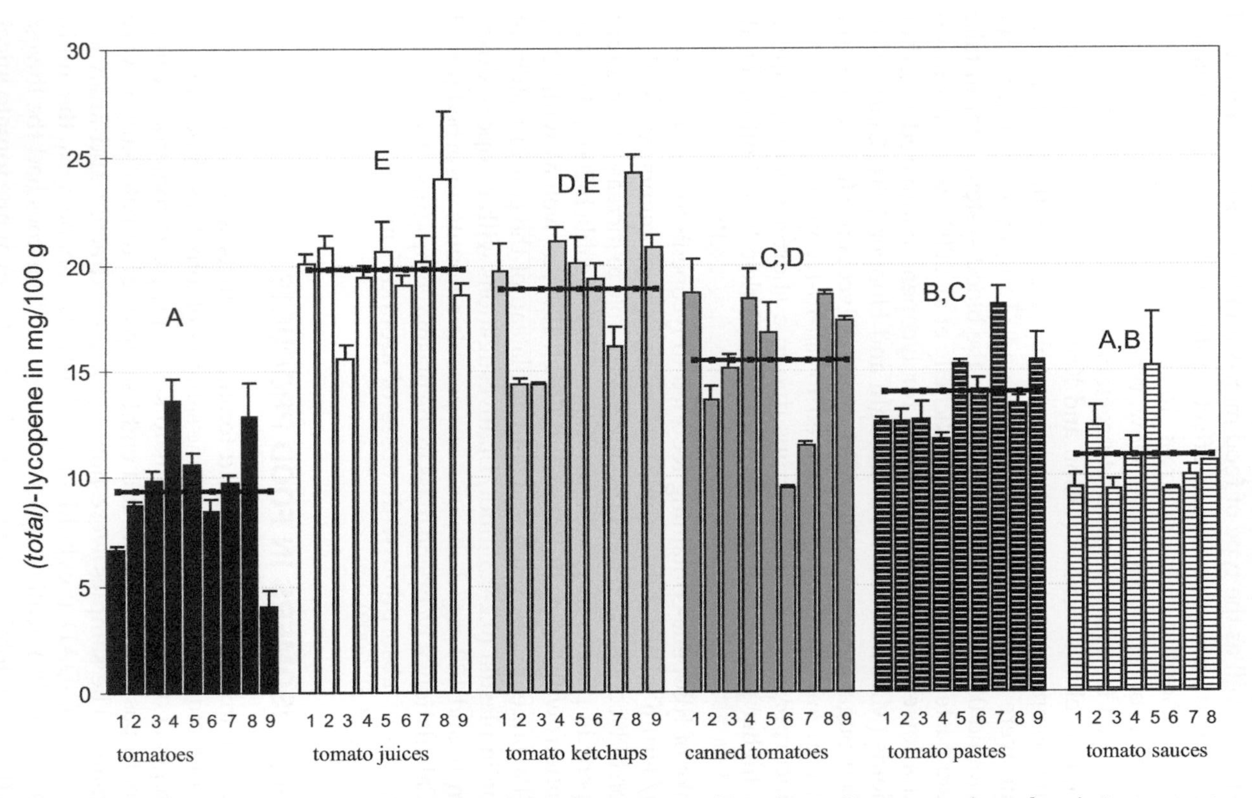

Fig. 3 Contents of *(total)*-lycopene [mg/100 g] of tomatoes and tomato products (M + s) as well as mean values of products groups; groups with different letter are significantly different ($p < 0.05$, one way ANOVA, Tukey-HSD).

isomer in all samples investigated was *(all-E)*-lycopene. The *(Z)*-isomers of lycopene were summarized to give the total content of *(Z)*-lycopenes. Tomatoes, tomato juices, tomato ketchups, canned tomatoes and tomato pastes showed comparable ($p > 0.05$) ratios between *(all-E)*-lycopene and *(Z)*-isomers of lycopene with 93-96% *(all-E)*-lycopene. In contrast, tomato sauces showed significantly ($p < 0.05$) lower relative contents of *(all-E)*-lycopene (about 82%). Thus, they contained around 18% *(Z)*-isomers of lycopene (Table 1). The analyses of commercial products showed higher contents of *(total)*-lycopene for all tomato products compared to the raw tomatoes. One reason for these differences is the loss of water during thermal processing of tomatoes. In addition, for tomato products, industry uses special tomato varieties with higher lycopene contents (Abushita et al. 2000). Many studies showed a high resistance of lycopene within the matrix against thermal degradation and isomerization (Khachik et al. 1992, Nguyen and Schwartz 1998, Abushita et al. 2000, Agarwal et al. 2001, Nguyen et al. 2001, Takeoka et al. 2001). In all samples presented here, *(all-E)*-, *(13Z)*-, *(15Z)*- and *(9Z)*-lycopene were detected. Processing of tomatoes can lead to isomerization. However, in all samples, except tomato sauces, more than 90% of the lycopene was found as *(all-E)*-lycopene. Longer heating times as well as addition of oil might be responsible for the higher content of *(Z)*-isomers of lycopene (18.3 ± 10.1%) in the tomato sauces.

Table 1 Mean relative portions of *(all-E)*-lycopene and of sum of *(Z)*-isomers of lycopene related to *(total)*-lycopene in tomatoes and tomato products (mean ± s), groups with different superscript letters are significantly different ($p < 0.05$, one way ANOVA, Tukey-HSD).

	tomatoes^A	*tomato juices*^{A}	*tomato ketchups*^{A}	*canned tomatoes*^{A}	*tomatoe pastes*^{A}	*tomato sauces*^{B}
(all-E)	95.2 ± 1.5	94.0 ± 1.0	93.6 ± 2.0	94.6 ± 1.9	93.1 ± 0.7	81.7 ± 10.1
Σ *(Z)*	4.8 ± 1.5	6.0 ± 1.0	6.4 ± 2.0	5.4 ± 1.9	6.9 ± 0.7	18.3 ± 10.1

HUMAN INTERVENTION STUDY

A human intervention study (Fröhlich et al. 2006) investigated the effect of ingestion of lycopene from three different matrices (tomatoes, tomato juice, tomato purée) on lycopene content in plasma as well as the lycopene isomer pattern. Seventeen non-smoking volunteers were divided randomly into three groups. After a 2 wk depletion period they received 12.5 mg/d of lycopene comprised in 145-320 g tomatoes or in 94-101 g tomato juice or in 25-28 g tomato purée for 4 wk. The volunteers were asked to eat the tomatoes or tomato products at breakfast together with a

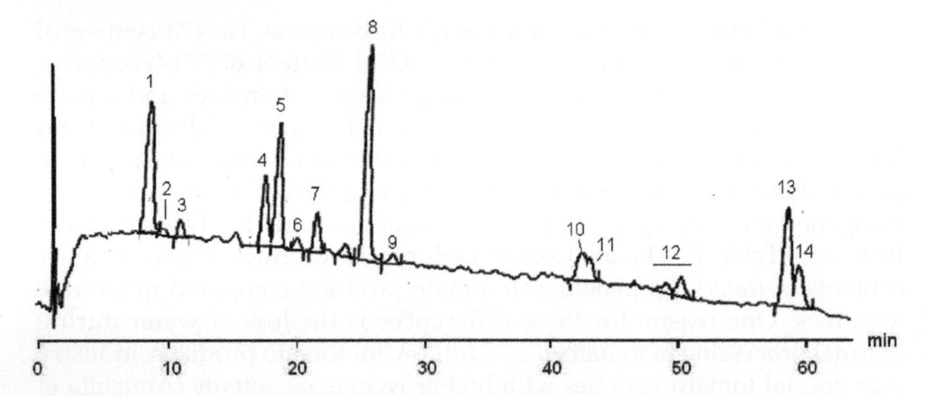

Fig. 4. HPLC chromatogram of a human plasma extract after 4 wk of intervention (T4) with tomatoes.

HPLC conditions YMC C_{30} (250 × 4.6 mm, 5 μm), 1.3 mL/min, gradient of methanol and methyl tert butyl ether, 23 ± 1°C, 450 nm.

1 = *(all-E)*-lutein, 2 = *(all-E)*-zeaxanthin, 3 = *(all-E)*-canthaxanthin, 4 = *(all-E)*-β-cryptoxanthin, 5 = echinenone (IS), 6 = *(13Z)*-β-carotene, 7 = *(all-E)*-α-carotene, 8 = *(all-E)*-β-carotene, 9 = *(9Z)*-β-carotene, 10 = *(13Z)*-lycopene, 11 = unidentified *(Z)*-lycopene, 12 = *(5Z,9'Z)*-+*(9Z)*-+*(5Z,9Z)*-lycopene, 13 = *(all-E)*-lycopene, 14 = *(5Z)*-lycopene.

portion of fat. During the whole intervention study, they had to follow a diet poor in lycopene. Blood samples were drawn before the depletion period (T-2), prior to start of intervention (T0) and weekly during the 4 wk of intervention (T1-T4). Figure 4 shows a representative HPLC chromatogram of a plasma extract after 4 wk intervention with tomatoes. Concerning the lycopene isomers, besides *(all-E)*-lycopene different *(Z)*-isomers were detected: *(13Z)-, (5Z,9'Z)-, (9Z)-,* (5Z,9Z), *(5Z)*-lycopene (for characterization of most of these *(Z)*-isomers see Fröhlich et al. (2005)). In addition, one yet unknown *(Z)*-isomer of lycopene was detected. Baseline concentrations (prior to depletion period) of *(total)*-lycopene were between 0.57 and 0.78 μmol/L and were comparable ($p > 0.05$) for the three intervention groups. The depletion period led to a significant decrease of *(total)*-lycopene by 45-62% in all groups. The intervention with tomatoes significantly increased the *(total)*-lycopene content in plasma from 0.25 ± 0.14 μmol/L to 0.39 ± 0.23 μmol/L after 1 wk of intervention. One week of intervention with tomato juice significantly increased the values from 0.43 ± 0.14 μmol/L to 0.61 ± 0.17 μmol/L. In contrast, intervention with tomato purée led to significantly increased contents of *(total)*-lycopene after 2 wk of intervention (0.40 ± 0.17 μmol/L to 0.72 ± 0.28 μmol/L). After 4 wk of intervention, the contents of lycopene in plasma were comparable ($p > 0.05$) for the three groups (0.53-0.81 μmol/L). The main lycopene isomers

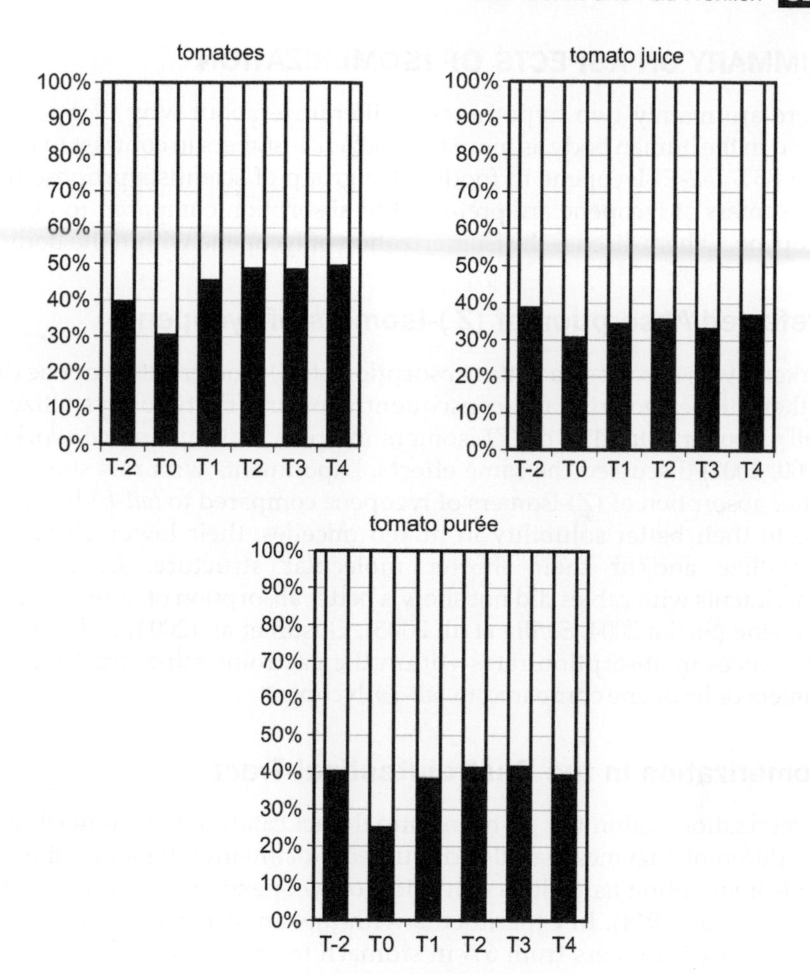

Fig. 5. Changes in relative portions [%] of *(all-E)*-lycopene (▪) and of sum of *(Z)*-isomers of lycopene (▫) related to *(total)*-lycopene in human plasma during the tomato intervention study.

in plasma were *(all-E)*-lycopene and *(5Z)*-lycopene. Figure 5 shows the changes in ratios of *(all-E)*-lycopene to total *(Z)*-isomers of lycopene for all three groups. Prior to the depletion period, the *(all-E)*:*(Z)*-ratio was around 40:60. During the depletion period, it changed to 25-30:70-75. Thus, the depletion period led to increased relative contents of *(Z)*-isomers of lycopene. After 4 wk of intervention, the ratio returned approximately to the baseline values for the groups ingesting tomato juice or tomato purée. After 4 wk of ingestion of tomatoes the *(all-E)*:*(Z)*-ratio was 50:50.

SUMMARY ON ASPECTS OF ISOMERIZATION

There are mainly two hypotheses in literature about why lycopene is found in the human body as more than 50% *(Z)*-isomers in contrast to more than 85% *(all-E)*-lycopene in foods. One group of scientists propose that *(Z)*-isomers of lycopene are preferred in absorption compared to *(all-E)*-lycopene. Others discuss the isomerization of lycopene within the body.

Preferred Absorption of (*Z*)-isomers of Lycopene

Parker (1997) described a better absorption of *(Z)*-isomers of lycopene due to their higher polarity and consequently lower affinity to crystallize as well as better solubility of *(Z)*-isomers in lipids. Boileau and co-workers (2000, 2002) discussed the same effects. Experiments with rats showed a better absorption of *(Z)*-isomers of lycopene compared to *(all-E)*-lycopene due to their better solubility in mixed micelles, their lower affinity to crystallize and/or their shorter molecular structure. In contrast, experiments with calves did not show a better absorption of *(Z)*-isomers of lycopene (Sicilia 2004, Sicilia et al. 2005). Gustin et al. (2004) did not find differences in absorption rates within the gastrointestinal tract for *(Z)*-isomers of lycopene compared to *(all-E)*-lycopene.

Isomerization in the Gastrointestinal Tract

Isomerization within the gastrointestinal tract catalysed by stomach acid and different enzymes was also discussed. *In vitro* investigations showed the isomerization as well as oxidation of lycopene with different acids (Zechmeister 1944). In experiments with rats an increase of the relative portion of *(Z)*-isomers from 6% in stomach to 18% in colon was shown. However, this increase is too low to explain 60% *(Z)*-isomers of lycopene in colon mucosa cells (Boileau et al. 1999). Investigations using model systems with human stomach juice showed also an isomerization (Re et al. 2001). Studies from Tyssandier et al. (2003) showed contradictory results. They investigated the isomer pattern of lycopene in human stomach juice, duodenal liquid and blood after ingestion of tomato paste. The isomer pattern did not change significantly during digestion.

Isomerization during Crossing of Mucosa Membrane or Formation of Chylomicrons

Investigation of different lipoproteins in human plasma on their lycopene isomer ratio showed a ratio of 57% *(Z)*-isomers and 43% *(all-E)*-lycopene in chylomicrons (Holloway et al. 2000). Those results support isomerization while passing the mucosa or during uptake into the chylomicrons. In contrast, another study showed a ratio of 65% *(all-E)*-lycopene and 35% *(Z)*-isomers in chylomicrons (Gärtner et al. 1997), supporting a later isomerization.

The human intervention study presented here (Fröhlich et al. 2006) showed a shift of the isomer ratio to higher relative contents of *(Z)*-isomers during the depletion period. One explanation could be a systematic isomerization of *(all-E)*-lycopene to *(Z)*-isomers within the body. A longer half-life of the *(Z)*-isomers compared to *(all-E)*-lycopene might also be possible or a transfer of *(Z)*-isomers from tissues to plasma. Pilot kinetic studies showed an increase in contents of *(all-E)*-lycopene as well as increasing relative contents of that isomer for up to 12 hr after ingestion of the tomato product. After 12 hr, the *(all-E)*:*(Z)*-ratio changed to higher relative contents of *(Z)*-isomers, supporting isomerization within the body (liver or other tissues).

One single explanation might not be enough to explain the differences in the lycopene isomer pattern of foods and tissue as well as plasma samples. Besides the isomerization of *(all-E)*-lycopene in the body a preferred metabolism of *(all-E)*-lycopene in blood as well as a mobilization of *(Z)*-isomers from body stores have to be discussed (Fröhlich 2007). Continuing human intervention studies can further elucidate the mechanism of isomerization of lycopene within the body. Especially experiments with isotopically labelled lycopene might give more insight into the isomerization process.

ACKNOWLEDGEMENTS

The authors acknowledge the supply of tomatoes and tomato products by the TV station "Hessischer Rundfunk", Frankfurt, Germany, as well as the generous gift of *(all-E)*-lycopene by BASF AG, Ludwigshafen, Germany. Special thanks are given to all volunteers of the human intervention studies for their participation as well as to the physicians for blood withdrawals.

REFERENCES

Abushita, A.A. and H.G. Daood, and P.A. Biacs. 2000. Change in carotenoids and antioxidant vitamins in tomato as a function of varietal and technological factors. J. Agric. Food Chem. 48: 2075-2081.

Agarwal, A. and H. Shen, S. Agarwal, and A.V. Rao. 2001. Lycopene content of tomato products: its stability, bioavailability and in vivo antioxidant properties. J. Med. Food 4: 9-15.

Böhm, V. and R. Bitsch. 1999. Intestinal absorption of lycopene from different matrices and interactions to other carotenoids, the lipid status, and the antioxidant capacity of human plasma. Eur. J. Nutr. 38: 118-125.

Böhm, V. and N.L. Puspitasari-Nienaber, M.G. Ferruzzi, and S.J. Schwartz. 2002. Trolox equivalent antioxidant capacity of different geometrical isomers of alpha-carotene, beta-carotene, lycopene, and zeaxanthin. J. Agric. Food Chem. 50: 221-226.

Boileau, A.C. and N.R. Merchen, K. Wasson, et al. 1999. Cis-lycopene is more bioavailable than trans-lycopene in vitro and in vivo in lymph-cannulated ferrets. J. Nutr. 129: 1176-1181.

Boileau, T.W. and S.K. Clinton, and J.W. Erdman. 2000. Tissue lycopene concentrations and isomer patterns are affected by androgen status and dietary lycopene concentration in male F344 rats. J. Nutr. 130: 1613-1618.

Boileau, T.W. and A.C. Boileau, and J.W. Erdman. 2002. Bioavailability of all-trans and cis-isomers of lycopene. Exp. Biol. Med. 227: 914-919.

Breinholt, V. and S.T. Lauridsen, B. Daneshvar, and J. Jakobsen J. 2000. Dose-response effect of lycopene on selected drug-metabolizing and antioxidant enzymes in the rat. Cancer Lett. 154: 201-210.

Breitenbach, J. and G. Sandmann. 2005. zeta-Carotene cis isomers as products and substrates in the plant poly-cis carotenoid biosynthetic pathway to lycopene. Planta 220: 785-793.

Breitenbach, J. and M. Kuntz, S. Takaichi, and G. Sandmann. 1999. Catalytic properties of an expressed and purified higher plant type zeta-carotene desaturase from Capsicum annuum. Eur. J. Biochem. 265: 376-383.

Carughi, A. and F.G. Hooper. 1994. Plasma carotenoid concentrations before and after supplementation with a carotenoid mixture. Am. J. Clin. Nutr. 59: 896-899.

Clark, R.M. and L. Yao, L. She, and H.C. Furr. 1998. A comparison of lycopene and canthaxanthin absorption: using the rat to study the absorption of non-provitamin A carotenoids. Lipids 33: 159-163.

Clark, R.M. and L. Yao, L. She, et al. 2000. A comparison of lycopene and astaxanthin absorption from corn oil and olive oil emulsions. Lipids 35: 803-806.

Clinton, S.K. 1998. Lycopene: chemistry, biology, and implications for human health, and disease. Nutr. Rev. 56: 35-51.

Clinton, S.K. and C. Emenhiser, S.J. Schwartz, et al. 1996. Cis-trans lycopene isomers, carotenoids, and retinol in the human prostate. Cancer Epidemiol. Biomarkers Prev. 5: 823-833.

DiMascio, P. and S. Kaiser, and H. Sies. 1989. Lycopene as the most efficient biological carotenoid singlet oxygen quencher. Arch. Biochem. Biophys. 274: 532-538.

Diwadkar-Navsariwala, V. and J.A. Novotny, D.M. Gustin, et al. 2003. A physiological pharmacokinetic model describing the disposition of lycopene in healthy men. J. Lipid Res. 44: 1927-1939.

Erdman, J.W. and T.L. Bierer, and E.T. Gugger. 1993. Absorption and transport of carotenoids. Ann. N.Y. Acad. Sci. 691: 76-85.

Etminan, M. and B. Takkouche, and F. Caamano-Isorna. 2004. The role of tomato products and lycopene in the prevention of prostate cancer: a meta-analysis of observational studies. Cancer Epidemiol. Biomarkers Prev. 13: 340-345.

Ferruzzi, M.G. and M.L. Nguyen, L.C. Sander, et al. 2001. Analysis of lycopene geometrical isomers in biological microsamples by liquid chromatography with coulometric array detection. J. Chromatogr. B Biomed. Sci. Appl. 760: 289-299.

Fröhlich, K. 2007. Lycopin-Isomere in Lebensmitteln und Humanplasma – Strukturaufklärung, antioxidative Aktitvität, Gehalte und relative *(E)-/(Z)-* Verhältnisse, Dissertation, Friedrich-Schiller-Universität Jena, Germany.

Fröhlich, K. and K. Kaufmann, R. Bitsch, and V. Böhm. 2006. Effects of ingestion of tomatoes, tomato juice and tomato purée on contents of lycopene isomers, tocopherols and ascorbic acid in human plasma as well as on lycopene isomer pattern. Br. J. Nutr. 95: 734-741.

Fröhlich, K. and J. Conrad, A. Schmid, et al. 2007. Isolation and structural elucidation of different geometrical isomers of lycopene. Int. J. Vitam. Nutr. Res. 77: (in print).

Gärtner, C. and W. Stahl, and H. Sies. 1997. Lycopene is more bioavailable from tomato paste than from fresh tomatoes. Am. J. Clin. Nutr. 66: 116-122.

Giovannucci, E. and E.B. Rimm, Y. Liu, et al. 2002. A prospective study of tomato products, lycopene, and prostate cancer risk. J. Natl. Cancer Inst. 94: 391-398.

Giuliano, G. and L. Giliberto, and C. Rosati. 2002. Carotenoid isomerase: a tale of light and isomers. Trends Plant Sci. 7: 427-429.

Grünwald, J. and C. Jänicke, and J. Freder. 2002. Lycopin: Eine neue Wunderwaffe gegen Zivilisationskrankheiten? Deutsche Apothekerzeitung 142: 856-869.

Gustin, D.M. and K.A. Rodvold, J.A. Sosman, et al. 2004. Single-dose pharmacokinetic study of lycopene delivered in a well-defined food-based lycopene delivery system (tomato paste-oil mixture) in healthy adult male subjects. Cancer Epidemiol. Biomarkers Prev. 13: 850-860.

Holloway, D.E. and M. Yang, G. Paganga, et al. 2000. Isomerization of dietary lycopene during assimilation and transport in plasma. Free Radic. Res. 32: 93-102.

Isaacson, T. and G. Ronen, D. Zamir, and J. Hirschberg. 2002. Cloning of tangerine from tomato reveals a carotenoid isomerase essential for the production of b-carotene and xanthophylls in plants. Plant Cell 14: 333-342.

Kaplan, L.A. and J.M. Lau, and E.A. Stein. 1990. Carotenoid composition, concentrations, and relationships in various human organs. Clin. Physiol. Biochem. 8: 1-10.

Karas, M. and H. Amir, D. Fishman, et al. 2000. Lycopene interferes with cell cycle progression and insulin-like growth factor I signaling in mammary cancer cells. Nutr. Cancer 36: 101-111.

Khachik, F. and M.B. Goli, G.R. Beecher, et al. 1992. Effect of food preparation on qualitative and quantitative distribution of major carotenoid constituents of tomatoes and several green vegetables. J. Agric. Food Chem. 40: 390-398.

Klipstein-Grobusch, K. and L.J. Launer, J.M. Geleijnse, et al. 2000. Serum carotenoids and atherosclerosis: The Rotterdam Study. Atherosclerosis 148: 49-56.

Krinsky, N.I and M.D. Russett, G.J. Handelman, and D.M. Snodderly. 1990. Structural and geometrical isomers of carotenoids in human plasma. J. Nutr. 120: 1654-1662.

Lee, M.T. and B.H. Chen. 2001. Separation of lycopene and its cis isomers by liquid chromatography. Chromatographia 54: 613-617.

Levy, J. and E. Bosin, B. Feldman et al. 1995. Lycopene is a more potent inhibitor of human cancer cell proliferation than either α-carotene or β-carotene. Nutr. Cancer 24: 257-266.

Lewinsohn, E. and Y. Sitrit, E. Bar, et al. 2005. Carotenoid pigmentation affects the volatile composition of tomato and watermelon fruits, as revealed by comparative genetic analyses. J. Agric. Food Chem. 53: 3142-3148.

Micozzi, M.S. and E.D. Brown, B.K. Edwards, et al. 1992. Plasma carotenoid response to chronic intake of selected foods and β-carotene supplements in men. Am. J. Clin. Nutr. 55: 1120-1125.

Nguyen, M.L. and S.J. Schwartz. 1998. Lycopene stability during food processing. Proc. Soc. Exp. Biol. Med. 218: 101-105.

Nguyen, M.L. and D. Francis, and S.J. Schwartz. 2001. Thermal isomerization susceptibility of carotenoids in different tomato varieties. J. Sci. Food Agric. 81: 910-917.

Olson, J.A. and N.I. Krinsky. 1995. Introduction: the colorful, fascinating world of the carotenoids: important physiologic modulators. Faseb J. 9: 1547-1550.

Paetau, I. and F. Khachik, E.D. Brown, et al. 1998. Chronic ingestion of lycopene-rich tomato juice or lycopene supplements significantly increases plasma concentration of lycopene and related tomato carotenoids in humans. Am. J. Clin. Nutr. 68: 1187-1195.

Parker, R.S. 1997. Bioavailability of carotenoids. Eur. J. Clin. Nutr. 51: S86-S90.

Porrini, M. and P. Riso, and G. Testolin. 1998. Absorption of lycopene from single or daily portions of raw and processed tomato. Br. J. Nutr. 80: 353-361.

Rao, A.V. and S. Agarwal. 1999. Role of lycopene as antioxidant carotenoid in the prevention of chronic diseases: a review. Nutr. Res. 19: 305-323.

Re, R. and P.D. Fraser, M. Long, et al. 2001. Isomerization of lycopene in the gastric milieu. Biochem. Biophys. Res. Commun. 281: 576-581.

Richelle, M. and K. Bortlik, S. Liardet, et al. 2002. A food-based formulation provides lycopene with the same bioavailability to humans as that from tomato paste. J. Nutr. 132: 404-408.

Romanchik, J.E. and D.W. Morel, and E.H. Harrison. 1995. Distributions of carotenoids and α-tocopherol among lipoproteins do not change when human plasma is incubated in vitro. J. Nutr. 125: 2610-2617.

Schierle, J. and W. Bretzel, I. Bühler, et al. 1997. Content and isomeric ratio of lycopene in food and human blood plasma. Food Chem. 59: 459-465.

Schmitz, H.H. and C.L. Poor, R.B. Wellman, and J.W. Erdmann. 1991. Concentrations of selected carotenoids and vitamin A in human liver, kidney and lung tissue. J. Nutr. 121: 1613-1621.

Seybold, C. and K. Fröhlich. R. Bitsch, et al. 2004. Changes in contents of carotenoids and vitamin E during tomato processing. J. Agric. Food Chem. 52: 7005-7010.

Sharoni, Y. and M. Danilenko, N. Dubi, et al. 2004. Carotenoids and transcription. Arch. Biochem. Biophys. 430: 89-96.

Shi, J. and M. Le Maguer. 2000. Lycopene in tomatoes: chemical and physical properties affected by food processing. Crit. Rev. Food Sci. Nutr. 40: 1-42.

Sicilia, T. 2004. Vorkommen, Bioverfügbarkeit und Metabolismus von Lycopin und dessen Wirkung gegen oxidativen Stress. Dissertation, Universität Karlsruhe, Germany.

Sicilia, T. and A. Bub, G. Rechkemmer, et al. 2005. Novel lycopene metabolites are detectable in plasma of preruminant calves after lycopene supplementation. J. Nutr. 135: 2616-2621.

Stahl, W. and W. Schwarz, A.R. Sundquist, and H. Sies. 1992. cis-trans isomers of lycopene and b-carotene in human serum and tissues. Arch. Biochem. Biophys. 294: 173-177.

Stahl, W. and J. von Laar, H.D. Martin, et al. 2000. Stimulation of gap junctional communication: comparison of acyclo-retinoic acid and lycopene. Arch. Biochem. Biophys. 373: 271-274.

Sugawara, T. and M. Kushiro, H. Zhang, et al. 2001. Lysophosphatidylcholine enhances carotenoid uptake from mixed micelles by Caco-2 human intestinal cells. J. Nutr. 131: 2921-2927.

Takeoka, G.R. and L. Dao, S. Flessa, et al. 2001. Processing effects on lycopene content and antioxidant activity of tomatoes. J. Agric. Food Chem. 49: 3713-3717.

Tang, G. and A.L.A. Ferreira, M.A. Grusak, et al. 2005. Bioavailability of synthetic and biosynthetic deuterated lycopene in humans. J. Nutr. Biochem. 16: 229-235.

Tyssandier, V. and E. Reboul, J.F. Dumas, et al. 2003. Processing of vegetable-borne carotenoids in the human stomach and duodenum. Am. J. Physiol. Gastrointest. Liver Physiol. 284: 913-923.

Zaripheh, S. and J.W. Erdman. 2005. The biodistribution of a single oral dose of [14C]-lycopene in rats prefed either a control or lycopene-enriched diet. J. Nutr. 135: 2212-2218.

Zechmeister, L. 1944. Cis-trans isomerization and stereochemistry of carotenoids and diphenylpolyenes. Chem. Rev. 34: 267-344.

Zhang, L.-X. and R.V. Cooney, and J.S. Bertram. 1991. Carotenoids enhance gap junctional communication and inhibit lipid peroxidation in C3H/10T1/2 cells: relationship to their cancer chemopreventive action. Carcinogenesis 12: 2109-2114.

16

Antioxidant Compound Studies in Different Tomato Cultivars

Giuseppina Tommonaro, Barbara Nicolaus, Rocco De Prisco, Alfonso De Giulio, Giuseppe Strazzullo[#] and Annarita Poli

Istituto di Chimica Biomolecolare, Consiglio Nazionale delle Ricerche (C.N.R.)
Via Campi Flegrei, 34 80078 Pozzuoli, Napoli, Italy

ABSTRACT

In this chapter is illustrated the determination of the antioxidant activity of hydrophilic and lipophilic fractions of nine different cultivars of tomato (*Lycopersicon esculentum*). To assess the nutritional value of all varieties, the total content of principal carotenoids, lycopene and β-carotene was analysed. The antioxidant activity was determined by DMPD and ABTS methods for hydrophilic and lipophilic fractions, respectively. It has been verified that there is no correlation between high concentration of lycopene and β-carotene and total antioxidant activity. It is reasonable to suppose, indeed, that the total antioxidant activity may be due to synergistic action of different bioactive compounds. Additionally, cytotoxic activity was determined by brine shrimps test on all lipophilic extracts to evaluate potential anti-tumoral activity.

INTRODUCTION

Fruits and vegetables play a significant role in human nutrition (Goddard and Matthews 1979, Block et al. 1992, Cohen et al. 2000) for their richness in health-related food components. Tomato and tomato products

---

[#]Present address: ISPA-CNR, Sezione di Torino, Via Leonardo Da Vinci, 44, 10095 Grugliasco - Torino

A list of abbreviations is given before the references.

represent a major natural source of several nutrients with beneficial effect on human health and ability to prevent some major chronic diseases such as some types of cancer and cardiovascular diseases. Tomatoes represent an effective way to supply nutrients such as folate, vitamin C, and potassium, but the peculiar compounds of these vegetables are carotenoids, particularly β-carotene and lycopene (Akanbi and Oludemi 2004, Beecher 1997). About 600 carotenoids are known in plants, but only 14 are found in human tissues (Khachik et al. 1995). Nine of these 14 carotenoids have been recovered in tomato and tomato products, in particular lycopene, neurosporene, gamma-carotene, phytoene and phytofluene. It is well established that the antioxidant activity of these carotenoids and high consumption of these nutrients prevent cardiovascular diseases and cancer. Among the class of carotenoids, lycopene is the more abundant and represents more than 80% of total tomato carotenoid content in the fully ripened fruit. This compound has notable antioxidant activity and several studies have been done to evaluate its anticancer activity, in particular against prostate cancer (Pastori et al. 1998, Giovannucci 1999, Hadley et al. 2002, La Vecchia 1998). Some biological activities of lycopene are potentiated in the presence of other compounds, such as α-tocopherol, glabridin, rosmarinic acid and garlic (Balestrieri et al. 2004, Pastori et al. 1998, Fuhrman et al. 2000, Crawford et al. 1998). A positive effect on health associated with their consumption is exerted by the pool of antioxidants, with noticeable synergistic effects. Therefore, to assess the nutritional quality of fresh tomatoes, it is important to study the main compounds having antioxidant activity.

Besides the compounds beneficial for human health, tomatoes also contain tomatine and dehydrotomatine, glycoalkaloids having well-known toxic properties (Friedman and McDonald 1997). The content of these glycoalkaloids decreases during ripening, whereas that of carotenoids increases (Kozukue et al. 1994, Rick et al. 1994). Therefore, it can be concluded that the consumption of well-ripened tomatoes should ensure maximum health benefit, with a high level of carotenoids coupled with the absence of glycoalkaloids. However, efforts are in progress to elucidate the physiological process as well as the storage conditions that can control the phytonutrient content in foods (Goldman et al. 1999, Grusak et al. 1999).

Tomato is represented by several hundred cultivars and hybrids in response to the fresh consumption market, which demands fruits that have very different characteristics (Leonardi 1994). Therefore, tomato cultivars for fresh consumption show great differences in fruit size (from a few grams to some hundreds of grams), shape (from flattened to elongated), and colour (from yellow to dark red). Besides, according to consumer and market requirements, tomato fruits are harvested at different stages of ripening: from breaking to red colour. There are several studies describing the variation of the qualitative characteristics of tomatoes in relation to cultivars (Davies and Winsor 1969, Gormley et al. 1983, Stevens et al. 1977) and growing conditions (Blanc 1989, La Malfa et al. 1995, Mitchell et al. 1991). Most of these works have taken into consideration only some qualitative characteristics (e.g., dry matter and soluble solids), whereas the antioxidant ability and the carotenoid composition have not been considered. Recently the antioxidant ability in water-soluble and water-insoluble fractions of different cultivars of tomato and the level of β-carotene and lycopene were established Moreover, cytotoxic activity by brine shrimps assay was tested on all lipophilic extracts to evaluate potential anti-tumoral activity (Strazzullo et al. 2007)

EVALUATION OF ANTIOXIDANT ACTIVITY

Nine different cultivars were harvested at the same stage of ripening at which they are usually consumed or processed by food industries. The samples of cultivars San Marzano 823, Cirio 3, All Flash 900, Castiglione del Lago Perugia, and Roma (commonly used by Italian tomato processing industries or for fresh consumption) and the varieties Holland, Red Beefsteak and Finest Tesco (English varieties) were cultivated in the area of Agro Nocerino-Sarnese (Salerno, Italy); the sample of English variety Black Tomato, mainly used as fresh food, was cultivated in the area of Pachino (Siracusa, Italy). Seeds of all varieties were bought from MFM International S.r.l., Torre del Greco, Naples, and all varieties were cultivated outdoors with the same fertilizing and pedoclimatic conditions. The field showed fertility, structure and presence of organic components. Seeds were planted in greenhouse in April and seedlings were transplanted in the open field in May. The berries were harvested at the end of July. All the fruits appeared red inside and outside, with a good degree of ripening.

Each sample was homogenized in a blender and centrifuged at 9.500 rpm for 20 min, and supernatant and pellet were collected separately for analysis. Pellets were extracted with diethyl ether (w/v 1:2) under stirring in the dark, overnight. Lipophilic extracts were filtered and concentrated in a rotary evaporator in vacuum ($T < 35°C$) and dried under N_2 (Table 1). Supernatant collected from each sample was tested for hydrophilic antioxidant activity and the test was carried out in triplicate, according to

DMPD (N,N-dimethyl-*p*-phenylenediaminedihydrochloride) method developed by Fogliano et al. (1999); the antioxidant activity was expressed as percentage of the absorbance of the uninhibited radical cation solution.

Crude diethyl ether extract from each sample was dissolved in dichloromethane analytical grade (20 mg/ml) and prepared to assess lipophilic antioxidant activity in triplicate according to ABTS [2,2′-Azino-bis (3-ethylbenzothiazoline-6-sulphonic acid) diammonium salt] method as described by Miller et al. (1996) and Miller and Rice-Evans (1997); the antioxidant activity was expressed as percentage of the absorbance of the uninhibited radical cation solution.

We reported the data of antioxidant activity of hydrophilic and lipophilic fractions of all tomato cultivars in terms of percentage of inhibition of radical cations $DMPD+^{\bullet}$ and $ABTS+^{\bullet}$. The percentage of inhibition, calculated in terms of percentage of inhibition per milligram of fresh material, ranged from 7.72% to 25.85% and 0.25% to 2.12% for hydrophilic and lipophilic fractions, respectively, as shown in Fig. 1. The cultivars Holland and Finest Tesco, among hydrophilic fractions, were those with high activity (25.85% and 20.23%, respectively).

Table 1 Amounts of material used for analyses of nine cultivars of tomatoes.

Cultivars	*Fresh material (g)*	*Supernatant (ml)*[a]	*Pellet (g)*[b]	*Diethyl ether extract (mg)*[c]
San Marzano 823	341.5	220	35.0	324.0
Cirio 3	353.6	240	87.3	223.0
All Flash 900	300.4	210	37.0	345.0
Castiglione del Lago Perugia	383.4	285	44.0	259.0
Roma	396.0	290	56.7	210.0
Black Tomato	394.0	380	68.8	265.6
Red Beefsteak	248.0	165	55.0	178.8
Finest Tesco	250.0	140	79.8	154.1
Holland	290.0	170	85.2	411.3

[a] Volumes of supernatant measured after centrifugation of homogenate of all fresh material
[b] Weight of pellets collected after centrifugation of homogenate of all fresh material
[c] Weight of lipophilic fractions measured after filtration and drying of diethyl ether extraction of pellets.

Between lipophilic fractions of all cultivars, the major antioxidant activity was observed for Italian varieties; the cultivar San Marzano, commonly used by processing industries, showed highest activity. In Table 2 we report the antioxidant activity expressed as TEAC (Trolox equivalent antioxidant capacity).

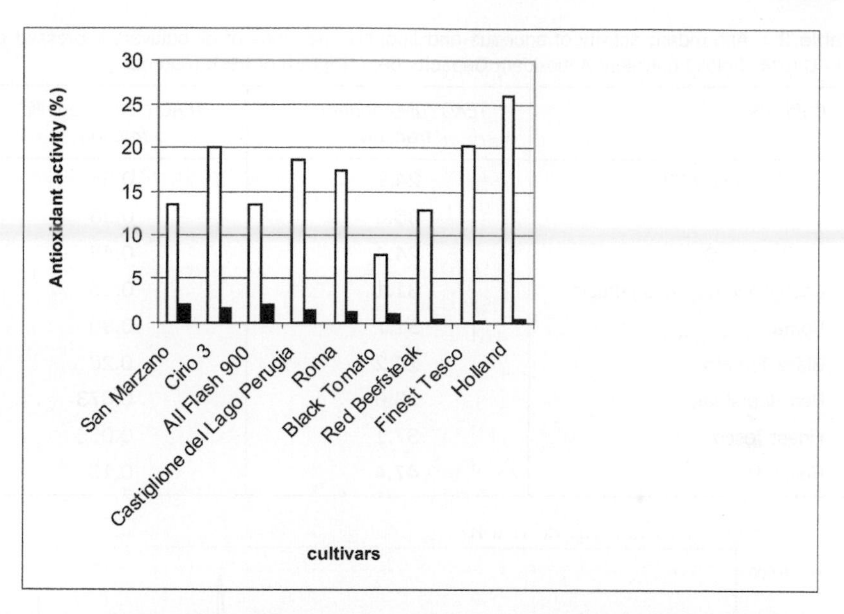

Fig. 1. Percentage of the absorbance of the uninhibited radical cation solution expressed per milligram of fresh material, obtained by using DMPD (N,N-Dimethyl-*p*-phenylenediaminedihydrochloride) and ABTS [2,2'-Azino-bis (3-ethylbenzothiazoline-6-sulfonic acid) diammonium salt] methods for aqueous and lipophilic fractions respectively of tomato cultivars. □ % inhibition radicalic cation DMPD+$^{\bullet}$ ■ % inhibition radicalic cation ABTS+$^{\bullet}$

CAROTENOID CONTENT

To assess the content of lycopene and β-carotene in all cultivars, the diethyl ether extracts of samples were purified by reversed-phase HPLC. The analysis of HPLC chromatograms allowed us to collect three fractions named FR1, FR2 and FR3. In Fig. 2 we report, as an example, the HPLC chromatogram of lipophilic extract of San Marzano cultivar. FR1 contained more unidentified polar compounds in low concentration; FR2 and FR3 were identified as lycopene and β-carotene, respectively, by retention time and co-injection with purchased authentic standards. The amounts in terms of percentage of lycopene and β-carotene calculated per unit weight of fruits in all cultivars are comparable except for the concentration of lycopene of Holland variety (2.5×10^{-3}%) and the content of β-carotene of San Marzano 823 and Black tomato (0.047% and 0.06%, respectively). In analysing the data reported in Fig. 3, we concluded that there is no correlation between high contents of lycopene and β-carotene and antioxidant activity. Indeed, in spite of the fact that the concentration values of lycopene and β-carotene in San Marzano 823 and Black tomato

Table 2 Antioxidant activity of aqueous and lipophilic fractions of all cultivars expressed as TEAC (μM)-Trolox Equivalent Antioxidant Capacity per milligram of fresh material

Cultivars	*TEAC (μM) aqueous fraction*	*TEAC (μM) lipophilic fraction*
San Marzano 823	24.9	0.55
Cirio 3	36.6	0.38
All Flash 900	24.4	0.48
Castiglione del Lago Perugia	31.1	0.35
Roma	31.7	0.30
Black Tomato	14.2	0.26
Red Beefsteak	23.6	0.073
Finest Tesco	37.1	0.065
Holland	47.4	0.10

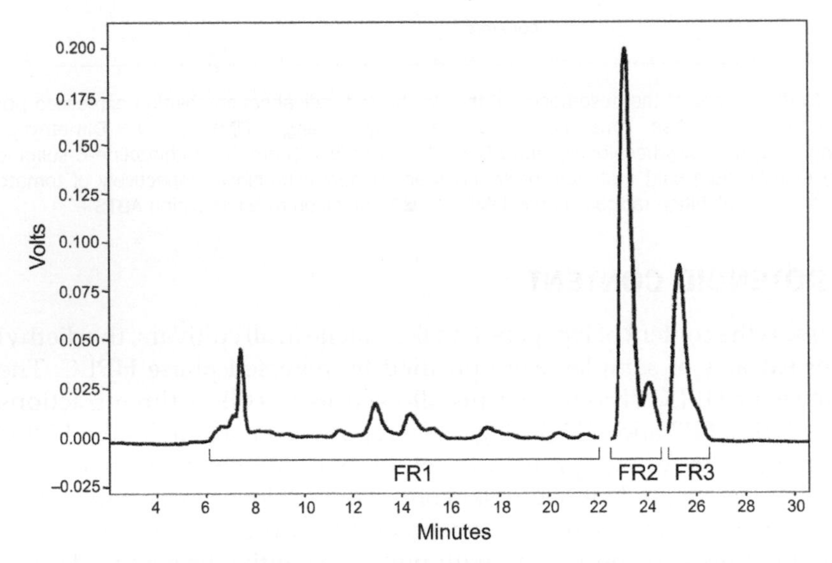

Fig. 2 Chromatogram of diethyl ether extract from San Marzano 823 cultivar obtained by reverse-phase HPLC on a Shimadzu LC 6A system with a Kromasil 100A C_{18} column, 5 μm, 250 × 10 mm (Phenomenex) with UV-VIS detector SPD 10A VP, CR 3A recorder, system controller SCL 10A VP, and Chemstation integration software Class–VP 5.0. Chromatographic conditions were: gradient elution 60:40 to 30:70, v/v, A/B [A was methanol/water 95/5 v/v (0.1% of butylated hydroxytoluene from Aldrich and 0.05% of triethylamin from Fluka) and B was dichloromethane (0.1% of butylated hydroxytoluene and 0.05% of triethylamin)], linear gradient changed over a period of 35 min and return to starting condition in 5 min before next injection; flow rate 1.5 ml min^{-1}; wavelength of UV detector, 450 nm, sensitivity adjusted to 0.04 AUFS; room temperature. FR1: more polar unidentified compounds; FR2: Lycopene; FR3: β-carotene.

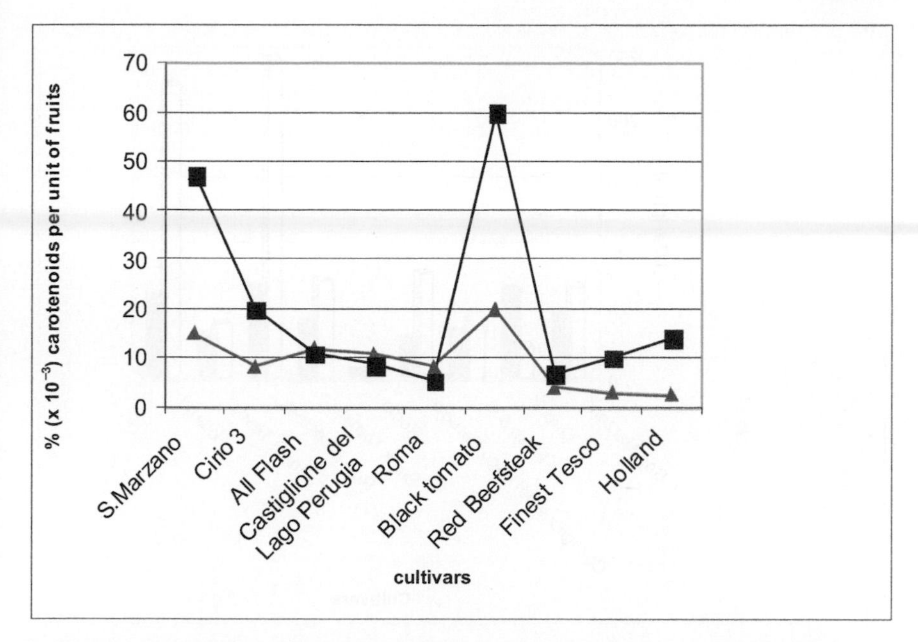

Fig. 3. Total concentration of lycopene and β-carotene recovered by reverse-phase HPLC from lipophilic extracts of tomato cultivars. Results were expressed as percentage ($x10^{-3}$) of carotenoids per unit weight of fruits. ■ β- carotene; ▲ lycopene.

type lipophilic extracts are comparable, the total antioxidant activities expressed as TEAC per milligram of fresh material are 0.55 μM and 0.26 μM, respectively.

BIOLOGICAL ACTIVITY BY BRINE SHRIMP ASSAY

The brine shrimp (*Artemia salina*) assay was performed in triplicate on lipophilic extracts and fractions recovered by HPLC of all cultivars of tomato, dissolved in DMSO (1% final volume) to reach final concentrations of 1000, 100 and 10 ppm, using 10 freshly hatched larvae suspended in 5 ml of artificial sea water (Meyer et al. 1982). Briefly, for each dose tested, surviving shrimps were counted after 24 hr, and the data statistically analysed by the Finney program (Finney 1971), which affords LD_{50} values with 95% confidence intervals.

In Fig. 4 are reported data concerning the cytotoxic activity of the diethyl ether extracts of all cultivars expressed in terms of lethal dose (LD_{50}). Among the crude extracts the Roma type exhibits the highest activity (lowest value of LD_{50}). We also performed brine shrimp assay on fractions FR1, FR2 and FR3 recovered by HPLC analyses of all lipophilic extracts. The data showed that FR2 (lycopene) and FR3 (β-carotene), for all

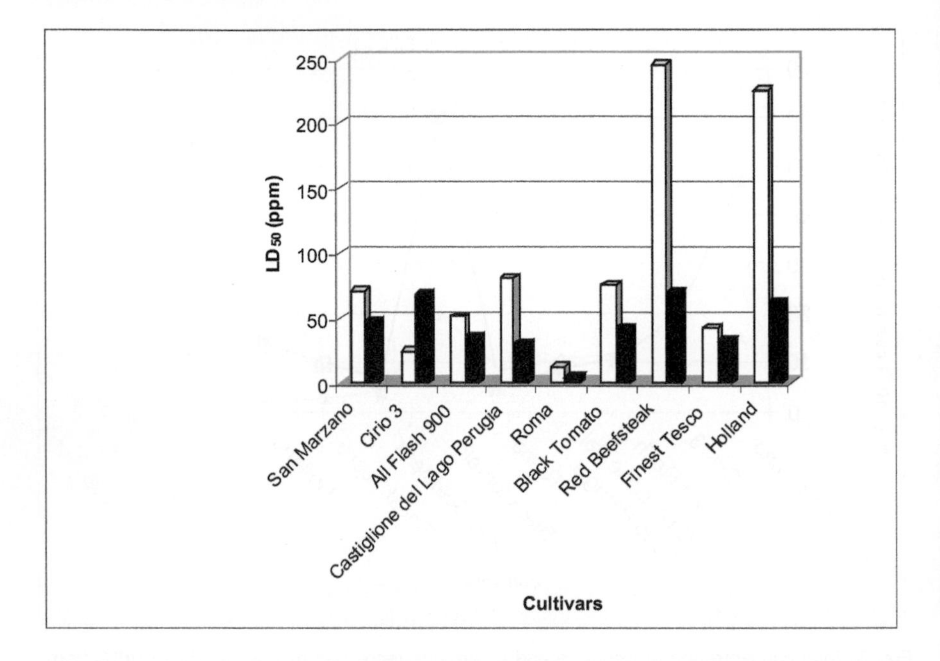

Fig. 4. Cytotoxic activities in brine shrimp assay of total lipophilic fractions and fraction FR1 (unidentified carotenoids) of nine tomato cultivars. Both fractions were tested at three concentrations, 1000, 100 and 10 ppm, using 10 freshly hatched larvae suspended in 5 ml of artificial sea water. For each dose tested, surviving shrimp were counted after 24 hr, and the data statistically analysed by the Finney program, which affords LD_{50} values. □ Diethyl Ether extract LD_{50} (ppm); ■ FR1 LD_{50} (ppm)

cultivars, presented low cytotoxic activity with values of LD_{50} (ppm) of 286.00 and 455.56 respectively. As shown in Fig. 4, where we have reported only the data of activity of FR1, the high cytotoxic activity of Roma variety detected in diethyl ether extract was mainly concentrated in FR1 and due probably to minor unidentified compounds.

CONCLUSION

We have assessed the total antioxidant activity of aqueous and lipophilic fractions of nine different cultivars of tomato, harvested at the stage at which they are usually consumed or processed by food industries. We have also quantified the total amount of lycopene and β-carotene (carotenoids) for each variety, because it is well known that carotenoids are natural lipophilic antioxidants. The lipophilic fraction of tomato, due to the high content of carotenoids, shows, obviously, antioxidant activity and the results are in agreement with those reported in scientific literature. Varietal, agricultural, technological, and environmental factors influence the carotenoid type and the content of lycopene and β-carotene in different cultivars of tomato. Our results show that lycopene and β-carotene content are not solely responsible for total antioxidant activity of tomato. Lycopene and β-carotene, in association with minor carotenoids (detected in FR1), play an important role in antioxidant activity, as reported above. Furthermore, these minor compounds are also responsible for the cytotoxic activity.

ACKNOWLEDGEMENTS

This work was supported by Provincia di Salerno.

ABBREVIATIONS

DMPD: N,N-Dimethyl-*p*-phenylenediaminedihydrochloride; ABTS: 2,2′-Azino-bis(3-ethylbenzothiazoline-6-sulphonic acid) diammonium salt; TEAC: Trolox equivalent antioxidant capacity; DMSO: Dimethylsulphoxide; HPLC: High performance liquid chromatography; LD_{50}: lethal dose at which 50% of mortality is observed

REFERENCES

Akanbi, C.T. and F.O. Oludemi. 2004. Effect of processing and packaging on the lycopene content of tomato products. Intl. J. Food Properties 1: 139-152.

Balestrieri, M.L. and R. De Prisco, B. Nicolaus, P. Pari, V. Schiano Moriello, G. Strazzullo, E.L. Iorio, L. Servillo, and C. Balestrieri. 2004. Lycopene in association with α-tocopherol or tomato lipophilic extracts enhances acyl-platelet-activating factor biosynthesis in endothelial cells during oxidative stress. Free Radic. Bio. Med. 36: 1058-1067.

Beecher, G.R. 1997. Nutrient content of tomatoes and tomato products. Proc. Soc. Exp. Biol. Med. 218: 98-100.

Blanc, D. 1989. The influence of cultural practices on the quality of the production in protected cultivation with special references to tomato production. Acta Hort. 191: 85-98.

Block, G. and B. Patterson, and A. Subar. 1992. Fruits, vegetables, and cancer prevention: a review of the epidemiological evidence. Nutr. Cancer 18: 1-29.

Cohen, J.H. and A.R. Kristal, and J.L. Stanford. 2000. Fruit and vegetable intakes and prostate cancer risk. J. Natl. Cancer Inst. 92: 61-69.

Crawford, R.S. and E.A. Lirk, M.E. Rosenfeld, R.C. LeBoeuf, and A. Chait. 1998. Dietary antioxidants inhibit development of fatty streak lesion in LDL receptor-deficient mouse. Arterioscl. Thromb. Vasc. Biol. 18: 1506-1513.

Davies, J.N. and G.W. Winsor. 1969. Some effects of variety on the composition and quality of tomato fruit. J. Hort. Sci. 44: 331-342.

Finney, D.J. 1971. Statistical logic in monitoring of reactions to therapeutic drugs. Meth. Inform. Med. 10(1): 1-8.

Fogliano, V. and V. Verde, G. Randazzo, and A. Ritieni. 1999. Method for measuring antioxidant activity and its application to monitoring the antioxidant capacity of wines. J. Agric. Food Chem. 47: 1035-1040.

Friedman, M. and G.M. McDonald. 1997. Potato glycoalkaloids: chemistry, analysis, safety, and plant physiology. Crit. Rev. Plant Sci. 16: 55-132.

Fuhrman, B. and N. Volkova, M. Rosenblat, and M. Aviram. 2000. Lycopene synergistically inhibits LDL oxidation in combination with vitamin E, glabridin, rosmarinic acid, carnosic acid, or garlic. Antioxid. Redox Signal. 2: 491-506.

Giovannucci, E. 1999. Tomatoes, tomato-based products, lycopene, and cancer: review of the epidemiological literature. J. Natl. Cancer Inst. 91: 317-331.

Goddard, M.S. and R.H. Matthews. 1979. Contribution of fruit and vegetables to human nutrition. Hort. Sci. 14: 245-247.

Goldman, I.L. and A.A. Kader, and C. Heintz. 1999. Influence of production, handling and storage on phytonutrients contents in foods. Nutr. Rev. 57: S46-S52.

Gormley, T.R. and M.J. Mather, and P.E. Walshe. 1983. Quality and performance of eight tomato cultivars in a nutrient film technique system. Crop Res. 23: 83-93.

Grusak, M.A. and D. Della Penna, and M.R. Welch. 1999. Physiologic processes affecting the content and distribution of phytonutrients in plants. Nutr. Rev. 57: S27-S33.

Hadley, C.W. and E.C. Miller, S.J. Schwartz, and S.K. Clinton. 2002. Tomatoes, lycopene, and prostate cancer: progress and promise. Exp. Biol. Med. 227: 869-880.

Khachik, F. and G.R. Beecher, and J.C. Smith Jr. 1995. Lutein, lycopene, and their oxidative metabolites in chemoprevention of cancer. J. Cell Biochem. Suppl. 22: 236-246.

Kozukue, N. and E. Kozukue, H. Yamashita, and S. Fujii. 1994. R-tomatine purification and quantification in tomatoes by HPLC. J. Food Sci. 59: 1211-1212.

La Malfa, G. and C. Leonardi, and D. Romano. 1995. Changes in some quality parameters of greenhouse tomatoes in relation to thermal levels and to auxin sprays. Agric. Med. 125: 404-412.

La Vecchia, C. 1998. Mediterranean epidemiological evidence on tomatoes and the prevention of digestive tract cancers. Proc. Soc. Exp. Biol. Med. 218: 125-128.

Leonardi, C. 1994. Studi su specie da orto ai fini della diversificazione colturale. Ph.D. Thesis. University of Catania, Italy.

Meyer, B.N. and N.R. Ferrigni, J.E. Putnam, L.B. Jacobsen, D.E. Nichols, and J.L. McLaughlin. 1982. Brine shrimp: a convenient general bioassay for active plant constituents. Planta Medica 45: 31-34.

Miller, J.N. and C.A. Rice-Evans. 1997. Factors influencing the antioxidant activity determined by the ABTS radical cation assay. Free Radic. Res. 26: 195-199.

Miller, J.N. and J. Sampson, L.P. Candeias, P.M. Bramley, and C.A. Rice-Evans. 1996. Antioxidant activities of carotenes and xanthophylls. FEBS Lett. 384: 240-242.

Mitchell, J.P. and C. Shennan, S.R. Grattan, and D.M. May. 1991. Tomato fruit yield and quality under water deficit and salinity. J. Amer. Soc. Hort. Sci. 116: 215-221.

Pastori, M. and H. Pfander, D. Boscoboinik, and A. Azzi. 1998. Lycopene in association with alfa-tocopherol inhibits at physiological concentration proliferation of prostate carcinoma cells. Biochem. Biophys. Res. Commun. 250: 582-585.

Rick, C.M. and J.W. Uhlig, and A.D. Jones. 1994. High alpha-tomatine content in ripe fruit of Andean *Lycopersicon esculentum* var. *cerasiforme*: developmental and genetic aspects. Proc. Natl. Acad. Sci. USA 91: 12877-12881.

Stevens, M.A. and A.A. Kader, M. Albright-Holton, and M. Algazi. 1977. Genotypic variation for flavor and composition in fresh market tomatoes. J. Amer. Soc. Hort. Sci. 102: 680-689.

Strazzullo, G. and A. De Giulio, G. Tommonaro, C. La Pastina, C. Saturnino, A. Poli, B. Nicolaus, and R. De Prisco. 2007. Antioxidative activity and lycopene and beta-carotene contents in different cultivars of tomato (*Lycopersicon esculentum*). Intl. J. Food Properties, DOI: 10.1080/10942910601052681.

17

Antioxidant Activity of Fresh-cut Tomatoes: Effects of Minimal Processing and Maturity Stage at Harvest

Milza Moreira Lana
Embrapa Hortaliças (Embrapa Vegetables)
Caixa Postal 218, 70359970, Brasília-DF, Brazil

ABSTRACT

The protective effect of a diet rich in fruits and vegetables against chronic diseases is well documented and at least partly ascribed to the action of antioxidant compounds such as carotenoids, ascorbic acid, tocopherols, and polyphenols. Because of that, campaigns to increase the consumption of fresh fruits and vegetables were launched in many countries. Minimally processed or fresh-cut fruits and vegetables (peeled, cut and ready to use) are presented as a convenient alternative to supply the dietary need for fresh food. However, the wounding stress resulting from processing operations can induce changes in the antioxidant content, composition and activity of these tissues that will affect their health-promoting properties. In the present review, changes in the antioxidant activity of fresh-cut tomato during refrigerated storage and during maturation are addressed. The potential impact of these changes in the value of fresh tomato as a source of bioactive compounds in the diet is discussed taking into consideration the characteristics and drawbacks of the available methodologies to measure antioxidant activity and current knowledge about other fresh-cut fruits and vegetables.

A list of abbreviations is given before the references.

INTRODUCTION

The consumption of a diet rich in fresh fruits and vegetables has been associated with a number of health benefits including the prevention of chronic diseases (Klerk et al. 1998, WHO 2003). This beneficial effect is believed to be due, at least partly, to the action of antioxidant compounds such as carotenoids, ascorbic acid, tocopherols, and polyphenols, which reduce oxidative damage in the body (Grassmann et al. 2002, Prior and Cao 2002, Wargovich 2000). However, the prescription of supplements containing antioxidants has resulted in contradictory results on human health, while epidemiological studies comparing populations with different diets show a clear trend in reduction of chronic diseases when there is an increase in the consumption of fruits and vegetables (Liu 2003). The joint FAO/WHO expert consultation on diet, nutrition and prevention of chronic diseases recommended the intake of a minimum of 400 g of fruits and vegetables per day (excluding potatoes and other starch tubers) for the prevention of chronic diseases (FAO/WHO 2005).

In urbanized societies, a series of barriers can result in less than optimal level of consumption of fresh fruits and vegetables (FAO/WHO 2005). These include lack of time for preparation and cooking as urbanization increases and more women work outside the home and rapid increase of fast food culture. Fresh-cut fruits and vegetables are presented as an alternative for healthy and convenient food. Tomato is a vegetable that presents great interest for this industry since it is a natural ingredient of fast food dishes such as sandwiches and salads. The consumption of tomato and tomato products is stimulated in view of its reported benefits on health (Caputo et al. 2004, Giovannucci 1999, Rao and Agarwal 2000, Willcox et al. 2003).

However, the impact of minimal processing on the amount and composition of nutrients and other bioactive components is not well known. There is a general perception that processed foods are of lower nutritional value and that preservation methods cause a depletion of naturally occurring antioxidants in food. However, food processing and preservation procedures can increase or decrease the overall antioxidant properties of food or promote no change, depending on the product and processes considered (Nicoli et al. 1999).

Most of the available information on the effect of dietary antioxidants from plant foods is related to operations such as canning, freezing, heating and blanching (Klein and Kurilich 2000) and much less is known about the effects of minimal processing operations such as peeling, cutting, shredding and slicing. These operations are expected to affect the content, activity and bioavailability of bioactive compounds and hence the health-promoting capacity of fruits and vegetables.

In this chapter, attention is given to the effects of minimal processing operations on the antioxidant activity of tomato fruit. Additionally, ways in which the antioxidant activity changes during maturation and ripening are discussed, as well as the possible interactions between the two factors.

EFFECTS OF MINIMAL PROCESSING ON ANTIOXIDANT ACTIVITY OF TOMATO FRUIT

Studies with fresh-cut tomato are quite scarce (Lana and Tijskens 2006) in contrast with a vast literature on the effects of heat processing on antioxidant activity and antioxidant content of tomato products (Dewanto et al. 2002, Lavelli and Giovanelli 2003, Sahlin et al. 2004, Sánchez-Moreno et al. 2006, Shi and Le Maguer, 2000, Takeoka et al. 2001). The effect of processing operations other than minimal is not considered here, and the interested reader should be addressed to the cited literature.

The scarce volume of literature on the effects of minimal processing operations in the chemical composition of tomato fruit makes it impossible at the moment to come to a final conclusion whether these operations alter the value of tomato as a source of bioactive compounds in the diet. In view of that, the results obtained for fresh-cut tomato are discussed together with results obtained for other fresh-cut fruits and vegetables and inferences are addressed about expected changes in the antioxidant activity of fresh-cut tissues.

Fresh-cut tissues are primarily subjected to oxidative stress (Brecht 1995), which presumably could induce changes in the composition and content of antioxidant compounds and consequently changes in the total antioxidant activity of the tissue. In view of that, it was first hypothesized that processed fruits and vegetables would have a lower health-protecting capacity than fresh ones because of depletion of antioxidant compounds in response to wounding stress. More recent studies show that this is not necessarily the case, depending on which product and which class of antioxidant is considered (Gil et al. 2006, Reyes et al. 2007).

In Table 1 is shown a summary of the most relevant results available in the literature about the effect of minimal processing on the antioxidant activity of fruits and vegetables. This subject is quite recent and has received less attention than other aspects of quality such as firmness, colour, microbial contamination and other visual or organoleptical aspects.

Table 1 Effects of minimal processing operations (peeling, cutting, slicing or shredding) on antioxidant activity of fruits and vegetables.

Fresh-cut produce	*Extract*	*Antioxidant method*	*Result*	*Comments*	*Reference*
Iceberg lettuce	Methanol	DPPH	Increased	Correlated with phenolics content	Kang and Saltveit 2002
	Methanol	FRAP	Increased	Correlated with phenolics content	
	Phosphate buffer	FRAP	Increased	Correlated with phenolics content	
Romaine lettuce	Methanol	DPPH	Increased	Correlated with phenolics content	Kang and Saltveit 2002
	Methanol	FRAP	Increased	Correlated with phenolics content	
	Phosphate buffer	FRAP	Increased	Correlated with phenolics content	
Celery, lettuce, parsnips, carrot, white cabbage, sweet potato	Methanol	DPPH	Increased	Correlated with soluble phenolics content	Reyes et al. 2007
Purple-flesh potatoes	Methanol	DPPH	Increased	Correlated with total phenolic content	Reyes and Cisneros-Zevallos 2003
Mandarin	Diluted juice	DPPH	Decreased	Paralleled by decrease in ascorbic acid	Piga et al. 2002
Spinach	Methanol	DPPH	Decreased	Related to a decrease in ascorbic acid and flavonoids	Gil et al. 1999

Contd.

Fresh-cut produce	*Extract*	*Antioxidant method*	*Result*	*Comments*	*Reference*
Zucchini, potato, red cabbage	Methanol	DPPH	Decreased	Correlated with soluble phenolics content	Reyes et al. 2007
Papaya	Phosphate buffer	ORAC	Decreased	Associated with depletion of ascorbic acid	Rivera-Lopez et al. 2005
Celery	Methanol	DPPH	Decreased initially, followed by increase and later decrease		Vina and Chaves 2006
Tomato	Na-phosphate buffer	ABTS	Initial decrease followed by small increase		Lana and Tijskens 2006
Lemon in slices 5 mm thick, wedges, ½ slices or ¼ slices	Methanol	DPPH	Did not change or slightly decreased	Related to ascorbic acid changes but not to phenolics changes	Artes-Hernandez et al. 2007
Orange carrot		ORAC	Did not change		Alasalvar et al. 2005
Purple carrot		ORAC	Did not change		
Tomato	Water	Lipid peroxidation inhibition	Did not change	Pro-oxidant activity	Lana and Tijskens 2006
	Methanol	Lipid peroxidation inhibition	Did not change		
	THF, hexane and ethyl acetate	Lipid peroxidation inhibition	Did not change		

Contd.

Tomato	Ethyl acetate	ABTS	Did not change		Lana and Tijskens 2006
Celery	Methanol	DPPH	Did not change	Short-term storage (24 hr)	Vina and Chaves 2007
Cactus fruit	Diluted juice	DPPH	Did not change	Paralleled by no changes in ascorbic acid	Piga et al. 2003
Radish	Methanol	DPPH	Did not change	Correlated with soluble phenolics content	Reyes et al. 2007

The results presented in Table 1 and Fig. 1 indicate that the antioxidant activity can increase, decrease or remain unchanged after processing. Reyes et al. (2007) proposed that the changes in antioxidant content and activity after wounding are tissue dependent. The antioxidant activity of methanol extracts measured by DPPH (2,2-diphenyl-1-picrylhydrazyl) assay increased in response to wounding in lettuce, celery, carrot, parsnips, and sweet potato, while it decreased for red cabbage, potato, and zucchini and did not change significantly for radish (Fig. 1). Other authors found increase in antioxidant activity after minimal processing for lettuce

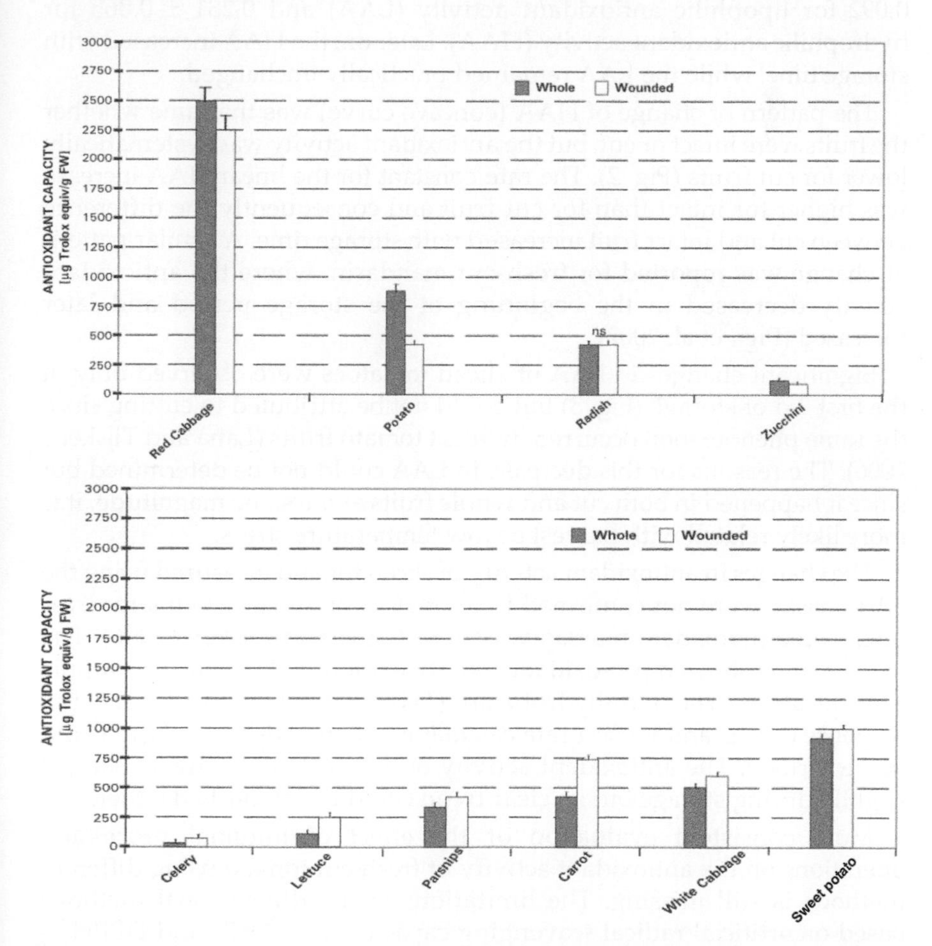

Fig. 1 Antioxidant capacity of unprocessed (whole) and minimally processed (wounded) vegetables after storage for 2 days at 15°C. Vertical lines represent one-sided standard deviation (n = 9). Non-significant differences in content after processing are marked with "ns" ($P > 0.05$). Reproduced from Reyes et al. (2007) with permission from Elsevier.

(Kang and Saltveit 2002) and potato (Reyes et al. 2007) and decrease for mandarin (Piga et al. 2002) and spinach (Gil et al. 1999). No changes were measured in the antioxidant activity of carrot (Alasalver et al. 2005), cactus (Piga et al. 2003), and lemon (Artes-Hernandez et al. 2007) after cutting.

The antioxidant activity of both the hydrophilic and the lipophilic extracts of sliced tomato stored at 5°C for 11 d was evaluated through ABTS (2,2'-azino-bis-(3-ethylbenzthiazoline-6-sulphonic acid) assay by Lana and Tijskens (2006). Both extracts showed the same exponential decay starting after processing, with a reference rate constant of 0.949 ± 0.092 for lipophilic antioxidant activity (LAA) and 0.281 ± 0.065 for hydrophilic antioxidant activity (HAA). Later on, the HAA increased with storage time, while the LAA remained practically unchanged.

The pattern of change of HAA (concave curve) was the same whether the fruits were intact or cut, but the antioxidant activity was systematically lower for cut fruits (Fig. 2). The rate constant for the linear HAA increase was higher for intact than for cut fruit and consequently the difference between cut and intact fruit increased with storage time. A similar pattern of change was reported for fresh-cut mandarin, where the antioxidant activity decreased in the beginning of the storage period and later increased (Piga et al. 2002).

Significant changes in LAA of sliced tomatoes were observed only in the first 3 d of storage (Fig. 3) but could not be attributed to cutting since the same phenomenon occurred in intact tomato fruits (Lana and Tijskens 2006). The reasons for this decrease in LAA could not be determined but since it happened in both cut and whole fruits at the same magnitude, it is more likely related with harvest or low temperature stress.

The changes in antioxidant activity of sliced tomato measured using the ABTS assay were not confirmed in another experiment when a rat liver microsome peroxidation assay was used (Lana and Tijskens 2006). The water extract showed pro-oxidant activity without a clear effect of sample concentration. The tetrahydrofuran (THF) extract showed very flat response curves and no accurate calculation of antioxidant activity could be performed. The antioxidant activity of the methanol extract changed slightly during storage but no clear trend could be established either.

More consistent evaluation of the effect of minimal processing operations on the antioxidant activity of fresh-cut tomato using different methods is still missing. The limitations of uni-dimensional methods based on artificial radical scavenging capacity such ABTS and DPPH to evaluate the health-promoting capacity of food as well as the difficulties in evaluating the antioxidant activity of lipophilic extracts are addressed in a later section. The importance of the antioxidant assay as much as the solvent used to extract the antioxidants can be gauged from the following

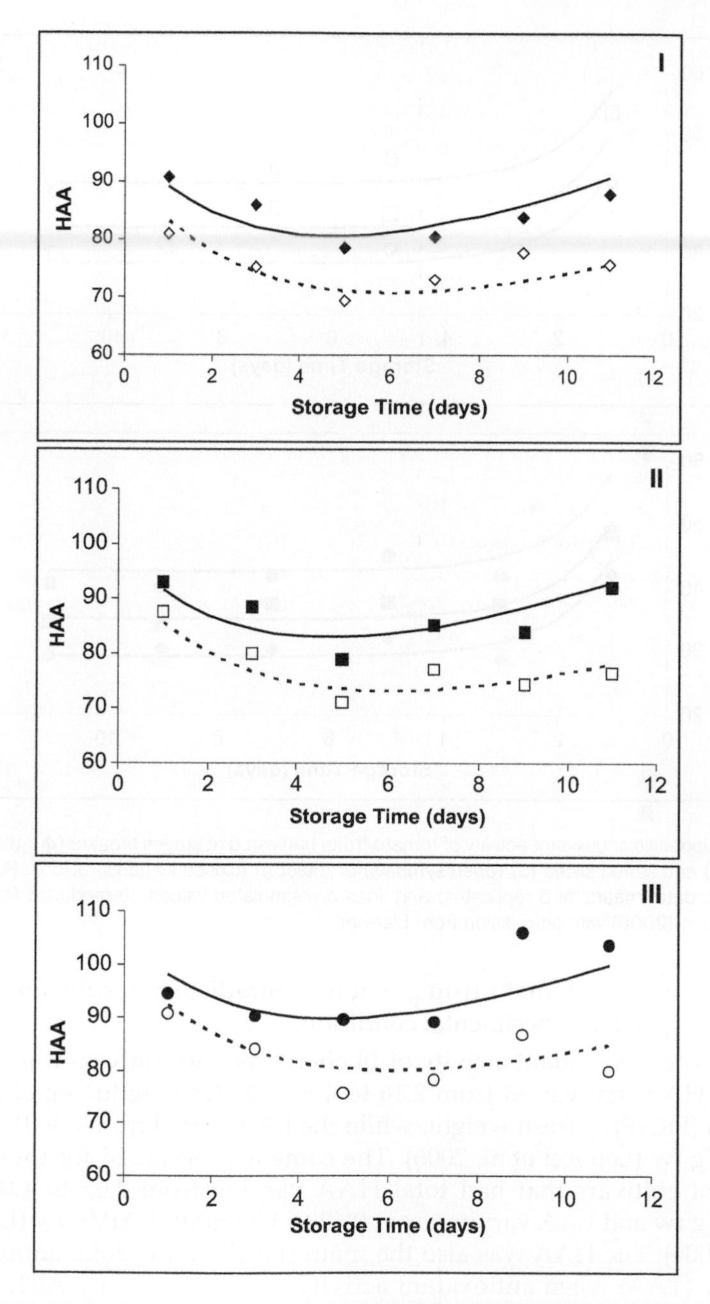

Fig. 2 Hydrophilic antioxidant activity of tomato fruits harvested at stages I (breaker), II (turning) or III (red) and stored sliced (open symbols and dotted line) or intact (closed symbols and solid lines) at 5°C. Points are measured data (means of 5 replicates) and lines are simulated values. Reproduced from Lana and Tijskens (2006) with permission from Elsevier.

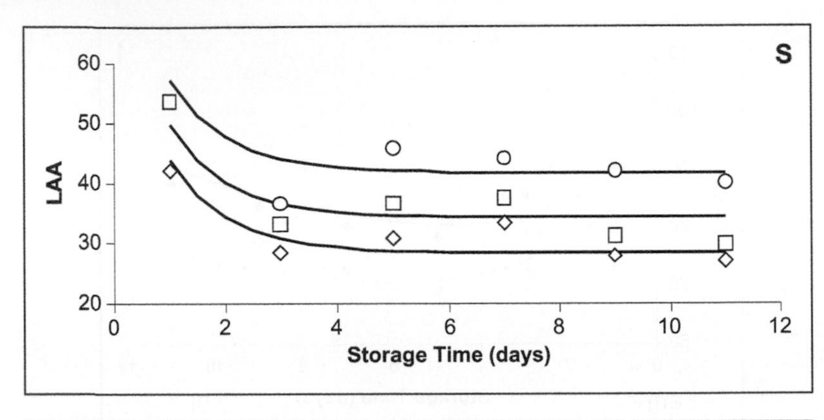

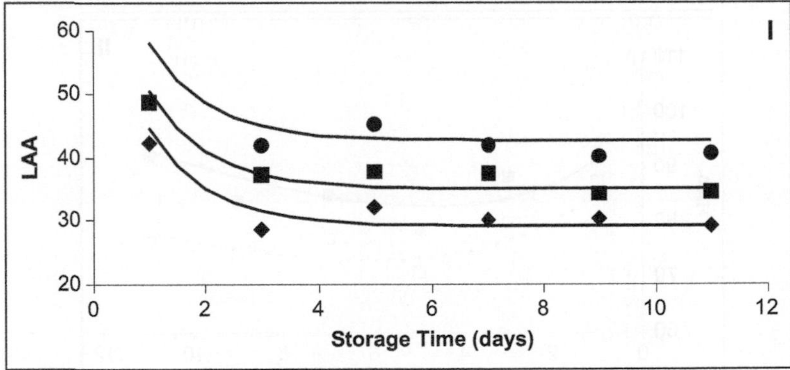

Fig. 3 Lipophilic antioxidant activity of tomato fruits harvested at stages breaker (◆), turning (■) or red (●) and stored sliced (S) (open symbols) or intact (I) (closed symbols) at 5°C. Points are measured data (means of 5 replicates) and lines are simulated values. Reproduced from Lana and Tijskens (2006) with permission from Elsevier.

reports for intact tomato fruits, where contradictory results are found depending on the experimental conditions.

The total antioxidant activity of 14 cherry tomato cultivars was mainly due to HAA that varied from 2.16 to 4.53 mM ferric reducing ability of plasma (FRAP)/g fresh weight, while the LAA varied from 0 to 0.71 mM FRAP/g fw (Lenucci et al. 2006). The same was observed for four high-pigment cultivars that had total HAA varying from 2.67 to 4.07 mM FRAP/g fw and LAA varying from 0.62 to 1.75 mM FRAP/g fw (Lenucci et al. 2006). The HAA was also the main contributor of total antioxidant activity (TAA) when antioxidant activity was evaluated by ABTS assay (Cano et al. 2003, Pellegrini et al. 2007, Toor and Savage 2005). This contribution varied from 71 to 93% in these reports.

With the Free Radical Quenching assay (FRQA) the activity of the lipophilic extract was much higher than that of the hydrophilic extract (George et al. 2004) and the ranking of several tomato genotypes was approximately the same in both extracts. With FRAP assay it was possible to measure the antioxidant activity of methanol but not of hexane extracts. Ranking tomato genotypes in the FRAP assay was different from ranking them in the FRQA (methanol extracts). Contrasting results depending on the model system used were also reported by Lavelli et al. (2000). The tomato lipophilic fraction, containing carotenoids, was not very effective in the xanthine oxidase-mediated reaction and in the neutrophil myeloperoxidase-mediated reaction systems but very effectively inhibited the copper-catalysed lipid peroxidation. This behaviour was contrary to that of the hydrophilic fraction.

Evaluation of the antioxidant activity of processed and stored products must include the evaluation of the intact product stored under the same conditions to avoid misinterpretation of data, since storage time and temperature can themselves induce changes in the antioxidant activity, as observed for intact tomato and apple fruit. The soluble phenolics and ascorbic acid contents of intact tomatoes harvested at light-red stage showed slight increases during storage parallel to an increase from 17 to 27% in the soluble antioxidant activity during storage regardless of temperature (Toor and Savage 2006). Long-term cold storage of apple fruit reduced the total phenolics concentration and TAA measured by DPPH assay (Tarozzi et al. 2004).

CHANGES IN ANTIOXIDANT CONTENT AND COMPOSITION IN RESPONSE TO WOUNDING STRESS AND ITS RELATION WITH ANTIOXIDANT ACTIVITY

Attempts to relate changes in antioxidant activity with changes in the content of individual antioxidant components are a common approach taken by many authors. In principle, this correlation will always be a partial one since the measurement of antioxidant activity takes into account interactions among individual components and the activity of "unknown compounds", both not measured in assays for the quantification of one individual known compound.

Studies on correlation content versus activity of antioxidants are then concentrated on those compounds believed to impart health benefits and to be of biological significance to humans, namely carotenoids, phenols and antioxidant vitamins such as C and E.

The results presented in Table 1 show that the changes in antioxidant activity of fresh-cut products are mainly related with changes in phenolic compounds or ascorbic acid. However, before concluding that in view of these results these are the two classes of compounds most affected by processing, one must take into consideration the limitations of the methods used (mainly radical scavenging methods) and consider that in the majority of the reports only aqueous or methanol extracts were used. This means that the activity of lipophilic antioxidants as important as carotenoids and α-tocopherol was not taken into account.

The biosynthesis of some phenols is clearly related to wound response (Kang and Saltveit, 2002, Lavelli et al. 2006, Tudela et al. 2002), while others can be degraded in response to stress (Vina and Chaves 2006). Changes in phenol composition can result in compounds with different antioxidant activities. With a similar increase in phenolics content (330% and 305% for Iceberg and Romaine lettuce respectively), there was an increase of 140% and 255% in the antioxidant activity. It appears that the wound-induced phenolics in Romaine have more antioxidant capacity than the phenolics in Iceberg, or that other compounds with different antioxidant activity at different concentrations in both lettuces influence the antioxidant activity (Kang and Saltveit 2002). With increased storage time, the antioxidant potential of a polyphenol-containing food can vary greatly because of progressive enzymatic or chemical oxidation. Polyphenols with an intermediate oxidation state can exhibit higher radical scavenging efficiency than non-oxidized ones (Piga et al. 2002).

Since phenols are part of a very large group of compounds, differences in individual phenols are expected (Table 2). Differences due to storage duration are also reported. Quercetin concentration was not significantly different between whole and chopped onions during storage for 11 d at 4°C, but higher quercetin concentration was found in chopped onions with 11-30 d storage (Martinéz et al. 2005). Chlorogenic acid content increased in cut celery in the first 24 hr after processing (Vina and Chaves 2007), while it decreased markedly after storage for 28 d (Vina and Chaves 2006).

Changes in total phenolics content was strongly correlated with the antioxidant capacity measured by DPPH (methanol extracts) and FRAP (phosphate buffered saline solution and methanol extracts) assays for many products but not for fresh-cut lemon (Table 1). The possibility was raised that other compounds with antioxidant activity could have influenced the antioxidant assay and/or that it would depend also upon synergistic effects among all water-soluble antioxidants and their interaction with other constituents of the fraction (Kang and Saltveit 2002, Lenucci et al. 2006).

Martinez-Valverde et al. (2002) used Pearson correlation to ascertain the relationship between individual antioxidants and antioxidant activity

Table 2 Effects of minimal processing operations (peeling, cutting, slicing or shredding) on antioxidant content of fruits and vegetables.

Product	*Antioxidant compound*	*Result*	*Comments*	*Reference*
Potato strips	Flavonol (Quercetin 3-glucosylrutinoside)	Increased	Lag time of 3 d	Tudela et al. 2002
	Flavonol (Quercetin 3-diglucoside)	Increased	Lag time of 3 d	
	Flavonol (Quercetin 3-rutinoside)	Increased	Lag time of 3 d	
	Total flavonols	Increased		
	Phenolic acids (Caffeic acid derivatives)	Increased		
Celery	Chlorogenic acid	Increased	Short-term storage (24 hr)	Vina and Chaves 2007
Shredded orange and purple carrot	Total phenolics	Increased	Purple carrots at a higher rate	Alasalvar et al. 2005
Lettuce	Total phenolics	Increased		Kang and Saltveit 2002
Potato	Total phenolics	Increased		Reyes and Cisneros-Zevallos 2003
Celery	Total flavonoids	Decreased		Vina and Chaves 2007
Celery	Chlorogenic acid	Decreased	Long-term storage (21 d)	Vina and Chaves 2006
Shredded red cabbage	Anthocyanin	Decreased		Reyes et al. 2007

Contd

Sliced lemon	Total phenols	Decreased after an initial increase	Decrease in wedges > slices > ½ slices, > ¼ slices	Artes-Hernandez et al. 2007
Sliced potato	Anthocyanin	Did not change or decreased	Decreased only in the peel of tubers exposed to light	Reyes and Cisneros-Zevallos 2003
Celery	Total phenols	Did not change		Vina and Chaves 2006, 2007
Spinach	Total flavonoids	Did not change		Gil et al. 1999
Shredded orange and purple carrot	Anthocyanins	Did not change	Only purple carrot	Alasalvar et al. 2005
Pumpkin – ½ slice, cubes and shredded	Ascorbic acid	Decreased	Decrease in shredded > cube > slice	Sasaki et al. 2006
Celery	Ascorbic acid	Transient increase at 14 d storage, decreasing to initial levels after 28 d		Vina and Chaves 2006
Shredded carrot, parsnips, celery, potato, zucchini, white cabbage	Reduced ascorbic acid	Decreased		Reyes et al. 2007
Shredded sweet potato, radish, red cabbage	Reduced ascorbic acid	Did not change		Reyes et al. 2007

Contd

Sliced lemon	Ascorbic acid and total vitamin C	Did not change		Artes-Hernández et al. 2007
Sliced lemon	Dehydroascorbic acid	Slight increase		Artes-Hernández et al. 2007
Sliced strawberry	Total ascorbic acid	Did not change		Wright and Kader 1997
	Reduced ascorbic acid	Initial decrease on day 1 followed by increase to levels close to initial		
	Dehydroascorbic acid	Increase on day 1 followed by decline		
Spinach	Vitamin C (AA + DHAA)	Decreased		Gil et al. 1999
Shredded orange and purple carrot	Total carotenoids	Decreased		Alasalvar et al. 2005, Martin-Diana et al. 2005
Cut pineapple, cantaloupe, mango and strawberry	Total carotenoids	Decreased	Spoilage occurred before significant losses of carotenoids	Gil et al. 2006
Pumpkin – ½ slice, cubes and shredded	Total carotenoids	Decreased	Decrease in shredded >> cube = slice	Sasaki et al. 2006
Shredded carrot, sweet potato	Total carotenoids	No changes		Reyes et al. 2007

Contd

Kiwi slices and watermelon cubes	Total carotenoids	No changes		Gil et al. 2006
Watermelon cubes	Lycopene	Decreased		Perkins-Veazie and Collins 2004
Sliced tomato	Lycopene	Increased or did not change depending on temperature		Lana et al. 2005
Watermelon cubes	β-carotene	No changes		Perkins-Veazie and Collins 2004
Papaya cubes and slices	β-carotene	Decreased	Decrease was significant after longer storage and dependent on temperature	Rivera-Lopez et al. 2005
Carrot sticks	β-carotene α-carotene	Initial small increase, followed by decrease	Returned to initial levels after 10 d	Lavelli et al. 2006

measured by DPPH and ABTS assay in tomato extracts. Lycopene, feluric acid and caffeic acid were found to be related with the antioxidant capacity, but not quercetin and chlorogenic acid. It must be considered, however, that only ripened fruits were used and the content of chlorogenic acid decreases with ripening so that its content in these fruits was low compared to that of other phenolic compounds.

The retention of ascorbic acid is dependent on the integrity of the tissue (Davey et al. 2000) and degradation of ascorbic acid is expected to contribute to a decrease in antioxidant activity. Differences between commodities in ascorbic acid degradation (Table 2) and in the relative importance of vitamin C in TAA (Table 1) could be due to its initial content, which varies greatly between and within species (Kalt 2005). It is also likely that some of the conflicting results are related with differences in extraction procedures and antioxidant assays.

The HAA of methanol extracts of tomato fruit (14 cherry cultivars and 4 high-lycopene cultivars) measured by FRAP assay showed a good linear correlation with ascorbic acid ($R^2 = 0.49$; $P < 0.001$) and was not correlated ($R^2 = 0.02$) with total phenolic content (Lenucci et al. 2006). This was not confirmed by Slimestad and Verheul (2005), who also used methanol extracts of cherry tomato and the FRAP assay. Changes in antioxidant activity were not correlated with changes in ascorbic acid and they seemed rather to be related with changes in the amount of phenols in general and especially that of chalconaringenin and chlorogenic acid in particular.

The antioxidant activity of tomato towards the ABTS radical using Na-phosphate buffer extract showed strong correlation with ascorbic acid ($R^2 = 0.90$) but no correlation with aqueous phenols (Cano et al. 2003). The water/methanol extracts using the same assay were not correlated with total flavonols (Caputo et al. 2004). When methanol extracts were used, the HAA was well correlated with total phenols ($R^2 = 0.7928$). When considering each polyphenol subclass, a good correlation was found for phenolic acids ($R^2 = 0.7796$), a low correlation for rutin and its analogue ($R^2 = 0.4685$), and no correlation for narigenin and its chalcone (Minoggio et al. 2003). Better extraction of phenols by methanol than by water (Yu et al. 2005) can partly explain the conflicting results reported by Cano et al. (2003) and Minoggio et al. (2003).

Decrease in the content of ascorbic acid of fresh-cut carrot, parsnips, celery, potato, zucchini and white cabbage did not affect the antioxidant activity measured by DPPH assay (Reyes et al. 2007), while for other products such as mandarin, spinach, lemon and cactus fruit the changes in antioxidant activity measured by the same DPPH assay reflected changes in ascorbic acid content (Table 1). The depletion of ascorbic acid in fresh-cut papaya was also associated with decrease in antioxidant activity by

ORAC (oxygen-radical absorbance capacity) assay (Rivera-Lopez et al. 2005).

Reyes et al. (2007) proposed that tissues with high levels of ascorbic acid are prepared to control the increase in radical oxygen species (ROS) induced by wounding; thus, the phenolics being synthesized are used for other purposes (e.g., lignin or suberin formation). Conversely, in tissues with low levels of reduced ascorbic acid, ascorbic acid is consumed readily and phenolics are possibly synthesized to partly control ROS. Since wounding could induce or decrease the synthesis of phenolic compounds in the tissues, it was hypothesized that the resulting antioxidant capacity of wounded produce would depend on the specific phenolic profile present. A change in the phenolic profile due to wounding would be reflected in changes in the antioxidant activity of the plant tissue.

The results reported by Lana and Tijskens (2006) are compatible with a situation in which minimal processing results in degradation of ascorbic acid and consequent reduction in HAA. The role of phenols in this equation is not clear at the moment, since Cano et al. (2003) reported that in the method used by both teams the correlation between aqueous phenols and HAA was not significant.

Changes in carotenoid content in fresh-cut tissues were expected in view of the instability of carotenoids when exposed to oxygen and light (Shi and Le Maguer 2000), a situation likely to occur when internal tissues are exposed after cutting. Wounding also promotes the production of wound ethylene (Brecht 1995), which hastens senescence, including the oxidation of fatty acids by lipoxygenase, during which carotenoids can be degraded by co-oxidation (Biacs and Daood 2000).

A summary of the most relevant results on retention of carotenoids in fresh-cut tissue is shown in Table 2. Decrease in total carotenoid content after minimal processing and during low-temperature storage was reported for carrot (Martin-Diana et al. 2005), watermelon (Perkins-Veazie and Collins 2004), papaya (Rivera-Lopez, 2005), pineapple, cantaloupe, mango and strawberry (Gil et al. 2006). Under different conditions no total carotenoid loss was observed during storage of fresh-cut carrots at 4 and 10°C (Lavelli et al. 2006). On the contrary, a small increase happened at 4°C after 3 d of storage, followed by a small decrease. During storage at 5°C for 9 d, no losses in carotenoid were found in kiwi slices (Gil et al. 2006).

Not all carotenoids are equally stable in fresh-cut tissues. In fresh-cut watermelon stored at 2°C for 10 d, reduction of lycopene content was observed but no changes in β-carotene (Perkins-Veazie and Collins 2004).

Changes in individual and total carotenoids of fresh-cut tissues are also expected to depend on storage temperature. Tomato slices stored at temperature varying from 8 to 16°C showed net increases of lycopene concentration during storage. At lower temperature down to 2°C, no increase or a small decrease was observed (Lana et al. 2005).

The LAA of hexane extracts measured by FRAP assay was correlated with lycopene content (R^2 = 0.45) and with the sum of all lipophilic antioxidants (lycopene, α-tocopherol, β-carotene) (R^2 = 0.46) (Lenucci et al. 2006). The LAA of tomato towards the ABTS radical paralleled the total amount of carotenoids, mainly lycopene, in THF extracts (Caputo et al. 2004), dichloromethane extracts (Raffo et al. 2002) and ethyl-acetate extracts (Cano et al. 2003). No clear trend was observed between LAA and α-tocopherol (Raffo et al. 2002) or between LAA and organic phenols (Cano et al. 2003).

The results reported by Lana and Tijskens (2006) are compatible with a situation in which minimal processing of tomato resulted in preservation of lycopene and no changes in LAA compared with intact fruit. A similar situation was reported for fresh-cut and intact watermelon that presented the same lycopene content, which decreased equally with increased storage time (Gil et al. 2006). As observed for sliced tomato, this decrease could not be ascribed to a processing effect.

STAGE OF MATURITY AND ANTIOXIDANT ACTIVITY

If minimal processing operations result in changes in antioxidant activity, it is expected that this will depend on the stage of maturity of the fruit, since its chemical composition will change with maturation and ripening.

Changes in antioxidant activity during ripening of tomato fruits were reported by Cano et al. (2003), Giovanelli et al. (1999), Jimenez et al. (2002), Lana and Tijskens (2006) and Raffo et al. (2002).

On the basis of results obtained for oxidative stress and antioxidant enzyme systems during maturation of two tomato cultivars, Mondal et al. (2004) proposed that tomato fruits respond to progressive increase in oxidative stress during earlier stages of maturation by increasing the activities of scavenging enzymes as well as the concentration of ROS scavenging compounds. However, at later stages, the ROS scavenging system does not cope with the production system, leading to the accumulation of ROS.

Antioxidant activity measured as radical scavenging capacity towards ABTS radical increased during ripening in both lipophilic (THF extracts) and hydrophilic (aqueous/methanol extracts) fractions (Caputo et al. 2004, Lana and Tijskens, 2006). The decrease in TAA from deep red stage to overripe stage was possibly associated with antioxidant depletion caused by fruit defence mechanisms against ROS, which are produced in large amounts during the climacteric (Caputo et al. 2004).

Similar results for LAA using ABTS assay were reported by Cano et al. (2003) and Raffo et al. (2002) and were related to changes in carotenoid content, mainly lycopene.

Contrary to what was reported by Lana and Tijskens (2006) and by Caputo et al. (2004), no change in HAA during ripening of tomato was observed by Cano et al. (2003) when using the same ABTS assay (Table 3) and the same extraction procedure as Lana and Tijskens (2006). In another assay, crocin bleaching inhibition test, water-soluble tomato extracts exhibited a slight decline during ripening (Raffo et al. 2002). This trend could be only partly attributed to changes in phenolics and ascorbic acid content, and it was not unequivocally related to any single component. It is plausible that the TAA depended also upon synergistic effects among all water-soluble antioxidants and their interactions with other constituents of the fraction.

Table 3 Total (TAA), hydrophilic (HAA) and lipophilic (LAA) antioxidant activities at successive ripening stages of on-vine tomatoes, expressed as μmol Trolox × 100 g^{-1} fw. Reproduced from Cano et al. (2003), with permission from Elsevier.

Ripening stage	*HAA measured*	*HAA[a] calculated*	*LAA measured*	*LAA[a] calculated*	*TAA[c]*
Green	195 a	129 a	33 a	12 a	228 a
Breaker	211 a	242 b	65 b	28 b	276 b
Pink	192 a	214 b	79 bc	57 c	271 b
Red-ripe	218 a	304 c	88 c	81 d	306 c

[a] HAA calculated as TEAC ascorbic acid (1) × [ascorbic acid] + TEAC gallic acid (3) × [amount of aqueous phenols].

[b] LAA calculated as TEAC β-carotene (2.5) × [β-carotene] + TEAC lycopene (3) × [lycopene + TEAC gallic acid (3) × [amount of organic phenols].

[c] Values followed by different letters within a column are significantly different at $P < 0.05$.

The effect of stage of maturity was more pronounced in LAA than HAA although TAA was mainly due to HAA at all stages (Table 3; Lana and Tijskens 2006).

Changes in antioxidant activity due to ripening are likely to be different whether the fruits ripen on vine or off vine (Giovanelli et al. 1999). Tomatoes that are picked at mature-green stage and ripened off vine developed greater levels of antioxidants (lycopene, β-carotene and ascorbic acid) than fruits picked at the full ripe stage. The synthesis of lycopene and β-carotene was linear with maturity for vine-ripened fruits, while for off-vine fruits it was exponential. In post-harvest tomatoes, ascorbic acid showed an initial decrease followed by a considerable increase in the last stages (upward concave curve), while in vine-ripened tomatoes, ascorbic acid accumulated during the first stages and then decreased. At later stages of maturity the ascorbic acid content was slightly higher for fruits ripened off vine. The two ripening conditions also gave different pattern of phenolic accumulation. Total phenolics on vine-ripened fruits showed a small linear decrease with ripening, while off vine it first increased, then decreased and increased again at later stages of maturity. The total phenolics content was higher for fruits ripened off vine.

When the antioxidant activity was measured as inhibition of lipid peroxidation in rat liver microsomes, the antioxidant activity of methanol extracts increased from mature-green stage to turning stage and decreased again at red stage (Lana and Tijskens 2006). No clear trend could be established for THF and water extracts in the same assay.

INTERACTION BETWEEN PROCESSING AND MATURITY STAGE IN THE CASE OF TOMATOES

An interaction between the effects of stage of maturity and minimal processing was expected in view of the changes that happen in the antioxidant content along maturation and ripening of the tomato fruit. However, differences due to stage of maturity, for both sliced and whole tomato, were mainly due to differences in the initial antioxidant activity when the fruits were processed, since the same pattern of changes during storage was observed for all stages (Lana and Tijskens 2006). The interaction between processing and maturity was not significant for LAA ($Pr > F = 0.8299$) and marginally significant for HAA ($Pr > F = 0.0732$). When the effects of time, processing and maturity were considered simultaneously by multiple non-linear regression, the parameters for HAA model could be estimated in common for all maturity stages in whole and cut fruit, confirming that a possible interaction between maturity stage and processing was not relevant in this case.

MEASUREMENT OF ANTIOXIDANT ACTIVITY—LIMITATIONS OF THE MOST COMMONLY USED METHODOLOGIES

The impact of processing on the amount and composition of antioxidant compounds of fruits and vegetables can be evaluated through the assessment of individual components known to be important antioxidants in each particular product (Table 2). This procedure can be cumbersome when there are many compounds of interest, each one evaluated by a different method. To measure TAA is a way to overcome this limitation with the additional benefit of taking into consideration synergistic interactions among antioxidants and the action of other compounds whose antioxidant activity is unknown. Total antioxidant activity is particularly useful in obtaining an overall picture of relative antioxidant activity in food and plant material and how it changes after processing or storage (Halliwell 2002).

There are numerous antioxidant assays to measure antioxidant activity but there are no approved standardized methods (Frankel and Meyer 2000). The advantages and drawbacks of the most commonly used assays have been the subject of numerous reviews (Arnao 2000, Aruoma, 2003, Frankel and Meyer, 2000, Halliwell 2002, Pellegrini et al. 2003, Roginsky and Lissi 2005).

Radical scavenging methods are the most widely used because they are sensitive and easy to perform. They involve the generation of radical species *in vitro* followed by their removal from the medium by the antioxidant present in the sample and include DPPH assay (Jimenéz-Escrig et al. 2000) ABTS assay (Cano et al. 2000), FRAP and TRAP assays (Pellegrini et al. 2003) and ORAC (Cao et al. 1993).

More complex methods involve testing of antioxidants in food and biological systems where a lipid or lipoprotein substrate is oxidized under standard conditions and the inhibition of oxidation after addition of the antioxidant is measured (Chang et al. 2000, Frankel and Meyer 2000, Sluis et al. 2000). Although closer to the situation experienced *in vivo*, these methods also present some drawbacks related to partitioning properties of the antioxidants between lipid and aqueous phase, oxidation conditions and physical state of the oxidizable substrate (Frankel and Meyer 2000).

As is shown in Table 1, the oxidant activity of fresh-cut products was mainly evaluated by radical scavenging methods. These results must be interpreted cautiously, since the ability of antioxidants to scavenge an artificial radical may not reflect the antioxidant activity *in vivo* or its action against physiologically relevant radicals (Frankel and Meyer 2000, Roginsky and Lissi 2005). In practice, these methods measure the H-donating activity, which does not always correlate with chain-breaking antioxidant activity (Roginsky and Lissi 2005). Although very sensitive, they neglect important compositional and interfacial phenomena concerning charge and solubility of multiple components in real food or biological systems that strongly affect antioxidant performance (Frankel and Meyer 2000). As a rule, the different antioxidant activity methods are poorly correlated with each other (Roginsky and Lissi 2005). Even methods based on the same principle such as DPPH and ABTS differ in their selectivity in the reaction with H-donors and consequently give different results for the same antioxidant (Martinez-Valverde et al. 2002, Wilborts 2003).

Another aspect that deserves attention refers to the antioxidant extraction procedure. Extraction of tomato antioxidants with a sequence of water and acetone was not effective in the case of foods with a high content of lipophilic compounds such as lycopene and β-carotene. An additional extraction with chloroform did not increase the recovery of these lipophilic compounds, which was equal to 20% of the total carotenoid extracted by THF (Pellegrini et al. 2007). Methanol extracts are predominant in the assays listed in Table 1 and a poor recovery of lipophilic antioxidants is expected in this situation.

Even within the hydrophilic antioxidant group, remarkable differences in antioxidant activity can be measured depending on the solvent or reaction medium used. For example, feluric acid, which showed high correlation with the antioxidant activity in aqueous tomato extracts, is 150% more efficient as antioxidant than caffeic and chlorogenic acid in aqueous phase (others by Martinez-Valverde et al. 2002).

The importance of the lipid substrate should also be taken into consideration, as shown from the results of Takeoka et al. (2001). The lycopene-containing hexane extract of fresh tomatoes presented greater antioxidant activity than the aqueous and methanol extracts when phosphatidylcholine liposome was used as substrate for oxidation but about half the activity of methanol extracts when linoleic acid was used as substrate.

Also, for HAA measurements, the substrate for oxidation was shown to interfere in the measured antioxidant activity. Hydrophilic extracts from tomato samples showed pro-oxidant action in a copper-catalysed linoleic acid oxidation assay (Lavelli et al. 2000). In another experiment, tomato aqueous extracts showed antioxidant action in a copper-catalysed phosphatidylcholine liposome oxidation assay but no activity in a linoleic acid emulsion oxidation induced by endoperoxide (Takeoka et al. 2001). The pro-oxidant action reported by Lavelli et al. (2000) was considered to be due to the presence of ascorbic acid and its action activating metal-ion catalysed reactions, which probably also happened in the conditions of the rat liver microsome peroxidation inhibition assay where ascorbic acid was used as oxidative inducer together with iron sulphate (Lana and Tijskens 2006).

The effect of cutting, fruit maturity stage and storage time on the antioxidant activity of tomato extracts was not the same in two different antioxidant assays (Lana and Tijskens 2006). Although different solvents were used in each assay, part of the compounds extracted by the Na-P buffer in the ABTS assay should be present in the water and methanol extracts used in the rat liver microsome peroxidation assay. Similarly, part of the compounds extracted by ethyl acetate in the ABTS assay should be present in methanol and THF extracts of the lipid peroxidation assay. The lack of antioxidant activity in the THF extract in the lipid peroxidation assay was highly unexpected, since this extract contains lycopene, which is known to be an efficient scavenger of peroxyl radicals (Woodall et al. 1997). However, Chen and Djuric (2001) questioned whether carotenoids can act as antioxidants in biological membranes and suggested that although carotenoids are sensitive to degradation by free radicals they do not protect against lipid peroxidation. The possibility that THF, a strong organic solvent, disturbed the biological system remains to be investigated.

Contrasting results between radical scavenging methods and *in vitro* lipid peroxidation methods were discussed in an extensive review by Frenkel and Meyer (2000) and reported by Garcia-Alonso et al. (2004). This apparently contradictory behaviour occurs because the effectiveness of antioxidants depends on the test system used (Frankel and Meyer 2000). Lipid peroxidation methods include mechanisms of antioxidant action besides radical scavenging. On the other hand, partitioning and interfacial properties are not taken into consideration in radical scavenging assays, while they represent an important factor in biological assays.

CHANGES IN ANTIOXIDANT ACTIVITY AS A RESULT OF PROCESSING—IMPLICATIONS FOR HUMAN HEALTH

Data on changes in antioxidant activity and their significance in diet and human health must be interpreted with care in view of the limitations of the methods for antioxidant activity measurement, as discussed in the previous section. Additionally, it is possible for an antioxidant to protect in one biological system, but fail to protect (or even sometimes to promote damage) in others (Halliwell 2002).

Measurement of TAA alone is probably not enough to conclude about the healthiness of the food. It is necessary to investigate whether an increase in HAA could, for example, be due to the increase in phenolic compounds with adverse effects on health (Ohshima et al. 1998). On the other hand, *in vitro* and animal studies indicate a potential for phenolics acids to have beneficial cardiovascular effects independent of their antioxidant activity (Morton et al. 2000).

An important point, generally neglected in the evaluation of the effect of minimal processing on the antioxidant activity and composition of fruits and vegetables, is whether the gain in one individual or in one class of antioxidants can compensate the loss of others. In fresh-cut carrot there was an increase in antioxidant activity of 77% and an increase in total phenol content of 191% (Reyes et al. 2007). For this same product there was a decrease in ascorbic acid of 82% in relation to initial content after 48 hr at 15°C. Is it right to conclude that the increase in phenolic content at the expense of a decrease in vitamin C is good for health? Vitamin C is an essential component of the diet, since humans are not able to synthesize ascorbic acid and virtually all the ascorbic acid in the western diet is derived from fruits and vegetables (Davey et al. 2000). On the other hand, the role of dietary antioxidants in the prevention of chronic and degenerative diseases is still a subject of debate (Astley 2003, Buttriss et al. 2002, Kaur and Kapoor 2001, Morton et al. 2000, Ohshima et al. 1998, Rietjens et al. 2002, Shvedova et al. 2000).

More consistent results showing that in fact the health-promoting activity of fruits can be reduced after harvest were shown not for fresh-cut but for intact apple fruit. Six-month storage markedly decreased the bioactivity of apples measured as the levels of intracellular antioxidant, cytoprotective and antiproliferative activity in Caco-2 cells (Tarozzi et al. 2004). In fresh-cut fruits that present a very short shelf-life the fruits commonly spoil visibly before significant nutrient loss occurs (Gil et al. 2006).

The results obtained for sliced tomato indicate that processing can induce a reduction of radical scavenging capacity, which is only one of the mechanisms by which antioxidants can prevent oxidative damage (Lana and Tijskens, 2006). The fact that an artificial radical was used should be kept in mind while extrapolating the results to an *in vivo* situation.

Although a minor decrease in antioxidant activity was measured in sliced tomato as a response to processing, at the moment it is not possible to know the impact of this decrease in the value of tomato as source of antioxidant in the diet.

CONCLUSIONS

The studies outlined in the present review show that inconsistent results have been obtained for different fresh-cut products and many times for a single product, depending on the system composition and analytical methods used to measure antioxidant activity. There is a predominance of *in vitro* radical scavenging methods, the results of which cannot be directly extrapolated to an *in vivo* situation.

It is clear that minimal processing operations induce changes in the content and composition of fresh-cut products, but it is not yet clear what is the impact of these changes in the total antioxidant activity of the tissue or in the value of this product as a source of antioxidants and other bioactive compounds in the human diet.

Studies with tomato fruit have concentrated on the impact of processing involving thermal operations other than minimal processing. The current knowledge on fresh-cut tomatoes indicates that processing tomato fruit into transversal slices induces a decrease in the radical scavenging capacity of hydrophilic extracts towards the ABTS radical. This indicates a potential effect of processing to decrease the antioxidant activity *in vivo*. This would represent a decrease in the value of cut tomatoes as a source of antioxidants in the diet compared with the fruit stored intact. The observed increase in hydrophilic antioxidant activity at the end of the storage period indicates that some repair or recycling mechanism is operative. This same mechanism is present in intact fruits, since they showed the same pattern of change as sliced fruits. The radical scavenging capacity of lipophilic extracts, rich in lycopene, was not affected by cutting. Since no relevant interaction was found between storage time and stage of maturity, the major factor determining the level of antioxidants in sliced tomatoes seems to be the initial level of antioxidant activity at the moment of harvest.

ABBREVIATIONS

ABTS: 2,2′-azino-bis-(3-ethylbenzthiazoline-6-sulphonic acid); DPPH: 2,2-diphenyl-1-picrylhydrazyl; FRAP: ferric reducing ability of plasma; FRQA: free radical quenching assay; FW: fresh weight; HAA: hydrophilic antioxidant activity; LAA: lipophilic antioxidant activity; ORAC: oxygen-radical absorbance capacity; ROS: radical oxygen species; TAA: total antioxidant activity; THF: tetrahydrofuran

REFERENCES

Alasalvar, C. and M. Al-Farsi, P.C. Quantick, F. Shahidi, and R. Wiktorowicz. 2005. Effect of chill storage and modified atmosphere packaging (MAP) on antioxidant activity, anthocyanins, carotenoids, phenolics and sensory quality of ready-to-eat shredded orange and purple carrots. Food Chem. 89: 69-76.

Arnao, M.B. 2000. Some methodological problems in the determination of antioxidant activity using chromogen radicals: a practical case. Trends Food Sci. Tech. 11: 419-421.

Artes-Hernandez, F. and F. Rivera-Cabrera, and A.A. Kader. 2007. Quality retention and potential shelf-life of fresh-cut lemons as affected by cut type and temperature. Postharvest Biol. Tech. 43: 245-254.

Aruoma, O.I. 2003. Methodological considerations for characterizing potential antioxidant actions of bioactive components in plant foods. Mutat. Res. Fund. Mol. M. 523-524: 9-20.

Astley, S.B. 2003. Dietary antioxidants—past, present and future? Trends Food Sci. Tech. 14: 93-98.

Biacs, P.A. and H.G. Daood. 2000. Lipoxygenase-catalysed degradation of carotenoids from tomato in the presence of antioxidant vitamins. Biochem. Soc. T. 28: 839-845.

Brecht, J.K. 1995. Physiology of lightly processed fruits and vegetables. HortScience 30: 18-22.

Buttriss, J.L. and J. Hughes, C.N.M. Kelly, and S. Stanner. 2002. Antioxidants in food: a summary of the review conducted for the Food Standards Agency. Nutr. Bull. 27: 227-236.

Cano, A. and M. Acosta, and M.B. Arnao. 2000. A method to measure antioxidant activity in organic media: application to lipophilic vitamins. Redox Rep. 5: 365-370.

Cano, A. and M. Acosta, and M.B. Arnao. 2003. Hydrophilic and lipophilic antioxidant activity changes during on-vine ripening of tomatoes (*Lycopersicon esculentum* Mill.). Postharvest Biol. Tech. 28: 59-65.

Cao, G. and H.M. Alessio, and R.G. Cutler. 1993. Oxygen-radical absorbance capacity assay for antioxidants. Free Radic. Bio. Med. 14: 303-311.

Caputo, M. and M.G. Sommella, G. Graziani, I. Giordano, V. Fogliano, R. Porta, and L. Mariniello. 2004. Antioxidant profiles of Corbara small tomatoes during ripening and effects of aqueous extracts on J774 cell antioxidant enzymes. J. Food Biochem. 28: 1-20.

Chang, S. and C. Tan, E.N. Frankel, and D.M. Barrett. 2000. Low-density lipoprotein antioxidant activity of phenolic compounds and polyphenol oxidase activity in selected clingstone peach cultivars. J. Agric. Food Chem. 48: 147-151.

Chen, G. and Z. Djuric. 2001. Carotenoids are degraded by free radicals but do not affect lipid peroxidation in unilamellar liposomes under different oxygen tensions. FEBS Lett. 505: 151-154.

Davey, M.W. and M.V. Montagu, D. Inzé, M. Sanmartin, A. Kanellis, N. Smirnoff, I.J.J. Benzie, J.J. Strain, D. Favell, and J. Fletcher. 2000. Plant L-ascorbic acid: chemistry, function, metabolism, bioavailability and effects of processing. J. Sci. Food Agric. 80: 825-860.

Dewanto, V. and X. Wu, K.K. Adom, and R.H. Liu. 2002. Thermal processing enhances the nutritional value of tomatoes by increasing total antioxidant activity. J. Agric. Food Chem. 50: 3010-3014.

[FAO/WHO] Workshop on Fruits and Vegetables for Health. 2005. Fruits and vegetables for health: Report of a Joint FAO/WHO Workshop, 2004, Kobe, Japan, pp 38.

Frankel, E.N. and A.S. Meyer. 2000. The problems of using one-dimensional methods to evaluate multifunctional food and biological antioxidants. J. Sci. Food Agric. 80: 1925-1941.

Garcia-Alonso, M. and S. de Pascual-Teresa, C. Santos-Buelga, and J.C. Rivas-Gonzalo. 2004. Evaluation of the antioxidant properties of fruits. Food Chem. 84: 13-18.

George, B. and C. Kaur, D.S. Khurdiya, and H.C. Kapoor. 2004. Antioxidants in tomato (*Lycopersicum esculentum*) as a function of genotype. Food Chem. 84: 45-51.

Gil, M.I. and F. Ferreres, and F.A. Tomas-Barberan. 1999. Effect of postharvest storage and processing on the antioxidant constituents (flavonoids and vitamin C) of fresh-cut spinach. J. Agric. Food Chem. 47: 2213-2217.

Gil, M.I. and E. Aguayo, and A.A. Kader. 2006. Quality changes and nutrient retention in fresh-cut versus whole fruits during storage. J. Agric. Food Chem. 54: 4284-4296.

Giovanelli, G. and V. Lavelli, C. Peri, and S. Nobili. 1999. Variation in antioxidant components of tomato during vine and post-harvest ripening. J. Sci. Food Agric. 79: 1583-1588.

Giovannucci, E. 1999. Tomatoes, tomato-based products, lycopene, and cancer: review of the epidemiologic literature. J. Natl. Cancer Inst. 91: 317-331.

Grassmann, J. and S. Hippeli, and E.F. Elstner. 2002. Plant's defence and its benefits for animals and medicine: role of phenolics and terpenoids in avoiding oxygen stress. Plant Physiol. Biochem. 40: 471-478.

Halliwell, B. Food Derived antioxidants: how to evaluate their importance in food and in vivo. pp. 145 *In*: Cadinas, E. and Packer, L. [ed.] 2002. Handbook of Antioxidants: M. Dekker, New York, USA.

Jiménez-Escrig, A. and I. Jiménez-Jiménez, C. Sánchez-Moreno, and F. Saura-Calixto. 2000. Evaluation of free radical scavenging of dietary carotenoids by the stable radical 2,2-diphenyl-1-picrylhydrazyl. J. Sci. Food Agric. 80: 1686-1690.

Jimenez, A. and G. Creissen, B. Kular, J. Firmin, S. Robinson, M. Verhoeyen, and P. Mullineaux. 2002. Changes in oxidative processes and components of the antioxidant system during tomato fruit ripening. Planta 214: 751-758.

Kalt, W. 2005. Effects of production and processing factors on major fruit and vegetable. J. Food Sci. 70: R11-R19.

Kang, H.M. and M.E. Saltveit. 2002. Antioxidant capacity of lettuce leaf tissue increases after wounding. J. Agric. Food Chem. 50: 7536-7541.

Kaur, C. and H.C. Kapoor. 2001. Antioxidants in fruits and vegetables—the millennium's health. Intl. J. Food Sci. Tech. 36: 703-725.

Klein, B.P. and A.C. Kurilich. 2000. Processing effects on dietary antioxidants from plant foods. HortScience 34: 580-584.

Klerk, M. and P. van't. Veer, and F.J. Kok. 1998. Fruits and Vegetables in Chronic Disease Prevention. LUW, Wageningen, the Netherlands, 86 pp.

Lana, M.M. and L.M.M. Tijskens. 2006. Effects of cutting and maturity on antioxidant activity of fresh-cut tomatoes. Food Chem. 97: 203-211.

Lana, M.M. and M. Dekker, P. Suurs, R.F.A. Linssen, and O. van Kooten. 2005. Effects of cutting and maturity on lycopene content of fresh-cut tomatoes during storage at different temperatures. Acta Hort. 682: 1871-1877.

Lavelli, V. and G. Giovanelli. 2003. Evaluation of heat and oxidative damage during storage of processed tomato products. II. Study of oxidative damage indices. J. Sci. Food Agric. 83: 966-971.

Lavelli, V. and C. Peri, and A. Rizzolo. 2000. Antioxidant activity of tomato products as studied by model reactions using xanthine oxidase, myeloperoxidase, and copper-induced lipid peroxidation. J. Agric. Food Chem. 48: 1442-1448.

Lavelli, V. and E. Pagliarini, R. Ambrosoli, J.L. Minati, and B. Zanoni. 2006. Physicochemical, microbial, and sensory parameters as indices to evaluate the quality of minimally-processed carrots. Postharvest Biol. Tech. 40: 34-40.

Lenucci, M.S. and D. Cadinu, M. Taurino, G. Piro, and G. Dalessandro. 2006. Antioxidant composition in cherry and high-pigment tomato cultivars. J. Agric. Food Chem. 54: 2606-2613.

Liu, R.H. 2003. Health benefits of fruit and vegetables are from additive and synergistic combinations of phytochemicals. Amer. J. Clin. Nutr. 78: 517S-520S.

Martin-Diana, A.B. and D. Rico, C. Barry-Ryan, J.M. Frias, J. Mulcahy, and G.T.M. Henehan. 2005 Comparison of calcium lactate with chlorine as a washing treatment for fresh-cut lettuce and carrots: quality and nutritional parameters. J. Sci. Food Agric. 85: 2260-2268.

Martinéz, J.A. and S. Sgroppo, C. Sanchéz-Moreno, B. De Ancos, and M.P. Cano. 2005. Effects of processing and storage of fresh-cut onion on quercetin. Acta Hort. 682: 1889-1894.

Martinez Valverde, I. and M.J. Periago, G. Provan, and A. Chesson. 2002. Phenolic compounds, lycopene and antioxidant activity in commercial varieties of tomato (*Lycopersicum esculentum*). J. Sci. Food Agric. 82: 323-330.

Minoggio, M. and L. Bramati, P. Simonetti, C. Gardana, L. Iemoli, E. Santangelo, P.L. Mauri, P. Spigno, G.P. Soressi, and P.G. Pietta. 2003. Polyphenol pattern and antioxidant activity of different tomato lines and cultivars. Ann. Nutr. Metab. 47: 64-69.

Mondal, K. and N.S. Sharma, S.P. Malhotra, K. Dhawan, and R. Suingh. 2004. Antioxidant systems in ripening tomato fruits. Biol. Plantarum 48: 49-53.

Morton, L.W. and R.A.A. Caccetta, I.B. Puddey, and K.D. Croft. 2000. Chemistry and biological effects of dietary phenolic compounds: relevance to cardiovascular disease. Clin. Exp. Pharmacol. P. 27: 152-159.

Nicoli, M.C. and M. Anese, and M. Parpinel. 1999. Influence of processing on the antioxidant properties of fruit and vegetables. Trends Food Sci. Tech. 10: 94-100.

Ohshima, H. and Y. Yoshie, S. Auriol, and I. Gilibert. 1998. Antioxidant and pro-oxidant actions of flavonoids: effects on DNA damage induced by nitric oxide, peroxynitrite and nitroxyl anion. Free Radic. Biol. Med. 25: 1057-1065.

Pellegrini, N. and M. Serafini, B. Colombi, D.de Rio, S. Salvatore, M. Bianchi, and F. Brighenti. 2003. Total antioxidant capacity of plant foods, beverages and oils consumed in Italy assessed by three different in vitro assays. J. Nutr. 133: 2812-2819.

Pellegrini, N. and B. Colombi, S. Salvatore, O.V. Brenna, G. Galaverna, D.D. Rio, M. Bianchi, R.N. Bennett, and F. Brighenti. 2007. Evaluation of antioxidant capacity of some fruit and vegetable foods: efficiency of extraction of a sequence of solvents. J. Sci. Food Agric. 87: 103-111.

Perkins-Veazie, P. and J.K. Collins. 2004. Flesh quality and lycopene stability of fresh-cut watermelon. Postharvest Biol. Tech. 31: 159-166.

Piga, A. and M. Agabbio, F. Gambella, and M.C. Nicoli. 2002. Retention of antioxidant activity in minimally processed mandarin and satsuma fruits. Lebensm. Wiss. Technol. 35: 344-347.

Piga, A. and A.D. Caro, I. Pinna, and M. Agabbio. 2003. Changes in ascorbic acid, polyphenol content and antioxidant activity in minimally processed cactus pear fruits. Lebensm. Wiss. Technol. 36: 257-262.

Prior, R.L. and G. Cao. 2000. Antioxidant phytochemicals in fruits and vegetables: diet and health implications. HortScience 35: 588-592.

Raffo, A. and C. Leonardi, V. Fogliano, P. Ambrosino, M. Salucci, L. Gennaro, R. Bugianesi, F. Giuffrida, and G. Quaglia. 2002. Nutritional value of cherry tomatoes (*Lycopersicon esculentum* Cv. Naomi F1) harvested at different ripening stages. J. Agric. Food Chem. 50: 6550-6556.

Rao, A.V. and S. Agarwal. 2000. Role of antioxidant lycopene in cancer and heart disease J. Amer. Coll. Nutr. 19: 563-569.

Reyes, L.F. and L. Cisneros-Zevallos. 2003. Wounding stress increases the phenolic content and antioxidant capacity of purple-flesh potatoes (*Solanum tuberosum* L.). J. Agr. Food Chem. 51: 5296-5300.

Reyes, L.F. and J.E. Villarreal, and L. Cisneros-Zevallos. 2007. The increase in antioxidant capacity after wounding depends on the type of fruit or vegetable tissue. Food Chem. 101: 1254-1262.

Rietjens, I.M.C.M. and M.G. Boersma, L.de Haan, B. Spenkelink, H.M. Awad, N.H.P. Cnubben, J.J. van Zanden, H. van der Woude, G.M. Alink, and J.H. Koeman. 2002. The pro-oxidant chemistry of the natural antioxidants vitamin C, vitamin E, carotenoids and flavonoids. Environ. Toxicol. Pharmacol. 11: 321-333.

Rivera-Lopez, J. and F.A. Vazquez-Ortiz, J.F. Ayala-Zavala, R.R. Sotelo-Mundo, and G.A. Gonzalez-Aguilar. 2005. Cutting shape and storage temperature affect overall quality of fresh-cut papaya cv. 'Maradol'. J. Food Sci. 70: S482-S489.

Roginsky, V. and E.A. Lissi. 2005. Review of methods to determine chain-breaking antioxidant activity in food. Food Chem. 92: 235.

Sahlin, E. and G.P. Savage, and C.E. Lister. 2004. Investigation of the antioxidant properties of tomatoes after processing. J. Food Compos. Anal. 17: 635-647.

Sánchez-Moreno, C. and L. Plaza, B. de Ancos, and M.P. Cano. 2006. Impact of high-pressure and traditional thermal processing of tomato purée on carotenoids, vitamin C and antioxidant activity. J. Sci. Food Agric. 86: 171-179.

Sasaki, F.F. and J.S. de Aguila, C.R. Gallo, E.M.M. Ortega, A.P. Jacomino, and R.A. Kluge. 2006. Alterações fisiológicas, qualitativas e microbiológicas durante o armazenamento de abóbora minimamente processada em diferentes tipos de corte. Hortic. Bras. 24: 170-174.

Shi, J. and M. Le Maguer. 2000. Lycopene in tomatoes: chemical and physical properties affected by food processing. Crit. Rev. Biotechnol. 20: 293-334.

Shvedova, A.A. and C. Kommineni, B.A. Jeffries, V. Castranova, Y.Y. Tyurina, V.A. Tyurin, E.A. Serbinova, J.P. Fabisiak, and V.E. Kagan. 2000. Redox cycling of phenol induces oxidative stress in human epidermal keratinocytes. J. Invest. Dermatol. 114: 354-363.

Slimestad, R. and M.J. Verheul. 2005. Seasonal variations in the level of plant constituents in greenhouse production of cherry tomatoes. J. Agric. Food Chem. 53: 3114-3119.

Sluis, A.A. van de and M. Dekker, R. Verkerk, and W.M.F. Jongen. 2000. An improved, rapid in vitro method to measure antioxidant activity. Application on selected flavonoids and apple juice. J. Agric. Food Chem. 48: 4116-4122.

Takeoka, G.R. and L. Dao, S. Flessa, D.M. Gillespie, W.T. Jewell, B. Huebner, D. Bertow, S.E. Ebeler, and L. Dao. 2001. Processing effects on lycopene content and antioxidant activity of tomatoes. J. Agric. Food Chem. 49: 3713-3717.

Tarozzi, A. and A. Marchesi, G. Cantelli-Forti, and P. Hrelia. 2004. Cold-storage affects antioxidant properties of apples in Caco-2 cells. J. Nutr. 134: 1105-1109.

Toor, R.K. and G.P. Savage. 2005. Antioxidant activity in different fractions of tomatoes. Food Res. Int. 38: 487-494.

Toor, R.K. and G.P. Savage. 2006. Changes in major antioxidant components of tomatoes during post-harvest storage. Food Chem. 99: 724-727.

Tudela, J.A. and E. Cantos, J.C. Espin, F.A. Tomas-Barberan, and M.I. Gil. 2002. Induction of antioxidant flavonol biosynthesis in fresh-cut potatoes. Effect of domestic cooking. J. Agric. Food Chem. 50: 5925-5931.

Vina, S.Z. and A.R. Chaves. 2006. Antioxidant responses in minimally processed celery during refrigerated storage. Food Chem. 94: 68.

Vina, S.Z. and A.R. Chaves. 2007. Respiratory activity and phenolic compounds in pre-cut celery. Food Chem. 100: 1654-1660.

Wargovich, M.J. 2000. Anticancer properties of fruits and vegetables. HortScience 35: 573-575.

[WHO]. 2003. Diet, Nutrition and the Prevention of Chronic Diseases. World Health Organization. Geneva, Switzerland, 149 pp.

Wilborts, M. 2003. Evaluation of different antioxidant assays. M.S. Thesis, Wageningen University, Wageningen, the Netherlands.

Willcox, J.K. and G.L. Catignani, and S. Lazarus. 2003. Tomatoes and cardiovascular health. Crit. Rev. Food Sci. 43: 1-18.

Woodall, A.A. and S.W. M. Lee, R.J. Weesie, M.J. Jackson, and G. Britton. 1997. Oxidation of carotenoids by free radicals: relationship between structure and reactivity. BBA Subjects 1336: 33-42.

Wright, K.P. and A.A. Kader. 1997. Effect of slicing and controlled-atmosphere storage on the ascorbate content and quality of strawberries and persimmons. Postharvest . Tech. 10: 39-48.

Yu, J. and M. Ahmedna, and I. Goktepe. 2005. Effects of processing methods and extraction solvents on concentration and antioxidant activity of peanut skin phenolics. Food Chem. 90: 199-206.

18

Tomato and Anticancer Properties of Saliva

Masahiro Toda and Kanehisa Morimoto

Department of Social and Environmental Medicine, Osaka University Graduate School of Medicine, 2-2 Yamada-oka, Suita, Osaka 565-0871, Japan

ABSTRACT

It has been suggested that intake of some nutrients may be related to the anticancer properties of saliva. Furthermore, epidemiological studies have suggested that tomatoes may protect against carcinogenesis. In this research, we investigated the effect of a period of daily tomato juice drinking on the anticancer properties of saliva. Of the 22 healthy male university students we recruited, 11 drank 3 cans (570 g) per day (morning, noon, and night) of tomato juice for 10 d (tomato-juice-drinking group). Saliva samples were collected immediately before and after this period of thrice daily consumption of tomato juice, and 3 d after the end. The anticancer properties of the saliva samples were determined using the umu test. The tomato-juice-drinking group showed a significant increase in the anticancer properties of saliva after drinking tomato juice for 10 d ($p < 0.05$). This increase, however, did not persist to 3 d later. The control group showed no such change. These findings suggest that consumption of tomato juice increases the anticancer properties of saliva and that a persistent effect of tomato consumption on the anticancer properties of saliva can be maintained only by regular consumption.

INTRODUCTION

In an earlier study (Toda et al. 2002a), we looked into the relationship between health-related lifestyle factors and the anticancer properties of saliva. Of the health-related lifestyle items, a nutritionally balanced diet was one of the main factors that promoted the anticancer properties of

A list of abbreviations is given before the references.

saliva. Thus, we conjectured that intake of some nutrients may be related to the anticancer properties of saliva. In this research, using the umu test, we investigated the effect of a period of daily tomato juice drinking on the anticancer properties of saliva.

Tomatoes seem to protect against carcinogenesis. Epidemiological studies have revealed an inverse relation between tomato intake, which increases serum lycopene, and the risk for several types of human cancers (Giovannucci 1999, Sengupta and Das 1999). Furthermore, lycopene, a carotenoid abundantly present in tomato, has been shown to be a powerful antioxidant (Di Mascio et al. 1989a, Khachik et al. 1995).

At present, the Ames test (Ames et al. 1975) and the umu test (Nakamura et al. 1987, Nakamura 1988, Oda et al. 1985) are widely used for assaying mutagenicity. In this study, we selected the umu test because of the many advantages it has over the Ames test: while the sensitivity is similar, only a single bacterial strain is required to test for various types of mutation. The assay is completed more quickly and is not influenced by histidine. Thus, the umu test seems more suitable for samples that can include histidine, such as components of the human body, and when large numbers of samples must be tested.

TESTING THE EFFICACY OF TOMATO JUICE

We recruited 22 healthy male university students to take part in the study. None were smokers or taking medication. These subjects were randomly assigned to either of two groups, a tomato-juice-drinking group (n = 11) and a control group (n = 11). Table 1 shows the basic characteristics of each group. We found no significant differences in age or physique between the two groups. In addition, responses to a questionnaire survey revealed no significant difference in their ordinary daily intake of vegetables or vegetable juice. In the protocol, the tomato-juice-drinking group drank 3 cans (570 g) per day (morning, noon, and night) of tomato juice (Yakult Co., Tokyo, Japan) for 10 d (Table 2).

Saliva samples of 10-15 ml were collected directly into test tubes (Toda et al. 2002b) immediately before and after this period of thrice daily consumption of tomato juice, and 3 d after the end (Fig. 1). To exclude the effects of possible circadian variation, samples were all collected at the same time (17:00). It has been reported that the presence of many components of saliva tends to be stable in the mid-afternoon (Ferguson and Botchway 1980). Subjects were requested to refrain from eating and drinking for at least 2 hr before each saliva sampling and not to consume similar beverages during the protocol period. The samples were stored at –80°C until the assay.

Table 1 Subject characteristics.

Group	Number	Values (mean ± SD)			
		Age (yr)	Height (cm)	Body weight (kg)	BMI (kg/m^2)
Tomato juice drinkers	11	23.5 ± 3.6	172.2 ± 4.1	60.6 ± 4.6	20.4 ± 1.2
Control group	11	22.6 ± 1.0	172.7 ± 4.5	62.7 ± 8.3	21.0 ± 1.9

The subjects were randomly assigned to either of two groups, a tomato-juice-drinking group and a control group. We found no significant differences in age or physique between the two groups.

Table 2 Nutrients in 190 g can of tomato juice.

Calories	36 kcal
Protein	1.3 g
Fat	0 g
Carbohydrates	7.8 g
Sodium	151 mg
Lycopene	19 mg

Source: Yakult Co.

The anticancer properties of the saliva samples were assayed using the umu test (Fig. 2). A volume of 0.1 ml (0.024 μg/ml) of AF-2 (furylfuramide) was used as a mutagen. The mutagenicity of AF-2 has been well confirmed by the umu test (Nakamura and Kosaka 1993, Okada et al. 1989, Uenobe et al. 1997). Bacteria were grown in either Luria's broth or TGA medium (1% bacto tryptone, 0.5% NaCl and 0.2% glucose) supplemented with ampicillin (20 mg/ml). Z-buffer was prepared as described previously (Miller 1972). Chemicals were of the purest grade available.

Using *Salmonella typhimurium* TA1535/pSK1002 as the tester strain, the umu test was carried out essentially as previously described (Nakamura et al. 1987, Nakamura 1988, Oda et al. 1985). It has been reported that 0.2 ml of saliva sufficiently inhibits the mutagenicity of AF-2 and that this amount has no effect on the proliferation of the tester strain (Okada et al. 1989). After adding 0.2 ml of saliva to the test system, using a previously described method (Miller 1972), SOS responses were evaluated according to β-galactosidase activity. The SOS response-inhibiting capacities of saliva, referred to here as the anticancer properties of saliva, were calculated using the following equation: $[1 - (A - C)/(B - D)] \times 100$ (%) (Yamamoto et al. 1994), where A is the assay result for SOS responses induced by the mutagen mixed with saliva, B is the result for the mutagen, C is the result for saliva, and D is the result without additions (baseline).

All results are displayed as mean values ± standard deviation. ANOVA with repeated measures was performed to detect inter-group and time-related differences. Bonferroni's test was used for multiple comparisons. Values were considered to be significantly different when $p < 0.05$.

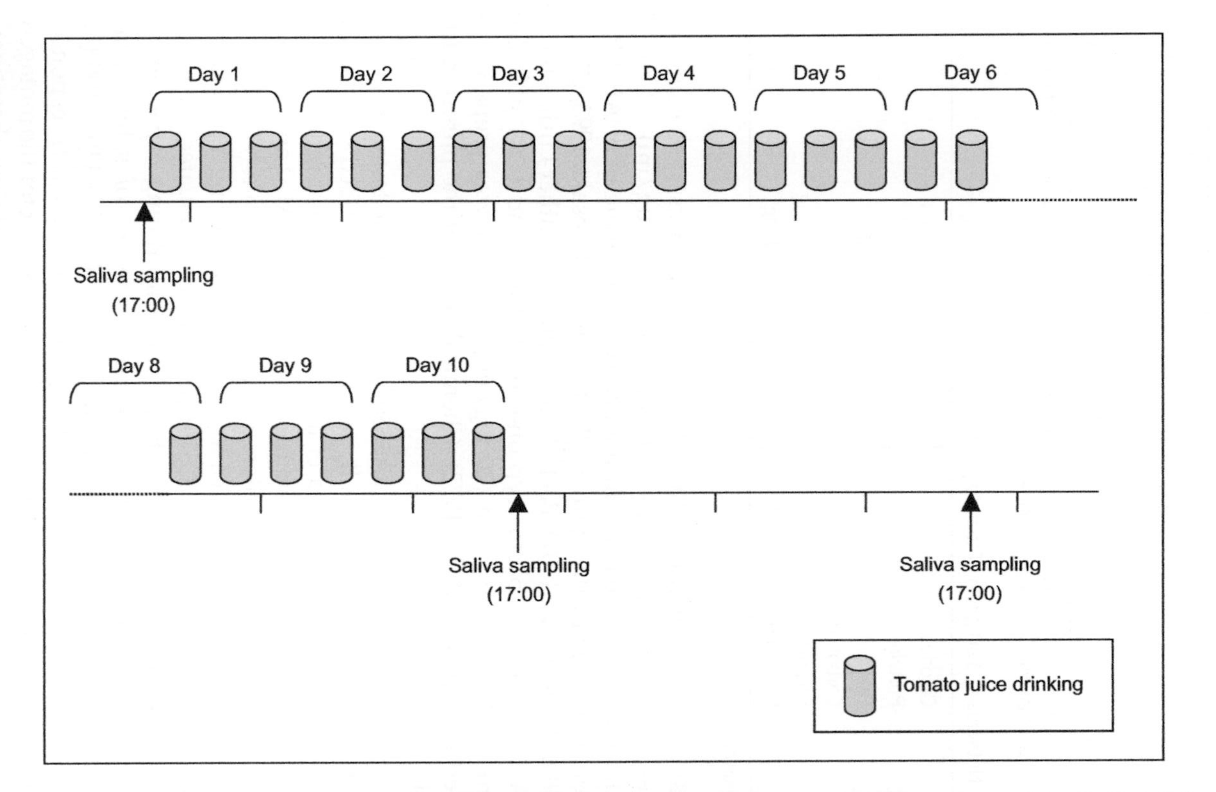

Fig. 1 Protocol of experiment. The tomato-juice-drinking group drank 3 cans per day (morning, noon, and night) of tomato juice for 10 d. Saliva samples were collected immediately before and after this period of thrice daily consumption of tomato juice, and 3 d after the end.

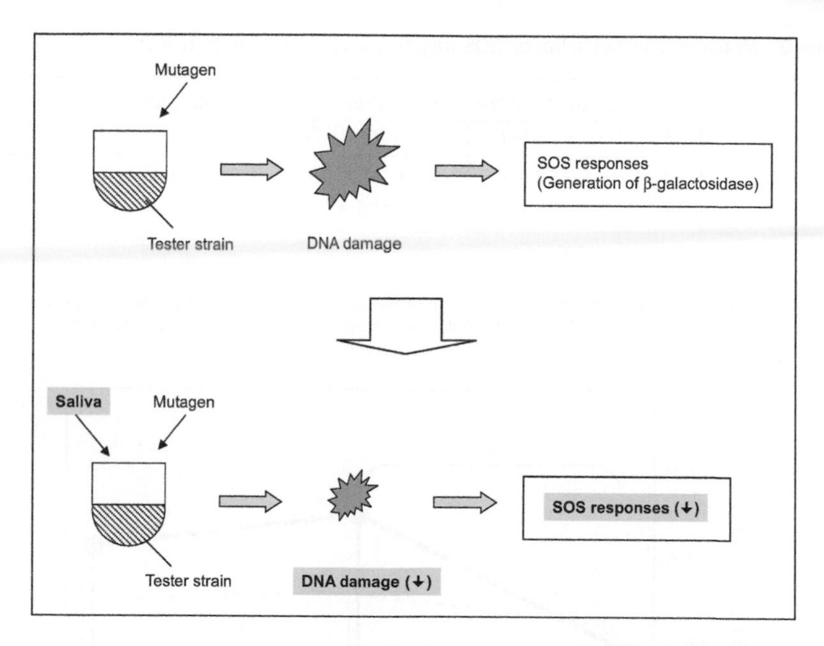

Fig. 2 Determination of anticancer properties of saliva using umu test. Tester strain: *Salmonella typhimurium* TA1535/pSK1002. Mutagen: furylfuramide (AF-2; 0.024 μg/ml). The anticancer properties of saliva were calculated using the following equation: $[1 - (A - C)/(B - D)] \times 100$ (%), where A is the assay result for the SOS responses induced by the mutagen mixed with saliva, B is the result for the mutagen alone, C is the result for saliva, and D is the result without additions (baseline).

FINDINGS

Table 3 shows the modifying effect of saliva on the mutagenicity of AF-2. Anticancer properties of 33.3 ± 25.5% were found.

The effect of drinking tomato juice on the anticancer properties of saliva is shown in Fig. 3. Tomato juice drinkers showed a significant increase in the anticancer properties of saliva after drinking tomato juice for 10 d ($p < 0.05$). This increase, however, did not persist to 3 d later. Meanwhile, in the control group, there was no such change.

IMPLICATIONS

It has been shown that active oxygen species such as singlet oxygen or free radicals can cause a wide spectrum of cell damage including lipid peroxidation, enzyme inactivation, and DNA damage (Cerutti 1985, Di Mascio et al. 1989b, Fridovich 1978, Meneghini 1988). This evidence suggests that the presence of active oxygen species may play an important role in carcinogenesis. Meanwhile, many antioxidants are antipromoters

Table 3 Modifying effect of saliva on SOS responses induced by AF-2 (n = 22).

	β-galactosidase activity (units)		*Anticancer properties (%)*
	AF-2 (0.024 μg/ml)	*Control (DMSO)*	
Saliva (+)	584.05 ± 165.99	171.29 ± 28.64	33.3 ± 25.5
Saliva (–)	829.53 ± 15.52	210.65 ± 5.64	

Values are expressed as mean ± SD.
AF-2 = furylfuramide; DMSO = dimethyl sulfoxide
0.2 ml of saliva was added to the test system in the presence of the mutagen (AF-2). SOS responses were evaluated according to β-galactosidase activity. Anticancer properties of 33.3 ± 25.5% were found.

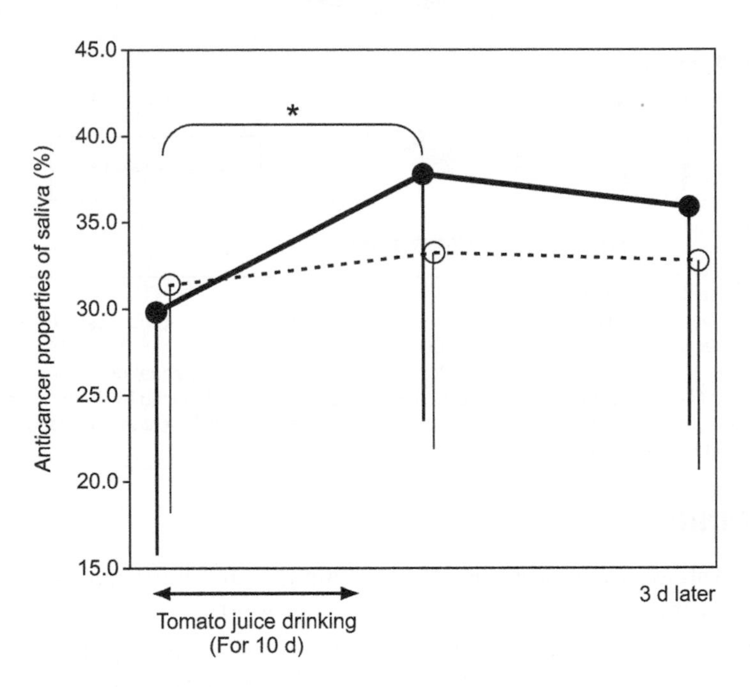

Fig. 3 Mean values (±SD) for anticancer properties of saliva in tomato juice drinkers (solid line; n = 11) and control group (broken line; n = 11) during the sampling period. *$p < 0.05$ (ANOVA with repeated measures and Bonferroni's test). Tomato juice drinkers showed a significant increase in the anticancer properties of saliva after drinking tomato juice for 10 d.

and anticarcinogens (Cerutti 1985): for example, in carotenoids, the conjugated polyene structure is involved in quenching harmful active oxygen species (Krinsky 1993). Furthermore, compared with other carotenoids, it has been reported that lycopene possesses exceptionally potent antioxidant capabilities (Di Mascio et al. 1989a). The tomato juice consumed in the present study included 19 mg of lycopene per can (Table 2). It is likely to have been a major contributor to the increased anticancer properties of saliva.

While the saliva of tomato juice drinkers showed a significant increase in anticancer properties, the increase did not persist to 3 d later. This finding suggests that a persistent effect of tomato consumption on the anticancer properties of saliva can be maintained only by regular consumption.

The optimal or minimal intake of lycopene or tomato to maintain efficacy has not yet been clearly determined and, at present, information on the absorption, distribution and metabolism of lycopene is sparse (Sengupta and Das 1999). In our protocol, for 10 d, subjects drank 3 cans per day of tomato juice, because we conjectured that this quantity might produce a measurable effect on the anticancer properties of saliva and that subjects could tolerate it without strain. As we expected, this level of consumption of tomato juice did have a significant effect on the anticancer properties of saliva. More detailed study is required, however, to better elucidate the relationship between intake of tomato juice and the anticancer properties of saliva.

In this study, we found that consumption of tomato juice did increase the anticancer properties of saliva. The research, however, has several limitations. The effect of substances other than lycopene in tomato juice cannot be excluded. This can be investigated by supplementing a regular diet with lycopene. Furthermore, salivary lycopene levels should be determined. To elucidate how dietary items affect the anticancer properties of saliva, it is necessary to carry out further studies with various nutrients.

ABBREVIATIONS

AF-2: Furylfuramide; TGA medium: Tryptone-glucose-ampicillin medium

REFERENCES

Ames, B.N. and J. McCann, and E. Yamasaki. 1975. Methods for detecting carcinogens and mutagens with the Salmonella/mammalian-microsome mutagenicity test. Mutat. Res. 31: 347-364.

Cerutti, P.A. 1985. Prooxidant states and tumor promotion. Science 227: 375-381.

Di Mascio, P. and S. Kaiser, and H. Sies. 1989a. Lycopene as the most efficient biological carotenoid singlet oxygen quencher. Arch. Biochem. Biophys. 274: 532-538.

Di Mascio, P. and H. Wefers, H.P. Do-Thi, M.V.M. Lafleur, and H. Sies. 1989b. Singlet molecular oxygen causes loss of biological activity in plasmid and bacteriophage DNA and induces single-strand breaks. Biochim. Biophys. Acta 1007: 151-157.

Ferguson, D.B. and C.A. Botchway. 1980. Circadian variations in the flow rate and composition of whole saliva stimulated by mastication. Arch. Oral Biol. 24: 877-881.

Fridovich, I. 1978. The biology of oxygen radicals. Science 201: 875-880.

Giovannucci, E. 1999. Tomatoes, tomato-based products, lycopene, and cancer: review of the epidemiologic literature. J. Natl. Cancer Inst. 91: 317-331.

Khachik, F. and G.R. Beecher, and J.C. Smith. 1995. Lutein, lycopene and their oxidative metabolites in chemoprevention of cancer. J. Cell. Biochem. Suppl. 22: 236-246.

Krinsky, N.I. 1993. Actions of carotenoids in biological systems. Annu. Rev. Nutr. 13: 561-587.

Meneghini, R. 1988. Genotoxicity of active oxygen species in mammalian cells. Mutat. Res. 195: 215-230.

Miller, J.H. 1972. Experiments in Molecular Genetics. Cold Spring Harbor Laboratory, New York, USA.

Nakamura, S. and Y. Oda, T. Shimada, I. Oki, and K. Sugimoto. 1987. SOS-inducing activity of chemical carcinogens and mutagens in Salmonella typhimurium TA1535/pSK1002: examination with 151 chemicals. Mutat. Res. 192: 239-246.

Nakamura, S. 1988. Development of screening test for mutagens. Kagaku to Kougyou 62: 142-148.

Nakamura, S. and H. Kosaka. 1993. Modification activity of human serum on SOS-inducing activity of chemical mutagens. Proc. Osaka Prefectural Institute of Public Health 31: 23-28.

Oda, Y. and S. Nakamura, I. Oki, T. Kato, and H. Shinagawa. 1985. Evaluation of the new system (umu-test) for the detection of environmental mutagens and carcinogens. Mutat. Res. 147: 219-229.

Okada, M. and S. Nakamura, K. Miura, and K. Morimoto. 1989. The anti-mutagenic effects of human saliva investigated by umu-test (part 1): effects of filtration and storage at low temperatures. Jpn. J. Hyg. 44: 1009-1013.

Sengupta, A. and S. Das. 1999. The anti-carcinogenic role of lycopene, abundantly present in tomato. Eur. J. Cancer Prev. 8: 325-330.

Toda, M. and K. Morimoto, S. Nakamura, and K. Hayakawa. 2002a. Daily lifestyles and anti-mutagenicity of saliva. Environ. Health Prev. Med. 7: 11-14.

Toda, M. and K. Morimoto, S. Nakamura, and K. Hayakawa. 2002b. A further note on the sampling device for the anti-mutagenicity of saliva. Environ. Health Prev. Med. 7: 27-29.

Uenobe, F. and S. Nakamura, and M. Miyazawa. 1997. Antimutagenic effect of resveratrol against Trp-P-1. Mutat. Res. 373: 197-200.

Yamamoto, N. and W. Sugiura, H. Kosaka, and S. Nakamura. 1994. The SOS-inhibition activity of human urine investigated by umu-test: diurnal and daily changes of SOS-inhibition activity of urine. Jpn. J. Hyg. 49: 791-796.

19

DNA Strand Breaks and Tomatoes

Karlis Briviba

Institute of Nutritional Physiology, Federal Research Centre for Nutrition and Food, Haid-und-Neu-Str. 9, 76131, Karlsruhe, Germany

ABSTRACT

Epidemiological and experimental data suggest that an increased intake of tomato products can reduce the risk of cancers, especially prostate and colon cancer. DNA damage is an important step in the initiation of carcinogenesis. Human intervention studies have shown that daily consumption of tomato products containing about 15 mg lycopene or more can decrease DNA damage and even those with about 6 mg or more can increase the antigenotoxic resistance of peripheral blood lymphocytes against strand breaks initiated by pro-oxidants such as hydrogen peroxide or iron ions. It would appear that the protective effects of tomatoes cannot be attributed solely to lycopene, which is, however, possibly a good marker for consumption of tomato products. Little is known about the protective effects of tomatoes on DNA damage in cells other than lymphocytes in human. Prostate and colon epithelial cells are of special interest because of well-known epidemiological links. Further studies are necessary to estimate the protective tomato compounds, molecular mechanisms of action, and subpopulations that are sensitive to the cancer-preventive effects of tomato.

DNA STRAND BREAKS, FORMATION, REPAIR AND BIOLOGICAL CONSEQUENCE

A DNA single strand break (SSB) is a discontinuity of one strand of a DNA duplex and can be caused by a number of reactive oxygen and nitrogen species (ROS/RNS) such as hydroxyl radical, singlet oxygen, hydrogen

A list of abbreviations is given before the references.

peroxide and peroxynitrite reacting directly with DNA (von Sonntag 1984, Epe et al. 1996). Reactive oxygen and nitrogen species are generated during numerous physiological processes such as oxidative phosphorylation in mitochondria or in a large number of enzymatic reactions. Increased formation of ROS/RNS is observed during pathological processes such as inflammation, ischemia/reperfusion, or tissue damage or induced by environmental factors such as ultraviolet irradiation from the sun, ozone, nitrogen oxides, tobacco smoke, asbestos particles, or medical drugs (Sies 1991). Many ROS/RNS can directly cause SSB by attacking the sugar-phosphate backbone (von Sonntag 1984), but they are also able to modify DNA bases. A number of modified DNA bases can be recognized by DNA base excitation and nucleotide-excitation repair enzymes and removed, causing indirect SSBs (Epe et al. 1996). Additionally, topoisomerases cause indirect SSBs. Most SSBs are rapidly repaired by a complex mechanism that includes removal of damaged bases and sugars, restoration of damaged 3' and 5' termini, gap filling and ligation (for review, see Caldecott 2007). More than 130 human DNA repair genes have been identified. Proteins such as X-ray repair cross-complementing protein 1 (XRCC1), 8-oxoguanosine DNA glycosylase (OGG1), excision-repair cross-complementing 1 (ERCC1) xeroderma pigmentosum complementation group (XPP), poly (ADP-ribose) polymerase-1 (PARP-1), DNA Ligase 3 (Lig3a), human polynucleotide kinase (hPNK,) apurinic/apyrimidinic endonuclease-1 (APE1), and DNA polymerase β (Pol β) play an important role in the repair of DNA base damage and SSB.

High frequency of SSBs can lead to formation of double strand breaks (DSBs), which are proposed to be the most hazardous lesions arising in the genome of eukaryotic organisms. The repair mechanisms of DSB are more complex than those of SSBs and efficient repair is very important for the maintenance of genomic integrity, cellular viability and prevention of tumorigenesis.

Many thousands of DNA lesions occur in each cell, every day. The steady-state level of strand breaks (SBs) is kept at a very low level and can be regulated by expression of many enzyme systems including antioxidant (e.g., glutathione peroxidase), detoxification of prooxidative xenobiotics (e.g., glutathione S-transferases), and the DNA repair system. There are a number of polymorphisms of some antioxidant, detoxification and DNA repair proteins, which can result in a large variety of possible combinations of activities of these enzymes and a large number of subsets of individuals with different resistance or susceptibility to genotoxic agents. There are some descriptive reports on the influence of nutrition in humans, but little information is available on how nutrition and, in particular, tomato consumption modulate this very complex system, which regulates the *in vivo* concentrations of genotoxic agents and DNA repair enzymes.

EFFECT OF TOMATO CONSUMPTION ON STRAND BREAK LEVELS IN HUMAN BLOOD CELLS

Pool-Zobel et al. (1997) first reported that consumption of heat-processed tomato juice (330 mL/d) by healthy subjects for 2 wk significantly reduced endogenous DNA strand breaks in lymphocytes (Pool-Zobel et al. 1997). The authors reasoned that the protective effect of tomato consumption could be attributed to the enhancement of a detoxification protein (cytosolic glutathione S-transferase P1) and DNA repair proteins (Pool-Zobel et al. 1998). Following this report a number of human intervention studies investigated the effect of daily consumption of tomato puree, sauce and extracts on DNA damage in humans (Riso et al. 1999, 2004, Porrini and Riso 2000, Briviba et al. 2004) (Table 1). Most of these studies showed a protective effect of consumption of various tomato products on DNA damage in lymphocytes. The most consistent effect seems to be an increased resistance of DNA against pro-oxidant-induced SB formation in lymphocytes in *ex vivo* experiments. Thus, three out of four studies and one out of one study showed a statistically significant increase in DNA resistance to H_2O_2 and Fe^{2+}, respectively (Pool-Zobel et al. 1997, Riso et al. 1999, 2004, Porrini and Riso 2000, Porrini et al. 2005). These data indicate that consumption of different processed tomato products increases the capacity of the cellular antioxidant network to more effectively inactivate H_2O_2 and/or prevent formation of the reactive species that can be formed from H_2O_2 such as hydroxyl, ferryl, oxo-ferryl radicals.

The effect on endogenous SSB formation is not so consistent; only two (Pool-Zobel et al. 1997, Briviba et al. 2004) out of six studies showed statistically significant reductions in endogenous levels of SB in lymphocytes (Table 1). Also, our data using a study design similar to that of Pool-Zobel et al. (1997) (tomato juice 330 mL; intervention time 2 wk) did not show statistically significant reduction in SBs in lymphocytes, in spite of a significant increase in plasma lycopene concentration (Fig. 1). The explanation for these discrepancies in the effects could be differences in the level of SB at the beginning of the experiments. In the study by Pool-Zobel et al. (1997), this level was about twice as high as ours. Also, the plasma concentrations of lycopene, which reflect the consumption of tomato products, were different at the beginning of the studies. Furthermore, the content of bioactive substances such as polyphenols and carotenoids differs significantly between tomato varieties and maturity

Table 1 Effect of intervention with tomato products on the levels of DNA strand breaks (SB) and/or on the resistance to H_2O_2-induced formation of DNA strand break in human lymphocytes.

Tomato product	*Lycopene content (mg/day)*	*Duration of intervention*	*Effect*	*References*
Tomato juice, 330 mL/day	40	2 wk	Significant decrease in SB; no effect on H_2O_2-induced SB formation	Pool-Zobel et al. 1997
Tomato puree, 60 g/day	16.5	3 wk	No effect on SB; significant increase in resistance to H_2O_2-induced SB formation	Riso et al. 1999
Tomato puree, 25 g/day	7	2 wk	No effect on SB; significant increase in DNA resistance H_2O_2-induced SB formation	Porrini et al. 2000
Tomato oleoresin extract (capsules)	14.64	2 wk	Significant decrease in SB	Briviba et al. 2004
Mix of tomato products	8	3 wk	Significant increase in DNA resistance to Fe^{2+}-induced SB formation	Riso et al. 2004
Tomato oleoresin extract (soft drink)	5.7	26 d	Significant increase in DNA resistance to H_2O_2-induced SB formation	Porrini et al. 2005
Tomato oleoresin extract (soft drink)	5.7	26 d	No effect on SB	Riso et al. 2006
Tomato juice, 330 mL/d	37	2 wk	No effect on SB	This work

Daily consumption of heat-processed tomato products containing 5.7 mg lycopene or more significantly increases the resistance of lymphocytes and possibly other cells against DNA strand break formation caused by pro-oxidants such as hydrogen peroxide or ferrous ion. Oleoresin extracts of tomatoes containing in addition to lycopene a number of other lipophilic tomato compounds are able to mimic these effects. The effect on the formation of DNA strand breaks is not consistent.

SB, DNA strand break; H_2O_2, hydrogen peroxide; Fe^{2+}, ferrous ion.

stages. Processing and storage also affect the concentrations and a large difference is reported between subjects in bioavailability and metabolism of carotenoids, polyphenols and other bioactive substances. These two studies indicate that not every tomato product (juice) can decrease DNA damage in every population subset. Some populations may be more sensitive to the protective effects of tomato products than others.

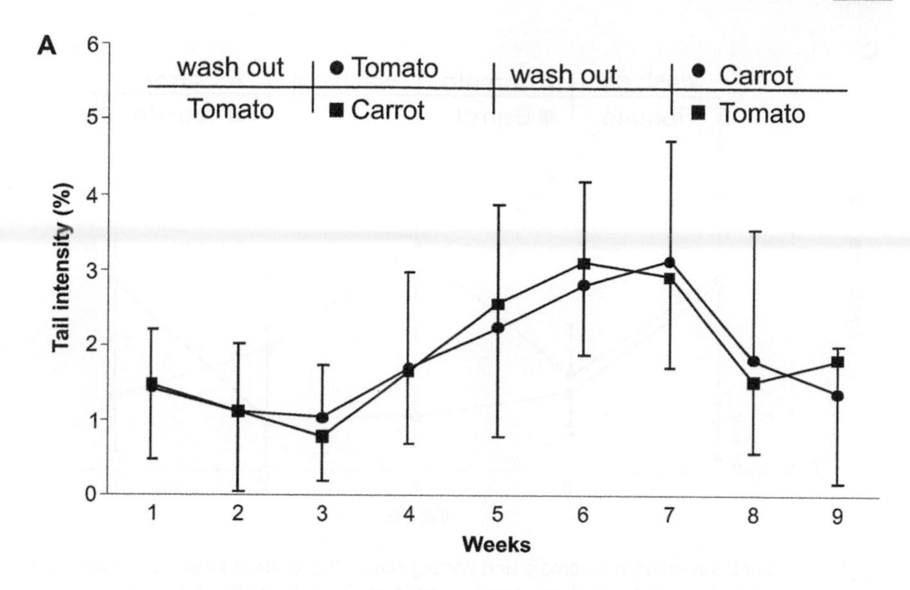
A
wash out
Tomato
● Tomato
■ Carrot
wash out
● Carrot
■ Tomato
Tail intensity (%)
Weeks

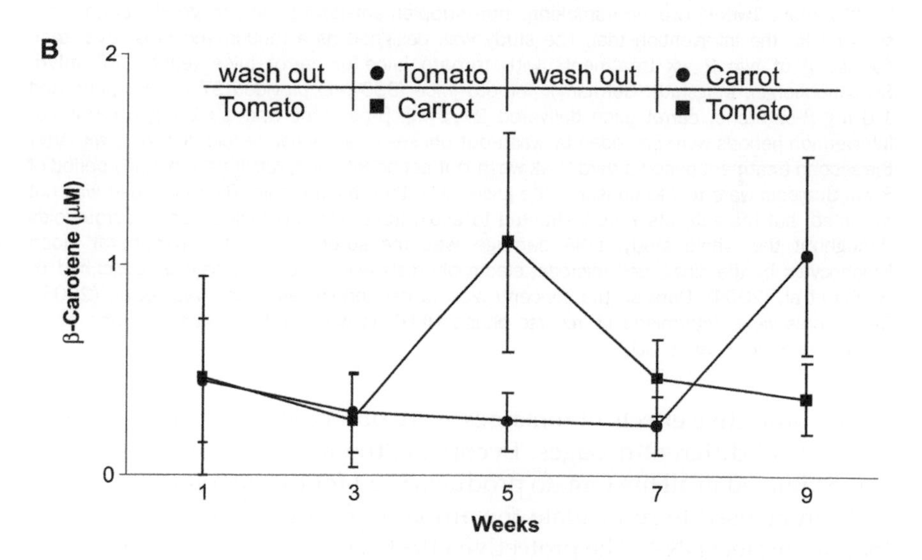
B
wash out
Tomato
● Tomato
■ Carrot
wash out
● Carrot
■ Tomato
β-Carotene (μM)
Weeks

Fig. 1 (Contd.)

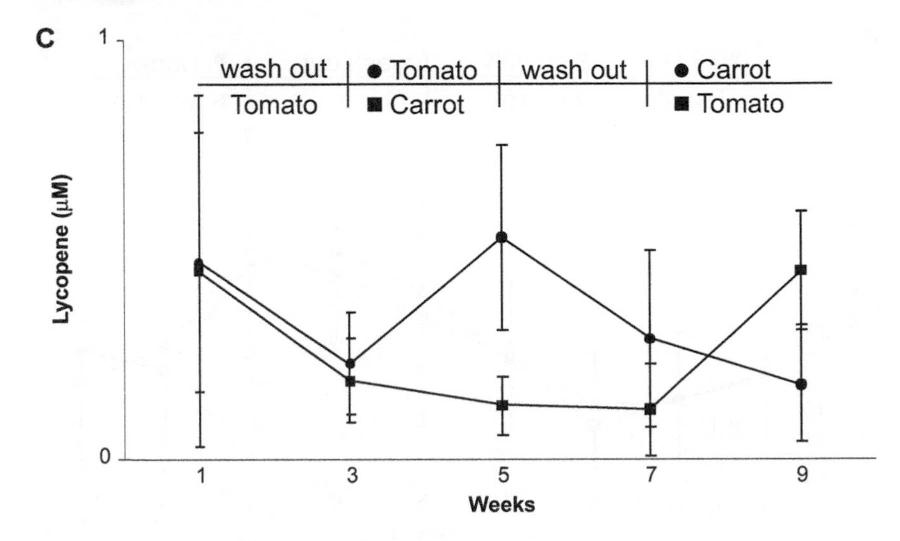

Fig. 1 Effect of intervention with tomato and carrot juice on the levels of single strand breaks in human blood lymphocytes (A) and the plasma levels of β-carotene (B) and lycopene (C) in healthy men. Twenty-two non-smoking, non-supplement-taking men in good health were selected for the intervention trial. The study was designed as a randomized cross-over trial, consisting of two 2 wk treatments with tomato juice or carrot juice (each 330 mL/d, Schoenenberger, Magstadt, Germany). Tomato juice (330 mL) provided 37 mg lycopene and 1.6 mg β-carotene, carrot juice delivered 27.1 mg β-carotene and 13.1 mg α-carotene. Intervention periods were preceded by wash-out phases of a low-carotenoid diet for 2 wk. After the second treatment period a third 2 wk wash-out period followed, resulting in a study period of 8 wk. Subjects were told to consume the juices with their main meals. Their daily diet was not restricted, but the subjects were instructed to avoid fruits and vegetables high in carotenoids throughout the whole study. DNA damage was measured in isolated peripheral blood lymphocytes by the single cell microgel electrophoresis assay (comet assay) as described by Briviba et al. (2004). Data on the concentration of carotenoids are from Watzl et al. (2003). Carotenoids were determined by reverse-phase HPLC as described previously (Briviba et al. 2001). Data are mean ± SD.

The protective effects of tomatoes were observed with various tomato products at different dosages. Lycopene, the major tomato carotenoid, was estimated in all the tomato products used for the intervention studies and can be used to recalculate the amount of different tomato products that can protect DNA. The protective effects were observed after intake of tomato products providing 5.7 mg lycopene or more (Table 1). This lycopene amount corresponds to about 25 g tomato puree, 60 mL tomato juice, 30 g ketchup or 210 g fresh tomatoes. The mean daily dietary intake of lycopene is close to 5.7 mg in some European countries such as the Netherlands (4.9 mg) and France (4.8 mg) but considerably lower in Spain (1.7 mg) (O'Neill et al. 2001). Thus, for at least a part of the European population the consumption of tomato products is close to that which can protect DNA from damage.

WHICH TOMATO COMPOUNDS ARE RESPONSIBLE FOR THE PROTECTIVE EFFECT?

An oleoresin tomato extract, which contains mostly lipophilic compounds, was able to mimic the effect of tomato products in at least two out of three studies. These protective effects of extract indicate that fat-soluble tomato ingredients are mostly responsible for the observed effects. Tomatoes are the most important dietary source of lycopene (about 80% in the United States). In addition to lycopene, tomatoes contain other carotenoids such as β-carotene, phytoene and phytofluene and bioactive compounds such as ascorbate, folate and polyphenols such as chlorogenic acid, rutin, naringenin, and chalconaringenin with potential antigenotoxic compounds. *In vitro* experiments show that these compounds can protect DNA from damage. Which of these compounds, or mixtures thereof, are responsible for the observed anti-genotoxic effects is not yet known.

The antigenotoxic effects of synthetic or isolated carotenoids such as β-carotene and lycopene have been investigated in a number of intervention studies. Recently, the study by Zhao et al. (2006) demonstrated a decrease in SSB by synthetic lycopene (12 mg/d, 56 d) or β-carotene (12 mg/d, 56 d), but there was no effect on DNA resistance in lymphocytes to H_2O_2 that seems to be typical for the tomato products. There are also reports that demonstrate no effect of lycopene on DNA SSB and/or resistance to H_2O_2 in lymphocytes in human intervention studies (Torbergsen and Collins 2000, Astley et al. 2004). No significant correlation was observed between plasma lycopene concentration and oxidative DNA damage in lymphocytes either (Collins et al. 1998). It appears that lycopene alone cannot completely mimic the antigenotoxic effects of tomatoes and the compounds responsible for this effect have not yet been identified.

The effect of tomatoes on DNA damage is not unique and the daily consumption of 1 kiwi fruit, blood orange juice (600 mL, 3 wk) or spinach (150 g, 3 wk) also caused decreased DNA damage in humans (Collins et al. 2003, Porrini et al. 2002, Riso et al. 2005). However, not all fruits and vegetables are able to protect DNA. Several studies report no effect of several types of fruits and vegetables such as blackcurrant juice, cooked carrots, tinned mandarins in orange juice and others (for a review see Moller and Loft, 2006) and about 600 g mixed fruits and vegetables also failed to show significant effect (van den Berg et al. 2001, Moller et al. 2003). Consumption of 12 servings (1200 g/d) of mixed vegetables and fruits a day was necessary to significantly decrease genotoxicity in lymphocytes (Thompson et al. 1999, 2005). Thus, it is possible to increase the antigenotoxic effect by selectively choosing foods such as tomato products.

CONCLUSIONS

Daily consumption of heat-processed tomato products containing 6 mg lycopene or more significantly increases the resistance of lymphocytes and possibly other cells against DNA strand break formation caused by pro-oxidants such as hydrogen peroxide or iron ions. Oleoresin extracts of tomatoes containing in addition to lycopene a number of other lipophilic tomato compounds seem to be able to mimic these effects. The effect of isolated or synthetic lycopene on DNA damage is not consistent. Taking all these experimental and epidemiological data into consideration, it appears that a daily consumption of tomato products may decrease the risk of cancer by protecting DNA against damage.

ACKNOWLEDGEMENTS

Supported by the Federal Ministry of Food, Agriculture and Consumer Protection, Germany.

ABBREVIATIONS

SSB: DNA single strand break; DSB: DNA double strand break; ROS/RNS: reactive oxygen and nitrogen species; H_2O_2: hydrogen peroxide; Fe^{2+}: ferrous ion

REFERENCES

Astley, S.B. and D.A. Hughes, A.J. Wright, R.M. Elliott, and S. Southon. 2004. DNA damage and susceptibility to oxidative damage in lymphocytes: effects of carotenoids in vitro and in vivo. Br. J. Nutr. 91: 53-61.

Briviba, K. and K. Schnabele, E. Schwertle, M. Blockhaus, and G. Rechkemmer. 2001. Beta-carotene inhibits growth of human colon carcinoma cells in vitro by induction of apoptosis. Biol. Chem. 382: 1663-1668.

Briviba, K. and S.E. Kulling, J. Moseneder, B. Watzl, B.G. Rechkemmer, and A. Bub. 2004. Effects of supplementing a low-carotenoid diet with a tomato extract for 2 weeks on endogenous levels of DNA single strand breaks and immune functions in healthy nonsmokers and smokers. Carcinogenesis 25: 2373-2378.

Caldecott, K.W. 2007. Mammalian single-strand break repair: Mechanisms and links with chromatin. DNA Repair (Amst.) 6: 443-453.

Collins, A.R. and B. Olmedilla, S. Southon, F. Granado, and S.J. Duthie. 1998. Serum carotenoids and oxidative DNA damage in human lymphocytes. Carcinogenesis 19: 2159-2162.

Collins, A.R. and V. Harrington, J. Drew, and R. Melvin. 2003. Nutritional modulation of DNA repair in a human intervention study. Carcinogenesis 24: 511-515.

Deeble, D.J. and D. Schulz, and C. von Sonntag. 1986. Reactions of OH radicals with poly(U) in deoxygenated solutions: sites of OH radical attack and the kinetics of base release. Intl. J. Radiat. Biol. Relat. Stud. Phys. Chem. Med. 49: 915-926.

Epe, B. and D. Ballmaier, I. Roussyn, K. Briviba, and H. Sies. 1996. DNA damage by peroxynitrite characterized with DNA repair enzymes. Nucl. Acid Res. 24: 4105-4110.

Moller, P. and S. Loft. 2006. Dietary antioxidants and beneficial effect on oxidatively damaged DNA. Free Radic. Biol. Med. 41: 388-415.

Moller, P. and Vogel, U., A. Pedersen, L.O. Dragsted, B. Sandstrom, and S. Loft. 2003. No effect of 600 grams fruit and vegetables per day on oxidative DNA damage and repair in healthy nonsmokers. Cancer Epidemiol. Biomarkers Prev. 12: 1016-1022.

O'Neill, M.E. and Y. Carroll, B. Corridan, B. Olmedilla, F. Granado, I. Blanco, H. van den Berg, I. Hininger, A.M. Rousell, M. Chopra, S. Southon, and D.I. Thurnham. 2001. A European carotenoid database to assess carotenoid intakes and its use in a five-country comparative study. Br. J. Nutr. 85: 499-507.

Pool-Zobel, B.L. and A. Bub, H. Muller, I. Wollowski, and G. Rechkemmer. 1997. Consumption of vegetables reduces genetic damage in humans: first results of a human intervention trial with carotenoid-rich foods. Carcinogenesis 18: 1847-1850.

Pool-Zobel, B.L. and A. Bub, U.M. Liegibel, S. Treptow-van Lishaut, and G. Rechkemmer. 1998. Mechanisms by which vegetable consumption reduces genetic damage in humans. Cancer Epidemiol. Biomarkers Prev. 7: 891-899.

Porrini, M. and P. Riso. 2000. Lymphocyte lycopene concentration and DNA protection from oxidative damage is increased in women after a short period of tomato consumption. J. Nutr. 130: 189-192.

Porrini, M. and P. Riso, and G. Oriani. 2002. Spinach and tomato consumption increases lymphocyte DNA resistance to oxidative stress but this is not related to cell carotenoid concentrations. Eur. J. Nutr. 41: 95-100.

Porrini, M. and P. Riso, A. Brusamolino, C. Berti, S. Guarnieri, and F. Visioli. 2005. Daily intake of a formulated tomato drink affects carotenoid plasma and lymphocyte concentrations and improves cellular antioxidant protection. Br. J. Nutr. 93: 93-99

Riso, P. and A. Pinder, A. Santangelo, and M. Porrini. 1999. Does tomato consumption effectively increase the resistance of lymphocyte DNA to oxidative damage? Amer. J. Clin. Nutr. 69: 712-718.

Riso, P. and F. Visioli, D. Erba, G. Testolin, and M. Porrini. 2004. Lycopene and vitamin C concentrations increase in plasma and lymphocytes after tomato intake. Effects on cellular antioxidant protection. Eur. J. Clin. Nutr. 58: 1350-1358.

Riso, P. and F. Visioli, C. Gardana, S. Grande, A. Brusamolino, F. Galvano, G. Galvano, and M. Porrini. 2005. Effects of blood orange juice intake on antioxidant bioavailability and on different markers related to oxidative stress. J. Agric. Food Chem. 53: 941-947.

Sies, H. [ed.] 1991. Oxidative Stress Oxidants and Antioxidants. Academic Press, London.

Thompson, H.J. and J. Heimendinger, A. Haegele, S.M. Sedlacek, C. Gillette, C. O'Neill, P. Wolfe, and C. Conry. 1999. Effect of increased vegetable and fruit consumption on markers of oxidative cellular damage. Carcinogenesis 20: 2261-2266.

Thompson, H.J. and J. Heimendinger, S. Sedlacek, A. Haegele, A. Diker, C. O'Neill, B. Meinecke, P. Wolfe, Z. Zhu, and W. Jiang. 2005. 8-Isoprostane F2alpha excretion is reduced in women by increased vegetable and fruit intake. Amer. J. Clin. Nutr. 82: 768-776.

Torbergsen, A.C. and A.R. Collins. 2000. Recovery of human lymphocytes from oxidative DNA damage; the apparent enhancement of DNA repair by carotenoids is probably simply an antioxidant effect. Eur. J. Nutr. 39: 80-85.

van den Berg, R. and T. van Vliet, W.M. Broekmans, N.H. Cnubben, W.H. Vaes, L. Roza, G.R. Haenen, A. Bast, and H. van den Berg. 2001. A vegetable/fruit concentrate with high antioxidant capacity has no effect on biomarkers of antioxidant status in male smokers. J. Nutr. 131: 1714-1722.

von Sonntag, C. 1984. Carbohydrate radicals: from ethylene glycol to DNA strand breakage. Intl. J. Radiat. Biol. Relat. Stud. Phys. Chem. Med. 46: 507-519.

Watzl, B. and A. Bub, K. Briviba, and G. Rechkemmer. 2003. Supplementation of a low-carotenoid diet with tomato or carrot juice modulates immune functions in healthy men. Ann. Nutr. Metab. 47: 255-261.

Zhao, X. and G. Aldini, E.J. Johnson, H. Rasmussen, K. Kraemer, H. Woolf, N. Musaeus, N.I. Krinsky, R.M. Russell, and K.J. Yeum. 2006. Modification of lymphocyte DNA damage by carotenoid supplementation in postmenopausal women. Amer. J. Clin. Nutr. 83: 163-169.

Tomato Carotenoids and the IGF System in Cancer

Joseph Levy[1*], Shlomo Walfisch[2], Yossi Walfisch[1], Amit Nahum[1], Keren Hirsch[1], Michael Danilenko[1] and Yoav Sharoni[1]

[1]Department of Clinical Biochemistry, Faculty of Health Sciences, Ben-Gurion University of the Negev, P.O. Box 653, Beer-Sheva, Israel

[2]Colorectal Unit, Faculty of Health Sciences, Ben-Gurion University of the Negev, P.O. Box 653, Beer-Sheva, Israel

ABSTRACT

Our interest on the subject of dietary carotenoids and the insulin-like growth factor (IGF) system stems from two basic observations. First, IGF-I is an important risk factor in several major cancers and second, the activity of this complex growth factor system can be modulated by various dietary regimes. In this short review we will examine the evidence on the interference of lycopene and other carotenoids with the activity of the IGF system and describe the data supporting further human studies of the long-term effects of dietary carotenoids that may provide a basis for a preventive strategy for some malignancies.

*Corresponding author

A list of abbreviations is given before the references.

INTRODUCTION

Insulin-like growth factor I (IGF-I) is important in normal growth and development. In addition, the IGF system plays a prominent role in cancer. Prospective cohort studies have observed that subjects with elevated blood levels of IGF-I , reduced levels of IGF-binding protein-3 (IGFBP-3) or an increased ratio of IGF-I to IGFBP-3 are at increased risk of developing several types of cancer, including prostate, premenopausal breast, lung and colorectal cancer (Giovannucci et al. 2003) (Fig. 1). Thus, attenuation of the IGF signaling may result in cancer prevention.

The IGF system comprises the IGF ligands (IGF-I and IGF-II), cell surface receptors that mediate the biological effects of the IGFs, including the IGF-I receptor (IGF-IR), the IGF-II receptor (IGF-IIR), and the insulin receptor (IR), as well as a family of IGF-binding proteins (IGFBPs) (Fig. 2). These binding proteins (IGFBP-1 to -6) affect the half-lives and availability of the IGFs in the circulation and in extracellular fluids. Thus, IGF action is determined by the availability of IGF-I to interact with the IGF-I receptor, which is regulated by the relative concentrations of IGFBPs (LeRoith and Roberts 2003). For example, a high concentration of IGFBP-3, the main IGFBP in plasma, reduces IGF-I action. IGF-IR is a transmembrane protein tyrosine kinase that mediates the biological effects of IGF-I and most of the actions of IGF-II (LeRoith and Roberts 2003). Binding of IGF-I to the IGF-IR results in receptor autophosphorylation, phosphorylation of intracellular substrates, and activation of specific signaling processes that promote growth (LeRoith and Roberts 2003) (Fig. 3).

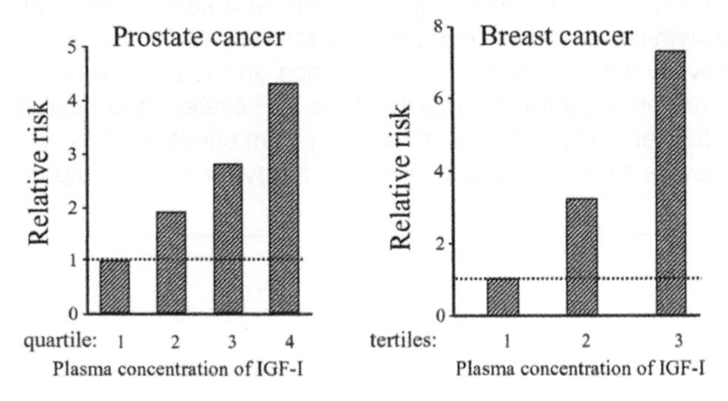

Fig. 1 Positive association between IGF-I levels and cancer risk. Prospective cohort studies have observed that subjects with elevated blood levels of IGF-I are at increased risk of developing several types of cancer, including prostate and premenopausal breast cancer. Based on results from Chan et al. (1998) and Hankinson et al. (1998).

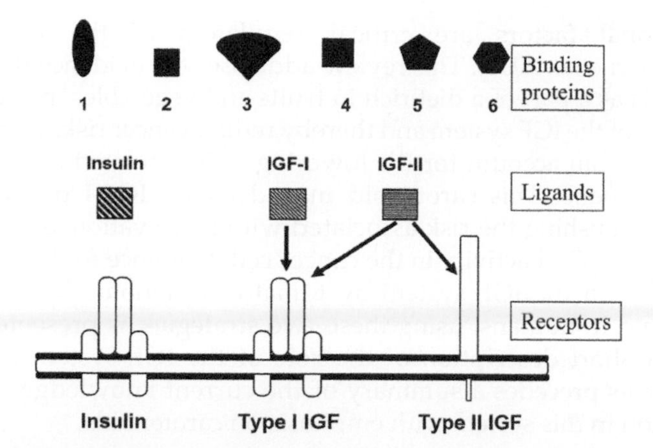

Fig. 2 The IGF axis. The IGF/insulin system comprises the ligands, IGF-I, IGF-II and insulin, cell surface receptors that mediate their biological effects including the IGF-I receptor (IGF-IR), the IGF-II receptor (IGF-IIR), and the insulin receptor (IR), as well as a family of six IGF-binding proteins (IGFBPs), which bind mainly IGF-I. IGF-IR mediates all the growth stimulatory effects of both IGF-I and IGF-II.

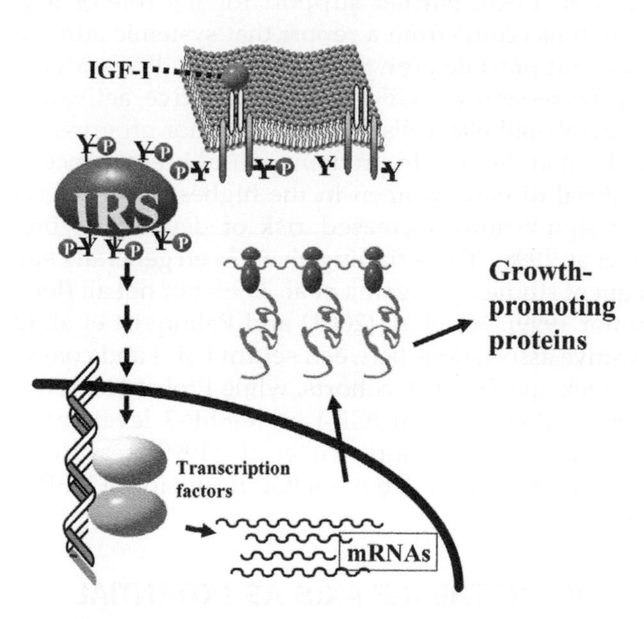

Fig. 3 A model for activation of transcription and cell growth by IGF-I. Binding of IGF-I to the IGF-IR results in receptor autophosphorylation on tyrosine (P-Y), phosphorylation of intracellular substrates (IRS), and activation of specific transcription processes that induce cell growth-promoting proteins.

Nutritional factors are critical regulators of the IGF system (Giovannucci et al. 2003). This review addresses the evidence that tomato carotenoids as a part of a diet rich in fruits and vegetables may attenuate the activity of the IGF system and thereby reduce cancer risk. Two possible mechanisms can account for the lowering of IGF-related cancer risk by tomato lycopene. This carotenoid may decrease IGF-I plasma levels, thereby diminishing the risk associated with its elevation, and/or it can interfere with IGF-I activity in the cancer cell. Evidence for the successful intervention in the IGF system by tomato carotenoids both in model systems and in humans, using these two strategies, is presented in this review. A short description of the role of the IGF system in specific malignancies precedes a summary of the current knowledge of dietary intervention in this system with emphasis on carotenoids.

IGF AND SPECIFIC CANCERS

Several prospective studies suggested that high circulating levels of IGF-I are associated with an increased risk of developing prostate cancer (Cohen 1998, Wolk et al. 1998). Further support for the role of IGF action in prostate growth has come from a report that systemic administration of IGF-I increases rat prostate growth (Torring et al. 1997). Mechanistically, deregulated expression of IGF-I and constitutive activation of IGF-I receptors in basal epithelial cells resulted in tumor progression similar to that seen in human disease. In premenopausal breast cancer, but not in post-menopausal disease, women in the highest tertile of serum IGF-I levels had a significantly increased risk of developing breast cancer (Hankinson et al. 1998). These findings have been generally supported by most subsequent studies (Vadgama et al. 1999) but not all (Jernstrom and Barrett-Connor 1999). Ma et al. (2001) and Palmqvist et al. (2002) have reported positive associations between serum IGF-I and colorectal cancer risk in US, Greek and Swedish cohorts, while Probst-Hensch et al. (2003) found an association between IGF-I or IGFBP-3 levels and colorectal cancer risk in a Chinese cohort. Yu et al. (1999) reported a positive association between high IGF-I, but not IGF-II, and low IGFBP-3 levels and lung cancer risk.

INTERFERENCE IN THE IGF AXIS AS POTENTIAL THERAPEUTIC APPROACH

The results of preclinical studies indicate that a diversity of factors that antagonize IGF-IR signaling or augment IGFBP-3 function can inhibit tumor cell growth in models of human cancers. For example, reducing the number of functional IGF-IRs through the use of antisense oligonucleotides (Resnicoff et al. 1996) or an IGF-I receptor dominant negative mutant (Scotlandi et al. 2002) is associated with decreased cell growth *in vitro* and delayed tumor formation with increased survival *in vivo*. Another experimental approach to interrupt IGF association with the IGF-IR uses αIR3, a monoclonal antibody that associates with the ligand-binding domain of the IGF-IR and inhibits IGF-I-mediated effects. *In vivo* administration of αIR3 has been shown to induce a complete regression of established tumors in nearly half of the mice inoculated with Ewing's sarcoma cells (Scotlandi et al. 1996). The clinical success of small molecule tyrosine kinase inhibitors has sparked interest that resulted in the development of compounds that specifically target the IGF-IR tyrosine kinase (Blum et al. 2000). The ultimate success of such inhibitors in the clinics remains to be determined.

Attenuation of IGF-IR signaling can also result from increased levels of specific IGF-binding proteins. Increased expression of wild-type IGFBP-3 or treatment with recombinant human IGFBP-3 has been shown to inhibit cancer cell growth and to induce cancer cell death in a variety of experimental systems (Lee et al. 2002). IGFBP-3 has also been shown to have potent antitumor activity *in vivo* (Devi et al. 2002), either alone or in combination with standard chemotherapeutic agents.

Clinical studies have demonstrated that reduced IGF-I levels (Pollak et al. 1990) and increased IGFBP-3 levels (Campbell et al. 2001), which result in decreased IGF-I/IGFBP-3 ratio, can be achieved by the antiestrogen tamoxifen. These effects are important in the treatment and prevention of breast cancer. Another estrogen receptor modulator, raloxifene (used for the treatment and prevention of osteoporosis in postmenopausal women), was shown to decrease the IGF-I/IGFBP-3 ratio and has been proposed for future studies in breast cancer prevention (Torrisi et al. 2001).

DIET AND THE IGF SYSTEM

Nutritional factors are critical regulators of IGFs (Giovannucci et al. 2003). In particular, under-nutrition of either protein or energy substantially lowers IGF-I levels (Thissen et al. 1994). For example, fasting for 10 d causes a 4-fold reduction in IGF-I to levels that are usually associated with growth hormone deficiency (Clemmons et al. 1981). Over-eating may increase IGF-I somewhat, but excess calories are not nearly as strong a stimulus as nutritional restriction (Giovannucci et al. 2003). Short-term studies of protein deprivation demonstrate a potent and independent role of protein consumption in the regulation of IGF-I levels. Deficiency of essential amino acids, in particular, has a severe depressing effect on IGF-

I levels (Clemmons et al. 1985). A high carbohydrate diet increases IGF-I levels relative to a high fat diet, possibly by maintaining hepatic sensitivity to growth hormone (Clemmons et al. 1981). Reports from the Nurses' Health Study showed that higher energy, protein, and milk intakes were associated with higher levels of IGF-I (Holmes et al. 2002).

CAROTENOIDS AND THE IGF SYSTEM

The central role of the IGF system in controlling cell proliferation, differentiation, apoptosis and malignant transformation, and the fact that this system has been implicated as a major cancer risk factor make it a suitable target for cancer prevention in particular for high-risk populations. Prevention strategies based on nutritional changes have obvious advantages over drug-based intervention. As discussed below, carotenoids, as part of a diet rich in vegetables and fruits, are promising candidates.

Epidemiological Studies

Several epidemiological studies looked for an association between tomato products or lycopene consumption and components of the IGF system (Signorello et al. 2000, Vrieling et al. 2004), but only one found a possible positive association between lycopene intake and IGFBP-3 (Holmes et al. 2002). In other studies, tomato consumption was inversely associated with IGF-I levels and/or its molar ratio with IGFBP-3 in disease-free men (Gunnell et al. 2003, Mucci et al. 2001). Among foods, the consumption of cooked tomatoes was significantly inversely associated with IGF-I levels. Mucci et al. (2001) suggested that the mechanism for the anticancer effects of tomato lycopene to prevent prostate cancer may be related to its ability to decrease IGF-I blood level.

A more recent study (Tran et al. 2006) investigated whether intakes of fruits, vegetables, and antioxidants (beta-carotene, lycopene, and vitamin C) are associated with plasma IGF-I and IGFBP-3 concentrations. Although higher intake of citrus fruit and vitamin C were associated with higher concentrations of IGF-I and lower concentrations of IGFBP-3, total intakes of beta-carotene and lycopene were not related to either IGF-I or IGFBP-3 concentrations.

Mechanistic Studies in Cell Cultures

A considerable amount of data from epidemiological studies have demonstrated that carotenoids, and lycopene in particular, decrease the risk of human malignancies (Giovannucci 1999). However, the molecular mechanisms of carotenoid inhibition of cancer cell growth as well as the mode of interaction between carotenoids and the IGF system at the cellular level remain largely unclear. We have previously demonstrated that lycopene inhibits the growth of breast, endometrial, and lung cancer (Levy et al. 1995) and leukemic (Amir et al. 1999) cells. Furthermore, growth stimulation of MCF-7 human breast cancer cells by IGF-I was markedly reduced by lycopene (Karas et al. 2000, Levy et al. 1995). Interestingly, these effects were not accompanied by either necrotic or apoptotic cell death but were associated with a marked inhibition of serum- and IGF-I-stimulated cell cycle transition from G1 to S phase (Amir et al. 1999, Karas et al. 2000, Nahum et al. 2001, 2006). Key proteins participating in this cell cycle transition are shown in Fig. 4. Lycopene treatment markedly reduced the IGF-I stimulation of tyrosine phosphorylation of IRS-1. These effects were not associated with changes in the number or affinity of IGF-I receptors, but with an increase in membrane-associated IGF-binding proteins (Fig. 5). Inhibition of cell cycle progression was evident also by other carotenoids such as beta-carotene (Murakoshi et al. 1989) and fucoxanthin, a carotenoid prepared from brown algae (Okuzumi et al. 1990).

Deregulated cell cycle is one of the major hallmarks of the cancer cell. Thus, elucidation of the mode by which carotenoids inhibit cell cycle progression would provide a mechanistic basis for the anti-cancer effect of

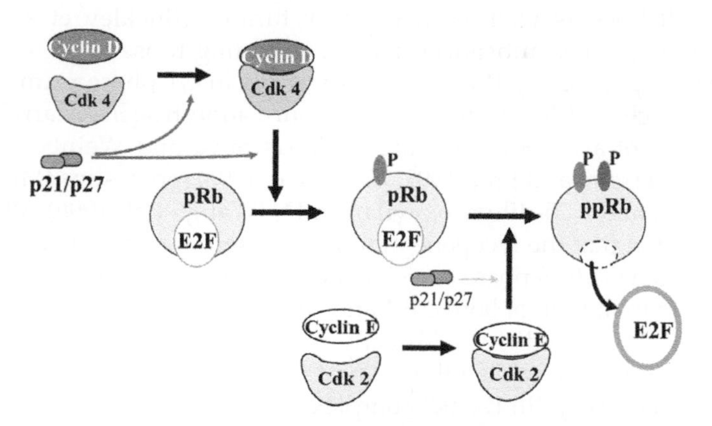

Fig. 4 Events during G1-S phase. Cell cycle transition through a late G1 checkpoint is governed by a mechanism known as the "pRb pathway". The central element of this pathway, retinoblasoma protein (pRb), is a tumor suppressor that prevents premature G1-S transition via physical interaction with transcription factors of the E2F family. The activity of pRb is regulated by an assembly of cyclins, cyclin-dependent kinases (Cdks) and Cdk inhibitors (p21/p27). Phosphorylation of pRb by Cdks results in the release of E2F, which leads to the synthesis of various cell growth-related proteins. Cdk activity is modulated in both a positive and a negative manner by cyclins and Cdk inhibitors, respectively.

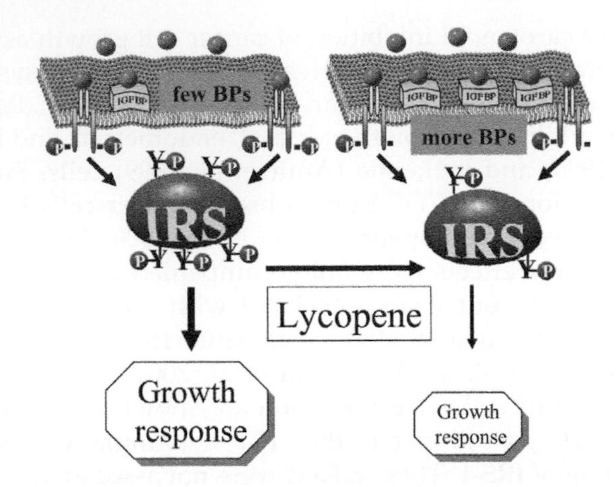

Fig. 5 Mechanism for lycopene inhibition of IGF-I induced cancer cell growth. Lycopene treatment markedly reduced the IGF-I stimulation of tyrosine phosphorylation (Y-p) of IRS. These effects were not associated with changes in the number or affinity of IGF-I receptors, but with an increase in membrane-associated IGF-binding proteins.

these micronutrients. Cell cycle progression is activated by growth factors primarily during G1 phase. The main components of the cell cycle machinery that act as growth factor sensors are the D type cyclins (Sherr 1995). Cyclin D1 is an oncogene that is overexpressed in many breast cancer cell lines as well as in primary tumors (Buckley et al. 1993). Interestingly, many anticancer agents, including those used for breast cancer therapy, convey their inhibitory effect in G1 phase primarily by reducing cyclin D1 levels, for example, pure antiestrogens (Carroll et al. 2000, Watts et al. 1994), tamoxifen (Planas Silva and Weinberg 1997), retinoids (Teixeira and Pratt 1997, Zhou et al. 1997), progestins (Musgrove et al. 1998), and 1,25-dihydroxyvitamin D_3 (Wang et al. 1996). We have demonstrated that the lycopene inhibition of serum-stimulated cell cycle traverse from G1 to S phase correlates with reduction in cyclin D1 levels, resulting in inhibition of both cdk4 and cdk2 kinase activity (Nahum et al. 2001). Inhibition of cdk4 was directly related to lower amount of cyclin D1-cdk4 complexes, while inhibition of cdk2 was related to retention of p27 molecules in cyclin E-cdk2 complexes due to the reduction in cyclin D1 level. These results, together with the fact that neither cyclin E nor cdk2 or cdk4 levels were changed by lycopene treatment, suggest that the inhibitory effect of this carotenoid on cell cycle progression is mediated primarily by down-regulation of cyclin D1 (Fig. 6). To further support this suggestion we determined the effect of lycopene on IGF-I-stimulated cell cycle progression in a clone of MCF-7 cells capable of exogenously

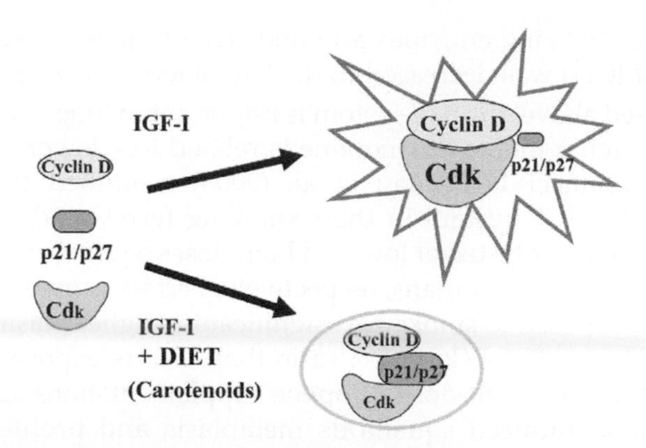

Fig. 6 Inhibitory effect of carotenoids on cell cycle progression induced by IGF-I is mediated primarily by down-regulation of cyclin D1. Growth factors affect the cell cycle apparatus primarily during G1, and the main components acting as growth factor sensors are the D-type cyclins. By increasing the level of cyclin D1 and thus the activity of Cdk2, IGF-I induce cell cycle progression and cancer cell growth. Lycopene and other carotenoids reduce the level of cyclin-D1 and thus attenuate cell cycle progression.

controlling cyclin D1 expression. We found that ectopic expression of cyclin D1 can overcome cell cycle inhibition caused by lycopene and all-trans retinoic acid (atRA) suggesting that attenuation of cyclin D1 levels by these compounds impairs the mitogenic action of IGF-I (Nahum et al. 2006).

More recently (Kanagaraj et al. 2007), the action of lycopene on components of the IGF system was examined in PC-3 androgen-independent prostate cancer cells. The cells treated with lycopene showed a significant decrease in proliferation accompanied by increased level of IGFBP-3 and a decrease in the IGF-IR expression. These changes were associated with increased apoptotic cell death. The authors suggest that the components of the IGF system may act as a positive regulator of lycopene-induced apoptosis in PC-3 cells.

Mechanistic Studies in Animal Models

The effect of lycopene on normal prostate tissue was examined to gain insight into the mechanisms by which lycopene can contribute to primary prostate cancer prevention (Herzog et al. 2005). Young rats were supplemented with 200 ppm lycopene for up to 8 wk. Lycopene accumulated predominantly in the lateral prostate lobe, which is primarily affected by malignancy. Transcriptomics analysis revealed that lycopene treatment mildly but significantly reduced gene expression of androgen-metabolizing enzymes and androgen targets. Moreover, local expression of IGF-I was decreased in the lateral lobe (Herzog et al. 2005).

As discussed above, the IGF system is important in lung cancer as well. In addition, higher intake of lycopene is related to a lower risk of lung cancer (Giovannucci 1999). Liu et al. (2003) examined the effect of lycopene on the IGF system in their smoking ferret model. Lycopene supplementation was tested at low and high doses equivalent to an intake of 15 and 60 mg/day in humans, respectively. Ferrets supplemented with lycopene and exposed to smoke had significantly higher plasma IGFBP-3 levels and a lower IGF-I/IGFBP-3 ratio than ferrets exposed to smoke alone. Both low- and high-dose lycopene supplementations substantially inhibited smoke-induced squamous metaplasia and proliferating cell nuclear antigen expression in the lungs of the ferrets. The authors concluded that lycopene may mediate its protective effects against smoke-induced lung carcinogenesis in ferrets through up-regulating IGFBP-3, which promote apoptosis and inhibit cell proliferation.

Human Intervention Studies

In a small clinical trial, a reduction in plasma IGF-I and IGFBP-3 levels was observed after tomato lycopene supplementation (Kucuk et al. 2001). However, a similar change was observed in the control group, suggesting that the reduction was not related to the lycopene treatment.

To determine whether short intervention with dietary tomato lycopene extract will affect serum levels of the IGF system components, a double-blind, randomized, placebo-controlled study was conducted in colon cancer patients (Walfisch et al. 2007). Patients, candidates for colectomy (n = 56), were recruited from the local community a few days to a few weeks before surgery. Personal and medical data were recorded. Plasma concentrations of lycopene and IGF-I, IGF-II and IGFBP-3 were measured. Plasma lycopene levels increased two-fold after supplementation with tomato lycopene extract. In the placebo-treated group, there was a small non-significant increase in lycopene plasma levels. The plasma concentration of IGF-I decreased significantly by about 25% after tomato lycopene extract supplementation, as compared to the placebo-treated group. No significant change was observed in IGFBP-3 or IGF-II, whereas the IGF-I/IGFBP-3 molar ratio decreased significantly. These results support the hypothesis that tomato lycopene extract has a role in the prevention of colon and possibly other types of cancer by modulating the plasma levels of IGF-I, which has been suggested as a risk factor for those malignancies.

In another study, a 4 wk randomized, double-blind, placebo-controlled lycopene supplementation was carried out with healthy male volunteers (Graydon et al. 2007). Plasma samples were analyzed for lycopene, IGF-I and IGFBP-3. Median change in lycopene was higher in subjects in the intervention than those on the placebo group. There was no difference in median change in IGF-I concentrations or in IGFBP-3 concentrations between intervention and control groups. However, the change in lycopene concentration was associated with the change in IGFBP-3 in the intervention group. The authors concluded that the association between change in lycopene concentration and change in IGFBP-3 in the intervention group suggests a potential effect of lycopene supplementation on IGFBP-3.

Riso et al. (2006) supplemented the diets of healthy subjects with a tomato-based drink. Twenty healthy young subjects participated in a double-blind, cross-over design study. Subjects consumed a tomato drink or a placebo drink for 26 d separated by 26 d wash-out. The tomato drink intake increased plasma lycopene, phytoene, phytofluene, and β-carotene concentrations. No significant effect of the tomato drink intake on IGF-I levels was observed. However, changes in lycopene before and after each experimental period were inversely and significantly correlated with those of IGF-I. No correlation was found with the other carotenoids. A significant reduction of IGF-I serum level was observed in subjects with the highest plasma lycopene response following the tomato drink intake. No effect was evident after the placebo treatment.

CONCLUSIONS

Growth factors, either in the blood or as part of autocrine or paracrine loops, are important for cancer cell growth. As discussed in this review, the IGF system plays a prominent role in cancer, and IGF-I have been implicated as a major cancer risk factor. Thus, attenuation of IGF signaling by dietary tomato lycopene may result in cancer prevention. The presented results of short intervention studies with healthy people or cancer patients suggest that a small but significant reduction in IGF-I activity (either decrease in IGF-I or increase in IGFBP-3) could be obtained with tomato lycopene. It should be noted that the source of the blood IGF-I and IGFBP-3 may be different in healthy people and in cancer patients. Thus, the decrease in IGF-I activity detected in healthy people is more relevant for cancer prevention. The accumulated evidence warrants further human studies of the long-term effects of tomato carotenoids and their combination with other phyto-nutrients. Such studies may provide the basis for a preventive strategy for some malignancies.

ABBREVIATIONS

IGF: insulin-like growth factor; IGFBP: IGF-binding protein; IGF-IR: Type I IGF-receptor; IGF-IIR: Type II IGF-receptor; IR: Insulin receptor; αIR3: Anti IGF Receptor, a monoclonal antibody to the ligand-binding domain of the IGF-IR; IRS-1: Insulin receptor substrate-1; Cdk: Cyclin-dependent kinase; atRA: All-trans retinoic acid

REFERENCES

Amir, H. and M. Karas, J. Giat, M. Danilenko, R. Levy, T. Yermiahu, J. Levy, and Y. Sharoni. 1999. Lycopene and 1,25-dihydroxyvitamin-D3 cooperate in the inhibition of cell cycle progression and induction of differentiation in HL-60 leukemic cells. Nutr. Cancer 33: 105-112.

Blum, G. and A. Gazit, and A. Levitzki. 2000. Substrate competitive inhibitors of IGF-1 receptor kinase. Biochemistry 39: 15705-12.

Buckley, M.F. and K.J. Sweeney, J.A. Hamilton, R.L. Sini, D.L. Manning, R.I. Nicholson, A. deFazio, C.K. Watts, E.A. Musgrove, and R.L. Sutherland. 1993. Expression and amplification of cyclin genes in human breast cancer. Oncogene 8: 2127-2133.

Campbell, M.J. and J.V. Woodside, J. Secker-Walker, A. Titcomb, and A.J. Leathem. 2001. IGF status is altered by tamoxifen in patients with breast cancer. Mol. Pathol. 54: 307-310.

Carroll, J.S. and O.W. Prall, E.A. Musgrove, and R.L. Sutherland. 2000. A pure estrogen antagonist inhibits cyclin E-Cdk2 activity in MCF-7 breast cancer cells and induces accumulation of p130-E2F4 complexes characteristic of quiescence. J. Biol. Chem. 275: 38221-38229.

Chan, J.M. and M.J. Stampfer, E. Giovannucci, P.H. Gann, J. Ma, P. Wilkinson, C.H. Hannekens, and M. Pollak. 1998. Plasma insulin-like growth factor-I and prostate cancer risk: A prospective study. Science. 279: 563-566.

Clemmons, D.R. and A. Klibanski, L.E. Underwood, J.W. McArthur, E.C. Ridgway, I.Z. Beitins, and J.J. Van Wyk. 1981. Reduction of plasma immunoreactive somatomedin C during fasting in humans. J. Clin. Endocrinol. Metab. 53: 1247-1250.

Clemmons, D.R. and M.M. Seek, and L.E. Underwood. 1985. Supplemental essential amino acids augment the somatomedin-C/insulin-like growth factor I response to refeeding after fasting. Metabolism 34: 391-395.

Cohen, P. 1998. Serum insulin-like growth factor-I levels and prostate cancer risk—interpreting the evidence. J. Natl. Cancer Inst. 90: 876-879.

Devi, G.R. and C.C. Sprenger, S.R. Plymate, and R.G. Rosenfeld. 2002. Insulin-like growth factor binding protein-3 induces early apoptosis in malignant prostate cancer cells and inhibits tumor formation in vivo. Prostate 51: 141-152.

Giovannucci, E. 1999. Tomatoes, tomato-based products, lycopene, and cancer: review of the epidemiologic literature. J. Natl. Cancer Inst. 91: 317-331.

Giovannucci, E. and M. Pollak, Y. Liu, E.A. Platz, N. Majeed, E.B. Rimm, and W.C. Willett. 2003. Nutritional predictors of insulin-like growth factor I and their relationships to cancer in men. Cancer Epidemiol. Biomarkers Prev. 12: 84-89.

Graydon, R. and S.E. Gilchrist, I.S. Young, U. Obermuller-Jevic, O. Hasselwander, and J.V. Woodside. 2007. Effect of lycopene supplementation on insulin-like growth factor-1 and insulin-like growth factor binding protein-3: a double-blind, placebo-controlled trial. Eur. J. Clin. Nutr. 61: 1196-1200.

Gunnell, D. and S.E. Oliver, T.J. Peters, J.L. Donovan, R. Persad, M. Maynard, D. Gillatt, A. Pearce, F.C. Hamdy, D.E. Neal, and J.M. Holly. 2003. Are diet-prostate cancer associations mediated by the IGF axis? A cross-sectional analysis of diet, IGF-I and IGFBP-3 in healthy middle-aged men. Br. J. Cancer 88: 1682-1686.

Hankinson, S.E. and W.C. Willett, G.A. Colditz, D.J. Hunter, D.S. Michaud, B. Deroo, B. Rosner, F.E. Speizer, and M. Pollak. 1998. Circulating concentrations of insulin-like growth factor I and risk of breast cancer. Lancet. 351: 1393-1396.

Herzog, A. and U. Siler, V. Spitzer, N. Seifert, A. Denelavas, P.B. Hunziker, W. Hunziker, R. Goralczyk, and K. Wertz. 2005. Lycopene reduced gene expression of steroid targets and inflammatory markers in normal rat prostate. Faseb. J. 19: 272-274.

Holmes, M.D. and M.N. Pollak, W.C. Willett, and S.E. Hankinson. 2002. Dietary correlates of plasma insulin-like growth factor I and insulin-like growth factor binding protein 3 concentrations. Cancer Epidemiol. Biomarkers Prev. 11: 852-861.

Jernstrom, H. and E. Barrett-Connor. 1999. Obesity, weight change, fasting insulin, proinsulin, C-peptide, and insulin-like growth factor-1 levels in women with and without breast cancer: the Rancho Bernardo Study. J. Women's Health Gend. Based. Med. 8: 1265-1272.

Kanagaraj, P. and M.R. Vijayababu, B. Ravisankar, J. Anbalagan, M.M. Aruldhas, and J. Arunakaran. 2007. Effect of lycopene on insulin-like growth factor-I, IGF binding protein-3 and IGF type-I receptor in prostate cancer cells. J. Cancer Res. Clin. Oncol. 133: 351-359.

Karas, M. and H. Amir, D. Fishman, M. Danilenko, S. Segal, A. Nahum, A. Koifmann, Y. Giat, J. Levy, and Y. Sharoni. 2000. Lycopene interferes with cell cycle progression and insulin-like growth factor I signaling in mammary cancer cells. Nutr. Cancer 36: 101-111.

Kucuk, O. and F.H. Sarkar, W. Sakr, Z. Djuric, M.N. Pollak, F. Khachik, Y.W. Li, M. Banerjee, D. Grignon, J.S. Bertram, J.D. Crissman, E.J. Pontes, and D.P. Wood Jr. 2001. Phase II randomized clinical trial of lycopene supplementation before radical prostatectomy. Cancer Epidemiol. Biomarkers Prev. 10: 861-868.

Lee, H.Y. and K.H. Chun, B. Liu, S.A. Wiehle, R.J. Cristiano, W.K. Hong, P. Cohen, and J.M. Kurie. 2002. Insulin-like growth factor binding protein-3 inhibits the growth of non-small cell lung cancer. Cancer Res. 62: 3530-3537.

LeRoith, D. and C.T. Roberts Jr. 2003. The insulin-like growth factor system and cancer. Cancer Lett. 195: 127-137.

Levy, J. and E. Bosin, B. Feldman, Y. Giat, A. Miinster, M. Danilenko, and Y. Sharoni. 1995. Lycopene is a more potent inhibitor of human cancer cell proliferation than either α-carotene or β-carotene, Nutr Cancer. 24: 257-267.

Liu, C. and F. Lian, D.E. Smith, R.M. Russell, and X.D. Wang. 2003. Lycopene supplementation inhibits lung squamous metaplasia and induces apoptosis via up-regulating insulin-like growth factor-binding protein 3 in cigarette smoke-exposed ferrets. Cancer Res. 63: 3138-3144.

Ma, J. and E. Giovannucci, M. Pollak, J.M. Chan, J.M. Gaziano, W. Willett, and M.J. Stampfer. 2001. Milk intake, circulating levels of insulin-like growth factor-I, and risk of colorectal cancer in men. J. Natl. Cancer Inst. 93: 1330-1336.

Mucci, L.A. and R. Tamimi, P. Lagiou, A. Trichopoulou, V. Benetou, E. Spanos, and D. Trichopoulos. 2001. Are dietary influences on the risk of prostate cancer mediated through the insulin-like growth factor system? BJU Intl. 87: 814-820.

Murakoshi, M. and J. Takayasu, O. Kimura, E. Kohmura, H. Nishino, J. Okuzumi, T. Sakai, T. Sugimoto, J. Imanishi, and R. Iwasaki. 1989. Inhibitory effects of α-Carotene on proliferation of the human neuroblastoma cell line GOTO. J. Natl. Cancer Inst. 81: 1649-1652.

Musgrove, E.A. and A. Swarbrick, C.S. Lee, A.L. Cornish, and R.L. Sutherland. 1998. Mechanisms of cyclin-dependent kinase inactivation by progestins. Mol. Cell. Biol. 18: 1812-1825.

Nahum, A. and K. Hirsch, M. Danilenko, C.K. Watts, O.W. Prall, J. Levy, and Y. Sharoni. 2001. Lycopene inhibition of cell cycle progression in breast and endometrial cancer cells is associated with reduction in cyclin D levels and retention of p27(Kip1) in the cyclin E-cdk2 complexes. Oncogene 20: 3428-3436.

Nahum, A. and L. Zeller, M. Danilenko, O.W. Prall, C.K. Watts, R.L. Sutherland, J. Levy, and Y. Sharoni. 2006. Lycopene inhibition of IGF-induced cancer cell growth depends on the level of cyclin D1. Eur. J. Nutr. 45: 275-282.

Okuzumi, J. and H. Nishino, M. Murakoshi, A. Iwashima, Y. Tanaka, T. Yamane, Y. Fujita, and T. Takahashi. 1990. Inhibitory effects of fucoxanthin, a natural carotenoid, on N-myc expression and cell cycle progression in human malignant tumor cells. Cancer Lett. 55: 75-81.

Palmqvist, R. and G. Hallmans, S. Rinaldi, C. Biessy, R. Stenling, E. Riboli, and R. Kaaks. 2002. Plasma insulin-like growth factor 1, insulin-like growth factor binding protein 3, and risk of colorectal cancer: a prospective study in northern Sweden. Gut 50: 642-646.

Planas Silva, M.D. and R.A. Weinberg. 1997. Estrogen-dependent cyclin E-cdk2 activation through p21 redistribution. Mol. Cell. Biol. 17: 4059-4069.

Pollak, M.N. and J. Costantino, and C. Polychronakos. 1990. Effect of tamoxifen on serum insulin-like growth factor I levels of stage I breast cancer patients. J. Natl. Cancer Inst. 82: 1693-1697.

Probst-Hensch, N.M. and H. Wang, V.H. Goh, A. Seow, H.P. Lee, and M.C. Yu. 2003. Determinants of circulating insulin-like growth factor I and insulin-like growth factor binding protein 3 concentrations in a cohort of Singapore men and women. Cancer Epidemiol. Biomarkers Prev. 12: 739-746.

Resnicoff, M. and J. Tjuvajev, H.L. Rotman, D. Abraham, M. Curtis, R. Aiken, and R. Baserga. 1996. Regression of C6 rat brain tumors by cells expressing an antisense insulin-like growth factor I receptor RNA. J. Exp. Ther. Oncol. 1: 385-389.

Riso, P. and A. Brusamolino, A. Martinetti, and M. Porrini. 2006. Effect of a tomato drink intervention on insulin-like growth factor (IGF)-1 serum levels in healthy subjects. Nutr. Cancer 55: 157-162.

Scotlandi, K. and S. Benini, M. Sarti, M. Serra, P.L. Lollini, D. Maurici, P. Picci, M.C. Manara, and N. Baldini. 1996. Insulin-like growth factor I receptor-mediated circuit in Ewing's sarcoma/peripheral neuroectodermal tumor: a possible therapeutic target. Cancer Res. 56: 4570-4574.

Scotlandi, K. and S. Avnet, S. Benini, M.C. Manara, M. Serra, V. Cerisano, S. Perdichizzi, P.L. Lollini, C. De Giovanni, L. Landuzzi, and P. Picci. 2002. Expression of an IGF-I receptor dominant negative mutant induces apoptosis, inhibits tumorigenesis and enhances chemosensitivity in Ewing's sarcoma cells. Intl. J. Cancer 101: 11-16.

Sherr, C.J. 1995. D-type cyclins. Trends Biochem. Sci. 20: 187-190.

Signorello, L.B. and H. Kuper, P. Lagiou, J. Wuu, L.A. Mucci, D. Trichopoulos, and H.O. Adami. 2000. Lifestyle factors and insulin-like growth factor 1 levels among elderly men. Eur. J. Cancer Prev. 9: 173-178.

Teixeira, C. and M.A.C. Pratt. 1997. CDK2 is a target for retinoic acid-mediated growth inhibition in MCF-7 human breast cancer cells. Mol. Endocrinol. 11: 1191-1202.

Thissen, J.P. and J.M. Ketelslegers, and L.E. Underwood. 1994. Nutritional regulation of the insulin-like growth factors. Endocr. Rev. 15: 80-101.

Torring, N. and L. Vinter-Jensen, S.B. Pedersen, F.B. Sorensen, A. Flyvbjerg, and E. Nexo. 1997. Systemic administration of insulin-like growth factor I (IGF-I) causes growth of the rat prostate. J. Urol. 158: 222-227.

Torrisi, R. and L. Baglietto, H. Johansson, G. Veronesi, B. Bonanni, A. Guerrieri-Gonzaga, B. Ballardini, and A. Decensi. 2001. Effect of raloxifene on IGF-I and IGFBP-3 in postmenopausal women with breast cancer. Br. J. Cancer 85: 1838-1841.

Tran, C.D. and C. Diorio, S. Berube, M. Pollak, and J. Brisson. 2006. Relation of insulin-like growth factor (IGF) I and IGF-binding protein 3 concentrations with intakes of fruit, vegetables, and antioxidants. Amer. J. Clin. Nutr. 84: 1518-1526.

Vadgama, J.V. and Y. Wu, G. Datta, H. Khan, and R. Chillar. 1999. Plasma insulin-like growth factor-I and serum IGF-binding protein 3 can be associated with the progression of breast cancer, and predict the risk of recurrence and the probability of survival in African-American and Hispanic women. Oncology 57: 330-340.

Vrieling, A. and D.W. Voskuil, H.B. Bueno de Mesquita, R. Kaaks, P.A. van Noord, L. Keinan-Boker, C.H. van Gils, and P.H. Peeters. 2004. Dietary determinants of circulating insulin-like growth factor (IGF)-I and IGF binding proteins 1, -2 and -3 in women in the Netherlands. Cancer Causes Control 15: 787-796.

Walfisch, S. and Y. Walfisch, E. Kirilov, N. Linde, H. Mnitentag, R. Agbaria, Y. Sharoni, and J. Levy. 2007. Tomato lycopene extract supplementation decreases insulin-like growth factor-I levels in colon cancer patients. Eur. J. Cancer Prevention 16: 298-303.

Wang, Q.M. and J.B. Jones, and G.P. Studzinski. 1996. Cyclin-dependent kinase inhibitor p27 as a mediator of the G1-S phase block induced by 1,25-dihydroxyvitamin D3 in HL60 cells. Cancer Res. 56: 264-267.

Watts, C.K. and K.J. Sweeney, A. Warlters, E.A. Musgrove, and R.L. Sutherland. 1994. Antiestrogen regulation of cell cycle progression and cyclin D1 gene expression in MCF-7 human breast cancer cells. Breast Cancer Res. Treat. 31: 95-105.

Wolk, A. and C.S. Mantzoros, S.O. Andersson, R. Bergstrom, L.B. Signorello, P. Lagiou, H.O. Adami, and D. Trichopoulos. 1998. Insulin-like growth factor 1 and prostate cancer risk: a population-based, case-control study. J. Natl. Cancer Inst. 90: 911-915.

Yu, H. and M.R. Spitz, J. Mistry, J. Gu, W.K. Hong, and X. Wu. 1999. Plasma levels of insulin-like growth factor-I and lung cancer risk: a case-control analysis. J. Natl. Cancer Inst. 91: 151-156.

Zhou, Q. and M. Stetler Stevenson, and P.S. Steeg. 1997. Inhibition of cyclin D expression in human breast carcinoma cells by retinoids in vitro. Oncogene 15: 107-115.

21

Tomato Paste and Benign Prostate Hyperplasia

Magda Edinger de Souza[1,2], Walter José Koff[1,2] and Tania Weber Furlanetto[1,3]

[1]Rua Demétrio Ribeiro, 244/603, Centro Porto Alegre, RS, Brazil, 90010-312

[2]Hospital de Clínicas de Porto Alegre Rua Ramiro Barcelos, 2350/835 Porto Alegre, RS, Brazil, 90035-903

[3]Hospital de Clínicas de Porto Alegre Rua Ramiro Barcelos, 2350/700 Porto Alegre, RS, Brazil, 90035-903

ABSTRACT

The prostate, a major male accessory gland, is a potential source of serious disorders affecting health and the quality of life in older men. Benign prostate hyperplasia (BPH) is the most prevalent disease in the prostate, and it represents a considerable health problem to aging men through its associated signs, symptoms, and complications. Although BPH is a major public health problem, little is known about its risk factors, including diet and other lifestyle factors. There seems to be a correlation between intake of tomato and its derivative products and the prevention of prostate diseases. Epidemiological studies have shown an inverse association between dietary intake of tomato products and prostate cancer risk. Nevertheless, there are only a few studies on the correlation between tomato and its byproducts and BPH. Tomato is an excellent source of nutrients and phytochemicals that may contribute to the health of the prostate. Lycopene is one of the main substances that might provoke such effects in prostate. Many mechanisms are proposed for the role of lycopene in the prevention of prostate diseases: antioxidant function, inhibition of cell cycle progression, increase in apoptotic index, increase of gap-junctional

A list of abbreviations is given before the references.

communication, inhibition of insulin-like growth factor-1 (IGF-1) signal transduction, inhibition of interleukin-6 expression, induction of phase II enzymes and inhibition of androgen activation and signaling. This carotenoid of red pigmentation is more bioavailable in tomato-processed products, such as tomato paste. Tomato paste is obtained by the concentration of ripened and healthy fruit pulp through an adequate technological process. Studies involving food have important limitations, such as the difficulty in performing a double-blind study. However, they should be encouraged because the bioactive substances, vitamins and minerals, are more bioavailable when consumed with foods than as supplements.

INTRODUCTION

The connection between diet and health has been recognized for many years. About 2,500 years ago Hippocrates used to say, "Let food be the medicine and medicine be the food." However, in the past few decades more and more has been discovered about how diet can prevent diseases (Andlauer et al. 1998).

The term *functional food* first appeared in 1993 in *Nature* magazine and was defined as "Any food or part of food with health-promoting and/or disease-preventing property, besides meeting the basic nutritional requirements" (Swinbanks and O'Brien 1993). The concept of functional foods had already been proposed in 1984, when a group of researchers in Japan started a nationwide project, sponsored by the government, to explore the interface between medicine and food (Arai 1996).

Currently, tomato and its byproducts are among the most studied functional foods. Several epidemiologic studies support the hypothesis that an increased intake of tomatoes and tomato-based products can influence prostate health, mainly reducing the risk of prostate cancer (Mills et al. 1989, Tzonou et al. 1999, Edinger and Koff 2006). These observations gave rise to the hypothesis that lycopene, the main carotenoid in tomatoes, may be the active component of tomatoes and their byproducts (Clinton 1998, Giovannucci et al. 2002).

This chapter specifically addresses the correlation between intake of tomato paste, which is the tomato byproduct with the highest lycopene concentration, and benign prostate hyperplasia (BPH), the most prevalent prostate disease (Lee et al. 1995).

TOMATO PASTE OR TOMATO CONCENTRATE

Tomatoes are important for worldwide agriculture, in terms of production and economic value, and tomato production is among the more industrialized in the world. In the United States, tomato consumption *per capita* is second only to that of potato (Garcia-Cruz et al. 1997, Gann and Khachik 2003). Over 80% of the processed tomatoes are consumed in the form of juice, pulp, sauce, puree and paste (Shi and Le Maguer 2000).

Tomato paste is produced from the concentration of ripened and healthy fruit pulp through an adequate technological process, with the addition of 1 to 3% of sodium chloride. It is low in calories and total fat. In Table 1 are described some nutritional facts of tomato paste used in a previous study (Edinger and Koff 2006). It is also referred to as "tomato mass" or "tomato concentrate", easily found in any supermarket all over the world, usually canned to avoid the addition of preservatives and therefore a natural and safe product.

Table 1 Nutritional information for 50 g of tomato paste.

Calories	18 Kcal
Carbohydrates	2.8 g
Proteins	1.3 g
Total fat	0.2 g
Saturated fats	0.1 g
Sodium	180 mg
Lycopene	13 mg

Source: Edinger and Koff (2006).

Tomato paste is used all over the world in sauces, pastas, lasagnas, pizzas or other dishes. The price of canned tomato paste is generally accessible to all social classes. In Brazil, a 350 g can of tomato paste costs around US$ 0.50; in the United States, the price varies from US$ 1.00 to 1.50.

TOMATO PASTE: A SOURCE OF LYCOPENE

The most frequent pigment in tomato is the carotenoid lycopene (ΨΨ-carotene), which gives the fruit its red coloring (Wertz et al. 2004). Lycopene corresponds to 80-90% of all tomato pigments. Other carotenoids, such as α-carotene, β-carotene, ξ-carotene, γ-carotene, phytofluene, neurosporene, lutein and β-cryptoxanthin, are found in small amounts (Curl 1961).

Other red fruits and vegetables can also be sources of lycopene (Hart and Scott 1995, Shi and Le Maguer 2000, Edinger 2005), but tomato and its derivative products are considered the largest sources of lycopene in the human diet (Takeoka et al. 2001), as illustrated in Table 2. In the United States, they correspond to over 80% of the lycopene intake (Arab and Steck 2000).

Table 2 Lycopene content in some red foods.

Food	*Lycopene content (mg/100 g)*
Tomato paste	18.4-26.8
Tomato	0.5-15.0
Red guava	5.2-5.5
Watermelon	2.3-7.2
Tomato pulp	8.6
Ketchup	6.8
Papaya	0.1-5.3
Pepper sauce	4.9
Grapefruit	0.4-3.4
Guava paste	2.5
Guava jam	1.9
Carrot	0.7-0.8
Guava juice	1.2
Guava jelly	0.9
Pumpkin	0.4-0.5
Strawberry jelly	0.3
Apple pulp	0.1-0.2
Sweet potato	0.0-0.1
Apricot	0.0-0.1

Source: Edinger (2005), Hart and Scott (1995), Shi and Le Maguer (2000).

The amount of lycopene in fresh tomatoes depends on the variety, degree of maturation, and environmental conditions during ripening. Typically, tomatoes contain 3-5 mg of lycopene in 100 g of raw matter. Some intense red varieties have over 15 mg lycopene in 100 g tomato, while the yellow varieties have 0.5 mg lycopene (Hart and Scott 1995). Thermal processing ruptures the tomato cell wall and allows lycopene extraction from chromoplasts (Gann et al. 1999), so processed or cooked products present the highest amount of bioavailable lycopene (Gartner et al. 1997).

A study (Pimentel 2003) conducted in Porto Alegre, Brazil, compared the content of lycopene in some tomato varieties and processed tomato products, in different parts of the world. The results are presented in Table 3.

Trans-lycopene is the most important geometric isomer in fresh tomatoes, but a change in structure results in processed tomato products with 1.7 to 10% of cis-isomers. Isomerization of trans to mono or poly-cis-lycopene can occur through light, through thermal energy or during chemical reactions (Agarwal and Rao 2000, Bramley 2000).

Table 3 Lycopene content of food in Porto Alegre, Brazil (Pimentel 2003), and other locations.

Products	*Porto Alegre (mg/100 g)*	*Other locations (mg/100 g)*	*References*
Fresh tomato	3.7		Tonucci et al. 1995
	3.5	1.6-4.8	Hart and Scott 1995
Longa Vida	8.2		
Super Marmante			
Santa Cruz Kada	6.6	6.5-19.4	Tonucci et al. 1995
Processed tomato	9.3	9.3	Nguyen and Schwartz 1999
Sauce	10.0	10.3-41.4	Tavares and Rodrigues-Amaya 1994
Pulp	27.2	30.1	Nguyen and Schwartz 1999
Ketchup			
Paste			

It was observed that blood concentration of lycopene was higher in people who consumed heat-processed tomato products than in those who consumed fresh tomato (Giovannucci et al. 1995, Gartner et al. 1997). It was also observed that tomato byproducts subjected to 100°C for one hour can have 20-30% of cis-isomers of lycopene (Stahl and Sies 1992). However, some studies suggest only a small increase (< 10%) in the cis-isomer content (Boileau et al. 2002). Thus, thermal processing, such as cooking, and destruction of mechanical texture, such as cutting and triturating, are useful methods to increase the bioavailability of lycopene by breaking the membrane of chromoplasts (Shi and Le Maguer 2000).

Boileau et al. (1999) demonstrated that cis-isomers of lycopene are more bioavailable than the trans structures, probably because cis-isomers are more soluble in the biliary acid micelles and can preferably be incorporated into the chylomicrons.

Lycopene is the most abundant carotenoid present in the prostate (Clinton et al. 1996), but a variety of other tomato carotenoids accumulate as well, including phytoene, phytofluene, β-carotene, ξ-carotene and γ-carotene (Khachik et al. 2002).

PROSTATE AND BENIGN PROSTATE HYPERPLASIA

The prostate is a gland that is part of the male reproductive system (Fig. 1). Its name derives from Greek *pro-histamai,* which means "being ahead", a reference to the distal structure from the bladder. Its shape is compared to a chestnut and it is crossed by the urethra. The prostate is responsible for producing prostatic liquid, a secretion that, combined with products of the

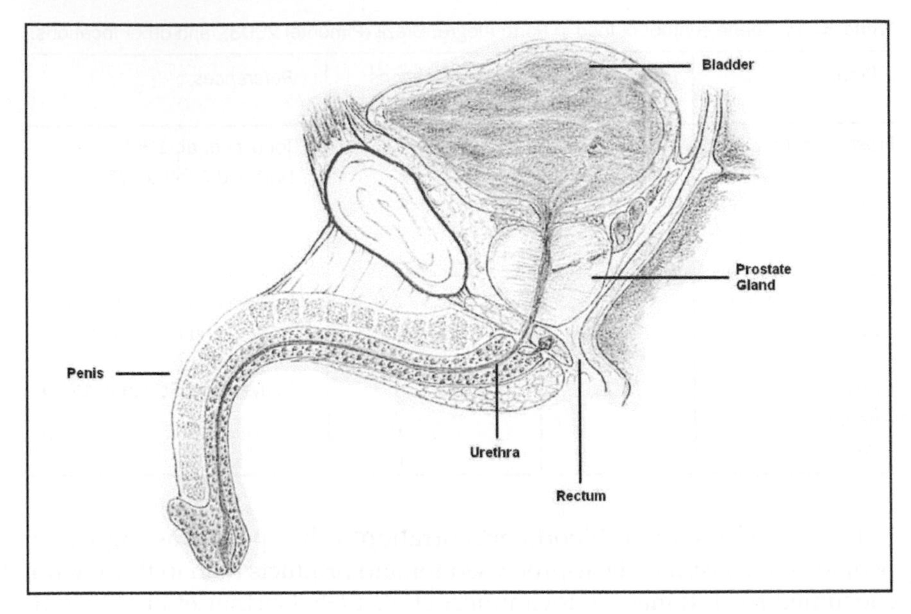

Fig. 1 Prostate location

seminal vesicle and paraurethral gland, constitutes the sperm, the liquid secreted during ejaculation. The prostate and seminal vesicle secretions represent 70% of the sperm. The prostatic secretion participates in the nutrition and preservation of spermatozoids and contains spermina, which acts on sperm liquefaction (Netto Júnior and Wroclawski 2000).

The prostate is affected by the two most frequent diseases in older men: BPH and prostate cancer (Berry et al. 1984, Brasil 2006). Although studies have shown that mortality rates due to BPH are low, 0 to 0.25%, this disease affects the quality of life (Furuya et al. 2006, Rassweiler et al. 2006) and increases health care costs. The cost of BPH treatment in Brazil is estimated to be US$ 2.26 to 3.83 billion a year (Suaid et al. 2003), and in the United States it represents the largest expense incurred in health care, over US$ 4 billion a year (Taub and Wei 2006).

Benign prostate hyperplasia is the most prevalent disease of the prostate, with about 50% of men presenting histological evidences of BPH at the age of 50 and 90% at the age of 80. It is considered a progressive disease, defined as a continuous prostate enlargement, leading to increasing symptoms and risk of complications with age, such as acute urinary retention (Lee et al. 1995).

Benign prostate hyperplasia starts in the prostate transition or paraurethral zone, under the hormonal action of 5-α-dihydrotestosterone (DHT), an active metabolite of testosterone transformed by 5-α-reductase, on the epithelium of ducts and acini. Substances such as epithelial growth factor, fibroblast growth factor, vascular growth factor, and insulin-like growth factor-1 (IGF-1) stimulate the prostate stroma hyperplasia (smooth and conjunctive muscle). The combined action of DHT and growth factors increases the infravesical resistance, determining instability of the detrusor and the progressive increase of vesical urine residue (Cury 2005).

Dietary and nutritional factors can change the risk of BPH through a number of mechanisms, but the scientific literature is not sufficiently clear on these topics (Gu 1997, Lagiou et al. 1999, Suzuki et al. 2002, Bravi et al. 2006).

Studies designed to evaluate the relationship between BPH and consumption of tomato or its byproducts, or lycopene supplementation, are rare. There are more studies about tomato and prostate cancer (Etminan et al. 2004, Edinger and Koff 2006).

STUDIES INVOLVING TOMATO AND ITS DERIVATIVE PRODUCTS OR LYCOPENE SUPPLEMENTATION IN PROSTATE HEALTH

As previously mentioned, studies on the correlation between tomato and its byproducts and BPH are rare. For this reason, most studies mentioned below are related to prostate cancer. However, we believe that they should be addressed in this section, as the carcinogenic process of the prostate has some similarity with benign prostate enlargement. For instance, DHT and IGF-1 have been implicated in benign and malignant tumor prostate enlargement (Furstenberger and Senn 2002, Oliver et al. 2004). Serum prostate-specific antigen (PSA) levels are significantly increased in prostate cancer patients, as well as in patients with infection or BPH (Brawer 2000, Balk et al. 2003).

Cohort Study of Diet, Lifestyle and Prostate Cancer

This cohort study started in 1976, and it was one of the first to establish an association between tomato consumption and the risk of prostate cancer. It included about 14,000 Adventist men, who answered a food frequency questionnaire for 7 d. Six years after that, 180 men had developed prostate cancer. Consumption of tomato, bean, lentil and pea and its byproducts was significantly associated with a reduced risk of prostate cancer. When comparing men who consumed tomato five times a week to those who consumed less than a portion a week, the relative risk (RR) of prostate cancer was 0.60 (95% confidence interval [CI] = 0.37-0.97, P = 0.02) (Mills et al. 1989).

Carotenoid and Retinol Intake in Relation to the Risk of Prostate Cancer

This study gave more strength to the hypothesis that foods with tomato can prevent prostate cancer. It was a prospective cohort study that started in 1986, with the purpose of investigating the relation between several carotenoids and the risk of prostate cancer. It involved approximately 47,000 men, all health professionals (Health Professionals Follow-up Study). They answered a detailed questionnaire, reporting their medical conditions, lifestyle, dietary habits and nutrition in 1988, 1990 and 1992. In 1992, 812 new cases of prostate cancer had been reported. Some fruits and vegetables were significantly associated with reduced risk of prostate cancer. The consumption of two to four portions of raw tomato a week was associated with a reduction of 26% in the risk of prostate cancer when compared to the consumption of one portion a week (RR = 0.74, 95% CI = 0.58-0.93, P = 0.03). In addition, the intake of tomato sauce was also significantly associated with a reduced risk of prostate cancer by 34%, when consumed two to four times a week, compared to no consumption at all (RR = 0.66, 95% CI = 0.49-0.90, P = 0.001). When all tomato sources were combined, the consumption of > 10 portions a week was associated with a significant reduction of 35% in the risk of prostate cancer, when compared to consumption of a little more than 1.5 portion a week (RR = 0.65, 95% CI = 0.44-0.95, P = 0.01) (Giovannucci et al. 1995).

Tomato Products, Lycopene and the Risk of Prostate Cancer

This study is a continuation of the Health Professionals Follow-up Study 1995 (Giovannucci et al. 1995). Since its first publication, 1,708 new cases of prostate cancer were reported, totaling 2,481 cases, in 1998. The later publication emphasized two hypotheses: the protective effect of tomato-based products can be higher against prostate cancer in older individuals and against prostate cancer in advanced stage. The relationship between tomato sauce intake and the risk of prostate cancer was not observed in men less than 65 years old (RR = 0.89, 95% CI = 0.67-1.17, P = 0.20), but there was a considerable risk reduction in men older than 65 (RR = 0.69, 95% CI = 0.56-0.84, P = 0.001). A reduction in metastasis risk was observed in those who consumed tomato or its byproducts twice a week, compared to those who consumed less than one portion a week (RR = 0.34, 95% CI = 0.13-0.90, P = 0.01) (Giovannucci et al. 2002).

Diet and Prostate Cancer: A Case-control Study in Greece

This study compared the dietary habits of 320 men with prostate cancer (cases) and 246 men without prostate cancer (controls), and reported that men with prostate cancer had a lower intake of cooked ($P < 0.005$) and raw tomato ($P = 0.12$). The authors concluded that the increased intake of cooked tomato was associated with a 15% reduced risk of prostate cancer (Tzonou et al. 1999).

High Serum Lycopene Levels and Reduced Risk of Prostate Cancer

This study was conducted to determine whether serum concentrations of different antioxidants, such as carotenoids, α-tocopherol or γ-tocopherol, have any connection with the risk of prostate cancer. It was a case-control study, and serum samples were obtained from individuals who participated in a double-blind randomized clinical trial with β-carotene (Hennekens et al. 1996). Thirteen years later, 578 men had developed prostate cancer (cases) and were compared to 1,294 controls sorted by age and smoking habit. Serum lycopene levels were defined as high (> 580 μg/L) and low (< 262 μg/L). Among all the antioxidants studied, only lycopene was significantly lower in prostate cancer patients ($P = 0.04$ for all cases). The odds ratio (OR) for prostate cancer was not significantly reduced in men with high serum lycopene levels (OR = 0.75, 95% CI = 0.54-1.06, $P = 0.12$), but prostate cancer was less aggressive in this group (OR = 0.56, 95% CI = 0.34-0.91, $P = 0.05$) (Gann et al. 1999).

Lycopene and the Modulation of Carcinogenesis Biomarkers

This study investigated 32 men with prostate cancer who received 30 mg/d of lycopene through tomato paste, distributed in food (pasta, lasagna, sauces) for 3 wk, before being subjected to radical prostatectomy. Both serum and prostate tissue were analyzed before and after tomato paste intake. Serum and prostate lycopene levels were higher by 1.97 and 2.92 times, respectively ($P < 0.001$). Serum PSA and 8-OH-deoxyguanosine levels decreased by 17.5% ($P < 0.002$) and 21% ($P < 0.005$), respectively, after lycopene administration.(Bowen et al. 2002). These results should be interpreted cautiously because there was no control group.

Powder Tomato, Lycopene and Energy Restriction in Mice

A very interesting study was performed with 194 mice that received three different diets: diet with 20% caloric restriction, diet with lycopene supplementation (0.025%) and diet with tomato powder (10%). When compared to the control group, mice fed tomato powder had a reduction of 26% in mortality related to prostate cancer (hazard ratio [HR] = 0.74, 95% CI = 0.59-0.93, P = 0.009). Lycopene was not associated with decreased mortality (HR = 0.91, 95% CI = 0.61-1.35, P = 0.630). However, caloric restriction diet reduced independently the specific mortality of prostate cancer (HR = 0.68, 95% CI = 0.49-0.96, P = 0.029). These data suggest that tomatoes have other substances, besides lycopene, that can modify prostate carcinogenesis. The mechanisms mediating the benefits associated with low-calorie diet and tomato phytochemicals are probably independent (Boileau et al. 2003).

Meta-analysis of Observational Studies

In total, 11 case-control studies and 10 cohort or nested case-control studies were included in this meta-analysis. The authors concluded that tomato-based products can play a role in prostate cancer prevention. There is also an indication that specific components of these foods, besides lycopene, can be related to benefits for prostate health (Etminan et al. 2004).

Tomato Paste Intake by Patients with BPH

This study was the first published study that specifically correlated a tomato-based product (in natural form and without supplementary lycopene addition) with BPH. This non-controlled study evaluated 43 BPH patients with serum PSA levels between 4 and 10 ng/mL, before and after the intake of 50 g tomato paste daily, mixed or not with other foods, for 10 wk. Mean serum PSA levels were 6.51 ± 1.48 ng/mL before tomato paste ingestion and decreased to 5.81 ± 1.58 ng/mL at the end of the observation period (P = 0.005). It was also observed that 88.3% of the patients considered the product intake good. The study concluded that tomato paste ingestion had some effect on prostate biology, but it cannot be asserted that such effect can improve BPH. Additional studies are required, mainly controlled clinical trials, with tomato and its byproducts, concerning the diseases that affect prostate (Edinger and Koff 2006).

Some Studies in Progress

Two studies are being developed to evaluate the effect of the intake of lycopene or tomato and its byproducts on BPH.

One study (van Breemen 2005) is already in the result analysis process. It is a controlled double-blind phase II clinical trial evaluating the effect of lycopene supplementation (30 mg/d) in a series of biological markers in men with prostate cancer or BPH (Fig. 2). It is a very interesting and important study, but it does not analyze the intake of tomato or its byproducts. As mentioned before, tomato is likely to have substances other than lycopene that can provide benefits to prostate health, or the lycopene present in tomato may be better utilized by the human body when combined with other tomato components. It is known that tomato ingestion had a better impact on prostate health than lycopene supplementation (Boileau et al. 2003).

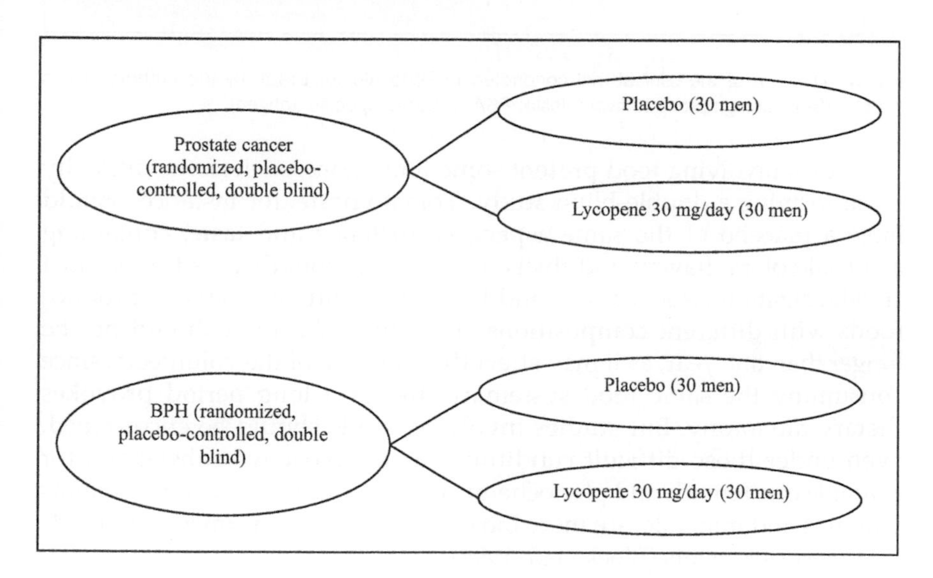

Fig. 2 Diagram of the placebo-controlled double-blind phase II study of the lycopene effects on intermediate endpoint markers in men with prostate cancer or benign prostate hyperplasia (BPH) (van Breemen 2005).

The second study is a double-blind randomized clinical trial, performed by the authors of this chapter, that is presently under way, comparing supplementation with lycopene and placebo, and tomato paste, non-blind, in BPH patients with serum PSA levels between 4 and 10 ng/mL (Fig. 3). The main purpose of this study is to verify what will happen with some biochemical and clinical parameters related to BPH. The results of this study are expected to be published in 2008.

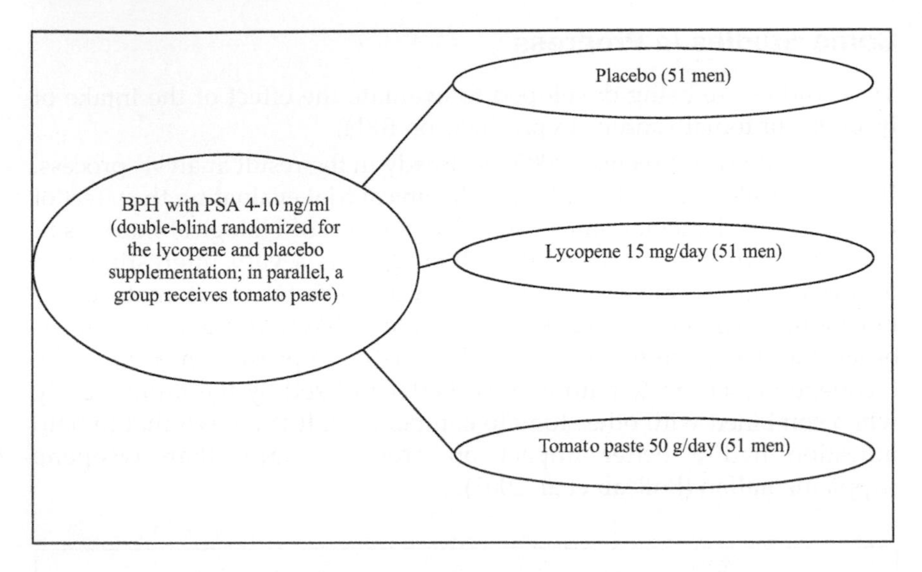

Fig. 3 Diagram of the clinical trial conducted in Porto Alegre, Brazil, by the authors of this chapter (BPH, benign prostate hyperplasia; PSA, prostate-specific antigen).

Studies involving food present some limitations, such as the difficulty in performing a double-blind study. Tomato paste, for instance, should have a placebo of the same aspect, consistency and taste, containing artificial colors, flavors and thickeners. In addition, there is the product standardization issue, as soils and harvests in different periods produce foods with different compositions. The study duration should not be longer than one year, as it may affect the adhesion of the volunteers, since consuming the same food systematically for a long period provokes dietary monotony. But studies involving foods should be encouraged, even under those difficult conditions, as the bioactive substances, for example carotenoids and phytochemicals, that are as essential nutrients as vitamins and minerals, are more bioavailable when consumed with foods than as supplements (Block et al. 1994).

PROPOSED MECHANISMS OF LYCOPENE ACTION ON THE PROSTATE

Although tomato and its byproducts have other substances besides the carotenoid lycopene, there is much evidence that this is one of the main substances responsible for the assumed ability of this food to prevent or mitigate prostate diseases. For this reason, possible mechanisms of the attributed effects of lycopene on prostate health have been widely studied.

Wertz et al. published an excellent revision in 2004 on the proposed lycopene action promoting prostate health. Several mechanisms are mentioned, such as antioxidant function, as it is the carotenoid with the highest free oxygen sequestration ability, inhibition of cell cycle progression, increase in apoptosis index, increase in gap junction communication, inhibition of IGF-1 sign transduction, inhibition of interleukin-6 expression, induction of phase II enzymes and inhibition of androgenic signaling and activation; the last may be more relevant, since the male hormones have a significant influence on BPH and prostate cancer development (Freire and Piovesan 1999, Brawley 2003, van Breemen 2005).

Studies suggest that DHT is the main androgen responsible for BPH and for the development of malignant tumors in prostate. This hormone is produced from testosterone, through the action of 5-α-reductase enzyme (Rizner et al. 2003). It was observed that men with 5-α-reductase deficiency have a hypoplastic prostate and do not develop prostate cancer (Bartsch et al. 2000). Then, the decrease in DHT synthesis might be one of the lycopene actions on the prostate, as it has already been observed that lycopene was associated with a decrease in the expression of 5-α-reductase in mice (Siler et al. 2004).

ACKNOWLEDGMENTS

Fundo de Incentivo à Pesquisa (FIPE) of the Hospital de Clínicas de Porto Alegre, RS, Brazil; Mr. Cláudio Oderich, of Conservas Oderich S/A, São Sebastião do Caí, RS, Brazil; Prof. Adriano Brandelli, of the Instituto de Ciências e Tecnologia de Alimentos, Universidade Federal do Rio Grande do Sul, RS, Brazil.

ABBREVIATIONS

BPH: Benign prostate hyperplasia; DHT: 5-α- Dihydrotestosterone; IGF-1: Insulin-like growth factor-1; PSA: Prostate-specific antigen

REFERENCES

Agarwal, S. and A.V. Rao. 2000. Tomato lycopene and its role in human health and chronic diseases. CMAJ 163: 739-44.

Andlauer, W. and P. Stehle, and P. Furst. 1998. Chemoprevention—a novel approach in dietetics. Curr. Opin. Clin. Nutr. Metab. Care 1: 539-547.

Arab, L. and S. Steck. 2000. Lycopene and cardiovascular disease. Amer. J. Clin. Nutr. 71: 1691S-5S; discussion 1696S-7S.

Arai, S. 1996. Studies on functional foods in Japan—state of the art. Biosci. Biotechnol. Biochem. 60: 9-15.

Balk, S.P. and Y.J. Ko, and G.J. Bubley. 2003. Biology of prostate-specific antigen. J. Clin. Oncol. 21: 383-391.

Bartsch, G. and R.S. Rittmaster, and H. Klocker. 2000. Dihydrotestosterone and the concept of 5alpha-reductase inhibition in human benign prostatic hyperplasia. Eur. Urol. 37: 367-380.

Berry, S.J. and D.S. Coffey, P.C. Walsh, and L.L. Ewing. 1984. The development of human benign prostatic hyperplasia with age. J. Urol. 132: 474-479.

Block, G. and R. Sinha, and G. Gridley. 1994. Collection of dietary-supplement data and implications for analysis. Amer. J. Clin. Nutr. 59: 232S-239S.

Boileau, A.C. and N.R. Merchen, K. Wasson, C.A. Atkinson, and J.W. Erdman Jr. 1999. Cis-lycopene is more bioavailable than trans-lycopene in vitro and in vivo in lymph-cannulated ferrets. J. Nutr. 129: 1176-1181.

Boileau, T.W. and A.C. Boileau, and J.W. Erdman Jr. 2002. Bioavailability of all-trans and cis-isomers of lycopene. Exp. Biol. Med. (Maywood) 227: 914-919.

Boileau, T.W. and Z. Liao, S. Kim, S. Lemeshow, J.W. Erdman Jr., and S.K. Clinton. 2003. Prostate carcinogenesis in N-methyl-N-nitrosourea (NMU)-testosterone-treated rats fed tomato powder, lycopene, or energy-restricted diets. J. Natl. Cancer Inst. 95: 1578-1586.

Bowen, P. and L. Chen, M. Stacewicz-Sapuntzakis, C. Duncan, R. Sharifi, L. Ghosh, H.S. Kim, K. Christov-Tzelkov, and R. van Breemen. 2002. Tomato sauce supplementation and prostate cancer: lycopene accumulation and modulation of biomarkers of carcinogenesis. Exp. Biol. Med. (Maywood) 227: 886-893.

Bramley, P.M. 2000. Is lycopene beneficial to human health? Phytochemistry 54: 233-236.

Brasil. 2006. Estimativa: Incidência de Câncer no Brasil. Rio de Janeiro: INCA, 2006. Disponível em <http://www.inca.org.br> Acesso em: 28 set. 2006.

Bravi, F. and C. Bosetti, L. Dal Maso, R. Talamini, M. Montella, E. Negri, V. Ramazzotti, S. Franceschi, and C. La Vecchia. 2006. Food groups and risk of benign prostatic hyperplasia. Urology 67: 73-79.

Brawer, M.K. 2000. Prostate-specific antigen. Semin. Surg. Oncol. 18: 3-9.

Brawley, O.W. 2003. Hormonal prevention of prostate cancer. Urol. Oncol. 21: 67-72.

Clinton, S.K. 1998. Lycopene: chemistry, biology, and implications for human health and disease. Nutr. Rev. 56: 35-51.

Clinton, S.K. and C. Emenhiser, S.J. Schwartz, D.G. Bostwick, A.W. Williams, B.J. Moore, and J.W. Erdman Jr. 1996. cis-trans lycopene isomers, carotenoids, and retinol in the human prostate. Cancer Epidemiol. Biomarkers Prev. 5: 823-833.

Curl, A.L. 1961. The xanthophylls of tomatoes. J. Food Sci. 26: 106-111.

Cury, J. Hiperplasia Prostática Benigna, p. 452. *In:* N. Schor. [ed.] 2005. Guia de Medicina Ambulatorial e Hospitalar de Urologia. Manole, Barueri - SP.

Edinger, M.S. 2005. Influência da ingestão dietética de extrato de tomate nos níveis plasmáticos do antígeno prostático específico (PSA) em pacientes com hiperplasia benigna da próstata, Programa de Pós Graduação em Medicina: Ciências Médicas. Universidade Federal do Rio Grande do Sul, Porto Alegre.

Edinger, M.S. and W.J. Koff. 2006. Effect of the consumption of tomato paste on plasma prostate-specific antigen levels in patients with benign prostate hyperplasia. Braz. J. Med. Biol. Res. 39: 1115-1119.

Etminan, M. and B. Takkouche, and F. Caamano-Isorna. 2004. The role of tomato products and lycopene in the prevention of prostate cancer: a meta-analysis of observational studies. Cancer Epidemiol. Biomarkers Prev. 13: 340-345.

Freire, G.C. and A.C. Piovesan. Prostatismo e HPB, p. 368. *In:* D. Bendhack and R. Damião. [eds.] 1999. Guia Prático de Urologia. BG Cultural, São Paulo.

Furstenberger, G. and H.J. Senn. 2002. Insulin-like growth factors and cancer. Lancet Oncol. 3: 298-302.

Furuya, S. and R. Furuya, H. Ogura, T. Araki, and T. Arita. 2006. A study of 4,031 patients of transurethral resection of the prostate performed by one surgeon: learning curve, surgical results and postoperative complications. Hinyokika Kiyo 52: 609-614.

Gann, P.H. and F. Khachik. 2003. Tomatoes or lycopene versus prostate cancer: is evolution anti-reductionist? J. Natl. Cancer Inst. 95: 1563-1565.

Gann, P.H. and J. Ma, E. Giovannucci, W. Willett, F.M. Sacks, C.H. Hennekens, and M.J. Stampfer. 1999. Lower prostate cancer risk in men with elevated plasma lycopene levels: results of a prospective analysis. Cancer Res. 59: 1225-1230.

Garcia-Cruz, C.H. and F.L. Hoffmann, S.M. Bueno, and T.M. Vinturim. 1997. Análise microbiológica de "catchup". Hig. Alim. 11: 43-46.

Gartner, C. and W. Stahl, and H. Sies. 1997. Lycopene is more bioavailable from tomato paste than from fresh tomatoes. Amer. J. Clin. Nutr. 66: 116-122.

Giovannucci, E. and A. Ascherio, E.B. Rimm, M.J. Stampfer, G.A. Colditz, and W.C. Willett. 1995. Intake of carotenoids and retinol in relation to risk of prostate cancer. J. Natl. Cancer Inst. 87: 1767-1776.

Giovannucci, E. and E.B. Rimm, Y. Liu, M.J. Stampfer, and W.C. Willett. 2002. A prospective study of tomato products, lycopene, and prostate cancer risk. J. Natl. Cancer Inst. 94: 391-398.

Gu, F. 1997. Changes in the prevalence of benign prostatic hyperplasia in China. Chin. Med. J. (Engl.) 110: 163-166.

Hart, D.J. and K.J. Scott. 1995. Development and evaluation of HPLC method for the analysis of carotenoid in foods, and the measurement of carotenoid content of vegetables and fruits commonly consumed in the UK. Food Chem. 54: 101-111.

Hennekens, C.H. and J.E. Buring, J.E. Manson, M. Stampfer, B. Rosner, N.R. Cook, C. Belanger, F. LaMotte, J.M. Gaziano, P.M. Ridker, W. Willett, and R. Peto. 1996. Lack of effect of long-term supplementation with beta carotene on the incidence of malignant neoplasms and cardiovascular disease. N. Engl. J. Med. 334: 1145-1149.

Khachik, F. and L. Carvalho, P.S. Bernstein, G.J. Muir, D.Y. Zhao, and N.B. Katz. 2002. Chemistry, distribution, and metabolism of tomato carotenoids and their impact on human health. Exp. Biol. Med. (Maywood) 227: 845-851.

Lagiou, P. and J. Wuu, A. Trichopoulou, C.C. Hsieh, H.O. Adami, and D. Trichopoulos. 1999. Diet and benign prostatic hyperplasia: a study in Greece. Urology 54: 284-290.

Lee, C. and J.M. Kozlowski, and J.T. Grayhack. 1995. Etiology of benign prostatic hyperplasia. Urol. Clin. N. Amer. 22: 237-246.

Mills, P.K. and W.L. Beeson, R.L. Phillips, and G.E. Fraser. 1989. Cohort study of diet, lifestyle, and prostate cancer in Adventist men. Cancer 64: 598-604.

Netto Júnior, N.R. and E.R. Wroclawski. 2000. Urologia: fundamentos para o clínico. Sarvier, São Paulo.

Nguyen, M.L. and S.L. Schwartz. 1999. Lycopene: chemical and biological properties. Food Techn. 53: 38-45.

Oliver, S.E. and B. Barrass, D.J. Gunnell, J.L. Donovan, T.J. Peters, R.A. Persad, D. Gillatt, D.E. Neal, F.C. Hamdy, and J.M. Holly. 2004. Serum insulin-like growth factor-I is positively associated with serum prostate-specific antigen in middle-aged men without evidence of prostate cancer. Cancer Epidemiol. Biomarkers Prev. 13: 163-165.

Pimentel, F.A. 2003. Efeito do processamento térmico da disponibilidade de licopeno em produtos processados de tomate, Curso de Engenharia de Alimentos. Universidade Federal do Rio Grande do Sul, Porto Alegre.

Rassweiler, J. and D. Teber, R. Kuntz, and R. Hofmann. 2006. Complications of Transurethral Resection of the Prostate (TURP)-Incidence, Management, and Prevention. Eur Urol.

Rizner, T.L. and H.K. Lin, and T.M. Penning. 2003. Role of human type 3 3alpha-hydroxysteroid dehydrogenase (AKR1C2) in androgen metabolism of prostate cancer cells. Chem. Biol. Interact. 143-144: 401-409.

Shi, J. and M. Le Maguer. 2000. Lycopene in tomatoes: chemical and physical properties affected by food processing. Crit. Rev. Food Sci. Nutr. 40: 1-42.

Siler, U. and L. Barella, V. Spitzer, J. Schnorr, M. Lein, R. Goralczyk, and K. Wertz. 2004. Lycopene and vitamin E interfere with autocrine/paracrine loops in the Dunning prostate cancer model. FASEB J. 18: 1019-1021.

Stahl, W. and H. Sies. 1992. Uptake of lycopene and its geometrical isomers is greater from heat-processed than from unprocessed tomato juice in humans. J. Nutr. 122: 2161-2166.

Suaid, H.J. and M.A. Goncalves, A.A. Rodrigues, Jr., J.P. Cunha, A.J. Cologna, and A.C. Martins. 2003. Estimated costs of treatment of benign prostate hyperplasia in Brazil. Intl. Braz. J. Urol. 29: 234-237.

Suzuki, S. and E.A. Platz, I. Kawachi, W.C. Willett, and E. Giovannucci. 2002. Intakes of energy and macronutrients and the risk of benign prostatic hyperplasia. Amer. J. Clin. Nutr. 75: 689-697.

Swinbanks, D. and J. O'Brien. 1993. Japan explores the boundary between food and medicine. Nature 364: 180.

Takeoka, G.R. and L. Dao, S. Flessa, D.M. Gillespie, W.T. Jewell, B. Huebner, D. Bertow, and S.E. Ebeler. 2001. Processing effects on lycopene content and antioxidant activity of tomatoes. J. Agric. Food Chem. 49: 3713-3717.

Taub, D.A. and J.T. Wei. 2006. The economics of benign prostatic hyperplasia and lower urinary tract symptoms in the United States. Curr. Urol. Rep. 7: 272-281.

Tavares, C.A. and D.B.R. Rodrigues-Amaya. 1994. Carotenoid composition of brazilian tomatoes and tomato products. Leb. Wiss. Techn. 27: 219-224.

Tonucci, L.H. and J.M. Holden, G.R. Beecher, F. Khachik, C. Davis, and G. Mulokozim. 1995. Carotenoid content of thermally processed tomato-based food productos. J. Agric. Food Chem. 43: 579-586.

Tzonou, A. and L.B. Signorello, P. Lagiou, J. Wuu, D. Trichopoulos, and A. Trichopoulou. 1999. Diet and cancer of the prostate: a case-control study in Greece. Intl. J. Cancer 80: 704-708.

van Breemen, R.B. 2005. How do intermediate endpoint markers respond to lycopene in men with prostate cancer or benign prostate hyperplasia? J. Nutr. 135: 2062S-2064S.

Wertz, K. and U. Siler, and R. Goralczyk. 2004. Lycopene: modes of action to promote prostate health. Arch. Biochem. Biophys. 430: 127-134.

22

Tomatoes and Components as Modulators of Experimental Prostate Carcinogenesis

Elizabeth Miller Grainger[1*], Kirstie Canene-Adams[2], John W. Erdman, Jr.[3] and Steven K. Clinton[1]

[1]A 434 Starling Loving Hall 320 West 10th Ave, The Ohio State University Columbus, Ohio 43210, USA

[2]905 S. Goodwin Ave. 448 Bevier Hall University of Illinois Urbana, Illinois, 61801, USA

[3]905 S. Goodwin Ave. 455 Bevier Hall University of Illinois Urbana, Illinois, 61801, USA

ABSTRACT

The accumulation of data from a variety of sources has led to a hypothesis that tomatoes, tomato products, or specific components derived from tomatoes may have the ability to inhibit certain phases of prostate carcinogenesis. However, not all studies support this relationship, and a definitive, randomized controlled trial of tomato products for the prevention of prostate cancer has never been undertaken. Thus, it is critical for readers to recognize that the relationship remains a hypothesis that requires continued reassessment as additional studies are completed and published in peer-reviewed scientific literature. Most important, readers should appreciate that no single study will provide the answers we need, although such an idea is often widely promoted by the public media. Only through a compilation of data derived from a variety of studies can we assess this relationship and provide evidence-based guidelines to enhance public health and patient care.

*Corresponding author

A list of abbreviations is given before the references.

An expanding array of *in vivo* rodent experimental models mimicking specific aspects of human prostate carcinogenesis and tumorigenesis have been developed. These models offer vital opportunities to conduct carefully controlled studies and assess dietary and nutritional hypotheses. Several laboratories have undertaken studies in experimental models targeting the relationship between tomatoes, tomato products and prostate cancer that provide data relevant to our interpretation of the epidemiologic, clinical, and *in vitro* data regarding tomato products or their components. These studies provide critical insight into efficacy, safety and biomarkers, which are relevant to the design of definitive intervention studies in humans. This chapter will focus primarily on a summary of accumulated knowledge derived from studies in rodent models relevant to the relationship between tomatoes and their constituents as modulators of prostate carcinogenesis.

INTRODUCTION

Prostate cancer is a common disease throughout the world, particularly in the economically developed nations of North America and northwestern Europe (Ferlay et al. 2001, World Cancer Research Fund 1997). In the United States in 2007, 29% of newly diagnosed male cancer cases will be from prostate cancer, which remains the second leading cause of cancer death (Jemal et al. 2007). As a result, one in every six men will develop prostate cancer over their lifetime (Jemal et al. 2007). Risk factors for prostate cancer that cannot be modified include gender, race, age, and family history. The etiology is certainly complex and multifactorial, and ways to define and quantify effective, safe, and acceptable interventions for prevention are not yet established.

Undoubtedly, we can conclude that environmental and lifestyle factors contribute to an increase in prostate cancer risk, and that a genetic component is also important. In addition, overall incidence may be influenced by governmental and economic policies influencing screening and diagnostic interventions. This issue is dramatically illustrated by the recently completed landmark study called the Prostate Cancer Prevention Trial, a randomized, double-blind and placebo-controlled study of finasteride (a drug that inhibits the enzyme 5-alpha reductase) (Thompson et al. 2003). The study involved over 18,000 men, aged 55 or older at randomization. Sixty percent of men in the finasteride group and 63% of men in the placebo group completed a prostate biopsy for clinical cause during the study, or as a planned end-of-study procedure on completion of seven years of study. In this cohort, 22% showed histological prostate cancer. This alarming rate of cancer diagnosis indicates how extensively the prostate carcinogenesis cascade has progressed in most American men over the age of 60. Yet, we believe that the majority of these cancers, which were diagnosed without meeting previously recommended screening criteria of serum prostate specific antigen (PSA) over 4.0 ng/mL or exam abnormalities, would probably follow an indolent biological course. Unfortunately, at the present time, we lack appropriate biomarkers to help us further refine the traditional predictive models for clinically significant disease based upon digital rectal exam findings, clinical staging, PSA, and Gleason scoring. We clearly need to define and validate additional biomarkers to discern which men with localized histopathological prostate cancer are most likely to benefit from therapeutic interventions. Research in this area is critical, so that we can determine with greater precision which cancers lack risk of progression, and predict which men will be able to avoid treatments and their associated side effects.

Another approach to solving the prostate cancer plague is to reduce the overall rate of prostate cancer progression, with a specific focus on the earliest steps in the transition from normal to premalignancy and early histological cancer. Indeed, prostate cancer prevention remains the ideal solution, yet investment in this approach, particularly translation of basic science into human clinical trials, has been woefully inadequate. It is often argued that the cost of prevention studies is prohibitive, yet the massive expenditures generated by the health care system for screening, diagnosis, and therapy of prostate cancer illustrates the ineffectiveness of the current strategy. The consumption of tomato products is only one of many active preventive hypotheses established through a range of investigations, including epidemiologic and clinical studies and preclinical laboratory investigations. Unfortunately, definitive controlled clinical trials remain rare.

THE EVOLUTION OF THE TOMATO, LYCOPENE, AND PROSTATE CANCER HYPOTHESIS

The landmark publication by Giovannucci et al. in 1995 stimulated great interest within the biomedical community as well as the popular press. The original findings in the Health Professionals Follow-up Study (HPFS) have been reinforced by subsequent publications as the prospectively evaluated cohort has aged (Giovannucci 2002, Giovannucci et al. 1995). These studies suggest a relationship between diets rich in tomato products and estimated lycopene intake and a lower risk of prostate cancer, particularly cancers with more aggressive features. However, not all epidemiologic studies support a relationship between tomato products and lower risk of prostate cancer, leading to divergent opinions within the field (Giovannucci 2005, Kristal and Cohen 2000, Peters et al. 2007). In such situations, reviews and meta-analyses can occasionally provide some additional insight simply by providing critical statistical power to detect modest effects within a heterogeneous disease. A meta-analysis conducted in 2004 examined relevant tomato and prostate cancer epidemiologic studies with 11 case-control and 10 cohort studies chosen for analysis (Etminan et al. 2004). It was estimated that compared to those rarely consuming raw tomatoes, the relative risk (RR) for prostate cancer in the highest quartile of intake was 0.89 (95%, confidence interval (CI) 0.80–1.00), and for those consuming cooked tomato products, the RR was 0.81 (95% CI 0.71–0.92). Although supportive of a benefit, it is now abundantly clear that epidemiologic studies alone are unlikely to resolve the dilemma for several reasons, including concerns regarding the precision of diet assessment tools, the well-known recall bias of case-control studies, and the likelihood that the relative risk reduction of 10-30% is simply difficult to define when superimposed on the complexity of prostate carcinogenesis. We should also appreciate the significant heterogeneity of human prostate cancer, which represents a family of diseases, each of which may have many interacting etiologic factors.

Closely intertwined within the tomato/prostate cancer relationship is the hypothesis that lycopene, the predominant carotenoid in tomatoes that provides the familiar red color, has anticancer properties and may mediate the protective effect against prostate cancer. However, in spite of the enormous emphasis on lycopene, investigators should recognize that tomatoes also contain a vast array of nutrients and phytochemicals that can be hypothesized either as single agents or as a collective entity, to modulate carcinogenesis. Nutrients such as the vitamins C, K, E, and folate are present, and their roles in cancer prevention are the subject of many reviews (World Cancer Research Fund 1997). It is our opinion that no single nutrient, using the classic definitions, can account for the hypothesized anti-prostate cancer properties of tomato products. We feel that the non-nutrient phytochemicals such as the polyphenols (Wang et al. 2003) (such as quercitin and rutin) and the carotenoids (including lycopene and its precursors, phytofluene and phytoene) are the more likely anticancer components (Canene-Adams et al. 2005).

Lycopene has received the bulk of research attention for several reasons. First, various *in vitro* and rodent studies suggest that lycopene may have unique antioxidant properties and perhaps additional biological activities that may alter carcinogenesis (Di Mascio et al. 1989, Herzog et al. 2005, Herzog and Wertz 2005, Kim et al. 2003). Second, serum or tissue lycopene can be readily measured by quality laboratories and may have some value as a surrogate biomarker of long-term exposure to tomato products, as few other foods are widely consumed in quantities sufficient to dramatically alter *in vivo* concentrations (Clinton 1998, 2005, Miller et al. 2002, Wu et al. 2003). The lycopene hypothesis gained support from the epidemiologic studies, as lycopene content of foods has been quantified in the USDA databases used to assess the nutrient and phytochemical content of foods. Predictably, estimated intake of lycopene was associated with a reduced risk of prostate cancer in epidemiologic studies using food frequency questionnaires (Giovannucci and Clinton 1998), as tomato products are the major source of lycopene intake (Clinton 2005, Clinton 1998, Giovannucci and Clinton 1998). The biological plausibility of lycopene mediating anti-prostate cancer activity was enhanced when lycopene and its isomers were found in significant concentrations in human prostate tissue (Clinton et al. 1996). In parallel, reports of "anti-cancer" activity for lycopene *in vitro*, using prostate cancer cells, have added to the growing enthusiasm.

A wave of additional epidemiologic studies of varying statistical power and precision have followed the original reports, focusing on the relationship of blood lycopene to prostate cancer risk. For example, in an analysis of plasma samples from men enrolled in the Physicians' Health Study, lycopene was found at significantly lower concentrations in the blood of men with prostate cancer than in matched controls ($P = 0.04$) (Gann et al. 1999). This inverse association was particularly evident for aggressive types of prostate cancer and for men who were not taking a beta-carotene supplement (highest v. lowest quintile $P = 0.006$). Another case-control study examined plasma carotenoids and risk for prostate cancer with a reported 83% reduction in prostate cancer risk observed in the group with the highest plasma lycopene concentration (0.40 :µ mol/L), in comparison to the lowest lycopene group (0.18 :µ mol/L) (Lu et al. 2001).

There is inconsistency among epidemiologic studies. Recently, two reports from the Prostate, Lung, Colorectal, and Ovarian Cancer Screening Trial cohort study, which have followed over 28,000 men, detected no relationship between estimated dietary lycopene consumption (Kirsh et al. 2006) or serum lycopene and prostate cancer incidence (P for trend = 0.433) (Peters et al. 2007). The inconsistency is perhaps predictable when estimates of intake are imprecise, the documentation of exposure during critical early time points is limited, and the benefit if modest (for example a 10-20% reduction in risk) may be difficult to detect. Furthermore, the epidemiologic studies are still limited in their ability to control for many potentially important variables regarding anticancer components with regard to absorption, distribution, and metabolism of specific phytochemicals that are defined by genetic variability and modulated by other dietary components. Additionally, prostate cancer is not a single disease, but rather a family of diseases affecting a single organ, and future efforts to define molecular signatures of prostate cancer subtypes showing specific biological patterns will help improve the data derived from epidemiologic studies.

The U.S. Food and Drug Administration (FDA) recently published a review of tomatoes, lycopene and risk of a number of different cancer subtypes, including ovarian, gastric, pancreatic and prostate cancers. The purpose of this review was to evaluate tomato, lycopene and cancer literature in the context of a qualified health claim. Health claims are granted by the FDA for use by food manufacturers, if a food or food component has met rigorous scientific criteria where there is a consistent and clear health benefit. The FDA report found that the accumulated evidence, which focused on human research, did not support the role of tomato products or lycopene for reduction in the risk for any of the cancer subtypes. Although this report garnered significant attention by the popular press, readers should understand that the approval of a health claim requires significant scientific agreement. The report indicated that studies of lycopene and tomatoes have produced some provocative, but inconsistent results, and more research is needed (Giovannucci 2007, Kavanaugh et al. 2007).

It is interesting to note that lycopene from tomatoes or tomato-containing food products is consumed primarily in the all-*trans* isomer form, but blood and tissues demonstrate a variety of *cis*-isomers. This observation has fueled speculation that lycopene isomer patterns may provide some insight into biological activity, but this has yet to be proven. Wu et al. (2003) investigated the relationship between plasma concentrations of *cis* and *trans*-lycopene isomers and total plasma lycopene in blood samples taken three to four years apart from 144 men enrolled in the HPFS. There was a strong correlation between concentrations of total lycopene and its major isomers over time, suggesting that a single plasma measurement of blood lycopene concentration is a reasonable estimate of long-term exposure (Wu et al. 2003).

Several groups have completed small human intervention studies with lycopene or tomato products and assessed biomarkers of biological activity. For example, we have observed that daily consumption of tomato juice, soup and sauce can quickly alter blood and tissue content of lycopene and its isomers (Allen et al. 2002, 2003, Clinton et al. 1996, Hadley et al. 2003, Jatoi et al. 2007). Several groups have reported an improvement in markers of oxidative stress and more favorable histopathological markers following intake of tomato products or lycopene (Chen et al. 2001, Hadley et al. 2003, Kim et al. 2003, Kucuk et al. 2001). None of the published studies have had the statistical power to fully address the ability of tomato products or components to influence prostate carcinogenesis, yet some of the data is supportive of moving forward with rigorously controlled multi-institutional clinical studies targeting specific biomarker outcomes.

Taken together, the epidemiologic and clinical research has provided enough circumstantial evidence to warrant continued investigations regarding the potential for tomato products and their components to prevent or slow the progression of prostate cancer. The remainder of this review focuses in greater detail on the contributions derived from important investigations of experimental prostate carcinogenesis and tumorigenesis where intake and exposure to tomato products or constituents can be precisely controlled.

EXPERIMENTAL PROSTATE CANCER

Dietary studies in rodent models of prostate carcinogenesis (*de novo* development of cancer) or tumorigenesis (growth of a tumor in a transplantable model) have a number of advantages that allow an investigator to control heterogeneity and provide estimates of efficacy and safety for a hypothesized intervention. These include (1) uniform age and genetic background of the subjects, (2) control of environmental variables, (3) precision in defining dietary exposures in terms of dose and source of components, (4) precise control of the timing and duration of exposure targeting critical stages of carcinogenesis, (5) uniform exposure to carcinogenic stimuli (chemical initiators and promoters, specific genetic alterations) or inoculation of tumor cells (defined number of cells), (6) the ability to examine combinations of chemopreventive agents or dietary patterns for additive or synergistic effects, (7) the ability to assess precisely critical hormones that influence human prostate cancer, which are similarly important in rodents, and (8) availability of biological specimens for cellular and mechanistic studies.

The interpretation of rodent prostate cancer studies must also consider the following issues when extrapolating experimental data to humans: (1) the carcinogenic stimuli may be irrelevant to humans, (2) absorption and distribution of nutrients or phytochemicals may differ among species, (3) metabolism of phytochemicals may involve different host enzymatic pathways or bacterial flora, and (4) the animal model may represent only one subtype of the "family" of human prostate cancer.

The perfect rodent model, one that mimics all characteristics of human prostate cancer, does not exist. The following characteristics are deemed highly desirable (Bosland 1999, Lucia et al. 1998, Tennant et al. 2000): (1) adenocarcinoma histology derived from the epithelium of the prostate, (2) availability of a large number of animals to all interested investigators, (3) androgen dependency during early stages of carcinogenesis,

(4) relatively slow-growing cancers, (5) tumors that respond to surgical or pharmacologic castration and show evolution of androgen independence over time, and (6) cancers that primarily metastasize to lymph nodes and bone. Efforts to define models that mimic many of these key characteristics continue and will provide important tools for those interested in studies on prevention by diet and nutrition, in addition to chemically defined chemopreventive agents.

Clearly, rodent experiments alone cannot serve as the basis for dietary guidelines or public health recommendations. However, *in vivo* rodent models provide a type of nutritional investigation that is far more convincing than *in vitro* cell culture studies. The cell-based approach, at best, can define potential molecular mechanisms of action in response to an agent and identify biomarkers, which of course is an important aspect of evaluating the tomato/prostate cancer hypothesis.

Although prostate tissue is nearly a ubiquitous accessory sex gland in the animal kingdom, spontaneous prostate cancer is remarkably rare among free-living mammals, except for humans and dogs, which present a similar disease. Thus, rodent models showing spontaneous prostate carcinogenesis are limited (Cohen 2002, Lamb and Zhang 2005, Shirai et al. 2000). The current rat prostate carcinogenesis models include several strains: Wistar Unilever, Fischer F344, Noble, ACI, and Lobund-Wistar with a variety of characteristics (Pollard and Suckow 2005). Murine models of prostate carcinogenesis continue to expand in number and are primarily based on genetic modification, often focusing on prostate specific expression of oncogenes, and have been reviewed (Klein 2005, Shappell et al. 2004). Transplantable models, such as the Dunning prostate cancer lines, are useful to evaluate tumorigenesis, as opposed to carcinogenesis (Bosland 1999).

No single animal model of prostate cancer can mimic the entire spectrum of characteristics of human prostate carcinogenesis. Indeed, if it existed, the model would be cumbersome, its tremendous heterogeneity making it difficult to employ, and would require large numbers of animals to detect effects of intervention and long periods of study. Thus, laboratory models are developed to be more homogeneous and easily reproducible among laboratories and investigators, and of acceptable cost to research teams and funding agencies.

EXPERIMENTAL RESULTS ON TOMATO AND PROSTATE CARCINOGENESIS

The results of the published studies are summarized in Table 1. An early rodent study by Imaida et al. (2001) employed the Fisher 344 rat treated

Table 1 Representative tomato and lycopene studies in rodent prostate carcinogenesis models.

Model used	*Dietary intervention*	*Results*	*Comments*
Fisher 344 (Imaida et al. 2001)	• Three 60-wk experiments with male rats 1) Carcinogen (DMAB) administration concurrent with diets containing either 15 ppm lycopene or 500 ppm curcumin 2) Animals pretreated with varying doses of DMAB followed by diets containing either 15 ppm lycopene or 500 ppm curcumin 3) Carcinogen (PhIP) administration followed by diets containing 45 ppm lycopene or 500 ppm curcumin or both for 50 wk.	• Lycopene given after DMAB significantly decreased incidence of PIN in the ventral prostate compared with DMAB alone (6% v. 70% respectively, $P < 0.05$). • None of the experiments suggested that lycopene could consistently inhibit rat prostate carcinogenesis.	Dietary concentrations of lycopene may be insufficient to provide significant tissue concentrations of lycopene
Muta™ mice (Guttenplan et al. 2001)	• 36 male mice randomized to one of three diets: 1) Control diet (AIN 76 diet) 2) Control diet + 7 g lycopene-rich tomato oleoresin/ kg diet (2.4% other carotenoids) 3) Control diet + 14 g lycopene-rich tomato oleoresin/ kg diet (2.4% other carotenoids)	• Lycopene tomato oleoresin resulted in a dose-dependent decrease in both spontaneous and BaP induced prostate mutagenesis, but not statistically significant.	This study also reported a significant increase in colon and lung mutagenesis in mice receiving the control diet + 14 mg lycopene/ kg diet
NMU in Wistar Rats (Boileau et al. 2003)	• 194 male rats, randomized to one of three diets: 1) Control diet 2) Tomato powder (10% of diet, 13 mg lycopene/ kg diet) 3) Lycopene beadlet (161 mg/kg diet)	• Diet with tomato powder significantly increased survival (HR = 0.74; 95% CI = 0.59-0.93; $P = 0.009$) • Diet with lycopene resulted in a non-significant reduction in survival from prostate cancer (HR = 0.91; 95% CI = 0.61-1.35; $P = 0.6$)	Tomato powder was more effective than lycopene in spite of much lower blood concentrations.
Lady mice (Venkateswaran et al. 2004)	• 76 mice, randomized to one of four diets: 1) Standard diet 2) Standard diet + antioxidant blend (lycopene, vit E and selenium) 3) 40% calorie restricted, high-fat diet 4) 40% calorie restricted, high-fat diet + antioxidant blend	• Animals receiving the antioxidant-supplemented diets developed less prostate cancer. • Cancer rates for each group were reported as follows:74% of animals in group 1 developed cancer, 11% of animals in group 2 ($P < 0.0001$), 100% of animals in group 3, and 16% of animals in group 4 ($P < 0.0001$)	Details regarding the diet composition and content of antioxidants are not reported. Individual contribution of the antioxidants employed are not known in this model.

DMAB, 3,2'-dimethyl-4-amino-biphenyl, NMU, *N*-methyl-*N*-nitrosourea; HR, hazard ratio; CI, confidence interval.

with the carcinogens 3,2″-dimethyl-4-aminobiphenol (DMAB) and 2-amino-1-methylimidazo[4,5-b]pyridine (PhIP). Lycopene was provided as the carotenoid-enriched tomato lipophilic extract called LycoRed at 5, 15, or 45 ppm (lycopene). In one of three studies, a slight trend towards a reduction in prostatic intraepithelial neoplasia (PIN) incidence was reported in the DMAB model but with no effect on prostate cancer in either the DMAB or PhIP studies. The study was compromised by the relatively low concentrations of lycopene reported in the serum and liver tissue of the rats. It is important for all investigators to appreciate that rats and mice typically show a lower absorptive capacity than humans for many carotenoids, including lycopene (Boileau et al. 2000, 2001). Thus, higher dietary concentrations must be fed to rodents, compared to humans, in order to achieve the blood and tissue concentrations that are typically associated with "anticancer" activity in human clinical or epidemiologic studies. As a result, the Imaida et al. (2001) study may not have achieved biologically relevant anti-cancer concentrations of the agent being tested. Additionally, the statistical power was limited by the small number of rats per dietary group (less than 20 in this study). This problem is common in the experimental carcinogenesis literature and limits the investigators' ability to detect real effects that may be in the range of 10-30%, a range of benefit that is hypothesized by epidemiologic studies.

The N-methyl-N-nitrosourea (NMU) androgen-induced rat model of prostate cancer has been used in a number dietary and chemoprevention studies (Bosland 1999, McCormick et al. 1998). Our laboratory (Boileau et al. 2003) compared prostate carcinogenesis in rats fed a semipurified American Institute of Nutrition (AIN) control diet or one prepared with 10% tomato powder (freeze-dried whole tomatoes) with 13 mg lycopene/kg diet, or lycopene beadlets at 0.0025 g/kg diet (161 mg lycopene/kg diet). The concentration of lycopene beadlets chosen was such that the rats accumulated lycopene in the prostate and blood in a range that approximates those found in humans (Boileau et al. 2000). The risk of death from prostate cancer was lower for rats fed the tomato powder diet than for rats fed the control diet with a hazard ratio of 0.74 (95% CI = 0.59 to 0.93, $P < 0.009$). Interestingly, the risk of prostate cancer-related death for rats fed the control diet or the lycopene beadlet diet was statistically similar. In addition, the tomato-fed rats showed an inhibition of prostate carcinogenesis even though the lycopene content of blood and tissues was much lower than the lycopene beadlet-fed rats. This study strongly suggests that tomatoes contain a variety of components that inhibit prostate carcinogenesis in the NMU model more effectively than lycopene alone. The study further suggests that blood and tissue lycopene may not serve as a biomarker for the anticancer activity of tomato products. It does not mean that lycopene may not be one of the contributing components, but that when fed alone it appears to be much less effective at this dose than when provided in the context of other tomato components. Perhaps only in combination with other tomato-derived phytochemicals can we see biologically relevant anticancer activity.

The first study evaluating a tomato component in murine prostate carcinogenesis employed the LADY transgenic model (Venkateswaran et al. 2004), which is based upon prostate expression of the viral large T antigen. The model progresses to androgen independence and a poorly differentiated neuroendocrine type of prostate cancer. Interestingly, the authors compared mice fed a "standard diet" to one containing a combination of supplemental selenium, lycopene, and vitamin E along with energy restriction. The authors report a significant reduction in prostate carcinogenesis for the experimental diet. However, it is impossible for another investigator to repeat the published work as the composition and formulation of the diets are not defined in the publication, and there is no documentation of the concentrations of the supplements in the diet or their sources. Furthermore, the design of the study does not allow one to determine the individual, additive, or synergistic effects of the components of the "intervention" diet. Thus, this report will remain an interesting testimony but is essentially non-interpretable with respect to the role of tomato components or lycopene as a preventive intervention for prostate carcinogenesis.

Another murine study employed the Muta™ mice given 7 or 14 g of lycopene-rich tomato oleoresin per kg diet (which also contained β-carotene, phytofluene, ξ-carotene, and 2,6-cyclolycopene-1,5-diol). The authors reported a dose-dependent but non-significant reduction in both spontaneous and benzo[*a*]pyrene-induced mutagenesis in the prostate (Guttenplan et al. 2001). Interestingly, this study also suggested a statistically significant increase in colon and lung mutagenesis in the animals receiving 14 g lycopene/kg diet compared with controls (Guttenplan et al. 2001). This interesting study, although not evaluating carcinogenesis, suggests that dietary tomato components may have tissue-specific effects that warrant further study.

EXPERIMENTAL TUMORIGENESIS

Models employing prostate cancer cell lines established by injection or transplantation, isolated tumor cells, or viable pieces of tumor allow investigators to examine tumor growth rates and biomarkers related to growth and progression. Several laboratories have conducted studies with lycopene or tomato components. These studies are summarized in Table 2. The MatLyLu subline of the Dunning rat model was derived from a rapidly growing metastatic prostate cancer that originally metastasized to the lymph nodes and lungs. When these poorly differentiated cells were injected directly into the ventral prostate of rats, it was found that neither dietary lycopene nor vitamin E caused a significant decrease in tumor size at 18 d (Siler et al. 2005). However, the authors did assess the tumor by magnetic resonance-imaging and on the basis of the examination of intratumor structure the authors suggest that there is an increase in the necrotic area of the tumors of rats fed lycopene at 200 ppm (Siler et al. 2005). Microarray analysis (mRNA) was conducted on these tumors, and lycopene treatment was associated with a down-regulation of 5-alpha reductase 1 suggesting an impact on androgen signaling. Other genes down-regulated in lycopene-fed prostate tumors were probasin, IGF-1, and IL-6.

Two studies have examined an effect of lycopene on human prostate cancer cell lines inoculated into immune-suppressed mice. The effects of lycopene on the growth of human DU145 tumor xenografts in BALB/c male immune-deficient nude mice were examined by Tang et al. (2005). The authors report that the tumor growth rate was reduced by 56 and 76% respectively in mice fed AIN-based diets containing lycopene at 100 and 300 mg/kg compared to those fed lycopene-free diets for 8 wk. The poorly differentiated PC-346C human prostate cancer cell line was employed in a study of orthotopic implantation in NMRI nu/nu mice (Limpens et al. 2006). Lycopene was provided daily at 5 and 50 mg/kg body weight. There was no significant effect on tumor growth observed for lycopene alone, but when both lycopene and vitamin E were provided at 5 mg/kg body weight a significant inhibition of tumor growth by 73% was reported over the 95 d duration of the study.

A recent study from our laboratory using the Dunning rat prostate cancer model employed the slow-growing and well-differentiated R3327H androgen-dependent subline (Canene-Adams et al. 2007). The rats were fed AIN-based semipurified diets containing various combinations of tomato and/or broccoli powder or lycopene for 22 wk (Figs. 1 and 2). The diet intervention groups were compared with rats that were castrated or received finasteride (5-alpha reductase inhibitor) in order to compare dietary effects with disruption of androgen signaling. The animals fed 10% whole tomato powder had a significant 33% reduction in tumor weights over the 18 wk study. In contrast, lycopene at 23 or 224 nM/g diet did not significantly reduce tumor weight, in spite of a slight positive trend of 7% and 18% smaller tumors. Of interest was the fact that castration at 16 wk caused dramatic tumor regression, while

Table 2 Representative tomato and lycopene studies in rodent prostate tumorigenesis models.

Model used	*Dietary intervention*	*Results*	*Comments*
Dunning MatLyLu (Siler et al. 2004)	30 male Copenhagen rats randomized to 1 of 5 groups: 1) Control diet 2) Control diet (40 ppm vitamin E) 3) Control diet + 540 ppm vitamin E 4) Control diet (40 ppm vitamin E) & 200 ppm lycopene 5) Control diet + 540 ppm vitamin E & 200 ppm lycopene	• No interventions decreased tumor size • Significant increase in tumor necrotic area with lycopene ($P = 0.02$) or vitamin E ($P = 0.006$)	Lycopene reduced 5-alpha reductase 1, IGF-1, & IL-6 gene expression measured by GeneChip analysis
Xenograft (BALB/c mice with DU145 cells) (Tang et al. 2005)	Male BALB/c mice injected with androgen-independent DU145 xenografts and randomized to 1 of 4 diet groups: 1) Control diet 2) Control diet + 10 mg/kg lycopene 3) Control diet + 100 mg/kg lycopene 4) Control diet + 300 mg/kg lycopene 5) DU145 cells pretreated with or without 20 μmol/L lycopene	• 100 mg/kg lycopene reduced tumor growth by 56% ($P < 0.05$) • 300 mg/kg lycopene reduced tumor growth by 76% ($P < 0.01$) • DU145 cells pretreated in vitro w/ 20 μmol/L lycopene did not form tumors	Anti-tumor effects correlated with parallel in vitro studies.
Xenograft (NMRI nu/nu mice with PC-346C cells) (Limpens et al. 2006)	54 male NMRI nu/nu mice inoculated with PC-346C tumor xenograft and randomized to 1 of 6 groups 1) Control diet 2) Control diet + 5 mg lycopene/kg body wt 3) Control diet + 50 mg lycopene/kg body wt 4) Control diet + 5 mg vit E/kg body wt 5) Control diet + 50 mg vit E/kg body wt 6) Control diet + 5 mg lycopene & 5 mg vit E/kg body wt	• No single treatment suppressed tumor growth • Lycopene + vitamin E suppressed orthotopic growth by 73% ($P < 0.05$) and increased median survival time by 40% ($P = 0.02$)	Plasma PSA levels were lowest in the mice fed the lycopene + vitamin E combination ($P = 0.06$ compared with controls)
Dunning R3327-H (Canene-Adams et al. 2007)	206 male Copenhagen rats randomized to 1 of 9 groups: 1) Control diet 2) Control diet + finasteride 3) Control diet + castration 4) Control diet + 23 nmol lycopene/g diet 5) Control diet + 224 nmol lycopene/g diet 6) Control diet with 10% tomato powder 7) Control diet with 10% broccoli powder 8) Control diet with 5% tomato & 5% broccoli powder 9) Control diet with 10% tomato & 10% broccoli powder	• Lycopene at 23 nmol/g insignificantly reduced tumor weights by 7% • 224 nmol lycopene/g insignificantly reduced tumor growth by 18% • Tomato powder reduced tumor weights by 34% ($P = 0.05$)	Rats consuming tomato powder exhibited lower proliferative index percentages (PCNA staining) and higher apoptosis rates (ApopTag staining).

PCNA, Proliferating Cell Nuclear Antigen

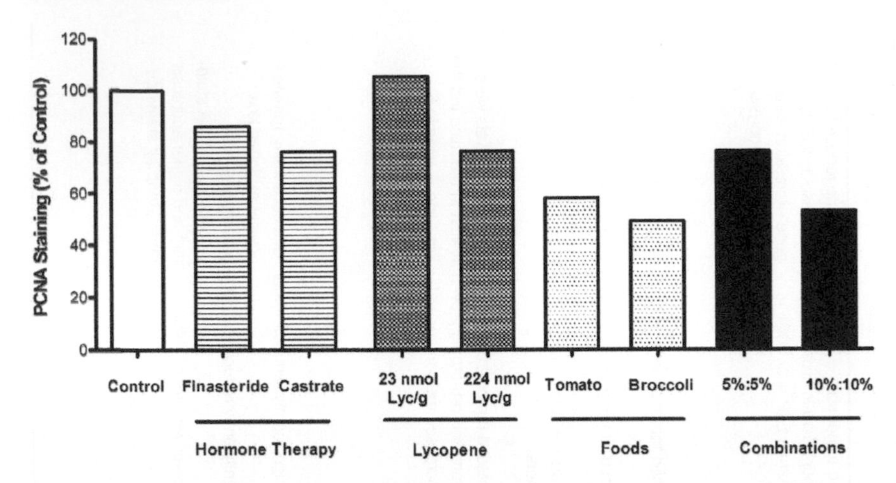

Fig. 1 The effects of various endocrine and dietary manipulations on the growth of the Dunning R3327H transplantable prostate adenocarcinoma in rats were evaluated (Canene-Adams et al. 2007). The data is presented as the percentage change compared to control. The tomato products were more effective than lycopene in this study and combinations of tomato and broccoli components were active and compared favorably to endocrine manipulations.

finasteride at 5 mg/kg body weight daily provided a 14% smaller tumor weight compared to controls, but this difference was not statistically significant. As seen in the NMU-induced carcinogenesis model, rats fed freeze-dried whole tomatoes had significantly less blood and tissue lycopene yet showed a more effective inhibition of tumor growth. Another key finding is that combinations of tomato and broccoli are more effective than individual food items alone and at least as effective as finasteride, an agent proven to serve as a human prostate cancer chemopreventive agent (Thompson et al. 2003). Among the biomarkers evaluated, tumor apoptotic index was increased and proliferative index reduced (Fig. 1) for those fed tomato components (Canene-Adams et al. 2007). Furthermore, tomato powder was more effective than finasteride for the inhibition of prostate cancer, which is of enormous interest when we consider that finasteride has been tested in a human prostate cancer prevention trial and shown to reduce the diagnosis of prostate cancer by approximately 25% over 7 yr of exposure (Thompson et al. 2003).

HYPOTHESIZED ANTI-CANCER MECHANISMS OF TOMATO PRODUCTS OR LYCOPENE

There are numerous anti-cancer mechanisms proposed for tomato products and lycopene (Bowen 2005), with the majority of efforts falling into several general categories, including reduced cellular oxidative damage reducing genetic damage, decreased tumor cell proliferation, enhanced sensitivity to apoptosis, anti-inflammatory activity, and inhibition of tumor-promoting hormone and growth factor action. The evidence for each of these pathways has been generated by a variety of approaches ranging from *in vitro* cell biology studies to *in vivo* rodent and human investigations, yet none has achieved the level of evidence necessary to be considered definitive.

The antioxidant activity of lycopene *in vivo* and its potential impact on carcinogenesis has been an experimental quagmire for decades. Although it is well established that carotenoids can quench specific types of reactive oxygen in highly controlled chemical studies, our ability to investigate the *in vivo* antioxidant activity is typically limited to indirect biomarker studies, and specific evidence for a linkage between antioxidant activity and reduced damage to critical genes or other macromolecules directly linked to carcinogenesis has been elusive. Lycopene is considered to be one of the most efficient of common dietary carotenoids in scavenging singlet oxygen and reactive oxygen species (Conn et al. 1991, Di Mascio et al. 1989). One *in vivo* study of 12-wk-old Wistar rats investigating the effect of lycopene on iron-induced oxidative DNA damage warrants particular interest. Rats pretreated with intraperitoneal injections of lycopene for 5 d prior to ferric nitrilotriacetate (Fe-NTA) injection showed a 70% reduction in prostate tissue 8-oxo-7,8-dihydro-2'-deoxyguanosine (8-oxoGuo) levels compared with animals that received only the Fe-NTA injection (Matos et al. 2006). Several small human clinical trials have supported the role of lycopene and tomato products in reducing oxidative damage. The first significant study, reported by Chen et al. (2001) in men fed tomato-based diets for several weeks, showed a reduction in DNA damage of the prostate in a pre-prostatectomy design. Another study, with non-smoking male subjects consuming 40 mg lycopene from tomato juice each day for 2 wk, documented a significant decrease in DNA strand breaks in peripheral blood lymphocyte cells (Pool-Zobel et al. 1997). In 2001, a clinical trial was reported involving 32 men with localized prostate adenocarcinoma consuming tomato sauce-based pasta dishes daily, providing approximately 30 mg lycopene per day, for 3 wk prior to a scheduled radical prostatectomy (Chen et al. 2001). Tomato product consumption resulted in decreased serum PSA from 10.9 to 8.7 ng/mL ($P < 0.001$) and leukocyte oxidative DNA damage was significantly lower (from 0.61 8-OHdG/10^5 dG with 95% CI = 0.45 to 0.77 8-OHdG/10^5 dG, compared to 0.48 8-OHdG/10^5 dG with 95% CI = 0.41 to 0.56 8-OHdG/10^5 dG)($P = 0.005$) after regular tomato consumption compared to before the dietary intervention. Additionally, 8-OHdG measurements of prostate tissue showed less oxidative DNA damage in men who consumed tomatoes. Hadley et al. (2003) investigated lycopene bioavailability and blood antioxidant capacity in 60 healthy adults who consumed between 23 and 35 mg lycopene from tomato products for 15 d. They reported that consumption of tomato products significantly enhanced the protection of lipoproteins to *ex vivo* oxidative stress (Hadley et al. 2003, Basu and Imrhan 2007). Although limited in number and scope, the above representative studies suggest that in cellular and *in vivo* models which measure biomarkers of oxidative damage, these biomarkers are improved with consumption of tomato products. Yet, definitive assessment of antioxidant activity focusing on targets directly affecting the carcinogenic cascade are lacking.

The balance between the number of proliferating cells and those undergoing apoptosis is critical with respect to normal tissue function and is dysregulated during tumor growth to favor survival of malignant cells. Several cell-culture studies report antiproliferative activity or increased apoptosis after treatment with lycopene. Lycopene has been known to inhibit the growth of human androgen-independent prostate cancer cells and Dunning AT6.3 cells by increasing apoptotic rate (Kim et al. 2003, Tang et al. 2005, Wang et al. 2003) and perhaps cause an inhibition of cell-cycle progression characterized by an increase in the number of cells in the G(2)/M phase of the cell cycle (Hwang and Bowen 2004, 2005, Hwang and Lee 2006, Tang et al. 2005). Additionally, lycopene has been shown in prostate cancer cell studies to induce the release of cytochrome c from the mitochondria, increase annexin V binding, and induce apoptosis (Hantz et al. 2005, Ivanov et al. 2007, Tang et al. 2005). Proliferating cell nuclear antigen (PCNA) was identified in 1978 and was so named because it was observed in the nucleus of dividing cells. Today PCNA is often used as a biomarker for tumor cell proliferation for *in vitro* and *in vivo* studies (Kelman 1997). Increasing lycopene intake and dietary tomato powder in the Dunning R3327-H rat model has been shown to reduce immunohistochemical staining for PCNA, reducing the proliferative index (Fig. 1), in parallel with a change in tumor volume indicating a decrease in cell proliferation (Fig. 2) (Canene-Adams et al. 2007). Similarly, immunohistochemical staining for apoptotic markers has been reported to be increased in BALB/c male nude mice that were inoculated with DU145 tumor xenografts and fed lycopene (100 or 300 mg/kg) and in the Dunning tumors of rats fed tomato powder (Canene-Adams et al. 2007). Flow cytometry analysis indicated a dose-dependent increase in apoptosis in the mice treated with lycopene (Tang et al. 2005). Although these studies suggest an increase in sensitivity to apoptosis and reduced proliferative rates, the precise mechanisms by which these components may act on prostate cancer cells remain to be defined.

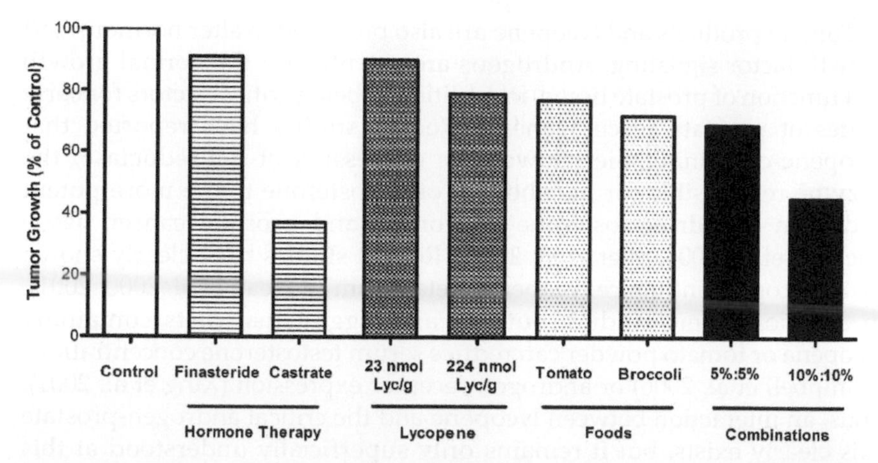

Fig. 2 The effects of various endocrine and dietary manipulations on the proliferative index (PCNA staining) of the Dunning R3327H transplantable prostate adenocarcinoma in rats were evaluated (Canene-Adams et al. 2007). The data is presented as the percentage change compared to control. The tomato products were more effective than lycopene in this study and combinations of tomato and broccoli components were active and compared favorably to endocrine manipulations.

It has been proposed that lycopene or other carotenoids may act on prostate cancer cells to modulate cell-cell communication and adhesion in a manner that reduces carcinogenesis or tumor growth (Vine et al. 2005). One small human study suggests that lycopene may increase the expression of connexin 43, a known tumor suppressor gene. In this study, 15 men were given either a placebo or Lyc-*O*-Mato (Beer-Sheva, Israel), containing 15 mg lycopene, 2.5 mg phytoene and phytofluene, and other minor carotenoids, twice a day for 3 wk before radical prostatectomy (Kucuk et al. 2001). Prostate tissue analysis revealed a suggestion of an increase in connexin 43 levels in the lycopene-supplemented groups compared with controls (mean absorbance: 0.63 ± 0.19 v. 0.25 ± 0.08, P = 0.13). However, larger studies with greater statistical power are need to investigate this potential mechanism in humans.

Inflammation within the prostate, perhaps initiated by a variety of pathways, is hypothesized to result in oxidative stress and macromolecule damage that enhances risk of carcinogenesis. The possibility that lycopene or other phytochemicals may inhibit this process is supported by a few studies. For example, the expression of inflammatory proteins such as proinflammatory cytokines, immunoglobulins and immunoglobulin receptors was found to be reduced in the prostate tissue of Copenhagen rats fed 200 ppm lycopene for 8 wk (Herzog et al. 2005). A few human studies have attempted to examine the relationship between lycopene and biomarkers of oxidative damage and inflammation, but no clear relationship for prostate cancer has yet emerged (Almushatat et al. 2006).

Tomato products and lycopene are also proposed to alter hormone and growth factor signaling. Androgens are essential for the normal growth and function of prostate tissue, in addition to being critical factors for early stages of prostate carcinogenesis. Rodent studies have reported that lycopene-containing diets lower the expression of 5-β-reductase, the enzyme responsible for metabolism of testosterone to the more potent androgen dihydrotestosterone in normal and prostate cancer tissue (Herzog et al. 2005, Siler et al. 2004). Rodent studies have clearly shown that androgens influence lycopene metabolism (Boileau et al. 2000, 2001). In addition, some studies, but not all, suggest that diets containing lycopene or tomato powder can reduce serum testosterone concentrations (Campbell et al. 2006) or androgen receptor expression (Xing et al. 2001). Thus, an interaction between lycopene and the critical androgen-prostate axis clearly exists, but it remains only superficially understood at this time.

Elevated levels of serum insulin-like growth factor-1 (IGF-1) have been correlated with an increased risk of benign prostate hyperplasia (BPH) and prostate cancer on the basis of several studies (Chen et al. 2004). Prostate cancer cell proliferation and survival is responsive to IGF-1 signaling (Wang et al. 2003). IGF-1 expression was reduced in Copenhagen rats as a result of lycopene-containing diets (Herzog et al. 2005, Siler et al. 2004). Our group has shown that several of the tomato polyphenols are potent inhibitors of IGF-1 signaling pathways in prostate cancer cells, resulting in anti-proliferative and pro-apoptotic signaling (Wang et al. 2003). Others have reported that lycopene disrupts IGF-I receptor expression in prostate cancer cells (Kanagaraj et al. 2007). A human case-control study investigating the relationship between dietary factors, serum IGF-1, and prostate cancer reports an inverse relationship between the consumption of cooked tomatoes and blood IGF-1 levels, with a mean reduction of 31.5% for each additional serving of cooked tomatoes consumed per day (95% CI, -49.4 to -7.9; $P = 0.014$) (Mucci et al. 2001). These representative examples from the literature suggest that additional studies of carotenoids, tomato polyphenols, and effects on autocrine, paracrine, and endocrine growth factor expression and signaling may provide key clues to a potential mechanism of action.

THE NEED FOR RANDOMIZED CONTROLLED CLINICAL TRIALS

Although evidence from epidemiology, cell biology, and rodent models provide enough support to establish the hypothesis that tomatoes or lycopene may have inhibitory activity for prostate carcinogenesis, the lack of randomized human intervention trials of substantial statistical power remains the major obstacle to adequate testing and procurement of definitive answers. Studies focusing on mechanisms and relevant biomarkers are needed, and, if supportive, definitive studies with a cancer endpoint will be necessary.

Although limited in number, there are several small human intervention studies which have investigated potential impact of tomato products and lycopene and specific biomarkers relevant to prostate carcinogenesis (Bowen 2005, Kim et al. 2003). In one study, 43 men with histologically diagnosed BPH consumed 50 g tomato paste daily for 10 wk. Plasma PSA values were analyzed at baseline, during and after intervention. The authors reported a significant reduction in mean plasma PSA levels between baseline and end-of-study ($P = 0.005$) (Edinger and Koff 2006). Prostatic premalignant lesions such as high-grade prostate intraepithelial neoplasia (HGPIN) are present in a large number of aging men; in a small proportion of these men, the lesions progress to clinically significant prostate cancer. The use of chemopreventive agents that are well tolerated and have minimal risk is of particular interest in this population. A recent study randomized 40 men with HGPIN to receive either 4 mg lycopene twice a day for one year (8 mg total per day, Lyc-O-Mato, Beer-Sheva, Israel), or routine monitoring (which included PSA analysis) every 3 mon. By the end of the intervention, men receiving the lycopene supplement had a lower serum PSA than at baseline (mean PSA 6.07 ng/mL at baseline, 3.5 ng/mL at end-of-study) compared with an increase reported in the control group (6.55 ng/ mL at baseline, 8.06 ng/mL at end-of-study) (Mohanty et al. 2005). Unfortunately, the authors did not provide information regarding statistical calculations or adequate information on randomization, diet, the type of supplement used or the use of other prescription medications that may affect serum PSA.

The role of tomatoes and lycopene in prevention of biochemical relapse after primary treatment for prostate cancer has also been investigated. Clark et al. (2006) conducted a dose-escalating trial of lycopene supplementation in 36 men with relapsed prostate cancer, based on PSA progression. The doses of lycopene ranged from 15 to 120 mg/d for one year. The median serum PSA at study entry was 4.4 ng/mL with a range of 0.8 to 24.9 mg/mL. Although no serum PSA responses (defined as a 50% decrease in baseline PSA) were observed, lycopene supplementation was well tolerated (Clark et al. 2006). It may be that the definition of a PSA response in this study (as typically used for chemotherapy) was too rigorous and other calculated outcomes such as PSA velocity and PSA doubling time may be more appropriate measures to assess impact of dietary interventions. PSA velocity was studied in a clinical trial of 37 men with localized prostate cancer who were receiving clinical surveillance only. All 37 men were given two lycopene tablets (5 mg lycopene per tablet plus vitamins C and E, phytoene, phytofluene, and beta carotene). The men had monthly clinical evaluations and serum PSA analyses for a mean intervention time of 10 mon. The velocity of PSA rise decreased in 26 of the 37 men in the study and in 8 men, the slope of PSA rise was negative after lycopene intervention (Barber et al. 2006). Although none of these studies provides definitive data regarding efficacy of tomato products or lycopene for the prevention of prostate cancer or prostate premalignancy, or as an adjunct to the treatment of prostate cancer, all suggest that studies are feasible with either carotenoid supplements or dietary interventions, and that no major toxicities are noted over a range of intake.

SUMMARY

The scientific evidence that tomatoes and tomato phytochemicals are important modulators of prostate carcinogenesis is provocative, yet further research is needed with a combination of well-designed studies in laboratory models, humans, and population groups (Clinton 2005, Gann and Khachik 2003). Thus far, on the basis of evidence from all available sources, our team hypothesizes a reduction in the risk of clinically significant prostate cancer, perhaps in the range of 10-20%, in response to daily tomato product consumption over a lifetime. We further propose that the data thus far accumulated are stronger for a benefit from tomato products than from pure lycopene. However, lycopene may be one of the components necessary for an optimal beneficial response. Mechanistic studies published to date also support a potential benefit, yet no unifying mechanism has been established. Additional studies should be undertaken, in hope of providing consistent data across the spectrum of studies from *in vitro* to rodent models, and humans. In addition, there are a number of critical issues that remain to be examined that will influence our understanding of the tomato, lycopene, and prostate cancer relationship. We are only beginning to understand the genetic polymorphisms that influence the absorption, metabolism, excretion, and biological activity of tomato-derived phytochemicals. For example, the role of genetic variation in carotenoid monooxygenase enzymes critically involved in carotenoid metabolism has not been examined with respect to anti-cancer activities of carotenoids (Lindshield et al. 2007). Additional efforts are needed to define ways in which the metabolism and biological activity of tomato phytochemicals are influenced by other components of the diet or pharmaceutical agents. Continued efforts to better understand prostate carcinogenesis and further classify prostate cancer into subtypes and families with specific biological behaviors and responses to interventions will allow a more precise assessment of how specific dietary components influence the disease risk and course.

ABBREVIATIONS

PSA: prostate specific antigen; ng: nanogram; HPFS: Health Professionals Follow-Up Study; RR: relative risk; CI: confidence interval; DMAB: 3,2"-dimethyl-4-aminobiphenol; PhIP: 2-amino-1-methylimidazo[4,5-b]pyridine; PIN: prostatic intraepithelial neoplasia; NMU: N-methyl-N-nitrosourea; AIN: American Institute of Nutrition; IGF-1: insulin-like growth factor 1; IL-6: interleukin 6; nM: nanomols; FeNTA: ferric nitrilotriacetate; 8-oxoGuo: 8-oxo-7,8-dihydro-2'deoxyguanosine; PCNA: proliferating cell nuclear antigen; ppm: parts per million; BPH: benign prostate hyperplasia; HG PIN: high grade prostate intraepithelial neoplasia; CMO: carotenoid monooxygenase

REFERENCES

Allen, C.M. and A.M. Smith, S.K. Clinton, and S.J. Schwartz. 2002. Tomato consumption increases lycopene isomer concentration in breast milk and plasma of lactating women. J. Amer. Diet. Assoc. 102: 1257-1262.

Allen, C.M. and S.J. Schwartz, N.E. Craft, E.L. Giovannucci, V.L. De Groff, and S.K. Clinton. 2003. Changes in plasma and oral mucosal lycopene isomer concentrations in healthy adults consuming standard servings of processed tomato products. Nutr. Cancer 47: 48-56.

Almushatat, A.S. and D. Talwar, P.A. McArdle, C. Williamson, N. Sattar, D.S. O'Reilly, M.A. Underwood, and D.C. McMillan. 2006. Vitamin antioxidants, lipid peroxidation and the systemic inflammatory response in patients with prostate cancer. Intl. J. Cancer 118: 1051-1053.

Barber, N.J. and X. Zhang, G. Zhu, R. Pramanik, J.A. Barber, F.L. Martin, J.D. Morris, and G.H. Muir. 2006. Lycopene inhibits DNA synthesis in primary prostate epithelial cells in vitro and its administration is associated with a reduced prostate-specific antigen velocity in a phase II clinical study. Prostate Cancer Prostatic Dis. 9: 407-413.

Basu, A. and V. Imrhan. 2007. Tomatoes versus lycopene in oxidative stress and carcinogenesis: conclusions from clinical trials. Eur. J. Clin. Nutr. 61: 295-303.

Boileau, T.W. and S.K. Clinton, and J.W. Erdman Jr. 2000. Tissue lycopene concentrations and isomer patterns are affected by androgen status and dietary lycopene concentration in male F344 rats. J. Nutr. 130: 1613-1618.

Boileau, T.W. and S.K. Clinton, S. Zaripheh, M.H. Monaco, S.M. Donovan, and J.W. Erdman Jr. 2001. Testosterone and food restriction modulate hepatic lycopene isomer concentrations in male F344 rats. J. Nutr. 131: 1746-1752.

Boileau, T.W. and Z. Liao, S. Kim, S. Lemeshow, J.W. Erdman Jr., and S.K. Clinton. 2003. Prostate carcinogenesis in N-methyl-N-nitrosourea (NMU)-testosterone-treated rats fed tomato powder, lycopene, or energy-restricted diets. J. Natl. Cancer Inst. 95: 1578-1586.

Bosland, M.C. 1999. Use of animal models in defining efficacy of chemoprevention agents against prostate cancer. Eur. Urol. 35: 459-463.

Bowen, P.E. 2005. Selection of surrogate endpoint biomarkers to evaluate the efficacy of lycopene/tomatoes for the prevention/progression of prostate cancer. J. Nutr. 135: 2068S-2070S.

Campbell, J.K. and C.K. Stroud, M.T. Nakamura, M.A. Lila, and J.W. Erdman Jr. 2006. Serum testosterone is reduced following short-term phytofluene, lycopene, or tomato powder consumption in F344 rats. J. Nutr. 136: 2813-2819.

Canene-Adams, K. and J.K. Campbell, S. Zaripheh, E.H. Jeffery, and J.W. Erdman Jr. 2005. The tomato as a functional food. J. Nutr. 135: 1226-1230.

Canene-Adams, K. and B.L. Lindshield, S. Wang, E.H. Jeffery, S.K. Clinton, and J.W. Erdman Jr. 2007. Combinations of tomato and broccoli enhance antitumor activity in dunning r3327-h prostate adenocarcinomas. Cancer Res. 67: 836-843.

Chen, L. and M. Stacewicz-Sapuntzakis, C. Duncan, R. Sharifi, L. Ghosh, R. van Breemen, D. Ashton, and P.E. Bowen. 2001. Oxidative DNA damage in prostate cancer patients consuming tomato sauce-based entrees as a whole-food intervention. J. Natl. Cancer Inst. 93: 1872-1879.

Chen, M.S. and D. Chen, and Q.P. Dou. 2004. Inhibition of proteasome activity by various fruits and vegetables is associated with cancer cell death. In Vivo 18: 73-80.

Clark, P.E. and M.C. Hall, L.S. Borden Jr., A.A. Miller, J.J. Hu, W.R. Lee, D. Stindt, R. D'Agostino Jr., J. Lovato, M. Harmon, and F.M. Torti. 2006. Phase I-II prospective dose-escalating trial of lycopene in patients with biochemical relapse of prostate cancer after definitive local therapy. Urology 67: 1257-1261.

Clinton, S.K. 1998. Lycopene: chemistry, biology, and implications for human health and disease. Nutr. Rev. 218: 140-143.

Clinton, S.K. 2005. Tomatoes or lycopene: a role in prostate carcinogenesis? J. Nutr. 135: 2057S-2059S.

Clinton, S.K. and C. Emenhiser, S.J. Schwartz, D.G. Bostwick, A.W. Williams, B.J. Moore, and J.W. Erdman Jr. 1996. cis-trans lycopene isomers, carotenoids, and retinol in the human prostate. Cancer Epidemiol. Biomarkers Prev. 5: 823-833.

Cohen, L.A. 2002. A review of animal model studies of tomato carotenoids, lycopene, and cancer chemoprevention. Exp. Biol. Med. (Maywood) 227: 864-868.

Conn, P.F. and W. Schalch, and T.G. Truscott. 1991. The singlet oxygen and carotenoid interaction. J. Photochem. Photobiol. B-Biol. 11: 41-47.

Di Mascio, P. and S. Kaiser, and H. Sies. 1989. Lycopene as the most efficient biological carotenoid singlet oxygen quencher. Arch. Biochem. Biophys. 274: 532-538.

Edinger, M.S. and W.J. Koff. 2006. Effect of the consumption of tomato paste on plasma prostate-specific antigen levels in patients with benign prostate hyperplasia. Braz. J. Med. Biol. Res. 39: 1115-1119.

Etminan, M. and B. Takkouche, and F. Caamano-Isorna. 2004. The role of tomato products and lycopene in the prevention of prostate cancer: a meta-analysis of observational studies. Cancer Epidemiol. Biomarkers Prev. 13: 340-345.

Ferlay, J. and F. Bray, and P. Pisani. 2001. GLOBOSCAN 2000: Cancer Incidence, Mortality and Prevalence Worldwide. International Agency for Research on Cancer. Lyon, France.

Gann, P.H. and F. Khachik. 2003. Tomatoes or lycopene versus prostate cancer: is evolution anti-reductionist? J. Natl. Cancer Inst. 95: 1563-1565.

Gann, P.H. and J. Ma, E. Giovannucci, W. Willett, F.M. Sacks, C.H. Hennekens, and M.J. Stampfer. 1999. Lower prostate cancer risk in men with elevated plasma lycopene levels: results of a prospective analysis. Cancer Res. 59: 1225-1230.

Giovannucci, E. 2002. A review of epidemiologic studies of tomatoes, lycopene, and prostate cancer. Exp. Biol. Med. (Maywood) 227: 852-859.

Giovannucci, E. 2005. Tomato products, lycopene, and prostate cancer: a review of the epidemiological literature. J. Nutr. 135: 2030S-2031S.

Giovannucci, E. 2007. Does prostate-specific antigen screening influence the results of studies of tomatoes, lycopene, and prostate cancer risk? J. Natl. Cancer Inst. 99: 1060-1062.

Giovannucci, E. and S. K. Clinton. 1998. Tomatoes, lycopene, and prostate cancer. Proc. Soc. Exp. Biol. Med. 218: 129-139.

Giovannucci, E., A. Ascherio, E.B. Rimm, M.J. Stampfer, G.A. Colditz, and W.C. Willett. 1995. Intake of carotenoids and retinol in relation to risk of prostate cancer. J. Natl. Cancer Inst. 87: 1767-1776.

Guttenplan, J.B. and M. Chen, W. Kosinska, S. Thompson, Z. Zhao, and L.A. Cohen. 2001. Effects of a lycopene-rich diet on spontaneous and benzo[a]pyrene-induced mutagenesis in prostate, colon and lungs of the lacZ mouse. Cancer Lett. 164: 1-6.

Hadley, C.W. and S.K. Clinton, and S.J. Schwartz. 2003. The consumption of processed tomato products enhances plasma lycopene concentrations in association with a reduced lipoprotein sensitivity to oxidative damage. J. Nutr. 133: 727-732.

Hantz, H.L. and L.F. Young, and K.R. Martin. 2005. Physiologically attainable concentrations of lycopene induce mitochondrial apoptosis in LNCaP human prostate cancer cells. Exp. Biol. Med. 230: 171-179.

Herzog, A. and K. Wertz. 2005. Lycopene effects for improved prostate health. Agro Food Industry Hi-Tech 16: 13-15.

Herzog, A. and U. Siler, V. Spitzer, N. Seifert, A. Denelavas, P.B. Hunziker, W. Hunziker, R. Goralczyk, and K. Wertz. 2005. Lycopene reduced gene expression of steroid targets and inflammatory markers in normal rat prostate. FASEB J. 19: 272-274.

Hwang, E.S. and P.E. Bowen. 2004. Cell cycle arrest and induction of apoptosis by lycopene in LNCaP human prostate cancer cells. J. Med. Food 7: 284-289.

Hwang, E.S. and P.E. Bowen. 2005. Effects of tomato paste extracts on cell proliferation, cell-cycle arrest and apoptosis in LNCaP human prostate cancer cells. Biofactors 23: 75-84.

Hwang, E.S. and H.J. Lee. 2006. Inhibitory effects of lycopene on the adhesion, invasion, and migration of SK-Hep1 human hepatoma cells. Exp. Biol. Med. 231: 322-327.

Imaida, K. and S. Tamano, K. Kato, Y. Ikeda, M. Asamoto, S. Takahashi, Z. Nir, M. Murakoshi, H. Nishino, and T. Shirai. 2001. Lack of chemopreventive effects of lycopene and curcumin on experimental rat prostate carcinogenesis. Carcinogenesis 22: 467-472.

Ivanov, N.I. and S.P. Cowell, P. Brown, P.S. Rennie, E.S. Guns, and M.E. Cox. 2007. Lycopene differentially induces quiescence and apoptosis in androgen-responsive and -independent prostate cancer cell lines. Clin. Nutr. 26: 252-263.

Jatoi, A. and P. Burch, D. Hillman, J.M. Vanyo, S. Dakhil, D. Nikcevich, K. Rowland, R. Morton, P.J. Flynn, C. Young, and W. Tan. 2007. A tomato-based, lycopene-containing intervention for androgen-independent prostate cancer: results of a Phase II study from the North Central Cancer Treatment Group. Urology 69: 289-294.

Jemal, A. and R. Siegel, E. Ward, T. Murray, J. Xu, and M.J. Thun. 2007. Cancer statistics, 2007. CA Cancer J. Clin. 57: 43-66.

Kanagaraj, P. and M.R. Vijayababu, B. Ravisankar, J. Anbalagan, M.M. Aruldhas, and J. Arunakaran. 2007. Effect of lycopene on insulin-like growth factor-I, IGF binding protein-3 and IGF type-I receptor in prostate cancer cells. J. Cancer Res. Clin. Oncol. 133: 351-359.

Kavanaugh, C.J. and P.R. Trumbo, and K.C. Ellwood. 2007. The U.S. Food and Drug Administration's evidence-based review for qualified health claims: tomatoes, lycopene, and cancer. J. Natl. Cancer Inst. 99: 1074-1085.

Kelman, Z. 1997. PCNA: Structure, functions and interactions. Oncogene 14: 629-640.

Kim, H.S. and P. Bowen, L. Chen, C. Duncan, L. Ghosh, R. Sharifi, and K. Christov. 2003. Effects of tomato sauce consumption on apoptotic cell death in prostate benign hyperplasia and carcinoma. Nutr. Cancer 47: 40-47.

Kirsh, V.A. and S.T. Mayne, U. Peters, N. Chatterjee, M.F. Leitzmann, L.B. Dixon, D.A. Urban, E. D. Crawford, and R.B. Hayes. 2006. A prospective study of lycopene and tomato product intake and risk of prostate cancer. Cancer Epidemiol. Biomarkers Prev. 15: 92-98.

Klein, R.D. 2005. The use of genetically engineered mouse models of prostate cancer for nutrition and cancer chemoprevention research. Mutat. Res. 576: 111-119.

Kristal, A.R. and J.H. Cohen. 2000. Invited commentary: tomatoes, lycopene, and prostate cancer. How strong is the evidence? Amer. J. Epidemiol. 151: 124-127; discussion 128-130.

Kucuk, O. and F.H. Sarkar, W. Sakr, Z. Djuric, M.N. Pollak, F. Khachik, Y.W. Li, M. Banerjee, D. Grignon, J.S. Bertram, J.D. Crissman, E.J. Pontes, and D.P. Wood, Jr. 2001. Phase II randomized clinical trial of lycopene supplementation before radical prostatectomy. Cancer Epidemiol. Biomarkers Prev. 10: 861-868.

Lamb, D.J. and L. Zhang. 2005. Challenges in prostate cancer research: animal models for nutritional studies of chemoprevention and disease progression. J. Nutr. 135: 3009S-3015S.

Limpens, J. and F.H. Schroder, C.M. de Ridder, C.A. Bolder, M.F. Wildhagen, U.C. Obermuller-Jevic, K. Kramer, and W.M. van Weerden. 2006. Combined lycopene and vitamin E treatment suppresses the growth of PC-346C human prostate cancer cells in nude mice. J. Nutr. 136: 1287-1293.

Lindshield, B.L. and K. Canene-Adams, and J.W. Erdman Jr. 2007. Lycopenoids: are lycopene metabolites bioactive? Arch. Biochem. Biophys. 458: 136-140.

Lu, Q.Y. and J.C. Hung, D. Heber, V.L.W. Go, V.E. Reuter, C. Cordon-Cardo, H.I. Scher, J.R. Marshall, and Z.F. Zhang. 2001. Inverse associations between plasma lycopene and other carotenoids and prostate cancer. Cancer Epidemiol. Biomarkers Prev. 10: 749-756.

Lucia, M.S. and D.G. Bostwick, M. Bosland, A.T.K. Cockett, D.W. Knapp, I. Leav, M. Pollard, C. Rinker-Schaeffer, T. Shirai, and B.A. Watkins. 1998. Workgroup I: Rodent models of prostate cancer. Prostate 36: 49-55.

Matos, H.R. and S.A. Marques, O.F. Gomes, A.A. Silva, J.C. Heimann, P. Di Mascio, and M.H. Medeiros. 2006. Lycopene and beta-carotene protect in vivo iron-induced oxidative stress damage in rat prostate. Braz. J. Med. Biol. Res. 39: 203-210.

McCormick, D.L. and K.V. Rao, L. Dooley, V.E. Steele, R.A. Lubet, G.J. Kelloff, and M.C. Bosland. 1998. Influence of N-methyl-N-nitrosourea, testosterone, and N-(4-hydroxyphenyl)-all-trans-retinamide on prostate cancer induction in Wistar-Unilever rats. Cancer Res. 58: 3282-3288.

Miller, E.C. and C.W. Hadley, S.J. Schwartz, J.W. Erdman Jr, T.W.-M. Boileau, and S.K. Clinton. 2002. Lycopene, tomato products, and prostate cancer prevention. Have we established causality? Pure Appl. Chem. 74: 1435-1441.

Mohanty, N.K. and S. Saxena, U.P. Singh, N.K. Goyal, and R.P. Arora. 2005. Lycopene as a chemopreventive agent in the treatment of high-grade prostate intraepithelial neoplasia. Urol. Oncol. 23: 383-385.

Mucci, L.A. and R. Tamimi, P. Lagiou, A. Trichopoulou, V. Benetou, E. Spanos, and D. Trichopoulos. 2001. Are dietary influences on the risk of prostate cancer mediated through the insulin-like growth factor system? BJU Intl. 87: 814-820.

Peters, U. and M.F. Leitzmann, N. Chatterjee, Y. Wang, D. Albanes, E.P. Gelmann, M.D. Friesen, E. Riboli, and R.B. Hayes. 2007. Serum lycopene, other carotenoids, and prostate cancer risk: a nested case-control study in the Prostate, Lung, Colorectal, and Ovarian Cancer Screening Trial. Cancer Epidemiol. Biomarkers Prev. 16: 962-968.

Pollard, M. and M.A. Suckow. 2005. Hormone-refractory prostate cancer in the Lobund-Wistar rat. Exp. Biol. Med. (Maywood) 230: 520-526.

Pool-Zobel, B.L. and A. Bub, H. Muller, I. Wollowski, and G. Rechkemmer. 1997. Consumption of vegetables reduces genetic damage in humans: first results of a human intervention trial with carotenoid-rich foods. Carcinogenesis 18: 1847-1850.

Shappell, S.B. and G.V. Thomas, R.L. Roberts, R. Herbert, M.M. Ittmann, M.A. Rubin, P.A. Humphrey, J.P. Sundberg, N. Rozengurt, R. Barrios, J.M. Ward, and R.D. Cardiff. 2004. Prostate pathology of genetically engineered mice: Definitions and classification. The consensus report from the bar harbor meeting of the mouse models of human cancer consortium prostate pathology committee. Cancer Res. 64: 2270-2305.

Shirai, T. and S. Takahashi, L. Cui, M. Futakuchi, K. Kato, S. Tamano, and K. Imaida. 2000. Experimental prostate carcinogenesis - rodent models. Mutat. Res.-Rev. Mutat. Res. 462: 219-226.

Siler, U. and L. Barella, V. Spitzer, J. Schnorr, M. Lein, R. Goralczyk, and K. Wertz. 2004. Lycopene and vitamin E interfere with autocrine/paracrine loops in the Dunning prostate cancer model. FASEB J. 18: 1019-1021.

Siler, U. and A. Herzog, V. Spitzer, N. Seifert, A. Denelavas, P.B. Hunziker, L. Barella, W. Hunziker, M. Lein, R. Goralczyk, and K. Wertz. 2005. Lycopene effects on rat normal prostate and prostate tumor tissue. J. Nutr. 135: 2050S-2052S.

Tang, L. and T. Jin, X. Zeng and J.S. Wang. 2005. Lycopene inhibits the growth of human androgen-independent prostate cancer cells in vitro and in BALB/c nude mice. J. Nutr. 135: 287-290.

Tennant, T.R. and H. Kim, M. Sokoloff, and C.W. Rinker-Schaeffer. 2000. The Dunning model. Prostate 43: 295-302.

Thompson, I.M. and P.J. Goodman, C.M. Tangen, M.S. Lucia, G.J. Miller, L.G. Ford, M.M. Lieber, R.D. Cespedes, J.N. Atkins, S.M. Lippman, S.M. Carlin, A. Ryan, C.M. Szczepanek, J.J. Crowley, and C.A. Coltman Jr. 2003. The influence of finasteride on the development of prostate cancer. N. Engl. J. Med. 349: 215-224.

Venkateswaran, V. and N.E. Fleshner, L.M. Sugar, and L.H. Klotz. 2004. Antioxidants block prostate cancer in lady transgenic mice. Cancer Res. 64: 5891-5896.

Vine, A.L. and Y.M. Leung, and J.S. Bertram. 2005. Transcriptional regulation of connexin 43 expression by retinoids and carotenoids: similarities and differences. Mol. Carcinog. 43: 75-85.

Wang, S. and V.L. DeGroff, and S.K. Clinton. 2003. Tomato and soy polyphenols reduce insulin-like growth factor-I-stimulated rat prostate cancer cell proliferation and apoptotic resistance in vitro via inhibition of intracellular signaling pathways involving tyrosine kinase. J. Nutr. 133: 2367-2376.

World Cancer Research Fund. 1997. Food, Nutrition and the Prevention of Cancer: A Global Perspective. American Institute for Cancer Research, Washington, DC.

Wu, K. and S.J. Schwartz, E.A. Platz, S.K. Clinton, J.W. Erdman Jr, M.G. Ferruzzi, W.C. Willett, and E.L. Giovannucci. 2003. Variations in plasma lycopene and specific isomers over time in a cohort of U.S. men. J. Nutr. 133: 1930-1936.

Xing, N. and Y. Chen, S.H. Mitchell, and C.Y. Young. 2001. Quercetin inhibits the expression and function of the androgen receptor in LNCaP prostate cancer cells. Carcinogenesis 22: 409-414.

23

Tomato Juice, Prostate Cancer and Adenosine Deaminase Enzyme

Aslihan Avci* **and Ilker Durak**

Ankara University, Faculty of Medicine, Department of Biochemistry, 06100 Sihhiye, Ankara, Turkey

ABSTRACT

Plants are taking an important place in the treatment of various diseases. Especially tomato and its products have been documented to play a significant role in the treatment of prostate disesases. One of the constituents of tomato is lycopene, which is a carotenoid found predominantly in tomato and tomato products. Lycopene is not destroyed during the processing of food; rather, its bioavailability improves. It plays an important role in reducing risk of prostate cancer because of its high antioxidant power. Cis-lycopene has been shown to predominate in both benign and malignant prostate tissues, suggesting a possible beneficial effect of high cis-isomer concentrations, and also the involvement of tissue isomerases in *in vivo* isomerization from all-trans to cis form. The plasma AD2 (adenosine deaminase isoform) is also increased in most cancers. Adenosine deaminase plays an important part in purine metabolism and DNA turnover. It irreversibly deaminates adenosine, converting it to the related nucleoside inosine, which then is deribosylated and converted to hypoxanthine. In some studies, it was shown that tomato extract significantly inhibits the adenosine deaminase of cancerous tissues, thereby eliminating this advantage of cancer cells, which might be the basis for the beneficial effect of tomato in prostate cancers.

*Corresponding author

A list of abbreviations is given before the references.

INTRODUCTION

The prostate, found only in men, is a walnut-sized gland located in front of the rectum and beneath the urinary bladder. It contains gland cells that make some of the fluid that protects and nourishes sperm cells in semen (Fig. 1). Male hormones, which are also called androgens, cause the prostate gland to develop in the fetus. The most common androgen is testosterone. The prostate continues to grow until a man reaches adulthood and then stays about the same size as long as male hormones are produced. In older men, the part of the prostate around the urethra may continue to grow, a condition called benign prostate hyperplasia (BPH). This can cause problems with urination because the urethra (the tube that carries the urine) is squeezed by the growing prostate tissue. If a cancer starts in glandular cells, it is named adenocarcinoma. Other types of prostate cancer are rare. Some prostate cancers can grow rapidly, whereas most prostate cancers grow slowly. Some doctors believe that prostate cancer begins with a condition called prostatic intraepithelial neoplasia (PIN), which begins to appear in some men in their twenties. Almost half of men have PIN by the time they reach 50. In this condition, there are changes in the microscopic appearance of prostate gland cells. These changes are classified as either low grade, meaning they appear almost normal, or high grade, meaning they look abnormal.

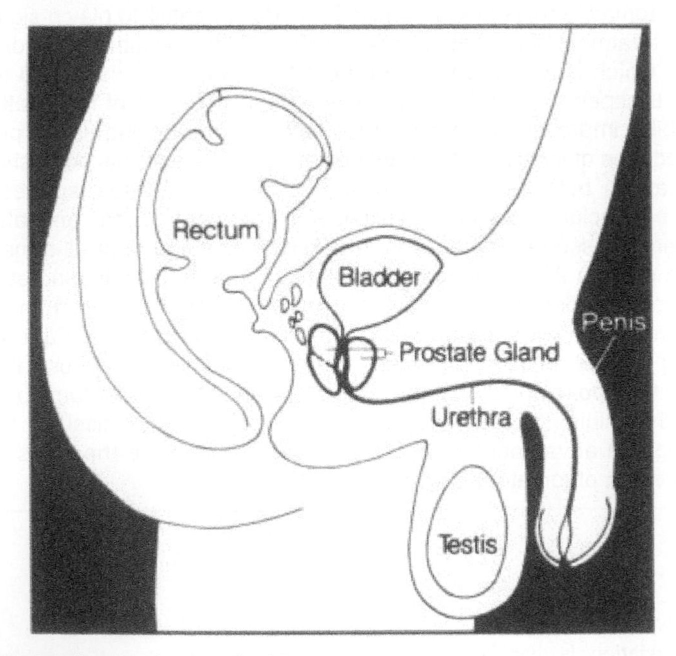

Fig. 1 Prostate is an endocrine gland in men.

Malign prostate diseases have become a major public health problem worldwide. Although the etiologic factors of prostate cancer remain largely unknown, dietary factors and physical activity might be important in the prevention of the disease. In some of the studies, it was observed that high consumption of meat and dairy products and high levels of circulating insulin-like growth factor 1 (IGF-1) have been linked to a greater risk, whereas consumption of fatty fish, vitamin E, selenium and tomato products has been associated with a reduced risk (Wolk 2005).

For more than 30 years, researchers have worked to identify the etiological factors that might explain the almost 40-fold difference in the reported incidence and the about 12-fold difference in mortality from prostate cancer between various geographic and ethnic populations (Hsing and Devesa 2001). Although it was estimated that 40% of prostate cancer can be explained by genetic factors (Lichtenstein et al. 2000), there is strong relation between lifestyle and environmental factors. It was shown that among migrants moving from countries with a low prostate cancer incidence (e.g., Japan) to North America, the incidence increases to almost equal that of white Americans within a generation. Large geographic differences were observed in the morbidity and mortality. Men from the same ethnic group but living in different environments have different risk ratio of prostate cancer (Kolonel 2004).

Tomatoes and tomato products are rich sources of folate, vitamin C, and potassium. The most abundant phytonutrients in tomatoes are the carotenoids (Fig. 2). Lycopene is the most prominent carotenoid, followed by beta-carotene, gamma-carotene and phytoene as well as several minor carotenoids. The antioxidant activity of lycopene as well as several other carotenoids and their abundance in tomatoes makes these foods rich sources of antioxidant activity (Fig. 3). The provitamin A activity of beta- and gamma-carotene, their modest levels in tomato products, and the high consumption of these foods results in a rich supply of vitamin A activity

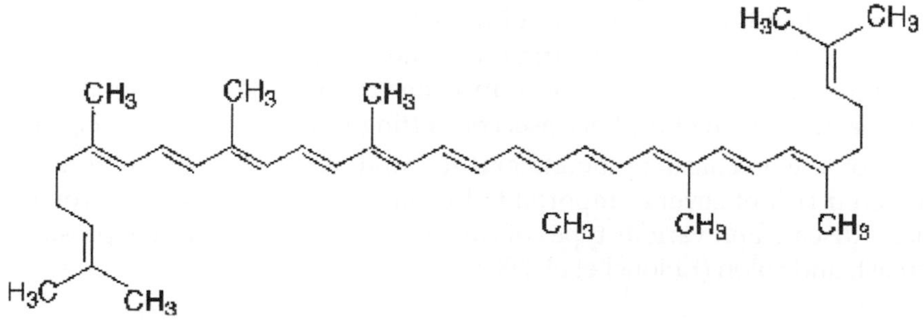

Fig. 2 Lycopene all-trans form.

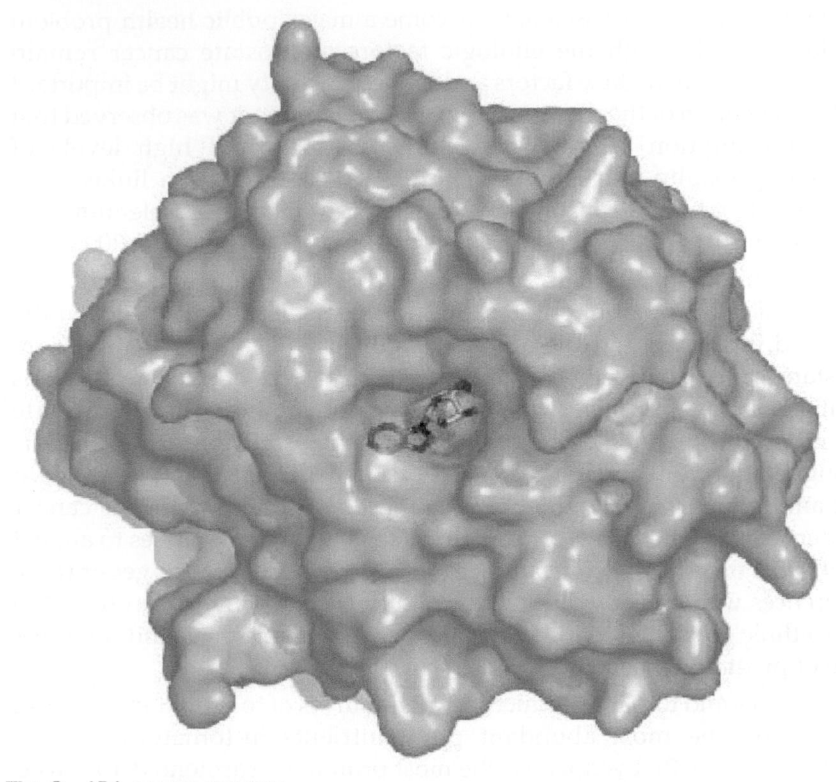

Fig. 3 ADA enzyme structure.

from tomato-based foods. Tomatoes also contain several other components that are beneficial to health, including vitamin E, trace elements, flavonoids, phytosterols, and several water-soluble vitamins (Beecker 1998).

Adenosine deaminase (ADA) is an enzyme (EC 3.5.4.4) involved in purine metabolism (Fig. 4). It is needed for the breakdown of adenosine from food and for the turnover of nucleic acids in tissues. It irreversibly deaminates adenosine, converting it to the related nucleoside inosine by the removal of an amine group. Inosine can then be deribosylated by purine nucleoside phosphorylase, converting it to hypoxanthine (Fig. 4).

People who consume tomato and its products in their diets present a lowered risk of several important chronic diseases, including coronary heart disease and various types of cancers such as cancers of the prostate, breast, and colon (Kolonel et al. 2004).

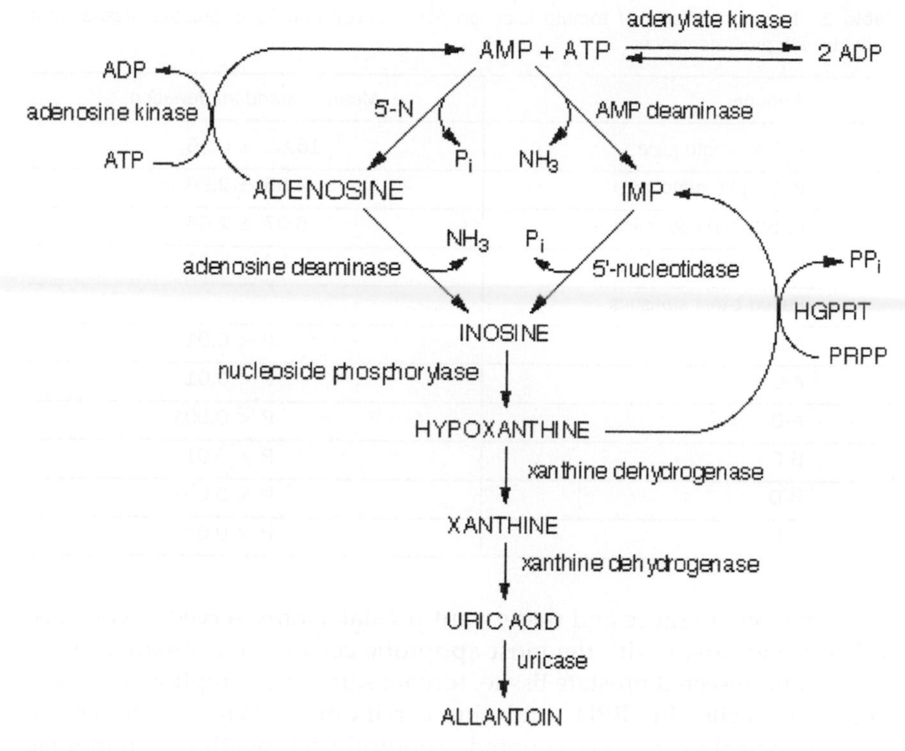

Fig. 4 Purine metabolism.

STUDIES

Tomato Consumption in Relation to Prostate Cancer

Stacewicz et al. showed that consuming tomato sauce and pasta significantly diminished oxidative DNA damage in leukocytes and prostate tissues in patients with prostate cancer 3 wk before their scheduled prostatectomy. Their explanations may show that lycopene affects the genes governing the androgen stimulation of prostate growth, cytokines and the enzymes producing reactive oxygen species (Stacewicz-Sapuntzakis and Bowen 2005). Population studies have suggested that lycopene, which is mostly found in tomato and tomato products, may reduce the risk of prostate cancer. A study by Kim et al. was designed to determine whether apoptotic cell death and associated Bcl-2 and Bax proteins were modulated by tomato sauce intervention (Kim et al. 2003). Thirty-two patients diagnosed by biopsy with prostate carcinoma were given tomato sauce (approximately 30 mg lycopene/day) for 3 wk before prostatectomy. Thirty-four patients with prostate cancer who did not

Table 1 Inhibitory effects of tomato juice on ADA activity (mIU/L) in prostate tissue from patients with prostate cancer.

Amounts	*Mean ± standard deviation*
A, No tomato juice	15.82 ± 6.36
B, 25 μl tomato juice	8.85 ± 2.96
C, 50 μl tomato juice	6.07 ± 2.64
D, 100 μl tomato juice	4.03 ± 1.93
Paired t-test statistics	
A-B	$P < 0.01$
A-C	$P < 0.01$
A-D	$P < 0.001$
B-C	$P < 0.01$
B-D	$P < 0.001$
C-D	$P < 0.05$

consume tomato sauce and underwent prostatectomy served as controls. When tumor areas with the most apoptotic cells were compared in the biopsy and resected prostate tissue, tomato sauce consumption increased apoptotic cells in BPH and in carcinomas. When comparable morphological areas were counted, apoptotic cell death in carcinomas increased significantly with treatment, from 0.84 ± 0.13% to 1.17 ± 0.19%, and apoptotic cell death in BPH showed a tendency toward an increase from 0.66 ± 0.10% to 1.20 ± 0.32%. When the values of apoptotic cells in BPH and carcinomas of patients who consume tomato sauce were compared with corresponding control lesions of the patients who did not consume tomato sauce in resected prostate tissue, the differences of values were not significant. They showed that tomato sauce consumption did not affect Bcl-2 expression but decreased Bax expression in carcinomas (Kim et al. 2003).

Epidemiological evidence associating the decreased risk of prostate cancer with frequent consumption of tomato products inspired us to conduct a small intervention trial among patients diagnosed with prostate adenocarcinoma. Tomato sauce pasta was consumed daily for 3 wk before their scheduled prostatectomy, and biomarkers of tomato intake, prostate cancer progression and oxidative DNA damage were followed in blood and the available prostate tissue. The whole food intervention was so well accepted by the subjects that the blood lycopene (the primary carotenoid in tomatoes responsible for their red color) doubled and the prostate

Table 2 Effects of different amounts of tomato juice on percentage inhibition of ADA activity.

Amounts	*% Inhibition*
A, No tomato juice	0
B, 25 µl tomato juice	32.58 ± 9.17
C, 50 µl tomato juice	50.27 ± 10.09
D, 100 µl l tomato juice	70.12 ± 5.15
Paired t-test statistics	
A-B	$P < 0.001$
A-C	$P < 0.001$
A-D	$P < 0.001$
B-C	$P < 0.01$
B-D	$P < 0.001$
C-D	$P < 0.01$

lycopene concentration tripled during this short period. They found that oxidative DNA damage in leukocytes and prostate tissues was significantly diminished, the latter mainly in the tumor cell nuclei, possibly because of the antioxidant properties of lycopene. Quite surprising was the decrease in blood prostate-specific antigen, which was explained by the increase in apoptotic death of prostate cells, especially in carcinoma regions. Prostate cancer cell cultures (LNCaP) were also sensitive to lycopene in growth medium, which caused an increased apoptosis and arrested the cell cycle. A possible explanation of these promising results may reside in lycopene effects on the genes governing the androgen stimulation of prostate growth, cytokines and the enzymes producing reactive oxygen species, all of which were recently discovered by nutrigenomic techniques. Other phytochemicals in tomato may act in synergy with lycopene to potentiate protective effects and to help in the maintenance of prostate health (Stacewicz-Sapuntzakis and Bowen 2005).

Adenosine Deaminase

Adenosine deaminase plays an important part in purine metabolism and DNA turnover. It irreversibly deaminates adenosine, converting it to the related nucleoside inosine, which then is deribosylated and converted to hypoxanthine. Mutations in the gene for ADA, which cause it not to be expressed, are one reason for severe combined immunodeficiency. Mutations causing it to be overexpressed are the causes of hemolytic anemia. There is some evidence that a different allele (ADA2) may lead to autism.

Table 3 Inhibitory effects of *urtica dioica* on ADA activity (mIU/mg) in prostate tissue from patients with prostate cancer.

Amounts	*(Mean ± standard deviation)*
A, No aqueous extract of *urtica dioica*	14.8 ± 6.8
B, 25 μl aqueous extract of *urtica dioica*	13.35 ± 4.24
C, 50 μl aqueous extract of *urtica dioica*	6.49 ± 2.38
D, 100 μl aqueous extract of *urtica dioica*	0.98 ± 0.76
Wilcoxon Signed Ranks Test	
A-B	$P > 0.05$
A-C	$P < 0.01$
A-D	$P < 0.02$
B-C	$P < 0.01$
B-D	$P < 0.02$
C-D	$P < 0.02$

There are two isoforms of ADA: ADA1 and ADA2. ADA is found in most body cells, such as lymphocytes and macrophages. ADA2 has been found only in the macrophage, where it co-exists with ADA1 and where the two isoforms regulate the ratio of adenosine to deoxyadenosine to potentiate the killing of parasites. ADA2 is the predominant form present in human blood plasma and is increased in many diseases, particularly those associated with the immune system: for example, rheumatoid arthritis, psoriasis and sarcoidosis. The plasma AD2 isoform is also increased in most cancers. Total plasma ADA can be measured using high performance liquid chromatography, enzymatic or colorimetric techniques. The simplest system is the measurement of the ammonia released from adenosine when broken down to inosine. After incubation of plasma with a buffered solution of adenosine the ammonia is reacted with a Berthelot reagent to form a blue color, which is proportionate to the amount of enzyme activity. To measure ADA2, erythro-9-(2-hydroxy-3-nonyl) adenine (EHNA) is then added to the reaction mixture so as to inhibit the ADA1. It is the absence of ADA1 that causes severe combined immunodeficiency (Aghaei et al. 2005).

Studies on ADA and Prostate Tissue

In one of our previous studies (Durak et al. 2004), we aimed to investigate possible effects of aqueous extract of *Urtica dioica* leaves on ADA activity in prostate tissue from patients with prostate cancer. In this study, 10

Table 4 Effects of different amounts of aqueous extract of *Urtica dioica* on percentage inhibition of ADA activities.

Amounts	*(Mean ± standard deviation)*
A, No aqueous extract of *urtica dioica*	0.0
B, 50 μl aqueous extract of *urtica dioica*	59.4 ± 9.2
C, 100 μl aqueous extract of *urtica dioica*	92.8 ± 5.3
Wilcoxon Signed Ranks Test	
A-B	$P < 0.02$
A-C	$P < 0.02$
B-C	$P < 0.02$

samples of prostate tissue from patients with pathologically proven localized prostate cancer (Gleason scores 4 to 7) were used. In the tissues, ADA activities with and without preincubation with different amounts of *U. dioica* extracts were measured. We found that aqueous extract of *U. dioica* results in significant inhibition of ADA activity of prostate tissue. We conclude that ADA inhibition might be one of the mechanisms in the observed beneficial effect of *U. dioica* in prostate cancer (Durak et al. 2004, Table 3 and Table 4).

In another study (Durak et al. 2003), we investigated possible effects of tomato juice on ADA activity in prostate tissue from cancer patients. Ten samples of prostate tissue from patients with pathologically proven localized prostate cancer (Gleason scores 4 to 7) were used in the study. In the tissues, ADA activities with and without preincubation with different amounts of tomato juice were measured. Tomato juice (pH 3.6) was prepared by homogenization and centrifugation (5000 rpm for 10 min) of Ayas tomato (Ayas is a city in Turkey). The upper clear tomato juice was used. We have seen that tomato juice causes significant inhibition of ADA activity in the prostate tissues. It has been suggested that the protective effect of tomato against prostate cancer may result from its potential to inhibit ADA activity (Durak et al. 2003, Table 1 and Table 2).

In yet another study (Durak et al. 2005), effects of extract of dried whole black grape including seed on ADA, 5' nucleotidase (5′NT) and xanthine oxidase (XO) enzymes were investigated in cancerous and non-cancerous human colon tissues. Enzyme activities were measured in 20 colon tissues, 10 from cancerous region and 10 from non-cancerous region with and without preincubation with black grape extract. ADA and 5′NT activities were found increased and that of the XO decreased in the cancerous tissues relative to non-cancerous ones. After incubation period with black grape extract for 12 hr, ADA and 5′NT activities were found to be significantly lowered but that of XO unchanged in both cancerous and

Table 5 Activities of ADA, XO and 5'NT enzymes in cancerous and non-cancerous human colon tissues incubated with and without whole black grape extract (mean ± SD).

	ADA (mIU/mg)	*XO (mIU/mg)*	*5' NT (mIU/mg)*
Group 1	8.67 ± 1.72	0.043 ± 0.020	10.72 ± 6.03
Group 2	1.53 ± 0.48*	0.038 ± 0.013	1.61 ± 1.31*
Group 3	12.66 ± 1.06**	0.021 ± 0.008	20.17 ± 14.16**
Group 4	3.03 ± 0.24**	0.039 ± 0.010	1.56 ± 0.23**

Group 1: Non-cancerous tissue without black grape extract. Group 2: Non-cancerous tissue with black grape extract. Group 3: Cancerous tissue without black grape extract. Group 4: Cancerous tissue with black grape extract.

*Group 1 vs. Group 2: $P < 0.05$; Mann-Whitney U test.

**Group 1 vs. Group 3: $P < 0.05$; Mann-Whitney U test.

***Group 3 vs. Group 4: $P < 0.05$; Mann-Whitney U test.

non-cancerous tissues. Results suggest that ADA and 5′NT activities increase but XO activity decreases in cancerous human colon tissues, which may provide advantage to the cancerous tissues in obtaining new nucleotides for rapid DNA synthesis through accelerated salvage pathway activity. Black grape extract significantly inhibits the ADA and 5′NT activities of cancerous and non-cancerous colon tissues, thereby eliminating this advantage of cancer cells, which might be the basis for the beneficial effect of black grape in some kinds of human cancers (Table 5) (Durak et al. 2005).

Table 6 Guisti Method.

	Reagent blank	*Standard*	*Sample blank*	*Sample*
Phosphate buffer	1.0 ml	-	-	-
Buffered adenosine solution	-	-	1.0 ml	1.0 ml
Ammonium sulphate standard solution	-	1.0 ml	-	-
Sample	-	-	-	0.05 ml
Distilled water	0.05 ml	0.05 ml	-	-
Incubate for 60 min in a 37°C water bath				
Phenol-nitroprusside solution	3.0 ml	3.0 ml	3.0 ml	3.0 ml
Sample	-	-	0.05 ml	-
Alkaline hypochloride solution	3.0 ml	3.0 ml	3.0 ml	3.0 ml

Incubate for 30 min in a 37°C water bath. Measure against distilled water.

ADA and Cancer

In a study by Aghaei et al. (2005), the potential relationship between ADA activity and cancer progression was examined by investigating the activity of total ADA and its isoenzymes in serum and simultaneously in the cancerous tissue of each patient with breast cancer. In that study, total ADA and its isoenzymes were measured using the Guisti method (Table 6). ADA2 activity was measured in the presence of a specific ADA1 inhibitor, EHNA. Their results indicated that ADA2 and total ADA activities were higher in serum and malignant tissues than those of corresponding controls ($P < 0.05$). Tumor ADA2 and total ADA activities were significantly ($P < 0.05$) correlated with lymph node involvement, histological grade and tumor size, whereas their levels in serum were significantly ($P < 0.05$) correlated with menopausal status and patient age. Although in serum and in tumoral tissues, total ADA activity and its ADA2 isoenzyme were both found to be increased, distinct correlation patterns were observed with some of the prognostic factors. It can be speculated that increased ADA and isoenzyme activities in serum originates from sources other than the breast tumors (Aghaei et al. 2005).

In a study by Walia et al. (1995), serum levels of ADA, 5′NT and alkaline phosphatase were studied in 25 patients with breast cancer and 25 normal subjects. Adenosine deaminase was found to be the better probable parameter for the detection of cancer and assessment of the development of various stages of cancer, whereas 5′NT had only diagnostic significance. After mastectomy, a significant decrease was found in the levels of serum ADA and 5′NT (Walia et al. 1995).

Prostate Cancer and Tomato

A study by Hodge et al. (2004) was designed to examine the risk of prostate cancer associated with foods and nutrients, including individual fatty acids and carotenoids. It was a population-based case-control study of 858 men aged < 70 yr and diagnosed with histologically confirmed prostate cancer of Gleason Grade 5 or greater, and 905 age-frequency-matched men, selected at random from the electoral rolls. Dietary intakes were assessed with a 121-item food frequency questionnaire. Results showed inverse associations between prostate cancer and the following foods (odds ratio, OR, 95% confidence intervals, 95% CI for tertile III compared with tertile I): allium vegetables 0.7, 0.5-0.9, p trend 0.01; tomato-based foods 0.8, 0.6-1.0, p trend 0.03; and total vegetables 0.7, 0.5-1.0, p trend 0.04. However, margarine intake was found to be positively associated with prostate cancer 1.3, 1.0-1.7, p trend 0.04. The only statistically significant associations observed with nutrients were weak inverse associations for palmitoleic acid (p trend 0.04), fatty acid 17:1 (p trend 0.04), and 20:5 n-6 (p trend 0.05), and a non-significant trend for oleic acid (p trend 0.09). Neither total nor beverage-specific intake of alcohol was associated with risk. On the basis of these findings, diets rich in olive oil (a source of oleic acid), tomatoes and allium vegetables might reduce the risk of prostate cancer (Hodge et al. 2004).

Carotenoids as well as their metabolites and oxidation products stimulate gap junctional communication between cells, which is thought to be one of the protective mechanisms related to cancer-preventive activities of these compounds. Increased intake of lycopene by consumption of tomatoes or tomato products has been epidemiologically associated with a diminished risk of prostate cancer. Aust et al. (2003) reported a stimulatory effect of a lycopene oxidation product on gap junctional communication in rat liver epithelial WB-F344 cells. The active compound was obtained by complete *in vitro* oxidation of lycopene with hydrogen peroxide/osmium tetroxide. For structural analysis, high performance liquid chromatography, gas chromatography coupled with mass spectrometry, ultraviolet/visible-, and infrared spectrophotometry were applied. The biologically active oxidation product was identified as 2,7,11-trimethyl-tetradecahexaene-1,14-dial. Their data indicate a potential role of lycopene degradation products in cell signaling and enhancing cell-to-cell communication via gap junctions (Aust et al. 2003).

In a study by van Breemen et al. (2002), men with clinical stage T1 or T2 prostate adenocarcinoma were recruited (n = 32) and consumed tomato sauce-based pasta dishes for 3 wk (equivalent to 30 mg lycopene per day) before radical prostectomy. Prostate tissue from needle biopsy just before intervention and prostectomy after supplementation from a subset of 11 subjects was evaluated for both total lycopene and lycopene geometrical isomer ratios. A gradient HPLC system using a C(18) column with UV-vis absorbance detection was used to measure total lycopene. Because the absorbance detector was insufficiently sensitive, HPLC with a C(30) column and positive ion atmospheric pressure chemical ionization mass spectrometric (LC-MS) detection was developed as a new assay to measure the ratio of lycopene cis/trans isomers in these samples. The limit of detection of the LC-MS method was determined to be 0.93 pmol of lycopene on-column, and a linear response was obtained over 3 orders of magnitude. Total lycopene in serum in the group supplemented with tomato sauce increased 2.0-fold from 35.6 to 69.9 µg/dL, whereas total lycopene in prostate tissue increased 3.0-fold from 0.196 to 0.582 ng/mg of tissue (from 0.365 to 1.09 pmol /mg). All-trans-lycopene and at least 14 cis-isomer peaks were detected in prostate tissue and serum. The mean proportion of all-trans-lycopene in prostate tissue was approximately 12.4% of total lycopene before supplementation but increased to 22.7% after dietary intervention with tomato sauce. In serum there was only a 2.8% but statistically significant increase in the proportion of all-trans-lycopene after intervention. These results indicate that short-term supplementation with tomato sauce containing primarily all-trans-lycopene (83% of total lycopene) results in substantial increases in total lycopene in serum and prostate, and a substantial increase in all-trans-lycopene in prostate but relatively less in serum (van Breemen et al. 2002).

Consumption of lycopene has been associated with reduced risk of prostate cancer. Guttenplan et al. (2001) investigated the effects of lycopene, fed as a lycopene-rich tomato oleoresin (LTO) at two doses, on *in vivo* mutagenesis in prostate, colon, and lungs of lacZ mice. Both short-term benzo[a]pyrene (BaP)-induced and long-term spontaneous mutagenesis were monitored. Non-significant inhibition of spontaneous mutagenesis in prostate and colon was observed at the higher dose of LTO, and the observation of inhibition in colon was facilitated by an unusually high spontaneous mutagenesis rate. BaP-induced mutagenesis was slightly inhibited by LTO in prostate. However, enhancement of BaP-induced mutagenesis was observed in colon and lung. These results indicate that any antimutagenic effects of LTO may be organospecific (Guttenplan et al. 2001).

In another study (Canene-Adams et al. 2007), combinations of tomato and broccoli in the Dunning R3327-H prostate adenocarcinoma model were evaluated. Male Copenhagen rats (n = 206) were fed diets containing 10% tomato, 10% broccoli, 5% tomato plus 5% broccoli (5:5 combination), 10% tomato plus 10% broccoli (10:10 combination) powders, or lycopene (23 or 224 nmol/g diet) for approximately 22 wk starting 1 mon prior to receiving s.c. tumor implants. Researchers compared the effects of diet to surgical castration (2 wk before termination) or finasteride (5 mg/kg body weight orally, 6 d/wk). It was found that castration reduced prostate weights, tumor areas, and tumor weight (62%, $P < 0.001$), whereas finasteride reduced prostate weights ($P < 0.0001$) but had no effect on tumor area or weight. Lycopene at 23 or 224 nmol/g of the diet insignificantly reduced tumor weights by 7% or 18%, respectively, whereas tomato reduced tumor weight by 34% ($P < 0.05$). Broccoli decreased tumor weights by 42% ($P < 0.01$), whereas the 10:10 combination (tomato plus broccoli) caused a 52% decrease ($P < 0.001$). Tumor growth reductions were associated with reduced proliferation and increased apoptosis, as quantified by proliferating cell nuclear antigen immunohistochemistry and the ApopTag assay (Canene-Adams et al. 2007).

Dietary lycopene has been reported to concentrate in the human prostate, so its uptake and subcellular localization was investigated in the controlled environment of cell culture using the human prostate cancer cell lines LNCaP, PC-3, and DU145. After 24 hr incubation with 1.48 µmol/L lycopene, LNCaP cells accumulated 126.6 pmol lycopene per million cells, which was 2.5 times higher than PC-3 cells and 4.5 times higher than DU145 cells. Among these cell lines, only LNCaP cells express prostate-specific antigen and fully functional androgen receptor. Levels of prostate-specific antigen secreted into the incubation medium by LNCaP cells were reduced 55% as a result of lycopene treatment (1.48 µmol / L). The binding of lycopene to the ligand-binding domain of the human androgen receptor was carried out, but lycopene was not found to be a ligand for this receptor. Next, subcellular fractionation of LNCaP cells exposed to lycopene was carried out using centrifugation and followed by liquid chromatography-tandem mass spectrometry quantitative analysis to determine the specific cellular locations of lycopene. The majority of lycopene (55%) was localized to the nuclear membranes, followed by 26% in nuclear matrix, and then 19% in microsomes. No lycopene was detected in the cytosol. These data suggest that the rapid uptake of lycopene by LNCaP cells might be facilitated by a receptor or binding protein and that lycopene is stored selectively in the nucleus of LNCaP cells (Liu et al. 2006).

It has been known that elevated serum androgens are associated with increased prostate cancer risk. Tomato consumption is also associated with reduced prostate cancer incidence, and the primary tomato carotenoid, lycopene, may modulate androgen activation in the prostate, yet little is known about other tomato carotenoids. In a study, animals were castrated or sham-operated to evaluate interrelations between phytofluene, lycopene, or tomato powder consumption and androgen status. For this aim, 8-wk-old male F344 rats were fed a control AIN 93G diet. Animals were provided with daily oral supplementation of phytofluene or lycopene (approximately 0.7 mg/d) or fed a diet supplemented with 10% tomato powder (AIN 93G) for 4 d. Sham-operated rats provided with either phytofluene, lycopene, or tomato powder had approximately 40-50% lower serum testosterone concentrations than the sham-operated, control-fed group. Tissue and serum phytofluene and lycopene concentrations were greater in castrated rats than in sham-operated rats, which might have been due in part to a decrease of hepatic CYP 3A1 mRNA expression and benzyloxyresorufin-O-dealkylase activity. Some changes in prostatic and testicular steroidogenic enzyme mRNA expression were found; in particular, prostate 17 beta-hydroxysteroid dehydrogenase 4 mRNA expression in castrated rats fed lycopene or tomato powder was 1.7-fold that of the sham-operated, control-fed group. Modest changes in mRNA expression of steroidogenic enzymes with short-term carotenoid intake may alter the flux of androgen synthesis to less potent compounds. Overall, results illustrate that short-term intake of tomato carotenoids significantly alters androgen status, which may partly be a mechanism by which tomato intake reduces prostate cancer risk (Campbell et al. 2006).

As seen from the studies, lycopene preferentially accumulates in androgen-sensitive tissues such as the prostate (Erdman 2005). When rats were fed diets containing tomato powder, the biodistribution of phytoene, phytofluene, and lycopene differed in various tissues. Interestingly, lycopene accumulated in higher amounts in the prostate even if tomato powder-fed animals received an oral dose of phytoene or phytofluene. Lycopene is the predominant carotenoid found in the prostate and its levels were significantly higher in malignant than in benign prostates (Clinton et al. 1996). Furthermore, in foods, lycopene is predominately found in the all-trans forms, but the prostate accumulated primarily cis-isomers of lycopene. This can be explained by a study in ferrets that demonstrated that cis-isomers of lycopene are more bioavailable than the all-trans form (Boileau et al. 1999).

There also appears to be differential tissue metabolism or accumulation of ^{14}C lycopene in rats fed lycopene. Liver and the androgen-sensitive seminal vesicles accumulated about 20% of radiolabeled lycopene dose as polar products. In contrast, the dorsolateral prostate, the region of the rat prostate most homologous to the peripheral region in which men develop prostate cancer, accumulated up to 80% of the radiolabeled dose as polar products of lycopene (40). It was believed that these polar products are lycopene metabolites and oxidation products that have yet to be identified. Further research determined that whether or not lycopene was prefed did not alter the amounts of polar products that accumulated in the prostate of rats after receiving a ^{14}C lycopene dose (Zaripheh et al. 2003). These studies together indicate that prostate tissue metabolizes or preferentially accumulates these polar lycopene products.

Lycopene Metabolites

The first *in vivo* lycopene metabolite reported was 5,6-dihydroxy-5,6-dihydrolycopene in human serum (Khachik et al. 1992a). The authors hypothesized that this metabolite resulted from oxidation of lycopene to form lycopene 5,6-epoxide, which then was metabolically reduced to 5,6-dihydroxy-5,6-dihydrolycopene (Khachik et al. 1992a, b, 1995). This same lab also identified epimeric 2,6-cyclolycopene-1,5-diols in human serum and milk. They hypothesized that they were produced in a similar fashion as 5,6-dihydroxy-5,6-dihydrolycopene.

CONCLUSION

As seen from the studies, the consumption of tomatoes has been demonstrated in *in vivo* and *in vitro* studies to have beneficial, protective effects with regard to prostate cancers. Scientific data suggest that use of tomato for the treatment of prostatic disorders including cancer and BPH might give some therapeutic advantage in addition to classic medicinal therapies. However, the subject needs further studies to elucidate possible mechanism(s) involved and adverse effects and interactions with some medicines used by the patients.

ABBREVIATIONS

BPH: Benign prostatic hyperplasia; neoplasia; Insulin-like growth factor 1; ADA: Adenosine deaminase; EC 3.5.4.4: Enzyme committee (Adenosine aminohydrolase); Low density lipoprotein; High performance liquid chromatography; N- methyl-N-nitrosurea; LNCaP: Prostate cell cultures; Erythro-9-(2-hydroxy-3-nonyl) adenine; 5′ NT: 5′ Nucleotidase; Liquid chromatography-mass spectrophotometry; LTO: Lycopene-rich tomato oleoresin.

REFERENCES

Aghaei, M. and F. Karami-Tehrani, S. Salami, and M. Atri. 2005. Adenosine deaminase activity in the serum and malignant tumors of breast cancer: The assessment of isoenzyme ADA1 and ADA2 activities. Clin. Biochem. 38: 887-891.

Aust, O. and N. Ale-Agha, L. Zhang, H. Wollersen, H. Sies, and W. Stahl. 2003. Lycopene oxidation product enhances gap junctional communication. Food Chem. Toxicol. 41: 1399-407.

Beecker, G.R. 1998. Nutrient content of tomatoes and tomato products. Proc. Soc. Exp. Biol. Med. 218: 98-100.

Boileau, A.C. and N.R. Merchen, K. Wasson, C.A. Atkinson, and J.W. Erdman Jr. 1999. Cis-lycopene is more bioavailable than trans-lycopene in vitro and in vivo in lymph-cannulated ferrets. Nutrition 129: 1176-1181.

Campbell, J.K. and C.K. Stroud, M.T. Nakamura, M.A. Lila, and J.W. Erdman Jr. 2006. Serum testosterone is reduced following short-term phytofluene, lycopene, or tomato powder consumption in F344 rats. J. Nutr. 136: 2813-2819.

Canene-Adams, K. and B.L. Lindshield, S. Wang, E.H. Jeffery, S.K. Clinton, and J.W. Erdman Jr. 2007. Combinations of tomato and broccoli enhance antitumor activity in dunning r3327-h prostate adenocarcinomas. Cancer Res. 15: 836-843.

Clinton, S.K. and C. Emenhiser, S.J. Schwartz, D.G. Bostwick, A.W. Williams, B.J. Moore, and J.W. Erdman Jr. 1996. Cis-trans lycopene isomers, carotenoids, and retinol in the human prostate. Cancer Epidemiol. Biomarkers Prev. 5: 823-833.

Durak, I. and H. Biri, A. Avci, S. Sozen, and E. Devrim. 2003. Tomato juice inhibits adenosine deaminase activity in human prostate tissue from patient with prostate tissue. Nutr. Res.. 23: 1183-1188

Durak, I. and H. Biri, E. Devrim, S. Sozen, and A. Avci. 2004. Aqueous extract of Urtica dioica makes significant inhibition on adenosine deaminase activity in prostate tissue from patients with prostate cancer. Cancer Biol. Ther. 3: 855-857.

Durak, I. and R. Cetin, E. Devrim, and I.B. Erguder. 2005. Effects of black grape extract on DNA turn-over enzymes in cancerous and noncancerous human kolon tissues. Life Sci. 6: 2995-3000

Erdman, J.W. Jr. 2005. How does nutritional and hormonal status modify the bioavailability, uptake, and distribution of different isomers of lycopene? J. Nutr. 135: 2046-2047.

Freeman, V.L. and M. Meydani, S. Yong, J. Pyle, Y. Wan, R. Arvizu-Durazo, and Y. Liao. 2000. Prostatic levels of tocopherols, carotenoids, and retinol in relation to plasma levels and self-reported usual dietary intake. Amer. J. Epidemiol. 151: 109-118.

Guttenplan, J.B. and M. Chen, W. Kosinska, S. Thompson, Z. Zhao, and L.A. Cohen. 2001. Effects of a lycopene-rich diet on spontaneous and benzo[a]pyrene-induced mutagenesis in prostate, colon and lungs of the lacZ mouse. Cancer Lett. 164: 1-6.

Hodge, A.M. and D.R. English, M.R. McCredie, G. Severi, P. Boyle, J.L. Hopper, and G.G. Giles. 2004. Foods, nutrients and prostate cancer. Cancer Causes Control 15: 11-20.

Hsing, A.W. and S.S. Devesa. 2001. Trends and patterns of prostate cancer: what do they suggest? Epidemiol. Rev. 23: 3-13.

Khachik, F. and G.R. Beecher, M.B. Goli, W.R. Lusby, 1992. Separation and quantitation of carotenoids in foods. Daitch, Methods Enzymol. 213: 205-219.

Khachik, F. and G.R. Beecher, M.B. Goli, W.R. Lusby, and J.C. Smith Jr. 1992b. Separation and identification of carotenoids and their oxidation products in the extracts of human plasma. Anal. Chem. 64: 2111-2122.

Khachik, F. and G.R. Beecher, and J.C. Smith Jr. 1995. Lutein, lycopene, and their oxidative metabolites in chemoprevention of cancer. J. Cell Biochem. 22: 236-246.

Kim, H.S. and P. Bowen, L. Chen, C. Duncan, L. Ghosh, R. Sharifi, and K. Christov. 2003. Effects of tomato sauce consumption on apoptotic cell death in prostate benign hyperplasia and carcinoma. Nutr. Cancer 47: 40-47.

Kolonel, L.N. and D. Altshuler, and E. Henderson. 2004. The multiethnic cohort study exploring genes, lifestyle and cancer risk. Nature Rev. 4: 1-9.

Lichtenstein, P. and N.V. Holm, P.K. Verkasalo, A. Iliadou, J. Kaprio, and M. Koskenvuo. 2000. Environmental and heritable factors in the causation of cancer-analyses of cohorts of twins from Sweden, Denmark, and Finland. N. Engl. J. Med. 343: 78-85.

Liu, A. and N. Pajkovic, Y. Pang, D. Zhu, B. Calamini, A.L. Mesecar, and R.B. van Breemen. 2006. Absorption and subcellular localization of lycopene in human prostate cancer cells. Mol. Cancer Ther. 5: 2879-2885.

Stacewicz-Sapuntzakis, M. and P.E. Bowen. 2005. Role of lycopene and tomato products in prostate health. Biochim. Biophys. Acta 30: 202-205.

van Breemen, R.B. and X. Xu, M.A. Viana, L. Chen , M. Stacewicz-Sapuntzakis, C. Duncan, P.E. Bowen, and R. Sharifi. 2002. Liquid chromatography-mass spectrometry of cis- and all-trans-lycopene in human serum and prostate tissue after dietary supplementation with tomato sauce. J. Agric. Food Chem. 10: 2214-2219.

Walia, M., and M. Mahajan, and K. Singh. 1995. Serum adenosine deaminase, 5'-nucleotidase and alkaline phosphatase in breast cancer patients. Indian J. Med. Res. 101: 247-249.

Wolk, A. 2005. Diet, lifestyle and risk of prostate cancer. Acta Oncol. 44: 277-281.

Zaripheh, S. and J.W. Erdman Jr. 2005. The biodistribution of a single oral dose of [14C]-lycopene in rats prefed either a control or lycopene-enriched diet. J. Nutr. 135: 2212-2218.

Zaripheh, S. and T.W Boileau, M.A. Lila, and J.W. Erdman Jr. 2003. [14C]-lycopene and [14C]-labeled polar products are differentially distributed in tissues of F344 rats prefed lycopene J. Nutr. 133: 4189-4195.

24

Effect of Tomato Juice on Prevention and Management of Lung Diseases: Cigarette Smoke-induced Emphysema in the Senescence-accelerated Mouse and Bronchial Asthma in Human

Kuniaki Seyama[1], Naoaki Tamura[2], Takahiro Inakuma[3] and Koichi Aizawa[3]

[1]Department of Respiratory Medicine, Juntendo University School of Medicine, 2-1-1 Hongo, Bunkyo-Ku, Tokyo 113-8421, Japan

[2]Department of Respiratory Medicine, Koto Hospital, 6-8-5 Ojima, Koto-ku, 136-0072, Tokyo, Japan

[3]Biogenics Research Department, KAGOME Co. Ltd., Research Institute, 17 Nishitomiyama, Nasusiobara-Shi Tochigi, 329-2762, Japan

ABSTRACT

Chronic obstructive pulmonary disease (COPD) and bronchial asthma are the most common inflammatory lung diseases and share various clinical manifestations in terms of their pathophysiology and treatment. Oxidative stress attracts increasing attention in aging, inflammatory diseases including COPD and asthma. The role of oxidative stress was investigated by the administration of tomato juice containing plenty of lycopene, a carotenoid with a powerful antioxidant activity. To investigate COPD, the senescence-accelerated mouse

A list of abbreviations is given before the references.

(SAM) was used. This is a naturally occurring animal model for accelerated aging after normal development and maturation. The SAMP1 strain shows age-related structural and functional changes in the lung and is considered to be a mouse model of the senile lung. Aging of the lung is postulated to be an important intrinsic process for the development of emphysema so that even a short period of cigarette smoke exposure may be able to generate emphysema in SAMP1 mice. When SAMP1 mice were exposed to cigarette smoke for 8 wk from the age of 12 to 20 wk, the mean liner intercepts (MLI) and destructive index (DI) of the lung were significantly increased (control air vs. smoke (mean ± SEM); MLI, 68.8 ± 0.7 vs. 75.3 ± 1.7 mm, $p < 0.05$ and DI, 8.6 ± 0.4 vs. 16.2 ± 1.5%, $p < 0.05$), whereas no significant changes were observed in SAMR1, control mice that show normal aging. In contrast, smoke-induced emphysema was completely prevented by concomitant ingestion of lycopene given as tomato juice (MLI: smoke with or without lycopene (mean ± SEM), 62.9 ± 0.8 vs. 66.9 ± 1.3 μm, $p < 0.05$). Smoke exposure increased the apoptosis of the airway and alveolar septal cells and reduced vascular endothelial cell growth factor (VEGF) in lung tissues, but tomato juice ingestion significantly reduced apoptosis and increased the tissue VEGF level. SAMP1 is a useful model for cigarette smoke-induced emphysema and a valuable tool to explore how aging influences the development of lung diseases, and both pathophysiologic mechanisms and the effect of therapeutic intervention on smoke-induced emphysema.

The antioxidative state was analyzed in both healthy subjects and asthma patients by determining the serum concentration of carotenoids. The results demonstrated that the serum lycopene concentration was significantly lower in asthma patients than in healthy subjects. On the other hand, no significant difference has been detected in other carotenoid concentrations. Accordingly, an additional study was conducted in patients with asthma who continued to take tomato juice every day for 12 mon to evaluate its effect on peak flow rate (PEFR), daily variation of PEFR, daily symptoms, and quality of life. After a year of daily tomato juice ingestion, the best PEFR value increased from 352 ± 19 to 399 ± 21 (L/min) after 12 mon ($p < 0.01$). Decrease of the daily variation of PEFR value was also observed ($p < 0.01$). Atopic patients with asthma responded better than non-atopic patients. Most patients had a positive impression on daily tomato juice intake since asthma patients realized the improvements of quality of life and asthmatic symptoms. No obvious adverse effect was observed throughout the study. Although the precise mechanism(s) and the active ingredient(s) responsible for these results must be elucidated in detail, the daily intake of tomato juice could therefore be a new candidate of controller for asthma patients.

INTRODUCTION

Chronic obstructive pulmonary disease (COPD) is defined as a preventable and treatable disease with significant extrapulmonary effects that may contribute to the severity in individual patients (Committee of

GOLD2006, 2006). Its pulmonary component is characterized by airflow limitation that is not fully reversible. The airflow limitation in COPD is usually progressive and associated with an abnormal inflammatory response of the lungs to noxious particles or gases such as tobacco smoke (Committee of GOLD2006, 2006). Accordingly, chronic cigarette-smoke exposure is the most important risk factor for COPD, evoking chronic inflammation in the airways and lung parenchyma. Asthma is another chronic inflammatory disorder of the lungs, especially of the airways, in which various inflammatory and allergic cells are involved. The chronic inflammation in asthma is associated with airway hyperresponsiveness leading to recurrent episodic respiratory symptoms (Committee of GINA2006, 2006). Therefore, both COPD and asthma are major chronic obstructive pulmonary diseases, involve underlying chronic inflammation, share many aspects in terms of pathophysiology and treatment, and tend to be difficult to differentiate from each other. This chapter describes the effect of tomato juice on the prevention of cigarette smoke-induced emphysema in a mouse model and the management of asthma in humans, and addresses the possible role of oxidative stress as an important component of the chronic inflammation-mediated disease process in both COPD and asthma.

TOMATO JUICE AND COPD

COPD and Its Pathogenesis—The Role of Oxidative Stress and Aging

The pathological changes characteristic of COPD can be identified in the entire structure of lungs including the proximal airways (bronchi more than 2 mm internal diameter), peripheral airways (bronchiole less than 2 mm internal diameter), lung parenchyma, and pulmonary vasculature. However, emphysema and inflammation in the peripheral airways (also referred to as small airways) are the principal pathological findings in COPD. Emphysema is defined as a condition of the lung characterized by abnormal, permanent enlargement of airspaces distal to the terminal bronchiole, accompanied by the destruction of their walls, and without any obvious fibrosis (Snider et al. 1985). Emphysema must be clearly distinguished from a senile lung or aging lung, which shows uniform respiratory airspace enlargement without apparent alveolar wall destruction. It is still unclear whether the airspace enlargement associated with aging is due to age alone or the combination of age and environmental history, but the occurrence of these changes in nearly all individuals suggests that the changes are "normal", thereby the term aging lung is preferable to "senile emphysema", which implies, by definition, the presence of alveolar wall destruction (Snider et al. 1985).

The prevailing theory for the pathogenesis of emphysema was originally assumed to be a proteinase/antiproteinase imbalance, which was deduced from the clinical observation of patients with alpha-1-antitrypsin deficiency who are at increased risk of early-onset emphysema (Eriksson 1999). Chronic cigarette smoke exposure induces airway and parenchymal inflammation by recruiting and activating inflammatory cells to release proteinases, particularly elastase from neutrophils and various metalloproteinases from alveolar macrophages, which will overwhelm the antiproteinase defense in the epithelial lining fluid and lung tissues. However, recent advances have revealed that not only various cells and mediators but also different biological processes are involved in the pathogenesis of emphysema, including an imbalance between oxidant and antioxidant defenses in the lungs (MacNee 2005a, Tuder et al. 2003b) or between cellular apoptosis and regeneration (Aoshiba et al. 2003, Tuder et al. 2003a, b) (Fig. 1).

COPD develops in only 10 to 15% of smokers, suggesting that some individuals have some intrinsic properties resulting in amplification of cigarette smoke-induced inflammation and will develop COPD. Since COPD usually develops in elderly individuals with a long smoking history, it may be associated with the aging process, especially the aging of the lungs. On the other hand, there are many studies concerned with the relationship between aging and oxidative stress (Irshad and Chaudhuri 2002, Junqueira et al. 2004). Moderate oxidative stress may gradually develop with advancement in age since plasma levels of lipoperoxidation products and antioxidant enzyme activities in red blood cells increase with age, while the plasma levels of nutritional antioxidants decrease (Junqueira et al. 2004). The lungs are persistently exposed to oxidants generated either endogenously from phagocytes and other cell types or exogenously from air pollutants or cigarette smoke (MacNee 2005b). Since cigarette smoke contains about 10^{17} oxidant molecules per puff and generates an oxidant/antioxidant imbalance in the lungs, oxidative stress is thus postulated to play an important role in the pathogenesis of emphysema (MacNee 2005b). In patients with COPD, biomarkers of oxidative stress, such as protein carbonyls and lipid peroxidation products, are reported to be elevated in the lungs (Rahman et al. 2002) and respiratory muscles (Barreiro et al. 2005).

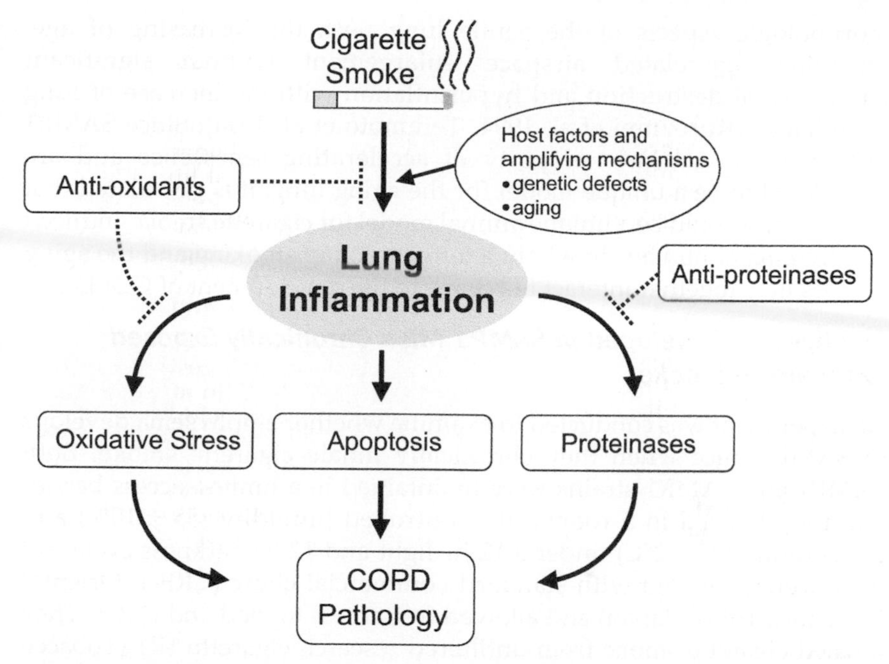

Fig. 1 Current understanding on pathogenesis of chronic obstructive pulmonary disease (COPD). COPD, which presents as emphysema and chronic inflammation of airways, is caused by chronic exposure to noxious gases or particles. Age-related cellular and molecular changes may be of significant pathophysiologic importance in the development of COPD. Accelerated aging may enhance susceptibility to COPD through amplification of a part or the entire pathophysiologic process presented in this figure modified from GOLD (Committee of GOLD 2006, 2006).

SAMP1 Strain of Mice: A Unique Animal Model for Cigarette Smoke-induced Emphysema

Senescence-accelerated Mice Strains

The senescence-accelerated mouse (SAM) strains, established by Takeda et al. (1981), are naturally occurring animal models for accelerated aging after normal development and maturation. The SAM strain of mice consists of series of SAMP (accelerated senescence-prone, short-lived) and SAMR (accelerated senescence-resistant, longer-lived) strains. While the SAMR strains show normal aging, the SAMP strains have been widely used as a model for investigating the aging process, including senile cataracts (Teramoto et al. 1992), amyloidosis (Takeshita et al. 1982), osteoporosis (Matsushita et al. 1986), senile lung (Kurozumi et al. 1994, Teramoto et al. 1994) and age-related impairment in memory and learning (Miyamoto et al. 1986). Among the several substrains of SAMP, both the SAMP1 and SAMP2 strains are reported to manifest the functional and morphologic aspects of the senile lung with the increasing of age, including age-related airspace enlargement without significant parenchymal destruction and hyperinflation with the increase of lung compliance (Kurozumi et al. 1994, Teramoto et al. 1994). Since SAMP1 mice have an intrinsic property of accelerating senescence and are considered to be a unique model for the aging lung, it is postulated that SAMP1 mice could be a unique animal model for cigarette smoke-induced emphysema of humans in whom a long period of smoking and the aging process may together interact in regard to the development of COPD.

Emphysema Developed in SAMP1 Mice Chronically Exposed to Cigarette Smoke

An experiment was conducted to examine whether emphysema develops in SAMP1 mice when they chronically inhale cigarette smoke. Both SAMP1 and SAMR1 strains were maintained in a limited access barrier and then housed in a room with controlled humidity (55 ± 10%) and temperature (24 ± 2°C) under a 12 hr light and 12 hr darkness cycle. All mice were provided with standard commercial chow (CRF-1, Oriental Kobo Inc., Tokyo, Japan) and allowed free access to food and water. They inhaled cigarette smoke from unfiltered research cigarette 1R1 (Tobacco Health Research Institute, Kentucky University, Lexington, Kentucky, USA). Researchers used the Tobacco Smoke Inhalation Experiment System for small animals (Model SIS-CS) (Shibata Scientific Technology Ltd., Tokyo, Japan), where mice are set in a body holder and inhale tobacco smoke through a nostril (Fig. 2). The precise protocol for smoke exposure is described elsewhere (Kasagi et al. 2006). The inhalation of either cigarette smoke or fresh air was initiated when mice were 12 wk old and was continued for 8 wk.

The body weights before and after exposure to cigarette smoke for 8 wk and body weight gain during the experiment did not change significantly in both SAMR1 and SAMP1 strains. However, there was an airspace enlargement in the SAMP1 mice as compared to the SAMR1 after 8 wk exposure to fresh air (age 5 mon), indicated by a significant increase in the mean linear intercept (MLI) (7.7%) (SAMP1 vs. SAMR1, 68.8 ± 0.7 vs. 63.8 ± 1.2 mm, mean ± SEM) ($p < 0.05$) (Fig. 3A). The destructive index (DI) was greater in SAMP1 than SAMR1 (Fig. 3B), but still less than a cut-off value of 10%, indicating the absence of significant alveolar destruction. These results confirmed the findings already reported by Kurozumi et al. (1994) that SAMP1 mice have an intrinsic accelerated aging process of the lungs and constitute a model for aging lung. On the other hand, after 8 wk exposure to cigarette smoke, the MLI was significantly increased in the SAMP1 mice (9.5%) in comparison to the air-exposed SAMP1 ($p < 0.01$)

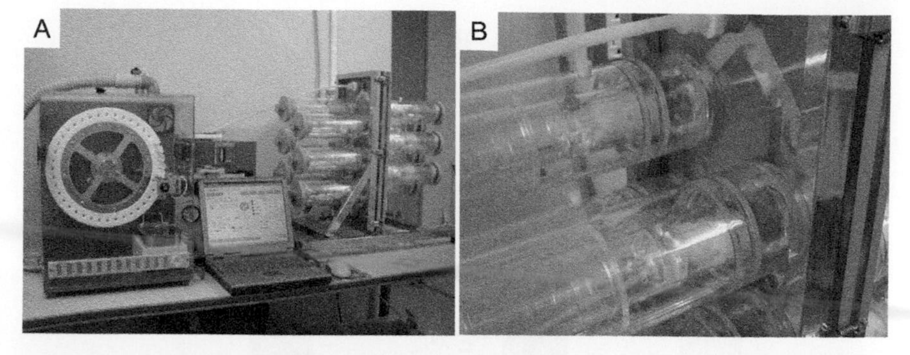

Fig. 2 A. Cigarette Smoke Inhalation Experiment System for small animals (Model SIS-CS) (Shibata Scientific Technology Ltd., Tokyo, Japan). Model SIS-CS consisted of both a cigarette smoke generator (Model SG-200) and the inhalation chamber. The smoke generator is controlled by a laptop computer and automatically generates cigarette smoke by setting a volume of syringe pump (10-50 cm^3 per puff) and a number of puffs per minute (1-12 puffs). Thereafter, the generated cigarette smoke is delivered to the inhalation chamber to which the mice body holders are set (a maximum of 12 body holders can be set at a time) and then the mice inhale cigarette smoke through their nostrils. B. Close-up view of the inhalation chamber. Mice are set in a holder and only their nostrils are exposed to the chamber through which the diluted cigarette smoke passes.

and smoke-exposed control strain SAMR1 ($p < 0.05$) (Fig. 3A). In contrast, there was no significant change in the MLI detected in the control SAMR1 strain exposed to cigarette smoke as compared with the air-exposed SAMR1. The DI was increased to greater than the cut-off value of 10% in smoke-exposed SAMP1 strain, while there was no relevant increase of DI recognized in SAMR1 mice (Fig. 3B). Taken together, these results indicate that emphysema developed in SAMP1 but not in SAMR1 after cigarette smoke exposure under these experimental conditions.

Oxidative Stress and Accelerated Senescence Rather than Inflammatory Cells-mediated Process Appears to Operate in the SAMP1 Strain

The SAMP1 and SAMR1 strains appear to be unable to evoke a TNFα-mediated inflammatory response and the associated influx of neutrophils into the lungs observed in other strains of mice such as C57B6 mice (Churg et al. 2002, 2003, 2004, Hautamaki et al. 1997, Shapiro 2000b). After 8 wk exposure to air or cigarette smoke, the total cell counts and cell populations in bronchoalveolar lavage fluid did not differ significantly in both SAMR1 and SAMP1 strains between air-exposed and smoke-exposed groups (Kasagi et al. 2006). Bronchoalveolar lavage cells comprised mainly macrophages and neutrophils were barely detected even after cigarette smoke exposure for 8 wk. Furthermore, TNFα did not

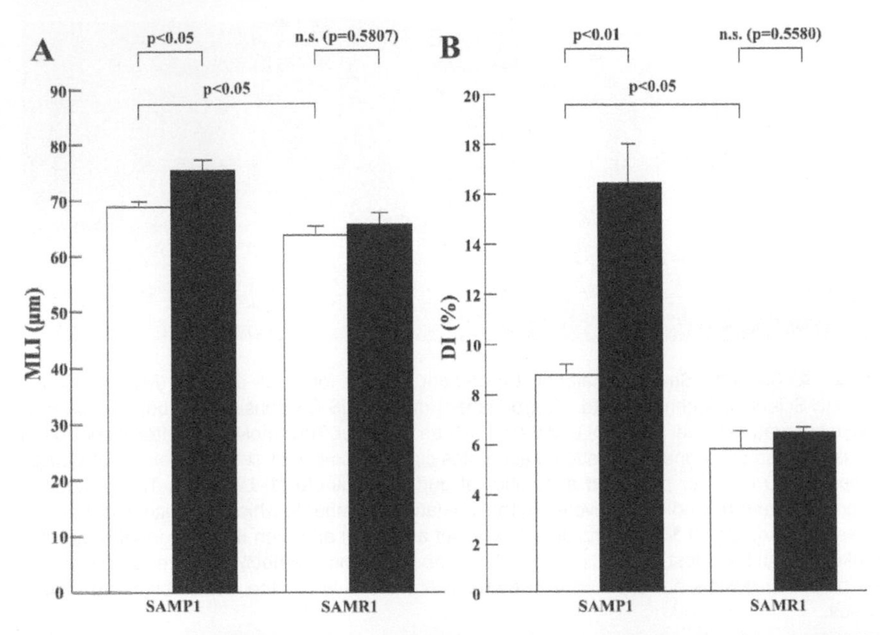

Fig. 3 Morphometric analysis of a lung specimen exposed to cigarette smoke from the SAMR1 and SAMP1 strains. A. MLI. Data are expressed as mean ± SEM (n = 6 in each strain). White and black bars indicate lung specimens exposed to air and cigarette smoke, respectively. B. DI. Data were presented in the same manner as in Fig. 3A (n = 6 in each strain). DI, destructive index; MLI, mean linear intercept; SAMP1, senescence-accelerated mouse, senescence-prone strain 1; SAMR1, a control strain for SAM and senescence-resistant strain 1. (Data from Kasagi et al. 2006. By copyright permission of American Physiological Society.)

increase in bronchoalveolar lavage fluid of smoke-exposed SAMP1 and SAMR1 after the 8 wk exposure. In other animal models of smoke-induced emphysema, TNFα-mediated inflammatory responses, the influx of inflammatory cells including neutrophils and macrophages and their subsequent release of various proteinases such as neutrophil elastase and metalloproteinases have been reported to contribute to pulmonary parenchymal destruction (Churg et al. 2002, 2003, 2004, Hautamaki et al. 1997, Shapiro 2000b). In the present model using SAMP1 mice, however, this does not appear to be the case. This may be due to specific characteristics of this strain or simply due to the limited period of smoke exposure, compared to others (Churg et al. 2002, 2003, 2004, Hautamaki et al. 1997, Shapiro 2000b).

Accordingly, another mechanism by which smoke-induced emphysema is generated must be occurring in the SAMP1 mice. It appears that the oxidant-antioxidant imbalance (MacNee 2005a, Tuder et al. 2003a, b) and the production of growth factors for lung cells (Kasahara et al. 2000,

2001) may play a role. Glutathione, a major antioxidant in the lung, was measured as well as VEGF, a growth factor important for maintaining the structural integrity of the lung parenchyma (Kasahara et al. 2000), in the lung homogenate of SAMP1 and SAMR1 (Fig. 4). The baseline level of lung glutathione (air-exposed mice) was significantly higher in SAMP1 than SAMR1, but the lung content significantly increased in SAMR1 in response to cigarette smoke exposure, while SAMP1 did not show a significant response (Fig. 4A). The experimental condition of chronic cigarette smoke exposure used in this study appeared to drive chronic oxidative stress appropriately in mice since the glutathione concentration was increased in the lung tissues after chronic exposure to cigarette smoke. This is consistent with the observation that an increase of the glutathione level in the epithelial lining fluid was associated with an increased oxidant burden in chronic smokers (Cantin et al. 1987) and that γ-glutamylcysteine synthetase was up-regulated by cigarette smoke in type-II alveolar epithelial cells (Rahman and MacNee 1999). Since higher oxidative stress is postulated to be one cause of the accelerated senescence

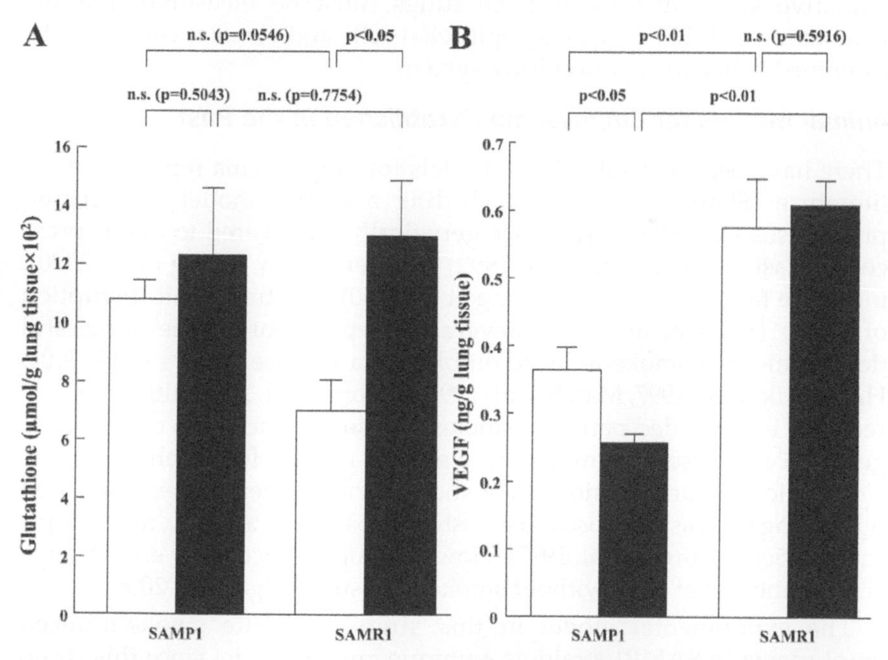

Fig. 4 Glutathione (A) and VEGF (B) contents in the SAMR1 and SAMP1 lung tissues specimens. Lung tissues were homogenized after exposure to air (white bar) or cigarette smoke (black bar) for 8 wk. Data are presented as mean ± SEM (n = 6 in each group). SAMP1, senescence-accelerated mouse, senescence-prone strain 1; SAMR1, a control strain for SAM and senescence-resistant strain 1; VEGF, vascular endothelial cell growth factor. (Data from Kasagi et al. 2006. By copyright permission of American Physiological Society.)

and age-dependent alterations in cellular structure and function (Hosokawa 2002), these observations may indicate that SAMP1 lungs have an intrinsic higher oxidative stress resulting in up-regulation of glutathione level, but cannot increase glutathione when additional exogenous oxidative stress is imposed on the lungs. However, it is also possible that the baseline level of glutathione content in the lung tissues is simply different between SAMP1 and SAMR1.

In contrast, the baseline level of VEGF in the lung tissues was lower in SAMP1 than SAMR1 and the response to cigarette smoke was different between SAMP1 and SAMR1. VEGF decreased in SAMP1 after cigarette smoke exposure, but it remained constant in SAMR1 (Fig. 4B). Taken together, these differences may be important factors contributing to both the phenotypical differences between SAMP1 and SAMR1 and the accelerated development of smoke-induced emphysema in SAMP1. It is possible that the senile lung observed in SAMP1 mice may result from a higher level of intrinsic oxidative stress and a decreased ability to produce growth factors to maintain lung cells. To prove this, the biomarkers of oxidative stress in SAMP1 mice lungs must be measured and the production of VEGF by airway epithelial cells and alveolar cells must be examined using an *in vitro* culture system.

Animal Models for Emphysema Established in the Past

There have been several animal models for emphysema reported in the literature (Shapiro 2000a, b) including a classic model introducing proteinases via the airways, mice genetically engineered to overexpress collagenase (D'Armiento et al. 1992) or interferon-γ (Wang et al. 2000), inducible targeting of IL-13 (Zheng et al. 2000), and transgenic disruption of *klotho* (Suga et al. 2000). Several groups reported the successful development of smoke-induced emphysema in mice (Churg et al. 2004, Hautamaki et al. 1997, March et al. 1999, Shapiro et al. 2003) although they required an extended period of smoke exposure (generally 6 or 7 mon) to generate emphysema. Among these animal models for emphysema, the *klotho* mice are unique since they show various phenotypes associated with aging such as arteriosclerosis, osteoporosis, skin atrophy, and ectopic calcifications (Kuro-o et al. 1997). However, *klotho* mice show emphysema early at the age of 4 wk without smoke exposure (Suga et al. 2000).

The experimental model in this study, cigarette smoke-induced emphysema in SAMP1 strain, is a unique animal model since this strain has intrinsically accelerated aging process of the lungs. It is possible that the age-related characteristics of the lungs in SAMP1 made it possible to generate smoke-induced emphysema in a short period while other groups required a longer period of smoke exposure (6-7 mon) to generate smoke-induced emphysema (Churg et al. 2004, Hautamaki et al. 1997, March et al. 1999, Shapiro et al. 2003). Therefore, this study suggests that aging is an important factor for the development of smoke-induced emphysema. However, it does not exclude the possibility that the different conditions of smoke generation and exposure employed here might have contributed to the development of emphysema within a relatively short period.

Daily Ingestion of Tomato Juice Inhibits Development of Cigarette Smoke-induced Emphysema in SAMP1 Mice

Carotenoids, common dietary constituents that are naturally contained in most fruits (orange-yellow) and vegetables (green leaves), are known to exert antioxidant activities and prevent free radical-induced cellular damage (Bendich 1993). Numerous studies have investigated the potential role of antioxidant nutrients in the prevention of chronic diseases and aging processes in human (Mayne 2003). Lycopene is a major carotenoid abundantly contained in tomatoes (Kaplan et al. 1990) and considered to be the most efficient biological carotenoid in quenching singlet oxygen (Di Mascio et al. 1989). Since cigarette smoke delivers an excess of oxidative stress into the lungs, it is conceivable that dietary carotenoid intake may influence the development of cigarette smoke-induced emphysema.

Accordingly, a preventive experiment was conducted to determine whether *ad libitum* ingestion of tomato juice prevents the development of smoke-induced emphysema in SAMP1. Tomato juice (lycopene 5 mg, vitamin A 52.6 mg, protein 0.68 g, lipid 0 g, fiber 0.68 g, sodium 110 mg, potassium 279 mg, calcium 6.8 mg, and 20 kcal in 100 g tomato juice) was diluted with equal volume of water to reduce its density and then administered to mice by replacing their drinking water with diluted tomato juice. There was no significant difference in the amount of commercial chow ingested whether water or diluted tomato juice was provided.

Histopathological and morphometrical analyses clearly demonstrated emphysematous change in smoke-exposed SAMP1 ingesting tap water but not in SAMP1 ingesting tomato juice (Fig. 5). The MLI was reproducibly increased in smoke-exposed SAMP1 as compared with air-exposed SAMP1 ($p < 0.05$) when tomato juice was not administered, but the increase of MLI was completely prevented by concomitant ingestion of tomato juice ($p < 0.05$) (Fig. 5A and C). In contrast, the increase of DI was significantly prevented by tomato juice but the DI did not return to the control level (Fig. 5B).

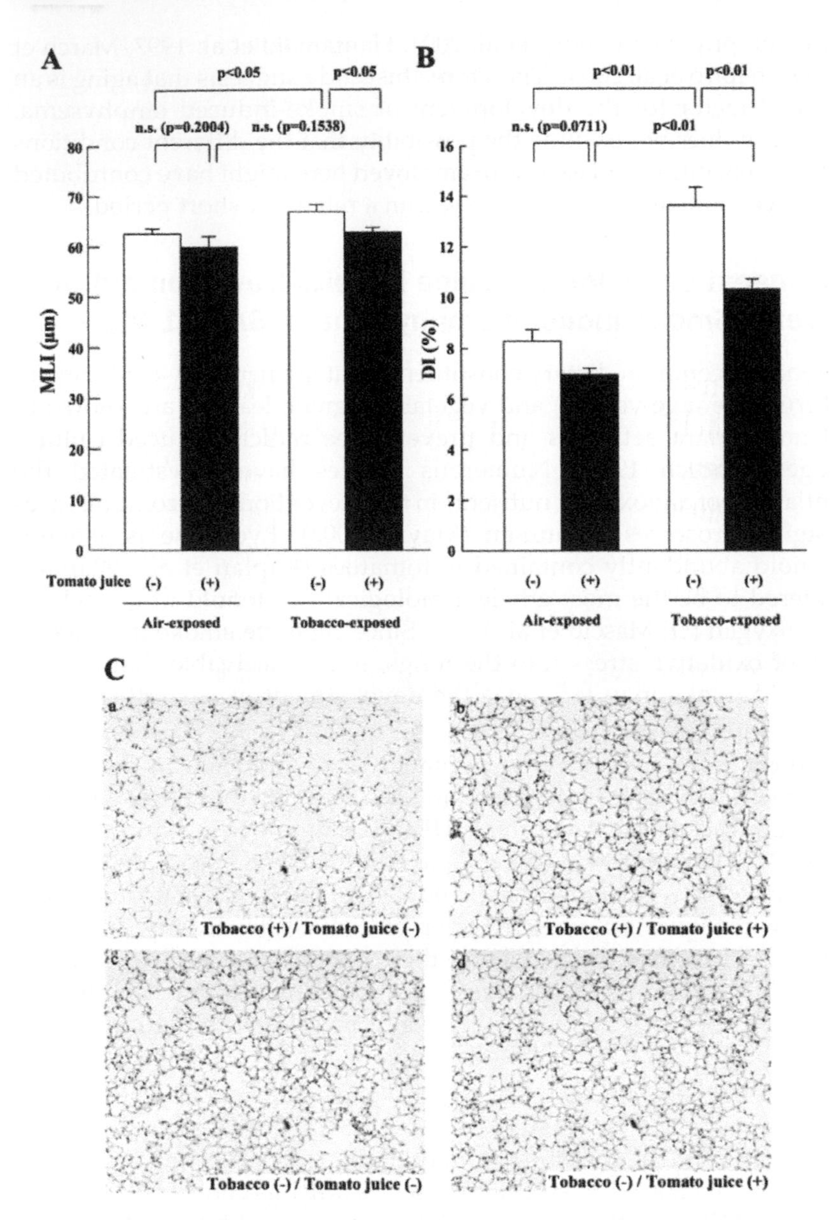

Fig. 5 Morphometric and histological observations of lung tissue specimens from SAMP1 after exposure to air or cigarette smoke with or without administration of tomato juice. A. MLI. Data are expressed as mean ± SEM (n = 6 in each group). White and black bars indicate without and with administration of tomato juice, respectively. B. DI. Data are presented in the same manner as in Fig. 5A (n = 6 in each strain). C. Representative histological views (n = 6 in each group) are presented (hematoxylin-eosin stain, original magnification 20×). Note that cigarette exposure for 8 wk generated airspace enlargement (a), but the administration of tomato juice prevented the development of emphysema (b). DI, destructive index; MLI, mean linear intercept; SAMP1, senescence-accelerated mouse, senescence-prone strain 1. (Data from Kasagi et al. 2006. By copyright permission of American Physiological Society.)

Lycopene was detected in both the serum and lung tissues when mice ingested tomato juice for 1 and 4 wk instead of tap water (Table 1). No significant difference in lycopene concentration was observed in either serum or lung tissue at 1 or 4 wk between air- and smoke-exposed SAMP1, although serum lycopene level tended to decrease in the smoke-exposed SAMP1 as compared with air-exposed SAMP1. The longer the mice ingested tomato juice, the more lycopene appeared to accumulate in the lungs and eventually reached a level similar to that in the serum. Lycopene was detected in neither serum nor lungs unless tomato juice was given to the SAMP1 mice. Taken together, these observations may indicate that lycopene was consumed at the sites of oxidative stress induced by cigarette smoke exposure and recruited from serum to the sites of consumption. In addition, lycopene recruited from serum to the lungs seemed enough to overcome the oxidative stress since no difference was noted between air- and smoke-exposed SAMP1 mice at both 1 and 4 wk exposure. It is possible that the prevention of emphysema in a smoke-exposed group with tomato juice ingestion was mediated by the antioxidant activity of lycopene since induction of glutathione synthesis tended to be suppressed and the lycopene level in both the serum and lung tissues decreased in the smoke-exposed as compared with the air-exposed SAMP1, probably because of its consumption by oxidant stress.

It is widely accepted that free radicals and reactive oxygen species delivered by cigarette smoke directly damage cells and induce apoptosis (Aoshiba et al. 2001, 2000, Church and Pryor 1985). Immunohistochemical examination using anti-ssDNA antibody demonstrates that the apoptosis of lung cells is enhanced by cigarette smoke and widely detected in

Table 1 Concentration of lycopene in serum and lungs of SAMP1 mice.

Period	*Serum (ng/mL)*	*p*	*Lung (ng/g tissue)*	*p*
1 wk				
air	13.25 ± 3.24	= 0.0569	1.11 ± 0.16	= 0.9716
smoke	4.49 ± 1.83		1.07 ± 1.07	
4 wk				
air	9.45 ± 2.30	= 0.2897	8.61 ± 2.09	= 0.4720
smoke	5.79 ± 2.05		6.84 ± 1.24	

Tomato juice was given instead of tap water for 1 or 4 wk with exposure to air or cigarette smoke. Data were presented as mean ± SEM (n = 4 in each group). Lycopene content in lung tissue significantly increased at 4 wk compared with that at 1 wk in both air-exposed ($p < 0.01$) and smoke-exposed ($p < 0.05$) SAMP1. However, there was no significant difference noted in serum lycopene concentration of the air-exposed or smoke-exposed SAMP1 between 1 wk and 4 wk. SAMP1, senescence-accelerated mouse, senescence-prone strain 1; and SAMR1, a control strain for SAM and senescence-resistant strain 1. (Data from Kasagi et al. 2006. By copyright permission of American Physiological Society.)

bronchial, bronchiolar, and alveolar septal cells. However, apoptosis is significantly reduced by concomitant ingestion of tomato juice in SAMP1 (Kasagi et al. 2006). VEGF in the lung homogenate reproducibly decreased after cigarette smoke exposure, but tomato juice prevented the decrease of VEGF associated with smoke exposure, while it even up-regulated VEGF in the smoke-exposed SAMP1 (Kasagi et al. 2006).

Recent studies clearly provide the evidence that apoptosis of lung cells, especially alveolar epithelial and endothelial cells, causes emphysema without apparent inflammation: the direct administration of activated caspase-3 generated emphysema (Aoshiba et al. 2003) and the blockade of vascular endothelial growth factor signaling by VEGF-receptor antagonist resulted in apoptosis of alveolar septal cells and generated emphysema (Kasahara et al. 2000). Several lines of evidence indicate an interaction between oxidative stress and apoptosis where a vicious cycle may be established: cells undergoing apoptosis show increased oxidative stress, which further contributes to apoptosis (Hockenbery et al. 1993, Tuder et al. 2003b). In the interaction between oxidative stress and apoptosis associated with emphysema, VEGF, a growth factor abundantly expressed in the lung that exerts a survival signal for endothelial cells, is considered a key molecule (Tuder et al. 2003b). Accordingly, the reduction of VEGF concentration could have some role in the apoptosis of alveolar septal cells in the present smoke-induced emphysema model. However, other mechanisms are likely to be involved in the apoptosis of lung cells, since tomato juice completely prevented the decrease of VEGF in the SAMP1 lung tissues by cigarette smoke exposure and appeared to up-regulate VEGF in the lungs, but apoptosis of the lung cells did not return to the control level. SAMP1 mice ingested tomato juice but were not given pure lycopene. As a result, it is possible that ingredients other than lycopene in tomato juice affected the results, including the increase of VEGF in smoke-exposed SAMP1 lung tissues.

In conclusion, these experiments demonstrated that cigarette smoke-induced emphysema in SAMP1 strain was completely prevented with the concomitant administration of lycopene given as tomato juice. This model appears to have a potential application for various *in vivo* experiments, not only to better understand the pathophysiology of cigarette smoke-induced emphysema but also to carry out interventional projects.

TOMATO JUICE AND BRONCHIAL ASTHMA

Bronchial Asthma and Oxidative Stress

Bronchial asthma is characterized by an allergic airway inflammation resulting in a reversible airflow limitation. However, it has been pointed out that the oxidative stress is likely to be involved in airflow limitation in bronchial asthma (Barnes et al. 1998, Bowler and Crapo 2002). Inflammatory cells such as eosinophils, neutrophils, and macrophages will migrate into the airway and generate reactive oxygen species to produce local airway damage by oxidative stress. Although there is a scavenger system involving various antioxidants to protect from oxidative stress, an imbalance of oxidative/antioxidative system may amplify the local inflammation (Misso et al. 2005). There have been several reports suggesting that increased production of reactive oxygen species and dysfunction of the scavenge systems due to viral infection and air pollution cause acute exacerbation of bronchial asthma (Caramori and Papi 2004). In addition, several epidemiologic studies have reported that decreased pulmonary function and various pulmonary symptoms are observed in individuals with lower intake of foods with active antioxidant activities (McKeever and Britton 2004, Ochs-Balcom et al. 2006, Paredi et al. 2002, Seaton et al. 1994). Accordingly, it is considered that some antioxidant compounds may play some role in the treatment and management of bronchial asthma.

Serum Concentration of Lycopene is Decreased in Patients with Bronchial Asthma

The serum concentration of various carotenoids including lycopene, α-carotene, β-carotene, zeaxanthin, lutein, β-cryptoxanthin and α-tocopherol and retinol were measured by HPLC (Fig. 6) in 266 healthy subjects (189 males and 77 females, age of 47 ± 1 (mean ± SEM) years old) and 53 patients with bronchial asthma (38 males and 15 females, age of 50 ± 1 years old). No significant difference was observed in α-carotene, β-carotene, zeaxanthin/lutein, β-cryptoxanthin and retinol between healthy volunteers and patients with bronchial asthma. Only the serum concentration of lycopene was significantly different between the healthy volunteers (0.26 ± 0.01 μg/mL, mean ± SEM) and patients with bronchial asthma (0.18 ± 0.02 μg/mL, $p < 0.001$) (Fig. 7). Compared to healthy subjects, serum lycopene concentration was about 30% lower in patients with bronchial asthma. Although patients with severe asthma often showed low values, a significant correlation between the severity of asthma and the serum lycopene concentration was not confirmed. It remains unclear, however, whether the lowered lycopene level is a reflection of an excessive consumption according to the increased oxidative stress in the airway and/or the clinical conditions of asthma.

In healthy individuals, there exists no correlation between serum levels of each of the carotenoids and lycopene. Gender, smoking habits, and the cholesterol level in the serum have been shown to influence the serum

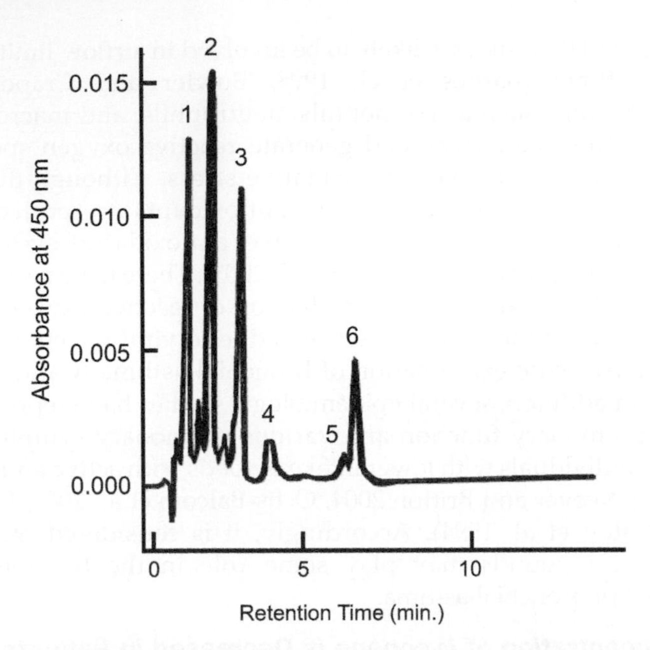

Fig. 6 Determination of carotenoids in the serum by HPLC. Carotenoids were extracted from 200 μl of serum with ethanol followed by hexane/dichloromethane (4:1) and then analyzed by HPLC (Oshima et al. 1997). Peaks indicated are: 1, zeaxanthin/lutein; 2, apocarotenal (internal control); 3, β-cryptoxanthin; 4, lycopene; 5, α-carotene; and 6, β-carotene. HPLC, high performance liquid chromatography.

lycopene concentration, but not age (Fig. 8). Although the study population was limited, the tendency observed in healthy individuals was also observed in patients with bronchial asthma.

Effect of Daily Intake of Tomato Juice for 12 Months on Asthma Control

Sixteen patients with bronchial asthma (11 males and 5 females, age 55 ± 4 (mean ± SEM)) drank at least 160 mL tomato juice a day (no salt added, Kagome Co., Ltd., Tochigi, Japan) for 3 mon. Changes in asthmatic symptoms and quality of life were measured. In addition, 46 patients with bronchial asthma (30 males and 16 females, age 53.1 ± 5.4 (mean ± SEM)) drank tomato juice (no salt added, Kagome Co., Ltd.) for 12 mon. The patients drank the tomato juice twice a day (at least 160 mL per day, basically at breakfast and dinner). Patients continued the medicines they had taken prior to the trial and did not change medication except in response to acute exacerbation.

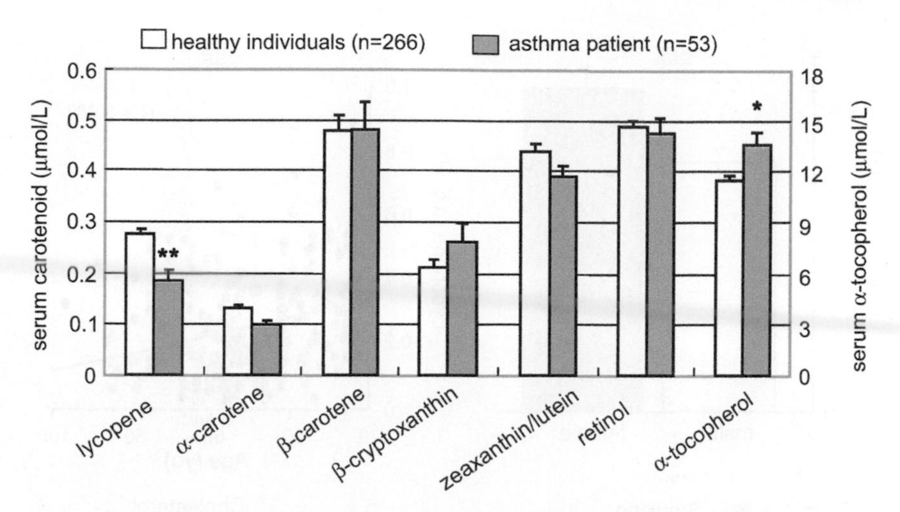

Fig. 7 Serum concentrations of carotenoids. The serum concentration of lycopene was significantly diminished in patients with bronchial asthma as compared with healthy individuals. Data are presented as mean ± SEM (*p < 0.05, **p < 0.001).

Before starting the daily intake of tomato juice, the maximum value of peak flow rate (PEFR) was 352 ± 19 L/min (mean ± SEM), but it gradually increased to 396 ± 19 L/min after 6 mon ($p < 0.01$, as compared with the baseline) and to 399 ± 21 L/min after one year (Fig. 9). In addition, the minimum value of PEFR before the trial was 259 ± 18 L/min and it increased up to 302 ± 20 L/min after 6 mon and up to 330 ± 18 L/min after one year ($p < 0.01$, see Fig. 9). Atopic patients and male patients showed greater improvement in PEFR than the non-atopic and female patients. The dairy variation of PFER was also improved from 27% to 16% (Fig. 9). Quality of life was improved regardless of gender, disease severity, age and atopic factors. The significant improvement on quality of life was already observed as early as 3 mon ($p < 0.05$) from starting the daily intake of tomato juice and continued until the end of the study. More than 80% of patients with bronchial asthma recognized the positive effect on their asthmatic symptoms while they were daily taking tomato juice. Blood examinations showed no significant change in the levels of eosinophils, IgE, liver function, and lipids while patients were taking tomato juice for a year. No adverse effect was observed.

Effect of Tomato Juice on Management of Bronchial Asthma

There have so far been many reports addressing tomatoes and lycopene that document their strong quenching capacity against cigarette smoke, increasing or controlling effect of vascular permeability, inhibition of histamine release, inhibition of leukotriene secretion by macrophages, and

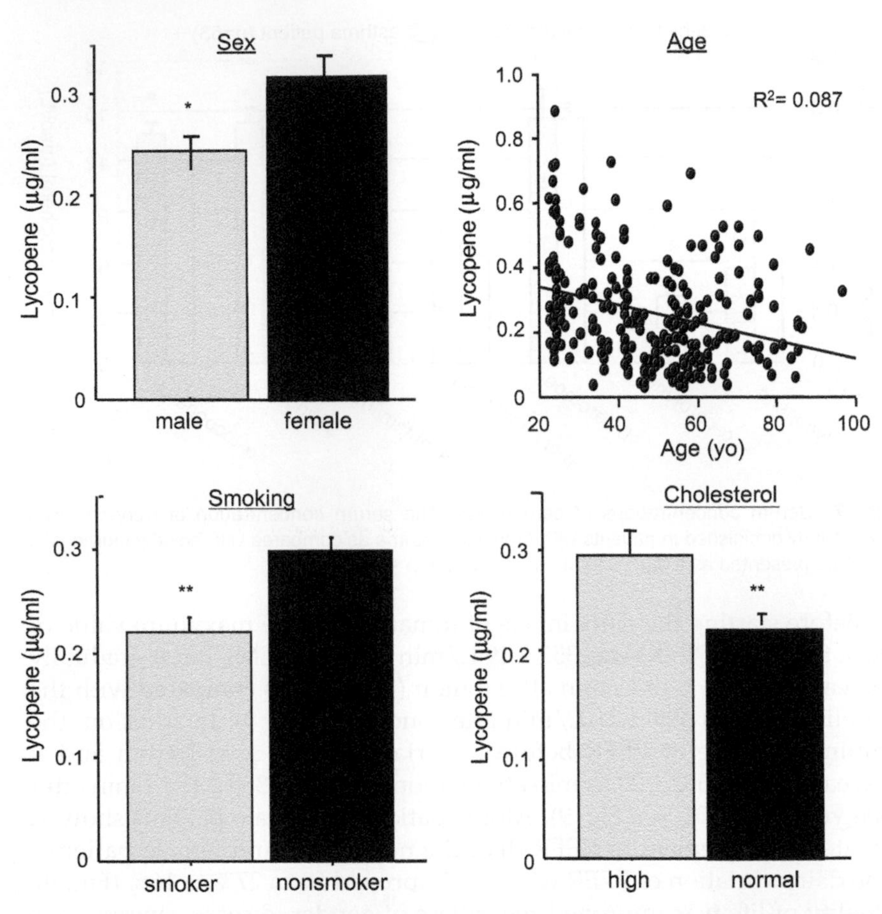

Fig. 8 Influence of clinical background on serum lycopene concentration in healthy volunteers. Data are presented as mean ± SEM (*$p < 0.05$, **$p < 0.001$) . The cholesterol in the serum is designated as high if it is greater than 220 mg/dL.

protection against anaphylactic reaction (Cantrell et al. 2003, Hsiao et al. 2005, Kaliora et al. 2006, McDevitt et al. 2005, Moore et al. 2004, Steck-Scott et al. 2004). These findings suggest that the components of tomatoes have the potential to improve allergenic diseases, but the possible prevention of asthmatic symptoms has not yet been documented. The observation that the PEFR, asthmatic symptoms, and quality of life were significantly improved after the daily intake of tomato juice for one year clearly demonstrates that tomato juice is a potential therapeutic for bronchial asthma. It appears that the improvement obtained during the one year study was completely attributed to the daily intake of tomato juice since the study population had been treated according to the standard

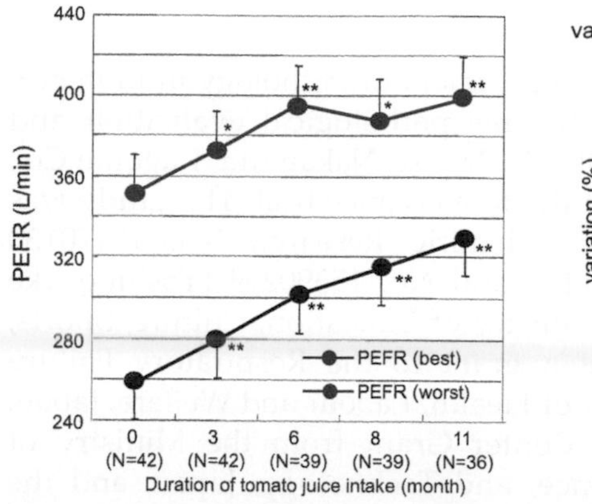

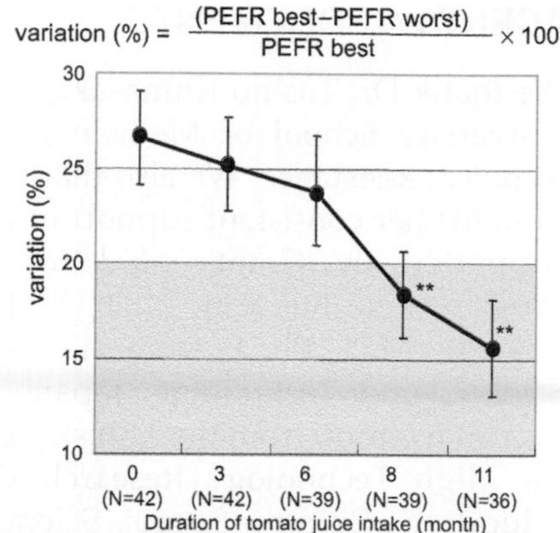

Fig. 9 Significant improvements of PEFR and its variation were observed over time. In the left panel, PEFR (best) or (worst) of the day during the given month was averaged per individual and then the value of PEFR (best) or (worst) of the given month was calculated (mean ± SEM, $*p < 0.05$, $**p < 0.01$). In the right panel, individual variations were calculated using the best or worst PEFR value of the given month, and then the value of variation of the given month was calculated (mean ± SEM, $**p < 0.01$). PEFR, peak flow rate.

guidelines prior to their enrollment and continued to take the same medications throughout the study period. Although this improvement was observed within a few months, the trial was designed to be long-term and to last one year to eliminate the effects of seasonal factors. However, further study is needed, including a control group treated only with a standard medication without tomato juice to demonstrate whether daily intake of tomato juice is truly effective in controlling asthma.

A daily variation in PEFR is characteristic of bronchial asthma and it shows the existence of airway hyperreactivity. It gradually decreased during the study period and the minimum value of PEFR showed more significant improvement than the maximum value of PEFR. These results suggest that the daily intake of tomato juice may therefore ameliorate allergic airway inflammation, thereby preventing the airway from being remodeled, and consequently improving airflow limitation. Tomato juice is less expensive than conventional medications and it acts via a different mechanism to manage bronchial asthma. It is necessary to evaluate the histopathological conditions in the airways, biomarkers of oxidative stress such as nitric oxide concentration in exhaled breath, hydrogen peroxide concentration or inflammatory cytokines in exhaled breath condensates before and after the use of this treatment in order to verify the efficacy of tomato juice for the management of bronchial asthma.

ACKNOWLEDGEMENTS

We thank Dr. Toshio Kumasaka, Department of Pathology in Juntendo University School of Medicine, for his pathological evaluation and technical assistance. We also thank Ms. Naoko Nakamura, Kagome Co., Ltd., for her consistent support of the tomato juice trial. This study was supported by Grant-in-Aid for Scientific Research No. 13470130 (Yoshinosuke Fukuchi, M.D., Ph.D.) and No. 15390259 (Yoshinosuke Fukuchi, M.D., Ph.D.) from the Ministry of Education, Culture, Sports, Science, and Technology, Japan, a grant to the Respiratory Failure Research Group from the Ministry of Health, Labour and Welfare, Japan, the High Technology Research Center Grant from the Ministry of Education, Culture, Sports, Science, and Technology, Japan, and the Institute for Environmental and Gender-Specific Medicine, Juntendo University, Graduate School of Medicine.

ABBREVIATIONS

COPD: chronic obstructive pulmonary disease; DI: destructive index; GINA: global strategy for asthma management and prevention; GOLD: global initiative for chronic obstructive lung disease; HPLC: high performance liquid chromatography; MLI: mean linear intercept; PEFR: peak flow rate; SAM: senescence-accelerated mouse; SAMP1: senescence-accelerated mouse, senescence-prone strain 1; SAMR1: a control strain for SAM and senescence-resistant strain 1; VEGF: vascular endothelial cell growth factor

REFERENCES

Aoshiba, K. and J. Tamaoki, and A. Nagai. 2001. Acute cigarette smoke exposure induces apoptosis of alveolar macrophages. Amer. J. Physiol. Lung Cell. Mol. Physiol. 281: L1392-1401.

Aoshiba, K. and S. Yasui, and A. Nagai. 2000. Apoptosis of alveolar macrophages by cigarette smoke. Chest 117: 320S.

Aoshiba, K. and N. Yokohori, and A. Nagai. 2003. Alveolar wall apoptosis causes lung destruction and emphysematous changes. Amer. J. Respir. Cell. Mol. Biol. 28: 555-562.

Barnes, P.J. and K.F. Chung, and C.P. Page. 1998. Inflammatory mediators of asthma: an update. Pharmacol. Rev. 50: 515-596.

Barreiro, E. and B. de la Puente, J. Minguella, J.M. Corominas, S. Serrano, S.N. Hussain, and J. Gea. 2005. Oxidative stress and respiratory muscle dysfunction in severe chronic obstructive pulmonary disease. Amer. J. Respir. Crit. Care Med. 171: 1116-1124.

Bendich, A. 1993. Biological functions of dietary carotenoids. Ann. NY Acad. Sci. 691: 61-67.

Bowler, R.P. and J.D. Crapo. 2002. Oxidative stress in allergic respiratory diseases. J. Allergy Clin. Immunol. 110: 349-356.

Cantin, A.M. and S.L. North, R.C. Hubbard, and R.G. Crystal. 1987. Normal alveolar epithelial lining fluid contains high levels of glutathione. J. Appl. Physiol. 63: 152-157.

Cantrell, A. and D.J. McGarvey, T.G. Truscott, F. Rancan, and F. Bohm. 2003. Singlet oxygen quenching by dietary carotenoids in a model membrane environment. Arch. Biochem. Biophys. 412: 47-54.

Caramori, G. and A. Papi. 2004. Oxidants and asthma. Thorax 59: 170-173.

Church, D.F. and W.A. Pryor. 1985. Free-radical chemistry of cigarette smoke and its toxicological implications. Environ. Health Perspect. 64: 111-126.

Churg, A. and J. Dai, H. Tai, C. Xie, and J.L. Wright. 2002. Tumor necrosis factor-alpha is central to acute cigarette smoke-induced inflammation and connective tissue breakdown. Amer. J. Respir. Crit. Care Med. 166: 849-854.

Churg, A. and R.D. Wang, H. Tai, X. Wang, C. Xie, J. Dai, S.D. Shapiro, and J.L. Wright. 2003. Macrophage metalloelastase mediates acute cigarette smoke-induced inflammation via tumor necrosis factor-alpha release. Amer. J. Respir. Crit. Care Med. 167: 1083-1089.

Churg, A. and R.D. Wang, H. Tai, X. Wang, C. Xie, and J.L. Wright. 2004. Tumor necrosis factor-alpha drives 70% of cigarette smoke-induced emphysema in the mouse. Amer. J. Respir. Crit. Care Med. 170: 492-498.

Committee of GINA2006. 2006. Global strategy for asthma management and prevention.

Committee of GOLD2006. 2006. Global strategy for the diagnosis, management, and prevention of chronic obstructive pulmonary disease 2006.

D'Armiento, J. and S.S. Dalal, Y. Okada, R.A. Berg, and K. Chada. 1992. Collagenase expression in the lungs of transgenic mice causes pulmonary emphysema. Cell 71: 955-961.

Di Mascio, P. and S. Kaiser, and H. Sies. 1989. Lycopene as the most efficient biological carotenoid singlet oxygen quencher. Arch. Biochem. Biophys. 274: 532-538.

Eriksson, S. 1999. Alpha 1-antitrypsin deficiency. J. Hepatol. 30 Suppl 1: 34-39.

Hautamaki, R.D. and D.K. Kobayashi, R.M. Senior, and S.D. Shapiro. 1997. Requirement for macrophage elastase for cigarette smoke-induced emphysema in mice. Science 277: 2002-2004.

Hockenbery, D.M. and Z.N. Oltvai, X.M. Yin, C.L. Milliman, and S.J. Korsmeyer. 1993. Bcl-2 functions in an antioxidant pathway to prevent apoptosis. Cell 75: 241-251.

Hosokawa, M. 2002. A higher oxidative status accelerates senescence and aggravates age-dependent disorders in SAMP strains of mice. Mech. Ageing Dev. 123: 1553-1561.

Hsiao, G. and Y. Wang, N.H. Tzu, T.H. Fong, M.Y. Shen, K.H. Lin, D.S. Chou, and J.R. Sheu. 2005. Inhibitory effects of lycopene on in vitro platelet activation and in vivo prevention of thrombus formation. J. Lab. Clin. Med. 146: 216-226.

Irshad, M. and P.S. Chaudhuri. 2002. Oxidant-antioxidant system: role and significance in human body. Indian J. Exp. Biol. 40: 1233-1239.

Junqueira, V.B. and S.B. Barros, S.S. Chan, L. Rodrigues, L. Giavarotti, R.L. Abud, and G.P. Deucher. 2004. Aging and oxidative stress. Mol. Aspects Med. 25: 5-16.

Kaliora, A.C. and G.V. Dedoussis, and H. Schmidt. 2006. Dietary antioxidants in preventing atherogenesis. Atherosclerosis 187: 1-17.

Kaplan, L.A. and J.M. Lau, and E.A. Stein. 1990. Carotenoid composition, concentrations, and relationships in various human organs. Clin. Physiol. Biochem. 8: 1-10.

Kasagi, S. and K. Seyama, H. Mori, S. Souma, T. Sato, T. Akiyoshi, H. Suganuma, and Y. Fukuchi. 2006. Tomato juice prevents senescence-accelerated mouse P1 strain from developing emphysema induced by chronic exposure to tobacco smoke. Amer. J. Physiol. Lung Cell. Mol. Physiol. 290: L396-404.

Kasahara, Y. and R.M. Tuder, L. Taraseviciene-Stewart, T.D. Le Cras, S. Abman, P.K. Hirth, J. Waltenberger, and N.F. Voelkel. 2000. Inhibition of VEGF receptors causes lung cell apoptosis and emphysema. J. Clin. Invest. 106: 1311-1319.

Kasahara, Y. and R.M. Tuder, C.D. Cool, D.A. Lynch, S.C. Flores, and N.F. Voelkel. 2001. Endothelial cell death and decreased expression of vascular endothelial growth factor and vascular endothelial growth factor receptor 2 in emphysema. Amer. J. Respir. Crit. Care Med. 163: 737-744.

Kuro-o, M. and Y. Matsumura, H. Aizawa, H. Kawaguchi, T. Suga, T. Utsugi, Y. Ohyama, M. Kurabayashi, T. Kaname, E. Kume, H. Iwasaki, A. Iida, T. Shiraki-Iida, S. Nishikawa, R. Nagai, and Y.I. Nabeshima. 1997. Mutation of the mouse klotho gene leads to a syndrome resembling ageing. Nature 390: 45-51.

Kurozumi, M. and T. Matsushita, M. Hosokawa, and T. Takeda. 1994. Age-related changes in lung structure and function in the senescence-accelerated mouse (SAM): SAM-P/1 as a new murine model of senile hyperinflation of lung. Amer. J. Respir. Crit. Care Med. 149: 776-782.

MacNee, W. 2005a. Oxidants and COPD. Curr. Drug Targets Inflamm. Allergy 4: 627-641.

MacNee, W. 2005b. Pulmonary and systemic oxidant/antioxidant imbalance in chronic obstructive pulmonary disease. Proc. Amer. Thorac. Soc. 2: 50-60.

March, T.H., E.B. Barr, G.L. Finch, F.F. Hahn, C.H. Hobbs, M G. Menache, and K.J. Nikula. 1999. Cigarette smoke exposure produces more evidence of emphysema in B6C3F1 mice than in F344 rats. Toxicol. Sci. 51: 289-299.

Matsushita, M. and T. Tsuboyama, R. Kasai, H. Okumura, T. Yamamuro, K. Higuchi, K. Higuchi, A. Kohno, T. Yonezu, A. Utani, M. Umezawa, and T. Takeda. 1986. Age-Related Changes in Bone Mass in the Senescence-Accelerated Mouse (SAM) SAM-R/3 and SAM-P/6 as New Murine Models for Senile Osteoporosis. Am. J. Pathol. 125: 276-283.

Mayne, S.T. 2003. Antioxidant nutrients and chronic disease: use of biomarkers of exposure and oxidative stress status in epidemiologic research. J. Nutr. 133 Suppl 3: 933S-940S.

McDevitt, T.M. and R. Tchao, E.H. Harrison, and D.W. Morel. 2005. Carotenoids normally present in serum inhibit proliferation and induce differentiation of a human monocyte/macrophage cell line (U937). J. Nutr. 135: 160-164.

McKeever, T.M. and J. Britton 2004. Diet and asthma, Amer. J. Respir. Crit. Care Med. 170: 725-729.

Misso, N.L. and J. Brooks-Wildhaber, S. Ray, H. Vally, and P.J. Thompson. 2005. Plasma concentrations of dietary and nondietary antioxidants are low in severe asthma. Eur. Respir. J. 26: 257-264.

Miyamoto, M. and Y. Kiyota, N. Yamazaki, A. Nagaoka, T. Matsuo, Y. Nagawa, and T. Takeda. 1986. Age-related changes in learning and memory in the senescence-accelerated mouse (SAM). Physiol. Behav. 38: 399-406.

Moore, E.H., M. Napolitano, M. Avella, F. Bejta, K.E. Suckling, E. Bravo, and K.M. Botham. 2004. Protection of chylomicron remnants from oxidation by incorporation of probucol into the particles enhances their uptake by human macrophages and increases lipid accumulation in the cells. Eur. J. Biochem. 271: 2417-2427.

Ochs-Balcom, H.M. and B.J. Grant, P. Muti, C.T. Sempos, J.L. Freudenheim, R.W. Browne, S.E. McCann, M. Trevisan, P.A. Cassano, L. Iacoviello, and H.J. Schunemann. 2006. Antioxidants, oxidative stress, and pulmonary function in individuals diagnosed with asthma or COPD. Eur. J. Clin. Nutr. 60: 991-999.

Oshima, S. and H. Sakamoto, Y. Ishiguro, and J. Terao. 1997. Accumulation and clearance of capsanthin in blood plasma after the ingestion of paprika juice in men. J. Nutr. 127: 1475-1479.

Paredi, P. and S.A. Kharitonov, and P.J. Barnes. 2002. Analysis of expired air for oxidation products. Amer. J. Respir. Crit. Care Med. 166: S31-37.

Rahman, I. and W. MacNee. 1999. Lung glutathione and oxidative stress: implications in cigarette smoke-induced airway disease. Amer. J. Physiol. 277: L1067-1088.

Rahman, I. and A.A. van Schadewijk, A.J. Crowther, P.S. Hiemstra, J. Stolk, W. MacNee, and W.I. De Boer. 2002. 4-Hydroxy-2-nonenal, a specific lipid peroxidation product, is elevated in lungs of patients with chronic obstructive pulmonary disease. Amer. J. Respir. Crit. Care Med. 166: 490-495.

Seaton, A. and D.J. Godden, and K. Brown. 1994. Increase in asthma: a more toxic environment or a more susceptible population? Thorax 49: 171-174.

Shapiro, S.D. 2000a. Animal models for chronic obstructive pulmonary disease: age of klotho and marlboro mice. Amer. J. Respir. Cell. Mol. Biol. 22: 4-7.

Shapiro, S.D. 2000b. Animal models for COPD. Chest 117: 223S-227S.

Shapiro, S.D. and N.M. Goldstein, A.M. Houghton, D.K. Kobayashi, D. Kelley, and A. Belaaouaj. 2003. Neutrophil elastase contributes to cigarette smoke-induced emphysema in mice. Amer. J. Pathol. 163: 2329-2335.

Snider, G.L. and J. Kleinerman, W.M. Thurlbeck, and Z.H. Bengali. 1985. The definition of emphysema. Amer. Rev. Respir. Dis. 132: 182-185.

Steck-Scott, S. and L. Arab, N.E. Craft, and J.M. Samet. 2004. Plasma and lung macrophage responsiveness to carotenoid supplementation and ozone exposure in humans. Eur. J. Clin. Nutr. 58: 1571-1579.

Suga, T. and M. Kurabayashi, Y. Sando, Y. Ohyama, T. Maeno, Y. Maeno, H. Aizawa, Y. Matsumura, T. Kuwaki, O.M. Kuro, Y. Nabeshima, and R. Nagai. 2000. Disruption of the klotho gene causes pulmonary emphysema in mice. Defect in maintenance of pulmonary integrity during postnatal life. Amer. J. Respir. Cell. Mol. Biol. 22: 26-33.

Takeda, T. and M. Hosokawa, S. Takeshita, M. Irino, K. Higuchi, T. Matsushita, Y. Tomita, K. Yasuhira, H. Hamamoto, K. Shimizu, M. Ishii, and T. Yamamuro. 1981. A new murine model of accelerated senescence. Mech Ageing Dev. Oct; 17(2): 183-194.

Takeda, T. and M. Hosokawa, K. Higuchi, M. Hosono, I. Akiguchi, and H. Katoh. 1994. A novel murine model of aging, Senescence-Accelerated Mouse (SAM). Arch. Gerontol. Geriatr. 19: 185-192.

Takeshita, S. and M. Hosokawa, M. Irino, K. Higuchi, K. Shimizu, K. Yasuhira, and T. Takeda. 1982. Spontaneous age-associated amyloidosis in senescence-accelerated mouse (SAM). Mech Ageing Dev. 20: 13-23.

Teramoto, S. and Y. Fukuchi, Y. Uejima, H. Ito, and H. Orimo. 1992. Age-related changes in GSH content of eyes in mice—a comparison of senescence-accelerated mouse (SAM) and C57BL/J mice. Comp. Biochem. Physiol. Comp. Physiol. 102: 693-696.

Teramoto, S. and Y. Fukuchi, Y. Uejima, K. Teramoto, T. Oka, and H. Orimo. 1994. A novel model of senile lung: senescence-accelerated mouse (SAM). Amer. J. Respir. Crit. Care Med. 150: 238-244.

Tuder, R.M. and I. Petrache, J.A. Elias, N.F. Voelkel, and P.M. Henson. 2003a. Apoptosis and emphysema: the missing link. Amer. J. Respir. Cell. Mol. Biol. 28: 551-554.

Tuder, R.M., L. Zhen, C.Y. Cho, L. Taraseviciene-Stewart, Y. Kasahara, D. Salvemini, N.F. Voelkel, and S.C. Flores. 2003b. Oxidative stress and apoptosis interact and cause emphysema due to vascular endothelial growth factor receptor blockade. Amer. J. Respir. Cell. Mol. Biol., 29: 88-97.

Wang, Z. and T. Zheng, Z. Zhu, R.J. Homer, R.J. Riese, H.A. Chapman, Jr., S.D. Shapiro, and J.A. Elias. 2000. Interferon gamma induction of pulmonary emphysema in the adult murine lung. J. Exp. Med. 192: 1587-1600.

Zheng, T. and Z. Zhu, Z. Wang, R.J. Homer, B. Ma, R.J. Riese, Jr., H.A. Chapman, Jr., S.D. Shapiro, and J.A. Elias. 2000. Inducible targeting of IL-13 to the adult lung causes matrix metalloproteinase- and cathepsin-dependent emphysema. J. Clin. Invest. 106: 1081-1093.

PART 3

Analysis and Methods

Proteomics of Tomato Seed and Pollen

Vipen K. Sawhney and Inder S. Sheoran

Department of Biology, 112 Science Place, University of Saskatchewan, Saskatoon, Saskatchewan S7N 5E2, Canada

ABSTRACT

Proteomic analyses of tomato seed, dry and during germination, and mature pollen were conducted using one- and two-dimensional gel electrophoresis and mass spectrometry. Both the tomato seed and pollen contain a number of defense-related proteins that could be part of the survival strategy of these small, free-floating structures. The tomato seed contains a high number of storage proteins, such as vicilins, legumins, albumins and prolamins, not identified in the pollen, which are potentially required for seed germination and growth of the multicellular embryo. The seed and pollen also possess cytoskeleton-related proteins, glycine-rich proteins, proteins involved in nucleic acid and general metabolism, and signaling proteins, all of which are designated to have important roles in germination and growth of these two structures.

INTRODUCTION

Proteins are the functional molecules of living cells that, either alone or in interaction with other proteins and metabolites, drive various metabolic processes in a biological system. An understanding of the nature of proteins and their interactions is, therefore, essential for determining how cells and their individual organelles function in an organism. With recent

A list of abbreviations is given before the references.

technological advancements in the areas of protein isolation, separations and molecular characterization, combined with the availability of genomic and proteomic databases and expressed sequence tags (ESTs) of various organisms, proteomic studies, that is, analyses of the nature of proteins, protein modifications and protein-protein interactions in a biological system, are now at the forefront of molecular biology (Hirano et al. 2004, Rose et al. 2004, Twyman 2004, Rossignol et al. 2006). An analysis of the role of a protein in a biological process is, however, complex because of the dynamic nature of these molecules in living cells. Proteins undergo several post-translational modifications, such as phosphorylation, acetylation, sulfation, ubiquitylation, and glycosylation, which affect both their activity and stability in a biological system. In addition, the localization of proteins in different cellular compartments adds to the complexity of protein analysis in cells and tissues. Although the proteomics of animal, human and yeast cells and tissues are well ahead of plants, in recent years there has been a surge of research activity in plant proteomics (Canovas et al. 2004, Park 2004, Peck 2005, Sheoran et al. 2005, 2006, Rossignol et al. 2006), especially because of the completion and availability of *Arabidopsis* and rice genome sequences, as well as an increase in the number of protein databases including that of tomato.

Tomato (*Lycopersicon esculentum*) is an important crop grown in nearly every part of the world. In the United States alone, it is the second major vegetable crop, next to maize, and is the center of a $2 billion industry of fresh and processed tomatoes (Decoteau 2000). Fruit production in tomato is dependent on successful sexual reproduction and an important structure in this process is the pollen grain (male gametophyte), which, after its release from an anther, lands on the stigma of a pistil (the female reproductive organ), germinates and forms a pollen tube through which sperm cells travel to the ovule and are delivered to the egg cell. After fertilization, fruit and seed develop, and seeds then produce the next generation of tomato plants. Since plants are sedentary organisms, pollen and seed serve as two important dispersal agents, the former for the transport of sperm cells and the latter for dissemination of offspring. The seed and pollen also have some common characteristics; they are both independent, dormant, and highly desiccated structures and have a tough protective coating, the seed coat and pollen wall, respectively. Thus, the proteome analysis of these two structures should provide not only insights into the survival strategies of these small free-floating structures, but also an understanding of proteins required for germination and subsequent growth of these two important plant organs.

This chapter will focus on our current understanding of tomato seed and pollen proteome and, wherever possible, a comparison will be made of proteins identified in similar structures in other species. First, a brief outline of the current procedures and technologies involved in proteome analysis is presented.

METHODOLOGY AND INSTRUMENTATION

The most direct approach for analyzing the proteome of a cell, organelle or tissue at any one stage or time of development is the separation of proteins from tissue extracts by one-dimensional gel electrophoresis (1-DE) based on molecular weight and two-dimensional gel electrophoresis (2-DE) based on charge and molecular weight. Protein bands and spots in 1-DE and 2-DE gels respectively are commonly visualized with stains such as Colloidal Coomassie Blue or the more sensitive silver stain, or fluorescent stains such as SYPRO Ruby (Miller et al. 2006). One of the limitations of 2-DE is the detection of low-abundance proteins, especially in crude extracts; therefore, some studies are restricted to either specific organs or cell components. Some non-gel techniques such as multidimensional protein identification technology (Mud-PIT), isotope labeling using isotope-coded affinity tags or isobaric tags for relative and absolute quantification, and protein-protein and protein-antibody micro-arrays are also used for proteome analyses (Paoletti et al. 2004, Roe and Griffin 2006, Wu et al. 2006). In Mud-PIT, the protein mixture is digested and the peptides are then separated on a strong cation exchange phase in the first dimension and reverse phase chromatography in the second dimension (Koller et al. 2002, Washburn 2004). Braconi et al. (2006) suggested using both 2-DE and Mud-PIT approaches, which would provide increased proteome coverage of a structure or an organelle. Another approach, especially for determining differences in proteins spots between samples from different treatments or from different genetic lines, is the differential in-gel electrophoresis. For this technique, proteins in tissue extracts are labeled with different fluorescent dyes and separated by 2-DE on a single gel, and the differential proteins are detected by scanning at dye specific wavelength and analyzed using 2-DE analysis software (Komatsu et al. 2006, Rossignol et al. 2006). This technique also eliminates gel-to-gel variation.

Protein analysis of spots from 2-DE gels involves excision of spots followed by digestion with proteases, e.g., trypsin, and the digests are then subjected to mass spectrometry (MS). There are variations in MS based on ionization of proteins and peptides; commonly matrix-assisted laser desorption ionization (MALDI) and electrospray ionization (ESI) is used. In many cases time-of-flight mass spectrometry (TOF MS) is combined with MALDI, or TOF is linked to quadrupole MS (Q-TOF MS). In other cases liquid chromatography-tandem mass spectrometry (LC-MS/MS) is used for the analysis and it may be combined with ESI and Q/TOF (LC-ESI-Q/TOF MS). The MALDI-TOF MS is generally used for protein identification by peptide mass fingerprinting, but for amino acid sequence of peptides and to identify post-translational modifications, Q-TOF MS is preferred (Hirano et al. 2004, Twyman 2004). For rapid analysis, Nano-LC MS/MS and Chip-LC MS/MS are used. The identification of proteins is achieved by using search engines such as MASCOT (www.matrixscience.com) or ExPASy (www.expasy.org). There are several plant and organelle databases, for example, the database of the National Center for Biotechnology Information or NCBI, available for protein identification, and for further information on these and other aspects of protein isolation, separation, identification and quantitation the reader is referred to various reviews (e.g., Hirano et al. 2004, Rose et al. 2004, Twyman 2004, Braconi et al. 2006, Rosssignol et al. 2006). A simplified flow chart of the steps involved in proteomic studies is presented in Fig. 1.

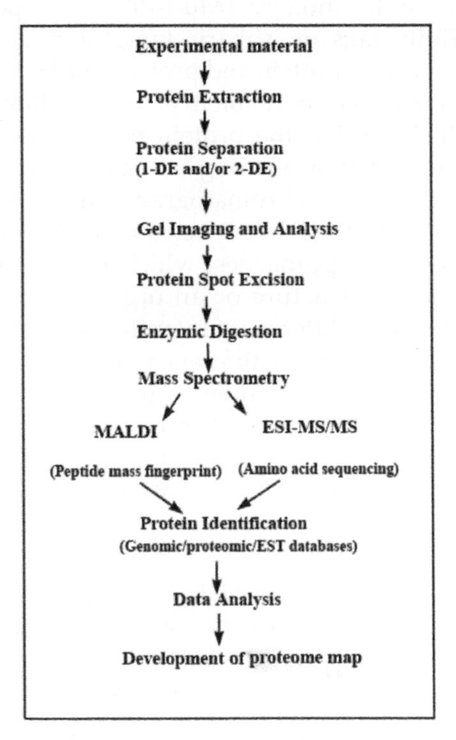

Fig. 1 Flow chart of steps involved in proteome analysis.

TOMATO SEED PROTEOME

Seeds are a rich source of proteins (Bewley 1997) and it has been estimated that tomato seeds, depending on the variety, have proteins in the range of 22-30% (Sogi et al. 2002). In tomato, the embryo occupies a major part of the seed and is very well embedded in the endosperm. By using the proteomic approach, we have identified a number of proteins in both the embryo and endosperm of dry seeds, as well as determined changes in major proteins in these two structures during seed germination.

The relative quantity of soluble protein in the embryo and endosperm each of dry seed (0 hr) was in the range of 60-100 μg/organ, and during germination, from 0 to 120 hr, the protein content increased gradually in the embryo to approximately 250 μg, but in the endosperm it declined rapidly after 48 hr (Sheoran et al. 2005). The 1-DE of embryo in the dry and germinating seed showed six major protein bands of 45, 36, 33, 26, 23 and 22 kDa (Fig. 2) and these constitute approximately 60% of total protein. These bands reflect major seed storage proteins (see details below); there

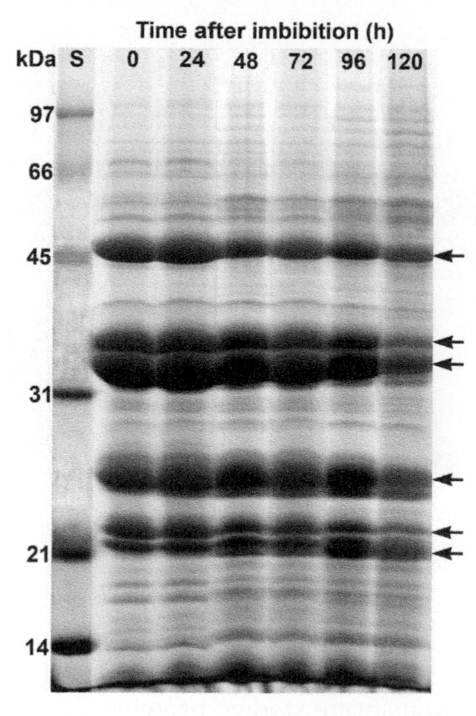

Fig. 2 1-DE gel of embryos from dry tomato seeds (0 hr) and during seed germination from 0 to 120 hr. The arrows indicate six major proteins, and the left lane contains standard molecular weight markers (from Sheoran et al. 2005, with permission from Wiley-VCH Verlag GmbH & Co KGaA).

was little change in the relative quantity of these bands during the early part of germination but later, especially at 120 hr, the quantity was reduced. A similar pattern of protein changes was observed in the endosperm, except that reduction in the protein level was earlier and greater than that in the embryo (Sheoran et al. 2005).

In 2-DE gels, 352 and 369 protein spots were detected in the embryo and endosperm respectively of dry seeds, and during germination from 0 to 120 hr the number of spots increased to 519 in the embryo (Fig. 3A and B) but declined to 298 in the endosperm (not shown). Also, in the embryo, whereas some groups of protein spots increased in intensity from 0 to 120 hr, others decreased and some were unchanged (see boxes in Fig. 3B). The relatively low number of total spots in these tomato tissues, relative to the proteomics of other seeds (Gallardo et al. 2001, Ostergaard et al. 2004), is attributed to the abundance of major storage proteins (Figs. 2 and 3). Protein spots (numbered on 2-DE gels) were analyzed both by MALDI-TOF MS and by LC-ESI-Q/TOF MS, and protein identification was achieved by searching various databases including tomato ESTs.

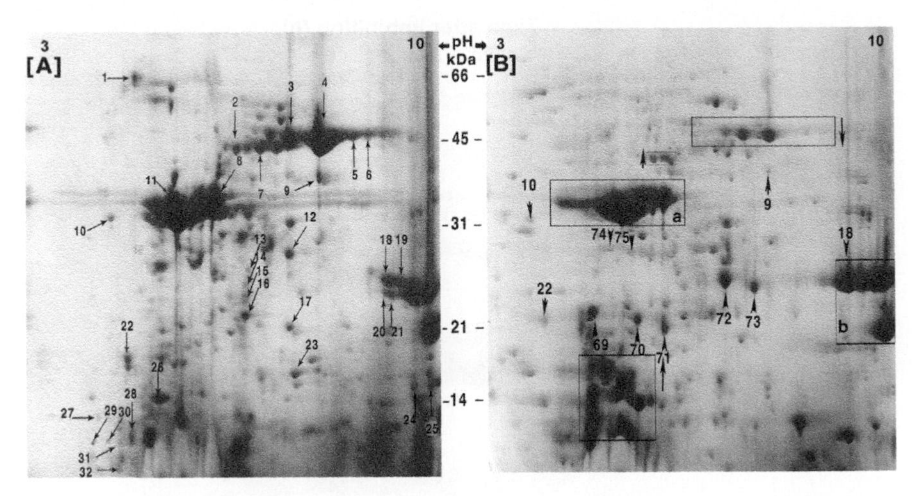

Fig. 3 2-DE gels of proteins extracted from tomato embryo at 0 hr (A) and at 120 hr (B). The numbered protein spots were identified by MS. In B the boxes indicate groups of proteins that were increased (↑), decreased (↓) or unchanged (a, b), and arrowheads indicate individual protein spots that increased or decreased from 0 to 120 hr (from Sheoran et al. 2005, with permission from Wiley-VCH Verlag GmbH & Co KGaA).

Of the proteins identified in both the embryo and endosperm of tomato, the two major groups were storage proteins (52%) and defense-related proteins (23%); others included proteins involved in nucleic acid metabolism, general metabolism, cytoskeleton formation and glycine-rich proteins (GRPs) (Fig. 4). The storage proteins included globulins (vicilins

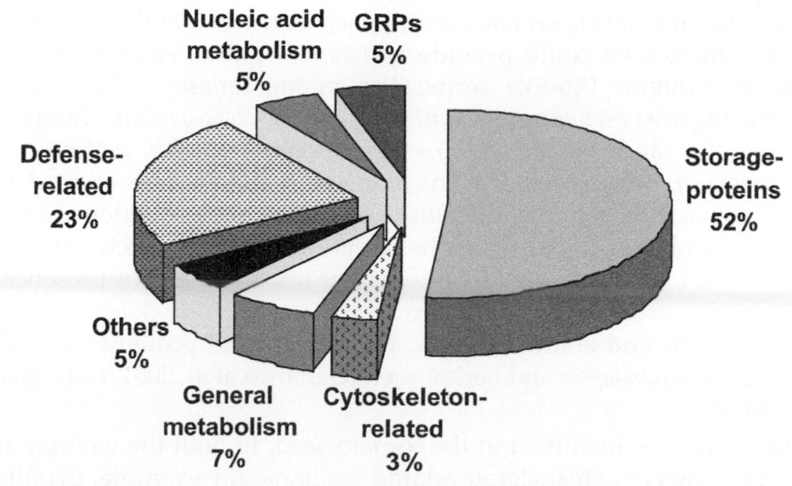

Fig. 4 The relative distribution of major proteins identified in the tomato seed.

and legumins), albumins and prolamins. Three forms of vicilin of approximately 44, 27 and 17 kDa, and α- and β-legumins were identified in both the embryo and endosperm (Sheoran et al. 2005). Globulins are the major storage proteins in the seeds of many dicotyledons (Bewley and Black 1994, Shewry et al. 1995) and legumins are especially reported in the legumes (Tiedemann et al. 2000). Both the vicilins and legumins are generally broken down during seed germination and their content also decreased during tomato seed germination. Two protein spots of 2S albumins were identified, which are cysteine-rich proteins and are present in the seed of a number of dicotyledons species, including tomato (Oguri et al. 2003). In addition, two spots of non-specific lipid transfer proteins (nsLTPs) were identified, which, along with 2S albumin, belong to a family of storage proteins called the prolamins (Shewry et al. 1995) and have roles in storage and in defense against pathogens (Blein et al. 2002, Maldonado et al. 2002). nsLTPs were especially high in the tomato endosperm and may also have a role in the mobilization of lipids from the endosperm to the embryo during seed germination. Proteomic analyses of the seed of other species have also reported various forms of storage proteins although the embryo and endosperm tissues were generally not analyzed separately (Gallardo et al. 2001, Koller at al. 2002, Mooney and Thelen 2004, Ostergaard et al. 2004, Hajduch et al. 2006).

A number of defense-related proteins, mainly in the endosperm, were identified in the tomato seed. These included coat proteins (CP) of tobacco mosaic virus (TMV) and tomato mosaic virus (ToMV). It has been shown that the constitutive expression of genes coding CP of TMV results in a virus-resistant plant (Koo et al. 2004), which suggests that the presence of CPs in tomato seed could provide resistance against viruses. Another group of proteins Pto-like serine/threonione kinases, which confer resistance against *Pseudomonas, Xanthomonas* and *Cladosporium* (Tang et al. 1999, Pedley and Martin 2003), were present in the embryo and endosperm of tomato seed. Vicilins along with 2S albumins and nsLTPS also play major roles in plant defense (Blein et al. 2002, Maldonado et al. 2002). A number of GRPs, which have various functions including their role in stress tolerance, were identified in the tomato seed. The presence of these various proteins in the tomato seed could provide defense against different biotic and abiotic stresses. Defense-related proteins were also identified in *Arabidopsis* and barley seed (Gallardo et al. 2001, Ostergaard et al. 2004).

Other proteins identified in the tomato seed, in both the embryo and endosperm, were cytoskeleton-related proteins, for example, profilins, which are actin-binding proteins, and their level increased during seed germination, especially in the embryo. The presence of profilins is reflective of their requirement for cell growth during radicle emergence and root growth. A number of housekeeping proteins involved in nucleic acid metabolism, such as KH domain-containing protein and MAR binding protein, and in general metabolism, such as glyceraldehydes-3-phosphate dehydrogenase, alcohol dehydrogenase and phenyl-ammonia lyase, were also present in the embryo and endosperm of tomato seed. The 14-3-3 protein, a highly conserved protein that acts as a signaling molecule in a number of processes including growth, apoptosis and cell cycle (Ferl 2004), was identified in both the tomato embryo and endosperm, and at various stages of seed germination (Sheoran et al. 2005).

The above represents only the partial proteome of tomato seed. A number of low-abundance proteins in 2-DE gels could not be identified and, although this is a major challenge in proteomic research (Hirano et al. 2004, Twyman 2004), our future efforts will be directed in this area.

TOMATO POLLEN PROTEOME

Mature pollen of tomato were collected from flowers at anthesis and clean preparations, checked under microscope, were used for protein extraction, separation and analysis by procedures similar to those for tomato seeds (as above). Of the 960 spots identified on 2-D gels, 190 were selected for analysis by MALDI-TOF MS, and 133 distinct proteins representing 158 spots were identified by searching various ESTs and databases.

The major groups of proteins in tomato pollen were those related with defense, energy, and protein synthesis and processing. The proteome profiles of rice and *Arabidopsis* also revealed the same major groups of proteins (Holmes-Davis et al. 2005, Noir et al. 2005, Dai et al. 2006, Sheoran et al. 2006). The defense-related proteins included Pto-serine/threonine kinases and ToMV coat protein, which are similar to those found in the tomato seed. Thus, tomato seed and pollen have similar defense mechanisms in place against pathogens. Tomato pollen also had a number of proteins related with abiotic stress resistance (Sheoran et al. 2007), such as heat shock proteins as protectants of heat stress, late embryogenesis abundant proteins for desiccation tolerance, and enzymes involved in the detoxification of reactive oxygen species, which, in high concentration, can damage cellular organelles.

Other proteins in tomato pollen included those involved in cytoskeleton formation, for example, profilins and actin-like proteins, as well as a number of proteins related with energy metabolism and protein synthesis and processing. All these proteins play potential roles in pollen germination and pollen tube growth. In addition, proteins involved with Ca^{2+} binding and signaling, such as calmodulin and calreticulin, and those with hormone signaling, such as G proteins, were identified in tomato pollen (Sheoran et al. 2007). Most of these proteins play important roles in pollen germination and growth of pollen tubes (Taylor and Hepler 1997) and many of them were also reported in *Arabidopsis* and rice pollen (Holmes-Davis et al. 2005, Noir et al. 2005, Dai et al. 2006, Sheoran et al. 2006). The tomato pollen also had 14-3-3 protein, as in the seed, and the LAT52 protein, which has a role in pollen development (Muschietti et al. 1994).

In conclusion, a comparison of the proteome of tomato seed and pollen indicates that both these free-floating structures are enriched with a number of defense-related proteins (Fig. 5) that could be part of their survival strategy. In addition, they contain cytoskeleton-related proteins, GRPs, 14-3-3 protein and a number of proteins and enzymes involved in nucleic acid metabolism and general metabolism (Fig. 4), all of which are important for the germination and growth of both the seed and pollen. The storage proteins were found only in the seed and this could be attributed to their requirement for germination and growth of the multicellular embryo with different organs. In contrast, in the pollen grain, germination and formation of the pollen tube involves the growth of a single vegetative cell, which may not necessitate the presence of stored proteins.

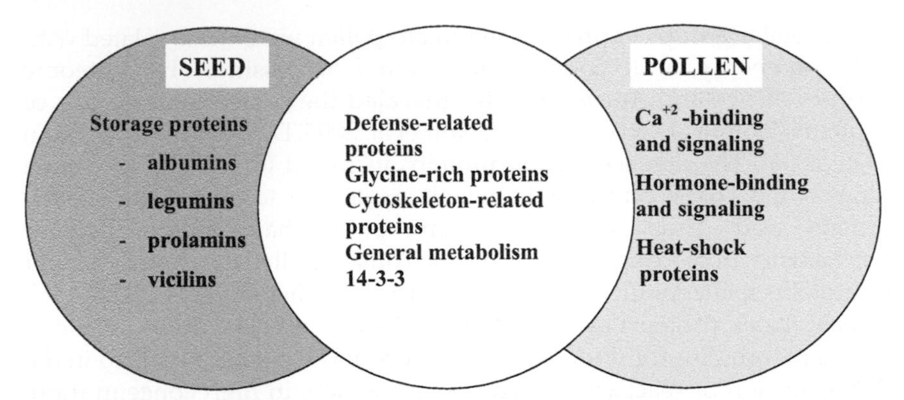

Fig. 5 Major proteins identified only in the seed (left) and pollen (right), and proteins common to seed and pollen (middle).

ACKNOWLEDGEMENTS

The authors gratefully acknowledge the help of Dr. Andrew Ross and Mr. Douglas Olson of the Plant Biotechnology Institute, National Research Council, Saskatoon, for mass spectrometric analysis. The research presented here was supported by a Natural Sciences and Engineering Research Council of Canada Discovery grant to VKS.

ABBREVIATIONS

1-DE: One-dimensional gel electrophoresis; 2-DE: Two-dimensional gel electrophoresis; CP: Coat protein; ESI: Electro-spray ionization; EST: Expressed sequence tag; GRP: Glycine-rich protein; LC: Liquid chromatography; MALDI: Matrix-assisted laser desorption ionization; MS: Mass spectrometry; Mud-PIT: Multidimensional protein identification technology; nsLTPs: non-specific lipid transfer proteins; Q-TOF: Quadrupole time-of-flight; TMV: Tobacco mosaic virus; ToMV: Tomato mosaic virus

REFERENCES

Bewley, J.D. 1997. Seed germination and dormancy. Plant Cell 9: 1055-1066.

Bewley, J.D. and M. Black. 1994. Seeds: Physiology of Development and Germination. Plenum Press, New York, USA.

Blein, J.P. and P. Coutos-Thevenot, D. Marion, and M. Ponchet. 2002. From elicitins to lipid-transfer proteins: a new insight in cell signalling involved in plant defence mechanisms. Trends Plant Sci. 7: 293-296.

Braconi, D. and R. Mini, and A. Santucci. Proteomics in plant research. pp. 119-138. *In*: A. Tiezzi and A. Santucci. [eds.] 2006. Plant Molecules: Basic and Applied Research. Gangemi editore, Rome, Italy.

Canovas, F.M. and E. Dumas-Gaudot, G. Recorbet, J. Jorrin, H.P. Mock, and M. Rossignol. 2004. Plant proteome analysis. Proteomics 4: 285-298.

Dai, S. and L. Li, T. Chen, K. Chong, Y. Xue, and T. Wang. 2006. Proteomic analysis of *Oryza sativa* mature pollen reveal novel proteins associated with pollen germination and tube growth. Proteomics 6: 2504-2529.

Decoteau, D.R. 2000. Vegetable Crops. Prentice-Hall Inc., New Jersey, USA.

Ferl, R.J. 2004. 14-3-3 proteins: regulation of signal-induced events. Physiol. Plant. 120: 173-178.

Gallardo, K. and C. Job, S.P.C. Groot, M. Puype, H. Demol, J. Vandekerckhove, and D. Job. 2001. Proteomic analysis of Arabidopsis seed germination and priming. Plant Physiol. 126: 835-848.

Hajduch, M. and J.E. Casteel, K.E. Hurrelmeyer, Z. Song, G.K. Agrawal, and J.J. Thelen. 2006. Proteomic analysis of seed filling in *Brassica napus*. Developmental characterization of metabolic isozymes using high-resolution two-dimensional gel electrophoresis. Plant Physiol. 141: 32-46.

Hirano, H. and N. Islam, and H. Kawasaki. 2004. Technical aspects of functional proteomics in plants. Phytochemistry 65: 1487-1498.

Holmes-Davis, R. and C.K. Tanaka, W.H. Vensel, W.J. Hurkman, and S. McCormick. 2005. Proteome mapping of mature pollen of *Arabidopsis thaliana*. Proteomics 5: 4864-4884.

Koller, A. and M.P. Washburn, B.M. Lange, N.L. Andon, C. Deciu, P.A. Haynes, L. Hays, D. Schieltz, R. Ulaszek, J. Wei, D. Wolters, and J.R. Yates. 2002. A proteomic survey of metabolic pathways in rice. Proc. Natl. Acad. Sci. USA 99: 520-526.

Komatsu, S. and X. Zang, and N. Tanaka. 2006. Comparison of two proteomics techniques used to identify proteins regulated by gibberellin in rice. J. Proteome Res. 5: 270-276.

Koo, J.C. and S. Asurmendi, J. Bick, T. Woodford-Thomas, and R.N. Beachy. 2004. Ecdysone agonist-inducible expression of a coat protein gene from tobacco mosaic virus confers viral resistance in transgenic Arabidopsis. Plant J. 37: 439-448.

Maldonado, A.M. and P. Doerner, R.A. Dixon, C.J. Lamb, and R.K. Cameron. 2002. A putative lipid transfer protein involved in systemic resistance signalling in Arabidopsis. Nature 419: 399-403.

Miller, I. and J. Crawford, and E. Gianazza. 2006. Protein stains for proteomic applications: Which, when and why? Proteomics 6: 5385-5408.

Mooney, B.P. and J.J. Thelen. 2004. High-throughput peptide mass fingerprinting of soybean seed proteins: automated workflow and utility of UniGene expressed sequence tag databases for protein identification. Phytochemistry 65: 1733-1744.

Muschietti, J. and L. Dircks, G. Vancanneyt, and S. McCormick. 1994. LAT52 protein is essential for tomato pollen development: pollen expressing antisense LAT52 RNA hydrates and germinates abnormally and cannot achieve fertilization. Plant J. 6: 321-338.

Noir, S. and A. Brautigam, T. Colby, J. Schmidt, and R. Panstruga. 2005. A reference map of the *Arabidopsis thaliana* mature pollen proteome. Biochem. Biophys. Res. Commun. 337: 1257-1266.

Oguri, S. and M. Kamoshida, Y. Nagata, Y.S. Momonoki, and H. Kamimura. 2003. Characterization and sequence of tomato 2S seed albumin: a storage protein with sequence similarities to the fruit lectin. Planta 216: 976-984.

Ostergaard, O. and C. Finnie, S. Laugesen, P. Roepstorff, and B. Svensson. 2004. Proteome analysis of barley seeds: Identification of major proteins from two-dimensional gels (pl 4-7). Proteomics 4: 2437-2447.

Paoletti, A.C., and B. Zybailov, and M.P. Washburn. 2004. Principles and applications of multidimensional protein identification technology. Expert Rev. Proteomics 1: 275-282.

Park, O.K. 2004. Proteomic studies in plants. J. Biochem. Mol. Biol. 37: 133-138.

Peck, S.C. 2005. Update on proteomics in Arabidopsis. Where do we go from here? Plant Physiol. 138: 591-599.

Pedley, K.F. and G.B. Martin. 2003. Molecular basis of *Pto*-mediated resistance to bacterial speck disease. Annu. Rev. Phytopathol. 41: 215-243.

Roe, M.R. and T.J. Griffin. 2006. Gel-free mass spectrometry-based high throughput proteomics: Tools for studying biological response of proteins and proteomes. Proteomics 6: 4678-4687.

Rose, J.K.C. and S. Bashir, J.J. Giovannoni, M.M. Jahn, and R.S. Saravanan. 2004. Tackling the plant proteome: practical approaches, hurdles and experimental tools. Plant J. 39: 715-733.

Rossignol, M. and J.B. Peltier, H.P. Mock, A. Matros, A.M. Maldonado, and J.V. Jorrin. 2006. Plant proteome analysis: A 2004-2006 update. Proteomics 6: 5529-5548.

Sheoran, I.S. and D.J.H. Olson, A.R.S. Ross, and V.K. Sawhney. 2005. Proteome analysis of embryo and endosperm from germinating tomato seeds. Proteomics 5: 3752-3764.

Sheoran, I.S. and A.R.S. Ross, D.J.H. Olson, and V.K. Sawhney. 2007. Proteomic analysis of tomato (*Lycopersicon esculentum*) pollen. J. Exp. Bot. 58: 3525-3535.

Sheoran, I.S. and K.A. Sproule, D.J.H. Olson, A.R.S. Ross, and V.K. Sawhney. 2006. Proteome profile and functional classification of proteins in *Arabidopsis thaliana* (Landsberg erecta) mature pollen. Sex. Plant Reprod. 19: 185-196.

Shewry, P.R. and J.A. Napier, and A.S. Tatham. 1995. Seed storage proteins—structure and biosynthesis. Plant Cell 7: 945-956.

Sogi, D.S. and M.S. Arora, S.K. Garg, and A.S. Bawa. 2002. Fractionation and electrophoresis of tomato waste seed proteins. Food Chem. 76: 449-454.

Tang, X.Y. and M.T. Xie, Y.J. Kim, J.M. Zhiu, D.F. Klessing, and G.B. Martin. 1999. Overexpresion of *Pto* activates defense responses and confers broad resistance. Plant Cell 11: 15-29.

Taylor, L.P. and P.K. Hepler. 1997. Pollen germination and tube growth. Annu. Rev. Plant Physiol. Mol. Biol. 48: 461-491.

Tiedemann, J. and B. Neubohn, and K. Muntz. 2000. Different functions of vicilin andlegumin are reflected in the histopattern of globulin mobilization during germination of vetch (*Vicia sativa* L.). Planta 211: 1-12.

Twyman, R.M. 2004. Principles of Proteomics. Garland Science Publishers, New York, USA.

Washburn, M.P. 2004. Utilization of proteomics datasets generated via multidimensional protein identification technology. Brief. Funct. Genomic. Proteomic. 3: 280-286.

Wu, W.W. and G.H. Wang, S.J. Baek, and R.F. Shen. 2006. Comparative study of three proteomic quantitative methods, DIGE, cICAT, and iTRAQ, using 2D gel- or LC-MALDI TOF/TOF. J. Proteome Res. 5: 651-658.

26

Gene Transfer in Tomato and Detection of Transgenic Tomato Products

Theodoros H. Varzakas[1*], Dimitris Argyropoulos[2] and Ioannis S. Arvanitoyannis[3]

[1]Department of Technology of Agricultural Products, School of Agricultural Technology, Technological Educational Institute of Kalamata, Antikalamos 24100, Kalamata, Hellas, Greece

[2]Institute of Biotechnology, National Agricultural Research Foundation, Sofokli Venizelou 1 Lykovrisi 14123, Attiki Greece

[3]Department of Agriculture, Ichthyology and Aquatic Environment, Agricultural Sciences, University of Thessaly, Fytokou Street, Nea Ionia Magnesias, 38446 Volos, Hellas, Greece

ABSTRACT

Transformation is the genetic alteration of a cell resulting from the uptake and expression of foreign genetic material. The development of improved transformation techniques with the immediate insertion of genes into parental lines of tomato cultivars facilitates the use of transgenes in advanced breeding programmes. The potential in the use of transgenes especially from plants in either sense or antisense origin is also quite high. Genes expressed during reproductive development or disease resistance have been manipulated in tomato plants, but concerns about consumption of genetically modified products still exist. *Agrobacterium tumefaciens* is more than the causative agent of crown gall disease affecting dicotyledonous plants. It is the most common natural instance for the introduction of foreign gene in plants allowing its genetic manipulation. Direct gene transfer also allows the insertion of foreign DNA into

*Corresponding author

A list of abbreviations is given before the references.

cells without the use of transfection vectors with future gene transfer methods targeting specific genes or DNA sequences. Green genetic engineering is still perceived in the first place as a risk. And although scientists have been conducting biological safety research on genetically modified plants for years, the belief that possible consequences for the environment have not been researched is widespread.

INTRODUCTION

The cultivated tomato, *Lycopersicon esculentum*, is one of the most important crops in the world. It is the number one vegetable crop consumed fresh worldwide, and in processed form it accounts for one third of the total world yield. The geographical distribution of the species was achieved by breeding for adaptation to diverse growth conditions and for resistance to plant diseases. Gene transfer for resistance to plant pathogens has acquired great importance nowadays in view of the various pesticides and fungicides and their residues as well as consumer interest in healthier foods through breeding for a high concentration of antioxidants and vitamins. Nowadays, different qualities are required from different varieties that will be used for diverse products ranging from canned tomatoes to juice, ketchup and pastes.

The development of genetic engineering techniques that modify fruit ripening (enhancement of total soluble solids) and other characteristics such as fruit quality and colour in conjunction with the development and improvement of tomato transformation protocols opens a new era in tomato breeding. Researchers have prepared technologies such as *Agrobacterium*-mediated transformation and antisense technology to develop new varieties.

GENE TRANSFER IN TOMATO

Gene Transfer Methods

One method of gene insertion makes use of a bacterial vector, *Agrobacterium tumefaciens*. To achieve genetic modification using this technique, a restriction enzyme is used to cut non-virulent plasmid DNA derived from *A. tumefaciens* and thus create an insertion point, into which the gene can be ligated. The engineered plasmid is then put into a strain of *A. tumefaciens*, which contains a "helper" plasmid, and plant cells are treated with the recombinant bacterium (Dandekar and Fisk 2005). The helper plasmid assists the expression of the new gene in plant as these grow in culture. The utility of *A. tumefaciens*-mediated gene transfer is illustrated by its use to develop potato plants that are resistant to Colorado beetle (Perlak et al. 1993). The inserted cryIIIa gene encodes for an insecticidal protein from soil bacterium *B. thuringiensis*. When expressed in potato plants this protein is toxic to Colorado beetles by combining with a protein found within their digestive epithelium to form a membrane channel that results in cell lysis (Fig. 1).

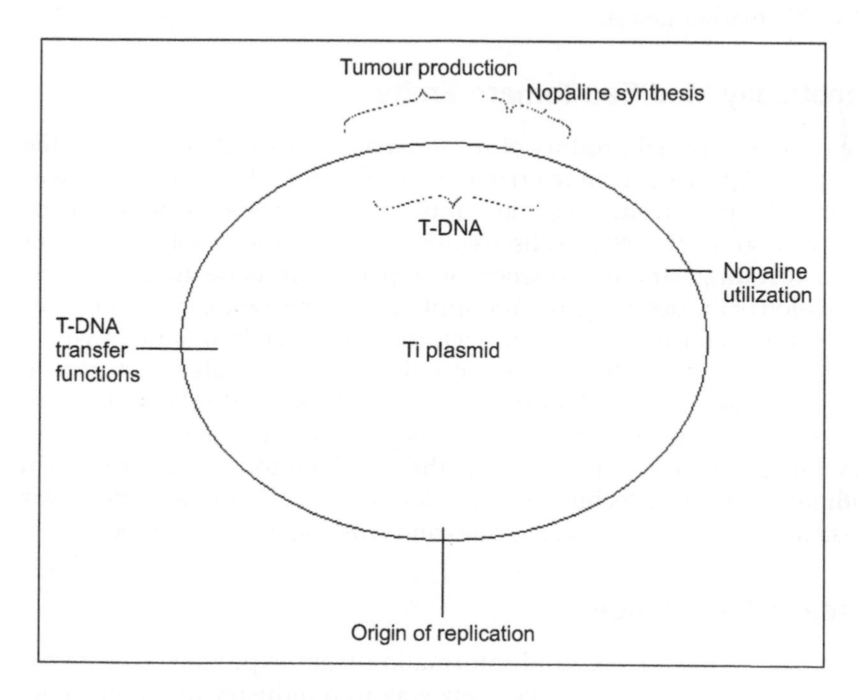

Fig. 1 *Agrobacterium tumefaciens* plasmid. Gene transfer with *A. tumefaciens* starts with activation of inner membrane protein VirA by acetosyringone and other plant cell wall intermediates. The subsequent steps are as follows: activation of cytoplasmic VirG protein by VirA; transcriptional activation of the vir gene operons by VirG; processing of T-DNA by vir gene products; transfer of T-DNA into the plant cell as single-stranded DNA; integration of T-DNA into plant nuclear genome; transcription of T-DNA genes; translation of T-DNA gene transcripts; production of auxin and cytokinin; stimulation of plant cell growth and division; production of opines catalysed by T-DNA gene products and uptake and catabolism of opines.

In microballistic impregnation or gene gun, the target gene is applied to minute particles of gold or tungsten, which are fired into plant tissue at high velocity. Cells that incorporate the target and marker genes can then be selected in culture and grown (Baum et al. 1997, Christou 1997). Electroporation is the application of a pulsed electric field that allows the transient formation of small pores in plant cells, through which genes can be taken up and in the form of naked DNA incorporated into the plant genome (Bates 1995, Terzaghi and Cashmore 1997).

Methods of direct DNA transfer include PEG-induced DNA uptake, microinjection of DNA into cultured cells, electroporation and microprojectile bombardment. Gene transfer is combined with efficient methods of plant regeneration, availability of new gene constructs, improved vector systems based on Ti and Ri plasmids of *Agrobacterium*, appropriate organ-specific promoters for gene expression and a series of selectable marker genes.

Genetically Modified Tomato Fruits

The first commercial product of recombinant DNA technology was the FlavrSavr™ tomato characterized by prolonged shelf life. This was achieved by introducing an antisense polygalacturonase gene (Redenbaugh et al. 1995). Antisense technology involves introduction of the non-coding strand sequence of a particular gene to inhibit the expression of the desired gene. The application of antisense technology has resulted in a consumer food product, the FlavrSavr™ tomato, in which shelf life is enhanced by suppression of the enzyme polygalacturonase (Sitrit and Bennett 1998). A similar increase in the shelf life of tomatoes can also be achieved using antisense technology against the enzyme 1-amino-1-cyclopropanecarboxylic acid synthase (Hamilton et al. 1995). In traditional breeding technology, the development of a new variety was possible only using particular species and crossing its known types.

Disease Resistance

Presented methodologies and information from experimental data for gene expression gradually find their way into industry for commercial production of tomato cultivars. Plants have evolved a large variety of sophisticated and efficient defence mechanisms to prevent the colonization of their tissues by microbial pathogens and parasites. The induced defence responses can be assigned to three major categories.

The incorporation of disease resistance genes into plants was initially initiated with classical breeding, which yielded high crop productivity. However, plant biology is currently oriented toward clone identification and characterization of genes involved in disease resistance. The following approaches are adopted toward developing disease-resistant transgenic plants (Punja 2004):

1. Expression of gene products that are directly toxic to or that reduce growth of the pathogen, including pathogenesis-related proteins such as hydrolytic enzymes, antifungal proteins, antimicrobial peptides and phytoalexins.

2. Expression of gene products that destroy or neutralize a component of the pathogen (lipase inhibition).
3. Expression of gene products that enhance the structural defences (high levels of peroxidase and lignin).
4. Expression of gene products that release signals regulating plant defences such as salicylic acid.
5. Expression of resistance gene products involved in race-specific avirulence interactions and hypersensitive response.

One of the most significant contributions of genetic engineering to agriculture has been the introduction of viral genes to plants, including tomato, as a means of conferring resistance to viral diseases (Barg et al. 2001). Apart from the introduction of *Bacillus thuringiensis* (Bt) toxin genes in tomato, characteristics related to the quality of fruit (total soluble solids, fruit colour) were also genetically engineered.

The site of gene integration into the genome also influences epitope and transgene accumulation in plants. *Agrobacterium tumefaciens* infection is most frequently used to achieve permanent integration into the nuclear DNA, where integration occurs at random chromosomal sites. A second promising approach is based on the integration of the gene or epitope into the circular chloroplast DNA that is present in multiple copies within defined plant cells. In this case transformation is usually achieved through the use of the "particle gun" and results in site-specific integration. These two approaches can be used to integrate genes into the plant genome and produce vaccines against infectious diseases.

Modified viruses such as derivatives of the tomato golden mosaic virus, in which an antibiotic resistance gene replaces the viral coat protein genes, can be used to enhance integration (Elmer and Rogers 1990). Once inside the genome, the virus has the ability to propagate, while the removal of the coat protein genes inhibits the production of new viable virus, thus leaving replicated copies of the viral DNA within the nucleus. Similar results have been obtained using other viral vector combinations such as the *Agrobacterium* geminivirus vectors and *Agrobacterium*-cosmid vectors (van den Eede et al. 2004).

Stable integration of a gene into the plant nuclear or chloroplast genome can transform higher plants (e.g., tobacco, potato, tomato, banana) into bioreactors for the production of subunit vaccines for oral or parental administration (Sala et al. 2003). This can also be achieved by using recombinant plant viruses as transient expression vectors in infected plants. The use of plant-derived vaccines may overcome some of the major problems encountered with traditional vaccination against infectious diseases, autoimmune diseases and tumours. They also offer a convenient tool against the threat of bio-terrorism (Table 1).

Table 1 Expressed gene product, and its effect on disease/composition.

Expressed gene product	*Effect on disease/composition*	*Reference*
Tomato chitinase	Fewer diseased plants due to *Cylindrosporium concentricum* and *Sclerotininia sclerotiorum*	Grison et al. 1996
Tomato *Cf9* gene	Delayed disease development due to *Leptosphaeria maculans* Resistance against *Cladosporium fulvum*	Hennin et al. 2001, Punja *2004*
1-aminocyclopropane-1-carboxylic acid deaminase	Increased tolerance to flooding stress and less subject to deleterious effects of root hypoxia on plant growth	Grichko and Glick 2001
Synthetic *cry1Ac* gene coding for an insecticidal crystal protein of *B. thuringiensis*	Expression resulted in high level of protection of transgenic plant leaves and fruits against larvae of tomato fruit borer (*Helicoverpa armigera*)	Mandaokar et al. 2000
Avr9 gene from *Cladosporium fulvum* LRR-TM structure	HR-associated defense responses in tomato plants that carry the matching *Cf-9* resistance gene	Joosten and De Wit 1999, Van den Ackerveken et al. 1992, Jones et al. 1994
Avr4 gene from *Cladosporium fulvum* LRR-TM structure	HR-associated defence responses in tomato plants that carry the matching *Cf-4* resistance gene	Joosten et al. 1994, Thomas et al. 1997
Avr2 gene from *Cladosporium fulvum* LRR-TM structure	HR-associated defence responses in tomato plants that carry the matching *Cf-2* resistance gene	Takken et al. 2000
Ecp1, *Ecp2* and *Ecp4*, 5 genes from *Cladosporium fulvum*	HR-associated defence responses in tomato plants that carry the matching *Cf-Ecp1* and *Cf-Ecp2* and *Cf-Ecp4*, 5 resistance genes	Van den Ackerveken et al. 1993, Lauge et al. 1997, 2000
Ve1 and Ve2 from *Verticillium dahliae* Coiled-coil structure LRR TM-Pro-Glu-Ser-Thr structure	Wilt resistance genes Unknown Avr matching genes	Kawchuk et al. 2001
	Pseudomonas syringae, *Xanthomonas campestris* bacterial resistance	
	Herbicide tolerance (Phospinothricin, glyphosate)	

Contd.

Contd.

	Fungal resistance such as powdery mildew resistance	(USDAAPHIS http://www.aphis.usda.gov/biotech/notday.html)
	Kanamycin	
Marker genes	PVY virus resistance	Hilbeck 2001
	Altering of seed colour and fruit ripening	Rosati et al. 2000
	Catalase level reduced, polyamine metabolism altered, polygalacturonase level reduced	Fraser et al. 2001
Lycopene cyclase (*Arabidopsis*)	Provitamin A and lycopene increase	Muir et al. 2001
Phytoene desaturase (*Erwinia*)	Provitamin A increase	Mehta et al. 2002
Chalcone isomerase (*Petunia*)	Flavonoids increase	Wang et al. 2006
Engineered polyamine accumulation	Lycopene increase	Siddiqui and Shaukat 2003
Coat protein gene transformed into tomato	Resistance to cucumber mosaic virus	Sanders et al. 1992
Expression of the reporter gene phl'-'lacZ reflects actual production of DAPG 2,4-diacetylphloroglucino, an antimicrobial metabolite	The ability of *Pseudomonas fluorescens* strain CHA0 to suppress root-knot plant disease caused by soil-borne pathogens.	Notz et al. 2002

Various expressed gene products are shown as well as their effects on disease/composition. Some very well-known genes are *Avr* and *Ecp* genes of *Cladosporium fulvum*, a biotrophic fungus causing leaf mould of tomato plants (Westerink et al. 2004). Analysis of the proteins present in the apoplast of colonized tomato leaves led to the cloning of seven genes that encode elicitor proteins. Four elicitor proteins (AVR2, AVR4, AVR4E and AVR9) are arre specific and trigger HR-associated defence responses in tomato plants that carry the matching Cf resistance gene. The other five elicitors, extracellular proteins ECP1, ECP2, ECP3, ECP4 and ECP5, are secreted by all strains of *C. fulvum*. The matching R genes, designated *Cf-ECPs*, have not yet been introduced into commercial cultivars.

Pest Resistance

A number of transgenes conferring resistance to insects and diseases and herbicide tolerance have been transferred into crop plants from a wide range of plant and bacterial systems. In the majority of cases, genes showing expression in transgenic plants are stably inherited into the progeny without detrimental effects on the recipient plant (Babu et al. 2003); transgenic plants under field conditions have also maintained increased levels of insect resistance. During the last 15 years, transformations have been produced in more than 100 plant species;

notable examples include maize, wheat, soybean, tomato, potato, cotton, and rice. Among these, herbicide-tolerant and insect-tolerant cotton, maize and soybean carrying Bt genes are grown on a commercial scale. Genetic transformation and gene transfer are routine in many laboratories. However, isolation of useful genes and their expression to the desired level to control insect pests still involves considerable experimentation and resources. Developing pest-resistant varieties by insertion of one or a few specific genes is becoming an important component of breeding. Use of endotoxin genes such as Bt and plant-derived genes (proteinase inhibitors) to the desired levels offers new opportunities to control insects and strategies involving combination of genes. Transgenic technology should be integrated in a total system approach for ecologically friendly and sustainable pest management. Issues related to intellectual property rights, regulatory concerns, and public perceptions for release of transgenics need to be considered (Babu et al. 2003). Providing a wealth of information on gene expression in higher plants by switching the gene on and off as and when required makes gene manipulation a more direct process for genetic improvement of crops.

To convert transgenics into an effective weapon in pest control, for example, by delaying the evolution of insect populations resistant to the target genes, it is important to deploy genes with different modes of action in the same plant. Pyramiding of different genes would reduce the likelihood of resistance development, since multiple mutations would be needed concurrently in individual insects. Integration of single transgenes to plant genomes has been valuable; however, multiple integration is essential to allow manipulation of complex metabolic pathways and to combine various agronomic characteristics in the next generation of plant molecular breeding. Sugita et al. (2000) reported that MAT-vector system used for generating marker-free transgenic plants is also an efficient and reliable transformation system for the repeated introduction of multiple transgenes independent of sexual crossing. Thus, pyramiding of single genes or multiple genes through new transformation procedures could be important for long-term durable resistance to insects.

Delannay et al. (1989) evaluated transgenic tomatoes expressing Bt insect control protein under field conditions in 1987 and 1988. The transgenic plants showed very limited feeding damage after infestation with tobacco hornworm (*Manduca sexta*), whereas control plants showed heavy feeding damage and were almost completely defoliated within 2 wk. Significant control of tomato fruit worm and tomato pinworm was also observed (Table 2).

Table 2. Gene insertion and expression in transgenic plants for insect resistance.

Crop target	*Inserted gene*	*Origin of transgene*	*Trait*	*Reference*
Lycopersicon esculentum L.	*Cry1A (c)*	*B. thuringiensis*	Resistance to tobacco hornworm (*Manduca sexta*)	Fischhoff et al. 1987
Lycopersicon esculentum L.	*Bt (k)*	*B. thuringiensis*	Resistance to tobacco hornworm (*Manduca sexta*), tomato pinworm (*Keiferia, Lycopersicella, Walshingham*) and tomato fruit worm *H. zea* Boddie	Delannay et al. 1989
Solanum tuberosum	*CpTi*	Cowpea	Tolerance to tomato moth	Gatehouse et al. 1997
Solanum tuberosum	*GNA*	Snowdrop lectin	Tolerance to tomato moth	Gatehouse et al. 1996

The R genes can be divided into six clases of proteins (Takken and Joosten, 2000). Three of these classes contain leucine-rich repeats (LRRs), of which the class of nucleotide-binding site proteins is the most abundant. The other two classes of LRR proteins involve the LRR-transmembrane anchored (LRR-TM) proteins providing resistance to fungi and nematodes and the LRR-TM-kinase proteins providing bacterial resistance. Members of the fourth class of R genes are protein kinases, which, in the case of Pto, confer resistance to *Pseudomonas syringae* pv. *tomato* carrying AvrPto. These genes also lead to fungal disease resistance. RPW8, a small putative membrane protein that confers powdery mildew resistance in *Arabidopsis*, represents the fifth class of R genes. Finally, the sixth class of R genes are the tomato verticillium wilt Ve resistance genes (Kawchuk et al. 2001). The two closely linked Ve1 and Ve2 genes, whose products might recognize different ligands, encode TM-surface glycoproteins having an extracellular LRR domain, endocytosis-like signals and leucine zipper or Pro-Glu-Ser-Thr sequences. The leucine zipper can facilitate dimerization of proteins through the formation of coiled-coil structures, while Pro-Glu-Ser-Thr sequences are often involved in internalization and protein degradation.

Field Effects

Parthenocarpy, development of the fruit in the absence of fertilization, has been employed to evaluate the equivalence of genetically modified (GM) and non-GM fruit and to evaluate the advantages of parthenocarpy produced by genetic engineering compared to traditional methods. In this work, we present an analysis of parthenocarpic tomato fruit obtained from field-grown GM plants to address some aspects of the equivalence of GM fruit. The trait of parthenocarpy is particularly important for crop plants whose commercial product is their fruit. During flowering, adverse environmental conditions may either prevent or reduce pollination and fertilization, decreasing fruit yield and quality. Moreover, parthenocarpic fruits are seedless, and seedlessness is highly valued by consumers in some fruit (e.g., table grape, citrus, eggplant, cucumber).

Thus, transgenic parthenocarpic plants have been obtained for horticultural crops (Carmi et al. 2003). In particular, the chimeric gene *DefH9-iaaM* has been used to drive parthenocarpic fruit development in several species belonging to different plant families (Ficcadenti et al. 1999). The *DefH9-iaaM* transgene promotes the synthesis of auxin (indole acetic acid) specifically in the placenta, ovules and tissues derived therefrom.

Rotino et al. (2005) have cultivated parthenocarpic tomato lines transgenic for the *DefH9-RI-iaaM* gene under open field conditions to address some aspects of the equivalence of GM fruit in comparison to controls (non-GM).

Under open field cultivation conditions, two tomato lines (UC 82) transgenic for the *DefH9-RI-iaaM* gene produced parthenocarpic fruits. *DefH9-RI-iaaM* fruits were either seedless or contained very few seeds. The quality of GM fruits, with the exception of a higher β-carotene level, was not different physically (colour, firmness, dry matter, °Brix, pH) or chemically (titratable acidity, organic acids, lycopene, tomatine, total polyphenols and antioxidant capacity) from that of fruits from the control line. Highly significant differences in quality traits exist between the tomato F1 commercial hybrid Allflesh and the three UC 82 genotypes tested, regardless of whether or not they are GM. Total yield per plant did not differ between GM and parental line UC 82. Fruit number was higher in GM lines, and GM fruit weight was lower.

DETECTION METHODS FOR GM TOMATO PRODUCTS

The food industry has long suffered loss of income due to extensive fraud. Since adulteration is always ahead of analysis and detection methods, there is an urgent need to invent faster, more accurate and cheaper technology and to get more reliable results. Biotechnological advances have paved the way for effective and reliable authenticity testing. The latter can be focused either on determination of variety and geographical origin or traceability testing of a wide range of food products, products of agricultural and animal origin and package counterfeiting. Apart from the widely employed DNA methods, other instrumental methods include Site-Specific Natural Isotope Fractionation–Nuclear Magnetic Resonance, and more updated technology micro-satellite marker, restricted fragment length polymorphism and Single Strand Conformation Polymorphism. Expensive fish and meat species in conjunction with orange juice and olive oil attracted the researchers' interest because of their adulteration with other species of lower quality and price. Another important issue is the detection of geographical origin of agricultural produce of added value and special properties such as rice, olive oil and wine. Post-commercialization tracking of GM crops requires three types of tests (Auer 2003):

1. A rapid detection assay to determine whether a GM crop is present in a sample of raw ingredients or food products.
2. An identification assay to determine which GM crop is present.
3. Quantitative methods to measure the amount of GM material in the sample.

The first stage can be accomplished by qualitative methods (presence or absence of transgene), whereas the third stage uses semi-quantitative methods (above or below a threshold level) or quantitative methods (weight/weight percentage or genome/genome ratio). Currently, the two most important approaches are immunological assays using antibodies that bind to the novel proteins and methods based on polymerase chain reaction (PCR) using primers that recognize DNA sequences unique to the GM crop. The two most common immunological assays are enzyme-linked immunosorbent assays (ELISA) and immunochromatographic assays (lateral flow strip tests). ELISA can produce qualitative, semi-quantitative and quantitative results in 1-4 hr of laboratory time (http://www.ers.usda.gov/publications/aib762/ and http://anrcatalog.ucdavis.edu/pdf/8077.pdf).

Nested PCR is used to confirm the PCR product, and it allows discrimination between specific and non-specific amplification signals. Therefore, the PCR product is re-amplified using another primer pair, located in the inner region of the original target sequence (Anklam et al. 2002). It increases PCR sensitivity, allowing low levels of GM organisms to be detected (Zimmermann et al. 1998b). In order to detect the presence of Roundup Ready soybean, a nested PCR method was applied to commercially available soy flour, infant formula containing soy protein isolate and soymilk powder samples. Greiner et al. (2005) analysed soy flour, infant foods and soy protein isolates.

Electrochemiluminescence (ECL), wherein light-emitting species are produced by reactions between electrogenerated intermediates, has become an important and powerful analytical tool in recent years. An ECL reaction using tri-propylamine (TPA) and tris (2,2-bipyridyl) ruthenium (II) (TBR) has been demonstrated to be a highly sensitive detection method for quantifying amplified DNA (Blackburn et al. 1991). TPA and TBR are oxidized at approximately the same voltage on the anode surface. After deprotonation, TPA chemically reacts with TBR and results in an electron transfer. The resulting TBR molecule relaxes to its ground state by emitting a photon. The TPA decomposes to dipropyl amine and is therefore consumed in this reaction. The TBR, on the other hand, is recycled. Since both reactants are produced at the anode, luminescence occurs there. Compared with other detection techniques, ECL has some advantages: no radioisotopes are used; detection limits are extremely low; the dynamic range for quantification extends over six orders of magnitude; the labels are extremely stable compared with those of most other chemiluminescence systems; and the measurement is simple and rapid, requiring only a few seconds. The ECL method is a chemiluminescent CL reaction of species generated electrochemically on an electrode surface. It is a highly efficient and accurate detection method. Liu et al. (2005) applied an ECL PCR combined with two types of nucleic acid probes hybridization to detect GM organisms. The presence of GM components in organisms was determined by detecting the cauliflower mosaic virus 35S (CaMV35S) promoter and nopaline synthase (NOS) terminator. The experimental results showed that the detection limit is 100 fmol of PCR products. The promoter and the terminator can be clearly detected in GM organisms. The method may provide a new means for the detection of GM organisms because of its simplicity and high efficiency. The instrument used was composed of an electrochemical reaction cell, a potentiostat, an ultra high sensitivity single photon counting module, a multi-function acquisition card, a computer and labview software. The electrochemical reaction cell contains a working electrode (platinum), a counter electrode (platinum) and a reference electrode ($Ag/AgCl_2$).

Peano et al. (2004) evaluated and compared four different methods for DNA extraction. To rank the different methods, the quality and quantity of DNA extracted from standards containing known percentages of GM material and from different food products were considered. The food products analysed, derived from both soybean and maize, were chosen on the basis of the mechanical, technological, and chemical treatment they had been subjected to during processing. Degree of DNA degradation at various stages of food production was evaluated through the amplification of different DNA fragments belonging to the endogenous genes of both maize and soybean.

From these results it is evident that the treatments of foodstuffs can affect DNA degradation level, and at the same time it is also evident that the method of extraction can greatly influence DNA degradation and/or yield and/or PCR amplification efficiency on degraded DNA. Among the analysed methods of extraction, the QIAamp DNA Stool Minikit gave good-quality DNA with a very low level of degradation from simple foodstuffs; the Nucleo Spin Food kit (Macherey-Nagel) proved to be the most efficient in recovering good-quality DNA with a low level of degradation from complex foodstuffs. Genomic DNA was extracted from Roundup Ready soybean and maize MON810 standard flours, according to four different methods, and quantified by real-time PCR with the aim of determining the influence of the extraction methods on DNA quantification through real-time PCR. It could be concluded that for each food matrix, the DNA extraction method that correlates best with performance of subsequent DNA analysis, such as real-time PCR testing, should be employed.

There are also in-process methods aimed at comparing the primal and GM plant. These techniques can be divided into targeted and non-targeted approaches. Targeted approaches monitor directly the consequence of novel gene product presence on the GM plant phenotype. Moreover, changes in chemical composition are detected. Non-targeted approaches consist of three basic levels: functional genomics, proteomics, metabolomics. Functional genomics contain methods such as mRNA fingerprinting and DNA microarray. Considering proteomics, the protein composition of original and GM plant is compared using methods such as two-dimensional electrophoresis (Gorg et al. 1999) and its modification or two-dimensional electrophoresis in connection with MALDI-TOF mass spectroscopy analysis (Andersen and Mann 2000). For testing two samples on the same gel, a method called difference gel electrophoresis (Unlu 1999) is used. The metabolomics level of analysis tries to identify and quantify the maximum amount of particular components. This involves separating methods such as gas chromatography, liquid chromatography, and high performance liquid chromatography combined with various detection methods such as nuclear magnetic resonance, Fourier transform impaired spectroscopy, mass spectroscopy, flame ionization detector (Celec et al. 2005).

The principle of direct detection of recombinant DNA in food by PCR is discussed following the three main steps: DNA extraction, PCR amplification, and verification of PCR products.

Suitable methods for genomic DNA isolation from homogeneous, heterogeneous, low DNA containing matrices (e.g., lecithin), gelatinizing material (e.g., starch), derivatives and finished products based on classical protocols and/or a combination with commercially available extraction kits are discussed. Various factors contribute to the degradation of DNA, such as hydrolysis due to prolonged heat treatment, nuclease activity and increased depurination and hydrolysis at low pH. The term "DNA quality" is defined as the degree of degradation of DNA (fragment size less than 400 bp in highly processed food) and by the presence or absence of potent inhibitors of the PCR; it is, therefore, a key criterion. In general, no DNA is detectable in highly heat-treated food products, hydrolysed plant proteins (e.g., soy sauce), purified lecithin, starch derivatives (e.g., maltodextrins, glucose syrup) and defined chemical substances such as refined soybean oil (Meyer 1999).

If the nucleotide sequence of a target gene or stretch of transgenic DNA is already known, specific primers can be synthesized and the segment of rDNA amplified. Detection limits are in the range 20 pg ± 10 ng target DNA and 0.0001 ± 1% mass fraction of GM organism. Amplification products are then separated by agarose gel electrophoresis and the expected fragment size estimated by comparison with a DNA molecular weight marker.

Several methods are used to verify PCR results and they vary in reliability, precision and cost. They include specific cleavage of the amplification products by restriction endonucleases or the more time-consuming, but also more specific, transfer of separated PCR products on to membranes (Southern Blot) followed by hybridization with a DNA probe specific for the target sequence. Alternatively, PCR products may be verified by direct sequencing. Nested PCR assays combine high specificity and sensitivity.

There are methods available for the screening of 35S-promoter, NOS terminator and other marker genes used in a wide range of GM organisms, the specific detection of approved products such as FlavrSavr tomatoes, Roundup Ready Soya, and Bt maize 176, and official validated methods for potatoes and genetically modified micro-organisms that have a model character. Methods to analyse new GM products are being validated by interlaboratory tests and new techniques are in development. However, these efforts may be hampered by the lack of availability of GM reference material as well as specific sequence information owned by the suppliers only.

The first detection method specifically developed for the identification of a commercialized genetically engineered plant was demonstrated for detection of the FlavrSavr tomato (Meyer 1995a, b). Specific detection of the FlavrSavr tomato (Calgene), the first approved genetically engineered crop, was demonstrated by a PCR assay targeting the combination of the sequence of the CaMV35S promoter followed by the antisense gene from the polygalacturonase. In addition, the presence of the nptII (neomycin phosphotransferase II) marker gene was determined.

Regarding Zeneca's GM tomato puree, genetic modification of a processing cultivar of tomato was carried out with a short sense construct; 210 individual transgenics were produced and these formed the basis for further selection, breeding and development. All the development work has been rigorously reviewed in the United States and United Kingdom by the regulatory bodies. It was alleged that the development of this product was due to "contamination or a chance turnaround" (Bright and Schuch 1999).

Genetically modified tomatoes intended for processing (especially for tomato paste) were analysed (Zeneca 1996) for their levels of a-tomatine, solanine and chaconine in both the GM fresh fruit and paste samples (Zeneca 1996). The results showed that the glycoalkaloid levels in the modified tomato paste fall well within the range of glycoalkaloid levels of commercially available pastes, and the genetic modification has not altered the levels of the glycoalkaloids in the paste made from modified tomatoes.

New methods and techniques have been developed within the framework of the European Research Project "Development of Methods to Identify Foods Produced by Means of Genetic Engineering", DMIF-GEN (project no. SMT4-CT96-2072). In the scope of this project, DNA-extraction methods have been compared (Zimmermann et al. 1998a), new primers and probes have been defined, ring tests with tomato, processed maize and soybean are being performed, and a database has been set up to record detailed information about food containing GM components on the market, sequences, primers and detection methods (Schreiber 1997).

In China, GM tomato Huafan No 1 with a character of long shelf life was the first GM plant approved for commercialization in 1996. To meet the requirement of the GM tomatoes labelling policy that has been implemented in China since 2001, screening and construct-specific PCR detection methods for detecting the universal elements transformed into tomato, such as CaMV35s promoter and the NOS terminator of *A. tumefaciens*, and the specifically inserted heterologous DNA sequence between CaMV35s promoter and antisense ethylene-forming enzyme gene were set up. To make the detection methods normative, a novel single copy tomato gene *LAT52* was also used as an endogenous reference gene in the PCR detection systems (Yang et al. 2005). The limit of detection of screening and construct-specific detection methods for Huafan No 1 was 68 haploid genome copies in conventional PCR detection, and three copies in TaqMan real-time PCR detection. The limit of quantitation of screening quantitative PCR assays for Huafan No 1 was three copies and for construct-specific quantitative PCR it was 25 copies. Two samples with known Huafan No 1 tomato content were detected using the established conventional and real-time PCR systems, and these results also indicated that the established Huafan No 1 screening and construct-specific PCR detection systems were reliable, sensitive and accurate. In addition to the official methods for GM tomato detection in the field or in food products, research methods have also been employed to evaluate possible toxic effects of GM products on other organisms. These methods are used by companies or government laboratories to evaluate possible toxic effects of GM products before their official release for human consumption (Table 3).

Table 3 Representative toxicity studies carried out on rats and mice with GM tomato (adapted from Kuiper et al. 2001, ILSI 2004).

Vegetable	*Trait*	*Duration*	*Parameters*	*Reference*
Tomato	Cry1 Ab endotoxin (*B. thuringiensis* var. *kurstaki*)	91 d	Feed consumption Body weight Blood chemistry Organ weights Histopathology	Noteborn et al. 1995
Tomato	Antisense polygalacturonase (tomato)	28 d	Feed consumption Body weight Blood chemistry Organ weights Histopathology	Hattan 1996

ACKNOWLEDGEMENTS

The authors T.H. Varzakas and I.S. Arvanitoyannis would like to thank their institutions for the funding provided.

ABBREVIATIONS

Bt: *Bacillus thuringiensis*; ECL: Electrochemiluminescence; GM: genetically modified; LRRs: leucine-rich repeats; LRR-TM: LRR-transmembrane; NOS: nopaline synthase; PCR: polymerase chain reaction; TBR: tris (2,2-bipyridyl) ruthenium (II); TPA: tri-propylamine

REFERENCES

Andersen J.S. and M. Mann. 2000. Functional genomics by mass spectrometry. FEBS Lett. 480: 25-31.

Anklam, E. and F. Gadani, P. Heinze, H. Pijnenburg, and G. Van Den Eede. 2002. Analytical methods for detection and determination of genetically modified organisms in agricultural crops and plant-derived food products. Eur. Food Res. Technol. 214: 3-26.

Auer, C.A. 2003. Tracking genes from seed to supermarket: techniques and trends. Trends Plant Sci. 18(12): 591-597.

Babu, R.M. and A. Sajeena, K. Seetharaman, and M.S. Reddy. 2003. Advances in genetically engineered (transgenic) plants in pest management—an overview. Crop Prot. 22: 1071-1086.

Barg, R. and S. Shabtai, and Y. Salts. Transgenic tomato. pp. 212-233. *In*: Y.P.S. Bajaj. [ed.] 2001. Biotechnology in Agriculture and Forestry 47. Transgenic Crops II, Springer, Berlin, Germany.

Bates, G.W. 1995. Electroporation of plant protoplasts and tissues. Methods Cell. Biol. 50: 363-373.

Baum, K. and B. Groning, and I. Meier. 1997. Improved ballistic transient transformation conditions for tomato fruit allow identification of organ-specific contributions of I-box and G-box to the RBCS2 promoter activity. Plant J. 12: 463-469.

Blackburn, G.F. and H.P. Shah, J.H. Kenten, J. Leland, R.A. Kamin, J. Link, J. Peterman, M.J. Powell, A. Shah, and D.B. Talley. 1991. Electrochemiluminescence detection for development of immunoassays and DNA probe assays for clinical diagnostics. Clinical Chemistry 37: 1534-1539.

Bright, S. and W. Schuch. 1999. Making sense of GM tomatoes. Nature 400: 14.

Burbidge, A. and T Grieve, A. Jackson, A. Thompson, D. McCarthy, and I. Taylor. 1999. Characterization of the ABA-deficient tomato mutant notabilis and its relationship with maize Vp14. Plant J. 17: 427-431.

Carmi, N. and Y. Salts, B. Dedicova, S. Shabtai, and R. Barg. 2003. Induction of parthenocarpy in tomato via specific expression of the *rolb* gene in the ovary. Planta 217: 726-735.

Celec, P. and M. Kukučková, V. Renczésová, S. Natarajan, R. Pálffy, R. Gardlík, J. Hodosy, M. Behuliak, B. Vlková, G. Minárik, T. Szemes, S. Stuchlík, and J. Turňa. 2005. Biological and biomedical aspects of genetically modified food. Biomed. Pharmacother. 59: 531-540.

Christou, P. Rice transformation: bombardment. 1997. Plant Mol. Biol. 35: 197-203.

Dai, N. and M.A. German, T. Matsevitz, R. Hanael, D. Swartzberg, Y. Yeselson, M. Petreikov, A.A. Schaffer, and D. Granot. 2002. *LeFRK* 2, the gene encoding the major fructokinase in tomato fruits, is not required for starch biosynthesis in developing fruits. Plant Sci. 162: 423-430.

Dandekar, A.M. and H.J. Fisk. 2005. Plant transformation: *Agrobacterium*-mediated gene transfer. Methods Mol. Biol. 286: 35-46.

Delannay, X. and B.J. La Vallee, R.K. Proksch, R.L. Fuchs, S.R Sims, J.T. Greenplate, P.G. Morrone, R.B. Dodson, J.J. Augustine, J.G. Layton, and D.A. Fischhoff. 1989. Field performance of transgenic tomato plants expressing the *Bacillus thuringiensis* var. *kurstaki* insect control protein. BioTechnology 7: 1265-1269.

Elmer, S. and S.G. Rogers. 1990. Selection for wild type derivatives of tomato golden mosaic virus during systemic infection. Nucleic Acids Res. 17: 2391-2403.

Else, M.A. and M.B. Jackson. 1998. Transport of 1-aminocyclopropane-1-carboxylic acid (ACC) in the transpiration stream of tomato (*Lycopersicon esculentum*) in relation to foliar ethylene production and petiole epinasty. Aust. J. Plant Physiol. 25: 453-458.

Ficcadenti, N. and S. Sestili, T. Pandolfini, C. Cirillo, G.L. Rotino, and A. Spena. 1999. Genetic engineering of parthenocarpic fruit development in tomato. Mol. Breed. 5: 463-470.

Fischhoff, D.A. and K.S. Bowdish, F.J. Perlak, P.G. Marrone, S.H. McCormick, J.G. Niedermeyer, D.A. Dean, K. Kusano-Kretzmer, E.J. Mayer, D.E. Rochester, S.G. Rogers, and R.T. Fraley. 1987. Insect tolerant transgenic tomato plants. Bio/Technology 5: 807-813.

Fraser, P.D. and S. Romer, J.W. Kiano, C.A. Shipton, P.B. Mills, R. Drake, W. Schuch, and PM. Bramley. 2001. Elevation of carotenoids in tomato by genetic manipulation. J. Sci. Food Agric. 81: 822-827.

Gatehouse, A.M.R. and R.E. Down, K.S. Powell, N. Sauvion, Y. Rahbe, C.A. Newell, A. Merryweather, W.D.O. Hamilton, and J.A. Gatehouse. 1996. Transgenic potato plants with enhanced resistance to the peach-potato aphid *Myzus persicae*. Entomol. Exp. Appl. 79: 295-307.

Gatehouse, A.M.R., G.M. Davison, C.A. Newell, A. Merryweather, W.D.O. Hamilton, and E.P.J. Burgess. 1997. Transgenic potato plants with enhanced resistance to the tomato moth, *Lacanobia oleracea*: growth room trials. Mol. Breed. 3: 1-15.

Giovannoni, J.J. 2005. Transcriptome and selected metabolite analyses reveal multiple points of ethylene control during tomato fruit development. Plant Cell 17: 2954-2965.

Gorg, A. and C. Obermaier, G. Boguth, and W. Weiss. 1999. Recent developments in two-dimensional gel electrophoresis with immobilized pH gradients: wide pH gradients up to pH 12, longer separation distances and simplified procedures. Electrophoresis 20: 712-717.

Greiner, R. and U. Konietzny, and A.L.C.H. Villavicencio. 2005. Qualitative and quantitative detection of genetically modified maize and soy in processed foods sold commercially in Brazil by PCR-based methods. Food Control 16: 753-759.

Grichko, V.P. and B.R. Glick. 2001. Flooding tolerance of transgenic tomato plants expressing the bacterial enzyme ACC deaminase controlled by the 35*S*, *rolD* or PRB-1*b* promoter. Plant Physiol. Biochem. 39: 19-25.

Grison, R. and B. Grezes-Besset, M. Schneider, N. Lucante, L. Olsen, Leguay J.-J., and A. Toppan. 1996. Field tolerance to fungal pathogens of Brassica napus constutively expressing a chimeric chitinase gene. Nature Biotechnol. 14: 643–646.

Hamilton, A.J. and R.G. Fray, and D. Grierson. 1995. Sense and antisense inactivation of fruit ripening genes in tomato. Curr. Top. Microbiol. Immunol. 197: 77-89.

Hattan, D. OECD, ed. 1996. Evaluation of toxicological studies on Flavr Savr tomato. pp. 58-60. *In*: Food Safety Evaluation. Organization for Economic Co-operation and Development, Paris, France.

Hennin, C. and M. Hofte, and E. Diederichsen. Functional expression of Cf9 and Avr9 genes in *Brassica napus* induces enhanced resistance to *Leptosphaeria maculans*. 2001. Molecular Plant-Microbe Interactions 14: 1075-1085.

Hilbeck, A. 2001. Implications of transgenic, insecticidal plants for insect and plant biodiversity. Persp. Plant Ecol. Evol. Syst. 4(1): 43-61.

ILSI. 2004. Chapter 3: Safety Assessment of Nutritionally Improved Foods and Feeds Developed through the Application of Modern Biotechnology, Vol. 3, Comprehensive Reviews in Food Science and Food Safety, pp. 63-71.

Jones, D.A. and C.M. Thomas, K.E. Hammond-Kosack, P.J. Balint-Kurti, and J.D.G. Jones. 1994. Isolation of the tomato Cf-9 gene for resistance to *Cladosporium fulvum* by transposon tagging. Science 266: 789-793.

Joosten, M.H.A.J. and T.J. Cozijnsen, and P.J.G.M. De Wit. 1994. Hot resistance to a fungal tomato pathogen host by a single base-pair change in an avirulence gene. Nature 367: 384-386.

Joosten, M.H.A.J. and P.J.G.M. De Wit. 1999. The tomato-*Cladosporium fulvum* interaction: A versatile experimental system to study plant-pathogen interactions. Annu. Rev. Phytopathol. 37: 335-367.

Kawchuk, L.M. and J. Hachey, D.R. Lynch, F. Kulcsar, G. Van Rooijen, D.R. Waterer, A. Robertson, E. Kokko, R. Byers, R.J. Howard, R. Fischer, and D. Prufer. 2001. Tomato Ve disease resistance genes encode cell surface-like receptors. Proc. Natl. Acad. Sci. USA 98: 6511-6515.

Kuiper, H.A. and G.A. Kleter, H.P.J.M. Noteborn, and E.J. Kok. 2001. Assessment of the food safety issues related to genetically modified foods. Plant J. 27: 503-528.

Lauge, R., M.H.A.J. Joosten, G.F.J.M. Van den Ackerveken, H.W. Van den Broek, and P.J.G.M. De Wit. 1997. The in planta produced extracellular proteins ECP1 and ECP2 of *Cladosporium fulvum* are virulence factors. Mol. Plant-Microbe Interact. 10: 725-734.

Lauge, R. and P.H. Goodwin, P.J.G.M. De Wit, and M.H.A.J. Joosten. 2000. Specific HR-associated recognition of secreted proteins from *Cladosporium fulvum* occurs in both host and nonhost plants. Plant J. 23: 735-745.

Liu, J. and D. Da Xing, X. Shen, and D. Zhu. 2005. Electrochemiluminescence polymerase chain reaction detection of genetically modified organisms. Anal. Chim. Acta 537: 119-123.

Mandaokar, A.D. and R.K. Goyal, A. Shukla, S. Bisaria, R. Bhalla, V.S. Reddy, A. Chaurasia, R.P. Sharma, I. Altosaar, and P.A. Kumar. 2000. Transgenic tomato plants resistant to fruit borer (*Helicoverpa armigera* Hubner). Crop Prot. 19: 307-312.

Mehta, R.A. and T. Cassol, N. Li, N. Ali, A.K. Handa, and A.K. Mattoo. 2002. Engineered polyamine accumulation in tomato enhances phytonutrient content, juice quality and vine life. Nature Biotech. 20: 613-618.

Meyer, R. 1995a. Detection of genetically engineered plants by polymerase chain reaction PCR using the Flavr Savr tomato as an example Z. Lebensm. Unters. Forsch.. 201: 583-586.

Meyer, R. 1995b. Detection of genetically engineered food by the polymerase chain reaction (PCR) Mitt. Gebiete Lebensm. Hyg. 86: 648-656.

Meyer, R. 1999. Development and application of DNA analytical methods for the detection of GMOs in food. Food Control 10: 391-399.

Muir, S.R. and G.J. Collins, S. Robinson, S. Hughes, A. Bovy, C.H.R, De Vos, A.J. van Tunen, and M.E. Verhoeyen. 2001. Overexpression of petunia chalcone isomerase in tomato results in fruit containing increased levels of flavonols. Nat. Biotechnol. 19: 470-474.

Noteborn, H.P.J.M. and M.E. Bienenmann-Ploum, J.H.J. van den Berg, G.M. Alink, L. Zolla, A. Reynerts, M. Pensa and H.A. Kuiper. Safety assessment of the *Bacillus thuringiensis* insecticidal crystal protein CRY1A (b) expressed in transgenic tomatoes. pp. 134-147. *In:* K.H. Engel, G.R. Takeola and R. Teranishi. [eds.] 1995. Genetically Modified Foods. Safety Issues, ACS Symposium Series 605, Washington, DC.

Notz, R. and M. Maurhofer, H. Dubach, D. Haas, and G. Defago. 2002. Fusaric Acid-Producing Strains of Fusarium oxysporum Alter 2,4-Diacetylphloro-glucino Biosynthetic Gene Expression in Pseudomonas fluorescens CHA0 In Vitro and in the Rhizosphere of Wheat. Applied and Environmental Microbiology. 68(5): 2229-2235.

Peano, C. and M.C. Samson, L. Palmieri, M. Gulli, and N. Marmiroli. 2004. Qualitative and quantitative evaluation of the genomic DNA extracted from GMO and non-GMO foodstuffs with four different extraction methods. J. Agric. Food Chem. 52: 6962-6968.

Perlak, F.J. and T.B. Stone, Y.M. Muskopf, L.J. Petersen, G.B. Parker, S.A. McPherson, J. Wyman, S. Love, G. Reed, and D. Biever. 1993. Genetically improved potatoes: protection from damage by Colorado potato beetles. Plant Mol. Biol. 22: 313-321.

Punja, Z.K. Genetic engineering of plants to enhance resistance to fungal pathogens. pp. 207-258. *In:* Z.K. Punja. [ed.] 2004. Fungal Disease Resistance in Plant. Haworth Press, Inc., New York, USA.

Redenbaugh, K. and W. Hiuatt, B. Martineau, and D. Emlay. 1995. Determination of the safety of genetically engineered crops. ACS Symp Ser 605. Genetically Modified Foods. Safety Issues. American Chemical Society, Washington, DC, pp. 72-87.

Rotino, G.L. and N. Acciarri, E. Sabatini, G. Mennella, R. Lo Scalzo, A. Maestrelli, B. Molesini, T. Pandolfini, J. Scalzo, B. Mezzetti, and A. Spena. 2005. Open field trial of genetically modified parthenocarpic tomato: seedlessness and fruit quality. BMC Biotechnol. 5: 32-38.

Rosati, C. and R. Aquilani, S.R. Dharmapuri, P. Pallara, C. Marusic, R. Tavazza, F. Bouvier, B. Camara, and G.I. Giuliano. Metabolic engineering of beta-carotene and lycopene content in tomato fruit. 2000. The Plant Journal 24(3): 413-420.

Sachetto-Martins, G. and L.D. Fernandes, D.B. Felix, and D.D. Oliveira. 1995. Preferential transcriptional activity of a glycine-rich protein gene from *Arabidopsis thaliana* in protoderm-derived cells, Intl. J. Plant Sci. 156: 460-470.

Sala, F. and M.M. Rigano, A. Barbante, B. Basso, A.M. Walmsley, and S. Castiglione. 2003. Vaccine antigen production in transgenic plants: strategies, gene constructs and perspectives. Vaccine 21: 803-808.

Sanders, P.R. and B. Sammons, W. Kaniewski, L. Haley, J. Layton, B.J. Lavallee, X. Delannay, and N.E. Tumer. 1992. Field resistance of transgenic tomatoes expressing the tobacco mosaic virus or tomato mosaic virus coat protein genes. Phytopathology 82: 683-690.

Schreiber, G.A. 1997. The European Commission Research Project: Development of methods to identify foods produced by means of genetic engineering. *In:* G.A. Schreiber and K.W. Bogl. [Eds.] Food Produced by Means of Genetic Engeneering, 2nd status report. Bundesinstitut fur gesundheitlichen Verbraucherschutz und Veterinarmedizin, BgVV, Berlin, 01/1997.

Siddiqui, I.A. and S.S. Shaukat. 2003. Suppression of root-knot disease by *Pseudomonas fluorescens* CHA0 in tomato: importance of bacterial secondary metabolite, 2,4-diacetylpholoroglucinol. Soil Biol. Biochem. 35: 1615-1623.

Sitrit, Y. and A.B. Bennett. 1998. Regulation of tomato fruit polygalacturonase mRNA accumulation by ethylene: A re-examination. Plant Physiol 116: 1145-1150.

Sugita, K. and E. Matsunaga, T. Kasahara, and H. Ebinuma. 2000. Transgene stacking in plants in the absence of sexual crossing. Mol. Breed. 6: 529-536.

Suzuki, T. and P.J. Higgins, and D.R. Crawford. 2000. Control selection for RNA quantitation. Biotechniques 29: 332-337.

Takken, F.L. and R. Luderer, S.H. Gabriels, N. Westerink, R. Lu, P.J.G.M. De Wilt, and M.H.A.J. Joosten. 2000. A functional cloning strategy, based on a binary PVX-expression vector to isolate HR-inducing cDNAs of plant pathogens. Plant J. 24: 275-283.

Terzaghi, W.B. and A.R. Cashmore. 1997. Plant cell transfection by electroporation. Methods Mol. Biol. 62: 453-462.

Thomas, C.M. and D.A. Jones, M. Parniske, K. Harrison, P.J. Balint-Kurti, K. Hatzixanthanis, and J.D.G. Jones. 1997. Characterization of the tomato Cf-4 gene for resistance to *Cladosporium fulvum* identifies sequences that determine recognitional specificity in Cf-4 and Cf-9. Plant Cell 9: 2209-2224.

Thompson, A. and A.C. Jackson, R.C. Symonds, B.J. Mulholland, A.R. Dadswell, and P.S. Blake. 2000b. Ectopic expression of a tomato 9-cis-epoxycarotenoid dioxygenase gene causes over-production of abscisic acid. Plant J. 23: 363-374.

Unlu, M. 1999. Difference gel electrophoresis. Biochem. Soc. Trans. 27: 547-549.

USDA APHIS. 2001. (http://www.aphis.usda.gov/biotech/notday.html).

Van den Ackerveken, G.F.J.M. and J.A.L. Van Kan, and P.J.G.M. De Wit. 1992. Molecular analysis of the avirulence gene Avr9 of the fungal tomato pathogen *Cladosporium fulvum* fully supports the gene-for-gene hypothesis. Plant J. 2: 359-366.

Van den Ackerveken, G.F.J.M. and J.A.L. Van Kan, M.H.A.J. Joosten, J.M. Muisers, H.M. Verbakel, and P.J.G.M. De Wit. 1993. Characterization of two putative pathogenicity genes of the fungal tomato pathogen *Cladosporium fulvum*. Mol. Plant-Microbe Interact. 6: 210-215.

Van den Eede, G. and H. Aarts, H.-J. Buhk, G. Corthier, H.J. Flint, W. Hammes, B. Jacobsen, T. Midtvedt, J. van der Vossen, A. von Wright, W. Wackernagel, and A. Wilcks. 2004. The relevance of gene transfer to the safety of food and feed derived from genetically modified (GM) plants. Food Chem. Toxicol. 42: 1127-1156.

Wang, Chen-Kuen and Po-Yen Chen, Hsin-Mei Wang, and Kin-Ying To. 2006. Cosuppression of tobacco chalcone synthase using Petunia chalcone synthase construct results in white flowers. Botanical Studies 47: 71-82.

Westerink, N. and M.H.A.J. Joosten, and P.J.G.M. De Wit. Fungal virulence factors at the crossroads of disease susceptibility and resistance, Chapter 4, pp. 93-137. *In*: Z.K. Punja. [ed.] 2004. Fungal Disease Resistance in Plants: Biochemistry, Molecular Biology and Genetic Engineering, Food Products Press, New York, London, Oxford.

Yang, L. and H. Shen, A. Pan, J. Chen, C. Huang, and D. Zhang. 2005. Screening and construct-specific detection methods of transgenic Huafan No. 1 tomato by conventional and real-time PCR. J. Sci. Food Agric. 85: 2159-2166.

Yi, S.Y. and A.Q. Sun, Y. Sun, J.Y. Yang, C.M. Zhao, and J. Liu. 2006. Differential regulation of Lehsp23.8 in tomato plants: Analysis of a multiple stress-inducible promoter. Plant Sci. 171: 398-407.

Zeneca. 1996. Genetically Modified Processing Tomatoes. Notification dossier C/ES/96/01.

Zimmermann, A. and J. Luthy, and U. Pauli. 1998a. Quantitative and qualitative evaluation of nine diferent extraction methods for nucleic acids on soybean food samples. Z. Lebensm. Unters. Forsch. A 207: 81-90.

Zimmermann, A. and W. Hemmer, M. Liniger, J. Luthy, and U. Pauli. 1998b. A sensitive detection method for genetically modified MaisGardTM corn using a nested PCR-system. LWT-Food Sci. Technol. 31: 664-667.

27

Assaying Vitamins and Micronutrients in Tomato

A.I. Olives[1], M.A. Martin[1], B. del Castillo[1] and M.E. Torija[2]

[1] S.D. Quimica Analitica, Facultad de Farmacia, Universidad Complutense de Madrid, Pza. Ramon y Cajal s/n, 28040 – Madrid, Spain

[2] Dept. Nutricion y Bromatologia II: Bromatologia, Facultad de Farmacia, Universidad Complutense de Madrid, Pza. Ramon y Cajal s/n, 28040 – Madrid, Spain

ABSTRACT

Recent epidemiological studies suggest a positive correlation between the ingestion of diets rich in vegetables and fruits and a reduced incidence of chronic diseases, such as cancer, cardiovascular disease, Alzheimer's disease and cataracts. This beneficial effect is primarily attributed to the occurrence of vitamins, minerals and secondary phytochemicals, for example, carotenoids, anthocyanins, flavonoids, and other phenolic compounds that are widely distributed throughout the plant kingdom. Tomatoes are a source of antioxidants and they contribute to the daily intake of a significant amount of these compounds. One hundred grams of tomato can contribute 24-48%, 0.5-0.8% and 2.7-4.0% of the recommended daily intake of vitamin C, vitamin E and β-carotene, respectively. Tomato fruits are also considered the major source of lycopene. Mineral content is normally between 0.60 and 1.80%; the major elements are K, Na, Ca, Mg, P, Fe and Mn.

The main problem of the analysis of micronutrients and bioactive compounds in tomatoes is the selection of an adequate pre-treatment and clean-up method. In the case of fat-soluble vitamins and carotenoids, the main analytical difficulty is the isolation of these compounds from a large excess of

A list of abbreviations is given before the references.

physically hydrosoluble materials, and hence a great deal of effort has been made to improve extraction and clean-up procedures. Liquid-liquid extraction is frequently used. An attractive alternative to alleviate the risk of isomerization and oxidation of these compounds is supercritical fluid extraction.

There are several methods for measuring total antioxidant activity, and in general all of them imply the generation of radical species and the subsequent evaluation of the presence of antioxidants by the disappearance of these free radicals.

Although there are also official methods based on spectrometric quantification, liquid chromatography has proved to be one of the most useful techniques for the analysis of vitamins. This is a consequence of the versatility in column technology and detection methods. Reverse phase high performance liquid chromatography is the method most widely employed for analysis of carotenoids because of their hydrophobic character. C-18 columns, which are less selective, are preferable for less detailed analysis, while C-30 columns in combination with gradient elution are suitable for the separation of carotenoids and their different isomers with a wide polarity range. High performance liquid chromatography coupled with mass spectrometry improves the sensitivity and the selectivity of the analyses of micronutrients and bioactive compounds, especially with the introduction of modern ionization systems such as atmospheric pressure chemical ionization.

For analysis of minerals, classical methods (Kjeldahl method for total nitrogen and colorimetric method for phosphate analyses) are still routinely employed. On the other hand, when several elements need to be analysed in a plant over a large range of concentration, the analytical tool of choice is usually inductively coupled plasma-mass spectrometry (ICP-MS), followed by ICP-atomic emission spectrometry (ICP-AES) and atomic absorption spectrometry. ICP-AES allows simultaneous measurement of metals together with some non-metals, such as P or B. Thus, this technique has rapidly replaced colorimetric procedures for P and B determinations.

This chapter also describes the different analytical methods that can be applied for the determination of micronutrients and bioactive compounds in tomato fruits and tomato products taking into account trends of the future in analytical methods.

INTRODUCTION

It is well known that a greater intake of fruits and vegetables can help prevent heart disease and mortality. Plant foods, such as fruits and vegetables, contain many components that are beneficial to human health and regular consumption of fruits and vegetables is associated with reduced risks of cancer, cardiovascular disease, stroke, Alzheimer's disease, cataracts, and some of the functional declines associated with aging. Their consumption has been reported as one of the safest and most effective strategies in the prevention of cardiovascular disease, and it is worth stressing that it is a more effective strategy than is treatment of chronic diseases (Greenwald et al. 2001, Liu 2003).

Epidemiological studies suggest a positive correlation between diets rich in vegetables and fruits and a reduced incidence of chronic diseases (Lasheras et al. 2000). This beneficial effect is primarily attributed to the occurrence of vitamins, minerals and secondary phytochemicals such as carotenoids, anthocyanins, flavonoids, and other phenolic compounds that are widely distributed throughout the plant kingdom.

Vegetables form an important part of the diet, supplying appreciable quantities of micronutrients in a relatively low-fat energy source food. Fruits have long been valued as foodstuffs owing to their pleasant flavour and aroma and their attractive appearance (Fig. 1).

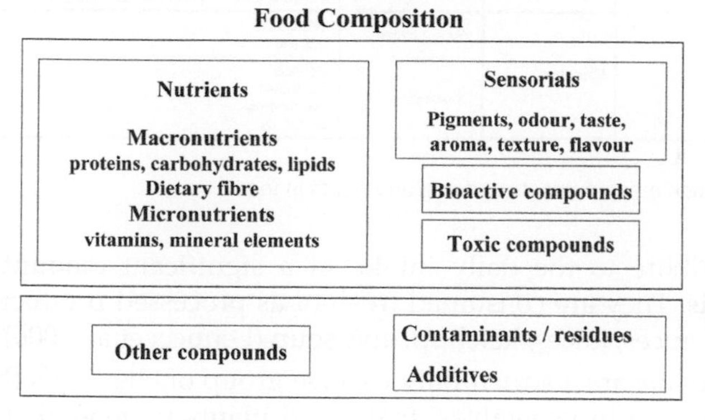

Fig. 1 General composition of foods.

The major component of fresh vegetables and fruits is water, with contents in the range of 80-95% in the edible portion. The nutrient composition of different vegetables and fruits varies considerably and can also be influenced by agronomic factors (Hallman and Renbialkowska 2007, Toor et al. 2006), storage, and processing or preservation techniques.

Vegetables are important sources of vitamin C (ascorbic acid), β-carotene (a vitamin A precursor), certain vitamins B, vitamin E, and minerals, including selenium, all of which have antioxidant potential. Fruits are an important low-fat energy source containing varying proportions of sugars and starches as well as being good sources of dietary fibre, mineral salts and vitamins, especially vitamin C (Fig. 2) (Arnao et al. 2001, Greenwald et al. 2001, Luthria et al. 2006, Pither 1995).

A large number of fresh fruits and vegetables are primary sources of bioactive antioxidants. Tomatoes are a major source of antioxidants and

Centesimal Composition					
	Water				
	Macronutrients	Proteins	Nitrogen Aminoacids Biogenic amines		
		Lipids	Total	Fatty acids Phospholipids Triglycerides	*Cis/trans*, PUFA, Essential ...
				Unsaponificable	Sterols, Triterpenes, Alcohols, Hydrocarbons, Antioxidants ...
		Carbohydrates	Sugars Starches Dextrins Dietary fibre		
	Ash				
	Micronutrients	Inorganics	Nonmetals	Halogens Sulphate Phosphate NO_2^- NO_3^-	
			Metals and metalloids	Nutritional	Ca, Co, Cr, Cu, Fe, K, Li, Mg, Mn, Na, Ni, Se, Zn ...
				Contaminants	As, Cd, Hg, Pb, Se ...
		Vitamins	Water-soluble	Vitamin C B group	
			Oil-soluble	Retinols Carotenes Tocopherols D_2 and D_3	

Fig. 2 Common macronutrients and micronutrients in foods.

they contribute to the daily intake of a significant amount of these compounds. They are consumed fresh or as processed products: canned tomatoes, sauces, juices, ketchup, and soup (Lenucci et al. 2006).

Antioxidants are a structurally diverse group of chemicals that can be naturally found in vegetables, fruits and plants in general. They have disease-fighting properties that protect cells from damage by substances known as free radicals. Antioxidants work by rendering ineffective harmful free radicals that are formed when our cells burn oxygen for energy. They may also help keep the immune system healthy and reduce the risk of some forms of cancers and other diseases. Antioxidant compounds found in fruits and vegetables may influence the risk of cardiovascular diseases by preventing the oxidation of cholesterol in arteries (Voutilainen et al. 2006).

BIOLOGICAL RELEVANCE OF MICRONUTRIENTS AND BIOACTIVE COMPOUNDS IN TOMATO

Carotenoids

Nowadays, the major interest of carotenoids is due not only to their provitamin A activity but also to their antioxidant action by scavenging oxygen radicals and reducing oxidative stress in the organism. Besides, they seem to help reduce the risk of some diseases, such as cancer or cardiovascular diseases or immunity system dysfunctions (Cantuti-Castelvetri et al. 2000, Clinton 1998, Levy and Sharoni 2004, Livny et al. 2002, Rao and Agarwal 2000, Rao and Honglei 2002). Tomato fruits and related products are important sources of lycopene and β-carotene. One hundred grams of tomato can contribute 24-48%, 0.5-0.8% and 2.7-4.0% of the recommended daily intake of vitamin C, vitamin E and β-carotene, respectively (Abushita et al. 1997).

The major importance of some of the carotenoids lies in their relationship to vitamin A. A molecule of orange β-carotene is converted into two molecules of colourless vitamin A in the animal body. Other carotenoids such as α-carotene, γ-carotene, and cryptoxanthin also are precursors of vitamin A, but because of minor differences in chemical structure one molecule of each of these compounds yields only one molecule of vitamin A (Oliver et al. 1998).

Lycopene is a carotenoid present in tomatoes, processed tomato products and other fruits. It is one of the most potent antioxidants among dietary carotenoids. Dietary intake of tomatoes and tomato products containing lycopene has been shown to be associated with a decreased risk of chronic diseases, such as cancer and cardiovascular disease. Serum and tissue lycopene levels have been found to be inversely related to the incidence of several types of cancer, including breast cancer and prostate cancer. Although the antioxidant properties of lycopene are thought to be primarily responsible for its beneficial effects, evidence is accumulating to suggest that other mechanisms may also be involved (Rao and Agarwal 2000, Rao and Honglei 2002).

Although lycopene has no provitamin A activity, it is now considered an important fruit component because of its ability to act as an antioxidant and to quench reactive oxygen species *in vitro*. Tomato fruits are considered to be important contributors of carotenoids to the human diet and the major source of lycopene.

The combination with other compounds of the diet has a synergic effect in this regard (Amir et al. 1999). This fact makes advisable the consumption of vegetables with high lycopene content rather than supplementation of diet with synthetic lycopene.

Lycopene is fairly stable after storage and cooking. In addition, heat processing such as cooking, required for the preparation of tomato sauces, is recommended because it increases the bioavailability of lycopene in the human body (Halliwell et al. 1995).

Agarwal and Rao (2000) have reported the lycopene content in tomato fruits and different tomato-derived products and these values are shown in Table 1.

Table 1 Daily intake of lycopene from tomato fruit and tomato products.

Product	*Serving size*	*Total daily lycopene intake (%)*
Tomato fruit	200 g	50.5
Tomato puree	60 mL	4.1
Tomato sauce	227 mL	6.0
Pizza sauce	60 mL	2.6
Tomato ketchup	15 mL	2.1
Tomato juice	250 mL	8.7

Data from Agarwal and Rao (2000).

Vitamins

Vitamins (hydrophilic and lipophilic), phenolic compounds, phytosterols, flavonoids and other organic micronutrients are naturally occurring plant substances. They are present in tomato and other foods of plant origin with significant biological relevance. Thus vitamins, in small amounts, are essential substances for normal metabolism of organisms and the other minor components and protect biological systems against potentially harmful effects. Epidemiological studies based on preclinical data show that the presence of foods with the above micronutrients in diets contributes to cancer prevention (Greenwald et al. 2001) and the alleviation of neurodegenerative diseases (Gilgun-Sherki et al. 2001) that appear as consequence of oxidative stress. The increased bioavailability of flavonoids and other compounds present in tomato puree (Simonetti et al. 2005) has positive consequences for human health.

Minerals

Minerals are present as salts of organic or inorganic acids or as complex organic combinations (e.g,. chlorophyll, lecithin); they are in many cases dissolved in cellular juice. Vegetables are richer in minerals than fruits. Mineral content is normally between 0.60 and 1.80% and more than 60 elements are present; the major elements are K, Na, Ca, Mg, Fe, Mn, Al, P, Cl and S (Belitz et al. 2004).

Among the vegetables that are especially rich in minerals are spinach, carrot, cabbage and tomato. Important quantities of potassium and absence of sodium chloride give a high dietetic value to fruits and to their processed products. Phosphorus is supplied mainly by vegetables.

Vegetables usually contain more calcium than fruits; green beans, cabbage, onions and beans contain more than 0.1% calcium. The calcium/phosphorus ratio is essential for calcium fixation in the human body; this value is considered normal at 0.7 for adults and at 1.0 for children (Belitz et al. 2004).

Some metals and non-metals (i.e., sodium, potassium, calcium, chloride, phosphorus) possess biological relevance because they play a role in maintaining the hydrosaline balance. Other minor elements (e.g., iron, copper, zinc) are part of prostetic groups in enzymes. Trace levels of these metals are essential in the diet, but at higher concentrations toxic properties become apparent. In light of the different range of concentration of these elements in food samples, the methodology and analytical technique should be selected with respect to the appropriate sensitivity required (Sumar and Coultate 1995).

The mean contents of vitamins, carotenoids and minerals most often found in tomato and tomato products are shown in Table 2.

Table 2 Comparative data of the micronutrient composition in tomato fruit and tomato products (values in mg/ 100 g tomato).

Composition	*Data from Souci et al. 1996 (tomato fruit)*	*Data from Mataix-Verdú et al. 1995 (tomato fruit)*	*Data from Mataix-Verdú et al. 1995 (tomato puree)*	*Data from Belitz et al. 2004 (tomato fruit)*	*Data from Salles et al. 2003 (tomato juice)**
Mineral components					
Sodium	6	6	240	6.3	12.6
Potassium	295	245	1150	297	3901
Magnesium	20	11	48	20	92.3
Calcium	14	10	48	14	9.3
Iron	0.500	0.600	1.600	0.5	–
Phosphate	25	25	94	26	746
Chloride	60	–	–	30	260
Organic components					
Carotenoids	0.820 g	0.036 (1)	0.217 (1)	0.600 (2)	
Vitamin E	0.800	1	5.370	0.800	
Vitamin K	0.008	–	–	–	
Vitamin C	25	16	38	23	
Vitamin B1	0.055	0.050	0.220	0.060	
Vitamin B2	0.035	0.030	0.120	0.040	
Vitamin B6	0.100	0.090	0.440	–	
Nicotinamide	0.530	0.080 (3)	4.100 (3)	0.500 (4)	
Panthotenic acid	0.310	–	–	–	
Folic acid	0.040	0.025	0.054	0.020	

* Values in mg/L.

(1) This amount corresponds to equivalent of retinol (vitamin A).

(2) This amount corresponds to equivalent of β-carotene.

(3) This amount corresponds to equivalent of niacin.

(4) This amount corresponds to equivalent of nicotinic acid.

In the following sections, the physical and chemical characteristics of the main micronutrients (vitamin C, tocopherols, vitamins of group B and minerals) and bioactive compounds (carotenoids) of the tomato, as well the different methods that can be applied for their determination in tomato fruits and tomato products are discussed.

ANTIOXIDANTS

Antioxidant activity (AOA) is related to the presence of vitamins A, C and E, carotenoids, flavonoids, and other simple phenolic compounds. It has been found in different amounts in cereals, fruits and vegetables. There are several methods for measuring total AOA; in general all of them imply the generation of radical species and subsequent evaluation of the presence of antioxidants by the disappearance of these free radicals (Arnao et al. 1999, 2001). The use of pre-treatment methods specific for lipid-soluble and water-soluble vitamins is necessary to ensure their determination in both extracts following different methods. A patented method for AOA determination based on the enzymatic system ABTS/ H_2O_2/HRP (2,2′-azino-bis-(3-ethyl benzothiazoline-6-sulphonic acid)/ H_2O_2/horseradish peroxidase) is easy, accurate and rapid, besides being free from interference due to peroxidase activity in the samples. The reactive cation $ABTS^{*+}$ is generated enzymatically from ABTS and then the sample is added to the reaction medium and the decrease in the absorbance due to $ABTS^{*+}$ can be related to the antioxidant concentration. This method has been successfully applied to hydrophilic (Cano et al. 1998) and lipophilic samples (Samaniego et al. 2007).

Other methods for evaluation of antioxidant levels in tomato employ the measurement of reducing power by means of the DPPH (2,2′-diphenyl-1-picryhydrazyl) scavenging ability. In the different tomato samples studied (fresh, freeze-dried and hot-air-dried tomatoes), the AOA found was the same as that observed using butylated hydroxyanisole as a reference antioxidant (Chang et al. 2006). This study reveals that the amounts of the different antioxidants vary depending on the drying process employed and the antioxidant determined. Thus, in hot-air-dried tomatoes the amounts of ascorbic acid (AA) were higher than the amounts of total flavonoids and also in the same type of sample the amounts of total phenolic compounds found were higher than lycopene. Hanson et al. (2004) suggest the evaluation of AOA by means of anti-radical power and inhibition of lipid peroxidation. Using these methods, the total phenolic compounds determined in tomato are principally responsible for the antioxidant behaviour of tomato compared to other compounds with antioxidant effect (lycopene, β-carotene and ascorbic acid). No significant differences were observed in the flavonoids and total phenolic content in different samples of tomato (cherry and high pigment) using the FRAP (ferric-reducing ability of plasma) reagent (1 mM 2,4,6-tripyridyl-1,3,5-triazine and 20 mM ferric chloride in 0.25 M sodium acetate buffer pH = 3.6) (Lenucci et al. 2006). The total phenolic compounds present in tomato were evaluated using the classical colorimetric method by Folin-Ciocalteu, but the individual phenolic acids were separated and examined by reverse phase (C_{18})-high performance liquid chromatography (HPLC) with diode array detection (DAD)-UV detection. The tomato samples were hydrolysed in alkaline media and the extracts were analysed in the chromatographic system. Caffeic acid was predominant among the phenolic compounds (Luthria et al. 2006).

CAROTENOIDS

Carotenoids are polyene hydrocarbons biosynthesized from eight isoprene units (tetraterpenes) and have a 40-C skeleton. The cyclization reaction can occur at one or both end groups.

They are highly soluble in apolar solvents but not in water. Hence, they are known as "lipochromes". Carotenoids are readily extracted from plant sources with ether, benzene or petroleum ether as well as ethanol or acetone (Belitz et al. 2004).

Regarding the presence of oxygen atoms in their chemical structure, carotenoids can be divided into two main groups: *carotenes* (pure polyene hydrocarbons, only composed of carbon and hydrogen atoms) and *xanthophylls* (oxygenated hydrocarbon derivatives that contain at least one oxygen function such as hydroxy, keto, epoxy or carboxylic acid groups) (Fig. 3A).

Their main structural characteristic is a conjugated double bond system, which influences their chemical, biochemical and physical properties. Because of the presence of this conjugated double bond system, carotenoids are responsible for the intense yellow, orange or red colour of some fruits and vegetables. For example, it is well known that the typically red colour of tomato is due to lycopene and the yellow colour in tomato cultivars is due to the mixture of lycopene precursors (such as phytoene or phytofluene) and β-carotene. Therefore, carotenoids are directly related to the perception of quality of fruits and vegetables, as colour does influence consumer preferences (Clydesdale 1993, Meléndez-Martínez et al. 2004).

Carotenoids exhibit a peculiar visible spectrum characterized by the presence of three maxima between 400 and 500 nm, whose wavelengths depend on the number of conjugated double bonds, the presence of methyl groups in the rings (hypsochromic shift), or oxo groups (bathochromic

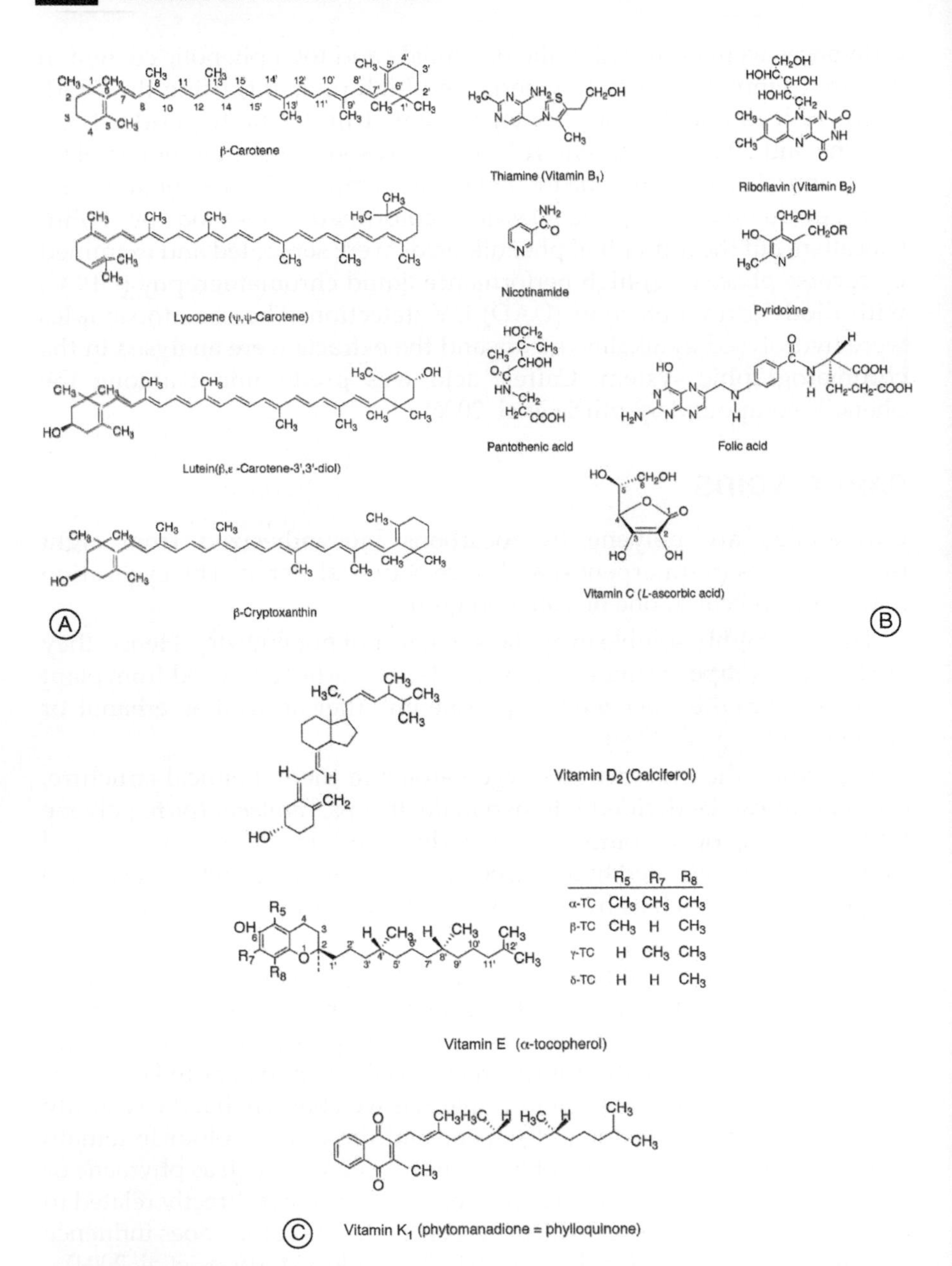

Fig. 3 Chemical structures of vitamins and carotenoids present in tomato. (A) Carotenoids. (B) Water-soluble vitamins. (C) Fat-soluble vitamins.

shift). Besides, the *cis-trans* isomerization produces a slight shift in the absorption bands with a new minor shoulder on the side of the shorter wavelength (Belitz et al. 2004, Britton 1992, 1995). The position of the absorption maxima and the shape of the spectrum are also strongly influenced by the environment of the pigment (e.g., solvent, temperature) (Schoefs 2002). In Table 3, the absorption maximum wavelengths and molar absorptivities of the main carotenoids found in tomato are summarized (Lin and Chen 2003, Rao et al. 1998, Scott 2001).

Carotenoids are highly sensitive to oxygen, light, heat and active surfaces. Therefore, determination of carotenoids must be carefully undertaken in order to avoid degradation, formation of stereoisomers, structural rearrangement or other physicochemical reactions. This is particularly important in the case of standard carotenoid solutions, especially lycopene, as is discussed below.

Table 3 Spectral characteristics of micronutrients found in tomato.

Compound	*λ (nm)*	*Log ε*	*Solvent*
Vitamins[a]			
α-Tocopherol	292	2.549	MeOH
β-Tocopherol	296	2.553	MeOH
γ-Tocopherol	298	2.554	MeOH
δ-Tocopherol	298	2.549	MeOH
Ascorbic acid	244	3.980	Water
Thiamine	235 (pH = 8)	4.100	Water
	268	3.900	Water
Riboflavin	267	4.510	Water
	373	4.020	Water
	447	4.090	Water
Niacin	255	3.400	EtOH
	262	3.400	EtOH
Calciferol	265	4.270	EtOH
Vitamin K	263	4.200	MeOH
	270	4.220	MeOH
	328	3.500	MeOH
Carotenoids[b]			
Trans-β-Carotene	425, 450, 478	5.143	Hexane
β-Cryptoxanthin	428, 450, 478	5.134	Hexane
Lutein	421, 445, 474	5.161	EtOH
Trans-lycopene	444, 470, 502	5.267	Hexane

[a]Lide 1992.
[b]Scott 2001.

In the literature there are several methods described for determining carotenoids in tomato and related products, most of them based on solvent extraction followed by HPLC separation (Cserháti and Forgács 2001, Rodríguez-Bernaldo de Quirós and Costa 2006, Shoefs 2002). The instability of carotenoids, especially lycopene, during the processes of extraction, handling, and elimination of organic solvents makes the preparation of a sample for analysis an extremely demanding task, often requiring successive and complex procedures to ensure that all the carotenoids are extracted. Therefore, the development of new methods that allow the analysis of these compounds without preliminary preparation is of relevance.

In general, the following steps can be distinguished in the analysis of carotenoids:

- sampling,
- homogenization, extraction and clean-up,
- preparation of standards, and
- quantification.

Sampling

Sample preparation consists mainly of quartering, cutting and mixing followed by grinding or homogenizing in a blender or food processor. If the determination is not carried out immediately, the homogenized and freeze-dried samples can be stored for 3-6 mon at -20°C until analysis (Konings and Roomans 1997). If the samples are not freeze-dried, they can be stored in an air-tight bottle under nitrogen at -20°C for up to 3 d prior to analysis (Scott 2001).

Extraction and Clean-up

Extraction and clean-up are necessary when UV-vis spectrophotometry or chromatographic techniques are employed. For the extraction of carotenoids from the tomato or tomato products, different systems can be used, such as liquid-liquid extraction, solid phase extraction (SPE) or supercritical fluid extraction (SFE). Microwave-assisted extraction has been reported for the extraction of carotenoids from paprika (*Capsicum annuum*) but not from tomato (Cserháti and Forgács 2001).

Liquid-liquid Extraction

In order to prevent carotenoid losses, the extraction must be carried out very quickly, avoiding exposure to light, oxygen, high temperatures and pro-oxidant metals (iron or copper). Antioxidants, such as butylated hydroxytoluene (BHT) (0.05-2.5%), are often added to the extracting solvent mixture (Davis et al. 2003, Lenucci et al. 2006, Kozukue and Friedman 2003, Rao et al. 1998, Sadler et al. 1990). It is also advisable to add magnesium or calcium carbonate to the extracting solvent mixture to neutralize trace levels of organic acids (Ferruzzi et al. 2001, Granado et al. 1992, Hart and Scott 1995, Konings and Roomans 1997, Kozukue and Friedman 2003).

Saponification with methanolic potassium hydroxide (20%) can be performed to enhance the analysis of lycopene by eliminating chlorophyll and lipid materials, which can interfere with its chromatographic elution and detection. Goula et al. (2006) have included saponification among the pre-treatment steps for the quantification of lycopene in tomato pulp. However, Granado et al. (1992) have reported a loss of total carotenoid content during saponification.

Instead of saponification, hydrolytic enzymes can be employed to help release the intracellular content, since in tomatoes β-carotene is found in the globulous chromoplast situated in the jelly part of the pericarp, while lycopene is in the chromoplast at the outer part of the pericarp. Enzyme-aided extraction of lycopene from whole tomatoes, tomato peel and tomato waste has been reported by Choudhari and Ananthanarayan (2007), obtaining remarkable increases in the yield of lycopene: in the case of cellulose-treated sample, 198% for whole tomato and 107% for tomato peel and, in the case of pectinase-treated sample, 224% and 206% for whole tomato and tomato peel, respectively.

Regarding the solvents employed in the extraction of carotenoids, there are several organic solvents as well as solvent mixtures cited in the literature. AOAC (1993) recommends methanol/tetrahydrofurane (MeOH/THF) (50:50 v/v) for extracting the carotenoids, because of their great solubility in THF. Moreover, Lin and Chen (2003) and Taungbodhitham et al. (1998) studied different solvent mixtures to extract lycopene and β-carotene from tomato juice and obtained the best extraction efficiency employing ethanol (EtOH)/hexane (4:3 v/v). Other authors have used hexane/acetone/EtOH in different proportions (Fish et al. 2002, Lenucci et al. 2006, Olives Barba et al. 2006, Sadler et al. 1990), hexane/acetone/MeOH (2:1:1 v/v/v) (Ferruzzi et al. 2001, Rao et al. 1998), acetone/petroleum ether (50:50 v/v) (Choudhari and Ananthanarayan 2007, Goula et al. 2006, Niizu and Rodríguez-Amaya 2005), acetone/diethyl ether (Kozukue and Friedman 2003), or chloroform/MeOH (3:1 v/v) (Abushita et al. 1997).

In the liquid-liquid extraction procedure, carotenoids are partitioned into the organic solvent and water. The organic phase is removed after the sample and the extracting solvent mixture have been magnetically stirred and centrifuged or, in other cases, filtered. These operations should be repeated until the pulp or matrix becomes colourless. The combined organic phases are evaporated to dryness under N_2 flow and under vacuum. Finally, the residue is dissolved in an appropriate solvent depending on the technique chosen for quantification (Ferruzzi et al. 2001, Granado et al. 1992, Lin and Chen 2003, Niizu and Rodríguez-Amaya 2005, Olives Barba et al. 2006). A scheme of this procedure is shown in Fig. 4a.

Solid Phase Extraction

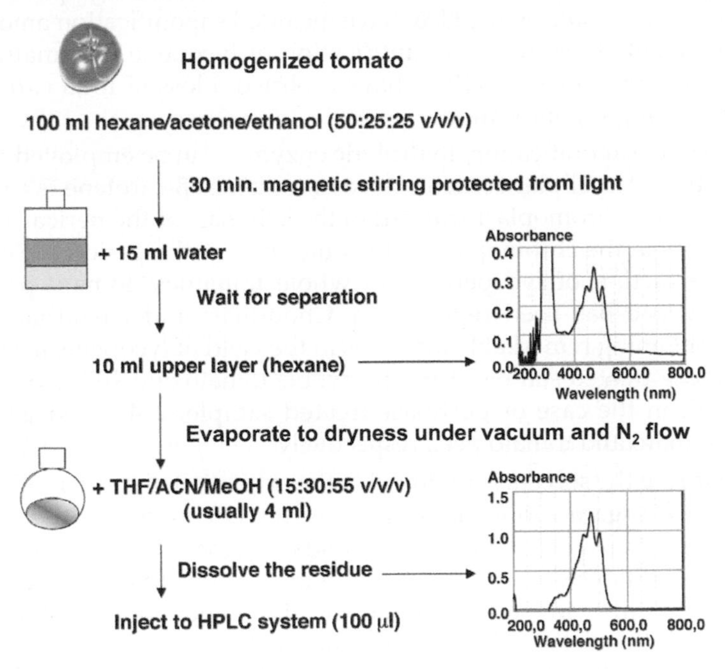

Fig. 4 Pre-treatment and clean-up procedures for extraction of carotenoids.

Solid phase extraction has been used by Baranska et al. (2006) for determining lycopene and β-carotene by HPLC in tomato fruits. An exactly weighed sample is dissolved in distilled water and the solution obtained is applied to Chromabond C18 cartridge. First, interfering substances such as sugars and organic acids are removed by washing the cartridge with water/MeOH (2:1 v/v). Second, carotenoid fraction is eluted with MeOH/dichloromethane (DCM) (1:1 v/v). Then the eluate is evaporated until dryness. The residue is redissolved in acetonitrile (ACN)/DCM (1:1 v/v), remaining ready to be injected in the HPLC system.

Solid phase extraction has also been employed for carotenoid extraction in orange juice (Fisher and Rouseff 1986) and in nutritional supplements (Iwase 2002).

Supercritical Fluid Extraction

Supercritical fluid extraction minimizes the risk of degradation via isomerization and oxidation of carotenoids, allowing an enhancement in the efficiency of the chromatographic separations. Besides, it provides a faster and more environmentally friendly extraction procedure. CO_2 has a low critical temperature (31°C), making it ideal for extraction of heat-labile compounds, e.g., carotenoids. Regarding sample pre-treatment, removal of water (freeze-drying or adding Hydromatrix) and reduction of particle size facilitate the SFE (Rodríguez-Bernaldo de Quirós and Costa 2006).

Gómez-Prieto et al. (2002, 2003) and Rozzi et al. (2002) extracted lycopene from tomato fruits and tomato processing by-products using CO_2 without modifier, while Baysal et al. (2000) employed EtOH, Vasapollo et al. (2004) used vegetable oil, and Pól et al. (2004) used MeOH as modifiers in the extraction of tomato paste waste and raw tomato. In Table 4, the SFE conditions employed by these authors are shown.

Moreover, Pól et al. (2004) developed an on-line SFE-HPLC system. The reliability and repeatability of the analysis were improved since the analysis and sample clean-up took place in a closed, automated system and the risks of sample loss and contamination decreased. Furthermore, the negative effects of light, atmospheric oxygen and moisture are eliminated.

Table 4 SFE conditions employed in carotenoid analysis.

Sample	*Supercritical fluid*	*Temperature*	*Pressure*	*Flow rate*
Tomato[a]	CO_2 without modifier	40°C	281 bar	4 mL/min
Tomato[b]	CO_2 + 10% vegetable oil	65-70°C	450 bar	18-20 kg/h
Tomato seeds and skins[c]	CO_2 without modifier	86°C	345 bar	2.5 mL/min
Tomato paste waste[d]	CO_2 + 5% EtOH	55°C (lycopene) 65°C (β-carotene)	300 bar	4 kg/h

[a] Gómez-Prieto et al. 2003.
[b] Vasapollo et al. 2004.
[c] Rozzi et al. 2002.
[d] Baysal et al. 2000.

Preparation of Standards

The preparation of adequate standards is a critical step in the determination of any substance, and especially in the case of carotenoids, because of the risk of photo-isomerization and oxidation during their manipulation. Besides, it is important not to assume that any carotenoid standard has remained stable during storage, even when kept in the dark, under nitrogen, at -20°C and stabilized with BHT. Therefore, it is very important that all standards are spectrophotometrically calibrated to calculate stock concentrations and their purity verified by HPLC (Craft 2001, Hart and Scott 1995, Scott et al. 1996, Scott 2001).

Individual stock standard solutions should be freshly prepared every day adding a suitable volume of hexane or another adequate solvent to the vial containing the carotenoid standard and mixing until complete dissolution; then the solutions are transferred to a volumetric flask and the concentration is determined spectrophotometrically using Beer's law and molar absorptivities shown in Table 3. An aliquot of each individual stock standard solution is evaporated to dryness under vacuum and nitrogen atmosphere to avoid oxidation and is redissolved in the phase mobile to check its purity. Thus, considering the purity obtained, the concentrations must be corrected according to the absorbance. Once the standard solution concentrations have been established, the individual stock solutions can be mixed to form calibration solutions of the required concentrations (Hart and Scott 1995, Konings and Roomans 1997, Olives Barba et al. 2006).

Quantification

The selection of the technique used to evaluate the carotenoid content depends on several factors, for example, on whether it is necessary to know the content of each individual carotenoid or, on the contrary, only the global content is required expressed as β-carotene or lycopene. On the other hand, sometimes it is interesting to use non-destructive techniques that avoid the pre-treatment steps and allow results to be obtained in the field.

UV-vis Absorption Spectrophotometry

This technique seems to be the simplest way to identify and quantify the major carotenoids in a mixture. The determination requires only pigment extraction with an adequate solvent in which specific or molar absorptivities have been determined followed by the absorbance measurement at its maximum wavelength. The precision of the method depends on the ability to determine the absorption maxima and on the accuracy of the absorptivity used for the calculation of the pigment concentration (Beer's law) (Schoefs 2002, Scott 2001).

The overlapping of the absorbance bands of the carotenoids present in an extract makes it difficult to estimate the individual carotenoid concentration. Another potential problem arises when the concentration of one of the carotenoids is much higher than that of the rest, and both limitations are present in the case of tomato. Olives Barba et al. (2006) have reported that the lycopene contents evaluated by HPLC or spectrophotometry at 501 nm were coincident in all samples analysed, but not for total carotenoids at 446 nm in tomato and watermelon, whose real content was lower than the content found by spectrophotometry. Therefore, UV-vis absorption spectrophotometry is not acceptable for the determination of total carotenoids expressed as β-carotene in lycopene-rich products, such as tomato, because of overestimation of β-carotene content as a consequence of the lycopene absorption.

Nevertheless, it can be considered a simple and quick alternative for routine analysis of lycopene content in products with lycopene as the major compound (Davis et al. 2003, Olives Barba et al. 2006, Rao et al. 1998).

A complete protocol to carry out the determination of carotenoids by UV-vis spectrophotometry is described by Scott (2001). Analyses of lycopene in tomato by this technique are reported by Choudhari and Ananthanarayan (2007) in petroleum ether extract from tomato tissues, by Goula et al. (2006) also in petroleum ether extract from tomato pulp, and by Davis et al. (2003) in hexane extract from tomato and tomato products such as juice, soup, ketchup, and different kinds of sauces.

Chromatographic Methods

Chromatographic techniques allow calculation of individual carotenoid contents, and even distinction among different isomers of a single carotenoid. They are the most commonly employed techniques in the determination of carotenoids in fruits and vegetables and their derived products. All of these methods require previous extraction of the analytes from the matrix followed by evaporation of the solvents under nitrogen and under vacuum and finally redissolution in the mobile phase or in a solvent mixture compatible with the mobile phase. On average, it takes 40-50 min to get the results of one sample if the preparation time and the chromatographic run are considered (Bhosale et al. 2004, Craft 2001, Pedro and Ferreira 2005). Figure 5A shows an example of a chromatographic separation of carotenoids present in tomato fruit.

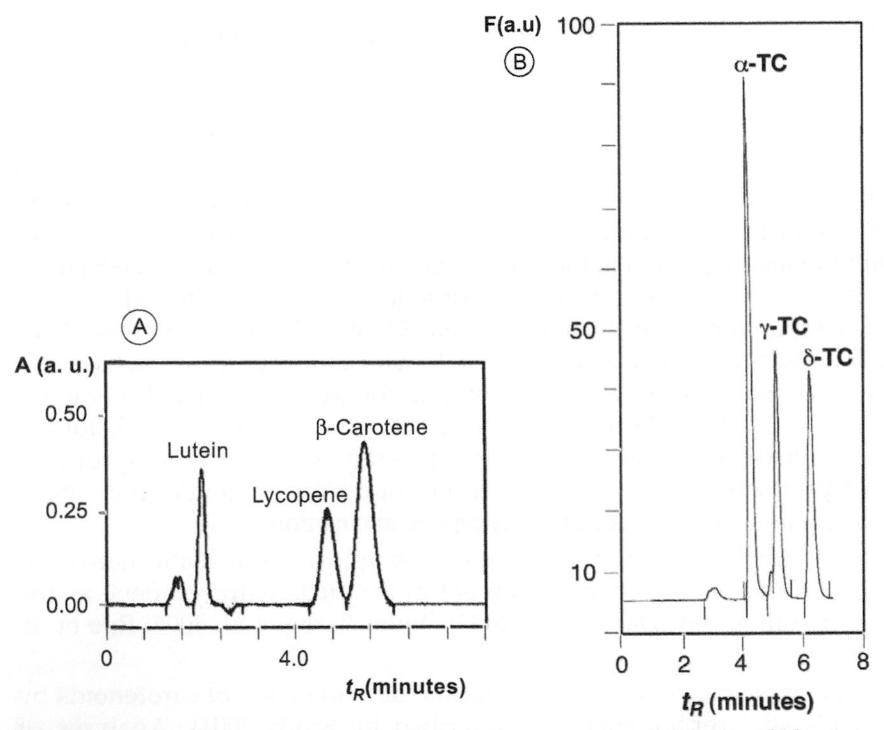

Fig. 5 Chromatograms of (A) carotenoids (isocratic reverse phase elution on C18 column, mobile phase: MeOH/ACN (90:10, v/v) + TEA 9 μM, UV-vis detection λ_{max} = 475 nm) and (B) tocopherols (isocratic normal phase elution on silica column, mobile phase: hexane/2-propanol (98:2, v/v), fluorescence detection λ_{ex} = 298 nm; λ_{em} = 330 nm) A (a.u)-absorbance in arbitrary units, F(a.u.)-Fluorescence intensity in arbitrary units.

Initially, the first analysis of carotenoids was carried out by open column chromatography, thin layer chromatography (TLC) or high performance TLC (HPTLC) (Almeida and Penteado, 1988, Cserháti and Forgács 2001, Rodriguez-Amaya 1996, Schiedt and Liaaen-Jensen 1995). Nowadays, open column chromatography or TLC is still employed for the isolation of standards (Meléndez-Martínez et al. 2007a, Niizu and Rodriguez-Amaya 2005). Gas chromatography (GC) is seldom used, because carotenoids typically decompose when exposed to the high temperatures of the GC process. Therefore, HPLC is the most widely used chromatography in carotenoid determination.

Concerning the type of stationary phase, normal or reverse phase HPLC can be used depending on the aim of the analysis. Normal phase HPLC is chosen when the interest is centred in the xanthophyll fraction, because the xanthophylls (more polar) are more strongly retained in the stationary phase than carotenes. The stationary phase could be silica or polar bond phases as alkylamine in combination with non-polar mobile phases, i.e., hexane/dioxane/indole-3-propionic acid (80:20:1.5, v/v/v) + triethylamine (TEA) 0.1% (Craft 2001).

The addition of TEA to the mobile phase is very common, because it helps to prevent both non-specific adsorption and oxidation of the analytes (Hart and Scott 1995, Niizu and Rodriguez-Amaya 2005, Olives Barba et al. 2006). It is also advisable to add BHT to the mobile phase in order to protect the carotenoids during the chromatographic analysis (Hart and Scott 1995, Meléndez-Martínez et al. 2003a). Other antioxidants such as ascorbic acid or ammonium acetate present more drawbacks (Rodríguez-Bernaldo de Quirós and Costa 2006).

The most widely employed method for carotenoids analysis, due to their hydrophobic character, is RP-HPLC. C-18 columns, which are less selective, are preferable for less detailed analysis, while C-30 columns in combination with gradient elution are suitable for the separation of carotenoids and their different isomers with a wide polarity range. Both isocratic and gradient elution systems are used in the case of C-18 columns. In general, the resolution is better with gradient elution, although the total analysis time is higher because of the need to re-equilibrate the column after each injection. Mobile phases are mixtures of MeOH/ACN in different proportions, with the occasional addition of other solvents such as THF, DCM, water, hexane, acetone, or propanol (Craft 2001, Cserháti and Forgács 2001, Rodríguez-Bernaldo de Quirós and Costa 2006, Schoefs 2002).

The literature contains numerous examples of the application of RP-HPLC to the analysis of carotenoids in tomato and tomato products; some of them are summarized below.

Using C-18 stationary phase:

- Abushita et al. (1997) separated 14 carotenoids, the major ones among them being lutein, lycopene and β-carotene, from tomato fruits using Chromasil C-18 (6 μm, 250 × 4.6 mm) as stationary phase and ACN/2-propanol/MeOH/water (39:52:5:4, v/v/v/v) as mobile phase at 1.0 mL/min.
- Garrido Frenich et al. (2005) analysed lycopene and β-carotene in four varieties of tomato (raf, cherry, rambo and daniela), dried tomato and canned tomato employing two coupled C-18 columns (Simmetry C-18 – 3.5 μm 75 × 4.6 mm + Atlantis dC-18 – 5 μm 150 × 2.0 mm, both from Waters) and a mobile phase composed by MeOH/THF/ACN (60:30:10, v/v/v) in isocratic mode.
- Hart and Scott (1995) found lutein, β-carotene and lycopene in tomato fruits (raw and cooked) and canned tomato using a column system consisting of ODS2 (5 μm, 10 mm) a metal-free guard column, with ODS2 (5 μm, 100 × 4.6 mm) a metal-free column and

Vydac 201 TP54 (5 μm, 250 × 4.6 mm) analytical column. The mobile phase consisted of ACN/MeOH/DCM (75:20:5, v/v/v) + BHT 0.1% + TEA 0.05% at 1.5 mL/min.

- Konings and Roomans (1997) separated lutein, β-carotene, *trans*-lycopene and *cis*-lycopene in tomatoes using a pre-column (Vydac 201 TP C-18, 10 μm, 10 × 4.6 mm) and an analytical C-18 column (Vydac 201 TP, 5 μm, 250 × 4.6 mm), eluted with a mixture of MeOH/THF (95:5, v/v) at 1.0 mL/min.
- Kozukue and Friedman (2003) determined β-carotene and lycopene from tomato fruits employing a C-18 Intersil ODS2 column (5 μm, 250 × 4.6 mm) and ACN/MeOH/DCM/hexane (50:40:5:5, v/v/v/v) as mobile phase at 1.0 mL/min.
- Lenucci et al. (2006) studied the contents of lycopene, β-carotene, α-tocopherol, vitamin C and total phenols and flavonoids in cherry tomatoes and four cultivars of high-pigment tomato hybrids. In the case of lycopene and β-carotene, the assay was carried out using an Acclaim column C-18 (5 μm, 250 × 4.6 mm) and a linear gradient of ACN (A), hexane (B) and MeOH (C), as follows: from 70% A, 7% B and 23% C to 70% A, 4% B and 26% C within 35 min at 1.5 mL/min.
- Niizu and Rodriguez-Amaya (2005) analysed the carotenoid composition of raw salad vegetables. They also found a high content of lycopene, together with lutein and β-carotene, in much smaller amounts, in tomato (*Lycopersicon esculentum*). The quantitative analysis was performed using a C-18 column (Spherisorb S3 ODS2, 3 μm, 150 × 4.6 mm) and a mobile phase consisting of ACN + TEA 0.05%/MeOH/ethyl acetate (60:20:20, v/v/v) at 0.8 mL/min.
- Olives Barba et al. (2006) quantified lycopene and β-carotene in four varieties of tomato (rambo, raf, cherry and canario), carrot, pepper, watermelon, persimmon and medlar by RP-HPLC employing a μBondapack C-18 guard column (10 μm, 20 × 3.9 mm) and a μBondapack C-18 analytical column (10 μm, 300 × 2 mm), both from Waters, with MeOH/ACN (90:10, v/v) + TEA 9 μM as mobile phase at 0.9 mL/min. The analysis is completed within 10 min.
- Vasapollo et al. (2004) determined lycopene content in tomato using a Zorbax column C-18 (5 μm, 250 × 4.6 mm) and a mixture of MeOH/THF/water (67:27:6, v/v/v) as mobile phase at 1.5 mL/min.

Using C-30 stationary phase:

- Ferruzzi et al. (2001) separated 13 lycopene isomers including prolycopene (a novel tetra-*cis*-lycopene) from common red Roma tomato and orange *Tangerine* tomato (a mutant variety that has the capacity to biosynthesize several *cis* isomers of lycopene that are not commonly found in other varieties). The separation was achieved using a polymeric C-30 column (3 μm, 250 × 4.6 mm, prepared by National Institute of Standards and Technology) and a gradient elution with a binary mobile phase of different concentrations of MeOH/methyl terc-butyl ether (MTBE)/ammonium acetate 1% in water within a run time of 50 min.
- Gómez-Prieto et al. (2003) obtained good or adequate selectivity and sensitivity for the determination of all-*trans*-lycopene from their *cis*-isomers as well as acceptable separations among the individual *cis* compounds themselves. The analyses were performed with Develosil UG C-30 column (250 × 4.6 mm) and gradient elution using as mobile phase MeOH/water (96:4 v/v) (eluent A) and MTBE (eluent B). A linear gradient was applied during 60 min from initial conditions (A/B, 83:17, v/v) to those maintained until the end of the analysis (A/B, 33:67, v/v). The samples analysed were tomato fruits (pear variety).
- Lin and Chen (2003) employed a C-30 column (YMC, 5 μm, 250 × 4.6 mm) and a mobile phase of ACN/1-butanol (7:3, v/v) (A) and methylene chloride (B) with the following gradient elution: A/B (99:1, v/v) initially, increased to 4% B in 20 min, 10% B in 50 min and returned to 1% B in 55 min. In these conditions, 16 carotenoids, including all-*trans*-lutein, all-*trans*-β-carotene, all-*trans*-lycopene and their 13 *cis* isomers could be separated, identified and resolved from tomato juices.

Another aspect to be taken into account is the type of detector coupled to the column, which influences the sensitivity of the method. The most widely employed is UV-vis absorption detector and, more recently, DAD, which allows a continuous collection of spectrophotometric data during the analysis contributing to the identification of the carotenoids (Abushita et al. 1997, Burns et al. 2003, Gómez Prieto et al. 2003, Hart and Scott 1995, Niizu and Rodriguez-Amaya 2005). When the chromatograms are not complex, the signals are measured at only one wavelength, at 450 nm (Craft 2001, Konings and Roomans 1997, Kozukue and Friedman 2003), at 475 nm (Lin and Chen 2003, Olives Barba et al. 2006, Sadler et al. 1990, Vasapollo et al. 2004) or at 503 nm (Lenucci et al. 2006, Rao et al. 1998).

When high sensitivity is required, electrochemical array detection is a good alternative (Ferruzzi et al. 1998, 2001, Rozzi et al. 2002). The advantage of coulometric array detection is its very low detection limits (picogram levels); the disadvantage is the need to add a supporting electrolyte to the mobile phase, which may alter the elution profiles (Ferruzzi et al. 1998).

In complex matrixes, when DAD is not enough to allow identification because of spectral interferences, mass spectrometry coupled with HPLC (HPLC-MS) has been successfully used. HPLC-MS is a powerful analytical tool for the identification of carotenoids, because it provides structural information and is highly sensitive (identifying even subpicogram levels). The most common interfaces used in carotenoid analysis are atmospheric pressure chemical ionization (APCI) (Garrido Frenich et al. 2005, Guil-Guerrero et al. 2003, Huck et al. 2000, Kurilich et al. 2003, van Breemen 2001, van Breemen et al. 2002) and electrospray ionization (ESI) (Careri et al. 1999, Hadden et al. 1999, van Breemen 2001).

In HPLC-ESI-MS, the typical mobile phase is a mixture of MeOH and MTBE (50:50, v/v), although other solvents such as water or THF can be used. Besides, volatile buffers such as ammonium acetate or ammonium carbonate can be used at concentrations < 40 mM. Electrospray ionization is compatible with a wide range of flow rates of the mobile phase, from 0.1 nL/min to 1 mL/min. The scan range is usually *m/z* 300 to 1000 in order to include known carotenoids and their esters, because ESI produces molecular ions, M^{+0}, with almost no fragmentation for carotenes and many xanthophylls. The optimum sensitivity for these compounds is obtained using the highest possible voltage (2000-7000 V) on the electrospray needle before corona discharge occurs. The dynamic range is relatively narrow, approximately two orders of magnitude (van Breemen 2001).

In HPLC-APCI-MS, the same mobile phases and scan range of *m/z* described in HPLC-ESI-MS can be employed. Carotenoids form molecular ions and protonated molecules during positive ion APCI or deprotonated molecules during negative ion analysis. The main advantage of APCI compared to ESI for carotenoid analysis is the higher dynamic range (four orders of magnitude) with similar sensitivity, which suggests that this tandem technique may become the standard for carotenoid quantification. However, APCI produces multiple molecular ion species, which might lead to ambiguous molecular weight determinations and, moreover, tends to produce more fragmentation in the ion source than ESI, although these fragment ions are often not abundant (van Breemen 2001).

With regard to the use of internal standard in the analysis of carotenoids, it is difficult to find a suitable substance with complex samples containing numerous peaks that can appear as a consequence of the photoisomerization of these compounds in spite of a careful manipulation. Craft (2001) states that compounds that may be used as internal standards for carotenoid analysis are unfortunately not available commercially and hence their use for quantification should be considered optional. For example, although the use of β-apo-8′-carotenal or echinenone or cholesterol as internal standard has been described (Gómez Prieto et al. 2003, Hart and Scott 1995, Konings and Roomans 1997, Pól et al. 2004), there are also recent papers in which the inner standard is not used (Garrido Frenich et al. 2005, Guil-Guerrero et al. 2003, Lenucci et al. 2006, Lin and Chen 2003, Niizu and Rodriguez-Amaya 2005, Olives Barba et al. 2006).

Mass Spectrometry

The high sensitivity and selectivity of MS facilitates the identification and structural analysis of carotenoids. In electron impact and chemical ionization MS, the carotenoid must be volatilized prior to ionization using thermal heating. Thus, it results in some pyrolysis before ionization and produces fragmentation after ionization. Therefore, electron impact and chemical ionization mass spectra show abundant fragment ions below *m/z* 300 and hardly any molecular ions. Fragments above *m/z* > 300 are useful in confirming the presence of specific ring system and functional groups (van Breemen 2001). Minimizing fragmentation has been achieved by means of matrix-mediated desorption techniques: fast atom bombardment, liquid secondary ion, matrix-assisted laser desorption/ ionization (MALDI), ESI or APCI.

Fast atom bombardment MS and liquid secondary ion MS employ 3-nitrobenzyl alcohol as a matrix for carotenoid ionization and the mass spectra obtained show molecular ions, M^{+0}, almost without fragmentation. If structurally significant fragment ions are desired, then collision-induced dissociation can be used to fragment the molecular ions followed by MS/MS to record the resulting product ions. These ions provide information about the presence of hydroxyl groups, esters, rings or the extent of conjugation of the polyene chain as well as the possibility of distinguishing between isomeric carotenoids. MALDI-TOF-MS (matrix-assisted laser desorption/ionization time of flight mass spectrometry) has been effective in the ionization of intact esterified carotenoids that would fragment too extensively using other ionization methods. The sample matrix employed is acetone saturated with 2,5-dihydroxybenzoic acid (van Breemen 2001). Electrospray ionization and APCI are combined with HPLC, as has been described above.

Non-destructive Methods

A rapid non-destructive estimation of carotenoid levels in intact fruits and vegetables and their juices could have great value in selecting nutritionally valuable crops for further propagation and commercial use (Gould 1992).

A non-destructive and rapid technique is the evaluation of colour, which is a very useful tool for quality control of carotenoids in the industry. The objective measurement of colour by means of colorimeters, which measure reflected visible colour, is being increasingly used for the analysis of carotenoids (Arias et al. 2000, Bicanic et al. 2003, Meléndez-Martínez et al. 2003b, Schoefs 2002).

Each colour can be described by a set of three parameters: hue, saturation and lightness. In 1976 the Commission Internationale de 1'Eclairage adopted a standard method of calculating colour attributes (CIE Lab Colour Space). The lightness coefficient L^* ranges from black ($L^* = 0$) to white ($L^* = 100$), while the coordinates a^* and b^* designate the colour on a rectangular-coordinate grid perpendicular to the L^* axis. The colour at the grid origin is achromatic (grey: $a^* = 0$ and $b^* = 0$). On the horizontal axis, a^* values indicate the hue of redness ($a^* > 0$) or greenness ($a^* < 0$), whereas on the vertical axis, b^* values indicate the hue of yellowness ($b^* > 0$) or blueness ($b^* < 0$) (Schoefs 2002). Meléndez-Martínez et al. (2007a) have summarized the colour coordinates for 17 carotenoids, among them lycopene, β-carotene, lutein and β-cryptoxanthin, and have studied the relationship between their colour and their chemical structure.

D'Souza et al. (1992) showed that the lycopene concentration of tomato (*L. esculentum* Mill.) can be estimated using a tristimulus colorimeter set to read in the CIE $L^* a^* b^*$ colour scale using the equation $(a^* / b^*)^2$ ($R^2 = 0.75$). Arias et al. (2000) reported that a^* / b^* readings on tomato yielded a highly linear regression ($R^2 = 0.96$) when compared with lycopene quantity. In this study, the samples were green tomatoes and red tomatoes with relatively low lycopene content. Thompson et al. (2000) also found a good correlation ($R^2 = -0.85$) between CIE $L^* a^* b^*$ hue values and lycopene content in tomato homogenate, but the correlation got worse in the red ripe stages.

Baranska et al. (2006), Bhosale et al. (2004) and Pedro and Ferreira (2005) employed Fourier Transform-Raman Spectroscopy (FT-Raman), Attenuated Total Reflection-Infrared Spectroscopy (ATR-IR) and/or Near Infrared Spectroscopy (NIR) for the determination of lycopene and β-carotene in tomato fruits and related products. These spectroscopies allow identification and analysis of carotenoids in minutes directly in the plant tissue and food products without preliminary sample preparation except grinding. Their use has also been reported in other vegetables such as carrot (Baranska and Schultz 2005, Baranska et al. 2005), sugar beet and pepper (Baranski et al. 2005) and maize (Brenna and Berardo 2004).

Koyama (1995) reported that the carotenoid response is characterized by three strong, high-frequency, Stokes-shifted signals originating from carbon-carbon double-bond and single-bond stretches of the molecule's polyene backbone and from methyl bends with associated Raman peak positions at 1525, 1159 and 1008 cm^{-1} respectively. Baranska et al. (2006) distinguished the same peaks described by Koyama in the FT-Raman spectrum of tomato puree and assigned the peak at 1510 cm^{-1} to lycopene and the shoulder at 1520 cm^{-1} to β-carotene.

Among these techniques, FT-Raman spectroscopy seems to be better suited for simultaneous determination of both carotenoids in tomato products, because ATR-IR and NIR spectra showed mainly strong water signals and no bands characteristic for carotenoids (Baranska et al. 2006). Nevertheless, although there has been good correlation between Raman or NIR measurements and the total carotenoid content determined by HPLC (Baranska et al. 2006, Bhosale et al. 2004, Pedro and Ferreira 2005), these spectroscopies will never replace HPLC as the technique of choice for the analysis of these compounds.

VITAMINS

Vitamin C

Most of the biological properties of vitamin C (Fig. 3B) are a consequence of its chemical properties and therefore these characteristics determine the method used for its analysis. Ascorbic acid (AA) is a diprotic acid (pK_{a1} = 4.1 and pK_{a2} = 11.8) and a strongly reducing agent because it is easily oxidized to dehydroascorbic acid (DHAA) in aqueous solution. This oxidation reaction is reversible and is catalysed in basic media in the presence of oxygen, Cu^{2+} and Fe^{2+}. Epidemiological studies have shown that the risk of cancer (stomach, oesophagus, pharynx, lung, pancreas and cervix) decreases in persons with high levels of intake of vitamin C (Negri et al. 2000).

During the last decades, various chromatographic methods, such as RP-HPLC, have been developed and applied to the quantification of AA and DHAA. An example of a method for preparation and purification of samples for the analysis of vitamin C is shown in Fig. 6.

More "classic" procedures involve the determination of AA by titration with 2,6-dichloroindophenol (DCIPh) and the colorimetric detection of the equivalence point. Ascorbic acid is in excess in the sample solution and then the solution remains colourless because the resulting reaction product and DHAA are colourless; however, the titration endpoint is indicated by the pink colour of the solution due to the absence of reducing agent (AA).

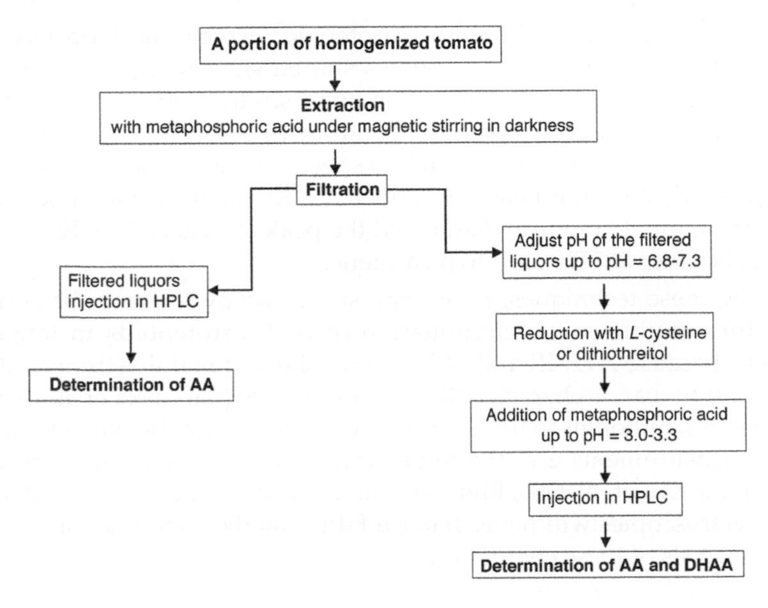

Fig. 6 Pre-treatment and clean-up procedures for extraction of vitamin C.

Fluorimetric detection affords higher sensitivity to quantitative analysis of vitamin C, besides being suitable for the quantification of vitamin C in coloured samples such as tomato samples. Ascorbic acid is oxidized to DHAA and then o-phenylenediamine (PDA) is added to the sample solution to produce a fluorescent quinoxaline derivative (Seiler 1993). This procedure allows the evaluation of AA and DHAA simultaneously in the sample. Both procedures, together with a semiautomated fluorimetric method based on the fluorescent reaction of vitamin C and PDA in a flow injection system, are reference AOAC methods. Sample preparation requires the maceration or extraction of vitamin C from the tomato samples and its stabilization with metaphosphoric acid or trifluoroacetic acid (AOAC 1996).

Ascorbic acid was determined spectrophotometrically in papaya and tomato pulp and the values obtained varied in the range of 40.5-41.4 mg/100 g; the effects of hydrocolloids (pectin, starch and ethyl cellulose) on the physicochemical, sensory and texture characteristics were studied. The presence of hydrocolloids does not affect the levels of vitamin C during 4 mon storage (Ahmad et al. 2005). Ascorbic acid and DHAA have been quantified in high-pigment tomato hybrids and cherry tomatoes. Differences in the AA/DHAA ratio were observed in different tomato cultivars. Considering that the environmental conditions affect this redox system, the tomato cultivars perceive the surrounding environment according to their genotype (Lenucci et al. 2006). The levels of vitamin C change with the storage and processing of the samples, as a consequence of the low stability of AA associated with its easy oxidation; this can be employed as a good parameter for detection of problems during the storage. The levels of vitamin C have been determined spectrophotometrically in different samples with the aim of studying the effect of storage (Meléndez-Martínez et al. 2007b).

Liquid chromatography (LC)-UV detection has been applied to the analysis of vitamin C in foods (Arnao et al. 2001), using UV detection (Table 3). Vitamin C was analysed in tomato fruits from different cultivars by RP-HPLC-UV detection, the selected wavelength being 225 nm; absorptiometric detection using DAD was also employed (Abushita et al. 1997). HPLC-UV detection has been employed for the determination of AA and total vitamin C (AA + DHAA) after reduction of DHAA with dithiothreitol in green beans (Sánchez-Mata et al. 2003), and the results obtained were compared to those from the reference spectrofluorimetic quantification of vitamin C. Spectrofluorimetric determination is adequate and more precise for the analysis of total vitamin C, while LC analysis shows better linearity and sensitivity (Sánchez-Mata et al. 2000). This method (HPLC-UV detection) has been applied to the quantification of vitamin C in tomato with the aim of characterizing *Lycopersicon* germplasm (Fernández-Ruiz et al. 2004).

Sensitivity and selectivity of analysis are improved when HPLC is coupled with MS. The ionization mode in MS was ESI at atmospheric pressure. The freezing, manipulation and preservation processes affect the levels of AA in tomato (Frenich et al. 2005).

More sophisticated and precise techniques such as capillary zone electrophoresis were employed for the determination of AA together with other organic acids such as oxalic, malic and citric acids (Galiana-Balaguer et al 2006). The levels of these compounds have been found to be a precise tool to classify and select *Lycopersicon* germplasm for internal quality attributes.

An attractive alternative to the conventional spectrofluorimetric analysis of vitamin C consists of a determination of total and reduced vitamin C in fruit using a colorimetric reaction developed in microplates and detection employing a plate reader (Stevens et al. 2006). For the evaluation of the total contents of vitamin C, the samples containing AA and DHAA are treated with dithiothreitol reducing the DHAA present in the samples. The plate containing the samples is then incubated at 37°C for 40 min in the presence of a coloured reagent (iron chloride) reduced from Fe^{3+} to Fe^{2+} by AA. The absorbance was read at 550 nm using a microplate reader and results obtained were compared with the results of the classical spectrofluorimetric determination of vitamin C. These methods are inexpensive and rapid and allow the determination of analytes in 96 samples at the same time and in the same reaction conditions.

Vitamin E

The name vitamin E comprises a group of compounds known as tocopherols (TCs), which are derived from 6-chromanol substituted by a saturated isoprenoid chain. The most abundant naturally occurring component is α-TC, while the dimethyl (β- and γ-TC) and monomethyl (δ-TC) derivatives are present in lower amounts (Fig. 3c). All these derivatives possess strong reducing properties due to the phenolic group and they are easily oxidized in the presence of trace amounts of metals and oxygen. Alkaline media favour the oxidation of TCs. Tocopherols absorb in the UV region (λ_{max} varies for each compound, Table 3), exhibit native fluorescence (λ_{em}= 330 nm), and can be electrochemically oxidized. Consequently TCs can be determined by LC with UV-DAD, fluorescence detection (FD) and electrochemical detection (ECD).

The main analytical problem for the analysis of vitamin E is related, as in the case of the other fat-soluble vitamins, to the isolation of these compounds from a large excess of physically hydrosoluble materials, and hence a great deal of effort has been taken to improve extraction and clean-up procedures. Liquid-liquid extraction is frequently used. Addition of antioxidants (AA or BHT) is necessary for the saponification step in the purification process. The influence of KOH concentration in the saponification procedure on the efficiency of the TC extraction from tomato was considered. The optimal conditions found were 60% KOH and a saponification time around 50 min at 70ºC (Lee et al. 2000). The use of SFE in fat-soluble vitamin analysis provides an interesting alternative to the use of organic solvents. Supercritical CO_2 can be employed for the extraction of carotenoids, lycopene and vitamins A, D, E and K from different vegetable samples (Turner et al. 2001).

According to the AOAC methods, vitamin E can be determined colorimetrically in foods after its isolation by TLC and also by GC-FID (flame ionization detector) (AOAC, 1996). The capillary GC analysis of vitamin E can be easily coupled with MS, and in such cases the conversion of the hydroxyl group of vitamin E into the corresponding acetate or butyrate esters enhances the sensitivity of the quantification of TCs.

Liquid chromatography separation of the TCs has been developed in normal phases (silica and polar bonded phases, i.e., amino) and also in reverse phase under isocratic elution conditions. The detection technique more frequently employed is UV-absorption or DAD, although the FD or

ECD enhance notably the detection limit for TCs. α-TC has been assayed by normal phase HPLC under gradient elution conditions. The solvents employed were ACN followed by hexane and MeOH, and α-TC was UV detected at 280 nm. The levels of α-TC change with the cultivar of tomato, being higher in the case of high-pigment than for cherry tomatoes (Lenucci et al. 2006). Normal phase HPLC with FD was also employed by Abushita et al. (1997) for the determination of TCs in tomato. α-TC and γ-TC predominate over β-TC and the total amount of TCs varies depending on the cultivars. The α-TC to γ-TC ratio changes for the 22 different vegetables studied. The clean-up procedure involves evaporation of the extraction liquors under a nitrogen atmosphere to preserve vitamin E. Among the different vegetables studied, the amount of α-TC detected in tomato by normal phase HPLC-FD was higher than that of γ-TC (Chun et al. 2006). RP-HPLC with ECD allows the separation and quantification of α-, γ- and δ-TC in trace amounts, no β-TC being found in the different genotypes of tomato (Sgherri et al. 2007). HPLC was the selected technique to evaluate the levels of vitamin E in different foods. The results obtained were compared with the corresponding data afforded by food composition tables and the differences found can be explained considering factors such as season, country of origin, ripeness and freshness (Sundl et al. 2007). New agricultural technologies may lead to changes in the nutrients of tomato and decrease the vitamin E amounts up to 12.69% (Hou and Mooneyham 1999). An extensive revision of the chromatographic analysis (GC and HPLC) of vitamin E in a great variety of samples including vegetables and plant samples discusses the experimental conditions and their advantages and drawbacks (Rupérez et al. 2001).

Vitamin D

Vitamin D, together with other micronutrients present in tomato fruit (vitamin E, lycopene and phenolic compounds), has been related in different epidemiological studies with a decrease of prostate cancer risk in men and these dietary substances are considered as chemopreventive agents against this disease (Bemis et al. 2006, Cohen 2002). Food composition tables usually give no vitamin D in tomato. Vitamin D_3 (cholecalciferol) is synthesized by vertebrates and is named "animal vitamin D" in contrast to vitamin D_2 (ergocalciferol) (Fig. 3C), which is known as "plant vitamin D", although vitamin D_3 and related compounds have been found in different plant species, e.g., potatoes and tomatoes (Aburjai et al. 1998, Prema and Raghuramulu 1996). Gradient elution in HPLC with UV detection (λ_{max} is shown in Table 3) was employed for identification and quantification of vitamin D_3 in tomato leaves (Aburjai et al. 1998). On the other hand, normal phase (propanol/hexane) and reverse phase (MeOH/ACN) in the mode of isocratic elution were employed by Prema and Raghuramulu (1996) for the characterization of vitamin D_3 in tomato plants.

The presence of the diene chain (the isoprenoid side chain) makes vitamin D sensitive to air and heat with isomerization and oxidation reactions being the main degradation pathways. This low stability must be taken into account in the clean-up of the samples to ensure the accurate quantification of vitamin D. The extraction procedures imply that fat components (carotenoids, vitamins D, E and K) remain in the same extract. Therefore, the colorimetric reaction of vitamin D with antimony trichloride (AOAC method) produces a coloured compound (λ_{max} = 500 nm) but vitamin A also produces the same coloured products. The derivatization of vitamin D with acyl chlorides facilitates its determination by GC-FID, GC-MS. HPLC-DAD and HPLC-MS have also been used for the determination of vitamin D with higher sensitivity (De Leenheer et al. 1995). The utility of HPLC with MS detection in the ionization mode ESI has been shown for the identification and quantification of vitamin D_2 and D_3 in different samples of vegetables (Wang et al. 2001). HPLC techniques have been employed for the separation of vitamin D_3 and their hydroxylated derivatives in various plant species belonging to Solanaceae family (Skliar et al. 2000). RP-HPLC using isocratic and gradient elution were employed for the separation of vitamin D, with UV and MS as detection techniques. The biosynthesis of vitamin D_3 in plant species (*Solanum glaucophyllum*) involves the same intermediates as in vertebrates. These compounds were isolated by organic solvent extraction and their identification and quantification were carried out by HPLC in the isocratic elution mode employing MeOH/water mixtures (Curino et al. 2001).

Vitamin K

There are different forms of vitamin K, among which vitamin K_1 (phylloquinone) (Fig. 3C) is naturally occurring in green plants. The levels of phylloquinone in tomato and tomato products are low. Vitamin K_2 is a group of substances named menaquinones and is produced by bacteria. Menadione is vitamin K_3 and exhibits vitamin K activity due to its conversion into menaquinones.

The problems for the extraction of vitamin K are close to those described for the other fat-soluble vitamins in a hydrophilic environment (fruits and vegetables). Therefore, besides liquid-liquid extraction, SPE on silica cartridges (Damon et al. 2005) and SFE have been successfully applied to vitamin K extraction (Turner et al. 2001).

The naphthoquinone ring is the chromophoric group in vitamin K, which is responsible for its UV-vis absorption (Table 3). The quinone form is not fluorescent but its reduction to the corresponding naphthalenediol leads to the appearance of a notable fluorescence that can be employed for its identification and quantification. Considering these properties, HPLC with UV detection and FD are widely employed. In the case of FD the post-column reactor is necessary to reduce the non-fluorescent quinone form of vitamin K to the corresponding quinolic fluorescent derivative. Thus, reverse phase and gradient elution HPLC-FD have been applied to the identification and quantification of vitamin K_1 in different food samples. The detection of the quinones was carried out in a post-column reactor packed with dry zinc (Dumont et al. 2003, Koivu-Tikkanen et al. 2000). This method has been applied to verify the levels of vitamin K and the values of the food composition data in Finland (Piironen and Koivu 2000). Vitamin K was also determined by RP-HPLC-FD with post-column derivatization with zinc in different samples of vegetables (broccoli, spinach, lettuce) including tomato. Tomatoes are typically low in phylloquinone, but since this compound is concentrated when water content decreases, the levels of phylloquinone found in tomato juice were 0.9 μg/100 g, while levels in tomato paste were 9.9 μg/100 g (Damon et al. 2005).

Group of B Vitamins

Vegetables are an important source of B vitamins. In the case of tomato, pyridoxine, nicotinamide and pantothenic acid present a special relevance. Several B vitamins are determined using microbiological tests, but other techniques can also be applied. A common characteristic of the vitamin B group is their participation in a number of enzymatic reactions and they also behave as coenzyme forms. The group of substances with vitamin B activity presents different chemical structures (Fig. 3B), different physicochemical behaviours (Table 3), and also different chemical properties and reactivity. Therefore, the analytical methods have to be considered individually for each compound.

Liquid chromatography has proved to be one of the most useful techniques for the analysis of B group of vitamins. Nicotinamide, pyridoxine, thiamine, folic acid and riboflavin can be separated in one single analysis. This is the consequence of the versatility in column technology and detection methods.

Thiamine

Thiamine is stable at pH values 2-4 but is labile at alkaline pH and it is also heat-labile. In alkaline media it is easily oxidized to the yellow-coloured and fluorescent compound known as thiochrome. Measurement of thiochrome fluorescence is a valuable method for quantification of thiamine and has been selected as the official method (AOAC 1996). The values obtained by Martín-Belloso and Llanos-Barriobero (2001) for thiamine in whole peeled canned tomatoes were 0.216 ± 0.008 mg/kg and they are lower than those described in Food Composition Tables (Souci et al. 1989). RP-HPLC facilitated the determination of thiamine in food samples, with thiamine usually reacting to produce thiochrome in a pre-column derivatization oxidation. The fluorescent compound is detected at λ_{ex} = 370 nm, λ_{em} = 430 nm. Pre-column derivatization reaction followed by HPLC separation with FD was employed for the determination of thiamine in a great variety of food samples (El-Arab et al. 2004). The levels for vitamin B_1 found were 0.0434 mg/100 g in raw tomato (moisture 93.2%), 0.0976 mg/100 g in canned tomato (moisture 74.4%), and 0.0326 mg/100 g in tomato salad (moisture 93.1%).

Riboflavin

Riboflavin is highly sensitive to UV light and its solutions are slowly degraded under daylight. It also shows a yellow-green fluorescence that decreases with time. Measurement of native fluorescence from riboflavin is a sensitive method that corresponds to the official method of AOAC (1996) for quantification of riboflavin in foods. This fluorimetric method can be automated employing a flow injection system with FD. According to Martín-Belloso and Llanos-Barriobero (2001), the content of riboflavin in tomatoes is 0.25 ± 0.04 mg/kg. The characteristic fluorescent emission of riboflavin (λ_{ex} = 370 nm, λ_{em} = 520 nm) is employed for the quantification of riboflavin by HPLC with FD (Esteve et al. 2001). The introduction of the new separation technique capillary zone electrophoresis with laser-induced fluorescence detection increases the sensitivity, selectivity and routine method for analysis of riboflavin in food samples (Cataldi et al. 2003). Riboflavin can be oxidized by $KMnO_4$ and the fluorescence emission (λ_{ex} = 440, λ_{em} = 565 nm) can be used for its quantitative determination in green beans (Sánchez-Mata et al. 2003).

Nicotinamide

Nicotinamide is stable in neutral solution, while nicotinic acid is stable in both acid and alkaline media. Nicotinamide adenine dinucleotide phosphate ($NADP^+$) and a reduced form of $NADP^+$ (NADPH) are determined spectrophotometrically at λ_{max} = 625 nm with DCIPh in the presence of glucose-6-phosphate, phenazine methosulfate, and glucose-6-phosphate hydrogenase. The concentrations were calculated on the basis of the absorbance values in the corresponding calibration curves for $NADP^+$ and NADPH (Sgherri et al. 2002, 2007). In different tomato genotypes the presence of NADPH was lower than that of $NADP^+$ and the ratio $NADPH/NADP^+$ increased during ripening.

Nicotinamide can also be colorimetrically quantified by reaction with sulfanilic acid and CNBr solution; the reaction produces a coloured compound absorbing at 470 nm. This method can be applied for the automated analysis of nicotinamide in a flow injection system with spectrophotometric detection (AOAC 1996). The influence of NaCl on the growth of tomato plants in controlled media has been proved. The levels of nincotinamide adenine dinucleotide (NAD^+) and its reduced form (NADH) in tomato plants change with the salinity of the media; under salinity 25 mM and 50 mM NaCl concentrations the NADH-GDH (glutamate dehydrogenase) activation increased by 64% and 89%, respectively. NAD-GDH activity is greatly decreased in the leaves and roots of tomato plants (Debouba et al. 2006).

Other B Vitamins

Pyridoxine is photosensitive. According to Martín-Belloso and Llanos-Barriobero (2001), the levels of pyridoxine found in whole peeled tomatoes were 0.47 ± 0.05 mg/kg. Pyridoxine can be determined spectrophotometrically (λ_{max} = 424 nm) by diazotation-diazo coupling of pyridoxine with $NaNO_2$ and sulfanilic acid (Sánchez-Mata et al. 2003).

Pantothenic acid is more stable in acid media than in alkaline media. Folic acid is easily oxidizable. When the samples are prepared for the determination of folic acid the use of antioxidants is required in order to avoid losses in the concentration of folic acid. Pantothenic acid is more stable in acid media than in alkaline media. Besides microbiological assays, ion exchange HPLC is used for the determination of folic acid and pantothenic acid. Radioimmunoassay and competitive protein binding assays have shown selective methods for the analysis of these acids (Parviainen 1995). The levels of folic acid change with the soil for different cultivars (Lester and Crosby 2002)

Factors such as soil composition, light and water affect the growth and ripeness of fruits. The cultivars and their varieties and new agricultural technologies also affect the composition of micronutrients and bioactive compounds in tomatoes. Some of these factors are considered below.

The effect of diluted sea water and ripening on the nutritional components of tomato has been studied (Sgherri et al. 2007); salinity induced oxidative stress and increased AA and α-TC, whereas total level of vitamin C (AA+DHAA) and total TC were decreased. These changes depended on the tomato cultivars and the breeding lines. Total content of vitamin C was determined by isocratic RP-HPLC with ECD exploiting the redox behaviour of vitamin C.

New agricultural technologies (Hou and Mooneyham 1999) affect the nutritional components in tomato. Thus, the levels of vitamin C and vitamin E in tomato decrease to 2.10% and 12.69% using agri-wave technology. However, the content of sugar, vitamin A and niacin is remarkably increased. There are no significant differences in the levels of vitamins D, B_1 and B_2. This procedure stimulates the growth of tomato and improves its quality.

The influence of agar-based substrates on tomato pollen germination has been studied. The presence of polyethylene glycol or mannitol affects the levels of vitamins (thiamine, riboflavin, pyridoxine, niacin and pantothenic acid) and produces an increase in pollen germination (Karapanos et al. 2006).

MINERALS

Minerals are the constituents that remain as ash after the combustion of a plant. Minerals are divided into main elements (Na, K, Ca, Mg, Cl and P), trace elements (Fe, I, F, Zn, Cu, Mn, Mo, Co), and ultra-trace elements (Cd, Pb, Sb, Sr, etc.). The mineral content in a vegetable or fruit can fluctuate greatly depending on genetic and climatic factors, agricultural procedures, soil composition and ripeness of the harvested crops, among other factors (Belitz et al. 2004, Bryson and Barker 2002). In tomato, ashes represent around 0.5% of the fresh edible portion. Potassium is by far the most abundant constituent, followed by magnesium, calcium and sodium. The major anions are phosphate, chloride and carbonate (Belitz et al. 2004). All other elements are present in much lower amounts (Table 2).

Ashes

Ashes were obtained by incineration in an oven at ≤ 450°C and dried until a constant dry weight was achieved (AOAC 1990, Bryson and Barker 2002, del Valle et al. 2006, Guil Guerrero et al. 1998a).

Total Nitrogen

Total nitrogen can be determined by the Kjeldahl method (AOAC 1996, Gálvez Mariscal et al. 2006).

Phosphorus

Phosphorus as phosphate is determined by the molybdate-vanadate method followed by spectrophotometric measurement at 470 nm (Bryson and Barker 2002, Guil-Gerrero et al. 1998a).

Mineral Nutrients

When several elements need to be analysed in a plant over a large range of concentration the analytical tool of choice is usually ICP-MS, followed by ICP-AES and atomic absorption spectrometry, but these techniques require chemical extraction or dissolution of the ashes with a mixture of HCl and HNO_3 (Bryson and Barker 2002, Chatterjee and Chatterjee 2003, Dong et al. 2004, Ebbs et al. 2006, Guil-Guerrero et al. 1998b, Henández et al. 2005, Martín-Belloso and Llanos-Barriobero 2001, Pohl and Prusisz 2007, Premuzic et al. 1998, Rouphael and Colla 2005, Ruiz et al. 2005, Salles et al. 2003). In the case of calcium, lanthanum salts must be added to avoid phosphorus interference (AOAC 1990). Salles et al. (2003) and Martín-Belloso and Llanos-Barriobero (2001) have used flame atomic emission spectrophotometry to evaluate sodium and potassium concentrations. ICP-AES allows the simultaneous measurement of metals together with some non-metals, such as P or B. Thus, this technique has rapidly replaced colorimetric procedures for P and B determinations (Taber 2004).

Dry ashing methods can lead to losses through volatilization of important heavy metals such as cadmium, lead or zinc, although digestion assisted by microwaves is acceptable for heavy metal analysis (Ebbs et al. 2006).

Another alternative for mineral determination is instrumental neutron activation analysis. This technique requires only simple sample preparation and its particularly sensitive for trace element analysis of plants and food; however, it is too slow for routine application. X-ray fluorescence spectrometry provides an alternative multielement analytical tool for routine non-destructive analysis with minimal sample preparation. Because of its poor sensitivity for many important elements, it is seldom applied to vegetables. Stephens and Calder (2004) described an X-ray fluorescence method for the rapid and non-destructive analysis of 30 elements (including P, Cl, S, Ca, K, Mg, Fe, Mn, Zn, Cu) in plant leaves for five orders of magnitude. The sample preparation consisted of pulverizing and briquetting of dried samples. Accuracy was evaluated using certified reference materials from NIST, among them NIST 1573 tomato leaves.

Tables 5 and 6 summarize the various methods (conventional, official, and others) and the relevant instrumental techniques most often

Table 5 Analytical methods for quantification of vitamins and carotenoids.

Organic compound	*Method/technique*
Carotenoids	Spectrophotometry RP-HPLC (UV-Vis detection and DAD. APCI-MS. ESI-MS)
Group of vitamin B	HPLC-UV. HPLC-FD (precolumn derivatization). HPLC (ion exchange). Microbiological assays. Radioimmunoassays
Vitamin C	Spectrophotometric titration with DCIPh Spectrofluorimetric determination with PDA Semiautomated spectofluorimetric determination with PDA RP-HPLC (UV-Vis detection and DAD. ESI-MS. ECD) Capillary zone electrophoresis Colorimetric reaction (Fe^{3+} to Fe^{2+}) in microplates
Vitamin D	Colorimetric reaction and detection HPLC (UV and MS detection)
Vitamin E	GC-FID. GC-MS HPLC (UV-Vis detection and DAD. FD. ECD)
Vitamin K	HPLC-UV detection and HPLC-FD

Table 6 Analytical methods for analysis of minerals.

Element	*Method/technique*
Na, K	Flame atomic emission spectrometry Ion selective electrodes ICP-AES
Ca, Mg, Fe, Cu, Zn, Mn, Se	Atomic absorption spectrometry ICP-AES Colorimetric reaction and detection
F, Cl, I	Direct titration with silver nitrate (for chloride and iodide). Potentiometric detection of the end point. Ion selective electrodes
P	Colorimetric reaction and detection (molybdic acid and vanadic acid) ICP-AES Instrumental neutron activation analysis

employed in the determination of vitamins, carotenoids and minerals. Non-routine techniques have been included as a future trend.

ACKNOWLEDGEMENTS

Financial support from the Spanish government, MEC through grant CTQ2006-13351/BQU, from UCM-CAM (Grupo nº 920234/2007), from

CICYT through grant AGF99-0602-C02-02 and from Universidad Complutense de Madrid through grant PR78/02-11037 is gratefully acknowledged.

ABBREVIATIONS

AA: Ascorbic acid; ABTS: 2,2′-azino-bis-(3-ethyl benzothiazoline-6-sulphonic acid); ACN: Acetonitrile; AOA: Antioxidant activity; APCI: Atmospheric pressure chemical ionization; ATR-IR: Attenuated Total Reflection Infrared Spectroscopy; BHT: Butylated hydroxytoluene; DAD: Diode array detection; DCIPh : 2,6-Dichloroindophenol; DCM: dichloromethane; DHAA: Dehydroascorbic acid; DPPH: 2,2′-Diphenyl-1-picryhydrazyl; ECD: Electrochemical detection; ESI: Electrospray ionization detection; EtOH: Ethanol; FD: Fluorescence detection; FID: Flame ionization detector; FRAP: Ferric-reducing ability of plasma; FT-Raman: Fourier Transform-Raman Spectroscopy; GC: Gas Chromatography; GDH: Glutamate dehydrogenase; HPLC: High performance liquid chromatography; HPTLC: High performance thin layer chromatography; HRP: Horseradish peroxidase; ICP-AES: Inductively coupled plasma atomic emission spectrometry; ICP-MS: Inductively coupled plasma mass spectrometry; λ_{em}: Fluorescence emission maximum; λ_{ex}: Fluorescence excitation maximum; λ_{max}: Maximum absorption wavelength; LC: Liquid Chromatography; MALDI: Matrix-assisted laser desorption/ionization; MALDI-TOF-MS: Matrix-assisted laser desorption/ionization time of flight mass spectrometry; MeOH: Methanol; MS: Mass spectrometry; MTBE: Methyl terc-butyl ether; NAD^+: Nicotinamide adenine dinucleotide; NADH: Reduced form of NAD^+; $NADP^+$: Nicotinamide adenine dinucleotide phosphate; NADPH: Reduced form of $NADP^+$; NIR Spectroscopy: Near Infrared Spectroscopy; PDA: *o*-Phenylenediamine; RP-HPLC: Reverse phase high performance liquid chromatography; SFE: Supercritical fluid extraction; SPE: Solid phase extraction; TCs: Tocopherols; TEA: Triethylamine; THF: Tetrahydrofurane; TLC: Thin layer chromatography; UV: Ultraviolet

REFERENCES

[AOAC] Association of Official Analytical Chemists. 1990. Official Methods of Analysis, 15th ed. AOAC, Washington, DC, USA.

[AOAC] Association of Official Analytical Chemists. 1993. Peer Verified Methods Program. AOAC, Manual on Policies and Procedures. AOAC, Arlington, Virginia, USA.

[AOAC] Association of Official Analytical Chemists. Supplement 1996. Official Methods of Analysis, 16th ed., AOAC International, Gaithersburg, Maryland, USA.

Aburjai, T. and S. Al-Khalil, and M. Abuirjeie. 1998. Vitamin D_3 and its metabolites in tomato, potato, egg plant and zucchini leaves. Phytochemistry 49: 2497-2499.

Abushita, A.A. and E.A. Hebshi, H.G. Daoodand, and P.A. Biacs. 1997. Determination of antioxidant vitamins in tomatoes. Food Chem. 60: 207-212.

Agarwal S. and A.V. Rao. 2000. Tomato lycopene and its role in human health and chronic diseases. Can. Med. Assoc. J. 163: 739-744.

Ahmad, S. and A.K. Vashney, and P.K. Srivasta. 2005. Quality attributes of fruit bar made from papaya and tomato by incorporating hydrocolloids. Int. J. Food Prop. 8: 89-99.

Almeida, L.B. and M.V.C. Penteado. 1988. Carotenoids and pro-vitamin A value of white fleshed Brazilian sweet potatoes (*Impomoea batatas* Lam.). J. Food Compos. Anal. 1: 341-352.

Amir H. and M. Karas, J. Giat, M. Danilenko, R. Levy, T. Yeniahn, J. Levy, and Y. Sharoni. 1999. Lycopene and 1,25-dihydroxy-vitamin D3 cooperate in the inhibition of cell cycle progression and induction of differentiation in HL-60 leukemic cells. Nutr. Cancer 33: 105-112.

Arias, R. and T.C. Lee, L. Logendra, and H. Janes. 2000. Correlation of lycopene measured by HPLC with the *L**, *a**, *b** color readings of a hydroponic tomato and the relationship of maturity with color and lycopene content. J. Agric. Food Chem. 48: 1697-1702.

Arnao, M.B. and A. Cano, and M. Acosta. 1999. Methods to measure the antioxidant activity in plant material: a comparative discussion. Free Radical Res. 31: S89-S96.

Arnao, M.B. and A. Cano, and M. Acosta. 2001. The hydrophilic and lipophilic contribution to total antioxidant activity. Food Chem. 73: 239-244.

Baranska, M. and H. Schulz. 2005. Spatial tissue distribution of polyacetylenes in carrot root. Analyst 130: 855-859.

Baranska, M. and H. Schulz, R. Baranski, T. Nothnagel, and L.P. Christensen. 2005. In situ simultaneous anlaysis of polyacetylenes, carotenoids and polysaccharides in carrot roots. J. Agric. Food Chem. 53: 6565-6571.

Baranska M. and W. Schüzte, and H. Schulz. 2006. Determination of lycopene and β-carotene content in tomato fruits and related products: comparison of FT-Raman, ATR-IR, and NIR Spectroscopy. Anal. Chem. 78: 8456-8461.

Baranski, R. and M. Baranska, and H. Schulz. 2005. Changes in carotenoid content and distribution in living plant tissue can be observed and mapped in situ using NIR-FT-Raman spectroscopy. Planta 222: 448-457.

Baysal, T. and S. Ersus, and D.A.J. Starmans. 2000. Supercritical CO_2 extraction of β-carotene and lycopene from tomato paste waste. J. Agric. Food Chem. 48: 5507-5511.

Belitz, H.-D. and W. Grosch, and P. Schieberle. 2004. Food Chemistry, 3rd ed. Springer, Berlin, Germany.

Bemis, D.L. and A.E. Katz, and R. Buttyan. 2006. Clinical trials of natural products as chemo-preventive agents for prostate cancer. Expert Opin. Invest. Drugs 15: 1191-1200.

Bhosale, P. and I.V. Ermakov, M.R. Ermakova, W. Gellermann, and P.S. Bernstein. 2004. Resonance Raman quantification of nutritionally important carotenoids in fruits, vegetables and their juices in comparison to high-pressure liquid chromatography analysis. J. Agric. Food Chem. 52: 3281-3285.

Bicanic, D. and M. Anese, S. Luterotti, D. Dadarlat, J. Gibkes, and M. Lubbers. 2003. Rapid, accurate and direct determination of total lycopene content in tomato paste. Rev. Sci. Instrum. 74: 687-689.

Brenna, O.V. and N. Berardo. 2004. Application of near-infrared reflectance spectroscopy (NIRS) to the evaluation of carotenoids content in maize. J. Agric. Food Chem. 52: 5577-5582.

Britton, G. Carotenoids. pp. 141-182. *In:* G.A.F. Hendry and J.D. Houghton. [eds.] 1992. Natural Foods Colorants. Blackie, Glasgow and London, UK.

Britton, G. UV/visible spectroscopy. pp. 13-62. *In:* G.Britton, S. Liaaen-Jensen and H. Pfander. [eds.] 1995. Carotenoids, Vol. 1B. Birkhäuser, Basel, Switzerland.

Bryson, G.M. and A.V. Barker. 2002. Determination of optimal fertilizer concentration range for tomatoes grown in peat-based medium. Commun. Soil Sci. Plant Anal. 33: 759-777.

Burns, J. and P.D. Fraser, and P.M. Bramley. 2003. Identification and quantification of carotenids, tocopherols and chlorophylls in commonly consumed fruits and vegetables. Phytochemistry 62: 939-947.

Cano, A. and J. Hernández-Ruiz, F. García-Canovas, M. Acosta, and M.B. Arnao. 1998. An end-point method for estimation of the total antioxidant activity in plant material. Phytochem. Anal. 9: 196-202.

Cantuti-Castelvetri, I. and B. Shukitt-Hale and J.A. Joseph. 2000. Neurobehavioral aspects of antioxidants in aging. Int. J. Dev. Neurosci. 18: 367-381.

Careri, M. and L. Elviri, and A. Mangia. 1999. Liquid chromatography-electrospray mass spectrometry of β-carotene and xanthophylls: validation of the analytical method. J. Chromatogr. A 762: 201-206.

Cataldi, T.R.I. and D. Nardiello, V. Carrara, R. Ciriello, and G.E. De Benedetto. 2003. Assessment of riboflavin and flavin content in common food samples by capillary electrophoresis with laser induced fluorescence detetion. Food Chem. 89: 309-314.

Chang, C.H. and H.Y. Lin, C.Y. Chang, and Y.C. Liu. 2006. Comparison of the antioxidants properties of fresh, freeze-dried and hot-air-dried tomatoes. J. Food Eng. 77: 478-485.

Chatterjee, J. and C. Chatterjee. 2003. Management of phytotoxicity of cobalt in tomato by chemical measures. Plant Sci. 164: 793-801.

Choudhari, S.M. and L. Ananthanarayan. 2007. Enzyme aided extraction of lycopene from tomato tissues. Food Chem. 102: 77-81.

Chun, J. and J. Lee, L. Ye, and R.R. Eitenmiller. 2006. Tocopherol and tocotrienol contents of raw and processed fruits and vegetables in the United States diet. J. Food Compos. Anal. 19: 196-204.

Clinton, S.K. 1998. Lycopene: chemistry, biology and implications for human health and disease. Nutr. Rev. 56: 35-51.

Clydesdale, F.M. 1993. Color as factor in food choice. Crit. Rev. Food Sci. Nutr. 33: 83-101.

Cohen, L.A. 2002. A review of animal model studies of tomato carotenoids, lycopene and cancer chemoprevention. Exp. Biol. Med. 277: 864-868.

Craft, N.E. Chromatographic techniques for carotenoid separation. pp. F2.3.1-F2.3.15. *In:* R. Wrolstad, T.E. Acree, H. An, E.A. Decker, M.H. Penner, D.S. Reid, S.J. Schwartz, C.F. Shoemaker and P. Sporns. [eds.] 2001. Current Protocols in Food Analytical Chemistry. John Wiley & Sons, Inc., New York, USA.

Cserháti, T. and E. Forgács. 2001. Liquid chromatographic separation of terpenoid pigments in foods and food products. J. Chromatogr. A 936: 119-137.

Curino, A. and L. Milanesi, S. Benassati, M. Skliar, and R. Boland. 2001. Effect of culture conditions on the synthesis of vitamin D_3 metabolites in *Solanum glaucophyllum* grown in vitro. Phytochemistry 58: 81-89.

D'Souza, M.C. and S. Singha and M. Ingle. 1992. Lycopene content of tomato fruit can be estimated from chromaticity values. Hortscience 27: 465-466.

Damon, M. and N.Z. Zhang, D.B. Haytowitz and S.L. Booth. 2005. Phylloquinone (vitamin K_1) content in vegetables, J. Food Compos. Anal. 18: 751-758.

Davis, A.R. and W.W. Fish, and P. Perkins-Veazie. 2003. A rapid spectrophotometric method for analyzing lycopene content in tomato and tomato products. Postharv. Biol. Tech. 28: 425-430.

De Leenheer, A. and H. Neils, and W. Lambert. Fat soluble vitamins. pp. 5382-5393. *In:* A. Townshend. [ed. in chief] 1995. Encyclopedia of Analytical Chemistry. Academic Press, London, UK.

Debouba, M. and H. Gouia, A. Suzuki, and M.H. Ghorbel. 2006. NaCl stress effects on enzyme involved in nitrogen assimilation pathway in tomato *"Lycopersicon esculentum"* seedlings. J. Plant Physiol. 163: 1247-1258.

Del Valle, M. and M. Cámara, and M.E. Torija. 2006. Chemical characterization of tomato pomace. J. Sci. Food Agric. 86: 1232-1236.

Dong, C.X. and J.M. Zhou, X.H. Fan, H.Y. Wang, Z.Q. Duan, and C. Tang. 2004. Application methods of calcium supplements affect nutrient levels and calcium forms in mature tomato fruits. J. Plant Nutr. 27: 1443-1455.

Dumont, J.F. and J. Peterson, D. Haytowitz, and S.L. Booth. 2003. Phylloquinone and dihydrophylloquinone contents of mixed dished, processed meats, soups and cheeses. J. Food Compos. Anal. 16: 595-603.

Ebbs, S. and J. Talbott, and R. Sankaran. 2006. Cultivation of garden vegetables in Peoria Pool sediments from the Illinois River: a case study in trace element accumulation and dietary exposures. Environ. Int. 32: 766-774.

El-Arab, A.E. and M. Ali, and L. Hussein. 2004. Vitamin B1 profile of the Egyptian core foods an adequacy of intake. J. Food Compos. Anal. 17: 81-97.

Esteve, M.J. and R. Farre, A. Frígola, and J.M. García-Cantabella. 2001. Simultaneous determination of thiamine and riboflavin in mushrooms by liquid chromatography. J. Agric. Food Chem. 49: 1450-1454.

Fernández-Ruiz, V. and M.C. Sánchez Mata, M. Cámara, M.E. Torija, C. Chaya, L. Galiana-Balaguer, S. Rosello, and F. Nuez. 2004. Internal quality characterization of fresh tomato fruits. Hortscience 39: 339-345.

Ferruzzi M.G. and L.C Sander, C.L. Rock, and S.J. Schwartz. 1998. Carotenoid determination in biological microsamples using liquid chromatography with a coulometric electrochemical array detector. Anal. Biochem. 256: 74-81.

Ferruzzi, M.G. and M.L. Nguyen, L.C. Sander, C.L. Rock, and S.J. Schwartz. 2001. Analysis of lycopene geometrical isomers in biological microsamples by liquid chromatography with coulometric array detection. J. Chromatogr. B 760: 289-299.

Fish, W.W. and P. Perkins-Veazie, and J.K. Collins. 2002. A quantitative assay for lycopene that utilizes reduced volumes of organic solvents. J. Food Compos. Anal. 15: 309-317.

Fisher, J.F. and R.L. Rouseff. 1986. Solid-phase extraction and HPLC determination of β-cryptoxanthin and α- and β-carotene in orange juice. J. Agric. Food Chem. 34: 985-989.

Frenich, A.G. and M.E.H. Torres, A.B. Vega, J.L.M. Vidal, and P.P Bolanos. 2005. Determination of ascorbic acid and carotenoids in food commodities by liquid chromatography with mass spectrometry detection. J. Agric. Food Chem. 53: 7371-7376.

Galiana-Balaguer, L. and S. Rosello, and F. Nuez. 2006. Characterization and selection of balanced sources of variability for breeding tomato (*Lycopersicon*) internal quality. Genet. Resour. Crop Evol. 53: 907-923.

Gálvez Mariscal, A. and I. Flores Argüello, and A. Farrés González Saravia. p. 147. *In*: S. Badui Dergal. [ed.] 2006. Química de los Alimentos. Pearson, Naucalpán de Juárez, México.

Garrido Frenich, A. and M.E. Hernández Torres, A. Belmonte Vega, J.L. Martínez Vidal, and P. Plaza Bolaños. 2005. Determination of ascorbic acid and carotenoids in food commodities by liquid chromatography with mass spectrometry detection. J. Agric. Food Chem. 53: 7371-7376.

Gilgun-Sherki, Y. and E. Melamed, and D. Offen. 2001. Oxidative stress induced-neurodegenerative diseases: the need for antioxidants that penetrate the blood brain barrier. Neuropharmacology 40: 959-975.

Gómez-Prieto, M.S. and M.M. Caja, M. Herraiz and G. Santa-María. 2002. Solubility in supercritical carbon dioxide of the predominant carotenes of tomato skin. J. Agric. Food Chem. 79: 897-902.

Gómez-Prieto, M.S. and M.M. Caja, M. Herraiz and G. Santa-María. 2003. Supercritical fluid extraction of all-*trans*-lycopene from tomato. J. Agric. Food Chem. 51: 3-7.

Goula, A. M. and K. G. Adamopoulos, P.C. Chatzitakis and V.A. Nikas. 2006. Prediction of lycopene degradation during a drying process of tomato pulp. J. Food Eng. 74: 37-46.

Gould, W.A. 1992. Tomato Production, Processing & Technology, 3rd ed. CTI Pub. Inc., Baltimore, USA.

Granado, F. and B. Olmedilla, I. Blanco, and E. Rojad-Hidalgo. 1992. Carotenoid composition in raw and cooked Spanish vegetables. J. Agric. Food Chem. 40: 2135-2140.

Greenwald, P. and C.K. Clifford, and J.A. Milner. 2001. Diet and cancer prevention. Eur. J. Cancer 37: 948-965.

Guil-Guerrero, J.L. and A. Giménez-Giménez, I. Rodríguez-García, and M.E. Torija-Isasa. 1998a. Nutritional composition of *Sonchus* species (*S. asper* L., *S. oleraceus* L. and *S. tenerrimus* L.). J. Sci. Food Agric. 76: 628-632.

Guil-Guerrero, J.L. and J.J. Giménez Martínez, and M.E. Torija Isasa. 1998b. Mineral nutrient composition of edible wild plants. J. Food Compos. Anal. 11: 322-328.

Guil-Guerrero, J.L. and M.M. Rebolloso-Fuentes, and M.E. Torija Isasa. 2003. Fatty acids and carotenoids from Stinging Nettle (*Urtica dioica* L.). J. Food Compos. Anal. 16: 111-119.

Hadden, W.L. and R.H. Watkins, L.W. Levy, E. Regalado, D.M. Rivadeneira, R.B. van Breemen, and S.J. Schwartz. 1999. Carotenoid composition of marigold (*Tagetes erecta*) flower extract used as nutritional supplement. J. Agric. Food Chem. 47: 4189-4194.

Halliwell, B. and M.A. Murcia, S.A. Chirico, and O.I. Aruoma. 1995. Free radicals and antioxidants in food and in vivo: what they do and how they work. Crit. Rev. Food Sci. Nutr. 35: 7-20.

Hallmann, E. and E. Rembialkowska. 2007. Comparison of the nutritive quality of tomato fruits from organic and conventional production in Poland. *In:* U. Niggli, C. Leifert, T. Alföndi, L. Lück and H. Willer. [eds.] The Proceedings of the 3rd QLIF Congress, Hohenheim, Germany. FiBL.

Hanson, P.L. and R.Y. Yang, J. Wu, J.T. Chen, D. Ledesma, S.C.S. Tsou, and T.C. Lee. 2004. Variation of antioxidants in tomato. J. Am. Soc. Hortic. Sci. 129: 704-711.

Hart, D.J. and K.J. Scott. 1995. Development and evaluation of an HPLC method for the carotenoids in foods, and the measurement of the carotenoid content of vegetables and fruits commonly consumed in the UK. Food Chem. 54: 101-111.

Hernández, M. and J. Rull, D. Ríos, E. Rodríguez, and C. Díaz. 2005. Chemical composition of cultivar of tomatoes resistant and non resistant against the tomato yellow leaf curl virus (TYLCV). Electron. J. Environ. Agric. Food Chem. (EJEAFChe—electronic journal) 4: 1049-1054.

Hou, T.Z. and R.E. Mooneyham. 1999. Applied studies of plant meridian system: I. The effect of the agri-wave technology on yield and quality of tomato. Amer. J. Chin. Med. 27: 1-10

Huck, C.W. and M. Popp, H. Scherz, and G.K. Bonn. 2000. Development and evaluation of a new method for the determination of the carotenoid content in selected vegetables by HPLC and HPLC-MS-MS. J. Chromatogr. Sci. 38: 441-449.

Iwase, H. 2002. Simultaneous sample preparation for high-performance liquid chromatographic determination of vitamin A and β-carotene in emulsified nutritional supplements after solid-phase extraction. Anal. Chim. Acta 463: 21-29.

Karapanos, I.C. and C. Fasseas, C.M. Olympios, and H.C. Passam. 2006. Factors affecting the efficacy of agar-based substrates for the study of tomato pollen germination. J. Hortic. Sci. Biotechnol. 81: 631-638.

Koivu-Tikkanen, T.J. and V. Ollilainen, and V.I. Piironen. 2000. Determination of phylloquinone in animal products with fluorescence detection after postcolumn reduction with metallic zinc. J. Agric. Food Chem. 54: 2606-2613.

Konings, E.J.M. and H.H.S. Roomans. 1997. Evaluation and validation of an LC method for the analysis of carotenoids in vegetables and fruit. Food Chem. 59: 599-603.

Koyama, Y. Resonance Raman spectroscopy. pp. 135-146. *In:* G. Britton, S. Liaaen-Jensen and H. Pfander. [eds.] 1995. Carotenoids, Vol. 1B. Birkhäuser, Basel, Switzerland.

Kozukue, N. and M. Friedman. 2003. Tomatine, chlorophyll, β-carotene and lycopene content in tomatoes during growth and maturation. J. Sci. Food Agric. 83: 195-200.

Kurilich, A.C. and S.J. Britz, B.A. Clevidence, and J.A. Novotny. 2003. Isotopic labelling and LC-APCI-MS quantification for investigating absorption of carotenoids and phylloquinone from kale (*Brassica oleracea*). J. Agric. Food Chem. 51: 4877-4883.

Lasheras, C. and S. Fernández, and A.M. Patterson. 2000. Mediterranean diet and age with respect to overall survival in institutionalised, non-smoking elderly people. Amer. J. Clin. Nutr. 71: 987-992.

Lee, J. and L. Ye, W.O. Landen, and R.R. Eitenmiller. 2000. Optimization of an extraction procedure for the quantification of vitamin E in tomato and broccoli using surface response methodology. J. Food Compos. Anal. 13: 45-57.

Lenucci, M.S. and D. Cadinu, M. Taurino, G. Piro, and G. Dalessandro. 2006. Antioxidant composition in Cherri and high-pigment tomato cultivars. J. Agric. Food Chem. 54: 2606-2613.

Lester, G.E. and K.M. Crosby. 2002. Ascorbic acid, folic acid and potassium content in postharvest green-flesh honeydew muskmelons: Influence of cultivar, fruit size, soil type and year. J. Amer. Soc. Hortic. Sci. 127: 843-847.

Levy, J. and Y. Sharoni. 2004. The functions of tomato lycopene and its role in human health. HerbalGram 62: 49-56.

Lide, D.R. [ed. in chief] 1992. CRC Handbook of Chemistry and Physics, 73rd ed., CRC Press Inc., Boca Raton, Florida, USA.

Lin, C.H. and B.H. Chen. 2003. Determination of carotenoids in tomato juice by liquid chromatography. J. Chromatogr. A 1012: 103-109.

Liu, R.H. 2003. Health benefits of fruits and vegetables are from additive and synergistic combinations of phytochemicals. Amer. J. Clin. Nutr. 78: 517S-520S.

Livny O. and I. Kaplan, R. Reifen, S. Polak-Charcon, Z. Madar, and B. Schwartz. 2002. Lycopene inhibits proliferation and enhances gap-junction communication of KB-1 human oral tumor cells. J. Nutr. 132: 3759-3854.

Luthria, D.L. and S. Mukhopadhyay, and D.T. Krizek. 2006. Contents of total phenolic acids in tomato (*Lycopersicon esculentum* Mill.) fruits as influenced by cultivar and solar UV radiation. J. Food Compos. Anal. 19: 771-776.

Martín-Belloso, O. and E. Llanos-Barriobero. 2001. Proximate composition, minerals and vitamins in selected canned vegetables. Eur. Food Res. Technol. 212: 182-187.

Mataix-Verdu, J. and M. Mañas-Almendros, J. Llopis-González, and E. Martínez de Victoria-Muñoz. 1995. Tabla de composición de alimentos españoles Instituto de nutrición y tecnología de alimentos. Universidad de Granada, Granada, Spain.

Mélendez-Martínez, A.J. and I.M. Vicario, and F.J. Heredia. 2003a. A routine high-performance liquid chromatography method for carotenoid determination in ultrafrozen orange juices. J. Agric. Food Chem. 51: 4219-4224.

Mélendez-Martínez, A.J. and I.M. Vicario and F.J. Heredia. 2003b. Application of tristimulus colorimetry to estimate the carotenoids content in ultrafrozen orange juices. J. Agric. Food Chem. 51: 7266-7270.

Mélendez-Martínez, A.J. and I.M. Vicario and F.J. Heredia. 2004. Correlation between visual and instrumental colour measurements of orange juice dilutions. Effect of the background. Food Qual. Prefer. 16: 471-478.

Mélendez-Martínez, A.J. and G. Britton, I.M. Vicario, and F.J. Heredia. 2007a. Relationship between the colour and the chemical structure of carotenoids pigments. Food Chem. 101: 1145-1150.

Meléndez-Martínez, A.J. and I.M. Vicario, and F.J. Heredia. 2007b. Provitamin A carotenoids and ascorbic acid contents of the different types of orange juices marketed in Spain. Food Chem. 101: 177-184.

Negri, E. and S. Franceschi, C. Bosetti, F. Levi, E. Conti, M. Parpinel, and C. La Vecchia. 2000. Selected micronutrients and oral and pharyngeal cancer. Int. J. Cancer 86: 122-127.

Niizu, P.Y. and D.B. Rodriguez-Amaya. 2005. New data on the carotenoid composition of raw salad vegetables. J. Food Compos. Anal. 18: 739-749.

Oliver, J. and A. Palou, and A. Pons. 1998. Semi-quantification of carotenoids by high-performance liquid chromatography saponification-induced losses in fatty foods. J. Chromatogr. A 829: 393-399.

Olives Barba, A.I. and M. Cámara Hurtado, M.C. Sánchez Mata, V. Fernández Ruiz, and M. López Sáenz de Tejada. 2006. Application of a UV-vis detection-HPLC method for a rapid determination of lycopene and β-carotene in vegetables. Food Chem. 95: 328-336.

Parviainen, M.T. Water-soluble vitamins. pp. 5393-5401. *In:* A. Townshend. [ed. in chief] 1995. Encyclopedia of Analytical Chemistry. Academic Press, London, UK.

Pedro, A.M.K. and M.M. Ferreira. 2005. Nondestructive determination of solids and carotenoids in tomato products by near-infrared spectroscopy and multivariate calibration. Anal. Chem. 77: 2505-2511.

Piironen, V. and T. Koivu. 2000. Quality of vitamin K analysis and food composition data in Finland. Food Chem. 68: 223-226.

Pither, R.J. Fruits and fruits products. pp. 1525-1531. *In:* A. Townshend. [ed. in chief] 1995. Encyclopedia of Analytical Chemistry. Academic Press, London, UK.

Pohl, P. and B. Prusisz. 2007. Fractionation analysis of manganese and zinc in tea infusions by two-column solid phase extraction and flame atomic absorption spectrometry. Food Chem. 102: 1415-1424.

Pól, J. and T. Hyötyläinen, O. Ranta-Aho, and M-L. Riekkola. 2004. Determination of lycopene in food by on-line SFE coupled to HPLC using a single monolithic column for trapping and separation. J. Chromatogr. A 1052: 25-31.

Prema, T.P. and N. Raghuramulu. 1996. Vitamin D_3 and its metabolites in tomato plant. Phytochemistry 42: 617-620.

Premuzic, Z. and M. Bargiela, A. Garcia, A. Rendina, and A. Iorio. 1998. Calcium, iron, potassium, phosphorus and vitamin C content of organic and hydroponic tomatoes. Hortscience 33: 255-257.

Rao, A.V. and S. Agarwal. 2000. Role of antioxidant lycopene in cancer and heart disease. J. Amer. College. Nutr. 19: 563-569.

Rao, A.V. and S. Honglei. 2002. Effect of low dose of lycopene intake on lycopene bioavailability and oxidative stress. Nutr. Res. 22: 1125-1131.

Rao A.V. and Z. Waseem, and S. Agarwal. 1998. Lycopene content of tomatoes and tomato products and their contribution to dietary lycopene. Food Res. Int. 31: 737-741.

Rodriguez-Amaya, D.B. 1996. Assessment of the provitamin A contents of foods— the Brazilian experience. J. Food Compos. Anal. 9: 196-230.

Rodríguez-Bernaldo de Quirós, A. and H.S. Costa. 2006. Analysis of carotenoids in vegetable and plasma samples: a review. J. Food Compos. Anal. 19: 97-111.

Rouphael, Y. and G. Colla. 2005. Growth, yield, fruit quality and nutrient uptake of hydroponically cultivated zucchini squash as affected by irrigation systems and growing seasons. Sci. Hortic. 105: 177-195.

Rozzi, N.L. and R.K. Singh, R.A. Vierling, and B.A. Watkins. 2002. Supercritical fluid extraction of lycopene from tomato processing by products. J. Agric. Food Chem. 50: 2638-2643.

Ruiz, J.J. and N. Martínez, S. García-Martínez, M. Serrano, M. Valero, and R. Moral. 2005. Micronutrient composition and quality characteristics of traditional tomato cultivars in southeast Spain. Commun. Soil Sci. Plant Anal. 36: 649-660.

Rupérez, F.J. and D. Martín, E. Herrera, and C. Barbas. 2001. Chromatographic analysis of α-tocopherol and related compounds in various matrices. J. Chromatogr. A 935: 45-69.

Sadler, G.D. and J.D. Davis, and D. Dezman. 1990. Rapid extraction of lycopene and β-carotene from reconstituted tomato paste and pink grapefruit homogenates. J. Food Sci. 55: 1460-1461.

Salles, C. and S. Nicklaus, and C. Septier. 2003. Determination and gustatory properties of taste-active compounds in tomato juice. Food Chem. 81: 395-402.

Samaniego, C. and A.M. Troncoso, M.C. García-Parrilla, J.J. Quesada, H. López, and M.C. López. 2007. Different radical scavenging tests in virgin olive oil and their relation to the total phenol content. Anal. Chim. Acta 593: 103-107.

Sánchez-Mata, M.C. and M. Cámara-Hurtado, C. Díez-Marqués, and M.E. Torija-Isasa. 2000. Comparison of high-performance liquid chromatography and spectrofluorimetry for vitamin C analysis of green beans (*Phaseolus vulgaris* L.). Eur. Food Res. Tech. 210: 220-225.

Sánchez-Mata, M.C. and M. Cámara-Hurtado, and C. Díez-Marqués. 2003. Extending shelf-life and nutritive value of green beans (*Phaseolus vulgaris* L.) by controlled atmosphere storage: micronutrients. Food Chem. 80: 317-322.

Schiedt, K. and S. Liaaen-Jensen. Isolation ana analysis. pp. 81-108. *In:* G. Britton, S. Liaaen-Jensen and H. Pfander. [eds.] 1995. Carotenoids, Vol. 1A. Birkhäuser, Basel, Switzerland.

Schoefs, B. 2002. Chlorophyll and carotenoid analysis in food products. Properties of the pigments and methods of analysis. Trends Food Sci. Technol. 13: 361-371.

Scott, K.J. Detection and measurement of carotenoids by UV/VIS spectrophotometry. pp. F2.2.1-F2.2.10. *In:* R. Wrolstad, T.E. Acree, H. An, E.A. Decker, M.H. Penner, D.S. Reid, S.J. Schwartz, C.F. Shoemaker and P. Sporns. [eds.] 2001. Current Protocols in Food Analytical Chemistry. John Wiley & Sons, Inc., New York, USA.

Scott, K.J. and P.M. Finglas, R. Seale, D.J. Hart, and I. de Froidmont-Görtz. 1996. Interlaboratory studies of HPLC procedures for the analysis of carotenoids in foods. Food Chem. 57: 85-90.

Seiler, N. Fluorescent derivatives. pp. 196-199. *In*: K. Blau and J.M. Halket. [eds.] 1993. Handbook of Derivatives for Chromatography, John Wiley, Chichester, UK.

Sgherri, C. and M.F. Quartacci, R. Izzo and F. Navari-Izzo. 2002. Relation between lipoic acid and cell redox status in wheat grown in excess copper. Plant Physiol. Biochem. 40: 591-597.

Sgherri, C. and F. Navari-Izzo, A. Pardosssi, G.P. Soressi and R. Izzo. 2007. The influence of the diluted seawater and ripening stage on the content of antioxidants in fruits of different tomato genotypes. J. Agric. Food Chem. 55: 2452-2458.

Simonetti, P. and P. Gardana, P. Riso, P. Mauri, P. Pietta and M. Porrini. 2005. Glycosylated flavonoids from tomato puree are bioavailable in humans. Nutr. Res. 25: 717-726.

Skliar, M. and A. Curino, E. Milanesi, S. Benassati, and R. Boland. 2000. *Nicotiana glauca* and other plant species containing vitamin D_3 metabolites. Plant Sci. 156: 193-199.

Souci, S.W. and W. Fachmann, and H. Kraut. 1989. Food Composition and Nutrition Tables. Wissenschaftliche verlagsgesellschaft, Stuttgart, Germany.

Souci, S.W. and W. Fachmann, and H. Kraut. 1996. Food Composition and Nutrition Tables. Wissenschaftliche verlagsgesellschaft, Stuttgart, Germany.

Stephens, W.E. and A. Calder. 2004. Analysis of non-organic elements in plant foliage using polarised X-ray fluorescence spectrometry. Anal. Chim. Acta 527: 89-96.

Stevens, R. and M. Buret, C. Garchery, Y. Carretero, and M. Causse. 2006. Technique for rapid, small scale analysis of vitamin C levels in fruits and application to a tomato mutant collection. J. Agric. Food Chem. 54: 6159-6165.

Sumar, S. and T.P. Coultate. Food and nutritional analysis. Introduction. pp. 1460-1468. *In*: A. Townshend. [Editor In-chief] 1995. Encyclopedia of Analytical Chemistry. Academic Press, London, UK.

Sundl, I. and M. Murkovic, D. Bandoniene, and B.M. Winklhofer. 2007. Vitamin E content of foods: Comparison of the results obtained from food composition tables and HPLC analysis. Clin. Nutr. 26: 145-153.

Taber, H.G. 2004. Boron analysis of mehlich n° 3 extractant with modified inductive coupled argon plasma techniques to eliminate iron interference. Commun. Soil Sci. Plant Anal. 35: 2957-2963.

Taungbodhitham, A.K. and G.P. Jones, M.L. Wahlqvist, and D.R. Briggs. 1998. Evaluation of extraction method for the analysis of carotenoids in fruits and vegetables. Food Chem. 63: 577-584.

Thompson, K.A. and M.R. Marshall, C.A. Sims, C.I. Wei, S.A. Sargent, and J.W. Scott. 2000. Cultivar, maturity, and heat treatment on lycopene content in tomatoes. J. Food Sci. 65: 791-795.

Toor, R.K. and G.P. Savage and A. Heeb. 2006. Influence of different types of fertilisers on the major antioxidant components of tomatoes. J. Food Compos. Anal. 19: 20-27.

Turner, C. and J.W. King, and L. Mathiasson. 2001. Supercritical fluid extraction and chromatography for fat-soluble vitamin analysis. J. Chromatogr. A 936: 215-237.

Van Breemen, R.B. Mass spectrometry of carotenoids. pp. F2.4.1-F2.4.13. *In:* R. Wrolstad, T.E. Acree, H. An, E.A. Decker, M.H. Penner, D.S. Reid, S.J. Schwartz, C.F. Shoemaker and P. Sporns. [eds.] 2001. Current Protocols in Food Analytical Chemistry. John Wiley & Sons, Inc., New York, USA.

Van Breemen, R.B. and X. Xu, MA. Viana, L. Chen, M. Stacewicz-Sapuntzakis, C. Duncan, P.E. Bowen, and R. Sharifi. 2002. Liquid chromatography-mass spectrometry of *cis*- and all-*trans*-lycopene in human serum and prostate tissue after dietary supplementation with tomato sauce. J. Agric. Food Chem. 50: 2214-2219.

Vasapollo, G. and L. Longo, L. Rescio, and L. Ciurlia. 2004. Innovative supercritical CO_2 extraction of lycopene from tomato in the presence of vegetable oil as co-solvent. J. Supercrit. Fluid 29: 87-96.

Voutilainen, S. and T. Nurmi, J. Mursu and T.H. Rissanen. 2006. Carotenoids and cardiovascular health. Amer. J. Clin. Nutr. 83: 1265-1271.

Wang, T. and G. Bengtsson, I. Kärnefelt, and L.O. Björn. 2001. Provitamins and vitamins D_2 and D_3 in *Cladina* spp. over a latitudinal gradient: possible correlation with UV levels. J. Photochem. Photobiol. B: Biology 62: 118-122.

28

Methods for PCR and Gene Expression Studies in Tomato Plants

Dimitris Argyropoulos[1*], Theodoros H. Varzakas[2], Charoula Psallida[1] and Ioannis S. Arvanitoyannis[3]

[1]Institute of Biotechnology, National Agricultural Research Foundation, Sofokli Venizelou 1 Lykovrisi 14123, Attiki Greece

[2]Department of Technology of Agricultural Products, School of Agricultural Technology, Technological Educational Institute of Kalamata, Antikalamos 24100, Kalamata, Hellas, Greece

[3]Department of Agriculture, Ichthyology and Aquatic Environment, Agricultural Sciences, University of Thessaly, Fytokou Street, Nea Ionia Magnesias, 38446 Volos, Hellas, Greece

ABSTRACT

In the last few decades, technological progress in molecular biology and information science has resulted in methods with extended capabilities in the study and analysis of genomes. Many countries are investing more and more in gene expression studies as a way to meet their needs for growth in food. Most economically important traits of crop plants follow a continuous distribution caused by the action and interaction of many genes and various environmental factors. Past techniques for genome analysis have been greatly improved and new methods have emerged, the latest based on micro-array technology. The core part in most of the methods for gene expression studies is the polymerase chain reaction technique. The commercial significance of tomato plants led to efforts of genetic modification and gene transfer from the early days of the molecular genetics revolution. As a result of these efforts, information data for

*Corresponding author

A list of abbreviations is given before the references.

the tomato genome are increasing constantly. Tomato microarrays for expression analysis are already useful to identify markers genes implicated in fruit ripening and development in the Solanaceae. Molecular marker analysis of tomato plants allows the identification of genome segments, so-called quantitative trait loci that help to select superior genotypes without uncertainties due to genotype by environment interaction and experimental error.

INTRODUCTION

Progress in molecular biology and bioformatics has offered new possibilities in the study of the plant genome. Since the time of Southern blot analysis, technological advances have led into new methods such as restriction fragment length polymorphism (RFLP), randomly amplified polymorphic DNA (RAPD), amplified fragment length polymorphism (AFLP) and more computerized methods such as expressed sequence tags—serial analysis of gene expression (EST-SAGE), real time polymerase chain reaction (PCR) and microarray analysis. The core part in all of these methods is PCR technology and, with the aid of information science, data resulting from these methods have already been stored in databases under the name of each method. Gene expression studies mainly include quantification of mRNAs being produced under different conditions such as stress, growth, development, and cell and tissue localization or as part of evaluation of the effects of gene transfection. A variety of techniques exist to quantify mRNAs and usually involve northern hybridization, ribonuclease protection or real-time PCR assays (Suzuki et al. 2000). An outline of these methods is presented, followed by applications in the analysis of tomato genome and gene expression profiling.

METHODS FOR GENOME ANALYSIS

Restriction Fragment Length Polymorphism

Using the RFLP technique, organisms may be differentiated by analysis of patterns derived from cleavage of their DNA. Organisms can differ in the distance between sites of cleavage of a particular restriction endonuclease; the length of the fragments produced will differ when the DNA is digested with a restriction enzyme. The restriction fragments are separated according to length by agarose gel electrophoresis. The resulting gel may be further analysed by Southern blotting using specific probes. The similarity of the patterns generated can be used to differentiate even

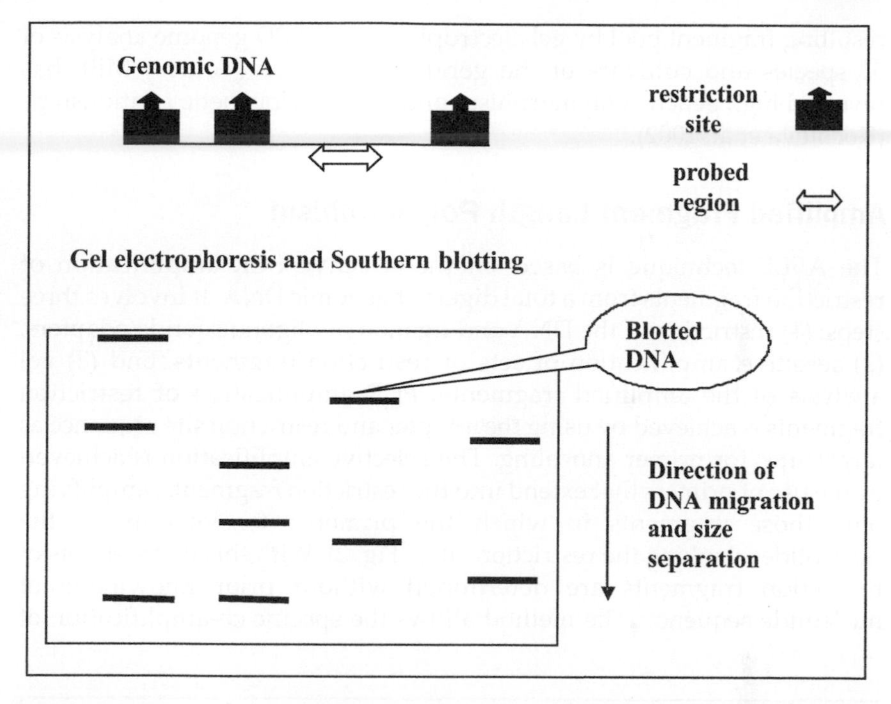

Fig. 1 Restriction Fragment Length Polymorphisms are molecular markers used in creating genetic maps of chromosomes. Combinations of enzymes and probes with digested DNA produce different profiles of DNA fragments, unique for each organism.

strains of the same organism (Fig. 1). The RFLP method has been combined with microsynteny analysis comparing tomato to *Arabidopsis* genome and new microsyntenic EST markers were rapidly identified (KwangChul et al. 2002)

Randomly Amplified Polymorphic DNA

The RAPD technique is a type of PCR reaction with DNA sequences amplified randomly. It is applied by using several arbitrary, short primers (8-12 nucleotides) to DNA in the expectation that some parts of DNA will amplify. By resolving the resulting patterns by agarose gel electrophoresis it is possible to differentiate strains of the same organism. The method has been used to assess populations of mRNA with differential display (Liang and Paedee 1992) or arbitrarily primed PCR (Welsh et al. 1992), amplification of random mRNA subsets and subsequent analysis of the resulting fragment pool by gel electrophoresis. RAPD genome analysis of 53 species and cultivars of the genus *Lycopersicon* (Tourn.) Mill. has revealed high genetic polymorphism and their phylogenetic relationships (Kochieva et al. 2002).

Amplified Fragment Length Polymorphism

The AFLP technique is based on the selective PCR amplification of restriction fragments from a total digest of genomic DNA. It involves three steps: (1) restriction of the DNA and ligation of oligonucleotide adapters, (2) selective amplification of sets of restriction fragments, and (3) gel analysis of the amplified fragments. PCR amplification of restriction fragments is achieved by using the adapter and restriction site sequence as target sites for primer annealing. The selective amplification is achieved by the use of primers that extend into the restriction fragments, amplifying only those fragments in which the primer extensions match the nucleotides flanking the restriction sites (Fig. 2). With this method, sets of restriction fragments are determined without prior knowledge of nucleotide sequence. The method allows the specific co-amplification of

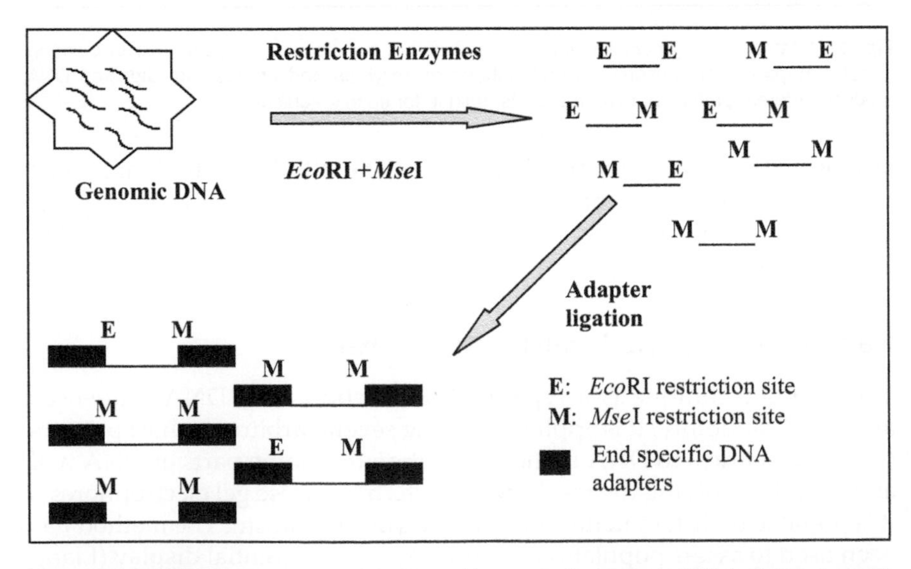

Fig. 2 In Amplified Fragment Length Polymorphism analysis, genomic DNA is digested usually with *Mse*I, and *Eco*RI enzymes. The resulting fragments are ligated to end-specific adapter molecules and are used in a preselective PCR with primers complementary to each of the two adaptor sequences having also an additional base at the 3′ end. Amplification of only 1/16 of *Eco*RI-*Mse*I fragments occurs.

high numbers of restriction fragments and is usually accompanied by automated capillary electrophoresis (Vos et al. 1995). The technique was originally described by Vos and Zabeau in 1993 in a patent submitted to the European Patent Office. AFLP, Simple Sequence Repeat and Single Nucleotide Polymorphism have been applied to the tomato genome for assessment of polymorphism and for genome mapping (Suliman-Pollatschek et al. 2002), while new AFLP sequences are constantly added at the Gene Bank database.

Serial Analysis of Gene Expression

Serial analysis of gene expression was developed as an elegant means to analyse mRNA populations by large-scale sequence determination of short identifying stretches of individual messengers (Velculescu et al. 1995), but the required depth of sequencing under different conditions makes it also labour intensive and time consuming (Fig. 3).

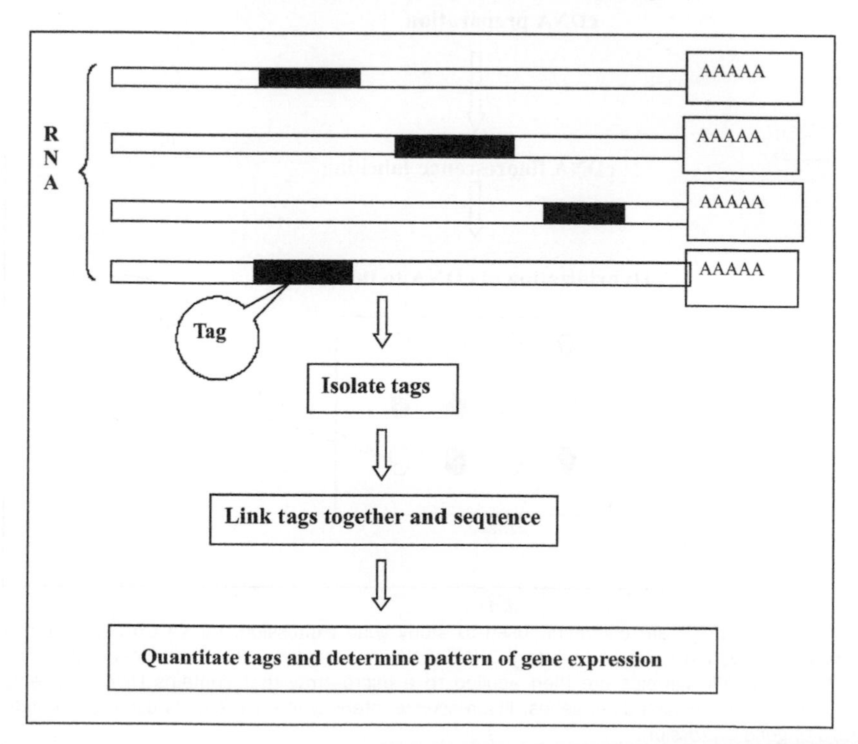

Fig. 3 In Serial Analysis of Gene Expression methodology, short sequence tags (10-14bp) obtained from a unique position within each transcript are used to uniquely identify transcripts. The number of times a particular tag is observed provides the expression level of the corresponding transcript.

Microarrays

More recent developments have made it possible to study altered gene expression in a more efficient and informative way by using DNA microarray technology (Lockhart and Winzeler 2000, Panda et al. 2003, Mockler et al. 2005). Using DNA microarrays, the expression of a large number of genes can be analysed simultaneously and in a semi-quantitative manner (Fig. 4). This allows for the analysis of different metabolic pathways in interaction and facilitates the identification of key

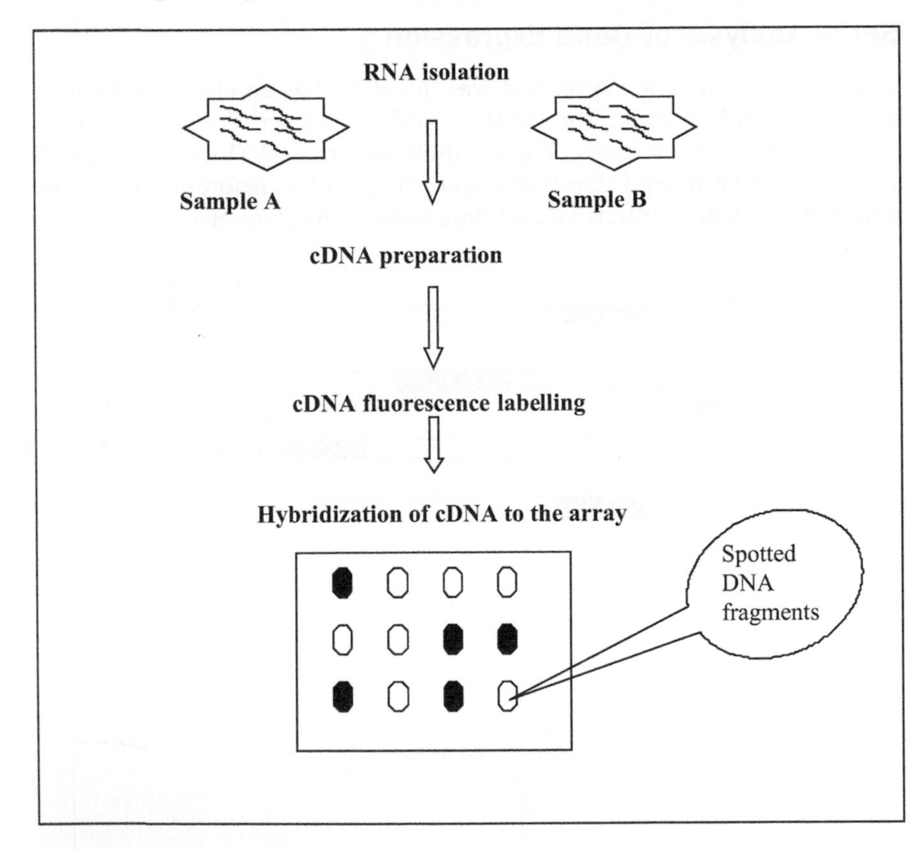

Fig. 4 Micro-arrays are commonly used to study gene expression. mRNA extracted from a sample is converted to complementary DNA (cDNA) and tagged with a fluorescent label. The fluorescent cDNA samples are then applied to a micro-array that contains DNA fragments corresponding to thousands of genes. Fluorescence intensity in each spot is used to estimate levels of gene expression.

responsive genes. For a limited number of species, microarrays have been constructed that represent all identified metabolic routes and genes active therein. These are the so-called whole genome arrays, oligo-arrays where all expressed gene sequences are represented by one or more short DNA sequences, usually up to 100 nucleotides (Mockler et al. 2005). Microarray analysis combined with suppression subtractive hybridization has been used in the identification of early salt stress response genes in tomato root (Ouyang et al. 2007).

Real Time PCR

In real time PCR, quantification of mRNA sequences is accomplished by absolute or relative analysis methods (Bustin 2002). Increasingly, the relative method of analysis is being used, as trends in gene expression can be better explained, but results depend on reference genes necessary to normalize sample variations (McMaugh and Lyon 2003, Weihong and Saint 2002).

A common technique in relative quantification is the choice of an endogenous control to normalize experimental variations, caused by differences in the amount of the RNA added in the reverse transcription (RT) PCR reactions. Specifications of reliable endogenous controls (i.e., housekeeping genes) are that they need to be abundant, remain constant in proportion to total RNA and be unaffected by the experimental treatments. The best choices proposed to be used as normalizers of isolated mRNA quantities are mainly RNAs produced from glyceraldehyde-3-phosphate dehydrogenase (GAPDH) (Wall and Edwards 2002, Bhatia et al. 1994), β-actin (Kreuzer et al. 1999), tubulin (Brunner et al. 2004) or rRNA (Bhatia et al. 1994). However, in general, depending on the developmental stage or environmental stimuli, the expression of certain reference genes is either up- or down-regulated.

The use of GAPDH mRNA as a normalizer is recommended with caution as it has been shown that its expression may be up-regulated in proliferating cells (Zhu et al. 2001). Use of 18S RNA as a normalizer is not always appropriate, as it does not have a polyA tail and thus prohibits synthesis of cDNA with oligo-dT. Expression of actin or tubulin often depends on the plant developmental stage (Diaz-Camino et al. 2005, Czechowski et al. 2005) and is affected by environmental stresses (Jin et al. 1999), making their use as normalizers inappropriate.

Argyropoulos et al. (2006) proposed an alternative method for an internal control that would be applicable to different organisms without prior knowledge of genomic sequences assuming that one wants to normalize against total mRNA. In the presented methodology, synthetic

DNA molecules (adapters) of known sequence tail cDNA during reverse transcription and are used instead of internal reference genes. Tailing sequences are further amplified with polymerase I or Klenow polymerase leading to second strand cDNA synthesis of the adapter molecules to be used as indicators of the total mRNA quantity (Fig. 5).

The method was applied in the study of transient expression pattern of the germination-specific endo-b-mannanase gene, in germinating tomato seeds (*Lycopersicon esculentum* Mill.), where there are no documented stably expressed genes to serve as normalizers. Endo-b-mannanase is expressed exclusively in the endosperm cap tissue, prior to radicle emergence. It has been shown that its activity develops prior to germination, specifically in the micropylar region of the endosperm opposite the radicle tip (Toorop et al. 1996, Still and Bradford 1997), and is involved in hydrolysis of the mannan-rich cell walls of the tomato endosperm during germination, leading to radicle protrusion (Nonogaki and Morohashi 1996, Still and Bradford 1997).

A common housekeeping gene, GAPDH, was tested in parallel to determine its efficiency as a normalizer of the sample variation. In

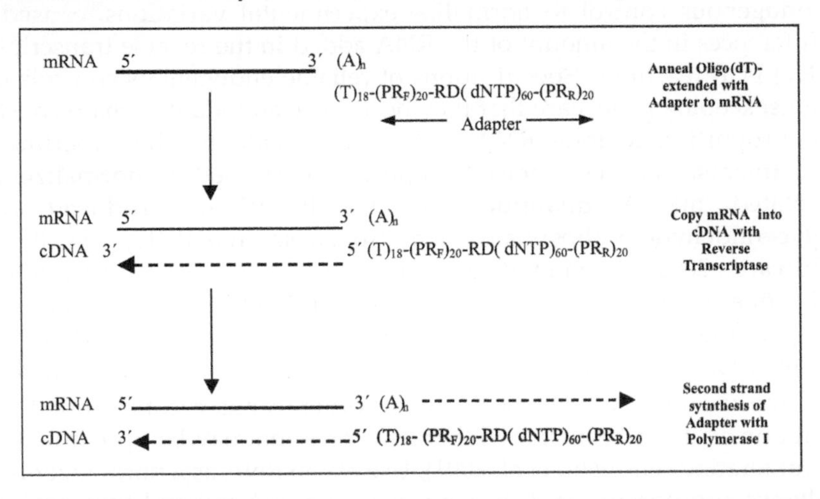

Fig. 5 Second strand synthesis of adapter molecules has been used to differentiate samples according to total mRNA quantity and to normalize sample variation. For each mRNA sample, reverse transcription was performed using Oligo $d(T)_{18}$ extended at the 5′ end with a synthetic DNA sequence of 100 bases length (adapter) forming a molecule of 118 bases. The adapter consisted of a random sequence of 60 bases RD $(dNTP)_{60}$ extended at both its ends with sequences of 20 bases length serving as the forward $(PRF)_{20}$ and the reverse $(PRR)_{20}$ primer of the adapter. (Argyropoulos et al. 2006).

conclusion, the proposed method with which to analyse mRNA helped to obtain a well-defined profile of the transiently expressed mannanase gene in germinating tomato seeds. Furthermore, it was shown that mannanase mRNA accumulated, in parallel with mannanase activity, to the same extent in seeds 5 hr prior to the completion of germination, irrespective of seed treatment, suggesting a close relationship between endo-β-mannanase accumulation, activity and tomato seed germination (Fig. 6).

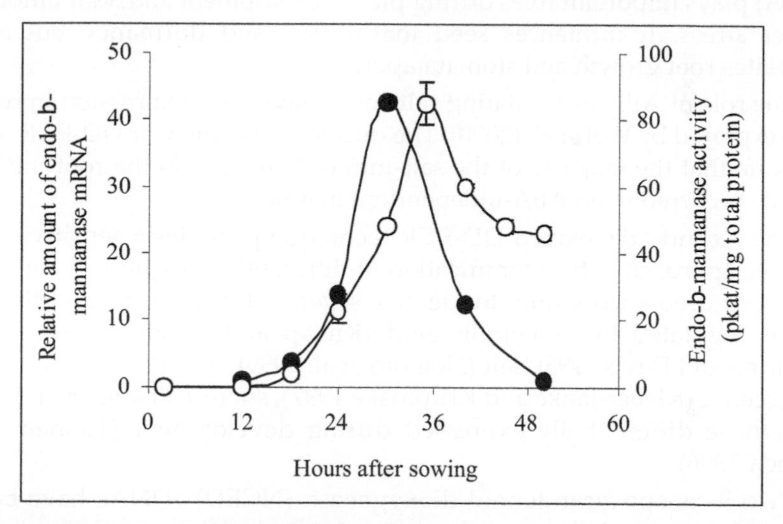

Fig. 6 Time course of mRNA production from the endo-b-mannanase gene (amounts expressed relative to that of 12h) and endo-b-mannanase activity (pkat/mg of total protein) during seed inbibition of the tomato cultivar ACE. Estimation of the mRNA levels was achieved by Real Time PCR normalizing sample variation using synthetic DNA molecules (adapters) of known sequence to tail cDNA during reverse transcription. Adapter molecules were used instead of reference genes (Argyropoulos et al. 2006).

GENE EXPRESSION STUDIES: EFFECTS IN MORPHOLOGY, PHYSIOLOGY AND AGRONOMIC TRAITS

Tomato Hormones

Representational difference analysis (RDA) is a hybridization selection procedure in which mRNAs that are present in one condition (wanted), and not in the other (unwanted), are selectively amplified. Tomato was chosen as a model system for a less well-documented species. cDNA libraries for two distinct physiological conditions of tomato fruits, red and green, were made (Kok et al. 2007). Tomato was used as a model because it is an important agricultural crop and a sufficient number of sequences have been annotated (NCBI/EMBL databases and TIGR tomato gene index database (http://www.tigr.org/tdb/tgi/plant.shtml). In addition, expression information is available, which allows for assessment of the success of the approach. The libraries were characterized by sequencing and hybridization analysis. The RDA procedure was shown to be effective in selecting for genes of relevance for the physiological conditions under investigation, and against constitutively expressed genes. Abscisic acid (ABA) plays important roles during plant development and adaptation to water stress. It influences seed maturation and dormancy but also regulates root growth and stomata aperture.

The role of ABA in regulating salt-responsive gene expression in roots was explored by Wei et al. (2000). The differential display or DD-PCR data indicate that the majority of the salt-induced changes in the root mRNA profile occurred in an ABA-independent manner.

The recently developed DD-PCR technique provides a sensitive and flexible approach to the identification of differentially expressed genes. It has been used successfully to identify several cDNAs corresponding to genes regulated by gibberellic acid (Knaap and Kende 1995), ozone (Sharma and Davis 1995), salt (Nemoto et al. 1999), heat (Joshi et al. 1996), senescence (Kleber-Janke and Krupinska 1997), sucrose (Tseng et al. 1995) and those differentially expressed during development (Tieman and Handa 1996).

Two 9-cis-epoxycarotenoid dioxygenase (NCED) cDNAs have been cloned from a petal library of *Gentiana lutea*. Both cDNAs carry a putative transit sequence for chloroplast import and differ mainly in their length and the 50-flanking regions. GlNCED1 was evolutionary closely related to *Arabidopsis thaliana* NCED6, whereas GlNCED2 showed highest homology to tomato LeNCED1 (Burbidge et al. 1997) and *A. thaliana* NCED3 (Zhu et al. 2007). In addition, genetic engineering has been applied to change the carotenoid cleavage capacity by over-expression or down-regulation of NCEDs. For example, constitutive over-expression of LeNCED1 from tomato resulted in a marked elevation of the ABA content (Thompson et al. 2000).

Temperature Resistance

Chilling tolerance is a multigenic complex trait (Tokuhisa and Browse 1999). Various mechanisms correlated with the acquisition of chilling tolerance have been identified including the 18:1 desaturase gene for the production of polyunsaturated membrane lipid (Miquel et al. 1993) and the catalase and alternative oxidase (AOX) genes in the oxidative stress defence mechanism (Kerdnaimongkol and Woodson 1999). Methyl salicylate (MeSA) vapour increased resistance against chilling injury in freshly harvested pink tomatoes. The expression patterns of AOX before and during the chilling period demonstrated that pre-treatment of tomato fruit with MeSA vapour increased the transcript levels of AOX. Methyl jasmonate (MeJA) and MeSA treatments induced chilling resistance in mature green tomatoes (Ding et al. 2002) and in green bell peppers (Fung et al. 2004).

However, MeJA is not as effective as MeSA in alleviating chilling injury of pink maturity stage tomatoes. This result confirms previous work demonstrating that tomato fruits at different maturity stages respond differently to plant growth regulator treatment (Ding and Wang 2003). Fung et al. (2006) used four EST tomato clones of AOX from the public database that belong to two distinctly related families, 1 and 2 defined in plants. Three clones were designated as LeAOX1a, 1b and 1c and the fourth clone as LeAOX2. Using RT-PCR, 1a and 1b genes were found to be expressed in leaf, root and fruit tissues, but 1c was expressed preferentially in roots. RNA transcript from LeAOX1a of AOX subfamily 1 was present in much greater abundance than 1b or 1c. Closely related LeAOX1a and 1b gene transcripts are expressed in chilled tomatoes (Holtzapffel et al. 2003). Enzymatically, LeAOX1a and 1b proteins were shown to be different in their regulatory properties.

The presence of longer AOX transcripts detected by RNA gel blot analysis in cold-stored tomato fruit was confirmed to be the un-spliced premRNA transcripts of LeAOX1a and LeAOX1b genes. Intron splicing of LeAOX1c gene was also affected by cold storage when it was detected in roots. This alternative splicing event in AOX pre-mRNAs molecules occurred, preferentially at low temperature, regardless of mRNA abundance. Fung et al. (2006) investigated the relationship of induced AOX gene expression and chilling tolerance by characterizing three closely related (Family 1) and one distinctly related (Family 2) LeAOX genes in chilled tomatoes.

A subset of SR family proteins (9G8) was shown to promote the nucleocytoplasmic export of mRNAs in mammalian cells (Huang and Steitz 2001). A gradual decrease in 9G8-SR transcript levels in cold-stored tomatoes was observed by Fung et al. (2006), who also suggested in their study that RNA processing steps may be important in the regulation of stress-related AOX expression at low temperature and are potentially related to chilling tolerance in tomatoes. High temperature is one of the most important restrictions on crop production. Plants respond to elevated temperatures by expressing several families of evolutionally conserved heat-shock proteins (HSPs). Heat-induced chilling tolerance, especially in fruits, has been related to the induction and maintenance of HSPs (Sabehat et al. 1996).

Small heat-shock proteins (sHSPs) are the major family of HSP induced by heat stress in plants. In this report, an approximately 1.9 kb of Lehsp23.8 50-flanking sequence was isolated from tomato genome. By using the b-glucuronidase (GUS) reporter gene system, the developmental and tissue-specific expression of the GUS gene controlled by the Lehsp23.8 promoter was characterized in transgenic tomato plants. Strong GUS staining was detected in the roots, leaves, flowers, fruits and germinated seeds after heat shock. The heat-induced GUS activity was different in the floral tissues at various developmental stages. Fluorometric GUS assay showed that the heat-induced GUS activity was higher in the pericarp than in the placenta. The heat-shock induction of the Lehsp23.8 promoter depended on the different stages of fruit development. The optimal heat-shock temperatures leading to the maximal GUS activity in the pericarp of green, breaker, pink and red fruits were 42, 36, 39 and 39.8°C, respectively. The heat-induced GUS activity in tomato fruits increased gradually within 48 hr of treatment and weakened during tomato fruit ripening. Obvious GUS activities under cold, exogenous ABA and heavy metal (Cd^{2+}, Cu^{2+}, Pb^{2+} or Zn^{2+}) stress conditions were also detected. These results show that the Lehsp23.8 promoter is characterized as strongly heat-inducible and multiple-stress responsive.

Structural and Developmental Proteins

Aquaporins facilitate water flux across biomembranes in plant cells and are involved in various physiological phenomena in several plant tissues (Baiges et al. 2002). Genome analysis has revealed that more than 30 aquaporins exist in higher plants (Johanson et al. 2001). The plant aquaporins are classified into plasma membrane intrinsic proteins (PIPs), tonoplast intrinsic proteins, small basic intrinsic proteins, and Nod26-like intrinsic proteins. The genes encoding plant aquaporins are expressed in several plant tissues and are regulated by certain environmental stress factors, such as cold, drought, and salinity (Tyerman et al. 2002, Jang et al. 2004). To determine whether aquaporins have physiological roles during seed development, Shiota et al. (2006) analysed the expression of genes encoding PIPs during seed and fruit development in tomato (*Lycopersicon esculentum* Mill.). Six genes encoding PIPs were detected in mature tomato seeds by RT-PCR using PCR primers corresponding to conserved transmembrane domains and asparagine-proline-alanine (NPA) motifs. The expression of these genes in developing seeds and fruit was analysed by RT-PCR using primers specific for nine tomato PIPs, including the six PIPs detected and an additional three PIPs from the tomato EST database. Increased expression of seven PIPs was detected during the earlier phase of seed development (12–32 days after flowering or DAF), and the expression of these genes decreased during the later phase (36–56 DAF). Each tomato PIP showed a distinct expression pattern during fruit development. In addition, the water content of the cells was calculated. The seed water content decreased gradually in the earlier phase of seed development (12–32 DAF) and was subsequently maintained at 44–50% from 36 to 56 DAF, whereas the water content of the fruit remained at 90% throughout fruit development. These results suggest that plasma membrane aquaporins play a physiological role during seed and fruit development in tomato. Previous studies have revealed the interaction of tomato aquaporins, tomato ripening-associated membrane protein and LeAqp2, but the expression of these genes in seeds has not been determined (Chen et al. 2001, Werner et al. 2001).

Glycine-rich proteins (GRPs) are a group of common proteins existing in various organisms. In plants, a wealth of GRPs has been identified and characterized (Sachetto-Martins et al. 2000). Plant GRPs represent a large group of proteins containing tandem glycine-rich (Gly-X)n domains and can be further classified into several sub-groups based on specific sequence domains. Furthermore, plant GRPs have been shown to localize in specific sub-cellular components; with the presence of certain specific sequence domains in some GRPs as suggestive of their sub-cellular localization, regulation of gene expression of plant GRPs varies considerably. Plant GRPs exhibit a wide range of tissue- or organ-specific expression patterns, even at a given developmental stage (de Oliveira et al. 1990). A number of plant GRPs are involved in the development of vascular and epidermal tissues (Parsons and Mattoo 1994). Some GRPs are preferentially expressed in specific organs or tissues such as fruit (Santino et al. 1997), flowers (Murphy and Ross 1998), roots (Matsuyama et al. 1999), root nodules (Kevei et al. 2002), or epidermis (Sachetto-Martins et al. 1995). Moreover, some GRPs are post-transcriptionally regulated (Carpenter et al. 1994) or post-translationally modified (Matsuyama et al. 1999).

To search for genes specifically or abundantly expressed in root, Lin et al. (2005) set out to identify tomato transcripts predominantly accumulated in roots, but not in fruit tissues. They reported the identification of a group of tomato genes (namely *LeGRP*), which are predominantly expressed in roots and exist as a gene family in plant genomes. The deduced amino acid sequences of two identified *LeGRP* genes share significant sequence homology to some plant structural GRPs that are targeted to the cell membrane/wall. *LeGRP* gene regulation is regulated by developmental cues as well as some environmental factors. The expression patterns of *LeGRPs* are quite unique and distinct from other plant GRPs that have been previously reported.

Lin et al. (2005) used Southern blot analysis to reveal the presence of multiple copies of *LeGRP* homologues in tomato, as well as in other dicot or monocot plant species tested. Comparative genomic Southern and PCR analyses indicated that some *LeGRP* genes could be cultivar-specific. An open-reading frame was identified in the sequence of one of the *LeGRP* genes (*LeGRP2*), with the predicted protein sharing good sequence homology with a few known and putative plant structural signal peptide-containing GRPs. However, the patterns and modulation of *LeGRP* gene expression were distinct from those of known plant GRP genes identified to date. *LeGRP* transcripts predominately accumulated in roots at different developmental stages, but not in leaves or ripe fruit tissues, and their levels declined gradually during plant development. Accumulation of *LeGRP* transcripts was effectively reduced by treatment with hydrogen peroxide and completely suppressed by salinity and application of sodium salicylate, but it was significantly enhanced with nutrient supplements such as HYPONeX solutions.

Chromosome Structure

Telomeres are specialized nucleoprotein structures located at the ends of linear eukaryotic chromosomes. They are essential for the maintenance of chromosome integrity and for the protection of chromosomes from exonucleolytic degradation or end-to-end fusion (Blackburn 1991). The functions of telomere-binding proteins that bind to these regions have been well described for humans and yeast and can be classified in two groups: control of telomere length by regulation of telomerase activity and maintenance of chromosome stability. Telomerase is a reverse transcriptase that uses its own RNA subunits as templates for the addition of telomere sequences to the ends of telomeres, compensating for the loss of telomeres in cell division.

Moriguchi et al. (2006) cloned and characterized the tomato telomere-binding protein LeTBP1 and transformed it into tobacco BY-2 cells to analyse its function *in vivo*. In addition, they discovered curious expression patterns of LeTBP1 during reproductive development. Sequence analysis revealed that LeTBP1 contained an open reading frame of 2067 bp and encoded a protein consisting of 689 amino acids with a predicted molecular mass of 77.0 kDa. The deduced amino sequence of LeTBP1 indicated that a myb-like motif, a structure particular to double-stranded telomere-binding proteins found in various organisms, existed in the C-terminal region. Southern blot analysis revealed that LeTBP1 was a single-copy gene in the tomato genome. Northern blot analyses showed that LeTBP1 was highly expressed in tissues with high cell division capacities as well as in fully differentiated tissues. The high level of expression of LeTBP1 in inflorescences was independent of flower formation, as shown with mutant inflorescences lacking flowers.

However, the level of LeTBP1 mRNA was greatly reduced in fruit. Gel shift assay revealed that LeTBP1509-689, which contained Gly509 to Ala689 of LeTBP1, including a myb-like motif, bound specifically to the double-stranded telomeric sequence, indicating that LeTBP1 is a double-stranded telomeric DNA-binding protein (Moriguchi et al. 2006).

Photoreceptors

The ability of plants to respond to light is achieved through a number of photoreceptor families, which include phytochromes that sense red and far-red light and blue-light-specific phototropins and cryptochromes (CRY) (Chen et al. 2004).

Cryptochromes are flavoproteins that share structural similarity to DNA photolyases but lack photolyase activity (Lin and Shalitin 2003). Although originally identified in *Arabidopsis*, cryptochromes have now been found in bacteria, plants and animals (Brudler et al. 2003). Most cryptochrome proteins, with the exception of CRY-DASH (or CRY3), are composed of two domains, an amino-terminal photolyase-related (PHR) region and a carboxy-terminal domain (DAS) of varying size. The PHR region appears to bind two chromophores: one is flavin adenine dinucleotide and the other 5,10-methenyltetrahydrofolate (pterin) (Lin et al. 1995). The carboxy-terminal domain of cryptochromes is generally less conserved than the PHR region; CRY-DASH proteins lack the DAS domain (Kleine et al. 2003). The CRY-DASH gene, recently characterized in Arabidopsis, shares little sequence homology with the other cryptochromes and carries an N-terminal sequence that mediates its import into chloroplasts and mitochondria. Furthermore, CRY-DASH lacks the C-terminal domain, which is present in most plant cryptochromes. In *Arabidopsis*, three cryptochrome genes (CRY1, CRY2 and CRY-DASH) have been described so far (Kleine et al. 2003). Plant cryptochromes play an important role in several blue-light-regulated developmental processes such as de-etiolation, flowering and flavonoid biosynthesis (Weller et al. 2001, Giliberto et al. 2005). CRY1 and CRY2 are intimately connected with the circadian clock machinery.

In tomato (*Solanum lycopersicum*), three cryptochrome genes have been discovered and analysed in detail so far: two CRY1-like (CRY1a and CRY1b) and one CRY2 gene [18,19]. The use of transgenic and mutant lines has shed light on the role of tomato cryptochromes in seedling photomorphogenesis, flavonoid and carotenoid accumulation, adult development, fruit pigmentation and flowering (Ninu et al. 1999).

A new member of the blue-light photoreceptor family, CRY-DASH, was reported in Arabidopsis, though its distinctive biological functions are still unclear. Facella et al. (2006) characterized the CRY-DASH gene of tomato and evidenced that its mRNA is expressed in both seeds and adult organs showing diurnal and circadian fluctuations. Moreover, the CRY-DASH transcription pattern is altered in both in a cry1a mutant and in a transgenic CRY2 over-expressor, suggesting that CRY-DASH regulation must be mediated at least partly by an interaction of CRY1a and CRY2 with the time-keeping mechanism.

Ripening

In climacteric fruit, including tomato, ripening is closely associated with an increase in respiration and the plant hormone ethylene stimulates both processes (Abeles et al. 1992). The ethylene-induced respiration has been related to the stimulation of the synthesis of nuclear-encoded proteins that are targeted to the mitochondria such as an alternative oxidase and uncoupling protein (Considine et al. 2001). Data of the screening of genes expressed during tomato ripening using cDNA microarrays (Alba et al. 2005) show that four genes putatively encoding mitochondrial proteins are up-regulated by ethylene: the 7 kDa subunit of a translocase of outer membrane, Nicotinamide adenine dinucleotide (NAD) aldehyde dehydrogenase, a protoporphyrinogen IX oxidase isozyme and a serine hydroxymethyltransferase. Since mitochondria are not a site for ethylene synthesis (Diolez et al. 1986) or for ethylene perception (Chen et al. 2002), stimulation of respiration appeared to be related to imported proteins only and not to the stimulation of its endogenous machinery. However, Zegzouti et al. (1999) showed that treating mature green tomato fruit with ethylene resulted in a transient enhanced accumulation of a transcript named ER49 encoded by a nuclear gene later denominated *LeEF-Tsmt* and corresponding to a mitochondrial elongation factor involved in protein synthesis in the mitochondria (Benichou et al. 2003). EF-Ts are nucleotide exchange factors that promote the exchange of guanosine 5′-diphosphate (GDP) for guanosine 5′-triphosphate (GTP) with elongation factors of the EF-Tu type (Merrick 1992). The elongation steps of mitochondrial translation are well documented in mammals and prokaryotes but not in plants. Girardi et al. (2006) have recently carried out a study aimed at assessing the function of the LeEF-Tsmt protein. It was demonstrated that the protein was targeted to the mitochondria rather than to the chloroplasts.

In order to further investigate the role of *LeEF-Tsmt*, Girardi et al. (2006) studied the spatio-temporal expression of the gene in different organs of tomato plants with special attention to the ripening fruit and to the effect of ethylene and abiotic stress conditions. Tomato plants in which the *LeEF-*

Tsmt gene has been over-expressed and down-regulated have been generated. Biochemical and physiological studies have shown that the level of LeEF-Tsmt RNA has no influence on fruit respiratory activity.

Hoeberichts et al. (2002) treated tomato fruit (*Lycopersicon esculentum* L. cv Prisca) with 1-methylcyclopropene (1-MCP), a potent inhibitor of ethylene action, delayed colour development, softening, and ethylene production in tomato fruit harvested at the mature green, breaker, and orange stages. 1-MCP treatment also decreased the mRNA abundance of phytoene synthase 1 (*PSY* 1), expansin 1 (*EXP* 1), and 1-aminocyclopropane-1-carboxylic acid (ACC) oxidase 1 (*ACO* 1), three ripening-related tomato genes, in mature green, breaker, orange, and red ripe fruit. These results demonstrate that the ripening process can be inhibited on a physiological and molecular level, even at very advanced stages of ripening. The effects of 1-MCP on ripening lasted 5–7 d and could be prolonged by renewed exposure.

Physiological changes associated with tomato fruit ripening can be halted or delayed by inhibiting ethylene perception, even when the fruit has reached advanced stages of ripening. Recent additional data showing that application of 1-MCP to ripe tomatoes results in an increase in postharvest life based on fruit appearance (Wills and Ku 2001) confirm these findings. Information is lacking on fruit genes that are specifically expressed at early developmental stages. Using a cDNA subtraction technique, Ohta et al. (2005) isolated fruit-specific genes that are expressed during the cell expansion phase of tomato (*Lycopersicon esculentum* Mill.) fruit development.

This phase is of physiological and horticultural importance, because the increase in cell volume during this time contributes greatly to the final fruit size (Gillaspy et al. 1993). One of the isolated cDNAs, LeODD, is transiently expressed 15 DAF in a nearly fruit-specific manner during the initial period of cell expansion. Southern blot analysis indicated that LeODD is encoded by a single gene. LeODD is homologous to 2-oxoglutarate-dependent dioxygenase genes, and the key amino acid residues in the binding sites for ferrous iron and 2-oxoglutarate are completely conserved. The amino acid sequence identity between LeODD and other 2-oxoglutarate-dependent dioxygenases is relatively low, suggesting that LeODD is a novel enzyme of this family. Another of the isolated, LeGLO2 cDNAs is also highly expressed at 15 DAF. LeGLO2 is thought to be a novel glycolate oxidase isoform that functions in fruit. 2-Oxoglutarate, the co-substrate of LeODD, could be supplied by a LeGLO2-mediated glycolate pathway in immature fruit. The coordinate expression of LeODD and LeGLO2 may play a role in the biosynthesis of a metabolite, such as a plant hormone or secondary metabolite that is required during the initial period of the cell expansion phase of fruit development.

Stress Resistance

Maskin et al. (2001) continued to dissect the *Asr* (ABA/water stress/ripening-induced) gene family originally described in tomato. Accumulated ABA is responsible for the expression of various target genes. Abscisic acid brings about the accumulation of different gene products during periods of water deficit. Tomato *Asr*1 was the first reported member of the *Asr* family, which subsequently drew the attention of other groups at the level of gene expression. *Asr*1 then led to the discovery of other members of the same family, *Asr*2 and *Asr*3, that map to chromosome 4 of tomato. RT-PCR-based strategy was developed to assess the organ (leaf, root and fruit) and developmental (immature and ripe fruit) specificity of expression of the three known members of the *Asr* gene family under normal and stress conditions. Maskin et al. (2001) concluded that, whereas *Asr*1 and *Asr*2 are the members of the family preferentially induced by desiccation in leaves, *Asr*2 is the only one activated in the roots from water-deficit-stressed plants. We also observed that expression of the three genes does not change significantly in fruit at different developmental stages, except for that of *Asr*2, which decreases after the breaker yellow stage. In addition, they identified a 72 amino acid polar peptide region, rich in His, Lys, Glu and Ala, which contains two internal imperfect repeats and is highly conserved in more recently discovered *Asr*-like proteins from other plant species exposed to different kinds of abiotic stress such as water deficit, salt, cold and/or limiting amount of light.

The molecular basis for the increase in ethylene production observed in flooded tomato plants is the induction of ACC synthase, a key enzyme in ethylene biosynthesis (Olson et al. 1995). ACC, which is synthesized in roots, is transported to shoots, where it is converted to ethylene (Else et al. 1998). Tomato plant roots express different ACC synthases at various times, each of which performs overlapping functions during flooding stress (Shiu et al. 1998). To determine the importance of the enzyme lipoxygenase (LOX) in the generation of volatile C6 aldehyde and alcohol flavour compounds, Griffiths et al. (1999) constructed two antisense LOX genes and transferred them to tomato plants. The first of these constructs (p2ALX) incorporated the fruit-specific 2A11 promoter. The second construct (pPGLX) consisted of the ripening-specific polygalacturonase (PG) promoter and terminator. High levels of ripening-specific reporter gene expression directed by tomato fruit PG gene-flanking regions were reported. Reduced levels of endogenous *TomloxA* and *TomloxB* mRNA (2–20% of wild-type) were detected in transgenic fruit containing the p2ALX construct compared to non-transformed plants, whereas the levels of mRNA for a distinct isoform, *TomloxC*, were either unaffected or even increased. The pPGLX construct was much less effective in reducing endogenous LOX mRNA levels. In the case of the p2ALX plants, LOX enzyme activity was also greatly reduced compared with wild-type plants. Analysis of flavour volatiles, however, indicated that there were no significant changes. These findings suggest that either very low levels of LOX are sufficient for the generation of C6 aldehydes and alcohols, or that a specific isoform such as *TomloxC*, in the absence of *TomloxA* and *TomloxB*, is responsible for the production of these compounds. To increase the soluble solid content of the tomato, Klann et al. (1996) used genetic engineering to switch off expression by adding a complement of the gene (antisense), without substantially altering other desirable traits of the fruit.

Antimacrobial Resistance

Plant defensins comprise a family of small cationic, cysteine-rich peptides (45-54 amino acid) that are mostly found to contribute to broad-spectrum host defence against pathogens and are widely distributed among plants, including wheat, barley, spinach, pea, and several members of the Brassicaceae family (Lay and Anderson 2005, Thomma et al. 2002).

Many plant defensins can inhibit the growth of a broad range of fungi at micromolar concentrations but are non-toxic to both mammalian and plant cells. The use of defensin genes in genetic engineering has resulted in broad spectrum resistance against fungal pathogens (Gao et al. 2000, Kanazaki et al. 2002). A recent reassessment of defensin-like sequence in the near complete genome sequence of *Arabidopsis thaliana* revealed that 317 homologous sequences could be identified (Silverstein et al. 2005). Defensins may have evolved into such a large multigene family in order to provide non-host resistance to numerous pathogens in many different tissues and in addition seem also to be involved in non-pathogen resistance mechanisms.

Solis et al. (2006) reported the isolation of a defensin gene, lm-def, from the Andean crop maca (*Lepidium meyenii*) with activity against the pathogen *Phytophthora infestans*, responsible for late blight disease of potato and tomato crops. The lm-def gene has been isolated by PCR using degenerate primers corresponding to conserved regions of 13 plant defensin genes of the Brassicaceae family assuming that defensin genes are highly conserved among cruciferous species. It belongs to a small multigene family of at least 10 members possibly including pseudogenes as assessed by genomic hybridization and nucleotide sequence analyses.

GENE EXPRESSION STUDIES: EFFECTS ON DIETARY PROPERTIES

Allergenic Properties

Tomato has been reported to cause allergic reactions after ingestion in patients with food allergy (Ortolani et al. 1989). Profilin, b-fructofuranosidase, and non-specific lipid transfer proteins (nsLTP) were recently identified as allergens in tomato (Westphal et al. 2004) and denominated as Lyc e 1, Lyc e 2, and Lyc e 3, respectively, by the International Union of Immunological Societies Allergen Nomenclature Subcommittee. Lyc e 3 belongs to a family of nsLTPs with a molecular size of 8 to 10 kd and is a member of the group of 14 pathogenesis-related proteins.

Lorenz et al. (2006) sought to achieve stable inhibition of expression of the allergenic nonspecific lipid transfer protein Lyc e 3 in tomato and to analyse the reduction of allergenicity *in vitro* by using histamine release assays and *in vivo* by using skin prick tests with transgenic tomato fruits. The aims of the study were to investigate the reduction of allergenicity of transgenic fruits showing a suppressed allergen expression *in vivo* by means of SPTs and to investigate the heritability of this effect on the next generation of plants. In addition, results of SPTs were confirmed by using *in vitro* mediator release assays. Their data showed a remarkable reduction of allergenicity in the majority of tested patients and stability of the inhibition of allergen expression in first-generation (T0) and second-generation (T1) transgenic fruits.

Prediction of potential allergenicity is one of the major issues in the safety assessment of genetically modified crops. During the assessment of the potential allergenicity of a protein it is recommended that its amino acid sequence be compared with those of known allergenic proteins (Gendel 2002). In a joint FAO/WHO expert consultation (2001), it was proposed that the amino acid sequence of a protein should be compared with all known allergenic proteins retrieved from protein databases to identify any stretches of 80 amino acids with more than 35% similarity or any small identical peptides of at least 6 amino acids; it was also proposed that the outcome of such comparisons should then be combined with other information on allergenicity, such as its digestibility and binding to IgE in sera from allergic patients.

SuSy and Invertase Pathways

The developing tomato (*Lycopersicon esculentum*) fruit metabolizes translocated sucrose either via sucrose-synthase (SuSy) that cleaves sucrose to UDP-glucose and fructose, or via invertase that cleaves sucrose to glucose and fructose. The SuSy pathway is associated with young green tomato fruits that undergo a period of transient starch accumulation (Wang et al. 1993), whereas the invertase pathway is associated with the starchless, later stages of fruit development (Ho and Hewitt 1986). Yet, the starch accumulation period is characterized by increased activities of SuSy and fructokinase together with ADP-glucose pyrophosphorylase and starch synthase, two key enzymes in starch biosynthesis (Schaffer and Petreikov 1997). Several fructokinase enzymes have been identified in tomato plants. Martinez-Barajas and Randall (1996) reported two isoforms of fructokinase in tomato fruit, both of which showed similar characteristics of inhibition by fructose, a phenomenon known as substrate inhibition. Petreikov et al. (2001) isolated three fructokinase isozymes from tomato fruit, FKI, FKII and FKIII, and found that FKI contributes most of the fructokinase activity in young fruits.

Three genes encoding different fructokinases have been isolated from tomato plants (Kanayama et al. 1998). One of them, *LeFRK* 1, is expressed throughout fruit development and its yeast-expressed gene product does not exhibit substrate inhibition. The second gene, *LeFRK* 2, is highly expressed at the early stages of fruit development and its product, when expressed in yeast, is inhibited by fructose (Kanayama et al. 1997). The third gene, *LeFRK* 3, is not expressed in fruits and does not exhibit substrate inhibition. It has been proposed that the *LeFRK* 2 gene product plays a role in transient starch accumulation in developing *L. esculentum* fruits. To analyse the role of *LeFRK* 2 in starch biosynthesis, Dai et al. (2002a) produced transgenic tomato plants with sense and antisense expression of *StFRK*, the potato homologue to *LeFRK* 2. Fruits of homozygous plants expressing sense or antisense *StFRK* exhibited suppression of *LeFRK* 2, concomitantly with specific elimination of the FKI isozyme, indicating that FKI is the gene product of *LeFRK* 2. The activity of fructokinase was reduced in antisense and sense homozygous fruits and increased in fruits of sense hemizygous plants. The modified activities of fructokinase led to small but significant changes both in the steady state levels of sugars and in the level of phosphorylated sugars in fruits. However, fruits lacking FKI had increased rather than decreased starch content. Hence, the authors concluded that *LeFRK* 2 is not required for transient starch accumulation in tomato fruits and might have a role in the regulation of sucrose import into tomato fruits.

Lycopene

Lycopene is the main carotene accumulated in ripe tomato fruits. This linear carotenoid is synthesized in plants through a pathway starting from geranylgeranyl diphosphate and is the biosynthetic precursor of most cyclic carotenoids including beta-carotene. Lycopene is converted into beta-carotene by the action of lycopene beta-cyclase (β-Lcy), an enzyme introducing beta ionone rings at both ends of the molecule (Cunningham et al. 1994). Moreover, the B gene entails the accumulation of beta-carotene in fruits. The genes controlling lycopene synthesis in tomato are Psy1 and Pds, respectively encoding the fruit-specific phytoene synthase and phytoene desaturase, which are up-regulated during fruit ripening (Corona et al. 1996), whereas those controlling lycopene cyclization such as β-Lcy or ε-Lcy are down-regulated. As a result, beta-carotene in ripe tomato fruits does not exceed 15% of the total carotenoids.

Lycopene may benefit the cardiovascular system by reducing the amount of oxidized low-density lipoprotein. Recent epidemiologic studies have suggested a potential benefit of this carotenoid in reducing the risk of prostate cancer, particularly the more lethal forms of this cancer. Five studies support a 30–40% reduction in risk associated with high tomato or lycopene consumption in the processed form in conjunction with lipid consumption, although other studies with raw tomatoes were not conclusive (Giovannucci 2002). In an intriguing paper, Mehta et al. (2002) used a genetic modification approach to modify polyamines in tomato fruit to retard the ripening process. These modified tomatoes had longer vine lives, suggesting that polyamines have a function in delaying the ripening process. There was also an unanticipated enrichment in lycopene content of the genetically modified tomato fruit. The lycopene levels were increased 2- to 3.5-fold compared to the conventional tomatoes. This is a substantial enrichment, exceeding that so far achieved by conventional means. This novel approach may work in other fruits and vegetables. Rosati et al. (2000) reported the over-expression and antisense repression of the β-Lcy under the control of the Pds promoter. The over-expression increased the levels of beta-carotene up to 7-fold, a level sufficient to cover the vitamin A RDA. Total fruit carotenoid levels were slightly increased.

CONCLUSIONS

The study of tomato genome and gene expression profiling targets mainly the production of genetic maps that are invaluable in marker-assisted selection for agronomical traits at the stage of cultivation as well as for the determination of useful properties for processed tomato products. As molecular biology, and to a greater extent biotechnology, enter the era of nanotechnology it will be possible for information of this kind to be used to create tomato products that have unique dietary or medicinal properties.

ABBREVIATIONS

ABA: Abscisic acid; AFLP: Amplified fragment length polymorphism; GAPDH: Glyceraldehyde-3-phosphate dehydrogenase; PCR: Polymerase chain reaction; RAPD: Randomly amplified polymorphic DNA; RDA: Representational difference analysis; RFLP: Restriction fragment length polymorphism; SAGE: Serial analysis of gene expression

REFERENCES

Abeles, F.B. and W.B. Morgan, and M.E. Saltveit Jr. 1992. Ethylene in Plant Biology. Academic Press, San Diego, USA.

Alba, R. and P. Payton, Z. Fei, R. McQuinn, P. Debbie, G.B. Martin, S.D. Tanksley, and J.J. Giovannoni. 2005. Transcriptome and selected metabolite analyses reveal multiple points of ethylene control during tomato fruit development. Plant Cell 17: 2954-2965.

Argyropoulos, D. and C. Psallida, and C. Spyropoulos. 2006. Generic normalization method for real-time PCR Application for the analysis of the mannanase gene expressed in germinating tomato seed. FEBS J. 273: 770-777.

Baiges, I. and A.R. Schaffner, M.J. Affenzeller, and A. Mas. 2002. Plant aquaporins. Physiol. Plant. 115: 175-182.

Benichou, M. and Z.-G. Li, B. Tournier, A. Chaves, H. Zegzouti, A. Jauneau, C. Delalande, A. Latche, M. Bouzayen, L. Spremulli, and J.C. Pech. 2003. Tomato EF-TSmt, a functional mitochondrial translation elongation factor from higher plants. Plant Mol. Biol. 53: 411-422.

Bhatia, P. and W.R. Taylor, A.H. Greenberg, and J.A. Wright. 1994. Comparison of glyceraldehyde-3-phosphate dehydrogenase and 28S-ribosomal RNA gene expression as RNA loading controls for northern blot analysis of cell lines of varying malignant potential. Anal. Biochem. 216: 223-226.

Blackburn, E.H. 1991. Structure and function of telomeres. Nature 350: 569-573.

Brudler, R. and K. Hitomi, H. Daiyasu, H. Toh, K. Kucho, M. Ishiura, M. Kanehisa, V.A. Roberts, T. Todo, J.A. Tainer, and E.D. Getzoff. 2003. Identification of a new cryptochrome class. Structure, function, and evolution. Mol. Cell. 11: 59-67.

Brunner, A.M. and I.A. Yakovlev, and S.H. Strauss. 2004. Validating internal controls for quantitative plant gene expression studies. BMC Plant Biol. 4: 14-19.

Burbidge, A. and T. Grieve, A. Jackson, A. Thompson, and T. Taylor. 1997. Structure and expression of a cDNA encoding a putative neoxanthin cleavage enzyme (NCE), isolated from a wilt-related tomato (*Lycopersicon esculentum* Mill.) library. J. Exp. Bot. 317: 2111-2112.

Bustin, S.A. 2002. Quantification of mRNA using realtime reverse transcription PCR (RT-PCR): trends and problems. J. Mol. Endocrinol. 29: 23-39.

Carpenter, C.D. and J.A. Kreps, and A.E. Simon. 1994. Genes encoding glycine-rich *Arabidopsis thaliana* proteins with RNA-binding motifs are influenced by cold treatment and an endogenous circadian rhythm. Plant Physiol. 104: 1015-1025.

Chen, G.P. and I.D. Wilson, S.H. Kim, and D. Grierson. 2001. Inhibiting expression of a tomato ripening-associated membrane protein increases organic acids and reduces sugar levels of fruit. Planta 212: 799-807.

Chen, M. and J. Chory, and C. Fankhauser. 2004. Light signal transduction in higher plants. Annu. Rev. Plant Biol. 38: 87-117.

Chen, Y.F. and M.D. Randlett, J.L. Findell, and G.E. Schaller. 2002. Localisation of ethylene receptor ETR1 to the endoplasmic reticulum of *Arabidopsis*. J. Biol. Chem. 277: 19861-19866.

Chen, Z.L. and H. Gua, Y. Li, Y. Su, P. Wu, Z. Jiang, X. Ming, J. Tian, N. Pan, and L-J. Qu. 2003. Safety assessment for genetically modified sweet pepper and tomato. Toxicology 188: 297-307.

Considine, M.J. and D.O. Daley, and J. Whelan. 2001. The expression of alternative oxidase and uncoupling protein during fruit ripening in mango. Plant Physiol. 126: 1619-1629.

Corona, V. and B. Aracri, G. Kosturkova, G.E. Bartley, L. Pitto, L. Giorgetti, P.A. Scolnik, and G. Giuliano. 1996. Regulation of a carotenoid biosynthesis gene promoter during plant development. Plant J. 9: 505-512.

Cunningham, F.X., Jr. and Z. Sun, D. Chamovitz, J. Hirschberg, and E. Gantt. 1994. Molecular structure and enzymatic function of lycopene cyclase from the cyanobacterium *Synechococcus* sp. strain PCC7942. Plant Cell 6: 1107-1121.

Czechowski, T. and M. Stitt, T. Altmann, M.K. Udvardi, and W.R. Scheible. 2005. Genome-wide identification and testing of superior reference genes for transcript normalization in arabidopsis. Plant Physiol. 139: 5-17.

Dai, N. and M.A. German, T. Matsevitz, R. Hanael, D. Swartzberg, Y. Yeselson, Petreikov M., A.A. Schaffer, and D. Granot. 2002a. LeFRK2, a gene encoding the major fructokinase in tomato fruits, is not required for starch accumulation in developing fruits. Plant Sci. 162: 423-430.

De Oliveira, D.E. and J. Seurinck, D. Inze, M. Van Montagu, and J. Botterman. 1990. Differential expression of five *Arabidopsis* genes encoding glycine-rich proteins. Plant Cell 2: 427-436.

Diaz-Camino, C. and R. Conde, N. Ovsenek, and M.A. Villanueva. 2005. Actin expression is induced and three isoforms are differentially expressed during germination of *Zea mays*. J Exp. Bot. 56: 557-565.

Ding, C.K. and C. Wang. 2003. The dual effects of methyl salicylate on ripening and expression of ethylene biosynthetic genes in tomato fruit. Plant Sci. 164: 589-596.

Ding, C.K. and C. Wang, K. Gross, and D. Smith. 2002. Jasmonate and salicylate induce the expression of pathogenesis related protein genes and increase resistance to chilling injury in tomato fruit. Planta 214: 895-901.

Diolez, P. and J. deVirville, A. Latche, J.C. Pech, F. Moreau, and M.S. Reid. 1986. Role of the mitochondria in the conversion of 1-aminocyclopropane-1-carboxylic acid to ethylene in plant tissues. Plant Sci. 43: 13-17.

Else, M.A. and M.B. Jackson. 1998. Transport of 1-aminocyclopropane-1-carboxylic acid (ACC) in the transpiration stream of tomato (*Lycopersicon esculentum*) in relation to foliar ethylene production and petiole epinasty. Australian Journal of Plant Physiology. 25: 453-458.

Facella, P. and L. Lopez, A. Chiappetta, M.B. Bitonti, G. Giuliano, and G. Perrotta. 2006. CRY-DASH gene expression is under the control of the circadian clock machinery in tomato. FEBS Lett. 580: 4618-4624.

FAO/WHO. 2001. Evaluation of Allergenicity of Genetically Modified Foods. Report of a Joint FAO/WHO Expert Consultation on Allergenicity of Foods Derived from Biotechnology, Rome, Italy, 22-25 January. (http://www.who.int/foodsafety/publications/biotech/en/ec_jan2001.pdf).

Fung, R.W. and C.Y. Wang, D.L. Smith, K.C. Gross, and M. Tian. 2004. MeSA and MeJA increase steady-state transcript levels of alternative oxidase and resistance against chilling injury in sweet peppers (*Capsicum annuum* L.). Plant Sci. 166: 711-719.

Fung, R.W.M. and C.Y. Wang, D.L. Smith, K.C. Grossa, Y. Tao, and M. Tian. 2006. Characterization of alternative oxidase (AOX) gene expression in response to methyl salicylate and methyl jasmonate pre-treatment and low temperature in tomatoes. J. Plant Physiol. 163: 1049-1060.

Gao, A.G. and S.M. Hakimi, C.A. Mittanck, Y. Wu, B.M. Woerner, and D.M. Stark. 2000. Fungal pathogen protection in potato by expression of a plant defensin peptide. Nat. Biotechnol. 8: 307-310.

Gendel, S.M. 2002. Sequence analysis for assessing potential allergenicity. Ann. NY Acad. Sci. 964: 87-98.

Giliberto, L. and G. Perrotta, P. Pallara, J.L. Weller, P.D. Fraser, P.M. Bramley, A. Fiore, M. Tavazza, and G. Giuliano. 2005. Manipulation of the blue light photoreceptor cryptochrome 2 in tomato affects vegetative development, flowering time, and fruit antioxidant content. Plant Physiol. 137: 199-208.

Gillaspy, G.H. and B-D.W. Gruissem. 1993. Fruits: a developmental perspective. Plant Cell 5: 1439-1451.

Giovannucci, E. 2002. A review of epidemiologic studies of tomatoes, lycopene, and prostate cancer. Exp. Biol. Med. 227: 852-859.

Girardi, C.L. and K. Bermudez, A. Bernadac, A. Chavez, M. Zouine, F.J. Miranda, M. Bouzayen, J.C. Pech, and A. Latche. 2006. The mitochondrial elongation factor LeEF-Tsmt is regulated during tomato fruit ripening and upon wounding and ethylene treatment. Postharvest Biol. Technol. 42: 1-7.

Griffiths, A. and S. Prestage, R. Linforth, J. Zhang, A. Taylor, and D. Grierson. 1999. Fruit-specific lipoxygenase suppression in antisense-transgenic tomatoes. Postharvest Biol. Technol. 17: 163-173.

Ho, L.C. and J.D. Hewitt. Fruit development. pp. 201-240. *In*: J.G. Atherton and J. Rudich. [Eds.] 1986. The Tomato Crop. Chapman and Hall, London, UK.

Hoeberichts, F.A. and L.H.W. Van Der Plas, and E.J. Woltering. 2002. Ethylene perception is required for the expression of tomato ripening-related genes and associated physiological changes even at advanced stages of ripening. Postharvest Biol. Technol. 26: 125-133.

Holtzapffel, R.C. and J. Castelli, P.M. Finnegan, A.H. Millar, J. Whelan, and D.A. Day. 2003. A tomato alternative oxidase protein with altered regulatory properties. Biochim Biophys Acta 1606: 153-162.

Huang, Y. and J.A. Steitz. 2001. Splicing factors SRp20 and 9G8 promote the nucleocytoplasmic export of mRNA. Mol. Cell. 7: 899-905.

Jang, J.Y. and D.G. Kim, Y.O. Kim, J.S. Kim, and H. Kang. 2004. An expression analysis of a gene family encoding plasma membrane aquaporins in response to abiotic stresses in *Arabidopsis thaliana.* Plant Mol. Biol. 54: 713-725.

Jin, S.M. and R.L. Xu, Y.D. Wei, and P.H. Goodwin. 1999. Increased expression of a plant actin gene during a biotrophic interaction between leaved mallow, *Maiva pusillia,* and *Colletotrichum gloeosporioides* f. sp. *malvae.* Planta 209: 487-494.

Johanson, U. and M. Karlsson, I. Johansson, S. Gustavsson, S. Sjovall, L. Fraysse, A.R. Weig, and P. Kjellbom. 2001. The complete set of genes encoding major intrinsic proteins in *Arabidopsis* provides a framework for a new nomenclature for major intrinsic proteins in plants. Plant Physiol. 126: 1358-1369.

Joshi, C.P. and S. Kumar, and H.T. Nguyen. 1996. Application of modified differential display technique for cloning the 3% region from three putative members of wheat HSP70 gene family. Plant Mol. Biol. 30: 641-646.

Kanayama, Y. and N. Dai, D. Granot, M. Petreikov, A. Schaffer, and A.B. Bennett. 1997. Divergent fructokinase genes are differentially expressed in tomato. Plant Physiol. 113: 1379-1384.

Kanayama, Y. and D. Granot, N. Dai, M. Petreikov, A. Schaffer, A. Powell, and A.B. Bennett. 1998. Tomato fructokinases exhibit differential expression and substrate regulation. Plant Physiol. 117: 85-90.

Kanazaki, H. and S. Nirasawa, H. Saitoh, M. Ito, M. Nishihara, and R. Terauchi. 2002. Overexpression of the wasabi defensin gene confers enhanced resistance to blast fungus (*Magnaporthe grisea*) in transgenic rice. Theor. Appl. Genet. 105: 809-814.

Kerdnaimongkol, K. and W.R. Woodson. 1999. Inhibition of catalase by antisense RNA increases susceptibility to oxidative stress and chilling injury in transgenic tomato plants. J. Amer. Soc. Hortic. Sci. 124: 330-336.

Kevei, Z. and J.M. Vinardell, G.B. Kiss, A. Kondorosi, and E. Kondorosi. 2002. Glycine-rich proteins encoded by a nodule-specific gene family are implicated in different stages of symbiotic nodule development in *Medicago* spp. Mol. Plant-Microb. Interact. 15: 922-931.

Klann, E.M. and B. Hall, and A.B. Bennett. 1996. Antisense acid invertase (TIV1) gene alters soluble sugar composition and size in transgenic tomato fruit. Plant Physiol. 112: 1321-1330.

Kleber-Janke, T. and K. Krupinska. 1997. Isolation of cDNA clones for genes showing enhanced expression in barley leaves during dark-induced senescence as well as during senescence under field conditions. Planta 203: 332-340.

Kleine, T. and P. Lockhart, and A. Batschauer. 2003. An *Arabidopsis* protein closely related to Synechocystis cryptochrome is targeted to organelles. Plant J. 35: 93-103.

Knaap, E. and H. Kende. 1995. Identification of a gibberellin-induced gene in deep water rice using differential display of mRNA. Plant Mol. Biol. 28: 589-592.

Kok, E.J. and N.L.W. Franssen-van Hal, L.N.W. Winnubst, E.H.M. Kramer, W.T.P. Dijksma, H.A. Kuiper, and J. Keijer. 2007. Assessment of representational difference analysis (RDA) to construct informative cDNA microarrays for gene expression analysis of species with limited transcriptome information, using red and green tomatoes as a model. J. Plant Physiol. 164: 337-349.

Kreuzer, K.A. and U. Lass, O. Landt, A. Nitsche, J. Laser, H. Ellebrok, G. Pauli, D. Huhn, and C.A. Schmidt. 1999. Highly sensitive and specific fluorescence reverse transcription-PCR assay for the pseudogene-free detection of beta-actin transcripts as quantitative reference. Clin. Chem. 45: 297-300.

KwangChul, Oh. and K. Hardeman, M.G. Ivanchenko, M.E. Ivey, A. Nebenführ, T.J. White, and T.L. Lomax. 2002. Fine mapping in tomato using microsynteny with the *Arabidopsis* genome: the *Diageotropica* (*Dgt*) locus. Genome Biol. 3: research 0049.1-0049.11.

Lay, F.T. and M.A. Anderson. 2005. Defensins – components of the innate immune system in plants. Curr. Protein Pept. Sci. 6: 85-101.

Liang, P. and A.B. Paedee. 1992. Differential display of eukaryotic messenger RNA by means of the polymerase chain reaction. Science 257: 967-971.

Lin, C. and D. Shalitin. 2003. Cryptochrome structure and signal transduction. Annu. Rev. Plant Biol. 54: 469-496.

Lin, C. and D.E. Robertson, M. Ahmad, A.A. Raibekas, M.S. Jorns, P.L. Dutton, and A.R. Cashmore. 1995. Association of flavin adenine dinucleotide with the Arabidopsis blue light receptor CRY1. Science 269: 968-970.

Lin, W.C. and M.L. Cheng, J.W. Wu, N.S. Yang, and C.P. Cheng. 2005. A glycine-rich protein gene family predominantly expressed in tomato roots, but not in leaves and ripe fruit. Plant Sci. 168: 283-295.

Lockhart, D.J. and E.A. Winzeler. 2000. Genomics, gene expression and DNA arrays. Nature 405: 827-836.

Lorenz, Y. and E. Enrique, L. LeQuynh, K. Fotisch, M. Retzek, S. Biemelt, U. Sonnewald, S. Vieths, and S. Scheurer. 2006. Skin prick tests reveal stable and heritable reduction of allergenic potency of gene-silenced tomato fruits. J. Allergy Clin. Immunol. 711-718.

Martinez-Barajas, E. and D.D. Randall. 1996. Purification and characterization of fructokinase from developing tomato (*Lycopersicon esculentum* Mill.) fruits. Planta 199: 451-458.

Maskin, L. and G.E. Gudesblat, J.E. Moreno, F.O. Carrari, N. Frankel, A. Sambade, M. Rossi, and N.D. Iusem. 2001. Differential expression of the members of the *Asr* gene family in tomato (*Lycopersicon esculentum*). Plant Sci. 161: 739-746.

Matsuyama, T. and H. Satoh, Y. Yamada, and T. Hashimoto. 1999. A maize glycine-rich protein is synthesized in the lateral root cap and accumulates in the mucilage. Plant Physiol. 120: 665-674.

McMaugh, S.J. and B.R. Lyon. 2003. Real-time quantitative RT-PCR assay of gene expression in plant roots during fungal pathogenesis. Biotechniques 34: 982-986.

Mehta, R.A. T. Cassol, N. Li, N. Ali, A.K. Handa, and A.K. Mattoo. 2002. Engineered polyamine accumulation in tomato enhances phytonutrient content, juice quality, and vine life. Nat. Biotechnol. 20: 613-668.

Merrick, W.C. 1992. Mechanism and regulation of eukaryotic protein synthesis. Microbiol. Mol. Biol. Rev. 56: 291-315.

Miquel, M. and D. James Jr., H. Dooner, and J. Browse. 1993. Arabidopsis requires polyunsaturated lipids for low temperature survival. Proc. Natl. Acad. Sci. USA 90: 6208-6212.

Mockler, T.C. and S. Chan, A. Sundaresan, H. Chen, S.E. Jacobsen, and J.R. Ecker. 2005. Applications of DNA tiling arrays for wholegenome analysis. Genomics 85(1): 1-15.

Moriguchi, R. and K. Kanahama, and Y. Kanayama. 2006. Characterization and expression analysis of the tomato telomere-binding protein LeTBP1. Plant Sci. 171: 166-174.

Murphy, D.J. and J.H. Ross. 1998. Biosynthesis, targeting and processing of oleosin-like proteins, which are major pollen coat components in *Brassica napus*. Plant J. 13: 1-16.

Nemoto, Y. and N. Kawakami, and T. Sasakuma. 1999. Isolation of early salt-responding genes from wheat (*Triticum aestivum* L.) by differential display. Theor. Appl. Genet. 98: 673-678.

Ninu, L. and M. Ahmad, C. Miarelli, A.R. Cashmore, and G. Giuliano. 1999. Cryptochrome 1 controls tomato development in response to blue light. Plant J. 18: 551-556.

Nonogaki, H. and Y. Morohashi. 1996. An endo-β-mannanase develops exclusively in the micropylar endosperm of tomato seeds prior to radicle emergence. Plant Physiol. 110: 555-559.

Ohta, K. and K. Kanahama, and Y. Kanayama. 2005. Enhanced expression of a novel dioxygenase during the early developmental stage of tomato fruit. J. Plant Physiol. 162: 697-702.

Olson, D.C. and J.H. Oetiker, and S.F. Yang. 1995. Analysis of *LE-ACS3*, a 1-aminocyclopropane-1-carboxylic acid synthase gene expressed during flooding in the roots of tomato plants. J. Biol. Chem. 270: 14056-14061.

Ortolani, C. and M. Ispano, E.A. Pastorello, R. Ansaloni, and G.C. Magri. 1989. Comparison of results of skin prick tests (with fresh foods and commercial food extracts) and RAST in 100 patients with oral allergy syndrome. J. Allergy Clin. Immunol. 83: 683-690.

Ouyang, B. and T. Yang, H. Li, L. Zhang, Y. Zhang, J. Zhang, Z. Fei, and Z.J. Ye. 2007. Identification of early salt stress response genes in tomato root by suppression subtractive hybridization and microarray analysis). Exp. Bot. 58(3): 507-520.

Panda, S. and T.K. Sato, G.M. Hampton, and J.B. Hogenesch. 2003. An array of insights: application of DNA chip technology in the study of cell biology. Trends Cell Biol. 3: 151-156.

Parsons, B.L. and A.K. Mattoo. 1994. A wound-repressible glycine-rich protein transcript is enriched in vascular bundles of tomato fruit and stem. Plant Cell Physiol. 35: 27-35.

Petreikov, M. and N. Dai, D. Granot, and A. Schaffer. 2001. Characterization of native and yeast expressed tomato fructokinases enzymes. Phytochemistry 58: 841-847.

Rosati, C. and R. Aquilani, S. Dharmapuri, P. Pallara, C. Marusic, R. Tavazza, F. Bouvier, B. Camara, and G. Giuliano. 2000. Metabolic engineering of beta-carotene and lycopene content in tomato fruit. Plant J. 24(3): 413-419.

Sabehat, A. and D. Weiss, and S. Lurie. 1996. The correlation between heat-shock protein accumulation and persistence and chilling tolerance in tomato fruit. Plant Physiol. 110: 531-547.

Sachetto-Martins, G. and L.O. Franco, and D.E. de Oliveira. 2000. Plant glycine-rich proteins: a family or just proteins with a common motif? Biochim. Biophys. Acta 1492: 1-14.

Santino, C.G. and G.L. Stanford, and T.W. Conner. 1997. Developmental and transgenic analysis of two tomato fruit enhanced genes. Plant Mol. Biol. 33: 405-416.

Schaffer, A.A. and M. Petreikov. 1997. Sucrose-to-starch metabolism in tomato fruit undergoing transient starch accumulation. Plant Physiol. 113: 739-746.

Sharma, Y.K. and K.R. Davis. 1995. Isolation of a novel ozone induced cDNA by differential display. Plant Mol. Biol. 29: 91-98.

Shiota, H. and T. Sudoh, and I. Tanaka. 2006. Expression analysis of genes encoding plasma membrane aquaporins during seed and fruit development in tomato. Plant Sci. 171: 277-285.

Shiu, O.Y. and J.H. Oetiker, W.K. Yip, and S.F. Yang. 1998. The promoter of *LE-ACS7*, an early flooding-induced 1-aminocyclopropane carboxylate synthase gene of the tomato, is tagged by a Sol3 transposon, Proc. Natl. Acad. Sci. USA 95: 10334-10339.

Silverstein, K.A.T. and M.A. Graham, T.D. Paape, and K.A. Vandenbosch. 2005. Genome organization of more than 300 defensin-like genes in *Arabidopsis.* Plant Physiol. 138: 600-610.

Solis, J. and G. Medrano, and M. Ghislain. 2006. Inhibitory effect of a defensin gene from the Andean crop maca (*Lepidium meyenii*) against *Phytophthora infestans.* J. Plant Physiol. 2007, Vol. 164, 8: 1071-1082.

Still, D.W. and K.J. Bradford. 1997. Endo-β-mannanase activity from individual endosperm caps and radicle tips in relation to germination rates. Plant Physiol. 113: 21-29.

Suliman-Pollatschek, S. and K. Kashkush, H. Shats, J. Hillel, and U. Lavi. 2002. Generation and mapping of AFLP, SSRs and SNPs in *Lycopersicon esculentum.* Cell Mol. Biol. Lett. 7 (2A): 583-597.

Suzuki, T. and P.J. Higgins, and D.R. Crawford. 2000. Control selection for RNA quantitation. Biotechniques. 29: 332-337.

Thomma, B.P. and B.P. Cammue, and K. Thevissen. 2002. Plant defensins. Planta 216: 193-202.

Thompson, A. and A.C. Jackson, R.A. Parker, D.R. Morpeth, A. Burbidge, and I.B. Taylor. 2000 a. Abscisic acid biosynthesis in tomato: regulation of zeaxanthin epoxidase and 9-cisepoxycarotenoid dioxygenase mRNAs by light/dark cycles, water stress and abscisic acid. Plant Mol. Biol. 42: 833-845.

Tieman, D.M. and A.K. Handa. 1996. Molecular cloning and characterization of genes expressed during early tomato (*Lycopersicon esculentum* Mill) fruit development by mRNA differential display, J. Amer. Soc. Hortic. Sci. 121: 52-56.

Tokuhisa, J. and J. Browse. Genetic engineering of plant chilling tolerance. pp. 79-93. *In*: J.K. Setlow. [ed.] 1999. Genetic Engineering. Kluwer Academic Publishers, New York, USA.

Toorop, P.E. and J.D. Bewley, and H.W.M. Hilhorst. 1996. Endo-b-mannanase isoforms are present in the endosperm and embryo of tomato seeds, but are not essentially linked to the completion of germination. Planta 200: 153-158.

Tseng, T.C. and T.H. Tsai, M.Y. Lue, and H. Lee. 1995. Identification of sucrose-regulated genes in cultured rice cells using mRNA differential display. Gene 161: 179-182.

Tyerman, S.D. and C.M. Niemietz, and H. Bramley. 2002. Plant aquaporins: multifunctional water and solute channels with expanding roles. Plant Cell Environ. 25: 173-194.

Velculescu, V.E. and L. Zhang, B. Vogelstein, and K.W. Kinzler 1995. Serial analysis of gene expression. Science Vol. 270, 5235: 484-487.

Vos, P. and R. Hogers, and M. Bleeker. 1995. AFLP: a new technique for DNA fingerprinting. Nucleic Acids Res. 23: 4407-4414.

Wall, S.J. and D.R. Edwards. 2002. Quantitative reversetranscription-polymerase chain reaction (RT-PCR): a comparison of primer-dropping, competitive, and real-time RT-PCRs. Anal. Biochem. 300: 269-273.

Wang, F. and A. Sanz, M.L. Brenner, and A. Smith. 1993. Sucrose synthase, starch accumulation, and tomato fruit sink strength. Plant Physiol. 101: 321-327.

Wei, J.-Z. and A. Tirajoh, J. Effendy, and A.L. Plant. 2000. Characterization of salt-induced changes in gene expression in tomato (*Lycopersicon esculentum*) roots and the role played by abscisic acid. Plant Sci. 159: 135-148.

Weihong, L. and D.A. Saint. 2002. Validation of a quantitative method for real time PCR kinetics. Biochem. Biophys. Res. Comm. 294: 347-353.

Weller, J.L. and G. Perrotta, M.E. Schreuder, A. van Tuinen, M. Koornneef, G. Giuliano, and R.E. Kendrick. 2001. Genetic dissection of blue-light sensing in tomato using mutants deficient in cryptochrome 1 and phytochromes A, B1 and B2. Plant J. 25: 427-440.

Welsh, J. and K. Chada, S.S. Dalal, D. Ralph, L. Cheng, and M. McClelland. 1992. Arbitrarily primed PCR fingerprinting of RNA. Nucleic Acids Res. 20: 4965-4970.

Werner, M. and N. Uehlein, P. Proksch, and R. Kaldenhoff. 2001. Characterization of two tomato aquaporins and expression during the incompatible interaction of tomato with the plant parasite *Cuscuta reflexa*. Planta 213: 550-555.

Westphal, S. and W. Kempf, K. Foetisch, M. Retzek, S. Vieths, and S. Scheurer. 2004. Tomato profilin Lyc e1: IgE cross-reactivity and allergenic potency. Allergy 59: 526-532.

Wills, R.B.H. and V.V.V. Ku. 2001. Use of 1-MCP to extend the time to ripen of green tomatoes and postharvest life of ripe tomatoes. Postharvest Biol. Technol. 26(1): 85-90, Aug. 2002.

Zegzouti, H. and B. Jones, P. Frasse, C. Marty, B. Maitre, A. Latche, J.C. Pech, and M. Bouzayen. 1999. Ethylene-regulated gene expression in tomato fruit: characterization of novel ethylene-responsive and ripening-related genes isolated by differential display. Plant J. 18: 589-600.

Zhu, C. and F. Kauder, S. Romer, and G. Sandmann. 2007. Cloning of two individual cDNAS encoding 9-cis-epoxycarotenoid dioxygenase from *Gentiana lutea*, their tissue-specific expression and physiological effect in transgenic tobacco. J. Plant Physiol. 164: 195-204.

Zhu, G.Y. and J. Chang Zuo, X. Dong, M. Zhang, C. Hu, and F. Fang. 2001. Fuderine, a C-terminal truncated rat homologue of mouse prominin, is blood glucose-regulated and can up-regulate the expression of GAPDH. Biochem. Biophys. Res. Comm. 281: 951-956.

29

DNA Analysis of Tomato Seeds in Forensic Evidence

Henry C. Lee[1*] and Cheng-Lung Lee[2]

[1]Division of Scientific Services, Department of Public Safety, 278 Colony Street, Meriden, Connecticut 06451, USA

[2]Head of Forensic Science Section, Hsin-Chu City Police Bureau, No. 1 Chung-Shan Rd., Hsin-chu, Taiwan 300

ABSTRACT

Vegetative materials such as stems, roots, and seeds found in a crime victim's stomach contents, on a suspect's clothing, or at the crime scene can be valuable evidence for providing investigative leads in criminal and civil litigations. Currently, the identification of vegetative materials still relies on microscopic and morphological methods. Most seeds from a particular species generally have the same microscopic appearance, especially those recovered from stomach content or in excretions. Therefore, the use of a DNA typing technique could provide a more reliable way for seed identification.

To further explore the use of tomato evidence in forensic investigations, we have developed a procedure to extract high quality DNA from tomato seeds for subsequent DNA- AFLP (Amplified Fragment Length Polymorphism) analysis. It was found that DNA in tomato seeds is often protected from acid degradation during stomach digestion by their tough exterior seed coat. The DNA-AFLP method was performed and we have found that DNA obtained from single seeds could be used for polymerase chain reaction analysis. From the AFLP results, several DNA markers for identifying seeds from tomato and other plant species were identified. These data on DNA analysis of tomato seeds indicate that AFLP analysis is a viable procedure for the individualization of seeds from stomach contents in medico-legal and forensic investigations.

*Corresponding author

A list of abbreviations is given before the references.

INTRODUCTION

Non-human materials such as plant and animal products not only provide important information for criminal investigations by linking a victim, suspect, witness or weapon to a crime, but also may verify an alibi or provide investigative leads during criminal investigations (Bock and Norris 1997, Mildenhall 2004). Tomato (*Lycopersicum esculentum*) seeds were selected as a model system since tomato is commonly found in a variety of cuisines around the world. Investigators may find different types of tomato evidence, such as tomato sauce, fresh tomato, tomato juice or tomato seeds in a vehicle, on clothing or at crime scene due to vomiting or defecation. In addition, tomato seed is often found in stomach contents collected during autopsy. The examination of ingested food from feces, stomach contents, and both small and large intestine are important steps in medico-legal investigation.

The results of such examination often provide valuable information for forensic purposes. Identification of seeds from stomach contents can verify the last meal of the deceased, provide linkages between victim and suspect, or help supply traceable seed evidence in death investigations (Bock and Norris 1997, Coyle 2005). In addition, stomach contents collected at autopsy may contain identifiable plants and seeds that can be used for estimating the time of death (Pathak 2002) or verifying an alibi (Bock and Norris 1997).

Several methods have been reported for the identification of plant species in gastric contents. It has been reported that the high performance liquid chromatography (HPLC) technique can be used to identify vegetable species in gastric contents (Hayashiba et al. 1996, Nagata et al. 1991). However, the majority of medical examiners and forensic scientists still rely on microscopic and morphological methods to identify plant and food materials from stomach contents (Bock and Norris 1997, Mertens 2003, Platek et al. 2001). Many edible fruits and vegetables contain seeds with tough, durable seed coats that remain intact after passing through the human digestive system. However, many of the materials are not recognizable after passing through the human digestive system. Therefore, alternative methods, such as DNA tests, must be developed. The DNA databases of genetic profiles with variety-specific markers can be established to identify and characterize different cultivars of edible plant and seed species for forensic application (Diaz et al. 2003, Suliman-Pollatschek et al. 2002, Vos et al. 1995).

DNA analysis of digested tomato seeds to determine the source of the seed sample is a novel area of research. The approximate time for the digestive process in humans has been established as follows: mouth 0-2 min, stomach 2-6 hr, small intestine 2-8 hr, and large intestine 6-9 hr, for a total time of 10-25 hr. Ingested seeds may acquire surface contaminants such as bacteria when they pass through this lengthy digestive system. In addition, seed components have different genetic inheritance patterns. The seed coat is maternal in origin, whereas the embryo is a unique (F1) hybrid of parental genetics due to recombination events. Limited DNA from a small seed may yield insufficient material for further testing.

Our study has shown that high quality DNA could be recovered from ingested tomato seeds and DNA extraction from a seed embryo will avoid the maternal component or other surface contaminants as sources of DNA. The amplified fragment length polymorphism (AFLP) technique was used to identify DNA from tomato seeds. It was first described by Vos et al. (1995) and represented a method for genotyping single source plant samples for a large number of markers to generate a highly discriminating DNA band pattern. The AFLP technique is based on the detection of genomic restriction fragments after polymerase chain reaction (PCR) amplification. Band patterns are produced without prior sequence knowledge, using a limited set of amplification primers. This technique has been used to identify fresh tomato samples but not previous to our work for ingested samples (Bonnema et al. 2002, He et al. 2003, Thomas et al. 1995). It permits the inspection of polymorphisms at a large number of loci within a short period of time and requires only a small amount of DNA (about 10 ng). Although other DNA typing techniques exist and have been developed for specific applications (Congiu et al. 2000, Jakse et al. 2001, Korpelainen and Virtanen 2003, Yoon 1993), AFLP analysis is the method of choice since it has general applicability to any single source plant sample regardless of the species (Miller Coyle et al. 2001, Coyle et al. 2002, Vos et al. 1995). Once tomato seeds have been identified that would yield sufficient quantity and quality of DNA for further testing, AFLP analysis may be performed. AFLP analysis was performed to generate special bands or peak patterns from tomato in this study. It is envisioned that AFLP can be performed to link a seed back to a parent plant or original fruit to aid in criminal investigations.

A SIMPLE METHOD FOR RECOVERY OF DNA FROM A SINGLE TOMATO SEED

Tomato seeds were collected from different varieties of fresh tomatoes obtained from the local grocery. Approximately 10-25 hr after each volunteer ingested a different tomato, excrement was collected. The seed samples were sieved and washed with 10% bleach. After air drying, seeds were examined under an Olympus microscope (model BH-2) at 40X magnification before dissection to avoid other adherent material that might later affect DNA quantitation on an agarose gel image. DNA extractions were performed as recommended by the manufacturer's protocol unless otherwise stated (DNeasy plant mini kit, QIAGEN, Valencia, California).

There are two kinds of seed preparation methods. In the first, samples were ground in the presence of liquid nitrogen using Mixer Mill MM300 (QIAGEN, and Retsch Technologies GmbH, Haan, Germany), following the manufacturer's protocol. In the second, a seed was ground with liquid nitrogen to a fine powder using a 2 mL collection tube and a disposable micropestle (Bel-Art Products, Pequannock, New Jersey) or a ceramic mortar and pestle ("hand grinding method"). For the final step, 50 μL of preheated (65°C) buffer AE was added to the DNeasy membrane to elute the extracted plant DNA. DNA yields were estimated by comparison with genomic DNA standards (K562, Gibco BRL) after electrophoresis on 1% agarose gels containing ethidium bromide for visualization. A portion of each DNA sample (10 μL) was loaded on the agarose gel for estimating yield. To test the ability of the DNA to amplify for AFLP analysis (Miller Coyle et al. 2003, Coyle 2005), DNA from a single representative seed was typed following the manufacturer's protocol (Plant Mapping kit, Applied Biosystems, Inc., Foster City, California).

The structure of the tomato seed (seed coat and embryo) is shown in Fig. 1. The tomato seeds were processed as dissected embryos (cut in half) and intact tomato seed (embryo plus seed coat, cut in half). The average yield from three DNA extractions of a single tomato seed was 100 ng and 62.5 ng for exterior seed coat and interior embryo respectively (Fig. 2). The yield of digested seed DNA was approximately equivalent to the fresh seed (Fig. 3). The above results showed that high quality DNA (40 ng total) could be extracted from only half a seed embryo. It means that the small amount of tomato sample is sufficient for further DNA testing.

For optimal DNA yield, hand grinding of seeds in liquid nitrogen to disrupt cells with a sterile disposable micropestle is recommended. The results indicate that hand grinding of samples in liquid nitrogen yielded high quality DNA and a greater quantity than grinding with the Mixer Mill. This result is probably due to the tube design used for the Mixer Mill and the small size of tomato seed samples. The major reason for the different yield between the two procedures was the sample tube design (Fig. 4) used in our experiments. With larger samples (e.g., leaf tissue), the tubes are efficient for optimal contact between the tissue and the beads used for tissue disruption; however, significantly smaller tissue samples a few millimeters in length (e.g., tomato seeds) become trapped in the space between the side of the sample tube and the bottom of the tube cap. This

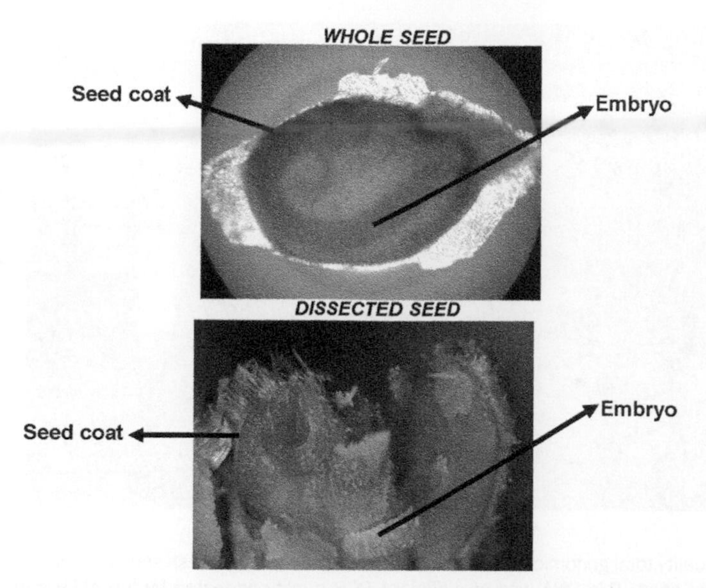

Fig. 1 The structure of a whole and dissected tomato seed viewed at 40 × magnification.

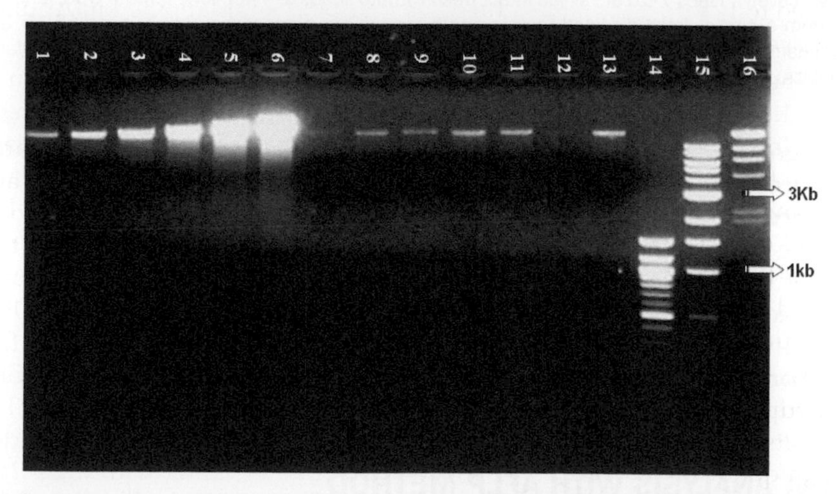

Fig. 2 High quality total genomic DNA was recovered from dissected embryos (cut in half) and intact tomato seeds (embryo plus seed coat, cut in half) from two of the five tomatoes. Lanes 1-6 are DNA quantitation standards, human K562 (12.5, 25, 50, 100, 200, 400 ng, respectively); lane 7 is a half embryo from tomato type 1; lane 8 is a half seed coat from tomato type 1; lane 9 is a half seed (embryo + seed coat combined) from tomato 1; lane 10 is a half seed (embryo + seed coat combined) from tomato 2; lane 11 is a half seed coat from tomato 2; lane 12 is a half embryo from tomato 2; lane 13 is an intact seed from tomato 3; lanes 14-16 are standard size markers 100 bp, 1 Kb and Lambda DNA-Hind III ladder, respectively (New England Biolabs, Beverly, Massachusetts). In all cases, the recovered DNA from the tomato seeds or dissected seed components was approximately 12.5 ng/10 μL or less.

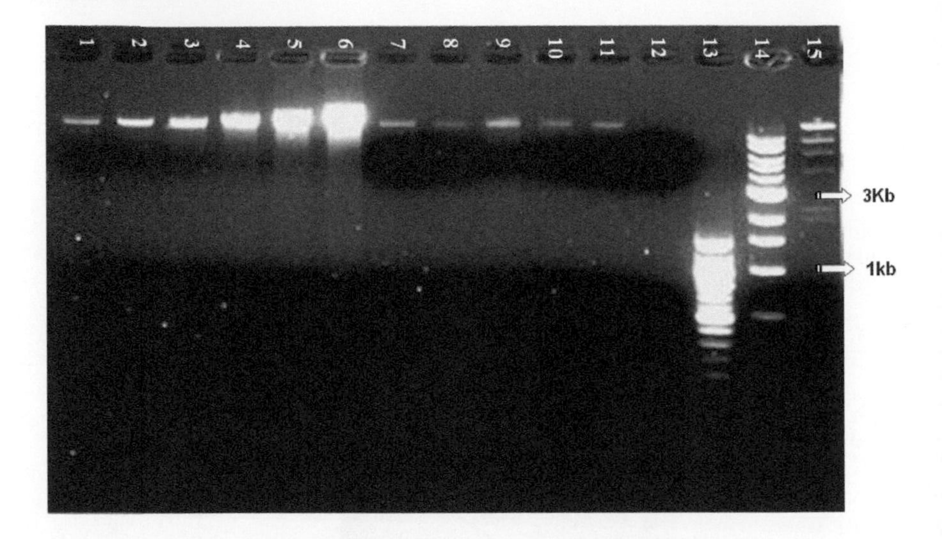

Fig. 3 High quality total genomic DNA was recoverable from a single fresh seed and single fresh tomato digested seed. This DNA was of sufficient PCR quality as tested by the AFLP method. Lanes 1-6 are DNA quantitation standards, human K562 (12.5, 25, 50, 100, 200, 400 ng, respectively); lanes 7-10 are single intact fresh tomato seeds 1-4 from one tomato type; lane 11 is from a single, fresh digested tomato seed; lane 12 is a negative control; lanes 13-15 are standard size markers 100 bp, 1 Kb and Lambda DNA-Hind III ladder respectively (New England Biolabs).

design results in limited contact between the bead and tissue sample and reduces the efficiency of the grinding procedure such that DNA yields are reduced. Our proposed modification of the tube design may improve single seed DNA yields with the Mixer Mill method. We found that the plant DNeasy kit from QIAGEN was used to efficiently process intact tomato seeds and dissected tomato seed embryos to obtain PCR-quality plant DNA. Fresh tomato seeds, after they have passed through the human digestive system, also yielded sufficient DNA for further PCR-based experiments (Fig. 3).

DNA ANALYSIS WITH AFLP METHOD

The Modified Process of AFLP

AFLP: Digestion-Ligation Reactions

Prepare amounts of reagents as with the AFLP Digesting-Ligation Worksheet:

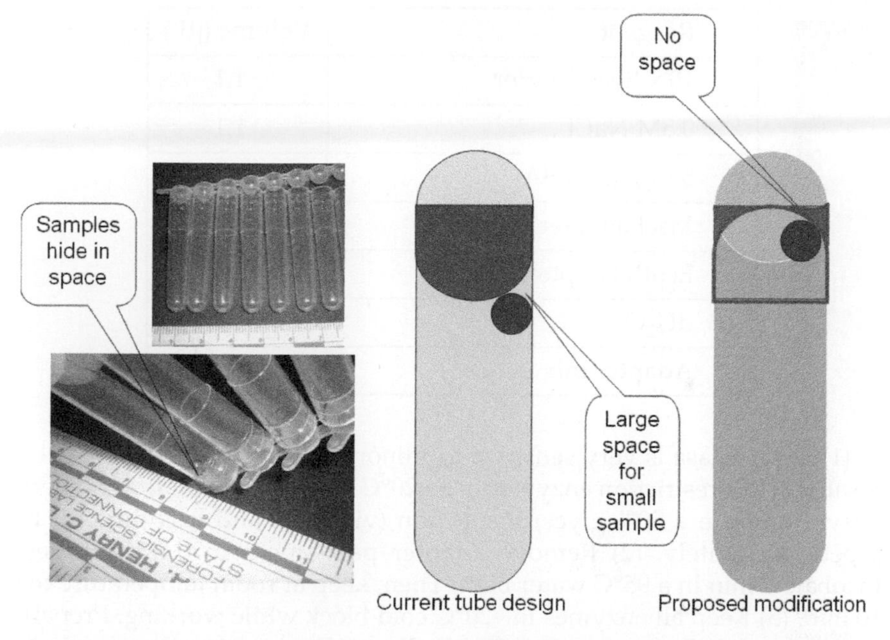

Fig. 4 The suggested improvement in tube design may increase single seed DNA yields with the Mixer Mill machine seed preparation method. The cap design is inverted to avoid the space in which the seed can become caught between the cap and the tube wall.

Enzyme/adapter mix	5.5 μL
DNA/dH2O	5.5 μL
Total volume/sample	**11 μL**

Reagent	Volume (μL)
MseI (50 U/μL)	0.02
EcoRI (50 U/μL)	0.1
T4 ligase (30 Weiss U/μL)	0.03
Enzyme mix	**0.153**

Reagent	Volume (μL)
10 × ligase buffer	1.1
0.5M NaCl	1.1
5 mg/mL BSA	0.11
MseI adapter	1
EcoRI adapter	1
dH_2O	1.037
Adapter mix	**5.347**

(1) DNA ligase is very sensitive to minor temperature changes. Keep MseI & EcoRI restriction enzymes in a -20°C cold block at all times. These enzymes are in a 50% glycerol solution (viscous), and it is difficult to pipette accurately. (2) Remove adapter pair stocks to 0.5 mL tubes. Incubate 5 min in a 95°C water bath. Then, keep at room temperature for 10 min. (3) Keep all enzymes in -20°C cold block while working. Prepare enzyme mix and adapter mix individually. (4) Prepare the adapter mix in 0.5 mL tube first, mix and centrifuge briefly. Then, transfer to the tube containing the enzyme mix. Mix carefully without forming bubbles. (5) Immediately add 5.5 mL of the enzyme/adapter mix to each of the DNA sample tubes. Carefully avoid forming bubbles. (6) Incubate at 37°C in a thermal cycler: GeneAmp PCR System 9700 Base Unit (PE Corporation, Foster City, California; P/N 805-0200) with a GeneAmp PCR 9700 interchangeable sample block (PE Corporation, P/N 4314443 or 4314445) with heated cover for 2 hr.

AFLP: Pre-selective Amplification

Keep those reagents in a -20°C non-frost freezer. Avoid exposing them to light for long periods of time. Thaw slowly, mix and centrifuge briefly before use. (1) Transfer 4 μL DL (Digestion-Ligation) to clean 0.2 mL PCR tubes. Prepare a negative control tube with 4 μL Tris EDTA (TE) buffer. (2) Prepare master mix by mixing the following: (n = number of samples)

(n + 1) × 1 μL AFLP EcoRI/MseI pre-amplification primers

(n + 1) × 15 μL AFLP core mix

(3) Quickly transfer 16 μL pre-amp mix to each DL tube and mix gently. (4) Using the GeneAmp PCR System 9700 thermal cycler according to the following parameters, with the ramp speed set to MAX. Unused DL can be stored at -20°C.

Pre-selective amplification (PSA) thermal-cycling temperature parameters:

72°C for 2 min.

20 cycles of 94°C for 2 sec, 56°C for 30 sec, 72°C for 2 min.

60°C for 30 min with a final 4°C hold.

AFLP: Selective Amplification

(1) Transfer 3 μL PSA product to 0.2 mL tubes. Prepare a negative control tube with 3 μL negative control PSA. (2) Prepare master mix by mixing the following and centrifuging briefly.

(n + 1) × 1 μL AFLP EcoRI (AAC) selective primer

(n + 1) × 1 μL AFLP MseI (CAA) selective primer

(n + 1) × 15 μL AFLP core mix

(3) Quickly transfer 17 μL selective amplification mix to each of 0.2 mL tubes and mix gently. (4) Use a GeneAmp PCR System 9700 thermal cycler according to the following parameters, with the ramp speed set to MAX. Unused DL can be stored at −20°C.

Selective amplification thermal-cycling temperature parameters:

94°C for 2 min.

1 cycle of: 94°C for 2 sec, 66°C for 30 sec, 72°C for 2 min.

1 cycle of: 94°C for 2 sec, 65°C for 30 sec, 72°C for 2 min.

1 cycle of: 94°C for 2 sec, 64°C for 30 sec, 72°C for 2 min.

1 cycle of: 94°C for 2 sec, 63°C for 30 sec, 72°C for 2 min.

1 cycle of: 94°C for 2 sec, 62°C for 30 sec, 72°C for 2 min.

1 cycle of: 94°C for 2 sec, 61°C for 30 sec, 72°C for 2 min.

1 cycle of: 94°C for 2 sec, 60°C for 30 sec, 72°C for 2 min.

1 cycle of: 94°C for 2 sec, 59°C for 30 sec, 72°C for 2 min.

1 cycle of: 94°C for 2 sec, 58°C for 30 sec, 72°C for 2 min.

1 cycle of: 94°C for 2 sec, 57°C for 30 sec, 72°C for 2 min.

20 cycles of: 94°C for 2 sec, 56°C for 30 sec, 72°C for 2 min.

60°C for 30 min with a 4°C hold.

Detecting AFLP Products with an ABI PrismTM 377 DNA Sequencer

The instrument and accompanying computer should be equipped with ABI PrismTM 377XL Firmware version 2.5. (1) Prepare acrylamide gel with a LongRanger® Singel pack® according to the manufacturer's instructions. Use 36 cm well-to-read distance glass plates, 0.2 mm spacers and a 34-well square-tooth comb. Polymerize at least 2 hr. After first 15 min of polymerization, wet and wrap the end of the gel with plastic wrap to prevent it from drying. (2) Plate check: Left Double click the Desktop "ABI Prism® 377 Collection Alias." Under the pull down file menu, select New, then select Genescan run. Close door, click plate check button. Wait for 1-2 min for "Scan Image" window to appear. All colors should be flat-lined at the bottom near 2000 units; if not, re-clean the region with dH2O, methanol, and/or canned air as needed. (3) Mix 500 μl deionized formamide with 100 μl blue dextran loading solution (FLS solution). (4) The number of samples to be loaded = n. In a separate tube, mix n × 4.75 μL formamide/blue dextran solution with n × 1 μL GeneScan-500 ROX size standard (red dye standard for sizing DNA fragments). (5) Mix 4 μL of each Selective Amplification product (designated SA) with 5 μL formamide/blue dextran/ROX solution. (6) Heat formamide/blue dextran/ROX/SA at 95°C for 2 min, then immediately transfer to ice or cold block. (7) Gently flush the wells of the gel with 1 × Tris, borate, EDTA (TBE) buffer. Pre-run the gel (Pre-Run Module: GS PR 36F-2400) with 1 × TBE buffer for 5 min, Hit Pause. (8) Gently flush the wells again. Load 2 μL of formamide/blue dextran/ROX/SA into each well. (9) To cancel the Pause Pre-Run, hit Resume, then Cancel. Run the following Run Module:

EP Voltage: 1680 volts

EP Current: 60.0 mA

EP Power: 200 watts

Gel Temp: 51°C

Laser Power: 40.0 mW

Run time: 5.50 hr

CCD Offset: 0

CCD Gain: 2

CCD X Pixel Position: 200

Genescan Analysis of AFLP Peak Profiles

(1) Open Run file and double click on the gel image. Allow the Genescan software to create a gel image. (2) Click on the Gel tab and scroll down to select Install New Gel Matrix. Choose an appropriate matrix and hit Open. (3) Click on the Gel tab and scroll down to select Track Lanes. When a pop-up menu appears, choose Auto-Track Lanes. (4) Click on the Gel tab and scroll down to select Adjust Gel Contrast to increase the contrast of the red

dye. (5) Click on the Gel tab and scroll down to select Regenerate Gel Image. When a pop-up menu appears, type in the smaller scan number as 2200 of the scan range and the larger number as the stop of the scan range. (6) Click on the Gel tab and scroll down to select Extract Lanes. When a pop-up menu appears, deselect Auto-Analysis New Sample Files and Over-Write Original Sample files. Select OK. (7) Choose the appropriate size standards and parameters for analysis of samples. In this case, to select a Size Standard, proceed to the top of the Size Standard column and use the right arrow to select threshold parameters as sample 50. Click Analyze. (8) View the analysis log. For AFLP samples, ROX size standards should cover the range from 50 to 500 bases. (9) Click on the Windows tab and scroll down to select Result Control. Change the number of panels to 8 by clicking and holding down the mouse button. To view each sample in its own panel, click on the number 1 on the left-hand side and do the same for the next 7 samples. Click on Display to view the allele peaks. (10) Check to make sure that all the negative and positive controls worked as expected.

A schematic diagram of the AFLP general procedure is shown in Fig. 5. The AFLP technique was performed according to the manufacturer's protocol and as modified in our previous published studies (Lee et al. 2005, 2006). To select optimal primer sets, different random selections of primers from 64 possible primers in the AFLP Plant kit were screened and used in further testing on samples.

Sixty-four possible primer combinations allow scoring for hundreds of markers (peaks). Peaks between 50 and 500 bases were sized using GeneScan software (version 3.1.2). AFLP patterns were determined for each seed sample and each DNA fragment within the pattern is represented as a "peak". If the pattern was the same for different seeds, the primer set was not useful for discriminating different seed samples. If each seed AFLP pattern was different, the primer set was suitable and used for distinguishing the seed sample. From the optimal primer combinations, the more different a pattern is, the better it will be for distinguishing the different seed types. In this purpose, many variable peaks were selected as markers to distinguish the different species or cultivar seed samples. Using customized Genotyper software (Applied Biosystems), AFLP polymorphic peaks were scored as present (1) and absent (0), and then the binary code for the different genotypes was compared. The criteria for scoring DNA fragments to binary code included selected peaks in the size range of 75-460 nucleotides (X-axis) and a relative fluorescence unit (RFU), Y-axis threshold of 50. There are 64 possible AFLP selective primers available from the kit. The sequence information of the adapters and PCR primers are presented in Table 1.

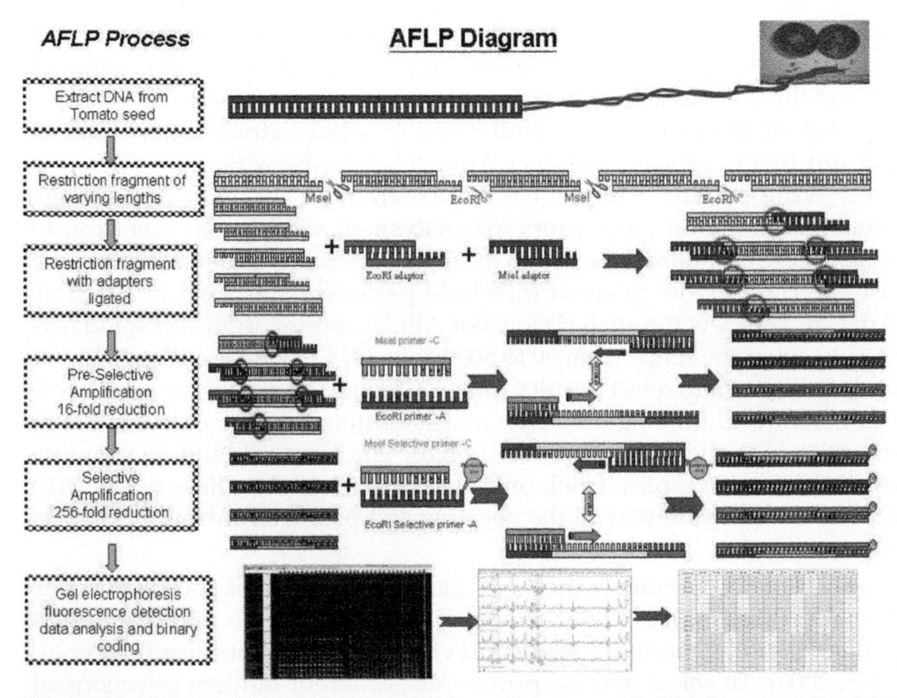

Fig. 5. The AFLP technique uses stringent PCR conditions to amplify DNA fragments. The AFLP amplification primers are generally 17-21 nucleotides in length and anneal perfectly to their target sequences, i.e., the adapter and restriction sites, and a small number of nucleotides adjacent to the restriction sites. This makes AFLP a reliable and robust technique, unaffected by small variations in amplification parameters (e.g., thermal cyclers, template concentration, PCR cycle profile). The high marker densities that can be obtained with AFLP are a characteristic of the technology: a typical AFLP "fingerprint" contains 50-100 amplified fragments. Moreover, the technology requires no sequence information or probe optimization prior to the generation of AFLP fingerprints. The frequency with which AFLP markers are detected depends on the level of sequence polymorphism between the tested DNA samples. Steps in AFLP analysis: 1. Restriction digestion of genomic DNA using MseI and EcoRI enzymes. 2. Ligation of adapters using EcoRI Adapter = 5'-CTCGTAGACTGCGTACC (EcoRI-oligo.1) and MseI Adapter = 5'-GACGATGAGTCCTGAG (MseI-oligo.1). 3. Pre-selective amplification using EcoRI + A oligo and Mse I + C oligo. 4. Selective amplification using EcoRI + ANN oligo and MseI + CNN oligo. 5. Gel electrophoresis, fluorescence detection, data analysis and binary coding.

The seed samples from different plants could be easily distinguished from each other using one of the 64 possible primer combinations (Table 1). In addition, with the AFLP patterns, many of the peaks differed between different species and cultivars; these were useful for discriminating different species and cultivars and may prove to be variety-specific markers. For example, some peaks were at the same position and common to all tomato types but different with all pepper types; those peaks were of no value for tracing the source of the tomato but may be the tomato-specific markers. We could recognize those peaks with

Table 1 PCR primer combinations used for AFLP analysis.

Primers used in pre-selective amplification	
EcoRI + 1-A	5′-GACTGCGTACCAATTC + A-3′
MseI + 1-C	5′-GATGAGTCCTGAGTAA + C-3′
Primer combinations used in selective AFLP amplification	
EcoRI + 3-**AAC*** 5′-GACTGCGTACCAATTC + **AAC**-3′	
MseI + 3-**CAA*** 5′-GATGAGTCCTGAGTAA + **CAA**-3′	
* *EcoRI* Primer + **AAC**/AAG/ACA/ACC/ACG/ACT/AGC/AGG	
* *MseI* Primer + **CAA**/CAC/CAG/CAT/CTA/CTC/CTG/CTT	

The PCR primer sequences used in this study are listed for the pre-selective and selective amplification steps for the AFLP analysis of tomato samples.

*Other possible primer combinations available in the AFLP kit but not used in this study.

different species (pepper). Some peaks were at a different position and varied with all tomato types; those peaks were valuable for tracing the tomato cultivar.

To make the experimental data easier to understand, the final analysis of AFLP pattern was converted to a binary Excel file (Microsoft Corp, Santa Rosa, California) (Table 2). The Excel file format provides a simple method to easily compare seed samples and types. With additional screening of genetically controlled seed populations, it may be possible to identify fragments that can be used as species/cultivar-specific markers.

AFLP Method to Identify Tomato with Other Plant Species

Ingested tomato and pepper seeds were used as a model system to test the feasibility of performing DNA testing for plant species identification. Seeds were collected from volunteers who had ingested different types of tomato and pepper. After approximately 10-25 hr, excrement was collected. The seed samples were sieved and washed from the feces to collect visible seeds that had passed through the digestive system (Fig. 6). When seeds of tomato and pepper pass through a human digestive system, it is difficult to differentiate them, because of similarities in size, shape and microscopic appearances. Fresh and ingested tomato and pepper seeds were selected as known control samples for this study. Additional seeds were selected for use in a blind test. This approach can be applied to other plant species in addition to tomato and pepper seeds.

From the tomato and pepper DNA extraction results, it is clear that high quality DNA could be extracted from either a single fresh seed or an ingested seed. The DNA quality and quantity were sufficient for the PCR methods using the DNA-AFLP technique. To select optimal primer sets,

Table 2 Binary coding for 12 variable AFLP markers for tomatoes A-E.

Sample	*01*	*02*	*03*	*04*	*05*	*06*	*07*	*08*	*09*	*10*	*11*	*12*
A1	**0**	**0**	**0**	**0**	**0**	**0**	**1**	**0**	**0**	**1**	**0**	**0**
A2	**0**	**1**	**0**	**0**	**0**	**0**	**1**	**0**	**0**	**1**	**0**	**0**
A3	**1**	**0**	**0**	**0**	**0**	**0**	**1**	**0**	**0**	**1**	**0**	**0**
A4	**0**	**1**	**0**	**0**	**0**	**0**	**1**	**0**	**1**	**1**	**0**	**0**
B1	**0**	**0**	**0**	**0**	**1**	**1**	**0**	**0**	**0**	**0**	**1**	**1**
B2	**1**	**0**	**0**	**0**	**1**	**1**	**0**	**0**	**0**	**0**	**1**	**0**
B3	**0**	**0**	**0**	**0**	**1**	**1**	**0**	**0**	**0**	**0**	**1**	**0**
B4	**0**	**0**	**0**	**1**	**1**	**1**	**0**	**1**	**0**	**0**	**1**	**1**
C1	**0**	**1**	**0**	**0**	**1**	**1**	**0**	**0**	**0**	**0**	**1**	**0**
C2	**0**	**0**	**0**	**0**	**1**	**1**	**0**	**1**	**0**	**1**	**1**	**1**
C3	**0**	**0**	**0**	**0**	**1**	**1**	**0**	**1**	**0**	**0**	**1**	**1**
C4	**0**	**0**	**0**	**0**	**1**	**1**	**0**	**1**	**0**	**0**	**1**	**1**
D1	**0**	**0**	**1**	**0**	**1**	**0**	**0**	**1**	**0**	**0**	**1**	**1**
D2	**0**	**0**	**0**	**0**	**1**	**1**	**0**	**1**	**0**	**0**	**1**	**0**
D3	**0**	**0**	**1**	**0**	**1**	**0**	**0**	**1**	**0**	**0**	**1**	**0**
D4	**0**	**0**	**1**	**0**	**1**	**1**	**0**	**1**	**0**	**0**	**1**	**1**
E1	**1**	**0**	**1**	**1**	**1**	**0**	**0**	**0**	**0**	**0**	**0**	**0**
E2	**1**	**0**	**0**	**1**	**1**	**1**	**0**	**0**	**0**	**0**	**0**	**0**
E3	**1**	**0**	**0**	**1**	**1**	**1**	**0**	**0**	**0**	**0**	**0**	**0**
E4	**1**	**0**	**0**	**1**	**1**	**1**	**0**	**0**	**1**	**1**	**0**	**0**

A comparison of embryos from each of the five morphologically different tomatoes showed that a single PCR primer combination could distinguish between samples. Some markers (peaks) were common to all of the tomato samples and may be useful for species identification. Others were highly variable and useful for individualizing samples from separate tomatoes. With a database of tomato markers, it should be easy to trace a source of a tomato sample for forensic purposes. (1) = peak present at bin location; (0) = peak absent at bin location. Twelve bins were scored for the selective primer set EcoRI-AAC and MseI-CAA. Letter designations for the sample column indicate tomato source (A-E); numbers 1-3 = embryos and 4 = intact seed.

64 possible primers in the AFLP plant kit were screened. We selected the AFLP pattern with some peaks at the same position and common to all tomato types but different with all pepper types; those peaks were valuable for recognizing the species and may be the tomato-specific markers. We could use those peaks to differentiate those species (pepper). The sequence information of the selected PCR primers are EcoRI + 3-AAC and MseI + 3-CAA as referenced in our published data (Lee et al. 2005, 2006). Using the primer combinations, it is easy to obtain AFLP patterns with many peaks that differed between tomato and pepper samples. Those peaks may be used as the species-specific markers (Fig. 7).

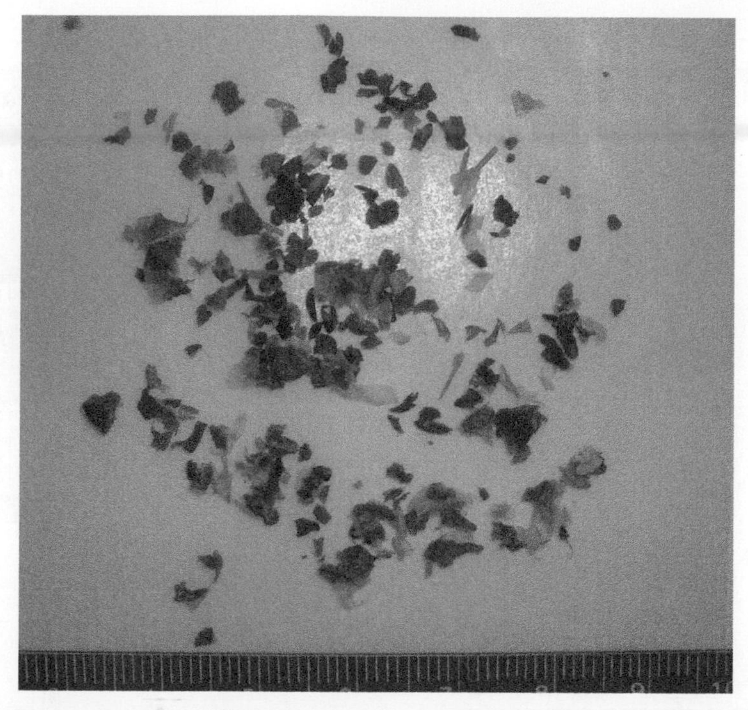

Fig. 6 Seeds were ingested and collected after they passed through the digestive system of volunteers. Different seed species from feces were sieved and mixed together. It is difficult to discern between tomato and pepper seeds visually.

From the above results, we showed that AFLP analysis can be performed to distinguish two plant seed samples at the species level, thus providing an additional tool for criminal investigations and other applications of forensic science. Using the AFLP method, it is possible to use a DNA-based system to discriminate different species and even closely related plant sources.

AFLP Method to Identify the Cultivar Tomatoes

Tomato seeds were used as a model system to test the feasibility of performing DNA testing for cultivar identification within species. Five phenotypically different fresh tomatoes were selected (Fig. 8, labeled A-E) from the local supermarket. No genetic variety names were available from the supermarket or its produce supplier. Digested seeds were obtained in the following manner: seeds were ingested and collected after they passed through the digestive system of volunteers. Seeds from feces were sieved, washed copiously with water to remove adherent material, and examined

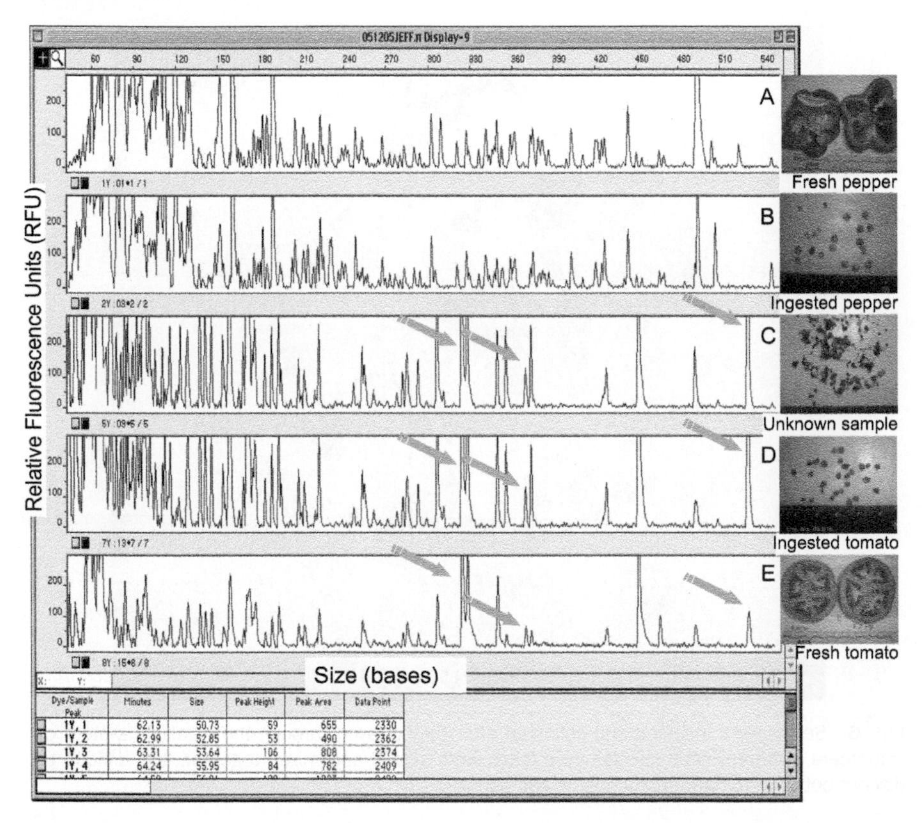

Fig. 7 There are five samples including A: Fresh pepper seed; B: Ingested pepper seed; C: Unknown seed sample recovered from feces; D: Ingested tomato seed; E: Fresh tomato seed. The AFLP patterns of A and B are similar, belonging to the pepper group. The AFLP patterns of D and E are similar, belonging to the tomato group. Patterns from each tomato and pepper AFLP profile was different (the arrows indicate marks present in tomato but absent in pepper).

microscopically. DNA extractions were performed as recommended by the above protocol. The identification of candidate markers for distinguishing phenotypically distinct tomatoes indicates that it is possible to use a biological method (DNA-based system) with sufficient markers to discriminate even closely related plant sources (e.g., two groups of seed from the same tomato plant as well as potentially for cultivar identification). The AFLP patterns showed us that many of the peaks differed between different cultivars of tomato; these were useful for discriminating different cultivars and may prove to be variety-specific markers. From the binary Excel file data, we can easily distinguish the different patterns for each tomato type. Here, we also show that AFLP analysis was performed to generate an individualizing band or peak

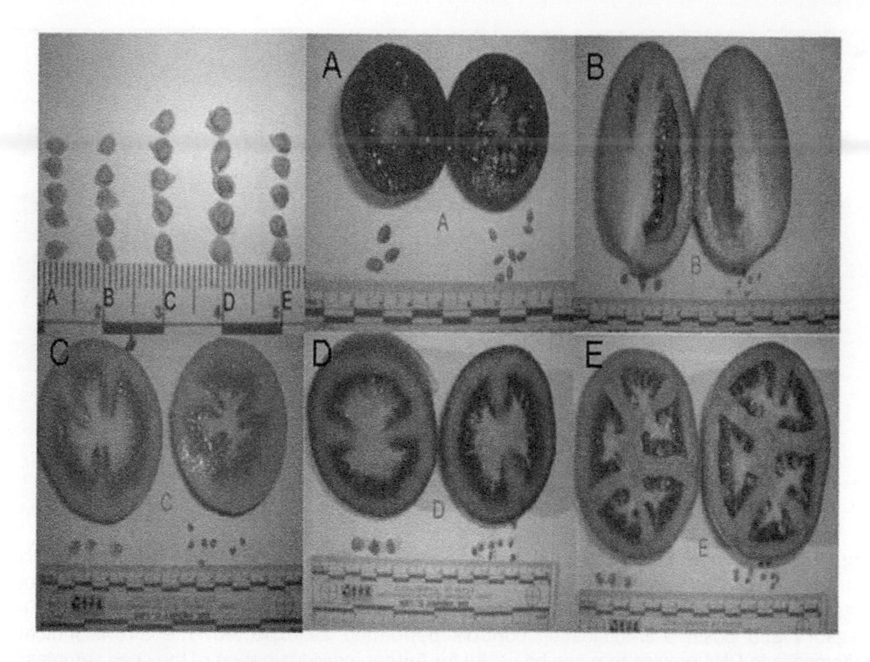

Fig. 8 Five different tomatoes were used in this study (designated A-E). The figure illustrates seeds and cross-sections of whole tomatoes (letter = color, diameter; A = red, 3.5 cm^2; B = red, 8.5 cm × 5.4 cm; C = yellow, 7 cm^2; D = red, 7.5 cm^2; E = red, 9 × 7.7 cm). The diameter of the tomato did not correlate to seed size and seed morphology was not a predictor of the tomato cultivar and thus not useful for cultivar identification.

pattern from tomato seeds. It is envisioned that AFLP can be performed to link a seed back to a parent plant or original fruit to aid in criminal investigations.

In Fig. 9, the arrows showed different peaks in the AFLP pattern and the data illustrates some representative DNA profiles generated from those tomato seeds. DNA typing with the AFLP method can be useful for further individualization of the tomato samples. Interestingly, these same markers were observed in whole seeds as well as from the three extracted embryos per tomato, which suggests that it may not be necessary to dissect the seeds for cultivar identification applications.

SUMMARY

Seeds from stomach contents can verify a last meal or provide traceable evidence in criminal investigations. Many seeds with tough, durable seed coats remain intact after passing through the human digestive system. These seeds can be retrieved from a crime scene where a suspect may have

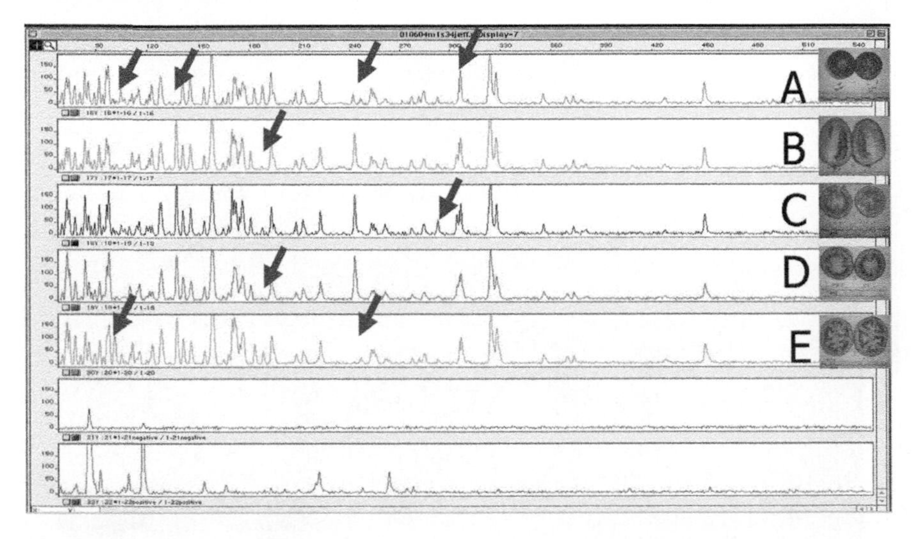

Fig. 9 Five phenotypically different tomato seed samples yielded AFLP profiles after DNA extraction using the hand grinding method. The DNA extracted from fresh tomato seeds, both digested and undigested, was of sufficient quality for PCR amplification by the AFLP method. The negative and positive amplification controls performed as expected. The arrows indicate differences in DNA profiles that can be useful for further individualization of the seed samples.

vomited or defecated. In addition, stomach contents collected at autopsy may contain identifiable plants and seeds that can be used for estimating the time of death or to verify an alibi. These pieces of information may become key evidence and very useful in forensic investigations. Our study shows DNA could be recovered from tomato seeds in fecal material, after passing through the stomach, small intestine and large intestine. Therefore, seeds collected from stomach contents, an earlier step in the digestive pathway, will yield equivalent or even better results.

The recovered tomato seed was useful for DNA typing by the AFLP method. From our discussion, a sufficient quality and quantity of DNA can be obtained to generate an individualizing DNA profile. The species/cultivar-specific markers could be easily distinguished by the AFLP method. By using the selected PCR primers EcoRI + 3-AAC and MseI + 3-CAA combinations, we could easily obtain AFLP patterns with many peaks that differed between species/cultivar samples. Those peaks were valuable and may be used as species-specific markers for further forensic applications. Once additional cultivar-specific markers have been identified and catalogued into a database, this method should be useful and provide a model system for tracing edible plant matter for criminal investigation.

ACKNOWLEDGEMENTS

The names of commercial manufacturers are provided for information only and do not imply an endorsement by any of the authors or their affiliated agencies. We want to thank the National Institute of Justice (NIJ grant #2001-IJ-CX-K011), the University of New Haven and the Henry C. Lee Forensic Science Institute for funding the research supplies. We also want to thank the National Science Council, Republic of China, for scholarship support (NSC92-2917-1-007-004) for Cheng-Lung Lee.

ABBREVIATIONS

AFLP: amplified fragment length polymorphism; HPLC: high performance liquid chromatography; PCR: polymerase chain reaction; TE: Tris EDTA buffer; DL: digestion ligation reaction; PSA: pre-selective amplification reaction; ROX: red dye standard for sizing DNA fragments; SA: selective amplification reaction; TBE: Tris, borate, EDTA buffer

REFERENCES

Bock, J. and D. Norris. 1997. Forensic botany: An under-utilized resource. J. Forensic Sci. 42: 364-367.

Bonnema, G. and P. van den Berg, and P. Lindhout. 2002. AFLPs mark different genomic regions compared with RFLPs: a case study in tomato. Genome/ National Research Council Canada = Genome/Conseil national de recherches Canada 45: 217-221.

Congiu, L. and M. Chicca, R. Cella, R. Rossi, G. Bernacchia. 2000. The use of random amplified polymorphic DNA (RAPD) markers to identify strawberry varieties: a forensic application. Mol. Ecol. 9: 229-232.

Coyle, H.M. 2004 Forensic Botany: Principles and Applications to Criminal Casework. CRC Press, Boca Raton, Florida, USA.

Coyle, H.M. 2005. Forensic Botany: Principles and Applications to Criminal Casework. CRC Press, Boca Raton, Florida, USA.

Coyle, H.M. and J. Germano-Presby, C. Ladd, T. Palmbach, and H.C. Lee. 2002. Tracking clonal marijuana using amplified fragment length polymorphism (AFLP) analysis: an overview. Conference proceedings: 13th International Symposium on Human Identification. Phoenix, Arizona, USA.

Diaz, S. and C. Pire, J. Ferrer, and M.J. Bonete. 2003. Identification of *Phoenix dactylifera* L. varieties based on amplified fragment length polymorphism (AFLP) markers. Cell. Mol. Biol. Lett. 8: 891-899.

Hayashiba, Y. and K. Kimura, S. Kashimura, T. Nagata, and T. Imamura. 1996. Identification of vegetable species in gastric contents using HPLC. Intl. J. Legal Med. 108: 206-209.

He, C. and V. Poysa, and K. Yu. 2003. Development and characterization of simple sequence repeat (SSR) markers and their use in determining relationships among *Lycopersicon esculentum* cultivars. TAG Theor. Appl. Genet. 106: 363-373.

Jakse, J. and K. Kindlhofer, and B. Javornik. 2001. Assessment of genetic variation and differentiation of hop genotypes by microsatellite and AFLP markers. Genome/National Research Council Canada = Genome/Conseil national de recherches Canada 44: 773-782.

Korpelainen, H. and V. Virtanen. 2003. DNA fingerprinting of mosses. J. Forensic Sci. 48: 804-807.

Lee, C.L. and H.M. Coyle, E. Carita, C. Ladd, N.C. Yang, T.M. Palmbach, I.C. Hsu, and H.C. Lee. 2006. DNA analysis of digested tomato seeds in stomach contents. Amer. J. Forensic Med. Pathol. 27: 121-125.

Lee, C.L. and H.M. Coyle, T.M. Palmbach, I.C. Hsu, and H.C. Lee. 2005. DNA analysis of ingested tomato and pepper seeds. Amer. J. Forensic Med. Pathol. 26: 330-333.

Mertens, J. 2003. Forensics follows foliage: botanists track plant evidence from crime scene to courtroom. Law Enforcement Technol. 30: 62-68.

Mildenhall, D.C. 2004. An example of the use of forensic palynology in assessing an alibi. J. Forensic Sci. 49: 312-316.

Miller Coyle, H. and C. Ladd, T. Palmbach, and H.C. Lee. 2001. The Green Revolution: botanical contributions to forensics and drug enforcement. Croatian Med. J. 42: 340-345.

Miller Coyle, H. and G. Shutler, S. Abrams, J. Hanniman, S. Neylon, C. Ladd, T. Palmbach, and H.C. Lee. 2003. A simple DNA extraction method for marijuana samples used in amplified fragment length polymorphism (AFLP) analysis. J. Forensic Sci. 48: 343-347.

Nagata, T. and Y. Hayashiba, and K. Kimura. 1991. Identification of species of weeds using high performance liquid chromatography in three crime cases. Intl. J. Legal Med. 104: 285-287.

Pathak, P.R, 2002. Role of autopsy surgeon in determination of time since death in criminal cases. J. Indian Med. Assoc. 100: 700-702.

Platek, S.F. and J.B. Crowe, N. Ranieri, and K.A. Wolnik. 2001. Scanning electron microscopy determination of string mozzarella cheese in gastric contents. J. Forensic Sci. 46: 131-134.

Suliman-Pollatschek, S. and K. Kashkush, H. Shats, J. Hillel, and U. Lavi. 2002. Generation and mapping of AFLP, SSRs and SNPs in *Lycopersicon esculentum*. Cell. Mol. Biol. Lett. 7: 583-597.

Thomas, C.M. and P. Vos, M. Zabeau, D.A. Jones, K.A. Norcott, B.P. Chadwick, and J.D. Jones. 1995. Identification of amplified restriction fragment polymorphism (AFLP) markers tightly linked to the tomato Cf-9 gene for resistance to *Cladosporium fulvum*. Plant J. 8: 785-794.

Vos, P. and R. Hogers, M. Bleeker, M. Reijans, T. van de Lee, M. Hornes, A. Frijters, J. Pot, J. Peleman, M. Kuiper and M. Zabeau. 1995. AFLP: a new technique for DNA fingerprinting. Nucleic Acids Res. 23: 4407-4414.

Yoon, C.K. 1993. Forensic science. Botanical witness for the prosecution. Science 260: 894-895.

Index

1-methylcyclopropene 67, 74, 75, 81
2,3-dihydrofarnesoic acid 287
2-pentadecanone 283
2-tridecanone 279
2-undecanone 279
5′ NT: 5′ Nucleotidase 472
5-α-dihydrotestosterone (DHT) 416
14-3-3 protein 508, 509
35S 526, 528, 532

A
Affects on human health 346
A proteinase/antiproteinase imbalance 478
ABTS 114-116, 121, 122, 125, 128, 544
Acid 3, 59, 71, 89-94, 98-103, 215-217, 220, 221, 286, 301, 310, 331, 334, 345, 346, 468, 506, 508, 544, 567, 568, 597, 601
Acid phosphatase 216, 217, 227, 229, 231
Acrylamide 259-265, 625
Actin-like proteins 509
ADA: Adenosine deaminase 472
AFLP 131, 586, 588, 589, 607, 613, 617, 619, 620, 622, 624-635
Aging 477, 478
Agrobacterium tumefaciens 515-517, 519
Albumins 507
ALP-Alkaline phosphatase 229, 467
Alternative oxidase 594, 600, 608
Ames text 378
Amino acid 71, 89, 90, 107, 110, 172, 173, 175, 178, 224, 225, 261, 504
Analysis 59, 60, 88, 96, 97, 103, 105-107, 109, 110, 165, 168, 170-172, 175, 177, 179, 184, 193, 205, 337, 403, 421, 444, 468, 482, 502, 524, 538, 585, 586, 617
Analysis of carotenoids 548
Analysis of vitamin C 561
Analytical methods 148
Animal models for emphysema 484
Anthocyanins 113, 114, 124, 358, 638
Anticancer properties of saliva 301, 377-379, 381-383
Antioxidant activity 41, 113, 115, 334, 443, 444, 459, 475, 487, 538, 544
and affects on human health 346
and antioxidant content 345, 347, 351, 357, 365
and ascorbic acid 345, 346, 348, 355, 356, 358, 359, 361, 362, 364, 365, 367, 369, 371, 374
and carotenoids 345, 346, 355, 356, 359, 360, 362, 363, 368, 370-372, 374, 375
and extraction procedure 364, 367
and lipid peroxidation methods 368
and lycopene 361-364, 367, 368, 370, 372-375
and maturation 345, 347, 363, 365
and microsome peroxidation assay 352, 368
and processing 345-348, 351, 352, 356, 357, 362, 363, 365, 366, 368-375
and radical scavenging methods 356, 366, 368, 370
and solvent 116, 117, 144, 148-150, 200, 204, 242, 262, 273, 279, 282, 288, 289, 354, 367, 368, 545, 547-550, 552, 553, 555, 558, 564-566

and storage 76-80, 98, 99, 103, 104, 133, 134, 142, 156, 157, 186, 235, 237, 242, 247, 281, 320, 334, 345, 350-352, 355-358, 360, 362, 363, 365, 366, 368-370, 388, 501, 505-507, 509, 539, 541, 552, 562, 563, 595
and substrate for oxidation 367
and wounding 345, 347, 351, 362
Antioxidants 537
Apoptosis 478, 487
Aquaporins 596, 597, 607, 610, 613
Arabidopsis 502
Aroma 71, 85-88
Ascorbic acid 19, 72, 73, 99, 115, 117, 119, 121, 122, 124, 237-239, 243, 247, 544, 555, 561, 562, 573, 577
and radical scavenging methods 356, 366, 368, 369
and storage 345, 350-352, 355-358, 360, 362, 363, 365-367, 369, 370, 372-375
and substrate for oxidation 367
and wounding 345, 347, 351, 362, 372, 374
Asparagine 261, 596
Assaying mutagenicity 378
Asthma 477
Auxin 51, 54, 55, 342, 517, 524

B

β-carotene 47, 51-54, 135, 328, 433, 453, 459, 537
Bacillus thuringiensis 12, 519, 530, 531
Benign prostate 411
Bioactive components 235
Bioavailability 146
Biochemical adaptation 215, 218, 219, 222, 231, 234
Biodegradable film 299, 300
Biogenesis 88
Biological activity 147
Biosynthesis of carotenoids 135
Bitter 103
BPH: Benign prostatic hyperplasia 472
Breast 148, 396, 398, 399, 401, 402, 460, 467, 541
Brine shrimp test 299, 300, 308
Bronchial asthma 475, 488

C

Calmodulin 509
Calreticulin 509
Carbohydrate-binding 166, 177-179, 187
Carotenoid(s) 72, 85, 92, 93, 96, 113, 114, 116, 118, 122, 125, 133-235, 246, 248-252, 319-321, 327, 333, 334, 340, 355, 356, 387, 388, 485, 537
- chemistry 134
- composition 138
Cases liquid chromatography-tandem mass spectrometry 504
Cell culture 226, 228, 231, 301, 318, 436, 470
Cell cycle 329, 401-403, 406, 407, 411, 423, 463, 508
Cherry tomatoes 31, 33, 39, 83, 84, 111, 118-121, 141, 169, 170, 181, 184, 562, 565
Chilling 74, 75, 80, 594, 595, 608, 609
Chitin-binding 165, 169-173, 177, 179, 186, 188
Chromatographic techniques 553
Chronic obstructive pulmonary disease (COPD) 476
Clean-up 548
Coat proteins (CP) 507
Colon cancer 385, 404, 409
Colour changes 69, 75
Consumer 29, 77, 85, 86, 99, 100, 103, 104, 125, 166, 237, 272, 335, 516, 518, 545
Cyclin D1 402, 403
Cysteine-rich proteins 507
Cytoskeleton-related proteins 508, 509

D

Defense-related proteins 506, 507, 509
Defensins 603, 611
Dehydrogenase 91-93, 220, 470, 508, 569, 573, 591, 607
Destructive index (DI) 480
DI 485
Differential 503
Differential display 587, 594, 610, 611

Differential in-gel electrophoresis 503
Dioxygenase 594, 601, 612
DNA damage 251, 379, 385, 387, 388, 390-392, 443, 450, 461-463
–quality 528, 629
–strand breaks 387, 388, 443
–Transfer 518
DMPD and ABTS methods 333

E

EC 3.5.4.4: Enzyme committee (Adenosine aminohydrolase 472
Edible coatings 67, 74, 78, 80
Effect of ripening and post-harvest storage 141
EHNA: Erythro-9-(2-hydroxy-3-nonyl) adenine 464, 467, 472
Electrochemiluminescence 525, 530, 533
Electrospray ionization (ESI) 503, 558
Embryo 505
Emphysema 475, 477, 478
Endo-β-mannanase 592, 613, 614
Endosperm 505, 507
Environmental factors 18, 33, 68, 72, 117, 125, 341, 386, 459, 585, 597
EST markers 587
Ethylene 37, 41, 61, 62, 67-69, 72-77, 80, 81, 98, 362, 600-602, 607
Exopolysaccharides 299, 301, 459
Exopolysaccharide from culture 299
Expansin 51, 60, 61, 63, 601
Expressed sequence tags (ESTs) 502
Extraction 548
Extraction of carotenoids 149

F

Fatty acid 90, 92, 245, 468
Ferrous ion 116, 388, 392
Field trial 51, 54, 57, 60, 65, 129, 534
Flavonoids 4, 113, 114, 119-125, 537, 539, 542, 544, 545, 556
Flavour 4, 15, 17-19, 29, 59, 69, 71, 85, 86, 91, 96, 98-100, 102-104, 539, 602, 603
FlavrSavr™ 518
Foreign DNA 515
FRAP 114-116, 120-122, 125, 545, 573
Fructokinases 605, 610
Fruit firmness 22, 47, 51, 58, 60, 61, 70, 98
Functional foods 411, 412

G

GAPDH 591, 592, 607, 615
Gene expression 42, 50, 52, 58-60, 65, 585, 586, 589-591, 593-595, 597, 598, 602, 604, 606, 607
Gene transfer 516-518, 522, 531, 585
Genetic engineering 47-50, 52, 57, 64, 594, 603, 614
Genetically modified 47-50, 52-54, 57, 59-63, 604, 606
Genetically modified plants 47, 49, 52, 60, 62, 516
Genotypes 51, 54, 55, 57, 58, 61, 63, 107, 111, 117, 118, 120, 122, 124, 125, 524, 565, 569, 586
Ginger 194
Ginger oil 205
Ginger rhizomes 205
GJC: Gap junctional communication 468, 472
Glandular trichomes 200
Globulins 506
Glucolipids 274
Glucose 4, 16, 17, 60, 71, 99-101, 240, 245, 253, 261, 303, 310, 314, 528, 568
Glutathione 483
Glycine 535, 597, 608
Glycine-rich proteins (GRPs) 506, 508
Greenhouse cultivation 54, 59
Greenhouse whitefly 275
Growth, environmental and technological factors affecting carotenoid content 138
Growth factor-1 (IGF-1) 417
GRPs 506, 508

H

Health effects 235
Heat shock proteins 509, 595, 596
Heat treatments 74, 75, 80, 83
High performance liquid chromatography (HPLC) 150

Histidine 378
HPLC 59, 73, 88, 104, 108, 150, 154, 158, 162, 257, 321, 324, 464, 468, 472, 490, 572, 618, 635
HPLC-MS 558
Hydrogen peroxide 385, 388, 392, 468, 598
Hyperplasia 411

I

IGF binding proteins 409
IGF-1 406, 412, 417, 423, 440, 441, 446, 449, 459
IGF receptor 406, 407
Insulin like growth factor 395, 396, 406-410, 459, 472
Intestinal absorption 321
IRS 397, 401, 402, 406
Isomerization 318, 323, 326, 327, 457, 538, 547, 551, 552, 566
Isomers 91, 113, 242, 253, 256, 317, 323, 414, 415, 433, 434, 468, 471, 473, 538, 553, 555, 557, 577
Italian market 259-261, 263-265

K

Klotho mice 484

L

LAT52 protein 509
Late embryogenesis abundant proteins 509
LCMS-Liquid chromatography-mass spectrophotometry 472
LDL: Low density lipoprotein 251
Leaf extracts 200
Leaf hairs 270
Lectin 165-188
Lectin-related protein 165, 168, 171, 172, 174, 177, 178, 181
Legumins 507
Liquid-liquid extraction 548
LNCaP: Prostate cell cultures 472
LTO: Lycopene rich tomato oleoresin 472
Lycopene 3, 4, 6, 16-19, 27, 30-32, 34, 69, 72, 73, 82, 85, 92, 93, 96, 104, 109, 111, 113, 116-123, 125, 134, 135, 411, 431-434, 438, 485, 487, 489, 537, 605, 606, 608
Lycopene beta-cyclase 51-53, 65, 606
Lymphocytes 180, 181, 187, 385, 387, 388, 390-392

M

Mass spectrometry (MS) 503, 559
MAT 522
MAT-vector 526
Matrix-assisted laser desorption ionization (MALDI) 503
Mean linear intercept (MLI) 480
Mediterranean diet 111, 259
Metabolism 89, 90, 96, 98, 145, 215, 216, 221, 222, 228, 251, 320, 327, 371, 382, 388, 433, 435, 446, 448
Methylketones 271
Microarrays 586, 590, 591, 600, 611
Microballistic 517
Micronutrients 537
Minerals 537
MLI 480, 485
Motifs 166, 596, 608
Mud-PIT 503
Multidimensional 503

N

Natural plant products 273
Nested PCR 525, 528, 536
Nicotinamide 568
N-NMU: N- methyl-N-nitrosurea 472
Non-destructive methods 559
Non-specific lipid transfer proteins 507
Nopaline synthase (NOS) 526
Nutritionally balanced diet 377

O

Odorant 87, 96
One-dimensional gel electrophoresis 503
ORAC 114-116, 120, 125
Oxidant/antioxidant imbalance 478
Oxidative stress 477, 478, 481, 488

P

P21 177, 401
P27 401, 402
Pantothenic acid 569

Parthenocarpy 47-49, 51-57, 124, 523, 524, 531
PCR 525-530, 532, 585-588, 591, 593-596, 598, 602, 603, 607, 622, 624, 625, 627-630, 634, 635
PEG-induced DNA 518
Perception 73, 81, 346, 545, 600, 601
Peripheral airways 477
Peroxidase 179, 216, 228, 386, 519, 544, 573
Phenolic compounds 72, 73, 93, 113, 116, 117, 119-121, 124, 126, 128, 235, 355, 356, 361, 362, 364, 365, 368, 370
Phosphate 35, 44, 51, 52, 65, 116, 215, 217, 228, 231, 232, 568, 570, 571, 573
Phosphate deficiency 234
Phytochemicals 235
Phytochromes 158, 599, 614
PIN: Prostatic intraepithelial neoplasia 458
Plasma 248-251, 321, 323, 324, 326-329, 387, 391, 396, 398, 434, 447, 464, 497, 573
PNP: Purine nucleoside phosphorilase 460
Pollen 502, 503, 508, 509
Pollen tube 502, 509
Polygalacturonase 51, 58, 60, 69, 502, 518, 521, 528, 530, 535, 602
Polymorphism 525, 586, 588, 589, 607, 617, 619, 628, 635
Polysaccharide extraction method 301
Post-harvest 47, 48, 50, 58, 67, 68, 71-74, 76-78, 80, 84, 98, 103, 111, 117, 119, 126, 134, 141, 142, 245, 365
Post-translational modifications 502, 504
Potato lectin 166, 167, 169-171, 176, 177, 186
Precursor 93, 94, 96, 135, 171, 172
Preparation of standards 548
Process 47-50, 334-336, 340-348, 351, 352, 355-357, 362, 363, 365, 366, 368-370, 386-388, 392, 396, 397, 539-542, 544, 548, 551, 554, 563, 564, 595, 599-601, 606
Products 411
Profilins 508, 509, 604, 614
Prolamins 507
Prostate 133, 148, 159, 235, 236, 248, 249, 385, 396, 398, 400, 403, 406, 411, 421, 422, 429-440, 442-449, 457-470
Prostate cancer 411
Prostate carcinogenesis 420, 429, 430, 432, 434-436, 438, 439, 446-448, 450
Prostate diseases 411
Prostate-specific antigen (PSA) 417
Protease 228, 231
Protein(s) 501-503
Protein identification 504
Protein identification technology (Mud-PIT) 503
Proteome 502, 503, 505, 508
Proteomics 501
Proteomic studies 502
Proximal airways 477
PSA 417, 419-421, 423, 431, 441, 443, 447-449, 625, 635
PSA: Prostate specific antigen 449
Pto-like serine/threonione kinases 508
Pto-serine/threonine kinases 509
Purple acid phosphatase 217, 227, 229, 231
Pyridoxine 569

Q

Quality 6, 8, 15, 17, 18, 22, 23, 47-49, 53, 54, 59, 61-63, 67-69, 73-78, 80, 81, 87, 98, 99, 103, 104, 111, 237, 273, 334, 347

R

RAPD 586, 587, 607, 635
RDA 593, 594, 606, 607, 611
Real Time PCR 591, 593
Resistance 13, 14, 21, 22, 56, 195, 213, 250, 269-274, 277, 281, 283, 284, 286, 291-294, 515, 516, 518, 519, 521-523, 532
Reverse phase HPLC 554
RFLP 586, 607
Riboflavin 568
Ripening 67-71, 140-142, 148, 149, 158, 159, 516, 521, 569, 586, 596, 597, 600-602, 606, 608
RNA interference 50, 58, 59, 61, 63
RNS 386, 392
Rodent experimental models 430
ROS 385, 386, 392

S

SA 626, 635

SAGE 586, 607

Salmonella typhimurium TA1535/pSK1002 378, 381, 384

Salt 37, 115, 116, 118, 125, 301, 490, 591, 602, 612

Seed 502, 503, 505, 509

-germination 505, 507

-storage proteins 505

Senescence-accelerated mice strains 479

Senescence-accelerated mouse (SAM) 475, 479

Senile lung or aging lung 477

Sesquiterpene hydrocarbons 271

Small airways 477

Softening 58, 60, 69, 70, 75, 77-79

Solanaceae 48, 167, 169, 170, 173, 179, 186, 566

Solid phase extraction 550

Soluble solids 17, 21, 52-54, 56, 61, 70, 71 73, 81, 99, 335, 516, 519

Sour 100, 101, 103

Spectrofluorimetric analysis 563

Spectroscopic analysis 299, 303

Spider mites 273

Stability of tomato carotenoids 156

Storage proteins 506, 509

Stress resistance 509, 602

Structural and biochemical properties 143

Sucrose-synthase 604

Sugar(s) 47, 52, 59, 70, 85, 86, 88, 98-103, 107, 180, 216, 245, 261, 272, 301, 303, 306, 310, 605

Sugar esters 284

Sugar-binding 165-167, 177-179, 182

Supercritical fluid extraction 551

Sweet 100, 101, 103, 270, 281, 282, 351, 359

T

Taste 15, 70, 85, 86, 88, 99, 100, 102, 103, 106, 107

Telomere 598, 612

Terpenoids 195

Thiamine 568

Tobacco mosaic virus (TMV) 507

Tomato 111-114, 116-125, 133-135, 138, 235, 411, 515, 516, 520

Tomato carotenoids 113, 127, 133-135, 143, 146-148, 150, 156, 157, 395, 398, 405, 415, 426, 470, 471, 576

Tomato concentrate 413

Tomato, cultivar

and antioxidant activity 345, 347, 348, 351-356, 361-375

Tomato cultivars 100, 545, 562, 570

Tomato, fresh-cut

and hydrophilic antioxidant activity 345, 347, 348, 351-356, 361-375

and lipophilic antioxidant activity 352, 370

Tomato juice 54, 100, 102, 103, 151, 152, 155-157, 235-240, 242, 243, 245-251, 323-325, 377-383, 387, 388, 390, 434, 443, 457, 462, 463, 465, 475-477, 485-488, 490-494, 542, 543, 549, 567, 618

Tomato juice drinking 377-380

Tomato lectin 165, 167-171, 173, 174, 176-178, 180-183, 186-188

Tomato mass 413

Tomato, maturation

and antioxidant activity 345, 347, 348, 351-356, 361-375

Tomato mosaic virus (ToMV) 507, 510

Tomato paste 412

Tomato products 27-29, 41, 44, 47, 103, 108, 130, 138, 235-237, 239, 242, 248-251, 317, 319, 321, 323, 327, 385, 387, 388, 390-392, 400, 411, 414, 429, 430, 543, 606

Tomato, ripening

and antioxidant activity 16, 73, 81, 111, 113-115, 125, 145, 247, 248, 333-341, 345, 347, 348, 351-356, 361-370, 443, 444, 459, 475, 487, 538, 544, 573

and ascorbic acid 345, 346, 348, 355, 356, 358, 359, 361, 362, 364, 365, 367, 369, 371, 374

and lycopene 361-364, 367, 368, 370, 372-375

and phenolic compounds 355, 361, 362, 368, 371, 373, 375

Tomato sauces 259-261, 263-265, 321, 323, 541

Tomato Seed 501
Tomato trichomes 273
Tomato waste 309, 314, 549
Tomato-processed products 412
Tomatoes, tomato products 429, 430
ToMV coat protein 509
Total acidity 17, 59, 60, 71, 73, 79
Total antioxidant 72, 73, 81, 250, 251, 538
Total antioxidant activity 538
Transformation 65, 69, 87, 104, 141, 147, 181, 262, 400, 515, 516, 519, 522
Transgenes 51, 58, 515, 521, 522
Transgenics 47, 49, 50, 52, 54, 56-60, 62-64, 112, 123-125, 231, 439, 484, 495, 515, 520, 523, 596, 599, 600, 602, 604, 605
Transgenic tomato products 515
Tumorigenesis 339, 386, 409, 430, 435, 436, 441
Two-dimensional gel electrophoresis 503

U

Umami 101, 102, 107
Umu test 378, 379, 381
UV-vis Absorption spectrophotometry 552

V

VEGF 483, 484
Vicilins 507
Vine life 51, 61, 65
Virus resistance 51, 57, 58, 62, 63, 521
Vitamin C 235, 537
Vitamin D 565
Vitamin E 235, 537
Vitamin K 566
Vitamins 537
Vitamins B 539
Volatile 34, 70, 71, 85-90, 92-94, 96, 98, 99, 104, 286, 558, 602

W

Weight loss 76, 77, 79, 83

Y

Yield 10, 17, 23, 29, 37, 42-45, 51, 53, 54, 65, 107, 117, 119, 122, 125, 149, 161, 171, 173, 184, 211, 242, 300, 619, 620, 634

Color Plate Section

Chapter 3

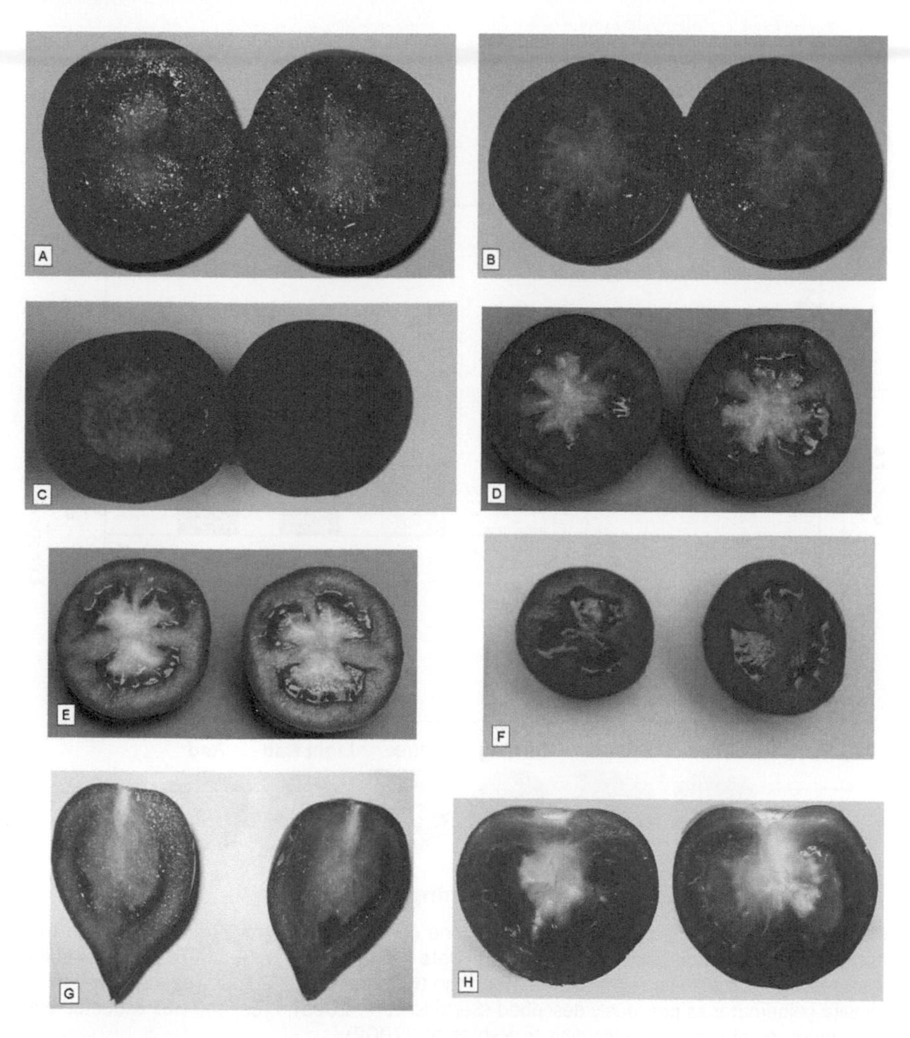

Fig. 1 Seedless parthenocarpic fruit from different tomato genotypes obtained by auxin-synthesis transgene expression. Cut fruits from different tomato genotypes engineered for the parthenocarpic trait. A, parental line L276 (Ficcadenti et al. 1999); B, parental line INB777 (our unpublished result); C hybrid "Giasone" (LH76PC × L276 iaaM 1-1) (Acciarri et al. 2000); D, an experimental hybrid using L276 iaaM 1-1 as male parent (Acciarri et al. 2000); E, cherry type line CM (Ficcadenti et al. 1999); F, cultivar UC 82 (Rotino et al. 2005); G, Italian tomato ecotype "Pizzutello" (our unpublished result); H, cultivar "Ailsa Craig" (our unpublished result). A, B, C, D, E and H tomato fruits transgenic for the auxin-synthesis gene *DefH9-iaaM*; F and G, tomato fruits transgenic for the *DefH9-RI-iaaM* gene (i.e., modified version of the *DefH9-iaaM* gene with reduced level of expression, see Pandolfini et al. 2002).

Chapter 4

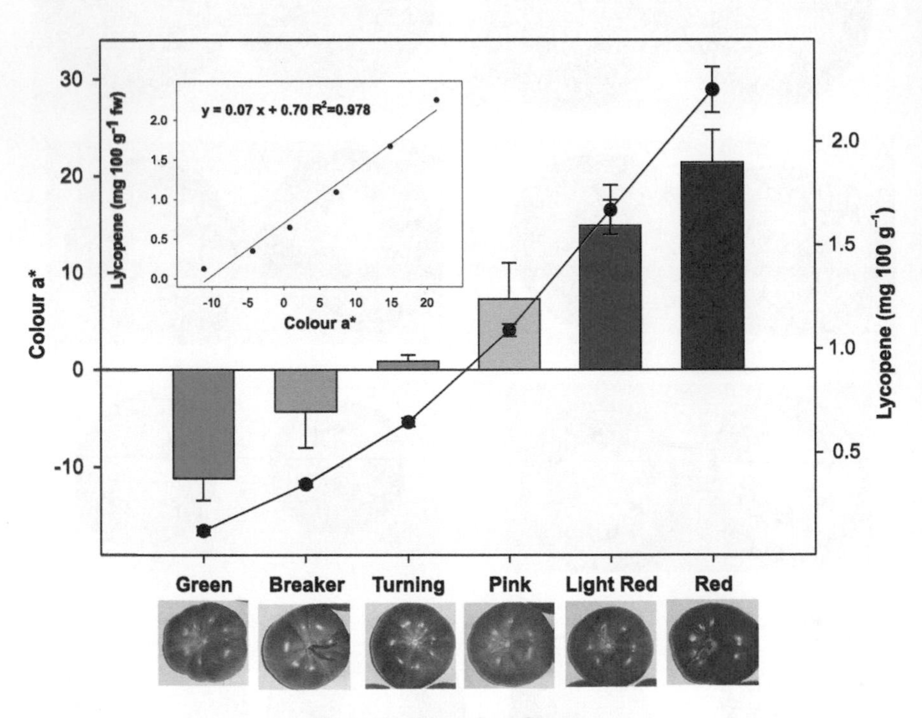

Fig. 1 Colour a* (bars) parameter and lycopene concentration (straight line) in tomato fruit from 'Raf' cultivar at different ripening stages. Data are the mean ± SE of 10 individual fruits. Inset plot shows the linear regression between the two parameters. Colour was measured using Minolta colorimeter as previously described (Serrano et al. 2005). Lycopene was extracted and determined by absorbance according to Fish et al. (2002).

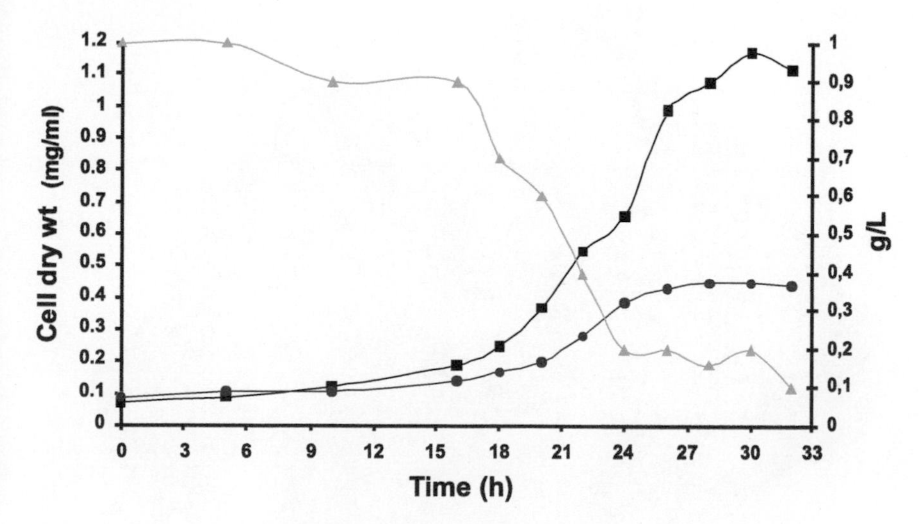

Chapter 14

Fig. 7 Strain Samu-SA1 cell growth (expressed as dry cell weight for ml of culture broth, mg/ml) on polysaccharide medium (■-■-■) and on TH medium (●-●-●) at 75°C. The growth was followed by measuring absorbance at 540 nm and converted to the cell dry weight by an appropriate calibration. The depletion of polysaccharide reported as g/L followed using Dubois method (▲-▲-▲).

Chapter 23

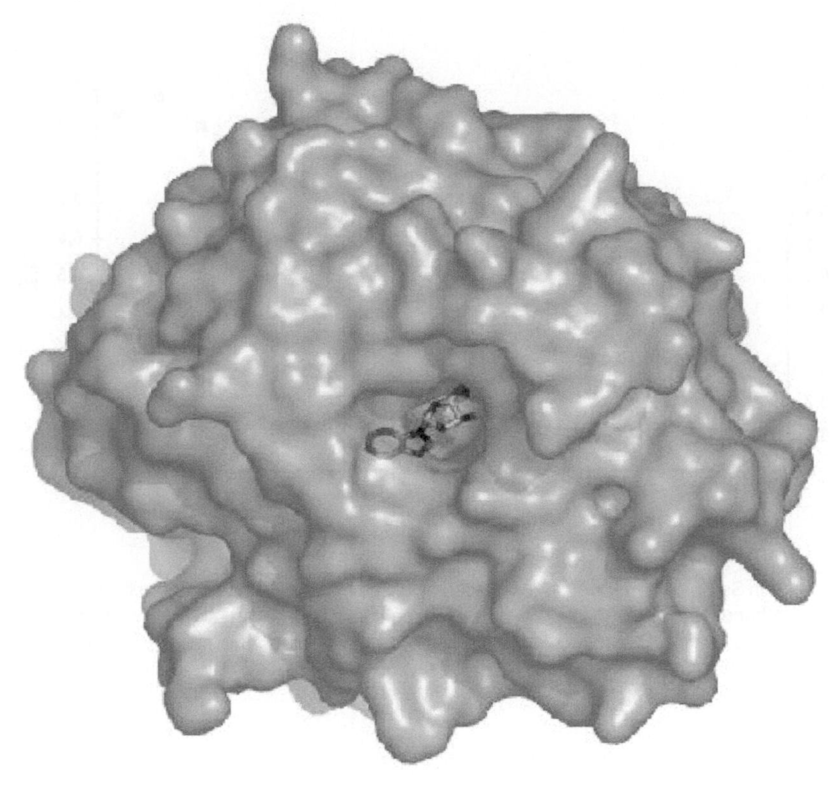

Fig. 3 ADA enzyme structure.

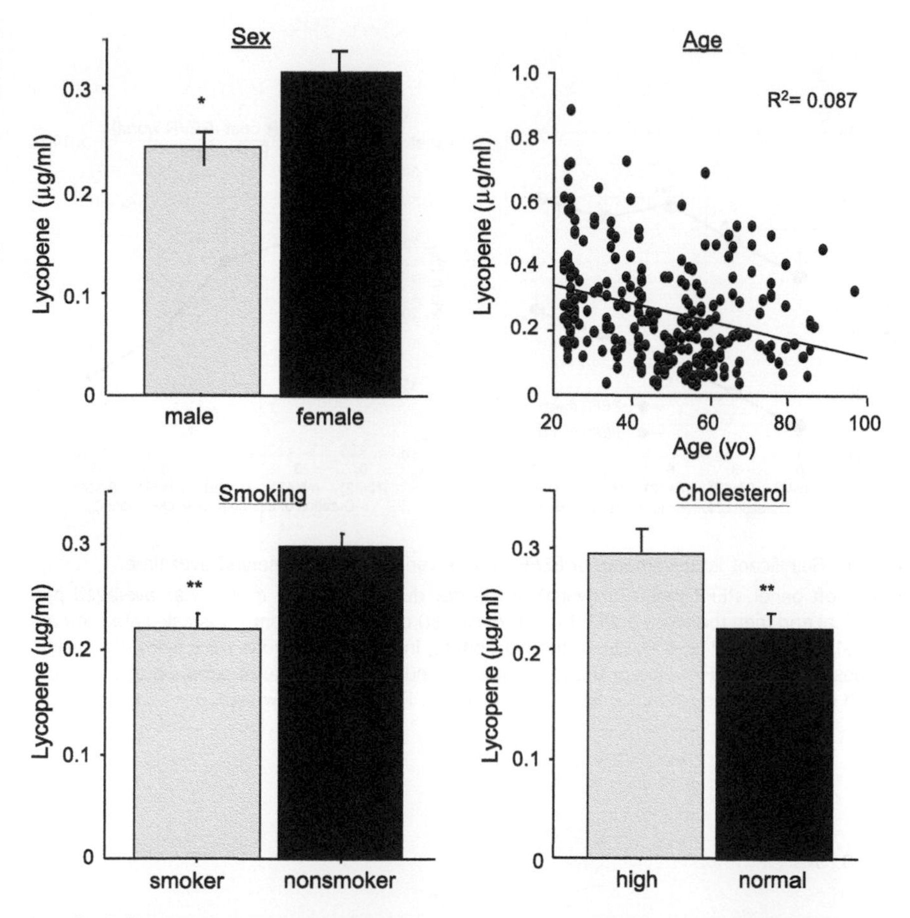

Chapter 24

Fig. 8 Influence of clinical background on serum lycopene concentration in healthy volunteers. Data are presented as mean ± SEM (*$p < 0.05$, **$p < 0.001$) . The cholesterol in the serum is designated as high if it is greater than 220 mg/dL.

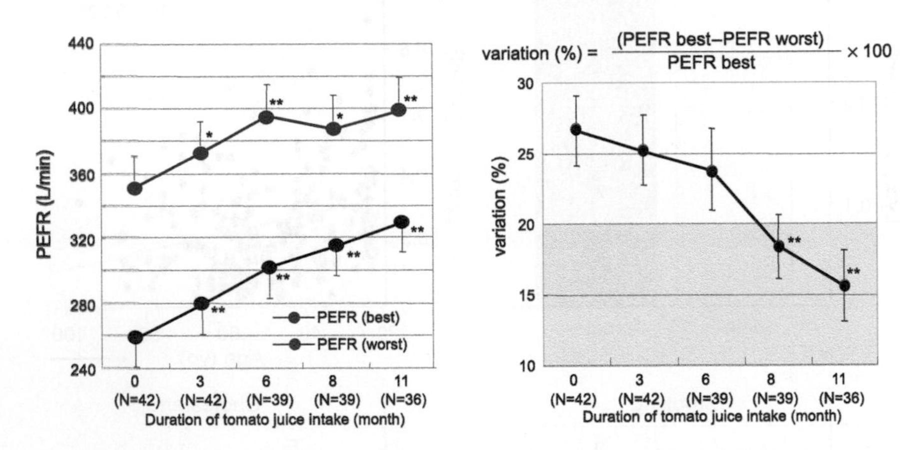

Fig. 9 Significant improvements of PEFR and its variation were observed over time.
In the left panel, PEFR (best) or (worst) of the day during the given month was averaged per individual and then the value of PEFR (best) or (worst) of the given month was calculated (mean ± SEM, $*p < 0.05$,$**p < 0.01$). In the right panel, individual variations were calculated using the best or worst PEFR value of the given month, and then the value of variation of the given month was calculated (mean ± SEM, $**p < 0.01$). PEFR, peak flow rate.

Printed and bound by CPI Group (UK) Ltd, Croydon, CR0 4YY
23/10/2024
01778227-0008